Cambridge
International AS and A Level

Chemistry

Peter Cann & Peter Hughes

HODDER
EDUCATION
AN HACHETTE UK COMPANY

Questions from Cambridge International AS and A Level Chemistry papers are reproduced by permission of Cambridge International Examinations.

Cambridge International Examinations bears no responsibility for the example answers to questions taken from its past question papers which are contained in this book/CD.

Questions from OCR past papers are reproduced by permission of OCR. OCR bears no responsibility for the example answers to questions taken from its past question papers which are contained in this book/CD.

Hachette UK's policy is to use papers that are natural, renewable and recyclable products and made from wood grown in sustainable forests. The logging and manufacturing processes are expected to conform to the environmental regulations of the country of origin.

Orders: please contact Bookpoint Ltd, 130 Milton Park, Abingdon, Oxon OX14 4SB. Telephone: (44) 01235 827720. Fax: (44) 01235 401. Lines are open 9.00–5.00, Monday to Saturday, with a 24-hour message answering service. Visit our website at www.hoddereducation.com

© Peter Cann and Peter Hughes 2015
First published in 2014 by
Hodder Education,
An Hachette UK Company
338 Euston Road
London NW1 3BH

Impression number	5	4	3	2	1
Year	2019	2018	2017	2016	2015

Cover photo by © kmiragaya – Fotolia
Illustrations by Barking Dog Art
Typeset in ITC Garamond Light 9/12 by Aptara Inc.
Printed in Dubai

A catalogue record for this title is available from the British Library
ISBN 978 1444 18133 3

Contents

A Level
Physical chemistry

Inorganic chemistry

Organic chemistry

Student's CD contents

Answers to 'Now try this' questions

Additional work

Interactive tests

Topic summaries

Revision checklists

Examination structure

Planning your revision

Examination technique

Glossary of command words

Mathematical background

Chemical data

Glossaries

Acknowledgements

We are grateful for the help given by Judy Potter in selecting and writing topic-oriented questions, and for the unstintingly professional support and cooperation given by the staff at Hodder Education: Nina Konrad, Emilie Kerton, Laurice Suess, Anne Trevillion and Anne Wanjie.

The Publishers would like to thank the following for permission to reproduce copyright material:

Photo credits: p.1 © The Granger Collection, NYC/TopFoto; **p.2** *tl* © Jeff Blackler/Rex Features; *tr* © Bicipici/Alamy; *bc* © Nils Jorgensen/Rex Features; *br* © Chris Lofty – Fotolia; **p.4** © Eye Of Science/Science Photo Library; **p.14** © lowefoto/Alamy; **p.21** © Rex Features; **p.23** © RGB Ventures/SuperStock/Alamy; **p.27** © Geoff Tompkinson/Science Photo Library; **p.73** © Zygimantas Cepaitis – Fotolia; **p.81** *bc* © Dr. Jeremy Burgess/Science Photo Library; *br* © Alfred Pasieka/Science Photo Library; **p.87** © David Hughes/Hemera/Thinkstock; **p.88** © Paul Fleet/Alamy; **p.90** *bc* © Charles D. Winters/Science Photo Library; *br* © Brian Cosgrove/Dorling Kindersley/Getty Images; **p.102** © Cozyta – Fotolia; **p.103** © fuyi – Fotolia; **p.104** © Martin, Custom Medical Stock Photo/Science Photo Library; **p.107** © NASA Marshall Space Flight Center (NASA-MSFC); **p.129** *l* © Martyn F. Chillmaid/Science Photo Library; **p.150** © ThyssenKrupp Uhde GmbH; **p.151** *tl* © Keystone-France/Gamma-Keystone via Getty Images; *tr* © The Granger Collection, NYC/TopFoto; **p.152** © Mr Korn Flakes – Fotolia; **p.164** *tl* © The Granger Collection, NYC/TopFoto; *tr* © The Granger Collection, NYC/TopFoto; **p.166** © Malcolm Fielding, Johnson Matthey Plc/ Science Photo Library; **p.169** © Martyn F. Chillmaid/Science Photo Library; **p.187** © Science Photo Library; **p.197** *tl* © Dr. Richard Roscoe/Visuals Unlimited, Inc./Science Photo Library; *bl* © Jiri Hamhalter/Alamy; **p.198** © Vanessa Miles/Alamy; **p.239** © Paul Rapson/Science Photo Library; **p.240** © Simon Fraser/Science Photo Library; **p.252** © Molymod; **p.264** © Jean Chung/Bloomberg News via Getty Images; **p.277** © Nasa/Science Photo Library; **p.278** © PUNIT PARANJPE/AFP/Getty Images; **p.289** © Andrew Lambert Photography/Science Photo Library; **p.291** *l* © Power And Syred/Science Photo Library; *c* © Roger Job/Reporters/Science Photo Library; *r* © Tom Parker/Rex Features; **p.333** *l* © maxximmm-Fotolia; *c* © NASA Johnson Space Center Media Archive (NASA JSCMA); *r* © William Arthur/Alamy, **p.337** © Science Photo Library; **p.340** *tl* © Christopher Bradshaw – Fotolia; *tr* © DEX Images Images/Photolibrary Group Ltd/Getty Images; **p.341** *tl* © Claude Nuridsany & Marie Perennou/Science Photo Library; *tr* © B.A.E. Inc./Alamy; *bl* © Imagestate Media (John Foxx); *br* © Charles D. Winters/Science Photo Library; **p.342** © Dr Kari Lounatmaa/Science Photo Library; **p.344** *tl* © David Ducros/Science Photo Library; *tc* © Ace Stock Limited/Alamy; *tr* © Peter Menzel/Science Photo Library; **p.352** © Andrew Lambert Photography/Science Photo Library; **p.377** © Andrew Lambert Photography/Science Photo Library; **p.383** © Andrew Lambert Photography/Science Photo Library; **p.388** *tl* © capude1957 – Fotolia; *bl* © Bygone Collection/Alamy; **p.390** © Georgios Kollidas/Alamy; **p.393** © Libby Welch/Alamy; **p.416** © INTERFOTO/Alamy; **p.431** © Roger Eritja/Alamy; **p.436** © INTERFOTO/Alamy; **p.455** *bl* © Don Fawcett/Science Photo Library; *br* © Wade Davis/Getty Images; **p.466** © J.C. Revy, Ism/Science Photo Library; **p.471** © yotrakbutda – Fotolia; **p.473** *tl* © kevinsday – Fotolia; *tc* © Antler; *tr* © Olivier DELAYE – Fotolia; **p.477** *tl* © Leslie Garland Picture Library/Alamy; *tr* © Dundee Photographics/Alamy; *br* © paramat1977 – Fotolia; **p.478** © Aberfeldy Golf Club; **p.482** © Fraunhofer IFAM; **p.483** © Harris Shiffman – Fotolia; **p.485** © helenlbuxton – Fotolia; **p.508** © James King-Holmes/Science Photo Library; **p.509** *tc* © Geoff Tompkinson/Science Photo Library; *bc* © James King-Holmes/Science Photo Library; **p.520** © TEK Image/Science Photo Library

© Martyn F. Chillmaid
p.6; **p.9**; **p.12**; **p.16**; **p.81**; **p.86**; **p.87**; **p.88**; **p.118**; **p.127**; **p.129**; **p.130**; **p.145**; **p.147**; **p.158**; **p.161**; **p.205**; **p.206**; **p.215**; **p.216**; **p.224**; **p.225**; **p.235**; **p.255**; **p.256**; **p.269**; **p.283**; **p.300**; **p.301**; **p.319**; **p.320**; **p.321**; **p.373**; **p.417**; **p.447**; **p.457**; **p.477**; **p.478**; **p.529**; **p.530**

Every effort has been made to trace all copyright holders, but if any have been inadvertently overlooked, the Publishers will be pleased to make the necessary arrangements at the first opportunity.

Introduction

Cambridge International AS and A Level Chemistry uses some of the content from *Chemistry for Advanced Level*, but has been completely revised by the original authors to cater for those students and teachers involved with the Cambridge International Examinations syllabus 9701.

The book has been fully endorsed by Cambridge International Examinations, and is listed as an endorsed textbook for students studying this syllabus. The syllabus content has been covered comprehensively, and has been separated into AS material, which comprises Topics 1–19, whilst the A Level material is dealt with in Topics 20–30.

All the Learning outcomes specified in the syllabus are included in the book. At the start of each Topic the specific Learning outcomes relevant to that Topic are clearly stated, using the same wording as in the syllabus, so that students can clearly see the syllabus areas covered by the Topic. The chart on the following page summarises the syllabus coverage in each Topic.

Throughout each Topic there are worked examples, with answers, to illustrate the concepts recently introduced. These are followed by a few 'Now try this' questions, allowing students to test themselves. Answers to these questions are on the accompanying Students' CD-ROM.

Each Topic ends with a summary of the key points covered, together with a list of key reactions where relevant. Finally, several past examination questions have been selected that illustrate how the subject matter of the Topic has been assessed in the past. Answers to these questions will be found on the Teachers' CD-ROM.

To allow students and teachers to locate easily the various aspects of the subject, the order of Topics is a logical one, starting with the essential basic principles of physical chemistry and then introducing the application of those principles firstly to inorganic chemistry and then to organic chemistry. No teaching order is implied by this, however. It has been found that mixing principles and applications with factual content throughout the course is often the best way to achieve a deeper and broader understanding of chemistry. Teachers are recommended to consult the schemes of work published by Cambridge International Examinations on their website for some suggested methods of delivering the subject material.

A feature of the new 2016 syllabus is the introduction of **Key concepts**. These are essential ideas, theories, principles or mental tools that help learners to develop a deep understanding of their subject and make links between the different topics. Although teachers are likely to have these in mind at all times when they are teaching the syllabus, we have included in the text the following icons at points where the Key concepts relate to the text.

Atoms and forces – Matter is built from atoms interacting and bonding through electrostatic forces. The structure of matter affects its physical and chemical properties, and influences how substances react chemically.

Experiments and evidence – Chemists use evidence gained from observations and experiments to build models and theories of the structure and reactivity of materials.

Patterns in chemical behaviour and reactions – By identifying patterns in chemical behaviour we can predict the properties of substances and how they can be transformed into new substances by chemical reactions. This allows us to design new materials of use to society.

Chemical bonds – The understanding of how chemical bonds are made and broken by the movement of electrons allows us to predict patterns of reactivity.

Energy changes – The energy changes that take place during chemical reactions can be used to predict both the extent and the rate of such reactions.

This book has been designed to be accessible to all AS and A Level students, but also attempts to go some way towards satisfying the curiosity of the able student, and to answering the questions of the inquisitive. Although based firmly on the AS and A Level syllabus of Cambridge International Examinations, teachers and students will find the subject matter and style of questions make it suitable for several other syllabuses. The subject matter has been extended in some areas where an application, or a more fundamental explanation, is deemed to be appropriate. These extensions are clearly delimited from the main text in panels, and can be bypassed on first reading.

The majority of students starting an AS course in chemistry come from a background of IGCSE Chemistry or Combined Science, and the initial chapters start at a level and a pace that is suited to all such students. Some students come to AS chemistry with the belief that they will find the mathematics difficult, although the mathematical concepts required for chemistry are simple in principle and few in number. We hope to demonstrate that, as long as the processes are understood, rather than learned by rote, the mathematics in both the AS and A Level Topics is well within the grasp of those who have gained a grade C at IGCSE®.

Students also sometimes consider that chemistry is a subject full of difficult concepts. This is not true. Most of

Introduction

chemistry is based on the very simplest idea of electrostatics – like charges repel, unlike charges attract. When the subtle ramifications of this generalisation are studied during the AS and A Level courses, students should constantly remind themselves of the inherent simplicity of this relationship.

Chemistry is the central science, at the crossroads of biology and its associated disciplines on the one hand, and physics on the other. Chemistry relies on physics for its understanding of the fundamental building blocks of matter, and biology relies on chemistry for an understanding of the structures of living organisms, and the processes that go on inside them that we call life. Standing at this crossroads, the chemist is uniquely positioned to understand, and make significant contributions to, many interdisciplinary areas of current and future importance. The chemistry-based sciences of biochemistry, genetic engineering, pharmacology, and polymer and material science will all make increasing contributions to our physical and material well-being in the future. Chemists are also playing a key role in the fight against industrial society's pollution of our environment.

We hope you enjoy discovering the secrets of chemistry during your AS/A Level course.

Peter Cann

Peter Hughes

® IGCSE is the registered trademark of Cambridge International Examinations.

How the Cambridge Learning outcomes are covered by the 30 Topics in this book

Topic	Learning outcomes	Topic	Learning outcomes
1	1.1–1.5	16	17.11
2	2.1–2.3	17	18.1
3	3.2, 3.3, 3.5c), 14. 3a)	18	19.1a),b), 19.3
4	3.1, 3.4, 3.5a),b), 4.1, 4.2, 4.3a),c),d)	19	–
5	5.1a)–c), 5.2a)	20	2.3g), 5.1b), d), 5.2a), 5.3, 5.4,10.1f),g)
6	1.5a)–c), 7.2a),b)	21	8.1c)–h), 8.2c), 8.3e)
7	4.3b), 6.1, 6.3g)	22	7.2c)–k), 7.3
8	5.2b), 8.1a),b), 8.2a),b), 8.3a)–d)	23	6.2, 6.3, 6.4
9	7.1	24	12.1, 12.2, 12.3, 12.4, 12.5
10	9.1a)–d), 9.2, 9.3, 10.1a)–e), 13.1, 13.2	25	14.3, 15.4, 17.2
11	11.1–11.5	26	17.1a), 19.1b)–e)
12	1.5b), 14.1a),b),d),e), 14.2, 14.4	27	20.1, 20.2, 20.3
13	15.1, 15.3a),b)	28	21.1, 21.2, 21.3, 21.4
14	14.3, 15.2	29	22.1, 22.2, 22.3, 22.4, 22.5
15	16.1, 16.2	30	23.1, 23.2

1 Chemical formulae and moles

In this topic we introduce chemical formulae, and show you how these can be worked out using the valencies (or combining powers) of atoms and ions. We also introduce the chemists' fundamental counting unit, the mole, and show you how it can be used to calculate both empirical formulae and the amounts of substances that react, or are formed, during chemical reactions.

As you read through the topic, study the worked examples, and then try the 'Now try this' questions that follow them.

Learning outcomes

By the end of this topic you should be able to:

1.1a) define and use the terms *relative atomic, isotopic, molecular* and *formula masses*, based on the ^{12}C scale
1.2a) define and use the term *mole* in terms of the Avogadro constant
1.4a) define and use the terms *empirical* and *molecular formula*
1.4b) calculate empirical and molecular formulae, using combustion data or composition by mass
1.5a) write and construct balanced equations
1.5b) perform calculations, including use of the mole concept, involving reacting masses (from formulae and equations), volumes of gases (e.g. in the burning of hydrocarbons), volumes and concentrations of solutions, and relate the number of significant figures in your answers to those given or asked for in the question
1.5c) deduce stoichiometric relationships from calculations such as those in **1.5b)**.

1.1 Introduction

What is chemistry?

Chemistry is the study of the properties of matter. By **matter**, we mean the substances that we can see, feel, touch, taste and smell – the stuff that makes up the material world. Passive observation forms only a small part of a chemist's interest in the world. Chemists are actively inquisitive scientists. We try to understand why matter has the properties it does, and how to modify these properties by changing one substance into another through chemical reactions.

Chemistry as a modern science began a few hundred years ago, when chemists started to relate the observations they made about the substances they were investigating to theories of the structure of matter. One of the most important of these theories was the Atomic Theory. It is just over 200 years since John Dalton put forward his idea that all matter was composed of **atoms**. His theory stated that:

● the atoms of different elements were different from each other
● the atoms of a particular element were identical to each other
● all atoms stayed the same over time and could be neither created nor destroyed
● all matter was made up from a relatively small number of elements (Dalton thought about 50) combined in various ways.

Although Dalton's theory has had to be modified slightly, it is still a useful starting point for the study of chemistry.

Figure 1.1 John Dalton, who first suggested the modern Atomic Theory

Since that time chemists have uncovered and explained many of the world's mysteries, from working out how elements are formed within stars to discovering how our genes replicate. On the way they have discovered thousands of new methods of converting one substance into another, and have made millions of new substances, many of which are of great economic and medical benefit to the human race (see Figure 1.2).

Figure 1.2 Some examples of the economic, medical and agricultural benefits of chemistry

Classifying matter – elements, compounds and mixtures

Chemists classify matter into one of three categories.

- **Elements** contain just one sort of atom. Although the atoms of a particular element may differ slightly in mass (see section 2.3), they all have identical chemical reactions. Examples of elements include hydrogen gas, copper metal and diamond crystals (which are carbon).
- **Compounds** are made up from the atoms of two or more different elements, bonded together chemically. The ratio of elements within a particular compound is fixed, and is given by its chemical formula (see page 7). The physical and chemical properties of a compound are always different from those of the elements that make it up. Examples of compounds include sodium chloride (containing sodium and chlorine ions), water (containing hydrogen and oxygen atoms) and penicillin (containing hydrogen, carbon, nitrogen, oxygen and sulfur atoms).

- **Mixtures** consist of more than one compound or element, mixed but not chemically combined. The components can be mixed in any proportion, and the properties of a mixture are often the sum of, or the average of, the properties of the individual components. Examples of mixtures include air, sea water and alloys such as brass.

1.2 Intensive and extensive properties

The properties of matter may be divided into two groups.

- The **extensive properties** depend on how much matter we are studying. Common examples are mass and volume – a cupful of water has less mass, and less volume, than a swimming pool.
- The other group are the **intensive properties**, which do not depend on how much matter we have. Examples include temperature, colour and density. A copper coin and a copper jug can both have the same intensive properties, although the jug will be many times heavier (and larger) than the coin.

The chemical properties of a substance are also **intensive**. A small or a large lump of sodium will react in the same way with either a cupful or a jugful of water. In each case it will fizz, give off steam and hydrogen gas, and produce an alkaline solution in the water.

Experiment

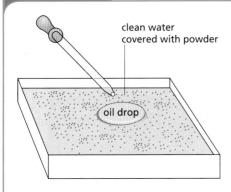

clean water covered with powder

oil drop

Figure 1.3 The oil-drop experiment

The oil-drop experiment

A bowl is filled with clean water and some fine powder, such as pollen grains or flour, is sprinkled over the surface. A small drop of oil is placed on the surface of the water, as shown in Figure 1.3. The oil spreads out. As it does so, it pushes the powder back, so that there is an approximately circular area clear of powder.

We can measure the volume of one drop by counting how many drops it takes to fill a micro measuring cylinder. (If we know the oil's density, an easier method would be to find the mass of, say, 20 drops.) We can calculate the area of the surface film by measuring its diameter. Assuming the volume of oil does not change when the drop spreads out, we can thus find the thickness of the film. This cannot be smaller than the length of one oil molecule (though it may be bigger, if the film is several molecules thick – there is no way of telling). The following are typical results:

$$\text{volume of drop} = 1.0 \times 10^{-4}\, \text{cm}^3$$
$$\text{diameter of oil film} = 30\, \text{cm}$$
so
$$\text{radius of film} = 15\, \text{cm}$$
$$\text{area of film} = \pi r^2$$
$$= 3.14 \times 15^2\, \text{cm}^2$$
$$= 707\, \text{cm}^2$$
$$\text{volume} = \text{area} \times \text{thickness}$$
so
$$\text{thickness} = \frac{\text{volume}}{\text{area}}$$
$$= \frac{1.0 \times 10^{-4}\, \text{cm}^3}{707\, \text{cm}^2}$$
$$= \mathbf{1.4 \times 10^{-7}\, cm}\ (1.4 \times 10^{-9}\, \text{m})$$

Distances this small are usually expressed in units of **nanometres**:
$$1 \text{ nanometre (nm)} = 1 \times 10^{-9} \text{ metres (m)}$$
So the film is 1.4 nm thick. Oil molecules cannot be larger than this.

1.3 The sizes of atoms and molecules

Just as we believe that elements are composed of identical atoms, so we also believe that compounds are made up of many identical units. These units are the smallest entities that still retain the chemical properties of the compound. They are called **molecules** or **ions**, depending on how the substance is bonded together (see Topics 3 and 4). **Molecules** contain two or more atoms bound together. The atoms may be of the same element (e.g. ozone, O_3, which contains three atoms of oxygen) or different elements (e.g. water, H_2O, which contains two atoms of hydrogen and one atom of oxygen). **Ions** are atoms, or groups of atoms, that carry an electrical charge.

Molecules are extremely small – but how small? Sometimes, a simple experiment, a short calculation and a little thought can lead to quite an amazing conclusion. The well-known oil-drop experiment is an example. It allows us to obtain an order-of-magnitude estimate of the size of a molecule using everyday apparatus (see the experiment on page 3).

Because molecules are made up of atoms, this means that atoms must be even smaller than the oil molecule. We can measure the sizes of atoms by various techniques, including X-ray crystallography. A carbon atom is found to have a diameter of 0.15 nm. That means it takes 6 million carbon atoms touching one another to reach a length of only 1 mm!

Figure 1.4 Coloured scanning tunnelling electron micrograph of carbon nanotubes, comprised of rolled sheets of carbon atoms. Individual atoms are seen as raised bumps on the surface of the tubes.

Standard form

The numbers that chemists deal with can often be very large, or very small. To make these more manageable, and to avoid having to write long lines of zeros (with the accompanying danger of miscounting them), we often express numbers in **standard form**.

A number in standard form consists of two parts, the first of which is a number between 1 and 10, and the second is the number 10 raised to a positive or negative power. Some examples, with their fully written-out equivalents, are given in Table 1.1.

Table 1.1 Standard form

Standard form	Fully written-out equivalent
6×10^2	600
7.142×10^7	71 420 000
2×10^{-6}	0.000 002
3.8521×10^{-4}	0.000 385 21

If the 10 is raised to a positive power, the superscript tells us how many digits to the right the decimal point moves. As in the examples in Table 1.1, we often need to write extra zeros to allow this to take place. If the 10 is raised to a negative power, the superscript number tells us how many digits to the left the decimal point moves. Here again, we often need to write extra zeros, but this time to the left of the original number.

Significant figures

In mathematics, numbers are exact quantities. In contrast, the numbers used in chemistry usually represent physical quantities which a chemist measures. The accuracy of the measurement is shown by the number of significant figures to which the quantity is quoted.

If we weigh a small coin on a digital kitchen scale, the machine may tell us that it has a mass of 4 g. A one-decimal-place balance will show its mass as 3.5 g, whereas on a two-decimal-place balance its mass will be shown as 3.50 g. We should interpret the reading on the kitchen scale as meaning that the mass of the coin lies between 3.5 g and 4.5 g. If it were just a little lighter than 3.5 g, the scale would have told us that its mass was 3 g. If it were a little heavier than 4.5 g, the read-out would have been 5 g. The one-decimal-place balance narrows the range, telling us the mass of the coin is between 3.45 g and 3.55 g. The two-decimal-place balance narrows it still further, to between 3.495 g and 3.505 g.

In this way the number of significant figures (1, 2 or 3 in the above examples) tells us the accuracy with which the quantity has been measured.

The same is true of volumes. Using a $100 \, cm^3$ measuring cylinder, we can measure a volume of $25 \, cm^3$ to an accuracy of $\pm 0.5 \, cm^3$, so we would quote the volume as $25 \, cm^3$. Using a pipette or burette, however, we can measure volumes to an accuracy of $\pm 0.05 \, cm^3$, and so we would quote the same volume as $25.0 \, cm^3$ (that is, somewhere between $24.95 \, cm^3$ and $25.05 \, cm^3$).

Most chemical balances and volumetric equipment will measure quantities to 3 or 4 significant figures. Allowing for the accumulation of errors when values are calculated using several measured quantities, we tend to quote values to 2 or 3 significant figures.

In the examples in Table 1.1, the number 6×10^2 has 1 significant figure, and the number 7.142×10^7 has 4 significant figures.

1.4 The masses of atoms and molecules

Being so small, atoms are also very light. Their masses range from 1×10^{-24} g to 1×10^{-22} g. It is impossible to weigh them out individually, but we can accurately measure their **relative masses**, that is, how heavy one atom is compared with another. The most accurate way of doing this is by using a mass spectrometer (see section 2.5).

Originally, the atomic masses of all the elements were compared with the mass of an atom of hydrogen:

$$\text{relative atomic mass of element E} = \frac{\text{mass of one atom of E}}{\text{mass of one atom of hydrogen}}$$

This is because hydrogen is the lightest element, so the relative atomic masses of all other elements are greater than 1, which is convenient.

Because of the existence of isotopes (see section 2.3), and the central importance of carbon in the masses of organic compounds, the modern definition uses the isotope carbon-12, ^{12}C, as the standard of reference:

relative atomic mass of element E $= \dfrac{\text{average mass of one atom of E}}{\frac{1}{12} \text{ the mass of one atom of } ^{12}C}$

The difference between the two definitions is small, since a carbon-12 atom has almost exactly 12 times the average mass of a hydrogen atom (the actual ratio is $11.91:1$). Relative atomic mass is given the symbol $\mathbf{A_r}$. Since it is the ratio of two masses, it is a dimensionless quantity – it has no units. We shall be looking at isotopes, and relative isotopic mass, in more detail in Topic 2, where we investigate the structure of atoms.

The masses of atoms and sub-atomic particles (see Topic 2) are often expressed in atomic mass units. An **atomic mass unit** (amu) is defined as 1/12 the mass of one atom of carbon-12. It has the value of 1.6606×10^{-24} g.

Although we cannot use a laboratory balance to weigh out individual atoms, we can use it to weigh out known ratios of atoms of various elements, as long as we know their relative atomic masses. For example, if we know that the relative atomic masses of carbon and magnesium are 12.0 and 24.0 respectively, we can be sure that 12.0 g of carbon will contain the same number of atoms as 24.0 g of magnesium. What is more, 24.0 g (12.0×2) of carbon will contain twice the number of atoms as 24.0 g of magnesium. Indeed, we can be certain that any mass of carbon will contain twice the number of atoms as the same mass of magnesium, since the mass of each carbon atom is only half the mass of a magnesium atom.

Similarly, if we know that the relative atomic mass of helium is 4.0 (which is one-third the relative atomic mass of carbon), we can deduce that identical masses of helium and carbon will always contain three times as many helium atoms as carbon atoms.

1.5 The mole

Chemists deal with real, measured quantities of substances. Rather than counting atoms individually, we prefer to count them in units that are easily measurable. The gram is a conveniently sized unit of mass to use for weighing out matter (a teaspoon of sugar, for example, weighs about 5 g). The chemist's unit of amount, the **mole** (symbol **mol**), is defined in terms of grams:

> One mole of an element is the amount that contains the same number of atoms as there are in 12.000 g of carbon-12.

Figure 1.5 One-tenth of a mole of each of the elements aluminium, sulfur, bromine and lead

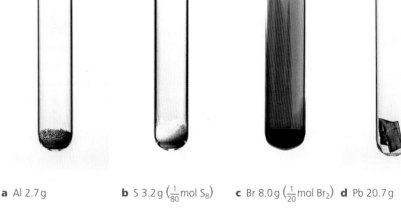

a Al 2.7 g   **b** S 3.2 g $\left(\frac{1}{80} \text{mol } S_8\right)$ **c** Br 8.0 g $\left(\frac{1}{20} \text{mol } Br_2\right)$ **d** Pb 20.7 g

The mass of one mole of an element is called its **molar mass** (symbol $\boldsymbol{M}$). It is numerically equal to its relative atomic mass, A_r, but is given in grams per mole:

relative atomic mass of carbon = 12.0
molar mass of carbon = 12.0 g mol^{-1}

relative atomic mass of magnesium = 24.0
molar mass of magnesium = 24.0 g mol^{-1}

It follows from the above definition that there is a clear relationship between the mass (m) of a sample of an element and the number of moles (n) it contains:

$$\text{amount (in moles)} = \frac{\text{mass}}{\text{molar mass}}$$

$$\text{or} \qquad n = \frac{m}{M} \text{ mol} \tag{1}$$

Now try this

Using the A_r values O = 16.0, Mg = 24.0, S = 32.0, calculate the amount of substance (in moles) in each of the following samples.

1 24.0 g of oxygen
2 24.0 g of sulfur
3 16.0 g of magnesium

Worked example

What is the amount (in moles) of carbon in 30 g of carbon?

Answer
Use the value $A_r(\text{carbon}) = 12.0$ to write its molar mass, and use equation (1) above:

$$m = 30\,\text{g} \quad \text{and} \quad M = 12.0\,\text{g mol}^{-1}$$

$$\text{so} \quad n = \frac{30\,\text{g}}{12.0\,\text{g mol}^{-1}}$$

$$= \mathbf{2.5\,mol}$$

As we saw on page 5, the actual masses of atoms are very small. We would therefore expect the number of atoms in a mole of an element to be very large. This is indeed the case. One mole of an element contains a staggering 6.022×10^{23} atoms (six hundred and two thousand two hundred million million million atoms). This value is called the **Avogadro constant**, symbol L.

$L = 6.022 \times 10^{23}\,\text{mol}^{-1}$

The approximate value of $L = 6.0 \times 10^{23}\,\text{mol}^{-1}$ is often adequate, and will be used in calculations in this book.

The relationship between the number of moles in a sample of an element and the number of atoms it contains is as follows:

$$\text{number of atoms} = L \times \text{number of moles}$$

$$\text{or} \qquad N = Ln \tag{2}$$

Now try this

Calculate the amount of substance (in moles) in

1 a sample of uranium that contains 1.0×10^{25} atoms
2 a sample of fluorine that contains 5×10^{21} atoms

Worked example

How many hydrogen atoms are there in 1.5 mol of hydrogen atoms?

Answer
Use equation (2), and the value of L given above:

$$L = 6.0 \times 10^{23}\,\text{mol}^{-1} \quad \text{and} \quad n = 1.5\,\text{mol}$$

$$\text{so} \quad N = 6.0 \times 10^{23}\,\text{mol}^{-1} \times 1.5\,\text{mol}$$

$$= \mathbf{9.0 \times 10^{23}}$$

1.6 Atomic symbols and formulae

Each element has a unique **symbol**. Symbols consist of either one or two letters. The first is always a capital letter and the second, if present, is always a lower-case letter. This rule avoids confusions and ambiguities when the symbols are combined to make the formulae of compounds. For example:

- the symbol for hydrogen is H
- the symbol for helium is He (not HE or hE)
- the symbol for cobalt is Co (not CO – this is the **formula** of carbon monoxide, which contains two atoms in its molecule, one of carbon and one of oxygen).

Symbols are combined to make up the **formulae** of compounds. If more than one atom of a particular element is present, its symbol is followed by a subscript giving the number of atoms of that element contained in one formula unit of the compound. For example:

- the formula of copper oxide is CuO (one atom of copper combined with one atom of oxygen)
- the formula of water is H_2O (two atoms of hydrogen combined with one atom of oxygen)
- the formula of phosphoric(V) acid is H_3PO_4 (three atoms of hydrogen combined with one of phosphorus and four of oxygen).

Sometimes, especially when the compound consists of ions rather than molecules (see Topic 4), groups of atoms in a formula are kept together by the use of brackets. If more than one of a particular group is present, the closing bracket is followed by a subscript giving the number of groups present. This practice makes the connections between similar compounds clearer. For example:

- the formula of sodium nitrate is $NaNO_3$ (one sodium ion, Na^+, combined with one nitrate ion, NO_3^-, which consists of one nitrogen atom combined with three oxygen atoms)
- the formula of calcium nitrate is $Ca(NO_3)_2$ (one calcium ion, Ca^{2+}, combined with two nitrate ions).

Note that in calcium nitrate, the formula unit consists of one calcium, two nitrogens and six oxygens, but it is not written as CaN_2O_6. This formula would not make clear the connection between $Ca(NO_3)_2$ and $NaNO_3$. Both compounds are nitrates, and both undergo similar reactions of the nitrate ions.

The formulae of many ionic compounds can be predicted if the valencies of the ions are known. (The valency of an ion is the electrical charge on the ion.) Similarly, the formulae of several of the simpler covalent (molecular) compounds can be predicted if the covalencies of the constituent atoms are known. (The covalency of an atom is the number of covalent bonds that the atom can form with adjacent atoms in a molecule.) Lists of covalencies and ionic valencies, and examples of how to use them, are given on pages 49 and 79.

Worked example

How many atoms of each element are present in one formula unit of each of the following compounds?

a $Al(OH)_3$ b $(NH_4)_2SO_4$

Answer

a The subscript after the closing bracket multiplies all the contents of the bracket by three. There are therefore three OH (hydroxide) groups, each containing one oxygen and one hydrogen atom, making a total of three oxygen atoms and three hydrogen atoms, together with one aluminium atom.

b Here there are two ammonium groups, each containing one nitrogen atom and four hydrogen atoms, and one sulfate group, containing one sulfur atom and four oxygen atoms. In total, therefore, there are:
- two nitrogen atoms
- eight hydrogen atoms
- one sulfur atom
- four oxygen atoms.

Now try this

How many atoms in total are present in one formula unit of each of the following compounds?

1 NH_4NO_3
2 $Na_2Cr_2O_7$
3 $KCr(SO_4)_2$
4 $C_6H_{12}O_6$
5 $Na_3Fe(C_2O_4)_3$

1.7 Moles and compounds

Relative molecular mass and relative formula mass

Just as we can weigh out a mole of carbon (12.0 g), so we can weigh out a mole of a compound such as ethanol (alcohol). We first need to calculate its **relative molecular mass**, M_r.

To calculate the relative molecular mass (M_r) of a compound, we add together the relative atomic masses (A_r) of all the elements present in one molecule of the compound (remembering to multiply the A_r values by the correct number if more than one atom of a particular element is present). So for ethanol, C_2H_6O, we have:

$$M_r = 2A_r(C) + 6A_r(H) + A_r(O)$$
$$= 2 \times 12.0 + 6 \times 1.0 + 16.0$$
$$= 46.0$$

Just as with relative atomic mass, values of relative molecular mass are ratios of masses, and have no units. The molar mass of ethanol is $46.0\,g\,mol^{-1}$.

For ionic and giant covalent compounds (see Topic 4), we cannot, strictly, refer to their relative molecular masses, as they do not consist of individual molecules. For these compounds, we add together the relative atomic masses of all the elements present in the simplest (empirical) formula. The result is called the relative formula mass, but is given the same symbol as relative molecular mass, M_r. Just as with molecules, the mass of one formula unit is called the molar mass, symbol M. For example, the relative formula mass of sodium chloride, NaCl, is calculated as follows:

$$M_r = A_r(Na) + A_r(Cl)$$
$$= 23.0 + 35.5$$
$$= 58.5$$

The molar mass of sodium chloride is $58.5\,g\,mol^{-1}$.

We can apply equation (1) (page 7) to compounds as well as to elements. Once the molar mass has been calculated, we can relate the mass of a sample of a compound to the number of moles it contains.

Figure 1.6 One-tenth of a mole of each of the compounds water, potassium dichromate(VI) ($K_2Cr_2O_7$) and copper(II) sulfate-5-water ($CuSO_4.5H_2O$)

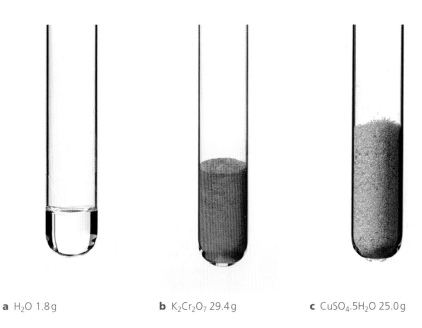

a H_2O 1.8g   b $K_2Cr_2O_7$ 29.4g   c $CuSO_4.5H_2O$ 25.0g

Worked example 1

Calculate the relative molecular mass of glucose, $C_6H_{12}O_6$.

Answer
$$M_r = 6A_r(C) + 12A_r(H) + 6A_r(O)$$
$$= 6 \times 12.0 + 12 \times 1.0 + 6 \times 16.0$$
$$= 72.0 + 12.0 + 96.0$$
$$= \mathbf{180.0}$$

Now try this

1 Calculate the relative formula mass of each of the following compounds. (Use the list of A_r values in the data section on the CD.)
   a iron(II) sulfate, $FeSO_4$
   b calcium hydrogencarbonate, $Ca(HCO_3)_2$
   c ethanoic acid, $C_2H_4O_2$
   d ammonium sulfate, $(NH_4)_2SO_4$
   e the complex with the formula $Na_3Fe(C_2O_4)_3$
2 How many moles of substance are there in each of the following samples?
   a 20 g of magnesium oxide, MgO
   b 40 g of methane, CH_4
   c 60 g of calcium carbonate, $CaCO_3$
   d 80 g of cyclopropene, C_3H_4
   e 100 g of sodium dichromate(VI), $Na_2Cr_2O_7$
3 What is the mass of each of the following samples?
   a 1.5 moles of magnesium sulfate, $MgSO_4$
   b 0.333 mole of aluminium chloride, $AlCl_3$

Worked example 2

How many moles are there in 60 g of glucose?

Answer

Convert the relative molecular mass calculated in Worked example 1 to the molar mass, M, and use the formula in equation (1):

$$n = \frac{m}{M}$$

$m = 60 \, g$
$M = 180 \, g\,mol^{-1}$

$$n = \frac{60 \, g}{180 \, g\,mol^{-1}}$$

$$= 0.33 \, mol$$

A mole of what?

When dealing with compounds, we need to define clearly what the word 'mole' refers to. A mole of water contains 6×10^{23} molecules of H_2O. But because each molecule contains two hydrogen atoms, a mole of H_2O molecules will contain two moles of hydrogen atoms, that is, 12×10^{23} hydrogen atoms. Likewise, a mole of sulfuric acid, H_2SO_4, will contain two moles of hydrogen atoms, one mole of sulfur atoms and four moles of oxygen atoms. A mole of calcium chloride, $CaCl_2$, contains twice the number of chloride ions as does a mole of sodium chloride, $NaCl$.

Sometimes this also applies to elements. The phrase 'one mole of chlorine' is ambiguous. One mole of chlorine molecules contains 6×10^{23} Cl_2 units, but it contains 12×10^{23} chlorine atoms (2 mol of Cl).

1.8 Empirical formulae and molecular formulae

The **empirical formula** is the simplest formula that shows the relative number of atoms of each element present in a compound.

If we know the percentage composition by mass of a compound, or the masses of the various elements that make it up, we can work out the ratios of atoms.

The steps in the calculation are as follows.

1 Divide the percentage (or mass) of each element by the element's relative atomic mass.

2 Divide each of the figures obtained in step **1** by the smallest of those figures.

3 If the results of the calculations do not approximate to whole numbers, multiply them all by 2 to obtain whole numbers. (In rare cases we might have to multiply by 3 or 4 to obtain whole numbers.)

Worked example

Calculate the empirical formula of an oxide of iron that contains 70% Fe by mass.

Answer

The oxide contains iron and oxygen only, so the percentage of oxygen is $100 - 70 = 30\%$. Following the steps above:

1 Fe: $\frac{70}{56} = 1.25$ O: $\frac{30}{16} = 1.875$

2 Fe: $\frac{1.25}{1.25} = 1.00$ O: $\frac{1.875}{1.25} = 1.50$

3 Multiply both numbers by 2: Fe = 2, O = 3.

Therefore the empirical formula is **Fe_2O_3**.

Now try this

Calculate the empirical formula of each of the following compounds.

1 a sulfide of copper containing 3.97 g of copper and 1.00 g of sulfur
2 a hydrocarbon containing 81.8% carbon and 18.2% hydrogen
3 a mixed oxide of iron and calcium which contains 51.9% iron and 18.5% calcium by mass (the rest being oxygen)

The molecular formula is either the same as, or a simple multiple of, the empirical formula. For example, the molecule of hydrogen peroxide contains two hydrogen atoms and two oxygen atoms. Its molecular formula is H_2O_2, but its empirical formula is HO.

> The **molecular formula** tells us the actual number of atoms of each element present in a molecule of the compound.

1.9 Equations

Mass is conserved

A chemical equation represents what happens during a chemical reaction. A key feature of chemical reactions is that they proceed with no measurable change in mass at all. Many obvious events can often be seen taking place – the evolution of heat, flashes of light, changes of colour, noise and evolution of gases. But despite these sometimes dramatic signs that a reaction is happening, the sum of the masses of all the various products is always found to be equal to the sum of the masses of the reactants.

This was one of the first quantitative laws of chemistry, and is known as the **Law of Conservation of Mass**. It can be illustrated simply but effectively by the following experiment.

Experiment

The conservation of mass

A small test tube has a length of cotton thread tied round its neck, and is half filled with lead(II) nitrate solution. It is carefully lowered into a conical flask containing potassium iodide solution, taking care not to spill its contents. A bung is placed in the neck of the conical flask, so that the cotton thread is trapped by its side, as shown in Figure 1.7. The whole apparatus is then weighed.

The conical flask is now shaken vigorously to mix the contents. A reaction takes place, and the bright yellow solid lead(II) iodide is formed. On re-weighing the conical flask with its contents, the mass is found to be identical to the initial mass.

Figure 1.7 During the formation of lead(II) iodide, mass is conserved.

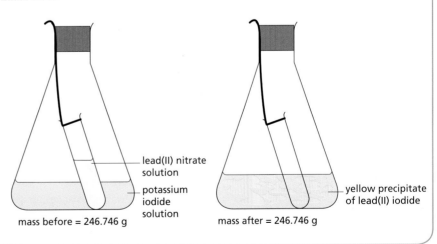

lead(II) nitrate solution

potassium iodide solution

yellow precipitate of lead(II) iodide

mass before = 246.746 g   mass after = 246.746 g

Balanced equations

The reason why the mass does not change during a chemical reaction is because no atoms are ever created or destroyed. The number of atoms of each element is the same at the end as at the beginning. All that has happened is that they have changed their chemical environment. In the example in the experiment above, the change can be represented in words as:

$$\begin{array}{c}\text{lead(II) nitrate} \\ \text{solution}\end{array} + \begin{array}{c}\text{potassium iodide} \\ \text{solution}\end{array} \rightarrow \begin{array}{c}\text{solid lead(II)} \\ \text{iodide}\end{array} + \begin{array}{c}\text{potassium nitrate} \\ \text{solution}\end{array}$$

Figure 1.8 The formation of lead(II) iodide

There are several steps we must carry out to convert this **word equation** into a **balanced chemical equation**.

1 Work out and write down the formula of each of the compounds in turn, and describe its physical state using the correct one of the following four **state symbols**:

(g) = gas   (l) = liquid   (s) = solid   (aq) = aqueous solution (dissolved in water)

For the above reaction:

lead(II) nitrate solution is $Pb(NO_3)_2(aq)$
potassium iodide solution is $KI(aq)$
solid lead(II) iodide is $PbI_2(s)$
potassium nitrate solution is $KNO_3(aq)$

The equation now becomes:

$$Pb(NO_3)_2(aq) + KI(aq) \rightarrow PbI_2(s) + KNO_3(aq)$$

2 The next step is to 'balance' the equation. That is, we must ensure that we have the same number of atoms of each element on the right-hand side as on the left-hand side.

a Looking at the equation in step **1** above, we notice that there are two iodine atoms on the right, in PbI_2, but only one on the left, in KI. Also, there are two nitrate groups on the left, in $Pb(NO_3)_2$, but only one on the right, in KNO_3.

b We can balance the iodine atoms by having two formula units of KI on the left, that is **2**KI. (Note that we cannot change the formula to KI_2 – that would not correctly represent potassium iodide, which always contains equal numbers of potassium and iodide ions.)

c We can balance the nitrates by having two formula units of KNO_3 on the right, that is **2**KNO_3. This also balances up the potassium atoms, which, although originally the same on both sides, became unbalanced when we changed KI to **2**KI in step **b**.

The fully balanced equation is now:

$$Pb(NO_3)_2(aq) + 2KI(aq) \rightarrow PbI_2(s) + 2KNO_3(aq)$$

It is clear that we have neither lost nor gained any atoms, but that they have swapped partners – the iodine was originally combined with potassium, but has ended up being combined with lead; the nitrate groups have changed their partner from lead to potassium.

Now try this

1 Copy the following equations and balance them.
   a $H_2(g) + O_2(g) \rightarrow H_2O(l)$
   b $I_2(s) + Cl_2(g) \rightarrow ICl_3(s)$
   c $NaOH(aq) + Al(OH)_3(s) \rightarrow$
      $NaAlO_2(aq) + H_2O(l)$
   d $H_2S(g) + SO_2(g) \rightarrow S(s) + H_2O(l)$
   e $NH_3(g) + O_2(g) \rightarrow N_2(g) + H_2O(l)$
2 Write balanced symbol equations for the following reactions.
   a magnesium carbonate → magnesium oxide + carbon dioxide
   b lead + silver nitrate solution → lead nitrate solution + silver
   c sodium oxide + water → sodium hydroxide solution
   d iron(II) chloride + chlorine (Cl_2) → iron(III) chloride
   e iron(III) sulfate + sodium hydroxide → iron(III) hydroxide + sodium sulfate

Worked example

Write the balanced chemical equation for the following reaction:

zinc metal + hydrochloric acid → zinc chloride solution + hydrogen gas

Answer

Following the steps given above:

1 Zinc metal is $Zn(s)$.
   Hydrochloric acid is $HCl(aq)$.
   Zinc chloride solution is $ZnCl_2(aq)$.
   Hydrogen gas is $H_2(g)$ (hydrogen, like many non-metallic elements, exists in molecules made up of two atoms).
   The equation now becomes:
   $Zn(s) + HCl(aq) \rightarrow ZnCl_2(aq) + H_2(g)$
2 a There are two hydrogen atoms and two chlorine atoms on the right, but only one of each of these on the left.
   b We can balance both of them by just one change – having two formula units of HCl on the left.
   The fully balanced equation is now:
   $Zn(s) + 2HCl(aq) \rightarrow ZnCl_2(aq) + H_2(g)$

1.10 Using the mole in mass calculations

We are now in a position to look at how the masses of the individual substances in a chemical equation are related. As an example, take the reaction between marble chips (calcium carbonate) and hydrochloric acid:

$$CaCO_3(s) + 2HCl(aq) \rightarrow CaCl_2(aq) + H_2O(l) + CO_2(g)$$

When this reaction is carried out in an open conical flask on a top-pan balance, the mass is observed to decrease. (Note that this is not due to the destruction of matter – as was mentioned on page 11, the overall number of atoms does not change during a chemical reaction. Rather, it is due to the fact that the gaseous carbon dioxide produced escapes into the air.) We can use the knowledge gained in this topic to calculate the answer to the following question:

● By how much would the mass decrease if 50 g of marble chips were completely reacted with an excess of hydrochloric acid?

We use the following steps:

1 We can use equation (1) (page 7) to calculate the number of moles of calcium carbonate in 50 g of marble chips:

$$n = \frac{m}{M} \qquad M_r(CaCO_3) = 40.1 + 12.0 + 3 \times 16.0 = 100.1$$
$$\text{so} \quad M = 100.1\,g\,mol^{-1}$$

$$n = \frac{50\ g}{100.1\ g\ mol^{-1}}$$

$$= \mathbf{0.50\,mol\ of\ CaCO_3}$$

2 From the balanced equation above, we see that one mole of calcium carbonate produces one mole of carbon dioxide. Therefore the number of moles of carbon dioxide produced is the same as the number of moles of calcium carbonate we started with, namely **0.50 mol of carbon dioxide**.
3 Lastly, we can use a rearranged form of equation (1) to calculate what mass of carbon dioxide this corresponds to.

$$n = \frac{m}{M} \quad \text{so} \quad m = n \times M \qquad M_r(CO_2) = 12.0 + 2 \times 16.0 = 44.0$$
$$\text{so} \quad M = 44.0\,\text{g}\,\text{mol}^{-1}$$
$$\text{also} \quad n = 0.50\,\text{mol}$$
$$m = 0.50\,\text{mol} \times 44.0\,\text{g}\,\text{mol}^{-1}$$
$$= 22.0\,\text{g}$$

The loss in mass (due to the carbon dioxide being evolved) is **22.0 g**.

The three steps can be summarised as shown in Figure 1.9.

Figure 1.9 Finding the mass of a product from the mass of reactant, or vice versa

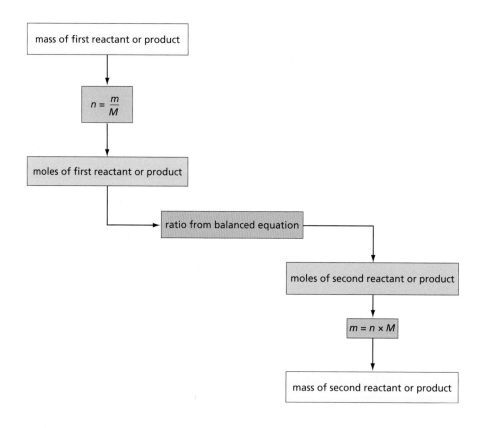

Worked example

The highly exothermic **thermit reaction** (see Figure 1.10) is used to weld together the steel rails of railway tracks. It involves the reduction of iron(III) oxide to iron by aluminium.

$$2Al(s) + Fe_2O_3(s) \rightarrow Al_2O_3(s) + 2Fe(s)$$

Figure 1.10 The thermit process is used to weld together the steel rails of railway tracks.

Use the chart in Figure 1.9 to calculate what mass of aluminium is needed to react completely with 10.0 g of iron(III) oxide.

Answer

1 $M_r(Fe_2O_3) = 2 \times 55.8 + 3 \times 16.0 = 111.6 + 48.0 = 159.6$
   so $M = 159.6\,g\,mol^{-1}$

   number of moles of iron(III) oxide $(n) = \dfrac{m}{M}$

   $= \dfrac{10.0}{159.6}$

   $= 0.0627\,mol$

2 From the balanced equation, one mole of iron(III) oxide reacts with two moles of aluminium, therefore:
   number of moles of aluminium $(n) = 0.0627 \times 2 = 0.125\,mol$

3 $A_r(Al) = 27.0$ so $M = 27.0\,g\,mol^{-1}$
   mass of aluminium $= n \times M$
   $= 0.125\,mol \times 27.0\,g\,mol^{-1}$
   $= \textbf{3.38\,g}$

Now try this

1 What mass of silver will be precipitated when 5.0 g of copper are reacted with an excess of silver nitrate solution?

   $Cu(s) + 2AgNO_3(aq)$
   $\rightarrow Cu(NO_3)_2(aq) + 2Ag(s)$

2 What mass of ammonia will be formed when 50.0 g of nitrogen are passed through the Haber process? (Assume 100% conversion.)

   $N_2(g) + 3H_2(g) \rightarrow 2NH_3(g)$

1.11 Moles of gases

The molar masses of most compounds are different. The molar volumes of most solid and liquid compounds are also different. But the molar volumes of gases (when measured at the same temperature and pressure) are all the same. This strange coincidence results from the fact that most of a gas is in fact empty space – the molecules take up less than a thousandth of its volume at normal temperatures (see section 4.13). The volume of the molecules is negligible compared with the total volume, and so any variation in their individual size will not affect the overall volume. At room temperature (25 °C, 298 K) and normal pressure (1 atm, 1.01×10^5 Pa):

the molar volume of any gas = 24.0 dm³ mol⁻¹

So we can say that:

amount of gas (in moles) $= \dfrac{\text{volume (in dm}^3)}{\text{molar volume}}$

$n = \dfrac{V}{24.0}$

or   volume of gas in dm³ = molar volume × moles of gas
$V = 24.0 \times n$

Worked example

What volume of hydrogen (measured at room temperature and pressure) will be produced when 7.0 g of iron are reacted with an excess of sulfuric acid?

Answer

The equation for the reaction is as follows:
$Fe(s) + H_2SO_4(aq) \rightarrow FeSO_4(aq) + H_2(g)$

1 $A_r(Fe) = 55.8$ so $M = 55.8\,g\,mol^{-1}$

   amount (in moles) of iron $= \dfrac{m}{M}$

   $= \dfrac{7.0\,g}{55.8\,g\,mol^{-1}}$

   $= 0.125\,mol$

2 From the balanced equation, one mole of iron produces one mole of hydrogen molecules, therefore:
   number of moles of $H_2 = \textbf{0.125\,mol}$

3 volume of H_2 (in dm³) = molar volume × moles of H_2
   $= 24.0\,dm^3\,mol^{-1} \times 0.125\,mol$
   $= \textbf{3.0\,dm}^3$

Now try this

1 What volume of carbon dioxide (measured at room temperature and pressure) will be produced when 5.0 g of calcium carbonate are decomposed by heating according to the following equation?

   $CaCO_3 \rightarrow CaO + CO_2$

2 Sulfur dioxide and hydrogen sulfide gases react according to the equation:
   $2H_2S + SO_2 \rightarrow 2H_2O + 3S$

   What volume of sulfur dioxide will be needed to react completely with 100 cm³ of hydrogen sulfide (both volumes measured at room temperature and pressure), and what mass of sulfur will be formed?

1.12 Moles and concentrations

Many chemical reactions are carried out in solution. Often it is convenient to dissolve a reactant in a solvent in advance, and to use portions as and when needed. A common example of this is the dilute sulfuric acid you find on the shelves in a laboratory. This has been made up in bulk, at a certain concentration. Solutions are most easily measured out by volume, using measuring cylinders, pipettes or burettes. Suppose we need a certain amount of sulfuric acid (that is, a certain number of moles) for a particular experiment. If we know how many moles of sulfuric acid each $1\,cm^3$ of solution contains, we can obtain the required number of moles by measuring out the correct volume. For example, if our sulfuric acid contains $0.001\,mol$ of H_2SO_4 per $1\,cm^3$, and we need $0.005\,mol$, we would measure out $5\,cm^3$ of solution.

In chemistry, the concentrations of solutions are normally stated in units of **moles per cubic decimetre** (= moles per litre). The customary abbreviation is **$mol\,dm^{-3}$**. Occasionally the older, and even shorter, abbreviation **M** is used.

If $1\,dm^3$ of a solution contains $1.0\,mol$ of solute, the solution's concentration is $1.0\,mol\,dm^{-3}$ (or $1.0\,M$, verbally described as 'a one molar solution').

If more moles are dissolved in the same volume of solution, the solution is a more concentrated one. Likewise, if the same number of moles is dissolved in a smaller volume of solution, the solution is also more concentrated. For example, we can produce a $2.0\,mol\,dm^{-3}$ solution ($2.0\,M$, 'two molar') by:

- either dissolving $2\,mol$ of solute in $1\,dm^3$ of solution
- or dissolving $1\,mol$ of solute in $0.5\,dm^3$ of solution.

$$\text{concentration} = \frac{\text{amount (in moles)}}{\text{volume of solution (in dm}^3)}$$

$$c = \frac{n}{V}$$

or   amount (in moles) = concentration × volume (in dm^3)

$$n = c \times V$$

Unlike the mass of a solid, or the volume of a gas, which are extensive properties of a substance (see page 3), the concentration of a solution is an intensive property. It does not depend on how much of the solution we have. Properties that depend on the concentration of a solution are also intensive. For example, the rate of reaction between sulfuric acid and magnesium ribbon, the colour of aqueous potassium manganate(VII), the sourness of vinegar and the density of a sugar solution do not depend on how much solution we have, but only on how much solute is dissolved in a given volume.

Figure 1.11 Colour is related to concentration, an intensive property that does not depend on how much solution there is.

Worked example 1

$200\,cm^3$ of a sugar solution contains $0.40\,mol$ of sugar. What is the concentration of the solution?

Answer
Use the first equation in the panel on page 16:

$$c = \frac{n}{V}$$

$$c = \frac{0.40\;mol}{0.20\;dm^3}$$

$$= \mathbf{2.0\,mol\,dm^{-3}}$$

Remember, $1000\,cm^3 = 1.0\,dm^3$
so $\quad 200\,cm^3 = 0.20\,dm^3$

Now try this

1 What are the concentrations of solute in the following solutions?
  a 2.0 mol of ethanol in 750 cm³ of solution
  b 5.3 g of sodium carbonate, Na₂CO₃, in 2.0 dm³ of solution
  c 40 g of ethanoic acid, C₂H₄O₂, in 800 cm³ of solution
2 How many moles of solute are in the following solutions?
  a 0.50 dm³ of a 1.5 mol dm⁻³ solution of sulfuric acid
  b 22 cm³ of a 2.0 mol dm⁻³ solution of sodium hydroxide
  c 50 cm³ of a solution containing 20 g of potassium hydrogencarbonate, KHCO₃, per dm³

Worked example 2

How many grams of salt, NaCl, need to be dissolved in $0.50\,dm^3$ of solution to make a $0.20\,mol\,dm^{-3}$ solution?

Answer
Use the second equation in the panel on page 16:

$n = c \times V$ $\qquad\qquad\qquad c = 0.20\,mol\,dm^{-3}$
$n = 0.20\,mol\,dm^{-3} \times 0.5\,dm^3$ $\qquad V = 0.50\,dm^3$
$\quad = 0.10\,mol$ of NaCl

Now use equation (1) (page 7) to convert moles into mass:

$n = \dfrac{m}{M}$ so $m = n \times M$ $\qquad M_r(NaCl) = 23.0 + 35.5 = 58.5$ so $M = 58.5\,g\,mol^{-1}$
$\qquad\qquad\qquad\qquad\qquad\qquad n = 0.10\,mol$

$m = 0.10\,mol \times 58.5\,g\,mol^{-1}$
$\quad = \mathbf{5.85\,g}$ of salt, NaCl

We shall come across the concentrations of solutions again in Topic 6, where we look at the technique of titration.

1.13 Calculations using a combination of methods

At the heart of most chemistry calculations is the balanced chemical equation. This shows us the ratios in which the reactants react to give the products, and the ratios in which the products are formed. This is called the **stoichiometry** of the reaction. Most calculations involving reactions can be broken down into a set of three steps (see Figure 1.12), similar to those described for mass calculations on page 14.

Figure 1.12 Calculations involving the stoichiometry of a reaction

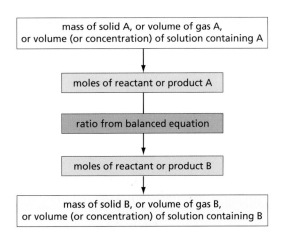

mass of solid A, or volume of gas A, or volume (or concentration) of solution containing A

moles of reactant or product A

ratio from balanced equation

moles of reactant or product B

mass of solid B, or volume of gas B, or volume (or concentration) of solution containing B

Now try this

Use Figure 1.12 and the A_r values in the data section on the CD to answer the following questions.

1 What volume of hydrogen gas will be given off when 2.3 g of sodium metal react with water?

$$2Na(s) + 2H_2O(l) \rightarrow 2NaOH(aq) + H_2(g)$$

2 The equation for the complete combustion of methane, CH_4, in oxygen is:

$$CH_4(g) + 2O_2(g) \rightarrow CO_2(g) + 2H_2O(g)$$

Calculate the volume of oxygen needed to burn 4.0 g of methane.

3 What volume of 0.50 mol dm^{-3} sulfuric acid, H_2SO_4, is needed to react exactly with 5.0 g of magnesium, and what volume of hydrogen will be evolved?

$$Mg(s) + H_2SO_4(aq) \rightarrow MgSO_4(aq) + H_2(g)$$

4 What mass of sulfur will be precipitated when an excess of hydrochloric acid is added to 100 cm^3 of 0.20 mol dm^{-3} sodium thiosulfate solution?

$$Na_2S_2O_3(aq) + 2HCl(aq) \rightarrow 2NaCl(aq) + SO_2(g) + H_2O(l) + S(s)$$

5 What would be the concentration of the hydrochloric acid produced if all the hydrogen chloride gas from the reaction between 50 g of pure sulfuric acid and an excess of sodium chloride was collected in water, and the solution made up to a volume of 400 cm^3 with water?

$$NaCl(s) + H_2SO_4(l) \rightarrow NaHSO_4(s) + HCl(g)$$

Summary

- **Atoms** and **molecules** are small and light – about 1×10^{-9} m in size, and about 1×10^{-22} g in mass.
- **Relative atomic mass** and **relative molecular mass** are defined in terms of the mass of an atom of carbon-12.
- One **mole** is the amount of substance that has the same number of particles (atoms, molecules, etc.) as there are atoms in 12.000 g of carbon-12.
- The relative molecular mass, M_r, of a compound is found by summing the relative atomic masses, A_r, of all the atoms present.
- The **empirical formula** of a compound is the simplest formula that shows the relative number of atoms of each element present in the compound.
- The **molecular formula** tells us the actual number of atoms of each element present in a molecule of the compound.
- Chemical reactions take place with no change in mass, and no change in the total number of atoms present.

- Chemical equations reflect this – when balanced, they contain the same numbers of atoms of each element on their left-hand and right-hand sides.
- The following equations allow us to calculate the number of moles present in a sample:

- amount (in moles) $= \dfrac{\text{mass}}{\text{molar mass}}$ or $n = \dfrac{m}{M}$

- amount (in moles) $= \dfrac{\text{volume of gas (in dm}^3)}{24.0}$ or $n = \dfrac{V}{24.0}$

- $\dfrac{\text{amount}}{\text{(in moles)}} = \dfrac{\text{concentration of}}{\text{solution (in mol dm}^{-3})} \times \dfrac{\text{volume}}{\text{(in dm}^3)}$

or $n = c \times V$

- $\dfrac{\text{amount}}{\text{(in moles)}} = \dfrac{\text{concentration of}}{\text{solution (in mol dm}^{-3})} \times \dfrac{\text{volume (in cm}^3)}{1000}$

or $n = c \times \dfrac{V}{1000}$

Examination practice questions

Please see the data section of the CD for any A_r values you may need.

1 Zinc is an essential trace element which is necessary for the healthy growth of animals and plants. Zinc deficiency in humans can be easily treated by using zinc salts as dietary supplements.

a One salt which is used as a dietary supplement is a hydrated zinc sulfate, $ZnSO_4.xH_2O$, which is a colourless crystalline solid. Crystals of zinc sulfate may be prepared in a school or college laboratory by reacting dilute sulfuric acid with a suitable compound of zinc. Give the formulae of **two** simple compounds of zinc that could **each** react with dilute sulfuric acid to produce zinc sulfate. [2]

b A simple experiment to determine the value of x in the formula $ZnSO_4.xH_2O$ is to heat it carefully to drive off the water.

$ZnSO_4.xH_2O(s) \rightarrow ZnSO_4(s) + xH_2O(g)$

A student placed a sample of the hydrated zinc sulfate in a weighed boiling tube and reweighed it. He then heated the tube for a short time, cooled it and reweighed it when cool. This process was repeated four times. The final results are shown below.

mass of empty tube/g	mass of tube + hydrated salt/g	mass of tube + salt after fourth heating/g
74.25	77.97	76.34

i Why was the boiling tube heated, cooled and reweighed four times?

ii Calculate the amount, **in moles**, of the anhydrous salt produced.

iii Calculate the amount, **in moles**, of water driven off by heating.

iv Use your results to **ii** and **iii** to calculate the value of x in $ZnSO_4.xH_2O$. [7]

c For many people, an intake of approximately 15 mg per day of zinc will be sufficient to prevent deficiencies. Zinc ethanoate crystals, $(CH_3CO_2)_2Zn.2H_2O$, may be used in this way.

i What mass of pure crystalline zinc ethanoate ($M_r = 219.4$) will need to be taken to obtain a dose of 15 mg of zinc?

ii If this dose is taken in solution as 5 cm^3 of aqueous zinc ethanoate, what would be the concentration of the solution used? Give your answer in mol dm^{-3}. [4]

[Cambridge International AS & A Level Chemistry 9701, Paper 21 Q1 November 2012]

2 Washing soda is hydrated sodium carbonate, $Na_2CO_3.xH_2O$.

A student wished to determine the value of x by carrying out a titration, with the following results.

5.13 g of washing soda crystals were dissolved in water and the solution was made up to 250 cm^3 in a standard volumetric flask.

25.0 cm^3 of this solution reacted exactly with 35.8 cm^3 of 0.100 mol dm^{-3} hydrochloric acid and carbon dioxide was produced.

a i Write a balanced equation for the reaction between Na_2CO_3 and HCl.

ii Calculate the amount, in moles, of HCl in the 35.8 cm^3 of solution used in the titration.

iii Use your answers to **i** and **ii** to calculate the amount, in moles, of Na_2CO_3 in the 25.0 cm^3 of solution used in the titration.

iv Use your answer to **iii** to calculate the amount, in moles, of Na_2CO_3 in the 250 cm^3 of solution in the standard volumetric flask.

v Hence calculate the mass of Na_2CO_3 present in 5.13 g of washing soda crystals. [6]

b Use your calculations in **a** to determine the value of x in $Na_2CO_3.xH_2O$. [2]

[Cambridge International AS & A Level Chemistry 9701, Paper 23 Q2 June 2012]

3 a A compound containing magnesium, silicon and oxygen is present in rock types in Italy. A sample of this compound weighing 5.27 g was found to have the following composition by mass: Mg, 1.82 g; Si, 1.05 g; O, 2.40 g. Calculate the empirical formula of the compound. Show your working. [2]

b Pharmacists sell tablets containing magnesium hydroxide, $Mg(OH)_2$, to combat indigestion. A student carried out an investigation to find the percentage by mass of $Mg(OH)_2$ in an indigestion tablet. The student reacted the tablet with dilute hydrochloric acid.

$Mg(OH)_2(s) + 2HCl(aq) \rightarrow MgCl_2(aq) + 2H_2O(l)$

The student found that 32.00 cm^3 of 0.500 mol dm^{-3} HCl was needed to react with the $Mg(OH)_2$ in a 500 mg tablet. [1 g = 1000 mg].

i Calculate the amount, in mol, of HCl used. [1]

ii Determine the amount, in mol, of $Mg(OH)_2$ present in the tablet. [1]

iii Determine the percentage by mass of $Mg(OH)_2$ present in the tablet. [3]

[OCR Chemistry A Unit F321 Q2 (part) May 2011]

2 The structure of the atom

In this topic we introduce the three sub-atomic particles – the electron, the proton and the neutron. We look at their properties, and how they are arranged inside the atom. We explain that some elements form isotopes, and describe how their relative abundances can be measured using the mass spectrometer. We outline the types of energy associated with the particles in chemistry. Against this background, we look at how the electrons are arranged around the nucleus and how this arrangement explains the positions of elements within the Periodic Table, their ionisation energies and the sizes of their atoms.

Learning outcomes

By the end of this topic you should be able to:

1.3a) analyse mass spectra in terms of isotopic abundances (part, see also Topic 29),

1.3b) calculate the relative atomic mass of an element given the relative abundances of its isotopes, or its mass spectrum

2.1a) identify and describe protons, neutrons and electrons in terms of their relative charges and relative masses

2.1b) deduce the behaviour of beams of protons, neutrons and electrons in electric fields

2.1c) describe the distribution of mass and charge within an atom

2.1d) deduce the numbers of protons, neutrons and electrons present in both atoms and ions given proton and nucleon numbers and charge

2.2a) describe the contribution of protons and neutrons to atomic nuclei in terms of proton number and nucleon number

2.2b) distinguish between isotopes on the basis of different numbers of neutrons present

2.2c) recognise and use the symbolism $^{x}_{y}A$ for isotopes, where x is the nucleon number and y is the proton number

2.3a) describe the number and relative energies of the s, p and d orbitals for the principal quantum numbers 1, 2 and 3 and also the 4s and 4p orbitals.

2.3b) describe and sketch the shapes of s and p orbitals

2.3c) state the electronic configuration of atoms and ions given the proton number and charge, using the convention $1s^2 2s^2 2p^6$, etc.

2.3d) explain and use the term ionisation energy, explain the factors influencing the ionisation energies of elements, and explain the trends in ionisation energies across a Period and down a Group of the Periodic Table (see also Topic 10)

2.3e) deduce the electronic configurations of elements from successive ionisation energy data

2.3f) interpret successive ionisation energy data of an element in terms of the position of that element within the Periodic Table.

2.1 The discovery of the sub-atomic particles

Our understanding of atoms is very much a nineteenth- and twentieth-century story. Although the Greek philosopher Democritus was the first to coin the term 'atom' (which is derived from the Greek word meaning 'cannot be cut'), in around 400 BCE, his idea of matter being composed of small particles was dismissed by Aristotle. Because Aristotle's reputation was so great, Democritus' atomic idea was ignored for centuries. It was not until 1808, when John Dalton published his Atomic Theory, that the idea of atoms as being indivisible constituents of matter was revived. Dalton suggested that all the atoms of a given element were identical to each other, but differed from the atoms of every other element. His atoms were the smallest parts of an element that could exist. They could not be broken down or destroyed, and were themselves without structure.

For most of the nineteenth century the idea of atoms being indivisible fitted in well with chemists' ideas of chemical reactions, and was readily accepted. Even today it is believed that atoms are never destroyed during a chemical reaction, but merely change their partners. However, in 1897 the physicist J.J. Thomson discovered the first **sub-atomic particle**, that is, a particle smaller than an atom. It was the **electron**.

Thomson found that the electron was much lighter than the lightest atom, and had a negative electrical charge. What is more, he found that under the conditions of his experiment, atoms of different elements produced identical electron particles. This suggested that all atoms contain at least one sub-atomic component in common.

Since atoms are electrically neutral objects, if they contain negatively charged electrons they must also contain particles with a positive charge. An important experiment carried out in 1911 by Ernest Rutherford (a New Zealander), Hans Geiger (a German) and Ernest Marsden (an Englishman) showed that the positive charge in the atom is concentrated into an incredibly small **nucleus** right in the middle of it. They estimated that the diameter of the nucleus could not be greater than 0.00001 times that of the atom itself. Eventually, Rutherford was able to chip away from this nucleus small positively charged particles. He showed that these were also identical to each other, no matter which element they came from. This positive particle is called the **proton**. It is much heavier than the electron, having nearly the mass of the hydrogen atom.

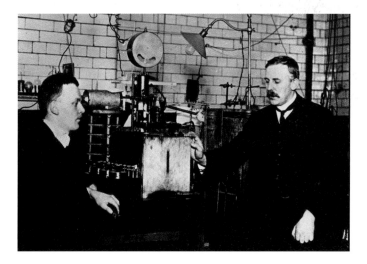

Figure 2.1 Ernest Rutherford (right) and Hans Geiger in their laboratory at Manchester University in about 1908. They are seen with the instrumentation they used to detect and count α-particles, which are the nuclei of helium atoms.

It was another 20 years before the last of the three sub-atomic particles, the **neutron**, was discovered. Although its existence was first suspected in 1919, it was not until 1932 that James Chadwick eventually pinned it down. As its name suggests, the neutron is electrically neutral, but it is relatively heavy, having about the same mass as a proton.

Scientists had therefore to change the earlier picture of the atom. In a sense the picture had become more complicated, showing that atoms had an internal structure, and were made up of other, smaller particles. But looked at in another way it had become simpler – the 90 or so different types of atoms that are needed to make up the various elements had been replaced by just three sub-atomic particles. It turns out that these, in different amounts, make up the atoms of all the different elements.

Figure 2.2 The picture of the atom in the early twentieth century

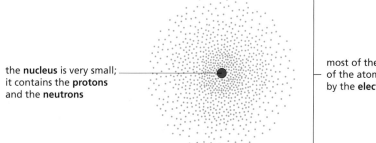

the **nucleus** is very small; it contains the **protons** and the **neutrons**

most of the volume of the atom is occupied by the **electrons**

2.2 The properties of the three sub-atomic particles

Table 2.1 lists some of the properties of the three sub-atomic particles.

Table 2.1 The properties of the sub-atomic particles. Note that the masses in the last row are given relative to $\frac{1}{12}$ the mass of an atom of carbon-12. These masses are often quoted relative to the mass of the proton instead, when the relative mass of the electron is $\frac{1}{1836}$, and the relative masses of the proton and neutron are both 1.

Property	Electron	Proton	Neutron
electrical charge/coulombs	-1.6×10^{-19}	$+1.6 \times 10^{-19}$	0
charge (relative to that of the proton)	-1	$+1$	0
mass/g	9.11×10^{-28}	1.673×10^{-24}	1.675×10^{-24}
mass/amu (see section 1.4)	5.485×10^{-4}	1.007	1.009

Because of their relative masses and charges, the three particles behave differently in an electric field, as shown in Figure 2.3. Neutrons are undeflected, being electrically neutral. Protons are attracted towards the negative pole, and electrons towards the positive pole. If their initial velocities are the same, electrons are deflected to a greater extent than protons because they are much lighter.

The picture of the atom assembled from these observations is as follows.

- Atoms are small, spherical structures with diameters ranging from 1×10^{-10} m to 3×10^{-10} m.
- The particles that contribute to the atom's mass (protons and neutrons) are contained within a very small central nucleus that has a diameter of about 1×10^{-15} m.
- The electrons occupy the region around the nucleus. They are to be found in the space inside the atom but outside the nucleus, which is almost the whole of the atom.
- All the atoms of a particular element contain the same number of protons. This also equals the number of electrons within those atoms.
- The atoms of all elements except hydrogen also contain neutrons. These are in the nucleus along with the protons. Almost the only effect they have on the properties of the atom is to increase its mass.

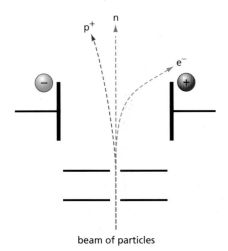

Figure 2.3 The behaviour of protons, neutrons and electrons in an electric field

2.3 Isotopes

At the same time as Rutherford and his team were finding out about the structure of the nucleus, it was discovered that some elements contained atoms that have different masses, but identical chemical properties.

These atoms were given the name **isotopes**, since they occupy the same (*iso*) place (*topos*) in the Periodic Table. The first isotopes to be discovered were those of the unstable radioactive element thorium. (Thorium is element number 90 in the Periodic Table.) In 1913, however, Thomson was able to show that a sample of neon obtained from liquid air contained atoms with a relative atomic mass of 22 as well as those with the usual relative atomic mass of 20. These heavier neon atoms were stable, unlike the thorium isotopes. Many other elements contain isotopes, some of which are listed in Table 2.2.

Table 2.2 The relative abundance of some isotopes

Isotope	Mass relative to hydrogen	Relative abundance
boron-10	10.0	20%
boron-11	11.0	80%
neon-20	20.0	91%
neon-22	22.0	9%
magnesium-24	24.0	79%
magnesium-25	25.0	10%
magnesium-26	26.0	11%

The **isotopes** of an element differ in their composition in only one respect – although they all contain the same numbers of electrons and protons, they have different numbers of neutrons.

Most of the naturally occurring isotopes are stable, but some, like those of uranium and also many artificially made ones, are unstable and emit radiation. These are called **radioactive** isotopes.

2.4 Extending atomic symbols to include isotopes

For most chemical purposes, the atomic symbols introduced in Topic 1 are adequate. If, however, we wish to refer to a particular isotope of an element, we need to specify its **mass number**, which is the number of protons and neutrons in the nucleus. We write this as a superscript before the atomic symbol. We often add the proton number as a subscript before the symbol. So for carbon:

- $^{12}_{6}C$ is the symbol for carbon-12, which is the most common isotope, containing 6 protons and 6 neutrons, with a mass number of 12
- $^{14}_{6}C$ is the symbol for carbon-14, which contains 6 protons and 8 neutrons, with a mass number of 14.

> The **proton number (atomic number)** of an atom is the number of protons in its nucleus. The **mass number** of an atom is the sum of the numbers of protons and neutrons.

Worked example

How many protons and neutrons are there in each of the following atoms?

a $^{18}_{8}O$ b $^{235}_{92}U$

Answer
The subscript gives the proton number, that is, the number of protons. So for oxygen, number of protons = **8**, and for uranium, number of protons = **92**.

Subtracting the proton number from the mass number gives the number of neutrons. So for oxygen, number of neutrons = 18 – 8 = **10**, and for uranium, number of neutrons = 235 – 92 = **143**.

Now try this

How many protons, electrons and neutrons are there in each of the following atoms?

1 $^{23}_{11}Na$

2 $^{127}_{53}I$

2.5 The mass spectrometer

The masses and the relative abundances of individual isotopes are easily measured in a **mass spectrometer**. This is a machine in which atoms that have been ionised by the loss of electrons are accelerated to a high velocity, and their trajectories (paths) are then deflected from a straight line by passing them through a magnetic field. A magnetic field has a similar bending effect on moving charged particles as the electric field in Figure 2.3 (see page 22).

Figure 2.4 Schematic diagram of a mass spectrometer

Figure 2.5 A modern mass spectrometer

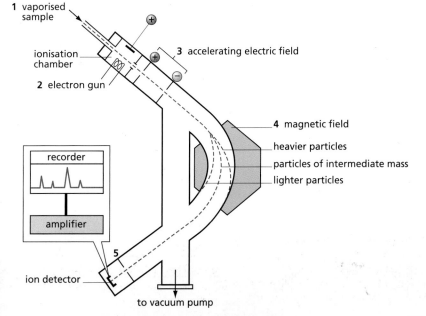

1 vaporised sample

ionisation chamber

2 electron gun

3 accelerating electric field

4 magnetic field
heavier particles
particles of intermediate mass
lighter particles

recorder

amplifier

5

ion detector

to vacuum pump

Five processes occur in a mass spectrometer.

1 If it is not already a gas, the element is vaporised in an oven.

2 Electrons are fired at the gaseous atoms. These knock off other electrons from the atoms:

$$M(g) + e^- \rightarrow M^+(g) + 2e^-$$

3 The gaseous ions are accelerated by passing them through an electric field (at a voltage of 5–10 kV).

4 The fast-moving ions are deflected by an electromagnet. The larger the charge on the ion, the larger is the deflection. On the other hand, the heavier the ion, the smaller is the deflection. Overall, the deflections are proportional to the ions' charge-to-mass ratios. If all ions have a +1 charge (which is usually the case), the extents of deflection will be inversely proportional to their masses.

5 The deflected ions pass through a narrow slit and are collected on a metallic plate connected to an amplifier. For a given strength of magnetic field, only ions of a certain mass pass through the slit and hit the collector plate. As the (positive) ions hit the plate, they cause a current to flow through the amplifier. The more ions there are, the larger the current.

The ions may travel a metre or so through the spectrometer. In order for them to do this without hitting too many air molecules (which would deflect them from their course), the inside of the spectrometer is evacuated to a very low pressure. When the situation is such that a steady stream of ions is being produced, the current through the electromagnet is changed at a steady rate. This causes the magnetic field to change in strength, and hence allows ions of different masses to pass successively through the slit. A **mass spectrum** is produced, which plots ion current against electromagnetic current. This is equivalent to relative abundance against mass number. Figure 2.6 shows the mass spectrum of krypton.

A mass spectrum enables us to analyse the proportions of the various isotopes in an element. However, by far the main use of the mass spectrometer nowadays is in analysing the formulae and structures of organic and inorganic molecules. We shall return to this application in Topic 29.

Figure 2.6 Mass spectrum of krypton

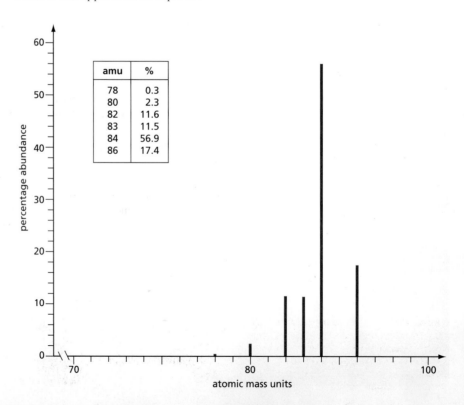

amu	%
78	0.3
80	2.3
82	11.6
83	11.5
84	56.9
86	17.4

Chlorine consists of two isotopes, with mass numbers 35 and 37, and with relative abundances 76% and 24% respectively. Calculate the average relative atomic mass of chlorine.

Answer
The percentages tell us that if we took 100 chlorine atoms at random, 76 of them would have a mass of 35 units, and 24 of them would have a mass of 37 units.

$$\text{total mass of the 100 random atoms} = (35 \times 76) + (37 \times 24)$$
$$= 3548 \text{ amu}$$

so   average mass of one atom $= \dfrac{3548}{100}$

$$= 35.5 \text{ amu}$$

that is, $A_r = \mathbf{35.5}$

1 Chromium has four stable isotopes, with mass numbers 50, 52, 53 and 54, and relative abundances 4.3%, 83.8%, 9.5% and 2.4% respectively. Calculate the average relative atomic mass of chromium.
2 Use the figures in Table 2.2 (page 22) to calculate the average relative atomic masses of boron, neon and magnesium to 1 decimal place.
3 (Harder) Iridium has two isotopes, with mass numbers 191 and 193, and its average relative atomic mass is 192.23. Calculate the relative abundances of the two isotopes.

Calculate the average relative atomic mass of krypton from the table in Figure 2.6.

Answer
We can extend the 100-random-atom idea from Worked example 1 to include fractions of atoms. Thus the average mass of one atom of krypton

$$= \frac{(78 \times 0.3) + (80 \times 2.3) + (82 \times 11.6) + (83 \times 11.5) + (84 \times 56.9) + (86 \times 17.4)}{100}$$
$$= \mathbf{83.9}$$

2.6 Chemical energy

The concept of energy is central to our understanding of how changes come about in the physical world. Our study of chemical reactions depends on energy concepts.

Chemical energy is made up of two components – **kinetic energy**, which is a measure of the motion of atoms, molecules and ions in a chemical substance, and **potential energy**, which is a measure of how strongly these particles attract one another.

Kinetic energy

Kinetic energy increases as the temperature increases. Chemists use a scale of temperature called the **absolute temperature scale**, and on this scale the kinetic energy is directly proportional to the temperature (see section 4.13).

Kinetic energy can be of three different types. The simplest is energy due to **translation**, that is, movement from place to place. For monatomic gases (gases made up of single atoms, for example helium and the other noble gases), all the kinetic energy is in the form of translational kinetic energy. For molecules containing two or more atoms, however, there is the possibility of **vibration** and **rotation** as well (see Figure 2.7). Both these forms of energy involve the movement of atoms, even though the molecule as a whole may stay still. In diatomic gases, the principal form of kinetic energy is translation, but in more complex molecules, such as ethane, vibration and rotation become the more important factors.

Figure 2.7 The two atoms in a diatomic gas molecule, such as nitrogen, behave as though they are joined by a spring, which lets them vibrate in and out. The two atoms can also rotate about the centre of the bond.

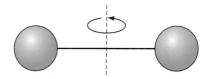

In solids, the particles are fixed in position, and the only form of kinetic energy is vibration. In liquids, the particles can move from place to place, though more slowly than in gases, and so liquid particles have translational, rotational and vibrational kinetic energy.

Potential energy

In studying chemical energy, we are usually more interested in the potential energy of the system than in its kinetic energy. This is because the potential energy gives us important quantitative information about the strengths of chemical bonds. At normal temperatures, potential energy is much larger than kinetic energy.

Potential energy arises because atoms, ions and molecules attract and repel one another. These attractions and repulsions follow from the basic principle of electrostatics, that unlike charges attract and like charges repel. Ionic compounds contain particles with clear positive and negative charges on them. Two positively charged ions repel each other, as do two negatively charged ions. A positive ion and a negative ion attract each other. Atoms and molecules, which have no overall charge on them, also attract one another (see section 3.17).

It requires energy to pull apart a sodium ion, Na^+, from a chloride ion, Cl^-. The potential energy of the system increases because we need to apply a force F (equal to the force of attraction between the two ions) for a distance d (see Figure 2.8). In a similar way, we increase the potential energy of a book if we pick the book up off the floor and put it on a desk (see Figure 2.9).

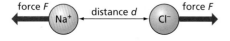

Figure 2.8 To separate oppositely charged ions through a distance d, a force F is required.

Figure 2.9 Energy is needed to lift a book from the floor and put it on the desk.

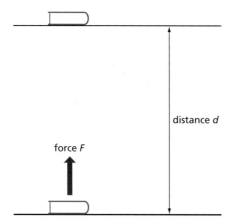

In contrast, if we start with a sodium ion and a chloride ion separated from each other and then bring them together, the potential energy decreases. We also decrease the potential energy of a book if we allow it to fall from the desk to the floor.

As we shall see in Topic 5, chemists are usually interested in the *change* in chemical energy that occurs during a reaction. This change is represented by the symbols ΔE – the Greek letter delta, Δ, being used to mean 'change'. If we look at energy changes that occur at constant pressure, which is normally the case in the laboratory, we use the symbol ΔH, which represents the **enthalpy change** of a reaction. The most commonly used unit of energy in chemistry is the **kJ mol^{-1}** (**kilojoule per mole**), where $1\,kJ = 1000\,J$.

2.7 The arrangement of the electrons – energy levels and orbitals

On page 21 we concluded that the atom is made up of a very dense, very small nucleus containing the protons and neutrons, and a much larger region of space around the nucleus that contains the electrons. We now turn our attention to these electrons. We shall see that they are not distributed randomly in this region of space. They occupy specific volume regions, called **orbitals**, which have specific energies associated with them.

Figure 2.10 An atomic absorption spectrometer

Energy levels and emission spectra

When gaseous atoms are given energy, either by heating them up to several hundred degrees, or by passing an electric current through them, the electrons become **excited** and move from lower energy levels to higher levels. Electrons in gaseous atoms can also move from lower to higher energy levels by absorbing specific frequencies of light. This leaves dark absorption lines in the spectrum of light transmitted through the gas. It was these absorption lines that provided the first evidence for a new element discovered in the outer regions of the Sun. It was named helium, after the Greek word for the Sun, *helios*. The technique of *atomic absorption spectroscopy* is widely used today to measure accurately, for example, the concentrations of calcium and sodium in a blood sample, or the elements contained in a sample of a steel alloy (see Figure 2.10).

Eventually, the excited electrons lose energy again by falling down to lower energy levels. During this process they radiate visible or ultraviolet light in specific amounts as 'packets' called **photons**. We can analyse the radiated light with a device called a spectroscope, shown in Figure 2.11. The **emission spectrum** shows that only a very few specific frequencies are emitted, and these are unique to an individual element. All the atoms of a particular element radiate at the same set of frequencies, which are usually different from those of all other elements (see Figure 2.12). This is the process that is responsible for the flame colours of the elements of Groups 1 and 2 (see Topics 10 and 19). The observation of line spectra, implying that photons of only specific energies are emitted or absorbed by atoms, constitutes strong evidence for the existence of discrete electronic energy levels within atoms.

Figure 2.11 Outline diagram showing how a spectroscope works

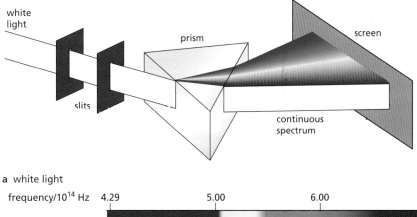

Figure 2.12 a White light spectrum
b Sodium emission spectrum
c Cadmium emission spectrum

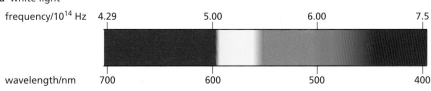

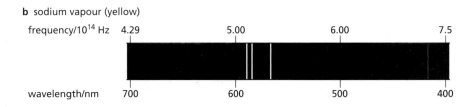

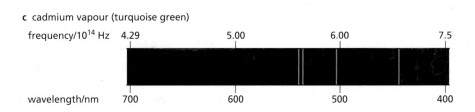

The energy E of a photon of light is related to its frequency, f, by Planck's equation:

$$E = hf$$

where h is the Planck constant.

If photons of a particular frequency are being emitted by an atom, this means that the atom is losing a particular amount of energy. This energy represents the difference between two states of the atom, one more energetic than the other (see Figure 2.13).

Figure 2.13 A photon is emitted as the atom moves between state E_1 and state E_2.

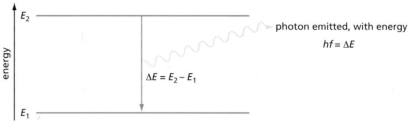

The spectrum of the simplest atom, hydrogen, shows a series of lines at different frequencies (see Figure 2.14). This suggests that the hydrogen atom can lose different amounts of energy, which in turn suggests that it can exist in a range of energy states. Transitions between the various energy states cause photons to be emitted at various frequencies. These energy states can be identified with situations where the single hydrogen electron is in certain orbitals, at specific distances from the nucleus, as we shall see.

Figure 2.14 Part of the hydrogen spectrum, showing the corresponding energy transitions

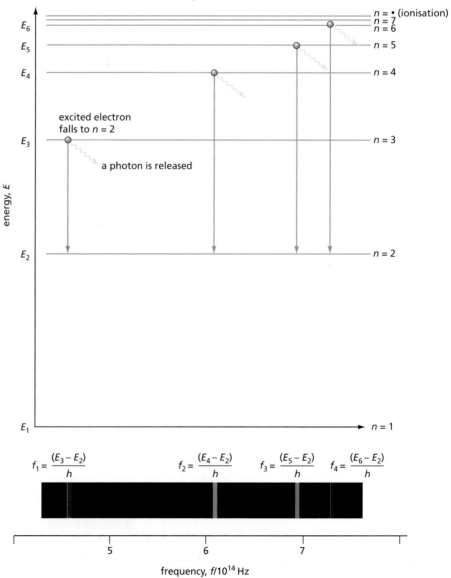

Quantum theory

The fact that only certain frequencies of light are emitted, rather than a continuous spectrum, is compelling evidence that the energy of the hydrogen atom can take only certain values, not a continuous range of values. This is the basic notion of the **quantum theory**. We say that the energy of the hydrogen atom is **quantised** (rather than continuous). It can lose (or gain) energy only by losing (or gaining) a **quantum** of energy.

A good analogy is a staircase. When you climb a staircase, you increase your height by certain fixed values (the height of each step). You can be four steps from the bottom, or five steps, but not four and a half steps up. By contrast, if you were walking up a ramp, you could choose to be at any height you liked from the bottom. It turns out that the energy of all objects is distributed in staircases rather than in ramps, but if the object is large enough, the height of each step is vanishingly small. The energy values then seem to be almost continuous, rather than stepped. It is only when we look at very small objects like atoms and molecules that the height of each step becomes significant.

The size of an energy quantum (the height of each step) is not fixed, however. It depends on the type of energy we are considering. We shall return to this point in Topic 29 when we study spectroscopy. We can use the methods of quantum mechanics to calculate not only the energies but also the probability distributions of orbitals (see Figures 2.16 and 2.18, page 30).

Energy levels in the atom

Hydrogen is a very small spherical atom with only two particles – a proton and an electron. Apart from energy of movement (translational kinetic energy), the only energy it can have is that associated with the electrostatic attraction between its two particles (see page 26). The different energy levels are therefore due to different electrostatic potential energy states, where the electron is at different distances from the proton (see Figure 2.15).

Figure 2.15 Energy against electron–proton distance for the hydrogen atom. The red curve represents how the potential energy of the hydrogen atom would vary with electron–proton distance if the energy could take on any value, as predicted by simple electrostatic theory. But, as we have seen above, the potential energy is quantised, and can only take on certain values, shown in blue. Therefore the electron can only be at certain (average) distances from the nucleus. These are the electronic orbitals, and are given the symbols 1s, 2s, etc. Note that the energy of the proton–electron system is usually defined as being zero when the two particles are an infinite distance apart. As soon as they start getting closer together, they attract each other and this causes the potential energy of the system to decrease. This is why the energy values on the y-axis are all negative.

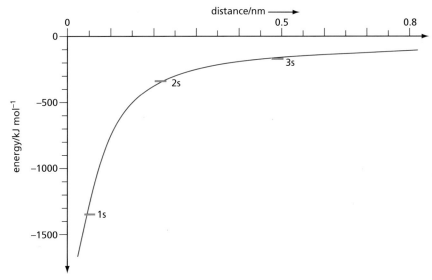

These energy levels, associated with different distances of the electron from the nucleus, are called **orbitals**, by analogy with the orbits of the planets at different distances from the Sun.

The orbitals are arranged in **shells**. Each shell contains orbitals of roughly the same energy. The shell with the lowest energy (the one closest to the nucleus) contains only one orbital. Shells with higher energies, further out from the nucleus, contain increasingly large numbers of orbitals, according to the formula:

number of orbitals in nth shell = n^2

Table 2.3 shows the number of orbitals in each shell, according to this equation.

Table 2.3 The number of orbitals in each shell

Principal shell number	Number of orbitals
1	1
2	4
3	9
4	16
5	25
6	36

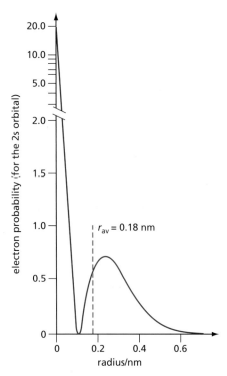

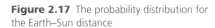

Figure 2.16 The electron probability distribution for a 2s orbital

Figure 2.17 The probability distribution for the Earth–Sun distance

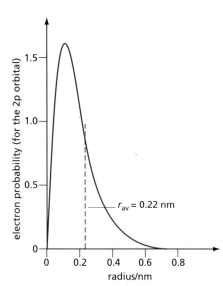

Figure 2.18 The electron probability distribution for a 2p orbital

2.8 Subshells and the shapes of orbitals

Electron probability and distance from the nucleus

Unlike a planet in a circular orbit, whose distance from the Sun does not change, an electron in an orbital does not remain at a fixed distance from the nucleus. Although we can calculate, and in some cases measure, the average electron–nucleus distance, if we were to take an instantaneous snapshot of the atom, we would be quite likely to find the electron either further away from or closer to the nucleus than this average distance. The graph of probability of finding the electron against its distance from the nucleus for a typical orbital is shown in Figure 2.16, and by contrast a similar one for the Earth–Sun distance is shown in Figure 2.17. Notice that, because of the gradual falling-off of the electron probability, there is a finite (but very small) chance of finding an electron a very long way from the centre of the atom.

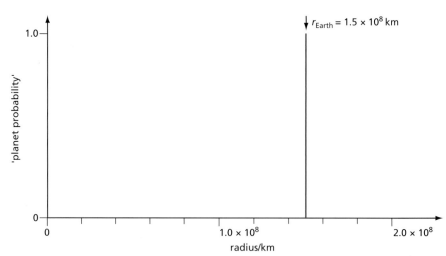

The first shell – s orbital

The single orbital in the first shell is spherically symmetrical. This means that the probability of finding the electron at a given distance from the nucleus is the same no matter what direction from the nucleus is chosen. This orbital is called the **1s orbital**.

The second shell – s and p orbitals

Of the four orbitals in the second shell, one is spherically symmetrical, like the orbital in the first shell. This is called the **2s orbital**. Its probability curve is shown in Figure 2.16. The other three second-shell orbitals point along the three mutually perpendicular x-, y- and z-axes. These are the **2p orbitals**. They are called, respectively, the $2p_x$, $2p_y$ and $2p_z$ orbitals. They do not overlap with one another. For example, an electron in the $2p_x$ orbital has a high probability of being found on or near to the x-axis, but a zero probability of being found on either the y-axis or the z-axis. These two different types of orbitals – the 2s and the 2p – are of slightly different energies. They make up the two **subshells** in the second shell of orbitals.

When an electron is located in an s orbital, there is a fair chance of finding it right at the centre of the atom, at the nucleus. But the distribution curve for the 2p orbital in Figure 2.18 shows that there is a zero probability of finding a p electron in the centre of the atom. This is general for all p orbitals. Electrons in these orbitals tend to occupy the outer reaches of atoms.

In three dimensions, an s orbital can be likened to a soft sponge ball, whereas a p orbital is like a long spongy solid cylinder, constricted around its centre to form two lobes, as shown in Figure 2.19.

a the 2s orbital

b the three
2p orbitals

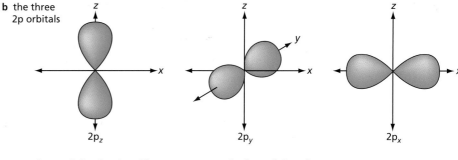

Figure 2.19 The shapes of **a** the 2s orbital and **b** the three 2p orbitals

The third shell – s, p and d orbitals

From the n^2 formula on page 29 we can predict that there will be nine orbitals in the third shell. One of these, the 3s orbital, is spherically symmetrical, just like the 1s and 2s orbitals. Three **3p orbitals**, the $3p_x$, $3p_y$ and $3p_z$ orbitals, each have two lobes, pointing along the axes in a similar fashion to the 2p orbitals. The other five have a different shape. The most common interpretation of the mathematical equations that describe their shape suggests that four of these orbitals each consist of four lobes in the same plane as one another, and pointing mutually at right angles, whereas the fifth is best represented as a two-lobed orbital surrounded by a 'doughnut' of electron density around its middle. They are called the **3d orbitals**, illustrated in Figure 2.20. So the third shell consists of three subshells – the 3s, the 3p and the 3d subshells.

Figure 2.20 The shapes of the five 3d orbitals. Four of them have the same shape, but in different orientations. The fifth has a different shape, as shown in **e**.

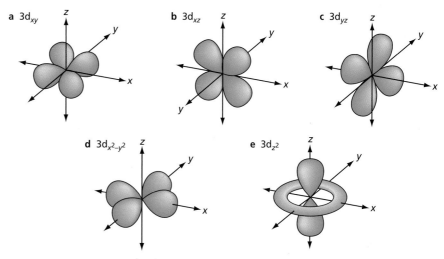

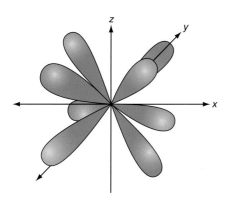

Figure 2.21 The shape of one 4f orbital. There are eight lobes pointing out from the centre.

The fourth shell – s, p, d and f orbitals

The process can be continued. In the fourth shell, we predict that there will be $4^2 = 16$ orbitals. One of these will be the 4s orbital, three will be the 4p orbitals, five will be the 4d orbitals, leaving seven orbitals of a new type. They are called the **4f orbitals**. Each consists of many lobes pointing away from one another, as shown in Figure 2.21.

Shells and orbitals in summary

The number and type of orbitals within the different shells are summarised in Table 2.4.

Table 2.4 The number and type of orbitals in each shell

Shell number, n	Number of orbitals within the shell, n^2	Number of orbitals of each type (that is, number of orbitals in each subshell)			
		s	p	d	f
1	1	1			
2	4	1	3		
3	9	1	3	5	
4	16	1	3	5	7

Now try this

How many orbitals will be in:
a the 5p subshell
b the 5f subshell?

Worked example

What is the total number of orbitals in the fifth shell? How many of these are d orbitals?

Answer

Total number of orbitals = 5^2 = 25. Of these, **5** are d orbitals. (There are five d orbitals in every shell above the third shell.)

2.9 Putting electrons into the orbitals

The atoms of different elements contain different numbers of electrons. A hydrogen atom contains just one electron. An atom of uranium contains 92 electrons. How these electrons are arranged in the various orbitals is called the atom's **electronic configuration**. This is a key feature in determining the chemical reactions of an element.

The electrons around the nucleus of an atom are most likely to be found in the situation of lowest possible energy. That is, they will occupy orbitals as close to the nucleus as possible. The single electron in hydrogen will therefore be in the 1s orbital. The electronic configuration of hydrogen is written as $1s^1$ (and spoken as 'one ess one'). The next element, helium, has two electrons, so its electronic configuration is $1s^2$ ('one ess two').

For reasons that we shall look at later, we find that each orbital, no matter what shell or subshell it is in, cannot accommodate more than two electrons. The third electron in lithium, therefore, has to occupy the orbital of next lowest energy, the 2s. The electronic configuration of lithium is $1s^2 2s^1$. Beryllium ($1s^2 2s^2$) and boron ($1s^2 2s^2 2p^1$) follow predictably.

When we come to carbon ($1s^2 2s^2 2p^2$), we need to differentiate between the three 2p orbitals. Because electrons are all negatively charged, they repel one another electrostatically. The three 2p orbitals are all of the same energy. Therefore we would expect the two p electrons in carbon to occupy different 2p orbitals (for example, $2p_x$ and $2p_y$), as far away from one another as possible. This is in fact what happens. Likewise, the seven electrons in the nitrogen atom arrange themselves $1s^2 2s^2 2p_x^1 2p_y^1 2p_z^1$. It is only when we arrive at oxygen ($1s^2 2s^2 2p_x^2 2p_y^1 2p_z^1$) that the 2p orbitals start to become doubly occupied. For many purposes, however, there is no need to distinguish between the three 2p orbitals, and we can abbreviate the electronic configuration of oxygen to $1s^2 2s^2 2p^4$.

This process continues until element number 18, argon. With argon, the 3p subshell is filled, and we would expect that the next electron should start to occupy the 3d subshell. But here the expectation is not what happens. Instead, the nineteenth electron in the next element, potassium, occupies the 4s subshell. We shall now explain why this is the case.

2.10 Shielding by inner shells

In a single-electron atom like hydrogen, the potential energy is due entirely to the single electrostatic attraction between the electron and the proton. When we move to the two-electron atom helium, two changes have occurred.

1 The nucleus now contains two protons, and has a charge of +2. It will therefore attract the electrons more, and reduce the potential energy of the system, that is, make it more stable.

2 However, the second electron in the atom will repel the first (and vice versa). This makes the decrease in potential energy described in **1** above less than it would otherwise have been.

This has a clear effect on the ionisation energies of the two elements.

The **ionisation energy** of an atom (or ion) is defined as the energy required to remove completely a mole of electrons from a mole of gaseous atoms (or ions). That is, the ionisation energy is the energy change for the following process:

$$X(g) \rightarrow X^+(g) + e^-$$

For hydrogen and helium the ionisation energies are:

$$H(g) \rightarrow H^+(g) + e^- \qquad \Delta H = 1312\,kJ\,mol^{-1}$$
$$He(g) \rightarrow He^+(g) + e^- \qquad \Delta H = 2372\,kJ\,mol^{-1}$$

Note that ΔH is the symbol chemists use for enthalpy change. The enthalpy change of a process is the energy change that occurs when the process is carried out at constant pressure (see Topic 5).

It takes more energy to remove an electron from a helium atom than from a hydrogen atom. This shows that, compared with the energy when the electron is at infinity, the energy of the 1s orbital has become more negative in helium (see Figure 2.22) than in hydrogen (see Figure 2.15, page 29).

Figure 2.22 The energy needed to remove an electron from a hydrogen atom and a helium atom

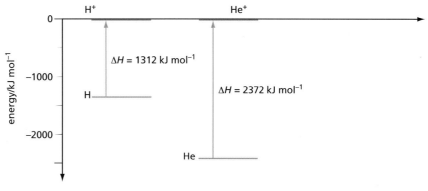

This decrease in energy for the 1s orbital continues as the proton number increases (see Figure 2.23). The decrease in energy is also true for other orbitals. The reason is that as the number of protons in the nucleus increases, the electrons in a particular orbital are attracted to it more. The decrease in energy is not regular, however, and is not the same for all orbitals. This is due to two factors.

Figure 2.23 Variation of the energy of the 1s orbital with proton number

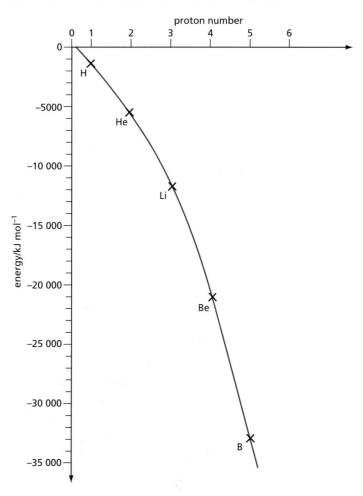

Figure 2.24 The lithium nucleus and inner shell

Figure 2.25 Graph of orbital energy against proton number for shell numbers 1–5 (see Table 2.4, page 31)

- The average distance of a p orbital from the nucleus is larger than that of the corresponding s orbital in the same shell (see Figures 2.16 and 2.18, page 30). Electrons in p orbitals therefore experience less of the stabilising effect of increasing nuclear charge than electrons in s orbitals. For a similar reason, electrons in d orbitals experience even less of the increasing nuclear charge than electrons in p orbitals.
- All electrons in outer shells are to some extent shielded from the nuclear charge by the electrons in inner shells. This shielding has the effect of decreasing the effective nuclear charge. In the case of lithium, for example, the outermost electron in the 2s orbital does not experience the full nuclear charge of +3 (see Figure 2.24). The two electrons in the filled 1s orbital mask a good deal of this charge. Overall, the **effective nuclear charge** experienced by the 2s electron in lithium is calculated to be +1.3, which is considerably less than the actual nuclear charge of +3.

A plot of orbital energy against proton number for several orbitals is shown in Figure 2.25.

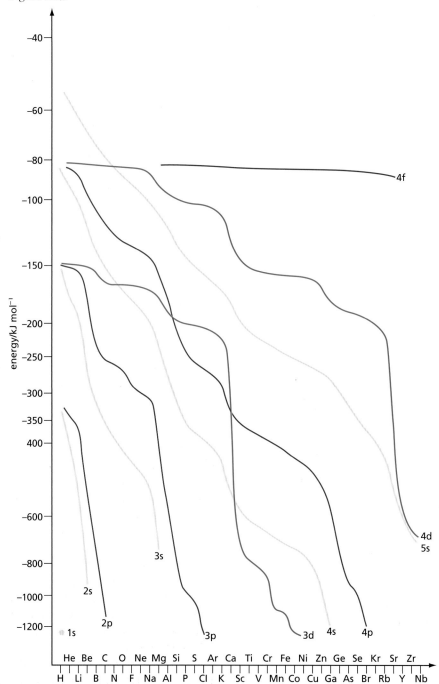

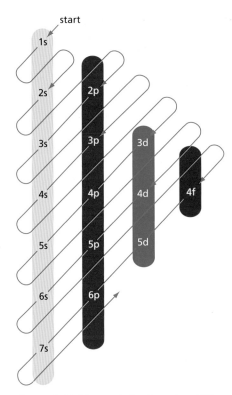

Figure 2.26 Mnemonic for the order of filling orbitals

It can be seen that, for each shell, the electrons in the s orbital decrease their energy faster than electrons in the p orbitals. Electrons in the p orbitals in turn decrease their energy faster than electrons in the d orbitals, and so on. In particular, electrons in the 4s and 4p orbitals decrease their energy faster than electrons in the 3d orbitals. So much so, that by the time 18 electrons have been added (it so happens), the next most stable orbital is the 4s rather than the 3d. The orbital filling continues after potassium in the order 4s–3d–4p–5s–4d–5p–6s–4f–5d–6p–7s, reflecting the 'lagging behind' of the d and f orbitals. This is due to the effective shielding, by filled inner electron shells, of these orbitals from the increasing nuclear charge.

The simple mnemonic diagram in Figure 2.26 will help you to remember the order in which the orbitals are filled.

2.11 Electron spin and the Pauli principle

Electrons are all identical. The only way of distinguishing them is by describing how their energies and spatial distributions differ. Thus an electron in a 1s orbital is different from an electron in a 2s orbital because it occupies a different region of space closer to the nucleus, causing it to have less potential energy. An electron in a $2p_x$ orbital differs from an electron in a $2p_y$ orbital because although they have exactly the same potential energy, they occupy different regions of space.

Two electrons with the same energy and occupying the same orbital must be distinguishable in some way, or else they would, in fact, be one and the same particle.

Experiments by Otto Stern and Walther Gerlach, in Germany in the 1920s, showed that an electron has a magnetic dipole moment. A spinning electrically charged sphere is predicted to produce a magnetic dipole – it acts like a tiny magnet, with a north and a south pole. Therefore the most common explanation for the results of the Stern–Gerlach experiment is that the electron is spinning on its axis, and the direction of spin can be either clockwise (let us say) or anticlockwise. These two directions of spin produce magnetic moments in opposite directions, often described as 'up' (given the symbol ↑) and 'down' (given the symbol ↓).

We could therefore distinguish between two electrons in exactly the same orbital if they had different directions of spin. All electrons spin at the same rate, and there are only two possible spin directions. Therefore there are only two possible ways of describing electrons in the same orbital (for example, 1s↑ and 1s↓). So there can only be two electrons in each orbital, and they must have opposite directions of spin. A third electron would need to have the same spin direction as one of the two already there, which would make it indistinguishable from the similarly spinning one.

The situation is neatly summarised by the **Pauli exclusion principle**, which states that:

No more than two electrons can occupy the same orbital, and if two electrons are in the same orbital, they must have opposite spins.

2.12 Filling the orbitals

We are now in a position to formulate the rules to use in order to predict the electronic configuration of the atoms of the elements, and also the ions derived from them. Collecting together the conclusions of sections 2.7–2.11, we arrive at the following.

1 Work out, from the element's proton number (and the charge, if an ion is being considered), the total number of electrons to be accommodated in the orbitals.

2 Taking the orbitals in order of their energies (see Figure 2.26), fill them from the lowest energy (1s) upwards, making sure that no orbital contains more than two electrons.

3 For subshells that contain more than one orbital with the same energy (the p, d and f subshells), place the electrons into different orbitals, until all are singly occupied. Only then should further electrons in that subshell start doubly occupying orbitals. For example, place one electron into each of $2p_x$, $2p_y$ and $2p_z$ before putting two electrons into any p orbital.

4 Two electrons sharing the same orbital must have opposite spins.

This procedure is known as the **Aufbau** or 'building-up' principle.

The electronic configurations of atoms and ions can be represented in a variety of ways. These are illustrated here by using sulfur (proton number 16) as an example. One way is as an 'electrons-in-boxes' diagram, as shown in Figure 2.27.

Figure 2.27 Electrons-in-boxes diagram showing the electronic configuration of sulfur. Note that the boxes can also be, and usually are, arranged in one row, with no 'steps' to indicate energy levels (as in Figure 2.28 below).

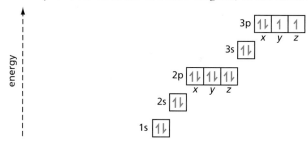

An alternative is the long linear form, specifying individual p orbitals:

$$1s^2 2s^2 2p_x^2 2p_y^2 2p_z^2 3s^2 3p_x^2 3p_y^1 3p_z^1$$

Below is the shortened linear form, which is the most usual representation:

$$1s^2 2s^2 2p^6 3s^2 3p^4$$

An even shorter form is:

$$[Ne]3s^2 3p^4$$

where [Ne] represents the filled shells in the neon atom, $1s^2 2s^2 2p^6$.

Sometimes, if we are only concerned with which shells are filled, rather than which sub-shells, the electron configuration of the sulfur atom can be described as 2.8.6, meaning 2 electrons in the first shell; 8 electrons in the second shell; and 6 electrons in the third shell.

Worked example

Write out:

a the 'electrons-in-boxes' representation of the silicon atom
b the shortened linear form of the electronic configuration of the magnesium atom
c the long linear form of the electronic configuration of the fluoride ion.

Answer

a Silicon has proton number 14, so there are 14 electrons to accommodate, as shown in Figure 2.28. Note that the last two electrons go into the $3p_x$ and $3p_y$ orbitals, with unpaired spins.

Figure 2.28

b Magnesium has proton number 12. The 12 electrons doubly occupy the six orbitals lowest in energy:

$$1s^2 2s^2 2p^6 3s^2$$

c Fluorine has proton number 9. The F⁻ ion will therefore have (9 + 1) = 10 electrons:

$$1s^2 2s^2 2p_x^2 2p_y^2 2p_z^2$$

Now try this

Write out the long linear form of the electronic configuration of each of the following atoms or ions.

1 N   2 Ca   3 Al³⁺

2.13 Experimental evidence for the electronic configurations of atoms – ionisation energies

A major difference between electrons in different types of orbital is their energy. We can investigate the electronic configurations of atoms by measuring experimentally the energies of the electrons within them. This can be done by measuring **ionisation energies** (see page 32).

Ionisation energies are used to probe electronic configurations in two ways:

- successive ionisation energies for the same atom
- first ionisation energies for different atoms.

We shall look at each in turn.

Successive ionisation energies

We can look at an atom of a particular element, and measure the energy required to remove each of its electrons, one by one:

$$X(g) \rightarrow X^+(g) + e^- \qquad \Delta H = IE_1$$
$$X^+(g) \rightarrow X^{2+}(g) + e^- \qquad \Delta H = IE_2$$
$$X^{2+}(g) \rightarrow X^{3+}(g) + e^- \qquad \Delta H = IE_3 \qquad \text{etc.}$$

These **successive ionisation energies** show clearly the arrangement of electrons in shells around the nucleus. If we take the magnesium atom as an example, and measure the energy required to remove successively the first electron, the second, the third, and so on, we obtain the plot shown in Figure 2.29.

Figure 2.29 Graph of the twelve ionisation energies of magnesium against electron number. The electronic configuration of magnesium is $1s^2 2s^2 2p^6 3s^2$.

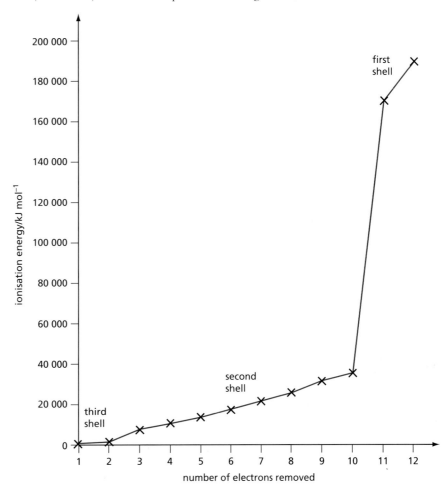

Successive ionisation energies are bound to increase because the remaining electrons are closer to, and less shielded from, the nucleus. But a larger increase occurs when the third electron is removed. This is because once the two electrons in the outer (third) shell have been removed, the next has to be stripped from a shell that is very much nearer to the nucleus (the second shell). A similar, but much more enormous, jump in ionisation energy occurs when the eleventh electron is removed. This has to come from the first, innermost shell, right next to the nucleus. These two large jumps in the series of successive ionisation energies are very good evidence that the electrons in the magnesium atom exist in three different shells.

The jumps in successive ionisation energies are more apparent if we plot the logarithm of the ionisation energy against proton number, as in Figure 2.30. (Taking the logarithm is a scaling device that has the effect of decreasing the differences between adjacent values for the larger ionisation energies, so the jumps between the shells become more obvious.)

Figure 2.30 Graph of logarithms of the twelve ionisation energies of magnesium against electron number

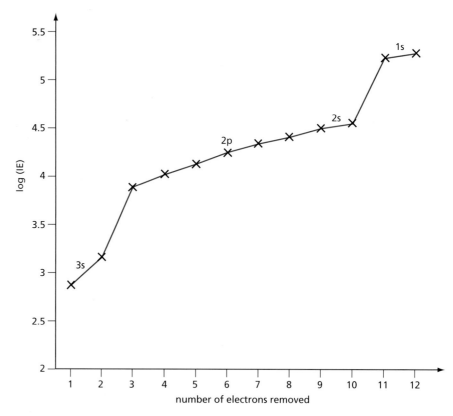

number of electrons removed

Worked example

The first five successive ionisation energies of element X are 631, 1235, 2389, 7089 and 8844 kJ mol^{-1}.

How many electrons are in the outer shell of element X?

Answer

The differences between the successive ionisation energies are as follows:

2 − 1: 1235 − 631 = 604 kJ mol^{-1}
3 − 2: 2389 − 1235 = 1154 kJ mol^{-1}
4 − 3: 7089 − 2389 = 4700 kJ mol^{-1}
5 − 4: 8844 − 7089 = 1755 kJ mol^{-1}

The largest jump comes between the third and the fourth ionisation energies, therefore X has three electrons in its outer shell.

Now try this

Decide which group element Y is in, based on the following successive ionisation energies:

590, 1145, 4912, 6474, 8144 kJ mol^{-1}

First ionisation energies

The second way that ionisation energies show us the details of electronic configuration is to look at how the first ionisation energies of elements vary with proton number. Figure 2.31 is a plot for the first 40 elements.

This graph shows us the following.

1 All ionisation energies are strongly endothermic – it takes energy to separate an electron from an atom.

2 As we go down a particular group, for example from helium to neon to argon, or from lithium to sodium to potassium, ionisation energies decrease. The larger the atom, the easier it is to separate an electron from it.

Figure 2.31 First ionisation energies for the first 40 elements

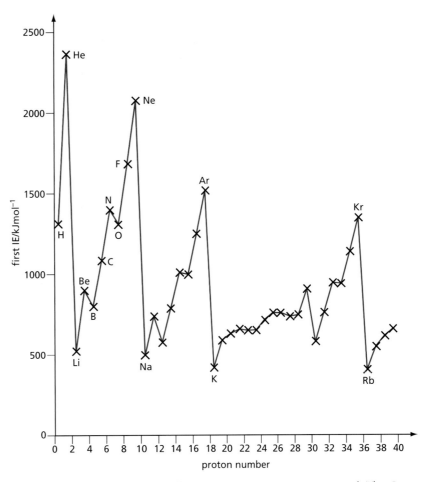

3 The ionisation energies generally increase on going across a period. The Group 1 elements (the alkali metals) have the lowest ionisation energy within each period, and the noble gases have the highest.

4 This general increase across a period has two exceptions. For the first two periods, these occur between Groups 2 and 13 and between Groups 15 and 16.
We shall comment on each of these features in turn.

1 The endothermic nature of ionisation energies

This is due to the electrostatic attraction between each electron in an atom (even the outermost one, which is always the easiest to remove) and the positive nucleus. It is worth remembering that this applies even to alkali metals like sodium, which we usually think of as 'wanting' to form ions. We must bear in mind, however, that ionisation energies as plotted in Figure 2.31 apply to the ionisation of isolated atoms in the gas phase. Ions are much more stable when in solid lattices or in solution. We shall be looking at this in detail in Topic 20.

2 The group trend

The nuclear charge experienced by an outer electron is the **effective nuclear charge**, Z_{eff}. It can roughly be equated to the number of protons in the nucleus, P, minus the number of electrons in the inner shells, E. As we explained when considering the ionisation energy of lithium on page 34, inner shells shield the effect of the increasing nuclear charge on the outer electrons. Because of this, the outer electrons of elements within the same group experience roughly the same effective nuclear charge no matter what period the element is in. What does change as we go down a group, however, is the atomic radius (see Figure 2.32). The larger the radius of the atom, the larger is the distance between the outer electron and the nucleus, so the electrostatic attraction between them is smaller.

Figure 2.32 The effective nuclear charge, Z_{eff}, and sizes of the cores of three alkali metals. The core, represented by the blue circles, comprises the nucleus plus all the inner shells of electrons. Despite the increase in Z_{eff}, the ionisation energies decrease from lithium to potassium, owing to the increased electron–nucleus distance.

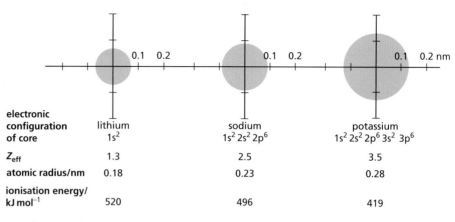

	lithium	sodium	potassium
electronic configuration of core	$1s^2$	$1s^2\,2s^2\,2p^6$	$1s^2\,2s^2\,2p^6\,3s^2\,3p^6$
Z_{eff}	1.3	2.5	3.5
atomic radius/nm	0.18	0.23	0.28
ionisation energy/ kJ mol^{-1}	520	496	419

3 The periodic trend

As we go across a period, we are, for each element, adding a proton to the nucleus, and an electron to the outermost shell. The extra proton will, of course, cause the nucleus to attract all the electrons more strongly. Electrons in the same shell are at (roughly) the same distance from the nucleus as one another. They are therefore not particularly good at shielding one another from the nuclear charge. As a result of this, the effective nuclear charge increases. This causes the electrostatic attraction between the ionising electron and the nucleus to increase too. Table 2.5 illustrates this for the second period.

Table 2.5 Comparing the values of $(P - E)$ and the effective nuclear charge, Z_{eff}, for Period 2. Z_{eff} is not exactly equal to $(P - E)$ for two reasons:
1 Electrons in s orbitals penetrate the inner orbitals to a certain extent, and are therefore less shielded by them than one might have predicted.
2 Electrons in the same shell do, to a certain extent, shield one another from the nucleus. This effect becomes larger as the outer shell becomes more full of electrons, and so the discrepancy between Z_{eff} and $(P - E)$ increases as we cross a period.

Element	Number of protons, P	Number of inner shell electrons, E	$P - E$	Effective nuclear charge, Z_{eff}
Li	3	2	1	1.3
Be	4	2	2	1.9
B	5	2	3	2.4
C	6	2	4	3.1
N	7	2	5	3.8
O	8	2	6	4.5
F	9	2	7	5.1
Ne	10	2	8	5.8

4 The exceptions

The two exceptions to the general increase in ionisation energy across a period (see Figure 2.31) arise from different causes.

In boron ($1s^2\,2s^2\,2p^1$), the outermost electron is in a 2p orbital. The average distance from the nucleus of a 2p orbital is slightly larger than that of a 2s orbital (see Figures 2.16 and 2.18). We would therefore expect the outermost electron in boron to experience less electrostatic attraction than the outermost 2s electron in beryllium. So the ionisation energy of boron is less.

The other exception is the decrease in ionisation energy from nitrogen to oxygen. This occurs when the fourth 2p electron is added, and is related to the fact that there are just three 2p orbitals. As we have seen before, because they are of the same electrical charge, electrons repel one another. The three successive electrons added to the series of atoms boron – carbon – nitrogen are therefore most likely to go into the three orbitals $2p_x$, $2p_y$ and $2p_z$. These orbitals are of equal energy and at right angles to one another, so allowing the electrons to be as far apart as possible. They will therefore experience the least electrostatic repulsion from one another, and so, overall, the atomic system will be the most stable. In oxygen, however, the fourth 2p electron has to be accommodated in an orbital that already contains an electron. These two electrons will be sharing the same region of space (by the Pauli principle, of course, their spins will have to be in opposite directions), and they will therefore repel each other quite strongly. This repulsion is larger than the extra attraction experienced by the new electron from the additional proton in oxygen's nucleus. So the energy

Figure 2.33 First ionisation energy against proton number for Period 2

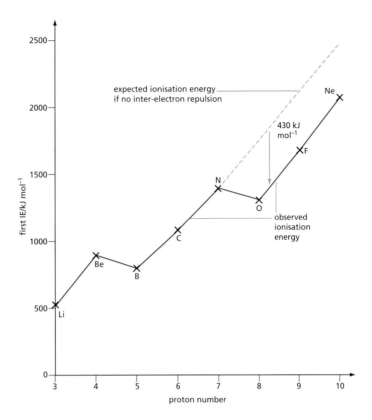

needed to remove the electron from the oxygen atom is less than the ionisation energy of the nitrogen atom. Similar repulsions are experienced by the additional electrons in the fluorine and neon atoms, so the ionisation energy of each of these elements is, like oxygen's, about $430\,kJ\,mol^{-1}$ less than one might have expected (see Figure 2.33).

Once the second shell has been filled, at neon, the next additional electron (in sodium) has to occupy the 3s orbital, and starts to fill the third shell. This is further out from the nucleus than the 2p electron in neon (remember, electrostatic attraction decreases with distance). It is also more shielded from the nucleus – by two inner shells, rather than the one in neon – hence Z_{eff} for sodium is only 2.5. On both accounts, we would expect a large decrease in ionisation energy from neon to sodium, which we indeed observe (see Figure 2.31).

2.14 The effect of electronic configuration on atomic radius

At first sight, the radius of an atom might seem an easy quantity to visualise. But, as we saw in Figures 2.16 and 2.18 (page 30), the outer reaches of atoms have an ever-decreasing electron probability, which is still greater than zero even at large distances. A filled orbital is a pretty elusive affair that can easily be squashed or polarised. We must therefore be prepared to accept that the value of the atomic radius will not be a fixed quantity, but will depend on the atom's environment.

Keeping this in mind, however, we can use the theories developed above to predict how the atomic radius might change with proton number. We have seen that we might expect two major influences on the size of an atom. One of these will be the number of shells – the more shells, the bigger the atom should be. We should see this effect as we go down a group. The other influence will be the effective nuclear charge – the larger the charge, the more the orbitals are pulled in towards the nucleus, and so the smaller the atom should be. We should see this effect as we go across a period. These two factors combine to produce a predictable pattern in the plot of atomic radius against proton number, which is borne out by experimental observations (see Figure 2.34). We shall return to the trends in atomic radius in section 10.3.

decreasing radius →

increasing radius

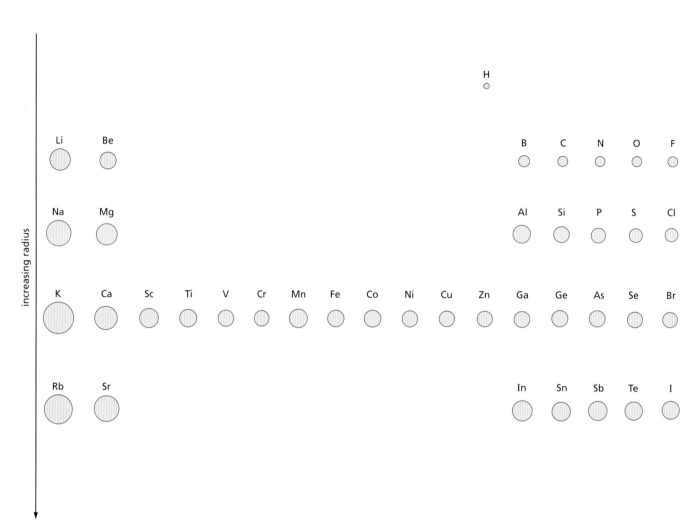

Figure 2.34 Atoms get larger down a group, as the number of shells increases. They get smaller across a period, as the effective nuclear charge pulls the electrons closer to the nucleus.

Summary

- All atoms are made of the same three **sub-atomic particles** – the **electron**, the **proton** and the **neutron**.
- Their relative electrical charges are, respectively, −1, +1 and 0.
- Their masses (relative to that of the proton) are, respectively, $\frac{1}{1836}$, 1 and 1.
- The **proton number** (**atomic number**) is the number of protons contained in the nucleus of an element's atoms. It tells us the order of the element in the Periodic Table.
- **Isotopes** are atoms of the same element (and therefore with the same proton number) but with different numbers of neutrons.
- The **mass number** of an atom is the sum of the numbers of protons and neutrons it contains.
- The full symbol for an atom shows its mass number as a superscript and its atomic (proton) number as a subscript. For example, $^{12}_{6}C$ shows a carbon atom with mass number 12 and proton number 6.
- Masses and relative abundances of isotopes can be measured using the **mass spectrometer**.

- **Chemical energy** has two components – the **kinetic energy** of moving particles, and the **potential energy** due to electrostatic attractions.
- The electrons are arranged around the nucleus of an atom in energy levels, or **orbitals**.
- When an electron moves from a higher to a lower energy level (orbital), a **quantum** of energy is released as a photon of light (sometimes visible, but often ultraviolet light).
- The number of possible orbitals in the nth shell is n^2. Each orbital can hold a maximum of two electrons.
- The first shell contains only one orbital, which is an s orbital. The second shell contains one s and three p orbitals, the third shell one s, three p and five d orbitals, and so on.
- The electrons in an atom occupy the lowest energy orbitals first. Orbitals of equal energy are occupied singly whenever possible.
- Successive ionisation energies of a single atom, and the trends in the first ionisation energies of the element across periods and down groups, give us information about the electronic configurations of atoms.

Examination practice questions

Please see the data section of the CD for any A_r values you may need.

1 Although the actual size of an atom cannot be measured exactly, it is possible to measure the distance between the nuclei of two atoms. For example, the 'covalent radius' of the Cl atom is assumed to be half of the distance between the nuclei in a Cl_2 molecule. Similarly, the 'metallic radius' is half of the distance between two metal atoms in the crystal lattice of a metal. These two types of radius are generally known as 'atomic radii'.

The table below contains the resulting atomic radii for the elements of Period 3 of the Periodic Table, Na to Cl.

Element	Na	Mg	Al	Si	P	S	Cl
Atomic radius/nm	0.186	0.160	0.143	0.117	0.110	0.104	0.099

   a i Explain qualitatively this variation in atomic radius.

     ii Suggest why it is not possible to use the same type of measurement for argon, Ar.   [4]

   b i Use the data section on the CD to complete the following table of radii of the cations and anions formed by some of the period three elements.

Radius of cation/nm			Radius of anion/nm		
Na^+	Mg^{2+}	Al^{3+}	P^{3-}	S^{2-}	Cl^-

     ii Explain the differences in size between the cations and the corresponding atoms.

     iii Explain the differences in size between the anions and the corresponding atoms.   [5]

[Cambridge International AS & A Level Chemistry 9701, Paper 23 Q1 a & b June 2012]

2 This question is about a model of the structure of the atom.

   a A model used by chemists includes the relative charges, the relative masses and the distribution of the sub-atomic particles making up the atom.
Copy and complete the table below.   [1]

Particle	Relative charge	Relative mass	Position within the atom
proton			
neutron			
electron		1/2000	shell

   b Early studies of ionisation energies helped scientists to develop a model for the electron structure of the atom. Define the term *first ionisation energy*.   [3]

   c A modern model of the atom arranges electrons into orbitals, sub-shells and shells.

Copy and complete the following table showing the maximum number of electrons which can be found within each region.   [3]

Region	Number of electrons
a 2p orbital	
the 3s sub-shell	
the 4th shell	

   d The modern Periodic Table arranges the elements in order of their atomic number.
Explain what is meant by the term *periodicity*.   [1]

   e In this part, you need to refer to the *Periodic Table of the Elements* in the data section on the CD.
From the first 18 elements **only**, choose an element which fits the following descriptions.

     i An element with an isotope that can be represented as $^{14}_{6}X$.   [1]

     ii The element which forms a 3– ion with the same electron structure as Ne.   [1]

     iii The element which has the smallest third ionisation energy.   [1]

     iv The element with the first six successive ionisation energies shown below, in $kJ\,mol^{-1}$.   [1]

      738   1451   7733   10541   13629   17995

[OCR Chemistry A Unit F321 Q1 May 2011]

3 Tin mining was common practice on Dartmoor (UK) in pre-Roman times. Most of the tin extracted was mixed with copper to produce bronze.

   a The table below shows the sub-atomic particles of an isotope of tin.

Isotope	Protons	Neutrons	Electrons
^{118}Sn			

     i Copy and complete the table.   [1]

     ii In terms of sub-atomic particles, how would atoms of ^{120}Sn differ from atoms of ^{118}Sn?   [1]

   b The relative atomic mass of tin is 118.7. Define the term *relative atomic mass*.   [3]

   c A bronze-age shield found on Dartmoor contained 2.08 kg of tin.
Calculate the number of tin atoms in this bronze shield. Give your answer to **three** significant figures.   [2]

   d Tin ore, known as cassiterite, contains an oxide of tin. This oxide contains 78.8% tin by mass.
Calculate the empirical formula of this oxide. You must show your working.   [2]

[OCR Chemistry A Unit F321 Q1 May 2010]

3 Chemical bonding in simple molecules

This topic is the first of two in which we look at how atoms bond together to form molecules and compounds, and how those particles arrange themselves into larger structures to form all the matter we see around us. Here we describe the various types of covalent bond, and explain some of the properties of simple covalent molecules.

Learning outcomes

By the end of this topic you should be able to:

3.2a) describe, including the use of 'dot-and-cross' diagrams, covalent bonding (as in hydrogen, oxygen, chlorine, hydrogen chloride, carbon dioxide, methane, ethene) and co-ordinate (dative covalent) bonding (as in the formation of the ammonium ion and in the Al_2Cl_6 molecule)

3.2b) describe covalent bonding in terms of orbital overlap, giving σ and π bonds, including the concept of hybridisation to form sp, sp^2 and sp^3 orbitals (see also Topic 14)

3.2c) explain the shapes of, and bond angles in, molecules by using the qualitative model of electron-pair repulsion (including lone pairs), using as simple examples: BF_3 (trigonal), CO_2 (linear), CH_4 (tetrahedral), NH_3 (pyramidal), H_2O (non-linear), SF_6 (octahedral), PF_5 (trigonal bipyramidal)

3.2d) predict the shapes of, and bond angles in, molecules and ions analogous to those specified in **3.2b)** (see also Topic 14)

3.3a) describe hydrogen bonding, using ammonia and water as simple examples of molecules containing N—H and O—H groups

3.3b) understand, in simple terms, the concept of electronegativity and apply it to explain the properties of molecules such as bond polarity and the dipole moments of molecules (part, see also Topic 10)

3.3c) explain the terms *bond energy*, *bond length* and *bond polarity* and use them to compare the reactivities of covalent bonds (see also Topic 15)

3.3d) describe intermolecular forces (van der Waals' forces), based on permanent and induced dipoles, as in $CHCl_3(l)$, $Br_2(l)$ and the liquid Group 18 elements

3.5c) show understanding of chemical reactions in terms of energy transfers associated with the breaking and making of chemical bonds (see also Topic 14)

3.1 Introduction

Of the total number of individual, chemically-pure substances known to exist, several million are **compounds**, formed when two or more elements are chemically bonded together. Less than 100 substances are **elements**. Only six of these elements consist of free, unbonded atoms at room temperature. These are the noble gases in Group 18. All other elements exist as individual molecules, or giant molecules, or metallic lattices. In all three cases, the atoms of the element are chemically bonded to one another.

This shows that the natural state of atoms (that is, the state where they have the lowest energy) is the bonded state. Atoms 'prefer' to be bonded to one another, rather than to be floating free through space. They give out energy when they form bonds. On the other hand, it always requires an input of energy to break a chemical bond – bond-breaking is an endothermic process (see Figure 3.1).

Figure 3.1 Bond breaking is endothermic.

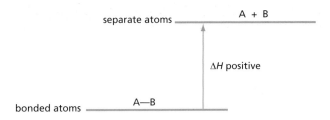

The bond strength is related to the value of ΔH in Figure 3.1, that is, the enthalpy change that occurs when one mole of bonds in a gaseous compound is broken, forming gas-phase atoms (see section 5.5). The larger ΔH is, the stronger is the bond.

We saw in Topic 2 that the negatively charged electrons in an atom are attracted to the positively charged nucleus: an electron in an atom has less potential energy than an electron on its own outside an atom. It therefore requires energy to remove an electron from an atom. In this topic we discover a similar reason why atoms bond together to form compounds. This is because the electrons on one atom are attracted to the nucleus of another, causing the bonded system to have less potential energy than in its unbonded state. (The three major types of bonding – covalent, ionic and metallic – differ from one another only in how far this attraction to another nucleus overcomes the attraction of the electron to its own nucleus.)

3.2 Covalent bonding – the hydrogen molecule

We shall look first at the simplest possible bond, that between the two hydrogen atoms in the hydrogen molecule. Imagine two hydrogen atoms, initially a large distance apart, approaching each other. As they get closer together, the first effect will be that the electron of one atom will experience a repulsion from the electron on the other atom, but this will be compensated by the attraction it will experience towards the other atom's nucleus (in addition to the attraction it always experiences from its own nucleus). Remember that the electron in an atom spends some of its time at quite a large distance from the nucleus (see Figure 2.16, page 30). As the hydrogen atoms get closer still, the two electrons will encounter an even greater attraction to the opposite nucleus, but will also continue to repel each other. Eventually, when the two nuclei become very close together, they in turn will start to repel each other, since they both have the same (positive) charge. The most stable situation will be when the attractions of the two electrons to the two nuclei are just balanced by the electron–electron and nucleus–nucleus repulsions. A **covalent bond** has formed. The decrease in potential energy at this point from the unbonded state is called the **bond energy**, and the nucleus–nucleus distance at which this occurs is known as the **bond length** (see Figure 3.2).

Figure 3.2 Potential energy against internuclear distance for the hydrogen molecule

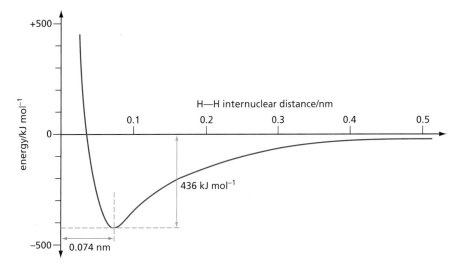

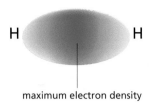

Figure 3.3 A molecular orbital is a region between two bonded atoms where the electron density is concentrated.

Being attracted to both the hydrogen nuclei, the two electrons spend most of their time in the region half-way between them, and on the axis that joins them. This is where the highest electron density (or electron probability) occurs in a single covalent bond (see Figure 3.3).

The two electrons are therefore *shared* between the two adjacent atoms. As we mentioned in section 2.7, we use the word (atomic) **orbital** to describe the region of space around the nucleus of an atom that is occupied by a particular electron. In a similar way, we can use the term **molecular orbital** to describe the region of space within a molecule where a particular electron is to be found.

3.3 Representing covalent bonds

Depending on the information we want to convey, we can use various ways to represent a covalent bond, as listed below.

● The dot-and-cross diagram:

H• + ×H → H ⦂ H

● The dot-and-cross diagram including Venn diagram boundaries – this is preferred for more complicated molecules, as it allows an easy check to be made of exactly which bonds the electrons are in:

(H•) (×H) → (H⦂H)

● The line diagram (one line for each bond) – this is not very informative for small simple molecules, but is often less confusing than drawing out individual electrons for larger, more complex molecules.

H• + •H → H — H

3.4 Covalent bonding with second-row elements

Bonding and valence-shell electrons

When an atom bonds with others, it is normally only the electrons in the outermost shell of the atom that take part in bonding. The outermost shell is called the **valence shell**. All electrons in the valence shell can be considered together as a group. When we look in simple terms at the bonding, we can ignore the distinction between the various types of orbitals (s, p, d, etc.) in a shell. The number of electrons in the valence shell is the same as the group number of the element in the Periodic Table (see Topic 10).

If an atom has more than one electron in its valence shell, it can form more than one covalent bond to other atoms. For example, the beryllium atom has a pair of 2s electrons. When forming a compound such as beryllium hydride, BeH_2, these can unpair themselves and form two bonds with two other atoms. The boron atom ($2s^2 2p^1$) can form three bonds. The bonding in beryllium hydride and boron hydride is shown in Figure 3.4.

Figure 3.4 Dot-and-cross diagrams showing beryllium hydride, BeH_2, and boron hydride, BH_3. (It is worth remembering that all electrons are identical, no matter which atom they came from. Our representing them as dots and crosses is merely for our own benefit, to make it clear to us which atom they were originally associated with.)

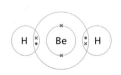

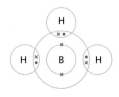

The carbon atom in methane has four electrons in its valence shell ($2s^2 2p^2$), and so forms four bonds.

Figure 3.5 Dot-and-cross diagram for methane, CH₄. (Often, for clarity, bonding diagrams show only the electrons in the outer shell, omitting the inner shells. For comparison, both diagrams for methane are shown here.)

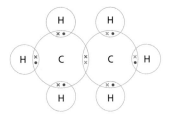

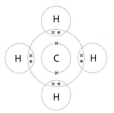 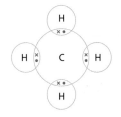

Figure 3.6 Dot-and-cross diagram for ethane, C₂H₆

The general rule is that:

Elements with **one to four** electrons in their valence shells form the same number of covalent bonds as the number of valence-shell electrons.

Importantly, carbon atoms form strong bonds to other carbon atoms (as do a few other elements). This allows the millions of organic compounds to be stable. The simplest molecule with a carbon–carbon bond is ethane (see Figure 3.6).

More than four valence-shell electrons

When the number of valence-shell electrons is greater than four, the maximum possible number of bonds is not always formed. This is because atoms of the elements of the second row of the Periodic Table have only four orbitals in their valence shells ($2s$, $2p_x$, $2p_y$, $2p_z$) and so cannot accommodate more than four pairs of electrons. The number of bonds they form is restricted by this overall maximum of eight electrons, because every new bond brings another electron into the valence shell. For example, nitrogen has five valence-shell electrons. In the molecule of ammonia (NH₃) there are three N—H bonds. These involve the sharing of three nitrogen electrons with three from the hydrogen atoms. These three additional electrons bring the valence shell total to eight. The outer shell is therefore filled, with a **full octet** of electrons. No further bonds can form. The remaining two of the five electrons in nitrogen's valence shell remain unbonded, as a **lone pair**, occupying an orbital associated with only the nitrogen atom (see Figure 3.7).
   The general rule is that:

Elements of the second period with **more than four** electrons in their outer shells form $(8 - n)$ covalent bonds, where n = the number of valence-shell electrons.

Similarly, oxygen ($2s^2 2p^4$), with six electrons in its valence shell, can form only two covalent bonds, with two lone pairs of electrons remaining. Fluorine ($2s^2 2p^5$), with seven valence-shell electrons, can form only one bond, leaving three lone pairs around the fluorine atom (see Figure 3.8).

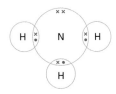

Figure 3.7 Dot-and-cross diagram for ammonia, NH₃

Figure 3.8 Dot-and-cross diagrams for water, H₂O, and hydrogen fluoride, HF. Oxygen can form two bonds (8 − 6 = 2) and fluorine just one (8 − 7 = 1).

Worked example

Draw diagrams showing the bonding in nitrogen trifluoride, NF₃, and hydrogen peroxide, H₂O₂.

Answer

Figure 3.9

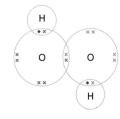

NF₃   H₂O₂

Now try this

Draw diagrams showing the bonding in the following molecules.

1 BF₃   2 N₂H₄
3 CH₃OH   4 CH₂F₂

3.5 Covalent bonding with third-row elements

Unlike elements in the second row of the Periodic Table, those in the third and subsequent rows can use their d orbitals in bonding, as well as their s and p orbitals. They can therefore form more than four covalent bonds to other atoms. Like nitrogen, phosphorus $(1s^2 2s^2 2p^6 3s^2 3p^3)$ has five electrons in its valence shell. But because it can make use of five orbitals (one 3s, three 3p and one 3d) it can use all five of its valence-shell electrons in bonding with fluorine. It therefore forms phosphorus pentachloride, PF_5, as well as phosphorus trifluoride, PF_3 (see Figure 3.10).

Figure 3.10 Dot-and-cross diagrams for phosphorus trifluoride, PF_3, and phosphorus pentafluoride, PF_5

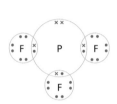 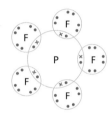

In a similar way, sulfur can use all its valence-shell electrons in six orbitals (one 3s, three 3p and two 3d) to form sulfur hexafluoride, SF_6. Chlorine does not form ClF_7, however. The chlorine atom is too small for seven fluorine atoms to assemble around it. Chlorine does, though, form ClF_3 and ClF_5 in addition to ClF.

Worked example

Draw a diagram to show the bonding in chlorine pentafluoride, ClF_5.

Answer

Figure 3.11

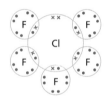

Now try this

1 Sulfur forms a tetrafluoride, SF_4. Draw a diagram to show its bonding.
2 a How many valence-shell electrons does chlorine use for bonding in chlorine trifluoride, ClF_3?
  b So how many electrons are left in the valence shell?
  c So how many lone pairs of electrons are there on the chlorine atom in chlorine trifluoride?

3.6 The covalency table

As explained in sections 3.4 and 3.5, the number of covalent bonds formed by an atom depends on the number of electrons available for bonding, and the number of valence-shell orbitals it can use to put the electrons into. For many elements, the number of bonds formed is a fixed quantity and is termed the **covalency** of the element. It is often related to its group number in the Periodic Table. As shown above, elements in Groups 15 to 17 in the third and subsequent rows of the Periodic Table (Period 3 and higher) can use a variable number of electrons in bonding. They can therefore display more than one covalency. Table 3.1 shows the most usual covalencies of some common elements.

Table 3.1 Covalencies of some common elements

Element	Symbol	Group	Covalency
hydrogen	H	1	1
beryllium	Be	2	2
boron	B	13	3
carbon	C	14	4
nitrogen	N	15	3
oxygen	O	16	2
fluorine	F	17	1
aluminium	Al	13	3
silicon	Si	14	4
phosphorus	P	15	3 or 5
sulfur	S	16	2 or 4 or 6
chlorine	Cl	17	1 or 3 or 5
bromine	Br	17	1 or 3 or 5
iodine	I	17	1 or 3 or 5 or 7

Now try this

1 What are the formulae of the three possible oxides of sulfur?
2 Use Table 3.1 to write the formulae of the simplest compounds formed between:
   a carbon and hydrogen
   b oxygen and fluorine
   c boron and chlorine
   d nitrogen and bromine
   e carbon and oxygen.

Worked example

What could be the formulae of compounds formed from the following pairs of elements?
a boron and nitrogen   b phosphorus and oxygen

Answer
The elements must combine in such a ratio that their total covalencies are equal to each other.
a The covalencies of boron and nitrogen are both 3, so one atom of boron will combine with one atom of nitrogen. The formula is therefore **BN** (3 for B = 3 for N).
b The covalency of oxygen is 2, and that of phosphorus can be either 3 or 5. We would therefore expect two possible phosphorus oxides: P_2O_3 (2 × 3 for P = 3 × 2 for O) and P_2O_5 (2 × 5 for P = 5 × 2 for O).

3.7 Dative bonding

So far, we have looked at covalent bonds where each atom provides one electron to form the bond. It is possible, however, for just one of the atoms to provide *both* bonding electrons. This atom is called the **donor atom**, and the two electrons it provides come from a lone pair on that atom. The other atom in the bond is called the **acceptor atom**. It must contain an empty orbital in its valence shell. This kind of bonding is called **dative bonding** or **co-ordinate bonding**. The apparent covalency of each atom can increase as a result of dative bonding.

For example, when gaseous ammonia and gaseous boron trifluoride react together, a white solid is formed, with the formula NH_3BF_3. The nitrogen atom in ammonia has a lone pair, and the boron atom in boron trifluoride has an empty 2p orbital. The nitrogen's lone pair can overlap with this empty orbital, as shown in Figure 3.12.

Figure 3.12 Forming a dative (co-ordinate) bond in NH_3BF_3

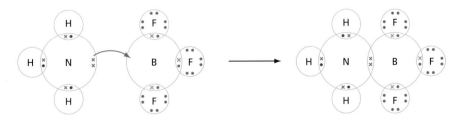

The dative bond, once formed, is no different from any other covalent bond. For example, when gaseous ammonia and gaseous hydrogen chloride react, the white solid ammonium chloride is formed. The lone pair of electrons on the nitrogen atom of ammonia has formed a dative bond with the hydrogen atom of the hydrogen

49

chloride molecule (see Figure 3.13). (At the same time, the H—Cl bond breaks, and the electrons that were in the bond form a fourth lone pair on the chlorine atom. With 18 electrons and only 17 protons, this now becomes the negatively charged chloride ion. Ionic bonding is covered in detail in Topic 4.)

Figure 3.13 Forming an ionic bond in ammonium chloride, NH_4Cl

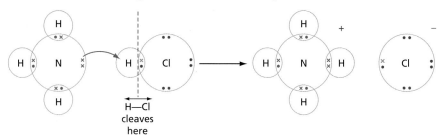

The shape of the ammonium ion is a regular tetrahedron, the same as that of methane molecule (see Figure 3.14). All four N—H bonds are exactly the same. It is not possible to tell which one was formed in a dative way.

Figure 3.14 The ammonium ion and methane are both tetrahedral.

3.8 Multiple bonding

Double and triple bonds

Atoms can share more than one electron pair with their neighbours. Sharing two electron pairs produces a double bond, and sharing three produces a triple bond. The covalencies still conform to those in Table 3.1. Examples of oxygen, nitrogen and carbon atoms forming multiple bonds are shown in Figure 3.15.

Figure 3.15 Dot-and-cross diagrams for O_2, N_2 and ethene, C_2H_4

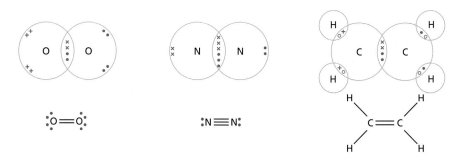

Figure 3.16

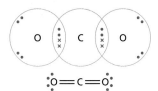

Worked example

Draw a dot-and-cross diagram and a line diagram of the carbon dioxide molecule, CO_2.

Answer
The covalencies of oxygen and carbon are 2 and 4 respectively. The only possible bonding arrangement is therefore O=C=O, and the electrons are shared as shown in Figure 3.16.

Now try this

Draw line diagrams and dot-and-cross diagrams to show the bonding in the following molecules.

1 HCN   2 H_2CO 3 C_2H_2 (ethyne)

3.9   The shapes of molecules

One of the major advances in chemistry occurred in 1874, when the Dutch chemist Jacobus van't Hoff suggested for the first time that molecules possessed a definite, unique, three-dimensional shape. The shapes of molecules are determined by the angles between the bonds within them. In turn, the bond angles are determined by the arrangement of the electrons around each atom.

VSEPR theory

Because of their similar (negative) charge, electron pairs repel each other. The electron pairs in the outer (valence) shell of an atom will experience the least repulsion when they are as far apart from one another as possible. This applies both to bonded pairs and to non-bonded (lone) pairs. This simple principle allows us to predict the shapes of simple molecules. The theory, developed by Nevil Sidgwick and Herbert Powell in 1940, is known as the **valence-shell electron-pair repulsion theory** (or **VSEPR** for short). In order to work out the shape of a molecule, the theory is applied as follows.

1 Draw the dot-and-cross structure of the molecule, and hence count the number of electron pairs around each atom.
2 These pairs will take up positions where they are as far apart from one another as possible. The angles between the pairs will depend on the number of pairs around the atom (see Figure 3.17).

Figure 3.17 The shapes in which *n* electron pairs around a central atom arrange themselves, with *n* = 2, 3, 4, 5 and 6

molecule shape	number of electron pairs	description
	2	linear
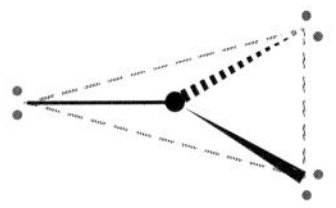	3	triangular planar (trigonal planar)
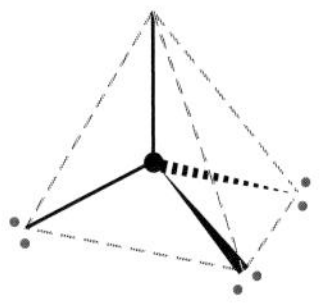	4	tetrahedral
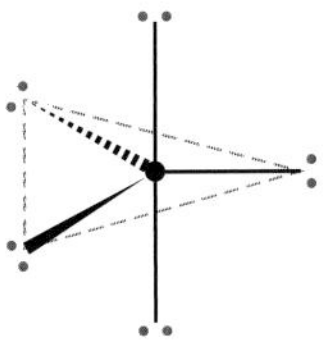	5	triangular bipyramidal (trigonal bipyramidal)
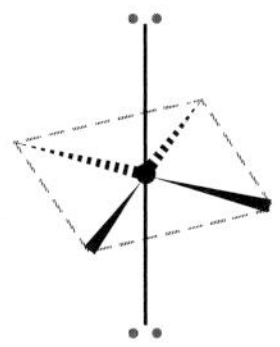	6	octahedral

3 Orbitals containing lone pairs are larger than those containing bonded pairs, and take up more space around the central atom. They therefore repel the other pairs that surround the atom more strongly. This causes the angle between two lone pairs to be larger than the angle between a lone pair and a bonded pair, which in turn is larger than the angle between two bonded pairs:

(LP–LP angle) > (LP–BP angle) > (BP–BP angle)

4 Although they are very important in determining the shape of a molecule, the lone pairs are not included when the molecule's shape is described. For example, the molecules CH_4, NH_3 and H_2O all have four pairs of electrons in the valence shell of the central atom (see Figure 3.18). These four pairs arrange themselves in a (roughly) tetrahedral fashion. But only the methane molecule is described as having a tetrahedral shape. Ammonia is pyramidal (it has only three bonds), and water is described as a bent molecule (it has only two bonds).

Figure 3.18 The shapes of methane, ammonia and water are as predicted from their four pairs of electrons around the central atom. Lone pairs repel other electron pairs more strongly than bonding pairs do, so the bond angles in ammonia and water are slightly smaller than the tetrahedral angle of 109.5°.

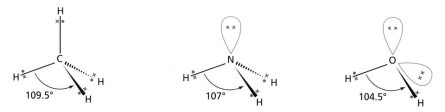

Worked example

Work out the shapes of the following molecules.

a CCl_4 **b** BF_3 **c** PF_5

Answer

a **Figure 3.19**

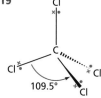

The valence shell of the carbon atom in CCl_4 contains eight electrons, arranged in four bonded pairs. It therefore takes up a regular tetrahedral shape, just like methane (see Figure 3.19).

b **Figure 3.20**

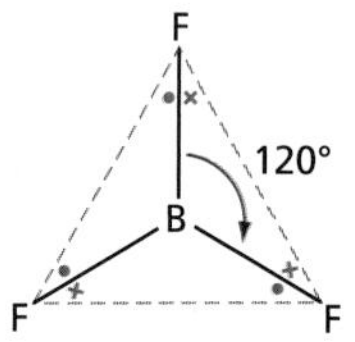

The valence shell of the boron atom in BF_3 contains six electrons, arranged in three bonded pairs (no lone pairs – see page 47). It thus takes up a regular triangular planar shape, with all bond angles being equal at 120° (see Figure 3.20).

c **Figure 3.21**

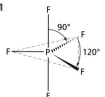

The valence shell of the phosphorus atom in PF_5 contains ten electrons: five from phosphorus, and one each from the five fluorine atoms (see page 48). The shape will be a regular trigonal bipyramid, with bond angles of 120° and 90° (see Figure 3.21).

Now try this

Work out the shapes of the following molecules.

1 BeH_2
2 ClF_3 (Explain why this is called a 'T-shaped' molecule.)
3 SF_6

Double bonds and VSEPR

When a molecule contains a double bond, the four electrons in this bond count as one group of electrons, as far as the VSEPR theory is concerned. For example, the carbon in carbon dioxide, which is doubly bonded to each of the oxygen atoms, is surrounded by just two electron groups. It is therefore predicted to be a linear molecule (see Figure 3.22).

180°

Ö═C═Ö

Figure 3.22 Carbon dioxide is a linear molecule.

Predict the shape of the sulfur dioxide molecule.

Answer

Sulfur shares two of its valence-shell electrons with each of the two oxygen atoms, forming two double bonds. It has two electrons left over, which form a lone pair. There are therefore three groups of electrons around the sulfur atom, and these will arrange themselves approximately into a triangular planar shape. The SO_2 molecule is therefore bent, as shown in Figure 3.23 (remember that lone pairs are ignored when the shape is described), with a O—S—O bond angle of about 120°. (In fact, the angle is very slightly less, at 119°, owing to the extra repulsion by the lone pair.)

Figure 3.23 Sulfur dioxide is a bent molecule.

Predict the shapes of the following molecules.

1 HCN   **2** H_2CO **3** C_2H_2 (ethyne)

(Note that a triple bond, like a double bond, can be considered as a single group of electrons.)

Later in this topic we shall see how we can predict the shapes of molecules such as ethene and carbon dioxide by studying their bonding (see section 3.15).

Summary: the shapes of molecules

Table 3.2 summarises how the numbers of bonded and non-bonded electron pairs determine the shapes of molecules.

Table 3.2 The shapes of molecules as determined by the numbers of bonding and non-bonding electron pairs

Total number of electron pairs	Number of bonding pairs	Number of lone pairs	Shape	Example
2	2	0	linear	BeF_2
3	3	0	triangular planar	BF_3 (see Figure 3.20)
4	1	3	linear	HF
4	2	2	bent	H_2O (see Figure 3.18)
4	3	1	triangular pyramidal	NH_3 (see Figure 3.18)
4	4	0	tetrahedral	CH_4 (see Figure 3.18)
5	2	3	linear	XeF_2 (see Figure 3.24)
5	3	2	T-shaped	ClF_3 (see Figure 3.24)
5	4	1	'see-saw'	SF_4 (see Figure 3.24)
5	5	0	trigonal bipyramidal	PF_5 (see Figure 3.21)

Figure 3.24 The shapes of XeF_2, ClF_3 and SF_4 are all determined by the five pairs of electrons around the central atom.

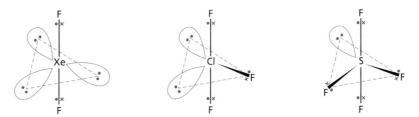

3.10 Electronegativity and bond polarity

The essential feature of covalent bonds is that the electrons forming the bond are shared between the two atoms. But that sharing does not have to be an equal one. Unless both atoms are the same (for example in the molecules of hydrogen, H_2, or oxygen, O_2), it is more than likely that they have different **electronegativities**, and this leads to unequal sharing.

The **electronegativity** of an atom is a measure of its ability to attract the electrons in a covalent bond to itself.

Several chemists have developed quantitative scales of electronegativity values. One of the most commonly used is the scale devised by the American chemist Linus Pauling in 1960. Numerical values of electronegativity on this scale range from fluorine, with a value of 4.0, to an alkali metal such as sodium, with a value of 0.9.

Figure 3.25 shows the electronegativity values of the elements in the first three rows of the Periodic Table. Notice that the larger the number, the more electronegative (electron attracting) the atom is.

Figure 3.25 Electronegativities of elements in the first three rows of the Periodic Table

H 2.1						
Li 1.0	Be 1.5	B 2.0	C 2.5	N 3.0	O 3.5	F 4.0
Na 0.9	Mg 1.2	Al 1.5	Si 1.8	P 2.1	S 2.5	Cl 3.0

The electronegativity value depends on the effective nuclear charge. This is similar to the way that ionisation energies depend on the effective nuclear charge, and the reason is also similar. The nucleus of a fluorine atom (electronegativity value = 4.0), for example, contains more protons and a higher effective nuclear charge than that of a carbon atom (electronegativity value = 2.5), and so is able to attract electrons more strongly than carbon. In the C—F bond, therefore, the electrons will be found, on average, nearer to fluorine than to carbon. This will cause a partial movement of charge towards the fluorine atom, resulting in a **polar bond** (see Figure 3.26).

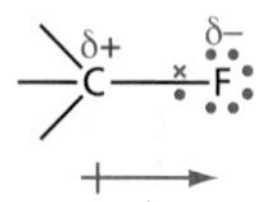

This **partial charge separation** is represented by the Greek letter δ (delta), followed by + or − as appropriate. The 'cross-and-arrow' symbol in Figure 3.26 is also sometimes used to show the direction in which the electrons are attracted more strongly.

Most covalent bonds between different atoms are polar. If the electronegativity difference is large enough, one atom can attract the bonded electron pair so much that it is completely transferred, and an ionic bond is formed (see section 4.6).

Figure 3.26 The carbon–fluorine bond is polar.

Worked example

Use the electronegativities in Figure 3.25 to predict the direction of polarisation in the following bonds.

a C—O   b O—H   c Al—Cl

Which of these three bonds is the most polar?

Answer

a $^{\delta+}$C—O$^{\delta-}$ electronegativity difference = 3.5 − 2.5 = 1.0
b $^{\delta-}$O—H$^{\delta+}$ electronegativity difference = 3.5 − 2.1 = 1.4
c $^{\delta+}$Al—Cl$^{\delta-}$ electronegativity difference = 3.0 − 1.5 = 1.5

Al—Cl is the most polar of these three bonds.

Now try this

Which is the most polar of the following three bonds?

Si—H   P—C   S—Cl

3.11 Polarity of molecules

Dipoles of molecules

The drift of bonded electrons to the more electronegative atom results in a separation of charge, termed a **dipole**. Each of the polar bonds in a molecule has its own dipole associated with it. The overall dipole of the molecule depends on its shape. Depending on the relative angles between the bonds, the individual bond dipoles can either reinforce or cancel each other. If cancellation is complete, the resulting molecule will have no dipole, and so will be non-polar. If the bond dipoles reinforce each other, molecules with very large dipoles can be formed.

For example, both hydrogen, H_2, and fluorine, F_2, are non-polar molecules, but hydrogen fluoride has a large dipole due to the much larger electronegativity of fluorine (see Figure 3.27).

Figure 3.27 The electronegativity difference gives hydrogen fluoride a large dipole.

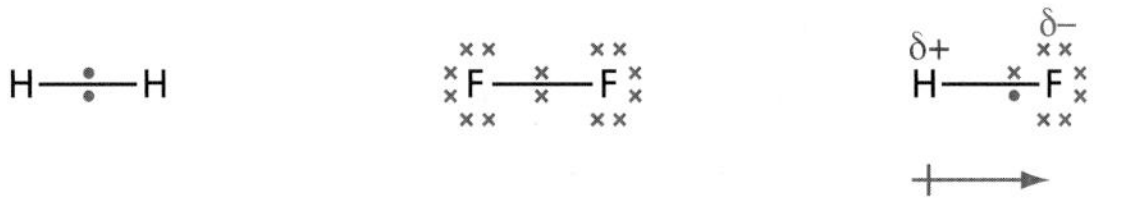

The two C=O bonds in carbon dioxide are polar, but because the angle between them is 180°, exact cancellation occurs, and the molecule of CO_2 is non-polar. A similar situation occurs in tetrachloromethane, but not in chloromethane (see Figure 3.28).

Figure 3.28 The dipoles cancel in carbon dioxide and tetrachloromethane, but not in chloromethane.

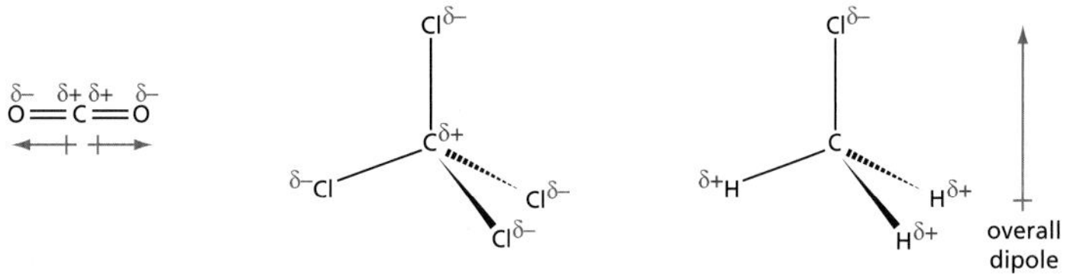

Worked example

Which of the molecules in Figure 3.29 has no dipole moment?

Figure 3.29 The dichlorobenzenes

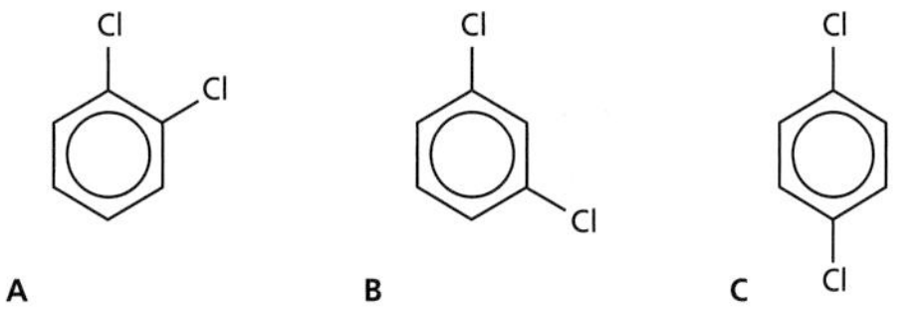

Answer
Compound **C**, 1,4-dichlorobenzene. In compound **C**, the C—Cl dipoles are on opposite sides of the benzene ring, so their dipoles oppose each other equally. In compounds **A** and **B**, the C—Cl dipoles are, more or less, on the same side of the ring.

Now try this

1 Which of the two compounds **A** and **B**, in the worked example above, will have the larger dipole moment, and why?
2 Work out the shapes of the following molecules and decide which of them are polar.
   a C_2H_6 b CH_2Cl_2 c CH_3OH d SF_4

Lone pairs and dipoles

Polarity of molecules is also caused by the presence of a lone pair of electrons on one side on the central atom. This is an area of negative charge and so it forms the δ− end of a dipole. In the ammonia molecule, this dipole reinforces the dipoles due to the N—H bonds, resulting in a highly polar molecule. A similar situation occurs with water (see Figure 3.30).

Figure 3.30 Ammonia and water are highly polar molecules.

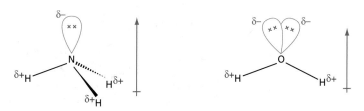

Table 3.3 The dipole moments of some common covalent molecules

Molecule	Dipole moment/D
HF	1.91
HCl	1.05
HBr	0.80
HI	0.42
H_2O	1.84
NH_3	1.48

Comparing dipoles

The strength of a dipole is measured by its **dipole moment**, which is the product of the separated electric charge, δ+ or δ−, and the distance between them. Dipole moments can be measured, and they can also be calculated. Table 3.3 lists the dipole moments of some common molecules. The debye unit, **D** (named after the chemist Peter Debye), has the dimensions of coulomb-metre. If, as in an ionic bond, a full electron charge were separated by a distance of 0.15 nm, the dipole moment would be 7.0 D. It can therefore be seen that even the most polar of covalent bonds, such as H—F, are still a long way from being ionic.

3.12 Induced polarisation of bonds

A bond's natural polarity, caused by the different electronegativities of the bonded atoms, can be increased by a nearby strong electric field. A nearby electric field can also produce a polarity in a bond that was originally non-polar. This **induced polarity** is often followed by the breaking of the bond. The nearby electric field can be due to an area of concentrated electron density, or an electron-deficient atom, or a small highly-charged cation. Examples of each of these cases are as follows:

- the polarisation of a bromine molecule by an ethene molecule, as shown in Figure 3.31 (see Topic 14)

Figure 3.31 Polarisation of bromine by ethene

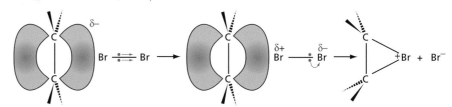

- the polarisation of a chlorine molecule by aluminium chloride, as shown in Figure 3.32 (see Topic 25)

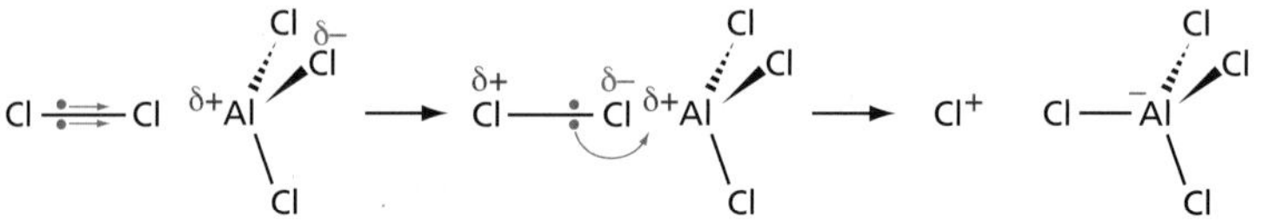

Figure 3.32 Polarisation of chlorine by aluminium chloride

- the polarisation of the carbonate ion by a Group 2 metal ion, as shown in Figure 3.33 (see Topic 20).

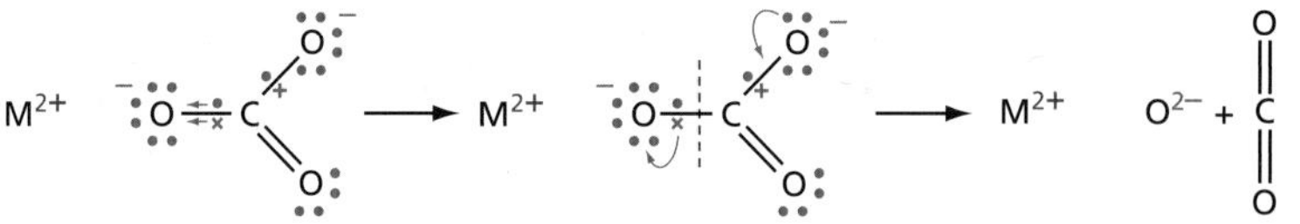

Figure 3.33 Polarisation of the carbonate ion by a Group 2 metal ion

The induced polarisation of **molecules,** and the effect this has on intermolecular attractions, is discussed later in this topic, in section 3.17.

3.13 Isoelectronic molecules and ions

Isoelectronic means 'having the same number of electrons'. The term can be applied to molecules or ions which have the same total number of electrons, or to molecules or ions that have the same number of electrons in their valence shells. Because the distribution of electrons in the valence shell determines the shape of a molecule, molecules or ions that are isoelectronic with one another have the same basic shape, often with identical bond angles too. For example:

- the tetrahydridoborate(III) ion, BH_4^-, methane, CH_4, and the ammonium ion, NH_4^+, are all regular tetrahedra, with H—X—H angles of 109.5°

- the borate ion, BO_3^{3-}, the carbonate ion, CO_3^{2-}, and the nitrate ion, NO_3^-, are all triangular planar, with O—X—O angles of 120°
- the oxonium ion, H_3O^+, and ammonia, NH_3, are both triangular pyramids with H—X—H angles of 107°.

Counting the electrons in the valence shell around an atom is therefore a useful way of predicting the shape of a molecule.

Worked example

Predict the shapes of the following ions:
a the methyl cation, CH_3^+
b the methyl anion, CH_3^-
c the nitronium cation, NO_2^+

Answer
a CH_3^+ is isoelectronic with BH_3 (six electrons in the outer shell), and so it is a triangular planar ion.
b CH_3^- is isoelectronic with NH_3, so it is a triangular pyramidal ion.
c NO_2^+ is isoelectronic with CO_2, so it is a linear ion.

Now try this

Given the information that sulfur trioxide is triangular planar, the sulfite anion, SO_3^{2-}, is a triangular pyramid, and the sulfate ion, SO_4^{2-} is tetrahedral, predict the shapes of the following ions.

1 ClO_3^- 2 PO_3^{3-} 3 PO_4^{3-}

3.14 The overlap of orbitals – σ bonds

A single bond is produced when a pair of electrons is shared between two atoms. The electron pair occupies a molecular orbital formed by the overlap of an atomic orbital on each atom. The atomic orbitals can be s orbitals or p orbitals, or a mixture of the two (see Figure 3.34).

Figure 3.34 The shapes of molecular orbitals formed by the overlap of atomic orbitals

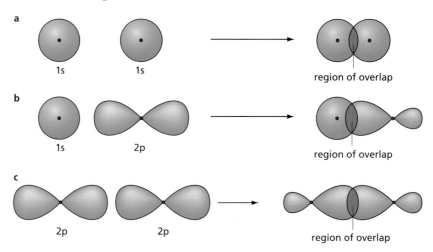

The resulting area of electron density is in the region between the bonded atoms, with the largest density on the axis joining the atom centres. Looked at 'end on', these orbitals have a circular symmetry (see Figure 3.35).

Figure 3.35 The C—C single bond (a σ bond)

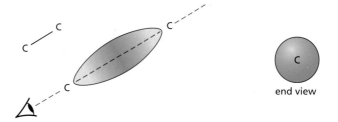

By analogy with the spherically symmetric s orbitals of atoms, these are termed σ (sigma) orbitals, σ being the Greek letter corresponding to s. The resulting bond is called a **σ bond**.

Hybridisation of orbitals

We saw on page 51 that the electron pairs which surround a central atom arrange themselves in such a way as to minimise the repulsion between them. If there are three electron pairs, they arrange themselves in a triangular planar arrangement, and if there are four electron pairs, they arrange themselves tetrahedrally. What molecular orbitals can be used by the electrons to point in these directions?

In Topic 2 we saw that, for second-row elements, the orbitals available for bonding are the spherical 2s orbitals and the three dumbbell-shaped 2p orbitals.

Figure 3.34 shows how two 2p orbitals can overlap to produce a σ molecular orbital. But, as can be seen in Figure 3.34, this molecular orbital points in the same direction as the original p atomic orbitals. So molecular orbitals formed from the $2p_x$, $2p_y$ and $2p_z$ orbitals would be pointing along the x-, y- and z-axes, that is, at right angles to one another. But for the triangular planar geometry of three electron pairs, there need to be three orbitals that are pointing at 120° from one another, and in the tetrahedral case of four electron pairs, there need to be four orbitals pointing at 109.5° from one another. Suitable orbitals that point in the right directions can be formed by mixing, or **hybridising**, the s and p atomic orbitals.

If the 2s, $2p_x$ and $2p_y$ orbitals are hybridised together, the result is three identical orbitals pointing at 120° from one another – which is exactly what we want! This type of hybridisation is called **sp²** (pronounced 'ess pee two'), and the three hybrid atomic orbitals are called sp² orbitals.

Similarly, if the 2s orbital is hybridised with all three of the 2p orbitals, four **sp³** ('ess pee three') hybrid orbitals are formed, which point at 109.5° from one another. Figure 3.36 shows what sp² and sp³ orbitals look like.

Figure 3.36 The sp² and sp³ orbitals

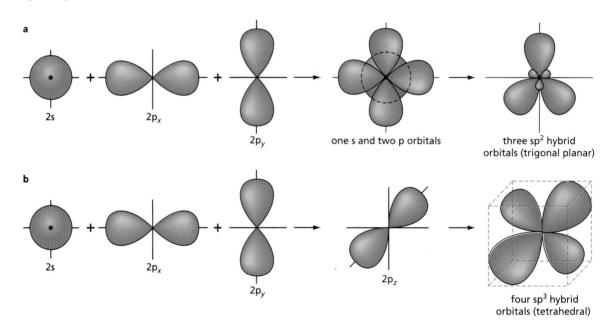

a

2s $2p_x$ $2p_y$ one s and two p orbitals   three sp² hybrid orbitals (trigonal planar)

b

2s $2p_x$ $2p_y$ $2p_z$ four sp³ hybrid orbitals (tetrahedral)

3.15 Localised π orbitals – the formation of π bonds

We now look at the orbitals that are used to form double bonds. We shall see that the way double bonds are formed can explain their shape and many of their properties.

In Figure 3.34 we have shown two p orbitals overlapping 'end-on' to form a σ bond. But it is also possible for two p orbitals to overlap sideways, to produce a molecular orbital which has two areas of electron density, one above and one below the axis joining the two atom centres (see Figure 3.37).

Figure 3.37 Sideways overlap of two p orbitals to form a π orbital

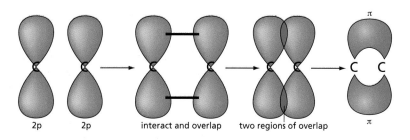

2p   2p   interact and overlap   two regions of overlap

The 'end-on' view of this orbital looks very like an atomic p orbital. It is called a π (pi) orbital, π being the Greek letter corresponding to p, and the resulting bond is called a **π bond**.

Figure 3.38 The π bond between two carbon atoms

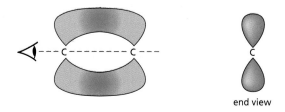

end view

This sideways overlap of two p orbitals is always accompanied by the end-on overlap of two other orbitals on the same atoms. So, although it is possible to join two atoms with only a σ bond, it is not possible to join two atoms with only a π bond. A **double bond** consists of a σ bond and a π bond between the same two atoms (see Figure 3.39).

Figure 3.39 A carbon–carbon double bond

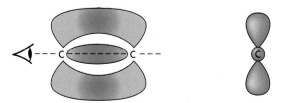

The bonding in ethene

The stronger of the two bonds in a double bond is usually the σ bond formed by end-on overlap, and the weaker bond is the π bond. The following data for ethane and ethene demonstrate this (assuming that the σ bond has the same strength in both compounds):

strength of C—C (single, σ) bond in ethane, CH_3—CH_3 = 376 kJ mol^{-1}
strength of C=C (double, σ + π) bond in ethene, CH_2=CH_2 = 720 kJ mol^{-1}

Therefore,

extra strength due to the π bond in ethene, CH_2=CH_2 = 344 kJ mol^{-1}

The full picture of the bonding in the ethene molecule can therefore be constructed as follows. Each carbon atom shares three electrons (one with each of the two hydrogen atoms, and one with the other carbon atom). These form five σ bonds (using sp^2 hybrid orbitals on carbon). The fourth electron in the valence shell on each carbon atom occupies an atomic p$_z$ orbital. These orbitals overlap with each other to form a π orbital. The π bond forms when the fourth electron on each atom is shared in this π orbital (see Figure 3.40).

Figure 3.40 The bonding in ethene

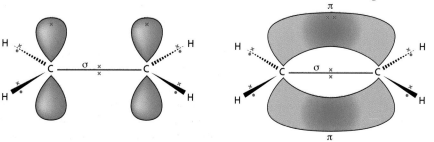

Mutual repulsion of the electrons in the three σ bonds around each carbon atom tends to place them as far apart from one another as possible (see page 51). The predicted angles (H—C—H = H—C—C = 120°) are very close to those observed (H—C—H = 117°, H—C—C = 121.5°).

The π bond confers various properties on the ethene molecule.

- Being the weaker of the two, the π bond is more reactive than the σ bond. Many reactions of ethene involve only the breaking of the π bond, the σ bond remaining intact.
- The two areas of overlap in the π bond (both above and below the plane of the molecule) cause ethene to be a rigid molecule, with no easy rotation of one end with respect to the other (see Figure 3.41). This is in contrast to ethane, and leads to the existence of *cis–trans* isomerism in alkenes.

These features are developed further in Topics 12 and 14.

Figure 3.41 The π bond prevents rotation around the carbon–carbon double bond.

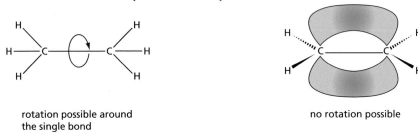

rotation possible around the single bond

no rotation possible

Triple bonds

Triple bonds are formed when two p orbitals from each of the bonding atoms overlap sideways, forming two π orbitals (in addition to the σ orbital formed by the end-on overlap of the third p orbital on each atom). This occurs in the ethyne molecule, C_2H_2, and in the nitrogen molecule, N_2 (see Figure 3.42).

Figure 3.42 Triple bonds in **a** nitrogen and **b** ethyne

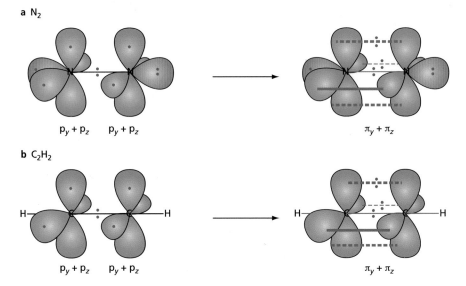

A single carbon atom can also use two of its p orbitals to form π orbitals with two separate bonding atoms. This occurs in the carbon dioxide molecule, CO_2 (see Figure 3.43).

Figure 3.43 The two π bonds in carbon dioxide

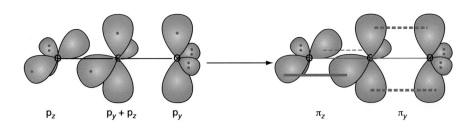

Worked example

Describe the bonding in and the shape of the molecules of:

a hydrogen cyanide, HCN

b methanal, CH_2O.

Answer

a The left-hand side of the HCN molecule is the same as one end of the ethyne molecule, and the right-hand side is the same as one end of the nitrogen molecule. HCN is linear, with one single bond and one triple bond. (One σ bond between H and C, one σ bond between C and N, and two π bonds between C and N.)

b The left-hand side of the CH_2O molecule is the same as one end of the ethene molecule, and the right-hand side is the same as one end of the carbon dioxide molecule. The molecule is triangular planar, with bond angles of approximately 120°.

Now try this

Describe the bonding in and the shape of each of the following molecules:

1 chloroethene, $CH_2{=}CHCl$
2 di-imide, $HN{=}NH$

3.16 Delocalised π orbitals

Delocalisation in benzene

Just as two p orbitals can overlap sideways to form a π bond, so two (or more) π bonds can overlap with each other to produce a more extensive π bond. The classic example is benzene, C_6H_6, whose molecule contains a planar, six-membered ring of sp^2-hybridised carbon atoms. The unhybridised p orbitals of the carbon atoms could overlap as shown in Figure 3.44a to produce three C=C double bonds between atoms 1 and 2, 3 and 4 and 5 and 6 of the ring, or just as likely, overlap as shown in Figure 3.44b to produce double bonds between atoms 2 and 3, 4 and 5, and 6 and 1 of the ring. In fact, both occur.

Figure 3.44 Alternative views of bonding between carbon pairs in benzene

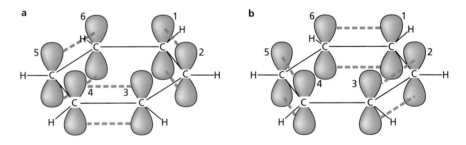

All six p orbitals overlap with each neighbour in an equal fashion, to produce a set of two doughnut-shaped ring orbitals (see Figure 3.45). This set of π orbitals can accommodate six electrons, none of which can be said to be localised between any two particular atoms. These π electrons are described as being **delocalised** around the ring. All six C—C bonds are the same length (shorter than a C—C single bond, but longer than a C=C double bond) and the shape of the ring is a regular hexagon, with every C—C—C angle being 120°. The interesting chemical properties that this ring of delocalised electrons gives benzene are discussed in Topic 25.

Figure 3.45 The delocalised bonding in benzene

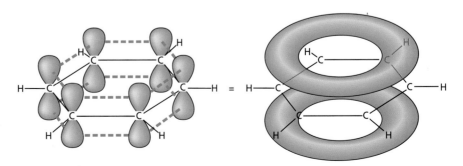

Delocalisation in the carboxylate ion

Another example of delocalisation occurs in the carboxylate anion (see Topic 18), which is sometimes represented as follows:

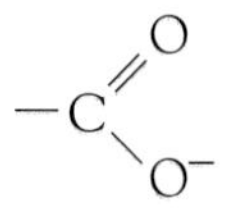

Here a lone pair of electrons in a p orbital on oxygen overlaps with the π orbital of an adjacent carbonyl group. Because of the symmetry of the system, the delocalisation is complete, resulting in each oxygen finishing with exactly half an electronic charge, on average. Also, each C—O bond has the same length, which is in between the lengths of the C—O single bond and the C=O double bond.

This structure can be derived from the simple localised model as follows. The C=O double bond consists of the usual σ bond + π bond pair. The singly bonded oxygen (which has an extra electron, making it anionic) contains three lone pairs of electrons. If one of the lone pairs is in a p orbital (see Figure 3.46a), this filled p orbital can overlap with the end of the π orbital on the carbon atom. The lone pair is thus partially donated, as a dative π bond (see Figure 3.46b).

Because carbon cannot have more than eight electrons in its valence shell, this partial donation by the lone pair on one oxygen will cause the π electrons in the C=O bond to drift towards the second oxygen atom. So the four π electrons (that is, the two in the C=O bond, and the two in the lone pair) become delocalised over the three atoms (see Figure 3.46c).

Figure 3.46 Delocalisation in the carboxylate ion

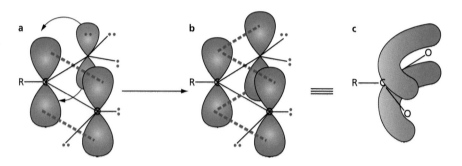

Delocalisation is sometimes represented by two (or more) localised structures joined by double-headed arrows, as in Figure 3.47.

Figure 3.47 The representation of delocalisation of charge in the carboxylate ion

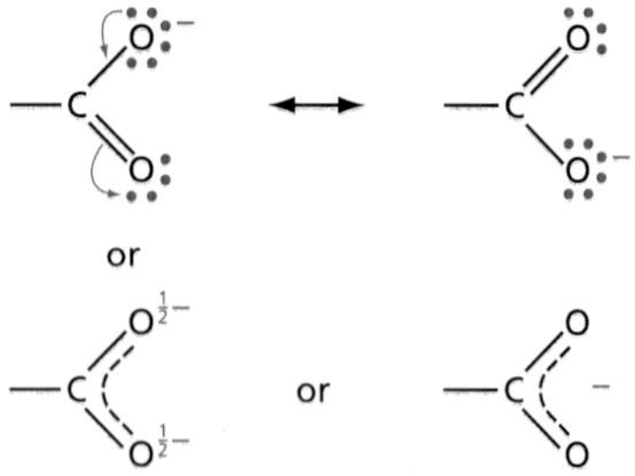

The double-headed arrow should be read as 'the actual structure is somewhere between the two extremes drawn here'. The curly arrows represent the movements of electron pairs that convert one extreme into the other.

The curly-arrow convention is often used when describing the movement of electrons, and the formation and breaking of bonds, during the reactions of organic compounds. See section 12.8 for a full description of what it means.

Worked example

All three C—O bonds in the carbonate ion are the same length. Use the idea of delocalisation to describe the bonding in the CO_3^{2-} anion, and to predict its shape.

Answer

If all the bonds in the carbonate ion were localised, it would consist of a carbonyl group to which are attached (by means of single σ covalent bonds) two oxygen atoms, each of which has its valence shell filled by an extra electron, giving it a negative charge (see Figure 3.48).

Figure 3.48 Localised bonding in CO_3^{2-}

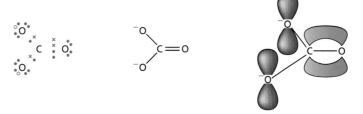

Just as in the carboxylate ion, each of the two singly bonded oxygen atoms has an extra electron, making it anionic. So each of these atoms contains three lone pairs of electrons. If one of the lone pairs is in a p orbital, this filled p orbital can overlap with the end of the π orbital on the central carbon atom. The lone pair is thus partially donated, as a dative π bond. The p orbital on each oxygen atom can overlap with the π orbital of the C=O group, creating a four-centre delocalised π orbital. There is nothing to distinguish one C—O bond from the other two, so all three are of equal length, and each oxygen atom carries, on average, $\frac{2}{3}$ of a negative charge (overall charge = $3 \times -\frac{2}{3} = -2$). The shape will be triangular planar, with all O—C—O bond angles 120° (see Figure 3.49).

Figure 3.49 Orbital overlap leading to delocalised bonding in CO_3^{2-}

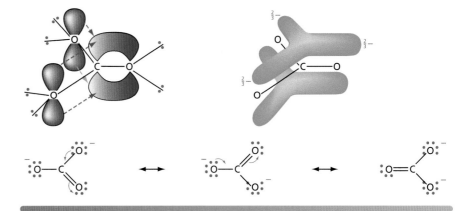

Now try this

All three N—O bonds in the nitrate ion, NO_3^-, are the same length, and the ion has a triangular planar shape. Describe the bonding in the nitrate ion. (Hint: there are three types of bond here – single (σ) bonds, dative single (σ) bonds, and delocalised π bonds.)

3.17 Intermolecular forces

Most of the covalently bonded particles we have looked at in this topic are simple covalent molecules. They differ from giant covalent, ionic and metallic lattices (see Topic 4) by having very low melting and boiling points. They are often gases or liquids at room temperature. This is because, although the covalent bonding within each molecule is very strong, the **intermolecular attractions** between one molecule and another are comparatively weak. When a substance consisting of simple covalent molecules melts or boils, no covalent bonds within the molecules are broken. The increased thermal energy merely overcomes the weak intermolecular forces. This requires little energy, and hence only a low temperature is required.

Why are these molecules attracted to one another at all? There are three main categories of intermolecular attraction, which we shall look at in turn. All are electrostatic in origin, as are all forces in chemistry. Strictly speaking, they can all be grouped under the umbrella term **van der Waals' forces,** although this term is now often used to describe only the instantaneous dipole force (see below).

Permanent dipole–dipole forces

As we saw on page 54, the uneven distribution of electronic charge within some molecules results in their having a permanent dipole. The positive end of one molecule's dipole can attract the negative end of another's, resulting in an intermolecular attraction in the region of $1–3\,kJ\,mol^{-1}$ (that is, at least 100 times weaker than a typical covalent bond). Such attractions are called **permanent dipole–dipole forces**, and an example is shown in Figure 3.50.

Figure 3.50 Dipole–dipole forces in hydrogen chloride

Instantaneous dipole forces

Being much lighter, the electrons within a molecule are much more mobile than the nuclei. This constant movement of negative charge produces a fluctuating dipole. If at any one instant a non-polar molecule possesses such an **instantaneous dipole** (owing to there being at that instant more electrons on one side of the molecule than the other), it will induce a corresponding dipole in a nearby molecule. This will result in an attraction between the molecules, called an **instantaneous dipole force** or an **induced dipole force** (also termed a **van der Waals' force**). The situation is similar to a magnet picking up a steel nail by inducing a magnetic moment in the nail, and hence attracting it.

An instant later, the first molecule might change its dipole through another movement of electrons within it, so the original dipole attraction might be destroyed. This changed dipole will induce new dipoles in molecules that surround it, however, and so intermolecular attraction will continue. The strength of the attraction decreases rapidly with distance ($F \propto 1/r^6$), and so it is only a very short-range force. Its magnitude depends on the number of electrons within a molecule, because the more electrons a molecule has, the larger the chance of an instantaneous dipole, and the larger that dipole is likely to be. Its magnitude also depends on the molecular shape, because the greater the 'area of close contact', the larger the force. Induced dipoles occur in all molecules, whether or not they have a permanent dipole, and whether or not they hydrogen bond with one another (see below). The magnitude of the instantaneous dipole force, as mentioned above,

Table 3.4 Boiling points of some molecular elements and compounds. None of these molecules has a permanent dipole, so the higher boiling points are due entirely to larger instantaneous dipoles.

Molecule	Boiling point/°C	Comments
CH_4	−162	These molecules are the same shape (tetrahedral), but the number of electrons increases going down the group, from 10 in CH_4 to 18 in SiH_4.
SiH_4	−112	
F_2	−188	All of these are linear diatomic molecules. Down the group the number of electrons increases, from 18 in F_2 to 34 in Cl_2 and 70 in Br_2.
Cl_2	−34	
Br_2	+58	
CH_4	−161	Along the alkane series the molecules are becoming longer (larger surface area of contact) and contain more electrons (an extra 8 for each CH_2 group).
C_2H_6	−89	
C_3H_8	−42	
C_4H_{10}	0	
$CH_3-\overset{\overset{\displaystyle CH_3}{\vert}}{\underset{\underset{\displaystyle CH_3}{\vert}}{C}}-CH_3$	+10	These two isomers have the same number of electrons, but the first is tetrahedral – almost spherical if the electron clouds around the atoms are taken into account – and the second is 'sausage shaped'. The area of close contact is therefore greater in the second one.
$CH_3CH_2CH_2CH_2CH_2CH_3$	+36	

depends on the size of the molecule, but is approximately 5–15 kJ mol^{-1}. The effect of this force is seen in the trends in boiling points shown in Table 3.4 – the greater the instantaneous dipole force, the higher the boiling point.

Hydrogen bonding

When a hydrogen atom bonds to another atom, its single 1s electron is taken up in bonding, and so it is located in a σ bonding orbital between it and the atom it is bonded with. Its other side, therefore, contains no electronic charge. It is a bare proton, with a significant δ+ charge. The positive charge is even greater if the hydrogen is bonded to an electronegative atom such as nitrogen, oxygen or fluorine. This highly δ+ hydrogen can experience a particularly strong attraction from a lone pair of electrons on an adjacent molecule, creating an intermolecular attractive force of about 20–100 kJ mol^{-1}. Such a force is called a **hydrogen bond**. It is roughly 10 times as strong as the dipole–dipole force or the instantaneous dipole force, and is only about 10 times weaker than a typical covalent bond. Atoms that contain lone pairs of electrons in orbitals that are small enough to interact with these δ+ hydrogen atoms also happen to be nitrogen, oxygen and fluorine. So it is between these elements that hydrogen bonding tends to occur (see Figure 3.51).

As a result of hydrogen bonding:

Figure 3.51 Hydrogen bonding in ammonia

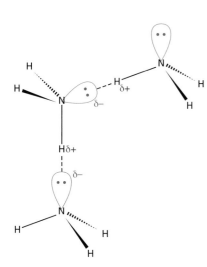

- the boiling points of ammonia, NH$_3$; water, H$_2$O; and hydrogen fluoride, HF; are considerably higher than that of methane, CH$_4$ (see Figure 3.52)
- the boiling points of ammonia, water and hydrogen fluoride are also considerably higher than those of other hydrides in their groups (see Figure 3.53)
- proteins fold in specific ways, giving them specific catalytic properties (see Topic 28, secondary and tertiary structure of proteins)
- in nucleic acids, the bases form bonds with specific partners, allowing the genetic code to be replicated without errors of transcription (see Figure 3.54) (see Topic 28, double helix)
- the boiling points of alcohols are considerably higher than those of alkanes with the same number of electrons, and hence roughly the same van der Waals' force (see Table 3.5).

Figure 3.52 The enhanced boiling points of NH₃, H₂O and HF

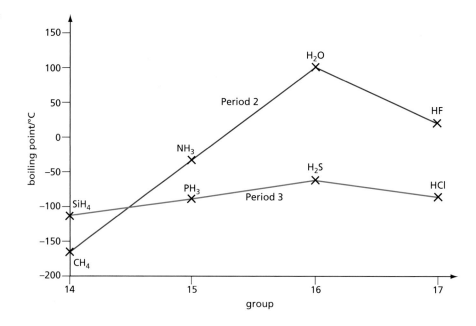

Figure 3.53 Boiling points of the hydrides of Groups 14 to 17

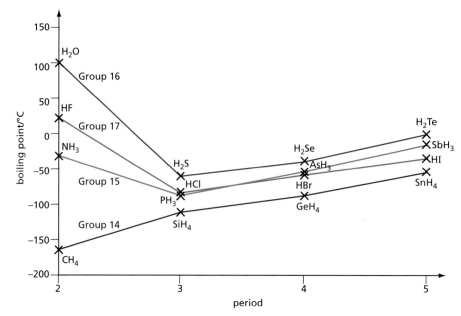

Figure 3.54 Thymine, adenine, cytosine and guanine are bases in DNA. The hydrogen bonding between them allows the DNA molecule to replicate (make new copies of itself) and also allows the genetic code to be expressed in the synthesis of proteins.

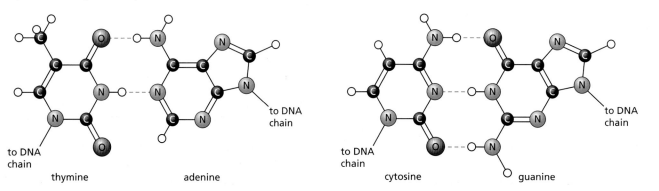

Table 3.5 Comparing the boiling points of alcohols and alkanes

Compound	Formula	Number of electrons	Boiling point/°C	Difference/°C
ethane	CH_3-CH_3	18	−89	154
methanol	CH_3-OH	18	+65	
propane	$CH_3-CH_2-CH_3$	26	−42	121
ethanol	CH_3-CH_2-OH	26	+79	
butane	$CH_3-CH_2-CH_2-CH_3$	34	0	97
propan-1-ol	$CH_3-CH_2-CH_2-OH$	34	+97	

Water shows several unusual properties owing to the large degree of hydrogen bonding within it. These are described in section 4.4.

Worked example

Describe all the intermolecular forces that can occur between the molecules in each of the following liquids, and hence predict which will have the lowest boiling point, and which will have the highest.

A $CH_3CH_2CH_2CH_3$ **B** CH_2ClCH_2OH **C** CH_3CHCl_2

Answer
Molecules of **A** will experience only van der Waals' (induced dipole) forces. Molecules of **B** will experience dipole–dipole forces (due to the C—Cl and C—O polar bonds), and hydrogen bonding (due to the –O—H groups), as well as van der Waals' forces. Molecules of **C** will experience dipole–dipole forces (due the dipoles of the C—Cl bonds not cancelling each other out) in addition to van der Waals' forces.

Therefore **A** will have the least intermolecular attraction, and hence the lowest boiling point, and **B** will have the highest.

Now try this

1 Describe the intermolecular forces in the following compounds, and place them in order of increasing boiling points:

   CH_2OH-CH_2OH CH_2Br-CH_2Br CH_2Cl-CH_2Cl.

2 Suggest a reason why the differences between the boiling points of the alkanes and alcohols in Table 3.5 become smaller as the chain length increases. (Hint: think of how the intermolecular forces other than hydrogen bonding are changing as chain length increases.)

Summary

- **Chemical bonding** results from the attraction of one atom's electrons by the nucleus of another atom.
- Atoms are more stable when they are bonded together than when they are free.
- A **covalent bond** is due to two atoms sharing a pair of electrons.
- An atom can often form as many bonds to other atoms as it has electrons in its **valence shell**, with the exception that, if the element is in Period 2, it cannot have more than eight electrons in its valence shell.
- Valence-shell electrons not involved in bonding arrange themselves into **lone pairs** around the atom.
- **Dative bonding** (**co-ordinate bonding**) occurs when one of the bonded atoms contributes both the bonding electrons.
- Double and triple bonds can occur if two atoms share two or three electron pairs between them.

- The shape of a covalent molecule is determined by the number of electron pairs there are in the valence shell of the central atom, according to **VSEPR theory**.
- If two bonded atoms differ in **electronegativity**, the shared electron pair is closer to the more electronegative atom. This results in the bond having a **dipole**.
- Complete molecules can have **dipole moments** if their bond dipoles do not cancel each other out.
- **Isoelectronic** molecules and ions usually have the same shape as one another.
- The overlap of atomic orbitals on adjacent atoms results in σ or π molecular orbitals.
- There are three types of **intermolecular forces**: permanent dipole, induced dipole and hydrogen bonding. Their strengths increase in this order.
- The strengths of intermolecular forces determine the physical properties of a substance, such as its boiling point.

Examination practice questions

Please see the data section of the CD for any A_r values you may need.

1 This question is about different models of bonding and molecular shapes.

  a Magnesium sulfide shows ionic bonding.

    i What is meant by the term *ionic bonding*?

    ii Draw a '*dot-and-cross*' diagram to show the bonding in magnesium sulfide. Show outer electron shells only.

  b '*Dot-and-cross*' diagrams can be used to predict the shape of covalent molecules.

    Fluorine has a covalent oxide called difluorine oxide, F_2O. The oxygen atom is covalently bonded to each fluorine atom.

    i Draw a '*dot-and-cross*' diagram of a molecule of F_2O. Show outer electron shells only.

    ii Predict the bond angle in an F_2O molecule. Explain your answer.

      [OCR Chemistry A Unit F321 Q3 (part) January 2010]

2 The structural formulae of water, methanol and methoxymethane, CH_3OCH_3, are given below.

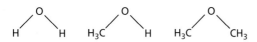

  a **i** How many lone pairs of electrons are there around the oxygen atom in methoxymethane?

    ii Suggest the size of the C—O—C bond angle in methoxymethane. [2]

The physical properties of a covalent compound, such as its melting point, boiling point, vapour pressure, or solubility, are related to the strength of attractive forces between the molecules of that compound.

These relatively weak attractive forces are called intermolecular forces. They differ in their strength and include the following:

A interactions involving permanent dipoles

B interactions involving temporary or induced dipoles

C hydrogen bonds

  b By using the letters **A**, **B**, or **C**, state the **strongest** intermolecular force present in **each** of the following compounds.

    i ethanal, CH_3CHO

    ii ethanol, CH_3CH_2OH

    iii methoxymethane, CH_3OCH_3

    iv 2-methylpropane, $(CH_3)_2CHCH_3$ [4]

  c Methanol and water are completely soluble in each other.

    i Which intermolecular force exists between methanol molecules and water molecules that makes these two liquids soluble in each other?

    ii Draw a diagram that clearly shows this intermolecular force. Your diagram should show any lone pairs or dipoles present on either molecule that you consider to be important. [4]

  d When equal volumes of ethoxyethane, $C_2H_5OC_2H_5$, and water are mixed, shaken, and then allowed to stand, two layers are formed.

    Suggest why ethoxyethane does not fully dissolve in water. Explain your answer. [2]

    [Cambridge International AS & A Level Chemistry 9701, Paper 2 Q1 June 2008]

3 Linus Pauling was a Nobel prize winning chemist who devised a scale of electronegativity. Some Pauling electronegativity values are shown in the table.

Element	Electronegativity
B	2.0
Br	2.8
N	3.0
F	4.0

  a What is meant by the term *electronegativity*? [2]

  b Show, using δ+ and δ− symbols, the permanent dipoles on each of the following bonds.

    N—F   N—Br [1]

  c Boron trifluoride, BF_3, ammonia, NH_3, and sulfur hexafluoride, SF_6, are all covalent compounds. The shapes of their molecules are different.

    i State the shape of a molecule of SF_6. [1]

    ii Using outer electron shells only, draw '*dot-and-cross*' diagrams for molecules of BF_3 and NH_3.

      Use your diagrams to explain why a molecule of BF_3 has bond angles of 120° and NH_3 has bond angles of 107°. [5]

    iii Molecules of BF_3 contain polar bonds, but the molecules are non-polar.

      Suggest an explanation for this difference. [2]

      [OCR Chemistry A Unit F321 Q3 January 2011]

4 Solids, liquids and gases

In Topic 3 we saw how atoms bond together covalently to form molecules. We finished that topic with a brief look at how the molecules themselves might attract one another. In this topic we examine how these attractions between molecules, and between other particles, can help to explain the differences between solids, liquids and gases. We explore the giant structures that atoms and ions can form, in which it is impossible to say where one structural unit finishes and the next one starts. We then take a closer look at gases, and see how the simple assumptions of the kinetic theory of gases lead us to the ideal gas equation, pV = nRT. Finally we see how the behaviour of real gases can differ markedly from that of an ideal gas.

Learning outcomes

By the end of this topic you should be able to:

3.1a) describe ionic (electrovalent) bonding, as in sodium chloride and magnesium oxide, including the use of 'dot-and-cross' diagrams

3.4a) describe metallic bonding in terms of a lattice of positive ions surrounded by delocalised electrons

3.5a) describe, interpret and predict the effect of different types of bonding (ionic bonding, covalent bonding, hydrogen bonding, other intermolecular interactions, metallic bonding) on the physical properties of substances

3.5b) deduce the type of bonding present from given information

4.1a) state the basic assumptions of the kinetic theory as applied to an ideal gas

4.1b) explain qualitatively in terms of intermolecular forces and molecular size the conditions necessary for a gas to approach ideal behaviour, and the limitations of ideality at very high pressures and very low temperatures

4.1c) state and use the general gas equation $pV = nRT$ in calculations, including the determination of M_r

4.2a) describe, using a kinetic-molecular model, the liquid state, melting, vaporisation, vapour pressure

4.3a) describe, in simple terms, the lattice structure of a crystalline solid which is ionic (as in sodium chloride, magnesium oxide), simple molecular (as in iodine and the fullerene allotropes of carbon – C_{60} and nanotubes only), giant molecular (as in silicon(IV) oxide and the graphite, diamond and graphene allotropes of carbon), hydrogen-bonded (as in ice) or metallic (as in copper)

4.3c) outline the importance of hydrogen bonding to the physical properties of substances, including ice and water (for example, boiling and melting points, viscosity and surface tension)

4.3d) suggest from quoted physical data the type of structure and bonding present in a substance.

4.1 The three states of matter

There are three states of matter – solid, liquid and gas (ignoring plasmas, which only form under extreme conditions). These are sometimes called **phases**. If we consider a substance that we experience every day in all three states, such as water, we can appreciate that the solid phase (ice) occurs at cold temperatures ($T < 0\,°C$); the liquid phase (water) occurs at intermediate temperatures ($0\,°C < T < 100\,°C$); and the gaseous phase (steam) occurs at high temperatures ($100\,°C < T$). We can see that in order to change a solid into a liquid, or a liquid into a gas, we need to give it energy, in the form of heat (thermal energy). This energy goes into overcoming the attractions between the particles. If we supply heat to a block of ice at a constant rate, and measure its temperature continuously, we obtain a graph like the one shown in Figure 4.1.

Figure 4.1 The change in temperature over time as a block of ice is heated

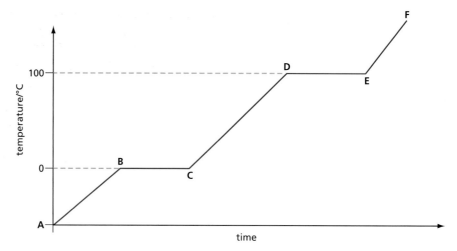

Changes in the molecular behaviour in the five regions of this graph can be described as follows.

- Between **A** and **B**, the ice is warming up. The H_2O molecules are fixed in the ice lattice (see page 71), but are vibrating more and more energetically from **A** to **B**.
- At **B**, the ice begins to melt. Here, the molecules are vibrating so strongly that they can begin to break away from their neighbours and move about independently.
- Between **B** and **C**, H_2O molecules continue to break away from the lattice. Although heat is still being absorbed by the ice–water mixture, the temperature does not rise. This is because any extra energy that the molecules in the liquid phase may pick up is soon transferred by collision to the ice lattice, where it causes another H_2O molecule to break away from the ice lattice. Not until all the ice has become liquid will the temperature start to rise again.
- From **C** to **D**, the liquid water is warming up. As the temperature increases, the H_2O molecules gain more kinetic energy, and move around faster. In cold water there are small closely knit groups of H_2O molecules, containing 5–12 molecules in each group, and, as the water warms, the extra energy also causes these to break up.
- At **D**, the H_2O molecules are moving so fast that they can overcome the remaining forces (mostly hydrogen bonds) that hold them together. The water starts to boil. The molecules evaporate to become single H_2O molecules in the gas phase, separated by distances of many molecular diameters. Just as in the region **B** to **C**, the temperature remains constant until the last H_2O molecule has boiled away.
- From **E** to **F**, the individual H_2O molecules in the steam gain more and more kinetic energy, so they travel faster and faster. If the steam is in an enclosed volume, this will cause the pressure to increase, as the molecules hit the walls of the container with increasing speed. If, on the other hand, the pressure is allowed to remain constant, the volume of gas will expand.

Worked example 1

A mole of liquid water has a volume of $18\,cm^3$, and a mole of steam at $100\,°C$ and room pressure has a volume of $33\,dm^3$. By how many times has the volume increased?

Answer
The volume has increased by $\dfrac{33 \times 10^3}{18} =$ **1833** times. (Remember, $1\,dm^3 = 10^3\,cm^3$.)

Worked example 2

Assuming that all the H_2O molecules in liquid water are touching one another, roughly how many molecular diameters are they apart in steam?

Answer
The increase in linear distance between the molecules is the cube root of the increase in volume. So in steam the molecules are $\sqrt[3]{1833} \approx$ **12** molecular diameters apart.

Table 4.1 summarises the differences between solids, liquids and gases.

Table 4.1 The positions and movement of the particles in solids, liquids and gases

State	Relative position of particles	Relative movement of particles
solid	touching one another	fixed in a regular three-dimensional lattice; can only vibrate about fixed positions
liquid	touching one another	moving randomly through the liquid, often in weakly bonded groups of several molecules at a time; vibration, rotation and translational motion are allowed.
gas	far apart from one another	individual molecules moving randomly at high speeds, colliding with, and bouncing off, one another and the walls of their container

4.2 Lattices

For most substances, the solid and liquid phases usually have about the same density. This is because in both solids and liquids the particles are touching one another, with little space in between them. As we have just mentioned, the major difference between solids and liquids is whether or not the particles have translational motion. In liquids, the particles move fairly randomly and are continually sliding over one another. In solids, on the other hand, the particles do not move around from one place to another. They can only vibrate around fixed positions.

There are two major types of solids – crystalline and non-crystalline solids.

In **non-crystalline** (or **amorphous**) solids, the particles are not arranged in any order, or pattern. They are fixed in random positions. If you took a video of a liquid, showing the particles moving around chaotically, and froze a frame from that video, the picture would be indistinguishable from that of a non-crystalline solid. Many non-crystalline solids, for example glass, are often called 'supercooled liquids' (see Figure 4.2).

In **crystalline** solids, the particles are arranged in a regular three-dimensional pattern or array. This is called a **lattice** (see Figure 4.3). Every lattice can be thought of as being made up from lots of subunits or building blocks stacked one on top of the other, in rows and columns. These repeating units are called **unit cells**. Once we have understood the geometry of the unit cell, and how the atoms are joined together in it, we are well on the way to understanding what gives crystalline solids their shapes, and their other physical properties.

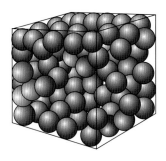

Figure 4.2 The structure of a non-crystalline (amorphous) solid. The particles are in a random arrangement.

Figure 4.3 The structure of a crystalline solid, zinc (left). The particles of sodium chloride (right) are in a regular lattice arrangement.

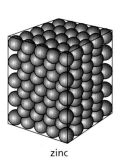

zinc

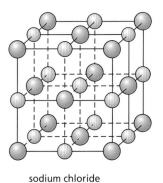

sodium chloride

4.3 The building blocks of lattices

The ultimate building blocks of matter are, of course, atoms. The lattices of most *elements* are composed simply of individual atoms. We can distinguish three types of **atomic lattice**:

- **monatomic molecular lattices**, such as in solid argon and the other noble gases
- **macromolecular covalent lattices**, such as in diamond and silicon
- **metallic lattices**, such as in iron and all other solid metals.

In addition, some elements that form polyatomic molecules, such as the non-metals in Groups 15, 16 and 17 (e.g. P_4, S_8 and I_2) exist as **simple molecular lattices**.

The lattice building blocks of solid *compounds* contain more than one element. There are three types:

- **simple molecular lattices**, such as in ice
- **giant molecular lattices**, such as in silicon(IV) oxide
- **ionic lattices**, such as in sodium chloride and calcium carbonate.

Each type of lattice confers different physical properties upon the substance that has it. We shall explore these properties in the sections that follow.

4.4 Simple molecular lattices

In simple molecular lattices, including molecular atomic lattices, there is no chemical bonding between the particles, just dipole–dipole forces, instantaneous dipole forces or hydrogen bonding attractions, which are comparatively weak. Figure 4.4 shows examples of simple atomic and molecular lattices. The molecules or atoms are packed tightly, but only a small amount of thermal energy is required to overcome the intermolecular forces and break the lattice. Substances containing this type of lattice therefore have low melting points (and low boiling points). As we saw in section 3.17, boiling points depend on the strength of the intermolecular bonding. The same is true for melting points (see Table 4.2).

Figure 4.4 In solid argon, a simple atomic lattice, the only intermolecular forces are weak instantaneous dipole forces. In solid hydrogen chloride, a simple molecular lattice, there are dipole–dipole forces. Solid iodine has strong instantaneous dipole forces.

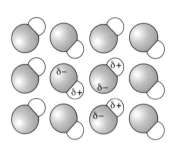

solid hydrogen chloride – dipole–dipole forces

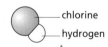

chlorine
hydrogen

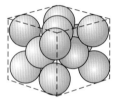

solid argon – weak instantaneous dipole forces

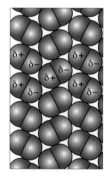

solid iodine – strong instantaneous dipole forces

Table 4.2 Intermolecular forces and melting points for some substances

Substance	Formula	Main type of intermolecular attraction	Melting point/°C
argon	Ar	weak instantaneous dipole	−189
hydrogen chloride	HCl	dipole–dipole	−115
water	H_2O	hydrogen bonding	0
iodine	I_2	strong instantaneous dipole	114
sucrose	$C_{12}H_{22}O_{11}$	strong hydrogen bonding	185

The properties of water are in several ways rather different from those of other simple molecular substances. Unlike the molecules of other compounds capable of forming hydrogen bonds, water has *two* lone pairs of electrons together with *two* δ+ hydrogen atoms. This means that it can form *two* hydrogen bonds per molecule on average, whereas alcohols, ammonia and hydrogen fluoride can only form *one*

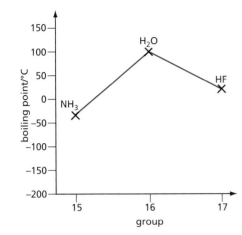

Figure 4.5 The boiling points of NH₃, H₂O and HF

hydrogen bond per molecule. This means that the hydrogen bonding in water is more extensive than in other compounds, which has the following effects on the properties of water.

- The boiling point of water is much higher than those of ammonia and hydrogen fluoride (see Figure 4.5).
- Liquid water has a high surface tension. The strongly hydrogen-bonded molecules form a lattice across the surface of water, allowing objects which you might expect to sink in water, such as pond skaters and even coins (if you are careful), to 'float' on water (see Figures 4.6 and 4.7).

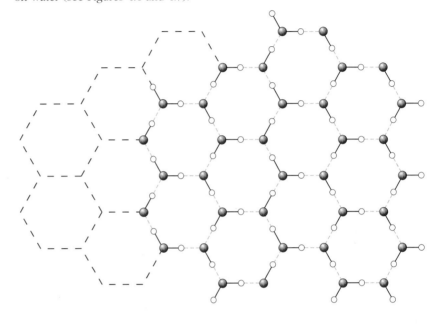

Figure 4.6 Hydrogen bonding on the surface of water forms a hexagonal array which provides a high surface tension.

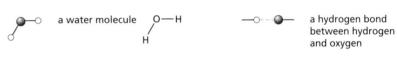

Figure 4.7 The surface tension of water supports the pond skater – the surface is depressed but the hydrogen bonds hold it together.

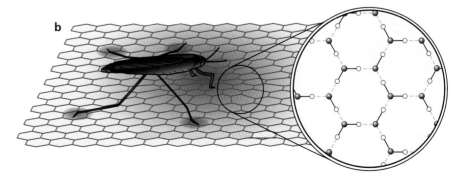

● Ice is less dense than liquid water. The hydrogen bonds between the water molecules in ice are positioned roughly tetrahedrally around each oxygen atom. This produces an open lattice, with empty spaces between some water molecules (see Figure 4.8). The more random arrangement of hydrogen bonds in liquid water takes up less space.

Figure 4.8 In ice, the molecules are hydrogen bonded in a tetrahedral arrangement, which makes ice less dense than liquid water.

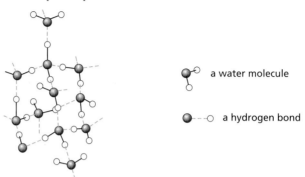

a water molecule

a hydrogen bond

For examples of carbon-containing simple lattices, see section 4.12, page 87.

4.5 Giant molecular lattices

These are sometimes referred to as macromolecular lattices. They consist of three-dimensional arrays of atoms. These atoms can be either all of the same type, as in the elements carbon (diamond, graphite or graphene, see page 84) and silicon, or of two different elements, such as in silicon(IV) oxide or boron(III) nitride. The atoms are all joined to one another by covalent bonds. So a single crystal of diamond or quartz is in fact a single molecule, containing maybe 1×10^{23} atoms. For atoms to become free from the lattice, these strong bonds have to be broken. Substances containing this type of lattice therefore have very high melting points and also high boiling points (see Table 4.3).

Table 4.3 Melting and boiling points are high for substances with a giant covalent lattice. (Note that boron(III) nitride sublimes – it changes straight from a solid to a gas when heated, so its melting and boiling points are the same.)

Substance	Formula	Type of interatomic attraction	Melting point/°C	Boiling point/°C
silicon	Si	giant covalent	1410	2355
silicon(IV) oxide	SiO_2	giant covalent	1610	2230
boron(III) nitride	BN	giant covalent	3027	3027
diamond	C	giant covalent	3550	4827

Figure 4.9 Giant molecular lattices have strong covalent bonds in three dimensions.

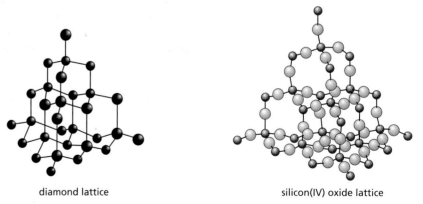

diamond lattice

silicon(IV) oxide lattice

The properties of giant covalent lattices (the terms 'giant molecular' and 'macrocovalent' are also used to describe these lattices) result from their structure and bonding.

● Their melting points are high, as described above.
● They are electrical insulators, because all their valence electrons are involved in localised covalent bonds, and so are unable to move when a potential difference is applied.
● They are hard, strong and non-malleable, because their covalent bonds are both strong and point in fixed directions, towards each adjacent atom.

Ceramics are inorganic compounds (usually oxides) that are hard and inert, with high melting points. They are often good electrical and thermal insulators, and find uses in furnaces, high voltage insulators and heat-resistant tiles (e.g. on the surface of the space shuttle). Common ceramics include the giant covalent silicon dioxide and the giant ionic magnesium oxide and aluminium oxide.

What happens to the valencies on the edge of a diamond crystal?

Although the carbon atoms inside a diamond crystal are all joined to four other atoms, and so have their full complement of covalent bonds, those on the flat surface of the side of a crystal have only three carbon atoms joined to them. They have a spare valency. Those atoms occupying positions along a crystal edge or at an apex are likely to have two spare valencies.

How the spare valencies are 'used up' was a mystery for many years. Some thought that they joined up in pairs to form double bonds. However, this would mean that the surface carbon atoms would have to be sp^2 hybridised, rather than sp^3 as they are in the inside of the crystal. The mystery was solved by studying the surface of very clean diamond crystals using sensitive techniques such as photoelectron spectroscopy and scanning tunnelling electron microscopy.

The spare valencies are used up by bonding with atoms other than carbon. It was discovered that under normal conditions the surface of a diamond crystal is covered with hydrogen atoms, each atom singly bonded to a carbon atom. Diamond is therefore not an element in the true sense, but a hydrocarbon (a polycyclic alkane) with a very high carbon-to-hydrogen ratio!

By heating a diamond crystal to a temperature of 1000 °C in an extremely high vacuum, it is possible to drive off the hydrogen atoms. The resulting surface is highly reactive, and can bond strongly to other atoms, such as oxygen, amine (—NH$_2$) groups, other carbon-containing molecules such as alkenes and even some large biological molecules such as DNA. There is hope that new molecular semiconducting devices might be constructed on wafer-thin diamond surfaces in the future, making computers even smaller and more powerful.

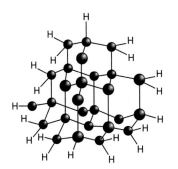

Figure 4.10 The carbon atoms on the edges of a diamond crystal have hydrogen atoms bonded to them.

 ## 4.6 Ionic bonding – monatomic ions

Ionic bonds are typically formed between a metallic element and a non-metallic element. Unlike covalent bonds, which involve the sharing of one or more pairs of electrons between two atoms, ionic bonding involves one atom (the metal) totally giving away one or more electrons to the non-metallic atom. This results in electrically charged atoms, called **ions**, being formed (see Figure 4.11).

Figure 4.11 Ionic lithium fluoride is formed by the transfer of an electron.

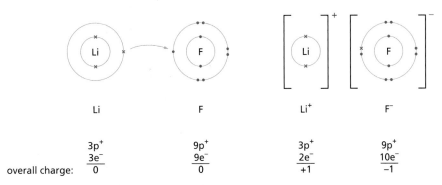

We shall delay taking a detailed look at the energetics of ionic bond formation until Topic 20 (though the panel on page 77 gives a brief account). It is instructive at this stage to compare the formation of an ionic bond with that of a covalent bond. We saw in section 3.11 that bonds between atoms of different electronegativities are polar, with an uneven distribution of electrons, and hence of electrical charge.

We can consider ionic bonding to be an extreme case: when the electronegativity difference between the two atoms is large enough, the bonding electrons will become *completely* transferred to the more electronegative atom, and an ionic bond results (see Figure 4.12).

Figure 4.12 Ionic bonds are formed between atoms of high electronegativity difference.

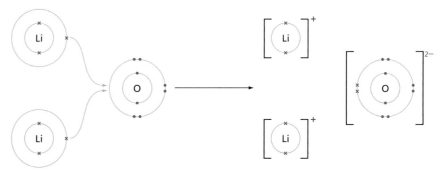

electronegativities: 4.0   4.0   2.2   4.0   1.0   4.0

difference: 0   1.8   3.0

When forming ionic bonds, metals usually lose all their outer-shell electrons. Their ionic valency therefore equals their group number. Likewise, non-metals usually accept a sufficient number of electrons to fill their outer shells. Their ionic valency therefore is $(18 - g)$ where g is the group number. (See Table 4.4, page 79, for a comprehensive list of ionic valencies.) The resulting ions combine in the correct proportions so as to cancel their charges. The compound is therefore electrically neutral overall.

In lithium oxide, for example, each lithium ion loses the electron in its second shell, and the oxygen atom accepts two electrons (one from each of two lithium atoms), forming a full octet in its second shell (see Figure 4.13). Its empirical formula is therefore Li_2O.

Figure 4.13 The formation of lithium oxide

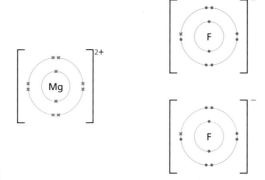

Worked example

Draw dot-and-cross diagrams showing the electronic configuration and charges in magnesium fluoride.

Answer
The 12 electrons in magnesium are arranged in the configuration 2.8.2. The Mg^{2+} ion has lost two outer-shell electrons, so it has the configuration 2.8.

The nine electrons in fluorine are in the configuration 2.7. The F^- ion has gained an electron, so it has the configuration 2.8. We need $2 \times F^-$ (see Figure 4.14).

Figure 4.14

Now try this

Draw dot-and-cross diagrams with charges showing the ionic bonding in magnesium oxide, MgO, lithium nitride, Li_3N, and aluminium oxide, Al_2O_3.

(Note the simplified representation of electronic configuration used here. Magnesium is $1s^2 2s^2 2p^6 3s^2$, shown as 2.8.2 by grouping together the electrons in each shell. Similarly, fluorine is $1s^2 2s^2 2p^5$, which becomes 2.7.)

Energetics of ionic bond formation – a brief summary

Ionic bonding involves a separation of charge, which is usually an unfavourable endothermic process. The reason why it becomes more favourable with metals is their comparatively low ionisation energies. In lithium fluoride, LiF, for example, the outer electron is fairly easily lost from the lithium atom, and becomes more attracted to the fluorine nucleus than to its original lithium nucleus. The benefits of total electron transfer are completed by the resulting electrostatic attraction between the cation and anion that are formed. Look carefully at the state symbols in the equations below. (For further descriptions of the terms used here, see Topic 20.)

$$Li(g) \rightarrow Li^+(g) + e^- \qquad \Delta H^\ominus = +513\,kJ\,mol^{-1} \text{ (ionisation energy)}$$
$$F(g) + e^- \rightarrow F^-(g) \qquad \Delta H^\ominus = -328\,kJ\,mol^{-1} \text{ (electron affinity)}$$

Therefore, in the gas phase:

$$Li(g) + F(g) \rightarrow Li^+(g) + F^-(g) \qquad \Delta H^\ominus = 513 - 328 = \mathbf{+185\,kJ\,mol^{-1}}$$

The fact that $\Delta H^\ominus$ is positive means that this is an energetically unfavourable process.

But when the ions form a solid lattice, the ions attract one another, and much energy is released:

$$Li^+(g) + F^-(g) \rightarrow Li^+F^-(s) \qquad \Delta H^\ominus = -1031\,kJ\,mol^{-1} \text{ (lattice energy)}$$

So the overall energy change, from gas-phase atoms to solid compound, is:

$$Li(g) + F(g) \rightarrow Li^+F^-(s) \qquad \Delta H^\ominus = 185 - 1031 = \mathbf{-846\,kJ\,mol^{-1}}$$

The complete process is seen to be highly exothermic, and therefore favourable.

4.7 Ionic bonding – compound ions

The ions that make up ionic compounds are not always monatomic (that is, they do not always contain *only one* atom). Several common ions contain *groups* of atoms which are covalently bonded together, and which have an overall positive or negative charge. Here are a few examples, some of which we came across in Topic 3.

The ammonium ion, NH_4^+

We first mentioned the ammonium ion in Topic 3 (pages 50 and 55). It is formed from one nitrogen atom and four hydrogen atoms, minus one electron (see Figure 4.15).

Figure 4.15 The formation of the ammonium ion

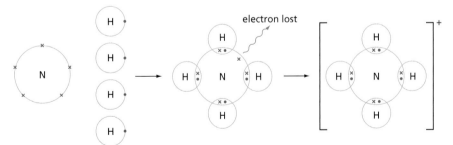

An alternative way of looking at its creation is by forming a dative bond from an ammonia molecule to a proton (see page 50). Figure 4.16 shows this approach.

Figure 4.16 The formation of the ammonium ion – a different view

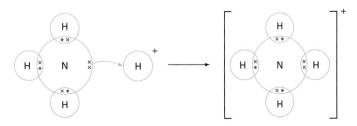

Either description can be used – the first method emphasises that all four bonds are exactly the same, while the second is a better representation of how the ammonium ion is usually formed, by reacting ammonia with an acid:

$$NH_3(g) + HCl(g) \rightarrow NH_4^+Cl^-(s)$$

The carbonate ion, CO_3^{2-}

To understand the bonding in polyatomic anions, it is often easier to look first at their corresponding acids. Chemically, carbonate ions are formed by reacting carbonic acid with a base, that is, by removing hydrogen ions from the acid. If this is done, we can see that the two hydrogen atoms leave their electrons on the carbonate ion, giving it a −2 charge (see Figure 4.17).

The delocalisation that occurs in this ion was discussed in section 3.16.

Figure 4.17 The carbonate ion

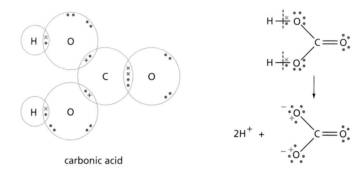

carbonic acid

The sulfate ion, SO_4^{2-}

To study the sulfate ion, let us look first at the corresponding acid. The bonding in sulfuric acid is best described as follows. The sulfur atom has six electrons in its outer shell. It can use four of these to form two double bonds to two oxygen atoms (just as sulfur dioxide, SO_2, see section 3.9). The other two electrons in the outer shell of sulfur can form two single bonds to the other two oxygen atoms. These singly-bonded oxygen atoms then in turn form single bonds to two hydrogen atoms (see Figure 4.18). (This bonding arrangement fits in with the respective covalencies given in Table 3.1, page 49.)

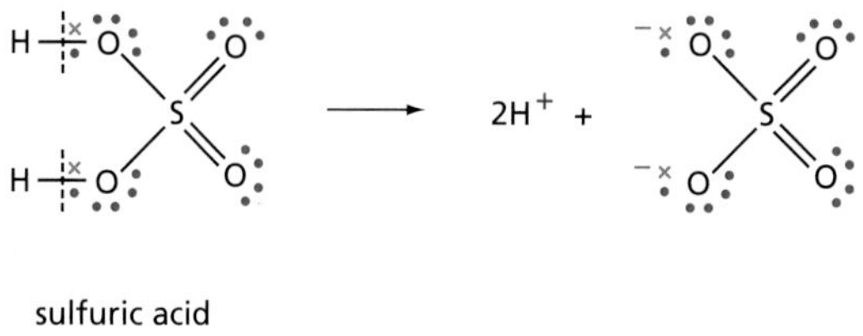

Figure 4.18 The sulfate ion

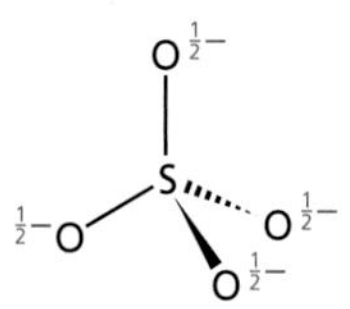

sulfuric acid

We can form sulfate ions by removing two hydrogen ions from sulfuric acid. Just as in the carbonate ion, delocalisation occurs, making all four S—O bonds equivalent. The shape of the SO_4^{2-} ion is a regular tetrahedron (see Figure 4.19).

Figure 4.19 In the sulfate ion, the charge is delocalised, and the bonds are all equivalent.

Worked example

a Deduce the formula of the ion formed by the loss of the H^+ ions from boric acid, H_3BO_3.
b What will be the overall charge on this ion, and hence the charge on each oxygen atom in it?
c Suggest a shape for the ion.

Answer
a If all three hydrogen atoms are lost as H^+ ions, the resulting ion will be **BO_3^{3-}**.
b The overall charge will therefore be **−3**, and each oxygen will have a charge of **−1**.
c Boron uses all its electrons in bonding with the oxygen atoms, so the shape will be triangular planar (see Figure 4.20).

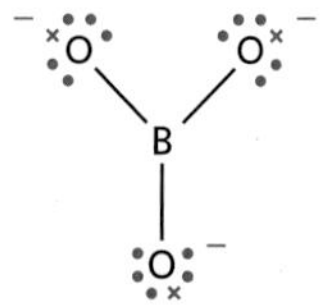

Figure 4.20

Now try this

Predict the shape of the ions derived from the following acids, and work out the fractional charge on each oxygen atom in each ion.

1 phosphonic acid, H_3PO_3

2 chloric(V) acid, $HClO_3$

3 chloric(VII) acid, $HClO_4$

4.8 The ionic valency table

The ionic valency of a single atom from Groups 1, 2 and 13–17 can easily be worked out from its group number. However, it is not so easy to predict the ionic valency of a transition metal, or of a compound ion. Table 4.4 includes all the valencies you will need to know.

Table 4.4 The ionic valencies of some common substances

Cations		Anions	
Name	**Formula**	**Name**	**Formula**
hydrogen	H^+	hydride	H^-
lithium	Li^+	fluoride	F^-
sodium	Na^+	chloride	Cl^-
potassium	K^+	bromide	Br^-
silver	Ag^+	iodide	I^-
copper(I)	Cu^+	oxide	O^{2-}
magnesium	Mg^{2+}	sulfide	S^{2-}
calcium	Ca^{2+}	nitride	N^{3-}
barium	Ba^{2+}	hydroxide	OH^-
copper(II)	Cu^{2+}	nitrite or nitrate(III)	NO_2^-
zinc	Zn^{2+}	nitrate or nitrate(V)	NO_3^-
iron(II)	Fe^{2+}	hydrogencarbonate	HCO_3^-
lead	Pb^{2+}	hydrogensulfate	HSO_4^-
aluminium	Al^{3+}	carbonate	CO_3^{2-}
iron(III)	Fe^{3+}	sulfite or sulfate(IV)	SO_3^{2-}
chromium	Cr^{3+}	sulfate or sulfate(VI)	SO_4^{2-}
ammonium	NH_4^+	phosphate	PO_4^{3-}

Ionic inorganic compounds are named according to the following general rules.

1 The name of the compound is usually made up of two words, the first of which is the name of the **cation** (the positive ion), and the second is the name of the **anion** (the negative ion).

2 The **oxidation state** (**oxidation number**) is the formal charge on a particular element in a compound or ion. Where elements can exist in different oxidation states (see section 7.2), the particular oxidation state it shows in the compound in question is represented by a Roman numeral. Therefore iron(III) represents iron with the oxidation number +3 (Fe^{3+}), and so iron(III) chloride is $FeCl_3$. This is particularly useful when referring to compound ions – the nitrogen atom in the nitrate(III) ion, NO_2^-, has an oxidation number of +3, whereas the nitrogen in the nitrate(V) ion, NO_3^-, has an oxidation number of +5.

3 Anions containing just a single atom have names ending in '-ide'. Anions containing oxygen as well as another element have names ending in '-ate'. If there is more than one anion containing a particular element combined with oxygen, the oxidation number of that element is indicated as in rule 2.

1 Use Table 4.4 to write the correct formula for:
a iron(II) fluoride
b magnesium nitride
c copper(I) oxide
d iron(III) hydroxide
e calcium phosphate
f ammonium sulfate
g copper(II) nitrate.

2 Write the names of the following ionic compounds.
a $FeSO_4$
b BaS
c $Mg(HCO_3)_2$
d KNO_2

3 Which of the following formulae are **incorrect**? For each incorrect formula, write the correct one.
a AgO
b $Ba(OH)_2$
c $Pb(NO_3)_3$
d AlI_2
e $Fe_2(SO_4)_3$

4 For some anions, the ending '-ite' is used in the traditional name to indicate that the particular element is combined with *some* oxygen, but not its maximum amount of oxygen. The recommended name uses the oxidation number instead. For example:

NO_2^- is either nitrite or nitrate(III)
NO_3^- is nitrate or nitrate(V)

SO_3^{2-} is either sulfite or sulfate(IV)
SO_4^{2-} is sulfate or sulfate(VI).

Use the ionic valency table (see Table 4.4) to predict the formula of:
a zinc nitrate
b aluminium sulfate.

Answer
a Zinc ions are Zn^{2+}, and nitrate ions are NO_3^-. To obtain a compound that is electrically neutral overall we need two nitrate ions to every one zinc ion:

$$Zn^{2+} + 2NO_3^- \rightarrow \mathbf{Zn(NO_3)_2}$$

b Aluminium ions are Al^{3+}, and sulfate ions are SO_4^{2-}, so we need two Al^{3+} (total charge = +6) for every three SO_4^{2-} (total charge = −6):

$$2Al^{3+} + 3SO_4^{2-} \rightarrow \mathbf{Al_2(SO_4)_3}$$

4.9 Ionic lattices

Co-ordination number

The sections above have described how individual ions form; we now consider how they collect together in a lattice. Clearly, oppositely charged ions will attract each other; also, ions of the same charge will repel each other. Each cation in an ionic lattice will therefore be surrounded by a number of anions as its closest neighbours, and each anion will be surrounded by a number of cations. We never find two cations adjacent to each other, nor can two anions be adjacent to each other. The number of ions that surround another of the opposite charge in an ionic lattice is called the **co-ordination number** of that central ion. The co-ordination number depends on two things – the relative sizes of the ions, and their relative charges.

The relative sizes of the ions

If one of the ions is very small, there will not be room for many oppositely charged ions around it. With Zn^{2+}, for example, the maximum co-ordination number is usually 4. If the cations and anions are nearly equal in size, one ion can be surrounded by eight others. The intermediate case of six neighbours occurs, as might be expected, when one ion is bigger than the other, but not by very much.

The relative charges of the ions

To gain electrical neutrality, a cation with a charge of +2 needs twice as many −1 ions as does a cation of charge +1. We therefore find that the Ca^{2+} ion in calcium chloride, $CaCl_2$, is surrounded by twice as many Cl^- ions as is the Li^+ ion in lithium chloride, $LiCl$.

These two factors are detailed in Table 4.5. Two of the ionic lattices from the table – those of sodium chloride, NaCl, and caesium chloride, CsCl – are shown in Figure 4.21, which clearly illustrates the differing co-ordination numbers of the ions in the structures.

Table 4.5 The co-ordination number depends on the relative sizes and the charges on the ions involved.

Compound	Cation radius/nm	Anion radius/nm	Anion radius / Cation radius	Co-ordination number of cation	Co-ordination number of anion
ZnS	0.08	0.19	2.4	4	4
NaCl	0.10	0.18	1.8	6	6
MgO	0.07	0.14	2.0	6	6
CsCl	0.17	0.18	1.1	8	8
CaF$_2$	0.10	0.13	1.3	8	4

Figure 4.21 Models of the lattices of sodium chloride, NaCl (**a** and **b**) and caesium chloride, CsCl (**c** and **d**); **b** and **d** show more clearly the alternating layers of ions.

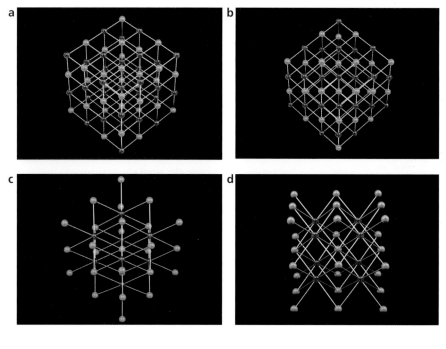

Figure 4.22 a Crystals of sodium chloride and **b** clear cubic crystals of calcium fluoride mixed with white dolomite

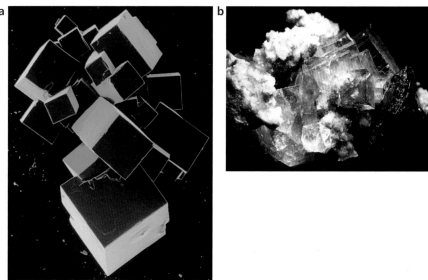

Ion polarisation and charge density

We saw in section 3.11 that many covalent bonds involve a slight separation of charge, due to the different electronegativities of the two atoms involved in the bond. This confers a small degree of ionic character on the covalent bond.

Starting from the other extreme, we find that a small degree of covalent character is apparent in some ionic bonds. This is due to polarisation of the (negative) anion by the (positive) cation. It occurs to the greatest extent when a small or highly charged cation is bonded to a large anion. The outer electrons around the large anion are not very firmly held by its nucleus, and can be attracted by the strong electric field around the small, highly charged cation. This causes an increase in electron density between the two ions, that is, a degree of localised covalent bonding (see Figure 4.23).

Figure 4.23 Ion polarisation between a small, highly charged cation and a large anion

The idea of ion polarisation is important in explaining the thermal decomposition of metal carbonates and nitrates (see section 10.4).

The strong electric field around small, highly charged cations attracts not only the outer-shell electrons in anions, but also the lone pairs on polar molecules such as water. We shall see in Topic 20 that the enthalpy changes of hydration of small ions are much more exothermic than those of larger ions.

The concept of charge density is a useful one for explaining the influence that an ion (usually a cation) has on the electrons in adjacent ions or molecules.

Imagine two +1 ions, one with a radius of 0.1 nm and one with a radius of 0.2 nm. For each ion, the +1 charge is spread over the surface of a sphere, for which the first ion has a surface area of $0.125 \, \text{nm}^2$ ($4\pi r^2$, with $r = 0.1$). The surface area of the second ion is $0.50 \, \text{nm}^2$, which is four times as large as the first (area $\propto r^2$).

The **charge density** (sometimes called **surface charge density**) is the amount of charge per unit area of surface. This is $\dfrac{1}{0.125} = 8$ electron units per nm^2 in the first case, and $\dfrac{1}{0.5} = 2$ electron units per nm^2 in the second. The extent of polarisation (and hence the attraction) of the electrons in adjacent anions, lone pairs and bonds depends on the charge density of the cation.

Small, highly charged ions (such as Al^{3+}) have a greater charge density than large, singly charged ions (such as Cs^+). If two ions have similar charge densities, this often results in their having similar properties.

4.10 Properties of ionic compounds

Melting and boiling points

In an ionic lattice, there are many strong electrostatic attractions between oppositely charged ions. We therefore expect that ionic solids will have high melting points. On melting, although the regular lattice is broken down, there will still be significant attractions between the ions in the liquid. This should result in high boiling points also. As we shall see in Topic 20, the ionic attractions are larger when the ions are smaller, or possess a larger charge. This is apparent from Table 4.6.

Table 4.6 Small, highly charged ions bond strongly, leading to high melting points.

Compound	Cation radius/nm	Anion radius/nm	Cation charge	Anion charge	Melting point/°C
NaCl	0.10	0.18	+1	−1	801
NaF	0.10	0.13	+1	−1	993
MgF₂	0.07	0.13	+2	−1	1261
MgO	0.07	0.14	+2	−2	2852

Strength and brittleness

If an ionic lattice is subjected to a shear or bending force, that is, a force that attempts to break up the regular array of ions, this will inevitably force ions of the same charge closer to each other (see Figure 4.24). The lattice resists this strongly. Ionic lattices are therefore quite hard and strong. If, however, the shearing force is strong enough, the lattice does not 'give', but breaks down catastrophically – the crystal shatters into tiny pieces. The strength of an ionic lattice is an 'all-or-nothing' strength.

Figure 4.24 A shearing force will shatter an ionic crystal.

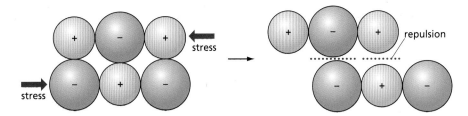

Electrical conductivity

The conduction of electricity is the movement of electrical charge. The individual ions that make up ionic compounds have a net overall electrical charge. If they are able to move, as they can when the compound is either molten or dissolved in water, then an electric current can flow. When in a solid lattice, however, the ions are fixed. Solid ionic compounds do not conduct electricity.

The electrical conductivity of a solution of an ionic compound increases as its concentration increases, because there are more ions to carry the current. Similarly, for the same concentration (say $1\,mol\,dm^{-3}$), a salt solution that contains highly charged ions, such as aluminium chloride, conducts better than a salt solution with ions of a small charge, such as sodium chloride.

The conduction of electricity through metals and other solid conductors is due to the movement of electrons (see below). When ions that are moving in a solution give up their charge to a solid conductor placed in the solution (an **electrode**), interesting chemical reactions take place. This process, called **electrolysis**, is described in Topic 23.

4.11 Metallic lattices

Table 4.7 Metallic bonding is much stronger than the instantaneous dipole forces in argon, but not as strong as bonding in the giant covalent lattice of carbon.

Element	Melting point/°C
argon	−189
silver	962
copper	1083
iron	1535
carbon	3550

All metals conduct electricity. They are also **malleable** (easily beaten or rolled into sheets), **ductile** (easily drawn into rods, wires and tubes) and **shiny**. These properties are due to how the atoms of the metal interact with their neighbours in the lattice.

In a pure metallic element, all the atoms are identical. If we look closely at most metallic lattices we find that the atoms are arranged in a very similar way to the argon atoms in solid argon, with each atom being surrounded by 12 others (see Figure 4.4, page 72). But clearly there are greater attractions between them, for the typical melting point of a metal is at least 1000 °C higher than that of argon (see Table 4.7).

On the other hand, an element whose atoms are strongly bonded in a macromolecular covalent lattice, such as carbon, has a melting point 2000 °C higher than the typical metal. So metallic bonding is strong, but not as strong as some covalent bonding.

A clue to the nature of metallic bonding comes from looking at the atomic properties of metallic elements – they all have only one, two or three electrons in their outer shells, and have low ionisation energies. The bulk property that is most characteristic of metals – their electrical conductivity – often increases as the number of ionisable outer-shell electrons increases. Electrical conductivity itself depends on the presence of mobile carriers of electric charge. We can therefore build up a picture of a metal structure consisting of an array of atoms, with at least some of their outer-shell electrons removed and free to move throughout the lattice (see Figure 4.25a). These **delocalised electrons** (see section 3.16) are responsible for the various characteristic physical properties of metals, as detailed below.

Figure 4.25 a The structure of a metallic lattice. **b** When a shearing force is applied, adjacent layers of cations can slide over one another, and the metal lattice can be deformed without shattering.

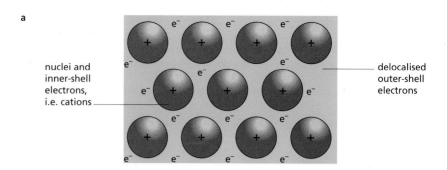

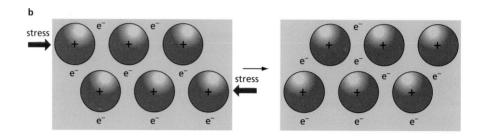

- After some electrons have been removed from the metal atoms, the atoms are left as positive ions. The attraction between the delocalised electrons and these positive ions is responsible for the strengths of the metallic lattices, and for their fairly high melting points.
- Electrons are very small, and move fast. When a potential difference is applied across the ends of a metallic conductor, the delocalised electrons will be attracted to, and move towards, the positive end. This movement of electrical charge is what we know as an **electric current**.
- The partially ionised atoms in a metal lattice are all positively charged. But they are shielded from each other's repulsion by the 'sea' of delocalised electrons in between them. This shielding is still present no matter how the lattice is distorted. As a result of this, and unlike ionic lattices, metal lattices can be deformed by bending and shearing forces without shattering (see Figure 4.25b). Metals are therefore malleable and ductile.
- The shininess and high reflectivity of the surfaces of metals is also a property associated with the delocalised electrons they contain. As a photon of light hits the surface of a metal, its oscillating electric field causes the electrons on the metal's surface to oscillate too. This allows the photon to bounce off the surface without any loss of momentum.

4.12 Graphite and the allotropy of carbon

Allotropes are two (or more) forms of the same element, in which the atoms or molecules are arranged in different ways.

On page 74 we looked at the macromolecular covalent structure of diamond. Carbon's other common allotrope, graphite, is very different in its properties (see Table 4.8).

Table 4.8 The properties of diamond and graphite

Property	Diamond	Graphite
colour	colourless and transparent	black and opaque
hardness	very hard – the hardest naturally occurring solid	very soft and slippery – over 500 times softer than diamond
electrical conductivity	very poor – a good insulator (resistivity $\approx 1 \times 10^{12}\,\Omega\,cm$)	very good along the layers (resistivity $\approx 5 \times 10^{-6}\,\Omega\,cm$)
density	$3.51\,g\,cm^{-3}$	$2.27\,g\,cm^{-3}$

These differences in properties are due to a major difference in the bonding between the carbon atoms in the two allotropes. In diamond, each carbon atom is tetrahedrally bonded to four others by single, localised covalent bonds. A three-dimensional network results. In graphite, on the other hand, each carbon atom is bonded to only three others (using planar sp^2 orbitals, see section 3.14). A two-dimensional sheet of carbon atoms is formed. Each carbon atom has a spare 2p orbital, containing one electron. These can overlap with one another (just as in benzene, see section 3.16) to form a two-dimensional delocalised π orbital spreading throughout the whole sheet of atoms. A graphite crystal is composed of many sheets or layers of atoms, stacked one on top of another (see Figure 4.26). Each sheet should be thought of as a single molecule. There is no bonding between one sheet and the next, but the instantaneous dipole attraction between them is quite substantial, because of the large surface area involved.

Figure 4.26 The structure of graphite

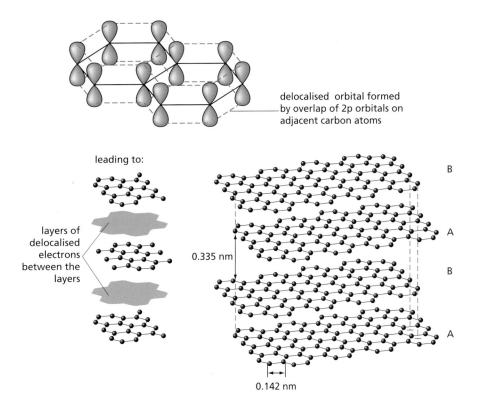

delocalised orbital formed by overlap of 2p orbitals on adjacent carbon atoms

leading to:

layers of delocalised electrons between the layers

0.335 nm

0.142 nm

B
A
B
A

Owing to its layered nature, graphite is an **anisotropic** material, which means that its properties are not the same in all directions. These properties of graphite can be clearly related to its structure, and result in a variety of uses, outlined below.

- The use of graphite as an electrical conductor – whether as electrodes for electrolysis or as brushes in electric motors – is well known. Owing to its delocalised electrons, graphite is a good conductor along the layers. But because electrons keep to their own layer, and cannot jump from one layer to the next, graphite is a poor conductor in the direction at right angles to the layers. The same is true for the conduction of heat. Graphite therefore finds a use in large crucibles, where the conduction of heat along the walls needs to be encouraged, but the transference of heat directly through the bottom is not wanted (see Figure 4.27).

Figure 4.27 Graphite conducts heat along its layers.

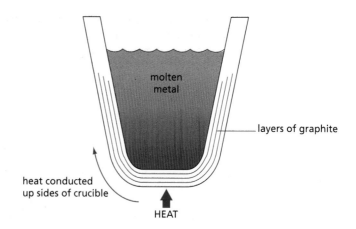

molten metal

layers of graphite

heat conducted up sides of crucible

HEAT

- The weak bonding between the layers allows them to slide over one another easily. One of the first uses of graphite (named from the Greek *grapho*, meaning 'to draw') made use of this property – layers of carbon atoms could easily be rubbed off a lump of graphite, leaving a black trace on paper or parchment. Apart from its continued use in pencils, graphite is also used today as a component of lubricating greases. The easy sliding of the layers over one another combines with a strong compression strength across the layers. This prevents a weight-bearing axle from grinding into the surface of its bearing (see Figure 4.28).

Figure 4.28 Graphite layers slide over one another, making it a useful lubricant.

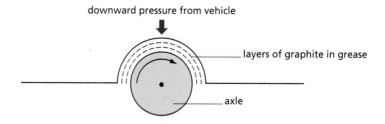

downward pressure from vehicle

layers of graphite in grease

axle

As well as diamond and graphite, carbon also forms other allotropes. The structure of one of these, buckminsterfullerene, is compared with those of diamond and graphite in Figure 4.29.

Figure 4.29 The structures of **a** diamond, **b** graphite and **c** buckminsterfullerene, C_{60}

a
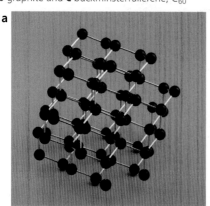

b
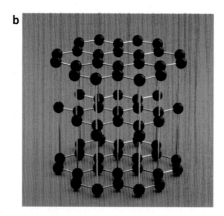

c
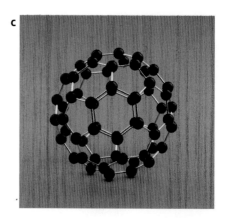

Single sheets of the graphite lattice are called **graphene**. Graphite, fullerenes and graphene all have one thing in common: delocalised electrons. Unlike diamond therefore, they are capable of conducting electricity, although with the smaller simple-molecular fullerenes this would only be around the surface of the molecule itself.

Fullerene is the generic name applied to simple molecular forms of carbon in the form of a hollow sphere, an ellipsoid or a tube. A whole series of fullerenes is now known. The first of these to be discovered was C_{60}. It was obtained by firing a powerful laser at a sample of graphite at a temperature of $10\,000\,°C$. Its discoverers, Harold Kroto, Robert Curl and Richard Smalley, were awarded the 1996 Nobel Prize in Chemistry for their work. They named C_{60} **buckminsterfullerene** in honour of the architect R. Buckminster Fuller, who used the principle of the geodesic dome in many of his buildings. (The alternating 5- and 6-membered rings in C_{60} give a bonding pattern similar to the struts in a geodesic dome – see Figure 4.30.) Other known fullerenes have the formulae C_{20}, C_{70}, C_{76}, C_{78}, C_{84}, C_{90} and C_{94}. More are likely to be synthesised as this exciting new area of chemistry develops.

Figure 4.30 Geodesic domes at the Eden Project in Cornwall, England

C_{60} itself is a highly symmetrical spherical football-shaped molecule (see Figure 4.31). Most of the higher fullerenes are derived from this basic shape by inserting rings of carbon atoms around the centre. Eventually, nanometre-sized tubes of carbon atoms are formed. Since the surface of a fullerene is covered by a cloud of delocalised electrons (as in graphite), these carbon nanotubes (CNTs) may find applications as small-scale conductors (see page 88).

Figure 4.31 The structure of buckminsterfullerene

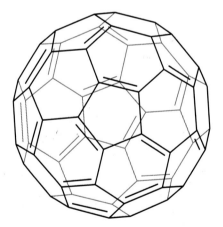

Adding functional groups to spherical fullerenes, such as C_{60}, allows them to interact with biological systems, and they are finding uses in cancer therapy and the targeted delivery of antibiotics. In the electrical field, they are showing promise as high temperature superconductors.

In graphene, as in graphite, the carbon atoms are arranged in a regular two-dimensional hexagonal lattice (see Figure 4.32). Graphene is the thinnest material known but is also extremely strong. It is an excellent conductor of both electricity and heat.

Single sheets of graphite were thought not to be stable on their own – it was assumed they would spontaneously clump together to re-form graphite, but in 2003 they were produced in quantity at the University of Manchester (UK) by Andre Geim and Kostya Novoselov. Geim and Novoselov were awarded the 2010 Nobel Prize in Physics for their work. Billions of dollars are presently being invested in graphene research: the material has many potential applications, such as flexible touch screens for mobile phones; electrodes in lightweight batteries; and single molecular sensors.

Graphene can be rolled into tubes, or two sheets can be combined to form *bilayer graphene*. The sheets can also be 'doped' with other atoms and ions, for example potassium, K^+. All of these forms have interesting electrical properties, and are now finding uses in nanosized electronic devices.

Figure 4.32 Graphene: a single sheet of carbon atoms

Carbon nanotubes (CNTs) are tubes whose walls are made of graphene sheets joined edge-to-edge. They could be considered either as forms of graphene, or as extended open-ended fullerenes, but their properties are sufficiently unique to class them separately from either. The longest nanotubes that have been made have a length approaching 20 cm, but with a diameter of only approximately 1 nm, their length-to-diameter ratio becomes an enormous 20 000 000. The walls of some CNTs are just one sheet thick, whereas others have walls composed of two, three or more graphene sheets, concentrically arranged. They are finding uses in electronics (as field-effect transistors, FETs) and optics, and their high thermal conductivity will encourage their use in other applications. They are also remarkably strong, as the data in Table 4.9 show.

Table 4.9 A comparison of the tensile strength of some materials

Material	Tensile strength (GPa)
multi-walled nanotube	70
single-walled nanotube	33
Kevlar	3.7
stainless steel	0.8

Now try this

In a particular carbon nanotube of length 20 cm, the wall consists of rings of six hexagons, so a particular slice through the wall contains 12 carbon atoms. If the distance along the tube between each ring of 12 carbon atoms is 0.10 nm, what is:

a   the relative molecular mass, M_r, of the carbon nanotube

b   its mass in grams?

Graphene and CNTs are being developed for several important energy applications.

- The use of CNTs in place of graphite at the positive electrode in lithium ion batteries. This will allow a much shorter recharge time.
- The use of CNTs for the storage of hydrogen gas in fuel cells, and in hydrogen fuelled vehicles.
- The use of CNTs, coupled with fullerenes and conducting polymers, to enhance the efficiency, and reduce the cost, of photovoltaic solar cells.
- The manufacture of ultracapacitors that can store electrical energy more effectively than batteries. These capacitors will not only have as much electrical storage capacity as lithium ion batteries, but they will be able to be recharged in only a few minutes.

4.13 The ideal gas equation

Experiments on gases – volume, pressure and temperature

Some of the first quantitative investigations in chemistry, in the seventeenth and eighteenth centuries, were into the behaviour of gases. Air in particular was readily available, and its volume was easily and accurately measurable. Scientists including Robert Boyle, Jacques Charles and Joseph-Louis Gay-Lussac studied how the volume of a fixed amount of air changed when either the pressure on the gas, or the temperature of the gas, was altered.

Their results are summarised below.

Boyle's Law

The volume V of a fixed mass of gas at a constant temperature is inversely proportional to the pressure p on the gas:

$$V \propto \frac{1}{p} \quad \text{or} \quad pV = \text{constant}$$

The graphs in Figure 4.33 show three ways of plotting the Boyle's Law relationship between the volume and the pressure of a gas.

Figure 4.33 These three graphs all show the Boyle's Law relationship.

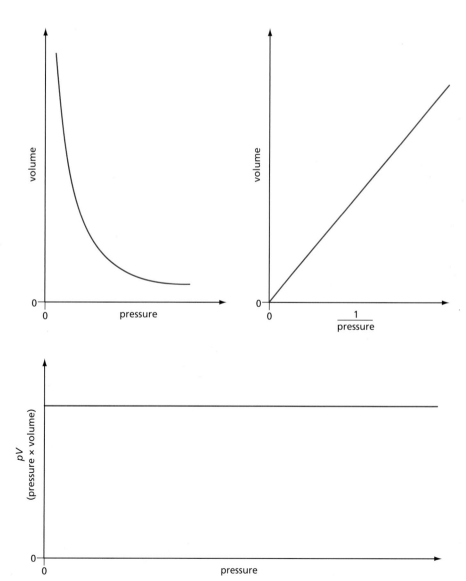

Charles's Law (also called Gay-Lussac's Law)
The volume V of a fixed mass of gas at a constant pressure is directly proportional to its temperature T:

$$V \propto T \quad \text{or} \quad \frac{V}{T} = \text{constant}$$

The units of pressure

The **pressure** of a gas is the force it exerts on a given area of its container. The SI unit of pressure is the **pascal**, Pa, which results from a force of one newton acting on an area of one square metre:

$$1\,\text{Pa} = 1\,\text{N m}^{-2}$$

On this scale, the pressure the atmosphere exerts on the surface of the Earth at sea level is about $1 \times 10^5\,\text{Pa}$. The atmospheric pressure at a particular point on the Earth's surface, however, changes with the weather. It also decreases with height above sea level. The variation with weather can be 10% or so, ranging from $0.95 \times 10^5\,\text{Pa}$ during a depression to $1.05 \times 10^5\,\text{Pa}$ on a fine, 'high-pressure' day.

Scientists have defined **standard atmosphere** as a pressure of $1.01325 \times 10^5\,\text{Pa}$. For most purposes, however, the approximation

$$1\,\text{atm} \approx 1 \times 10^5\,\text{Pa}$$

is adequate. The unit $1.00 \times 10^5\,\text{Pa}$ has been given the name **bar**. The bar is now replacing the standard atmosphere as the most convenient unit of pressure.

Boyle would have measured his pressures with a mercury barometer (see Figure 4.34). The pressure of the atmosphere can support a column of mercury about 0.75 m high. 'Standard atmosphere' was originally defined as the pressure that would support a column of mercury exactly 760 mm high. This is why the conversion factor from atmospheres to pascals is not an exact power of 10.

Figure 4.34 A simple barometer (left) and the vernier scale used to read the height to the nearest 0.1 mm

Atmospheres, bars and pascals are all used today as pressure units. The first is more often used by chemical engineers, whereas laboratory scientists usually use pascals. Two other scales are in common use: car tyre pressures are often measured in pounds per square inch (lb in^{-2}), or in kilograms per square metre (kg m^{-2}).

Absolute zero and the kelvin temperature scale

When the volume of a gas is plotted against its temperature using the Celsius temperature scale, a straight line of positive slope is obtained. If this line is extrapolated back to the point where it crosses the temperature axis, we find the temperature at which the volume of gas would be zero. Accurate measurements show that this point occurs at −273.15 °C (see Figure 4.35). The same temperature is found no matter what volume of gas is used, or at what pressure the experiment is carried out. What is more, the same extrapolated temperature is found no matter what gas we use.

Figure 4.35 This graph illustrates Charles's Law and shows the theoretical derivation of absolute zero.

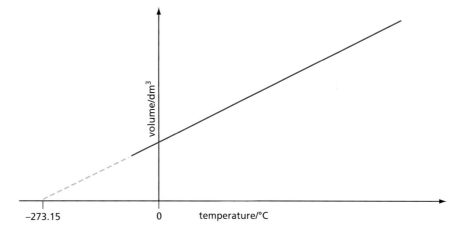

This is a universal and fundamental point on the temperature scale. Below this temperature, Charles's Law predicts that gases would have negative volumes. That is clearly impossible. So presumably it is impossible to attain temperatures lower than −273.15 °C. This point is known as the **absolute zero** of temperature, and experiments in many other branches of chemistry arrive at the same conclusion, and the same value for the absolute zero of temperature. Not only is it impossible to attain temperatures lower than absolute zero, it is impossible even to equal it.

Chemists use a temperature scale which starts at absolute zero. As mentioned in section 2.6, it is called the **absolute temperature scale**, and its unit is the kelvin (K). On this scale, each degree is the same size as a degree on the Celsius scale, so the conversion between the two is easy:

$$\text{temperature/K} = \text{temperature/°C} + 273.15$$

or, more usually:

$$\text{temperature/K} = \text{temperature/°C} + 273$$

Figure 4.36 shows the Charles's Law relationship of volume plotted against temperature in both °C and K.

Figure 4.36 The relationship between the absolute temperature scale and the Celsius scale

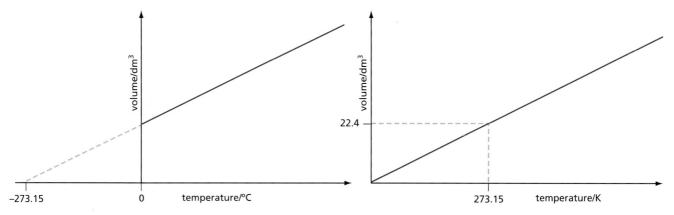

1 Convert the following temperatures from degrees Celsius to kelvin.
   a −197 °C
   b +273 °C
2 Convert the following from kelvin to degrees Celsius.
   a 198 K
   b 500 K

Worked example

What are the following temperatures on the absolute temperature scale?
a 25 °C
b −0.5 °C
c −15 °C

Answer

In each case, we add 273 to the temperature in °C.
a Absolute temperature = 273 + 25 = **298 K**
b Absolute temperature = 273 − 0.5 = **272.5 K**
c Absolute temperature = 273 − 15 = **258 K**

Amount of substance and volume

One other clear influence on the volume of a gas is the amount of gas we are studying. If we perform experiments on 2.0 g of gas rather than 1.0 g, we expect that under the same conditions of temperature and pressure the first sample will have twice the volume of the second. This is found to be the case:

The volume of gas, under identical conditions of temperature and pressure, is proportional to the amount (in moles) of gas present:

$$V \propto n$$

Arriving at the ideal gas equation

We can combine all three influences on the volume of a gas into one relationship:

$$V \propto \frac{1}{p} \qquad V \propto T \qquad V \propto n \qquad \text{so} \qquad V \propto \frac{nT}{p}$$

We can convert this relationship into an equation by introducing a constant of proportionality. This is the **gas constant**. Chemists have given it the symbol R.

$$V \propto \frac{nT}{p} \qquad \text{so} \qquad V = R\left(\frac{nT}{p}\right) \qquad \text{or} \qquad \boldsymbol{pV = nRT}$$

This equation is the **ideal gas equation**. The definition of an **ideal gas** is one which follows this equation exactly, under all conditions of pressure and temperature. The behaviours of many real gases, such as air, and its main components nitrogen and oxygen, together with hydrogen, helium and neon, do fit in with the equation fairly accurately. Some other gases, however, follow the equation only approximately when experiments are carried out at low temperatures or high pressures. The next section will look at why this is the case.

One surprising discovery was that all gases, no matter how they differ in their chemical reactions, or in the sizes or shapes of their molecules, obey the equation (at least approximately). Their points of zero volume ($T_{V=0}$) are all −273.15 °C, and R has the same numerical value for all of them, namely $8.31 \text{J K}^{-1} \text{mol}^{-1}$.

To understand why this is the case, we must return to the major difference between liquids and gases. We saw in section 4.1 that water expands 1833 times when it forms steam at 100 °C. This means that water molecules take up only one part in 1833 of steam (the molecules themselves do not expand when converting from the liquid to the gaseous state). The rest, which amounts to 99.95% of the volume ($= 100 \times \frac{1832}{1833}$), is empty space. This empty space is common to all gases, no matter what sort of molecules they contain. The properties of gases are not the properties of empty space, of course. They arise from the molecules moving about and colliding with one another within that empty space. But as long as the different molecules of various gases move and bump into one another in a similar way, the values of R and $T_{V=0}$ will be the same.

> ### Worked example
>
> Calculate the volume taken up by 1 mol of an ideal gas at room temperature and pressure (25 °C and 1.01×10^5 Pa).
>
> **Answer**
> We rearrange the ideal gas equation to:
>
> $$V = \frac{nRT}{p}$$
>
> $n = 1.0$ mol
> $R = 8.31\,\text{J}\,\text{K}^{-1}\,\text{mol}^{-1}$ [$= 8.31\,\text{N}\,\text{m}\,\text{K}^{-1}\,\text{mol}^{-1}$]
> $T = 273 + 25 = 298$ K
> $p = 1.01 \times 10^5$ Pa   [$= 1.01 \times 10^5\,\text{N}\,\text{m}^{-2}$]
>
> $$V = \frac{1.0 \times 8.31 \times 298}{1.01 \times 10^5}$$
>
> $$= 2.45 \times 10^{-2}\,\text{m}^3$$
> $$= \mathbf{24.5\,dm^3}$$
>
> Note that when carrying out calculations using the ideal gas equation, volumes must be in cubic metres. In calculations we often use the approximate value of 24 dm^3 ($2.4 \times 10^{-2}\,\text{m}^3$) for the volume of 1 mol of an ideal gas at room temperature and pressure.

> ### Now try this
>
> How many moles of ideal gas are there in the following volumes?
>
> 1 2.8 dm^3 of gas at a pressure of 1.01×10^5 Pa and a temperature of 10 °C
> 2 54 dm^3 of gas at a pressure of 5.0×10^5 Pa and a temperature of 600 °C
> 3 92 cm^3 of gas at a pressure of 9.5×10^4 Pa and a temperature of 100 °C

Avogadro's Law

One consequence of the fact that all gases consist mostly of empty space is that the volume of a gas does not depend on the type of molecules it contains. The Italian chemist Amedeo Avogadro was the first to realise this general property, and it can be summarised as follows:

> Under the same conditions of temperature and pressure, equal volumes of all gases will contain equal numbers of molecules.

4.14 The behaviour of real gases

Assumptions about ideal gases

It is possible to derive the ideal gas equation from the basic principles of mechanics. The **kinetic theory of gases** starts by making the following assumptions about an ideal gas.

- The molecules of an ideal gas behave as rigid spheres.
- There are no intermolecular forces between the molecules of an ideal gas.
- Collisions between molecules of an ideal gas are 'perfectly elastic' – that is, there is no loss of kinetic energy during collision.
- The molecules of an ideal gas have no volume.

The theory then considers that the pressure exerted by the gas is due to the bouncing of the gas molecules off the sides of the container. It calculates the magnitude of this pressure by assuming that the molecules are in constant random motion, and that no kinetic energy is lost during collisions with one another or with the walls of the container.

None of the above four assumptions is 100% true for real gases, however. We saw in Topic 2 that atoms, and hence molecules, are not rigid, but are rather fuzzy around the edges. And we saw in Topic 3 that there are various ways in which intermolecular forces can arise. Both of these factors will cause inelastic collisions. Finally, it is clearly the case that molecules do have a volume greater than zero.

How some real gases behave

The behaviour of some real gases is compared with that of an ideal gas in Figure 4.37. As we might expect, the ways in which real gases depart from ideal gas behaviour are different for each gas, because their molecules are different shapes and sizes. In general, though, we can see that the deviations become greater at high pressures.

Figure 4.37 Some real gases depart from ideal behaviour, particularly when the pressure is high.

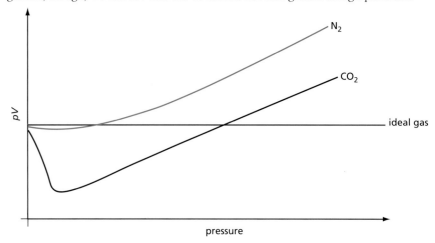

If we now look at how one particular gas behaves as a result of changing the temperature (see Figure 4.38) we see that the deviation is greatest at low temperatures.

Figure 4.38 The behaviour of nitrogen departs from ideal behaviour to a greater extent as the temperature is lowered.

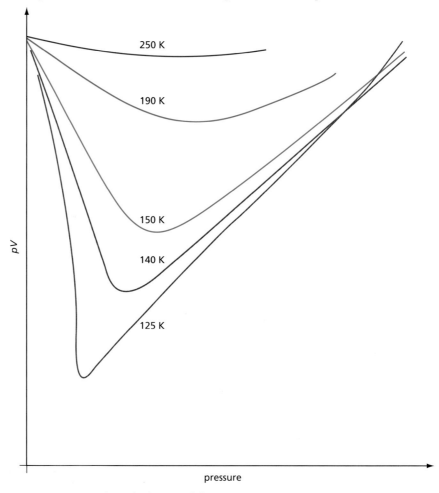

We can summarise these findings as follows:

The behaviour of real gases is **least** like that of an ideal gas at low temperatures and high pressures.

Explaining deviation from ideal behaviour

When we increase the pressure on a gas, we decrease its volume (Boyle's Law). We are forcing the molecules closer together, and the empty space between them is becoming less. Let us take our example of steam, which is 99.95% empty space at atmospheric pressure. 1 g of water occupies $1\,cm^3$, and this expands to $1833\,cm^3$ when it forms steam at $100\,°C$ (see section 4.1). If we compress $1.0\,g$ of steam to 20 atmospheres (which is about the pressure inside the boiler of a steam locomotive), the volume deceases to $\frac{1}{20}$ of its original $1833\,cm^3$, which is $92\,cm^3$. The volume of the molecules is still $1\,cm^3$, which as a percentage of the total volume is now $\frac{1}{92} \times 100 = 1.1\%$. When compressed, the molecules occupy a higher percentage of the gas than the 0.05% they occupy at atmospheric pressure.

> The main reason why gases behave less ideally at high pressures is because the volume of their molecules becomes an increasingly significant proportion of the overall volume of the gas.

Looking now at the effect of temperature, we know that decreasing the temperature of a gas also decreases its volume (Charles's Law) and hence reduces the empty space between the molecules. But, in addition, decreasing the temperature of a gas also decreases the kinetic energy the molecules possess. They travel more slowly, and bounce into one another with less force. The less the bouncing force of collision, the more significant will be the forces of attraction that exist between the molecules. If they are moving more slowly, they are likely to stick to one another more effectively. This causes the collisions to become less elastic. Eventually, on further cooling, the intermolecular forces become larger than the bouncing force, and the molecules spend more time sticking together than moving between one another. This is the molecular explanation of why a gas condenses to a liquid when it is cooled to below its boiling point.

> The main reason why gases behave less ideally at low temperatures is because the intermolecular forces of attraction become comparable in size to the bouncing forces the molecules experience. This causes the collisions between the molecules to be inelastic.

Now try this

1 Explain the following observation in terms of the sizes of the molecules and the intermolecular forces between them.
   - At room temperature, carbon dioxide can be liquefied by subjecting it to a pressure of 10 atm. Nitrogen, however, cannot be liquefied at room temperature, no matter how much pressure is applied.
2 Place the following gases in order of decreasing ideality, with the most ideal first. Explain your reasons for your order.
   CH_4 CH_3Br Cl_2 HCl H_2

Summary

- The major differences between the structures of solids, liquids and gases are whether or not their particles are touching one another, and whether they are fixed in position or moving.
- **Crystalline** solids are composed of **lattices**, which can be **simple molecular**, **macromolecular**, **metallic** or **ionic**. The major physical properties of a solid depend on the type of lattice it has.
- Simple molecular lattices have low melting points, and are electrical insulators under all conditions.
- Macromolecular lattices have high melting points, and are electrical insulators under all conditions.
- Ionic compounds are formed between (usually) metallic cations and non-metallic anions.
- Ionic lattices have high melting points, are brittle, and only conduct electricity when molten or in solution.

- Metals and graphite conduct electricity because of the delocalised electrons they contain.
- Metallic lattices are **malleable**, **ductile**, and have reasonably high melting points.
- Graphite is an **anisotropic** material, owing to the layer nature of its lattice.
- The **ideal gas equation**, $pV = nRT$, can be derived by applying the simple principles of mechanics to a collection of gas particles.
- Although the behaviour of many real gases approximates roughly to that of an ideal gas, real gases are least likely to behave like an ideal gas at low temperatures and at high pressures.

Examination practice questions

Please see the data section of the CD for any A_r values you may need.

1 A new method of making very light, flexible batteries using nanotechnology was announced in August 2007. Read the passage and answer the questions related to it.
Researchers have developed a new energy-storage device that could easily be mistaken for a simple sheet of black paper. The nano-engineered battery is lightweight, ultra-thin and completely flexible. It is geared towards meeting the difficult design and energy requirements of tomorrow's gadgets, such as implantable medical devices and even vehicles.
Researchers soaked 'paper' in an ionic liquid electrolyte which carries the charge. They then treated it with aligned carbon nanotubes, which give the device its black colour. The nanotubes act as electrodes and allow the storage devices to conduct electricity. The device, engineered to function as both a battery and a supercapacitor, can provide the long, steady power output comparable to a conventional battery, as well as a supercapacitor's quick burst of high energy. The device can be rolled, twisted, folded, or cut into shapes with no loss of strength or efficiency. The 'paper' batteries can also be stacked, like a pile of printer paper, to boost the total power output.

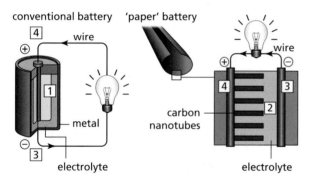

Conventional batteries produce electrons through a chemical reaction between electrolyte and metal.
Chemical reaction in the 'paper' battery is between electrolyte and carbon nanotubes.
Electrons collect on the negative terminal of a battery.
Electrons must flow from the negative terminal, through the external circuit to the positive terminal for the chemical reaction to continue.

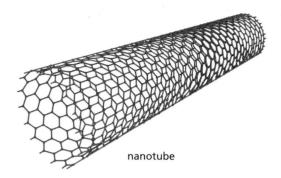

nanotube

a From your knowledge of the different structures of carbon, suggest which of these is used to make nanotubes. [1]

b Suggest a property of this structure that makes it suitable for making nanotubes. [1]

c Carbon in its bulk form is brittle like most non-metallic solids. Suggest why the energy storage device described can be rolled into a cylinder. [1]

d Name an example of an 'ionic *liquid* electrolyte' (not a solution). [1]

[Cambridge International AS & A Level Chemistry 9701, Paper 41 Q8 November 2009]

2 Solids exist as lattice structures.

a Giant metallic lattices conduct electricity. Giant ionic lattices do not. If a giant ionic lattice is melted, the molten ionic compound will conduct electricity.
Explain these observations in terms of bonding, structure and particles present. [3]

b The solid lattice structure of ammonia, NH_3, contains hydrogen bonds.

i Draw a diagram to show hydrogen bonding between **two** molecules of NH_3 in a solid lattice. Include relevant dipoles and lone pairs. [2]

ii Suggest why ice has a higher melting point than solid ammonia. [2]

c Solid SiO_2 melts at 1610 °C. Solid $SiCl_4$ melts at −70 °C. Neither of the liquids formed conducts electricity.
Suggest the type of lattice structure in solid SiO_2 and in solid $SiCl_4$ and explain the difference in melting points in terms of **bonding** and **structure**. [5]

[OCR Chemistry A Unit F321 Q5 May 2011]

5 Energy changes in chemistry

Chemistry is the study of how atoms combine and recombine to form different compounds. Energy changes take place during these reactions. In this topic, the importance of energy changes in chemistry is discussed. Measurement of them allows us to find out why chemical reactions take place in the way that they do, and also to discover the strengths of the bonds that hold atoms together.

Learning outcomes

By the end of this topic you should be able to:

5.1a) explain that some chemical reactions are accompanied by energy changes, principally in the form of heat energy; the energy changes can be exothermic (ΔH is negative) or endothermic (ΔH is positive)

5.1b) explain and use the terms *enthalpy change of reaction* and *standard conditions*, with particular reference to formation, combustion, hydration, solution, neutralisation, atomisation; and *bond energy* (ΔH positive, i.e. bond breaking) (part, see also Topic 20)

5.1c) calculate enthalpy changes from appropriate experimental results, including the use of the relationship: enthalpy change, $\Delta H = -mc\Delta T$

5.2a) apply Hess's Law to construct simple energy cycles, and carry out calculations involving such cycles and relevant energy terms, with particular reference to determining enthalpy changes that cannot be found by direct experiment, e.g. an enthalpy change of formation from enthalpy changes of combustion, and average bond energies (part, see also Topic 20)

5.1 Chemical energy revisited – introducing enthalpy

We saw in section 2.6 that chemical energy is a combination of kinetic energy and potential energy. The potential energy is made up of the electrostatic attractions between the particles and, under the usual conditions of temperature and pressure, is usually the larger of the two.

Enthalpy and enthalpy changes

The total chemical energy of a substance is called its **enthalpy** (or **heat content**). Values of enthalpy are large because the particles attract one another strongly, and it is measured for a very large number of particles, usually for one mole (6.0×10^{23}, see section 1.5). If this same number of particles is considered in each case, then a comparison of the enthalpies of different substances gives a valid comparison of the forces of attraction that exist between the particles in these substances.

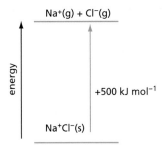

Figure 5.1 If energy is added to the system, ΔH is positive. The products have more energy than the reactants.

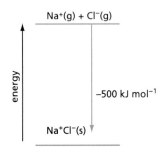

Figure 5.2 If energy is removed from the system, ΔH is negative. The products have less energy than the reactants.

In chemical reactions there are changes in chemical energy and therefore changes in enthalpy. An **enthalpy change** is given the symbol ΔH. The Greek letter delta, Δ, means 'change', and H is the symbol for enthalpy.

If the products of a reaction have greater energy than the reactants, then ΔH is positive. For example, we can represent the pulling apart of the ions in one mole of sodium chloride (see section 2.6) by the following equation:

$$Na^+Cl^-(g) \rightarrow Na^+(g) + Cl^-(g)$$

Because the potential energy of the system is increased, ΔH is large and positive (see Figure 5.1). It has a value of 500000 joules per mole. This is written as:

$$\Delta H = +500\,kJ\,mol^{-1}$$

If the energy of the system is decreased, ΔH is negative, for example when a mole of sodium ions and a mole of chloride ions react together in the reverse of the process above (see Figure 5.2).

This is shown by the following equation:

$$Na^+(g) + Cl^-(g) \rightarrow Na^+Cl^-(s) \qquad \Delta H = -500\,kJ\,mol^{-1}$$

When dealing with enthalpy changes, it is important to specify the physical state of each species in an equation: as the above equation shows, changes of state can be associated with very large changes of enthalpy.

Gaseous sodium and chloride ions exist only at extremely high temperatures, so they cannot be studied under normal laboratory conditions. They can be studied in solution, however. We can represent the dissolution of sodium chloride in water by the following equation:

$$Na^+Cl^-(s) \rightarrow Na^+(aq) + Cl^-(aq) \qquad \Delta H = +4\,kJ\,mol^{-1}$$

Two questions can be asked at this stage.

- How has this value of $+4\,kJ\,mol^{-1}$ been measured?
- Why is this enthalpy change so much smaller than the enthalpy change for the gas-phase reaction?

The first question is answered in section 5.2. The second question requires a knowledge of how sodium and chloride ions pack together in the solid and how they associate with water molecules when in solution. These points are discussed in Topic 20.

Worked example 1

What type of kinetic energy is present in ice? Explain your reasoning.

Answer
Ice is a solid, and the atoms are fixed in position (see section 2.6). The only motion is vibration, so the atoms have vibrational kinetic energy.

Worked example 2

State the sign of ΔH when ice changes to water. Explain your answer.

Answer
When ice melts, some intermolecular bonds are broken, so energy is taken in and ΔH is positive.

5.2 Measuring enthalpy changes directly

Exothermic and endothermic reactions

The **Law of Conservation of Energy** states that energy cannot be created or destroyed; it can only be converted into another form of energy.

The enthalpy change of a reaction usually appears as heat, which means that there is a temperature change. If the products have less enthalpy than the reactants, there is an enthalpy decrease during the reaction (ΔH is negative) and an equivalent amount of heat energy must be given out by the reaction; the reaction is said to be **exothermic**. The heat given out is passed to the **surroundings** – that is, the environment around the reaction – where it can be measured. Most chemical reactions are exothermic, but there are some in which the enthalpy increases (ΔH is positive), and the reaction is then said to be **endothermic**. These reactions take in heat from the surroundings because the enthalpy of the products is more than the enthalpy of the reactants.

It is not surprising that most chemical reactions are exothermic. In everyday life, the changes we observe are usually those in which the potential energy decreases and the kinetic energy increases in the form of heat. If we push a book to the side of a desk, we expect to see it fall from the desk to the floor (its potential energy therefore decreases). We would be very surprised if it suddenly rose back up again (its potential energy would increase). Both processes are possible in theory, as they do not break the Law of Conservation of Energy.

- For **exothermic reactions**, ΔH is negative. The temperature of the surroundings increases and the potential energy of the system (that is, the reacting chemicals) decreases.
- For **endothermic reactions**, ΔH is positive. The temperature of the surroundings decreases and the potential energy of the system (the reacting chemicals) increases.

The direction of a chemical change is determined by the relative energy levels of the reactants and products. If the enthalpy of the reactants is higher than that of the products (exothermic, ΔH negative), the reaction is **thermodynamically** possible. It might not, however, take place because the rate is too slow; it is then said to be **kinetically** controlled. These kinetic factors are considered in Topic 8).

Measuring temperature changes and calculating ΔH

If we measure the heat given out or taken in during a reaction, we can find this enthalpy change. The simplest way of measuring it is to use the energy to heat (or cool) some water or a solution. We need to make the following measurements:

The **enthalpy change of a reaction**, ΔH, is the change in enthalpy accompanying the complete conversion of one mole of reactants into products.

- the mass of the reactants
- the mass of water, m (or its volume, since its density is $1.00\,\mathrm{g\,cm^{-3}}$)
- the rise (or fall) in temperature of the water or solution, ΔT.

We also need to know the amount of energy needed to raise the temperature of water by one degree. This is known as the **specific heat capacity** of water and is given the symbol c. It has the value $4.18\,\mathrm{J\,g^{-1}\,K^{-1}}$. (Note that we can measure the temperature change using a thermometer marked in degrees Celsius, because a change in temperature is the same on either the Celsius or the kelvin scale.)

If we represent the heat change as q, we have the following equation:

$$q = mc\Delta T$$

If we are dealing with aqueous solutions rather than pure water, it is the mass of water in the solution that should be included as the 'm' in this equation. Since the solutions are often quite dilute, and since the volume and heat capacity of a dilute solution are about the same as those of the water it contains, we normally use the following equation:

$$q = 4.18 \times v \times \Delta T, \quad \text{where } v = \text{volume of the solution in cm}^3$$

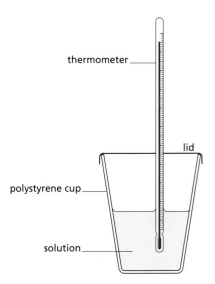

Figure 5.3 A basic calorimeter, used for simple heat experiments

Figure 5.3 shows a simple apparatus that can be used measure the heat change. The expanded polystyrene cup has a lid to keep heat losses to a minimum.

Under these conditions all the heat produced is used to raise the temperature of the contents of the plastic cup, which therefore behaves as both the system and the surroundings. An apparatus used for measuring heat changes in this way is called a **calorimeter**.

Experiment

To find the enthalpy change of solution of ammonium nitrate

The **enthalpy change of solution**, ΔH_{sol}, of a substance is the enthalpy change when one mole of the substance is dissolved in water. Ammonium nitrate is used in this experiment because the temperature change when it dissolves is quite large. A known volume of water is placed in the calorimeter and the initial temperature is measured. The finely powdered solid is weighed in a beaker and then tipped into the water. The water is stirred until all the solid has dissolved, and the lowest temperature is recorded. The worked example below shows how the enthalpy change of solution may be calculated from measuring a temperature change in this way.

Worked example

Some powdered ammonium nitrate was added to water in a plastic beaker and the following results were obtained. Calculate the **enthalpy change of solution**, ΔH_{sol}, of ammonium nitrate.

mass of water	$= 100\,g$
specific heat capacity of water	$= 4.18\,J\,g^{-1}\,K^{-1}$
mass of ammonium nitrate	$= 7.10\,g$
initial temperature	$= 18.2\,°C$
final temperature	$= 12.8\,°C$

Answer

$q = mc\Delta T$
$= 100 \times 4.18 \times (12.8 - 18.2)$ (remember that 'Δ' means 'final − initial')
$= -2260\,J$

$M_r(NH_4NO_3) = 14.0 + 4.0 + 14.0 + 48.0 = 80.0$

so $\qquad M = 80.0\,g\,mol^{-1}$

$n(NH_4NO_3) = \dfrac{m}{M} = \dfrac{7.10}{80.0} = 0.089\,mol$

0.089 mol takes in 2260 J

so 1.0 mol takes in $\dfrac{2260}{0.089} = 25 \times 10^3\,J$

Because heat is taken in, the reaction is endothermic and ΔH_{sol} is positive.

$\Delta H_{sol} = \textbf{+25 kJ mol}^{-1}$

It is usual to show the sign of ΔH, even when it is positive.

Experiment

Measuring other enthalpy changes

The **enthalpy change of neutralisation**, ΔH_{neut}, of an acid is the enthalpy change accompanying the neutralisation of an acid by a base to give one mole of water. To find it, a known amount of acid in solution is placed in a polystyrene cup and its temperature is recorded. An equivalent amount of base is added and the rise in temperature is measured.

Worked example

$50\,cm^3$ of $2.0\,mol\,dm^{-3}$ sodium hydroxide solution were added to $50\,cm^3$ of $2.0\,mol\,dm^{-3}$ hydrochloric acid in a polystyrene cup.

$$HCl(aq) + NaOH(aq) \rightarrow NaCl(aq) + H_2O(l)$$

The following results were obtained:

initial temperature of HCl $= 17.5\,°C$
initial temperature of NaOH $= 17.9\,°C$
final temperature $\qquad = 31.0\,°C$

Calculate the enthalpy change of neutralisation, ΔH_{neut}, for this reaction.

Answer

Average temperature of the HCl and NaOH $= 17.7\,°C$
 (We can take the average temperature because the volumes of acid and base solutions are equal. Because the solution is very dilute, its specific heat capacity is taken to be the same as that of water, namely $4.18\,J\,g^{-1}\,K^{-1}$. This approximation is always used in calculations involving reactions in dilute aqueous solution.)

$$q = mc\Delta T$$
$$= (50 + 50) \times 4.18 \times (31.0 - 17.7)$$
$$= 5560\,J$$

$n(HCl) = n(NaOH) = c \times V$ (in dm^3)   (see section 1.12)

$$= 2.0 \times \frac{50}{1000} = 0.10\,mol$$

$0.10\,mol$ gives out $5560\,J$

so $1.0\,mol$ gives out $\dfrac{5560}{0.10} = 56 \times 10^3\,J$

Because heat is evolved, we know that the reaction is exothermic and ΔH_{neut} is negative.
 $\Delta H_{neut} = \mathbf{-56\,kJ\,mol^{-1}}$ (that is, per mole of water formed)

In a similar way, the temperature change can be measured for a variety of reactions which take place on mixing, and ΔH can be calculated. If one of the reactants is in excess, it is then only necessary to know the exact amount of the reactant that is completely used up, as it is this reactant that determines the energy change. For example, if magnesium ribbon is dissolved in excess hydrochloric acid, it is the quantity of magnesium that determines the energy change. The actual amount of hydrochloric acid has no effect on the amount of heat given out. The mass of the solution must, however, be known in order to measure the heat evolved, using the formula $q = mc\Delta T$.

Now try this

1 $25\,cm^3$ of $1.0\,mol\,dm^{-3}$ nitric acid, HNO_3, were placed in a plastic cup. To this were added $25\,cm^3$ of $1.0\,mol\,dm^{-3}$ potassium hydroxide, KOH. The initial temperature of both solutions was $17.5\,°C$. The maximum temperature reached after mixing was $24.1\,°C$.
  a Calculate the heat given out in the reaction.
  b Calculate the number of moles of nitric acid.
  c Hence calculate ΔH for the neutralisation.
2 $75\,cm^3$ of $2.0\,mol\,dm^{-3}$ ethanoic acid, CH_3CO_2H, were placed in a plastic cup. The temperature was $18.2\,°C$. To this were added $75\,cm^3$ of $2.0\,mol\,dm^{-3}$ ammonium hydroxide, NH_4OH, whose temperature was $18.6\,°C$. After mixing, the highest temperature was $31.0\,°C$. Calculate ΔH for the neutralisation.
3 $0.48\,g$ of magnesium ribbon was added to $200\,cm^3$ of hydrochloric acid in a plastic beaker. The temperature at the start was $20.0\,°C$, and after the magnesium ribbon had dissolved the temperature rose to $21.2\,°C$.
  a Write an equation for the reaction of magnesium ribbon with dilute hydrochloric acid.
  b Calculate ΔH for this reaction.

5.3 Enthalpy changes of combustion

The **enthalpy change of combustion**, ΔH_c, of a substance is the enthalpy change accompanying the complete combustion of one mole of the substance in oxygen.

Most organic compounds burn readily and give off a lot of heat. This provides our main source of energy for homes and for industry.

Energy sources

Fossil fuels

Our main source of energy is the combustion of **fossil fuels**, namely coal, oil and natural gas. They give out large amounts of energy when burnt, but there are two main disadvantages to their use. The first is that supplies are finite and may run out in the foreseeable future. The second is that this combustion produces carbon dioxide, which contributes to the greenhouse effect (see section 13.4), and also other pollutants.

Coal, oil and natural gas are respectively solid, liquid and gaseous fuels. The following considerations are made when choosing which to use.

Figure 5.4 A coal-burning power station. Coal has the highest energy density of the fossil fuels, giving out 76 kJ of energy for each cm³ of coal burnt.

- Ease of combustion – solid fuels are difficult to ignite and do not burn at a constant rate. For this reason, coal is often powdered before being injected into a furnace.
- Storage and ease of transport – for local use, natural gas does not need to be stored, as it is carried by pipeline from the source. Oil is easily stored and can also be distributed by pipeline. In both cases, expensive pipelines must be laid. Coal has the disadvantage that it needs to be carried by road or rail to where it will be used. To transport methane without a pipeline, it must be liquefied, which requires expensive refrigeration. This makes natural gas of limited use as a vehicle fuel, though propane and butane are more useful as they can be liquefied by pressure alone (and are then called liquefied petroleum gas, LPG). LPG is an important fuel for cooking and heating in rural areas without piped natural gas. As increasing numbers of pipelines for natural gas are installed throughout the world, for example in India and China, the use of LPG in factories and homes is gradually being phased out. But it will still be the fuel of choice in most rural areas of the world. The use of coal for driving steam engines and trains was once important but is now of only historical interest throughout the world.
- Pollution – coal contains many impurities and, when burnt, gives off much sulfur dioxide (see section 10.6). The same is true of oil, unless it is carefully refined before use. Crude oil is notorious for the environmental damage caused by spillage from tankers or from the oil extraction and flaring of the associated natural gas.
- Economic and political considerations – the relative prices of coal, oil and gas must take into account the costs of transport, storage and pollution-reducing measures, as well as political expediency, and are subject to wide local variations. Fuel taxes raise revenue for governments, and the resulting increased price of fossil fuels may encourage people to reduce their energy consumption.

Because there are serious problems associated with burning fossil fuels, much research has been carried out to find alternative energy sources. These are increasingly important, but it is unlikely that they will provide more than a quarter of our energy demands during the next few years. Many alternative energy sources are **renewable** – they can be replaced as fast as they are used.

Figure 5.5 Wind farms provide pollution-free energy, but have a visual effect on the landscape.

Alternative energy sources

Alternative energy sources do not rely on the combustion of fuels. The following are some of those in current use.

- **Hydroelectric power** – there are many places in the world where rivers have been dammed and the resulting difference in water level used to provide energy to generate electricity, for example at Kariba on the Zambezi river between Zimbabwe and Zambia, at Itaipu on the Parana river between Brazil and Paraguay, and the Three Gorges dam on the Yangtze river in China. The last two examples alone generate over 6×10^{17} J of energy per year, the equivalent of burning 6×10^7 tonnes of coal in a coal-fired power station. This saves pumping millions of tonnes of carbon dioxide into the atmosphere. However, there can be resultant damage to the environment, which needs to be considered as it may be unacceptable.
- **Tidal and wave power** – these are little used, even though the large-scale tidal power generator at La Rance in northern France first produced electricity 50 years ago.
- **Wind power** – the use of wind turbines is increasing, although there is some objection to the construction of large windmills in isolated regions as they have a dramatic effect on the landscape. The wind speed is not always high enough to turn the blades in summer, but it usually is in winter, when the demand for electricity is highest. Another difficulty is that the places which are the most windy (and therefore the most suitable for the production of wind power) are often not the most populous, so the power has to be transported to where it is needed or power lines constructed to link to a national grid.
- **Geothermal power** – potentially this could be an easy way to provide energy for home heating, though it can only be carried out at sites with particular geological features. Water is pumped from deep under the ground and comes to the surface at a temperature near to boiling point. It is not usually possible to use this hot water to produce electricity, because superheated steam is needed in order to obtain a reasonable energy conversion. However, the Philippines, New Zealand and Costa Rica obtain a reasonable proportion of their energy from this source.
- **Solar panels** – the direct conversion of sunlight into electricity using solar cells used to be quite expensive, but the application of conducting polymers has reduced the cost of these, and in sunny places, for example in California, Spain and India, solar energy is now economically viable.
- **Nuclear power** – the use of nuclear reactors based on **fission reactions**, in which large nuclei split into two smaller nuclei and give out nuclear energy, has sharply declined as a result of accidents such as that at Chernobyl in 1986. There is much debate over the long-term implications of the disposal of nuclear waste, and most countries had cut back on their nuclear programmes until a few years ago, when concern for the rise in atmospheric carbon dioxide caused a rethink. There is the possibility that reactors using **fusion reactions**, in which lighter nuclei join to form a heavier one, may be developed in the next few decades. These provide the best hope for a long-term solution to the energy problem, because they are 'cleaner' and do not produce radioactive waste.

Renewable fuels

Most of the alternative energy sources listed above are renewable because they are powered by the Sun, and so are always available. Some fuels are made out of resources that can be replaced quickly. For example, tropical countries such as Brazil can grow fast-ripening crops such as corn or sugar cane which can be used to produce ethanol cheaply (see section 16.4) and this can be used mixed with petrol in motor vehicles.

Hydrogen is another renewable energy resource. It has the advantage of being pollution free, since its combustion produces water only, and can be used in fuel cells (see section 23.6). Its great disadvantage is the difficulty of transporting it (being much lighter than natural gas, it is difficult to contain it), though there is hope that the use of metal hydrides (compounds of transition metals with hydrogen) may overcome this problem.

For special applications, for example in space rockets (see page 107), special synthetic fuels are being developed.

Standard enthalpy changes

We can measure enthalpies of combustion very accurately, and they give us information about the forces of attraction that exist between the atoms in molecules. The conditions of the reaction need to be stated very precisely. For example, the following equation shows the combustion of methane:

$$CH_4(g) + 2O_2(g) \rightarrow CO_2(g) + 2H_2O(l)$$

The enthalpy change for this reaction at room temperature is about 10% greater than it is at 200 °C. By convention, it has been agreed that **standard conditions** of both reactants and products are at a pressure of either 1 bar (10^5 Pa) or 1 atm (1.013×10^5 Pa) and, unless otherwise specified, at a temperature of 25 °C. The bar and atm are so nearly the same that interchanging them usually makes no significant difference to quoted standard thermodynamic data. Under these conditions, the enthalpy change is known as the **standard enthalpy change of combustion** and is given the symbol $\Delta H_c^\ominus$, the 'c' indicating combustion and the '$\ominus$' indicating that it was measured and calculated under standard conditions.

It is also important to specify exactly the amount of substance involved in the change. Two common ways of writing the combustion of hydrogen are as follows:

$$H_2(g) + \tfrac{1}{2}O_2(g) \rightarrow H_2O(l)$$
$$2H_2(g) + O_2(g) \rightarrow 2H_2O(l)$$

The first equation, in which one mole of hydrogen gas undergoes combustion, represents $\Delta H_c^\ominus$. The enthalpy change for the second reaction is $2 \times \Delta H_c^\ominus$.

The **standard enthalpy change of combustion**, $\Delta H_c^\ominus$, is the enthalpy change when one mole of the substance is completely burnt in excess oxygen at 1 bar or 1 atm pressure and at a specified temperature (usually 25 °C).

The bomb calorimeter

Highly accurate values of $\Delta H_c^\ominus$ can be found only by using a specially constructed apparatus called a **bomb calorimeter**, shown in Figures 5.6 and 5.7. The 'bomb' is a sealed pressure vessel with steel walls. The fuel is placed in the crucible and the 'bomb' filled with oxygen at a pressure of 15 atm. The 'bomb' is then placed in an insulated calorimeter containing a known mass of water. The fuel is ignited by an electric current and the temperature change measured to within 0.01 K. To eliminate heat losses, the calorimeter is placed in another water bath whose temperature is raised with an electric heater so that it just matches the average temperature in the calorimeter. The apparatus is first calibrated using benzoic acid, the enthalpy of combustion of which is recognised as a standard and which can be readily obtained in a high state of purity. There are a number of small corrections which must be applied to the results, but values accurate to 0.1% can be obtained.

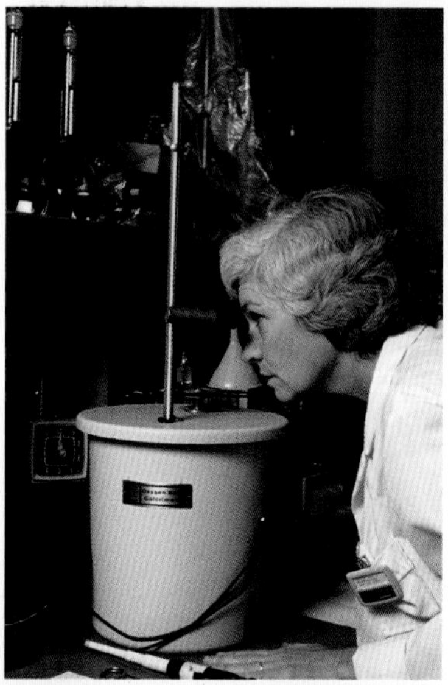

Figure 5.6 Bomb calorimeters are used to measure the energy content not only of fuels but also of foods.

Figure 5.7 In a bomb calorimeter, accurate values of the enthalpy change of combustion can be measured because heat losses to the air are minimised.

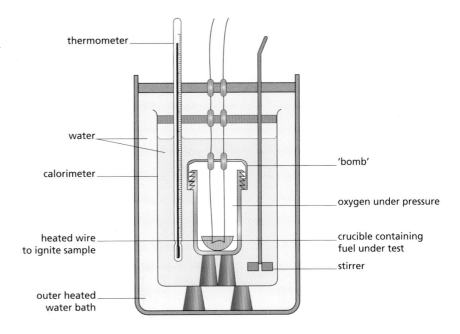

thermometer

water

calorimeter

heated wire to ignite sample

outer heated water bath

'bomb'

oxygen under pressure

crucible containing fuel under test

stirrer

To measure $\Delta H_c^\ominus$

A simple apparatus

Figure 5.8 shows a simple apparatus to measure the enthalpy change of combustion for a fuel such as methanol. A known volume of water is placed in a copper calorimeter and its temperature is measured. The calorimeter is clamped so that its base is just a few centimetres above a spirit burner, which contains the fuel. The spirit burner is weighed, placed under the calorimeter and lit. The water in the calorimeter is stirred with the thermometer. When the temperature has risen about 10 °C, the flame is put out, the temperature is noted and the spirit burner re-weighed.

Figure 5.8 A simple apparatus used to measure enthalpy changes of combustion

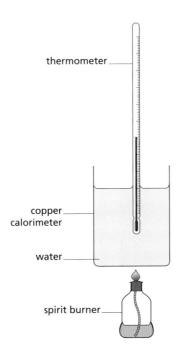

thermometer

copper calorimeter

water

spirit burner

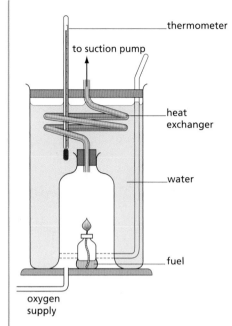

thermometer

to suction pump

heat exchanger

water

fuel

oxygen supply

Figure 5.9 In Thiemann's fuel calorimeter, a supply of oxygen ensures complete combustion, and heat loss to the air is reduced.

Thiemann's fuel calorimeter

When determining $\Delta H_c^{\ominus}$ using a copper calorimeter, there are two major sources of error:

- the methanol is not all completely burnt to carbon dioxide and water (some incomplete combustion takes place)
- not all the heat given off is passed to the water.

These errors can be reduced by means of **Thiemann's fuel calorimeter** (see Figure 5.9). The fuel is burnt in a stream of oxygen to ensure complete combustion, and the gases are sucked through a copper spiral placed in water (a heat exchanger) so that very little heat is lost to the air.

Measurements similar to those with the simple copper calorimeter experiment are made. The cap is replaced on the spirit burner after putting out the flame to reduce losses of fuel by evaporation before re-weighing. The oxygen should be supplied fast enough so that the fuel burns with a clear blue flame. The suction pump is usually fully on, but it may need to be turned down if the suction is so vigorous that the flame is pulled off the spirit burner. Thiemann's apparatus can give results to within 80% of quoted values.

Worked example

In an experiment to determine $\Delta H_c^{\ominus}$ for methanol, CH_3OH, the following readings were obtained. Calculate $\Delta H_c^{\ominus}$ for methanol.

mass of water in calorimeter	= 200 g
mass of methanol and burner at start	= 532.68 g
mass of methanol and burner at end	= 531.72 g
temperature of water at start	= 18.3 °C
temperature of water at end	= 29.6 °C

Answer

We shall ignore the heat taken in by the calorimeter.

temperature rise of water = 11.3 K

$q = mc\Delta T$

$= 200 \times 4.18 \times 11.3 = 9447\,J$

mass of methanol burnt = 0.96 g   and $M_r(CH_3OH) = 32.0$

so   amount of methanol burnt $= \dfrac{0.96}{32.0} = 0.030\,mol$

Because heat is evolved, we know that the reaction is exothermic and $\Delta H_c^{\ominus}$ is negative.

$$\Delta H_c^{\ominus} = \frac{-9447}{0.030} = -315 \times 10^3\,J\,mol^{-1} \text{ or } \mathbf{-315\,kJ\,mol^{-1}}$$

Now try this

1 a What is meant by 'standard conditions'?
   b Why is it necessary to specify the conditions of a reaction?

2 A burner containing hexanol, $C_6H_{13}OH$, had a mass of 325.68 g. It was lit and placed under a copper calorimeter containing 250 cm³ of water. The temperature of the water rose from 19.2 °C to 31.6 °C. Afterwards the burner's mass was 324.37 g. Calculate:
   a the heat evolved
   b ΔH_c for hexanol.

3 a State the two main sources of error in the experiment described in question **2**.
   b Explain how these two errors are made as small as possible in Thiemann's apparatus.

Rocket fuels

To launch a rocket into space, an explosive fuel called a **propellant** is needed. All propellants have two main components:

- a combustible fuel
- an oxidising agent (oxidant).

In the atmosphere, the oxidant is oxygen from the air, but since a rocket operates in space it needs to carry its own oxidising agent. When the fuel combines with the oxidising agent, it gives out a large amount of heat, which causes the gases produced to expand. This expansion provides thrust for the rocket. The efficiency of the rocket engine is determined by the temperature and volume of the gaseous products.

The fuel and oxidant of a rocket motor make up most of the mass of the rocket. The main function of the propellant is to produce as much energy for a given mass as possible. Cost may be of secondary importance to the heat evolved per kilogram. So, while cars that operate in the atmosphere use a cheap fuel (see Topic 13), rocket fuels are often expensive chemicals.

There are two types of rocket fuels – liquid and solid.

- Liquid fuels are often relatively cheap but may need refrigeration, so the rocket can only be fuelled immediately before lift-off.
- Solid fuels have the advantage that they can be stored in the rocket. They are used in small rockets, for example fireworks, as well as in large launcher rockets, for example the space shuttle. They have the disadvantage that they produce solid as well as gaseous products, and this limits their power output.

Figure 5.10 The solid fuel boosters of this spacecraft burn ammonium chlorate(VII) and aluminium. Clouds of aluminium oxide are produced as well as the gases that propel the craft into space.

Liquid propellants

Liquid propellants may contain relatively common chemicals, such as liquid oxygen with a hydrocarbon or liquid hydrogen fuel. They must be refrigerated and have a relatively poor power-to-weight ratio.

More exotic liquid fuels are also used. The lunar module used to transport astronauts from the Apollo command module to the Moon's surface was powered with liquid dinitrogen tetraoxide, N_2O_4, and dimethylhydrazine, $(CH_3)_2NNH_2$. These react as follows:

$$2N_2O_4(l) + (CH_3)_2NNH_2(l) \rightarrow 2CO_2(g) + 3N_2(g) + 4H_2O(g)$$
$$\Delta H^{\ominus} = -1800 \, kJ \, mol^{-1} = -6980 \, kJ \, (kg \, of \, reagents)^{-1}$$

Fuels with an even higher power-to-weight ratio have been investigated (for example, fluorine and boron hydride), but they are too toxic for use on the ground.

$$6F_2(g) + B_2H_6(g) \rightarrow 6HF(g) + 2BF_3(g)$$
$$\Delta H^{\ominus} = -2800 \, kJ \, mol^{-1} = -10\,900 \, kJ \, (kg \, of \, reagents)^{-1}$$

Solid propellants

The most familiar of solid propellants is the mixture of carbon, sulfur and potassium nitrate known as gunpowder or 'black powder' that is used in fireworks. The simplified chemical equation for the reaction is as follows:

$$3C(s) + S(s) + 2KNO_3(s) \rightarrow K_2S(s) + 3CO_2(g) + N_2(g)$$
$$\Delta H^{\ominus} = -280 \, kJ \, mol^{-1} = -1040 \, kJ \, kg^{-1}$$

The booster stage of the space shuttle used a mixture of ammonium chlorate(VII), NH_4ClO_4, and aluminium powder:

$$10Al(s) + 6NH_4ClO_4(s) \rightarrow 5Al_2O_3(s) + 3N_2(g) + 6HCl(g) + 9H_2O(g)$$
$$\Delta H^{\ominus} = -3250 \, kJ \, mol^{-1} = -9850 \, kJ \, kg^{-1}$$

5.4 Hess's Law and enthalpy change of formation

Introducing Hess's Law

There are very few reactions whose enthalpy change can be measured directly by measuring the change in temperature in a calorimeter. Fortunately, we can find enthalpy changes for other reactions indirectly. To do this we make use of **Hess's Law**, which states that the value of ΔH for a reaction is the same whether we carry out the reaction in one step or in many steps, provided that the initial and final states or conditions are the same.

Hess's Law states that the enthalpy change, ΔH, for a reaction is independent of the path taken.

This law is illustrated in Figure 5.11. If $(\Delta H_1 + \Delta H_2)$ was greater than ΔH, we would be able to get energy 'for nothing' by going round the cycle: reactants, intermediate, products, reactants. This would break the Law of Conservation of Energy (page 99). An example of the use of Hess's Law allows us to find ΔH for the decomposition of calcium carbonate:

$$CaCO_3(s) \rightarrow CaO(s) + CO_2(g)$$

This reaction is slow and requires a high temperature to bring it about. Direct measurement of the temperature change is therefore impracticable. We can, however, carry out two reactions that take place readily at room temperature and use their enthalpy changes to find the ΔH value that we want. These are the reactions of calcium carbonate, and of calcium oxide, with dilute hydrochloric acid:

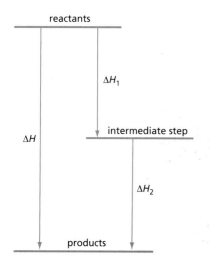

Figure 5.11 Hess's Law: ΔH is independent of the path taken, and $\Delta H = \Delta H_1 + \Delta H_2$.

$$CaCO_3(s) + 2HCl(aq) \rightarrow CaCl_2(aq) + H_2O(l) + CO_2(g)$$
$$\Delta H_1^{\ominus} = -17 \, \text{kJ mol}^{-1} \quad (1)$$
$$CaO(s) + 2HCl(aq) \rightarrow CaCl_2(aq) + H_2O(l) \qquad \Delta H_2^{\ominus} = -195 \, \text{kJ mol}^{-1} \quad (2)$$

There are two ways in which we can use these values to find the enthalpy change for the decomposition of calcium carbonate.

Method 1: Subtracting equations

If we subtract equation (2) from equation (1), we have:

$$CaCO_3(s) - CaO(s) \rightarrow CO_2(g) \qquad \Delta H^{\ominus} = -17 - (-195) = +178 \, \text{kJ mol}^{-1}$$

This equation is equivalent to $CaCO_3(s) \rightarrow CaO(s) + CO_2(g)$, which is the equation we want.

In this method we have taken away '2HCl(aq)' from the left-hand sides of equations (1) and (2) and '$CaCl_2(aq) + H_2O(l)$' from the right-hand sides. These terms are associated with a fixed amount of energy, and we are taking away this same amount of energy from both equations, so this has no effect on the final answer.

Method 2: Constructing a Hess's Law diagram

This method is preferred with more complicated examples. We draw a diagram like that in Figure 5.12 to indicate the two routes by which the reaction can be carried out. The diagram is not intended to give any indication of the actual energy levels (that is why Hess's Law diagrams are not drawn vertically).

In order to go from '$CaCO_3(s)$' to '$CaCl_2(aq) + H_2O(l) + CO_2(g)$', we can go either directly or via '$CaO(s) + CO_2(g)$'.

Considering the enthalpy change involved for both routes gives us ΔH:

$$\Delta H + (-195) = -17$$
$$\text{so} \quad \Delta H = -17 - (-195) = +178 \, \text{kJ mol}^{-1}$$

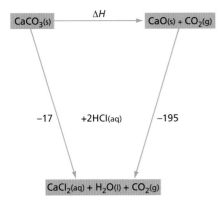

Figure 5.12 Hess's Law cycle for the decomposition of calcium carbonate. All figures are in kJ mol^{-1}.

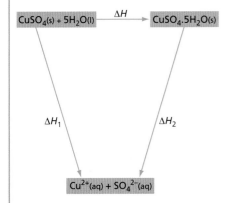

Figure 5.13 Hess's Law cycle to find the enthalpy change when anhydrous copper(II) sulfate crystals are hydrated

To determine the enthalpy change when copper(II) sulfate is hydrated

The equation for the hydration of copper(II) sulfate is as follows:

$$CuSO_4(s) + 5H_2O(l) \rightarrow CuSO_4.5H_2O(s)$$

The enthalpy change for this reaction cannot be found directly: if we add five moles of water to one mole of anhydrous copper(II) sulfate, we do not produce hydrated copper(II) sulfate crystals. These can only be made by crystallisation from a solution. The enthalpy change can, however, be found indirectly by determining the enthalpy change of solution of both anhydrous copper(II) sulfate and hydrated copper(II) sulfate (see Figure 5.13).

According to Hess's Law, $\Delta H_1 = \Delta H + \Delta H_2$. 0.10 mol of anhydrous copper(II) sulfate is added to 100 cm^3 of water in a plastic cup, fitted with a lid and a thermometer. When the solid has dissolved, the change in temperature can be used to calculate ΔH_1.

The experiment is repeated using 0.10 mol of powdered hydrated copper(II) sulfate, but using only 91 cm^3 of water (because the hydrated salt already contains 9 cm^3 (0.5 mol) of water). The change in temperature when the hydrated copper sulfate dissolves can be used to calculate ΔH_2.

The required enthalpy change of reaction ΔH is given by $\Delta H = \Delta H_1 - \Delta H_2$.

To find the enthalpy change of decomposition of sodium hydrogencarbonate

On heating, sodium hydrogencarbonate decomposes:

$$2NaHCO_3(s) \rightarrow Na_2CO_3(s) + H_2O(l) + CO_2(g)$$

The enthalpy change for this decomposition may be found by measuring the enthalpy change when sodium hydrogencarbonate and sodium carbonate react separately with hydrochloric acid (see Figure 5.14).

$$NaHCO_3(s) + HCl(aq) \rightarrow NaCl(aq) + H_2O(l) + CO_2(g) \qquad \Delta H_1$$
$$Na_2CO_3(s) + 2HCl(aq) \rightarrow 2NaCl(aq) + H_2O(l) + CO_2(g) \qquad \Delta H_2$$

Figure 5.14 Hess's Law cycle for the decomposition of sodium hydrogencarbonate

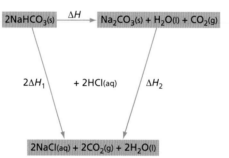

Note that ΔH_1 is multiplied by 2 in Figure 5.14. This is because the first equation must be multiplied by 2 in order for the equations to subtract correctly. $100\,cm^3$ of dilute hydrochloric acid are placed in a plastic cup, fitted with a lid and thermometer. $0.05\,mol$ of sodium hydrogencarbonate is added. The temperature rise is used to calculate ΔH_1. The value of ΔH_2 is found by repeating the experiment using $0.05\,mol$ of anhydrous sodium carbonate. The required enthalpy change can then be calculated, as $\Delta H = 2\Delta H_1 - \Delta H_2$.

Enthalpy change of formation

Although actual values of the enthalpy contained in individual substances are not known, it is possible to obtain accurate values of a quantity that is related to it, namely the **standard enthalpy change of formation**, $\Delta H_f^{\ominus}$.

The **standard enthalpy change of formation** of a substance, $\Delta H_f^{\ominus}$, is the enthalpy change when one mole of the substance is formed from its elements in their **standard states**. A substance in its standard state is at a pressure of 1.0 bar or 1.0 atm and at a specified temperature, often 298 K.

The values of $\Delta H_f^{\ominus}$ for a few compounds, such as oxides, can be determined experimentally but those for most other compounds must be calculated using Hess's Law.

The value of $\Delta H_f^{\ominus}$ of the oxide often has the same value as $\Delta H_c^{\ominus}$ of the element. For example, for carbon:

$$C(s) + O_2(g) \rightarrow CO_2(g) \qquad \Delta H_f^{\ominus}(CO_2(g)) = \Delta H_c^{\ominus}(C(s)) = -393.5\,kJ\,mol^{-1}$$

and for hydrogen:

$$H_2(g) + \tfrac{1}{2}O_2(g) \rightarrow H_2O(l) \qquad \Delta H_f^{\ominus}(H_2O(l)) = \Delta H_c^{\ominus}(H_2(g)) = -285.9\,kJ\,mol^{-1}$$

In other examples, the two are not the same because the two processes are represented by different equations. For example, for aluminium:

$$2Al(s) + 1\tfrac{1}{2}O_2(g) \rightarrow Al_2O_3(s) \qquad \Delta H_f^{\ominus} = -1675.7\,kJ\,mol^{-1}$$

$$Al(s) + \tfrac{3}{4}O_2(g) \rightarrow \tfrac{1}{2}Al_2O_3(s) \qquad \Delta H_c^{\ominus} = -837.8\,kJ\,mol^{-1}$$

The value of $\Delta H_f^\ominus(Al_2O_3(s))$ is twice that of $\Delta H_c^\ominus(Al(s))$.

We can use values of $\Delta H_f^\ominus$ to calculate $\Delta H^\ominus$ for a reaction. On page 109 we showed how to use Hess's Law to find the value of $\Delta H^\ominus$ for the reaction:

$$CaCO_3(s) \rightarrow CaO(s) + CO_2(g)$$

Another method uses standard enthalpy changes of formation in a different cycle to find this enthalpy change, $\Delta H^\ominus$, as shown in Figure 5.15.

Figure 5.15 Hess's Law cycle to find the enthalpy change for the decomposition of calcium carbonate using standard enthalpy changes of formation

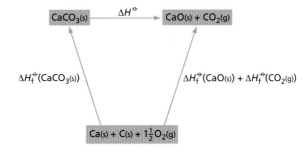

$$\Delta H_f^\ominus(CaCO_3(s)) + \Delta H^\ominus = \Delta H_f^\ominus(CaO(s)) + \Delta H_f^\ominus(CO_2(g))$$
$$-1206.9 + \Delta H^\ominus = -635.5 + (-393.5)$$
$$\Delta H^\ominus = -1029.0 - (-1206.9) = +177.9 \, kJ \, mol^{-1}$$

If an element in its standard state appears in the equation, it can be ignored in the calculation because its $\Delta H_f^\ominus$ value is zero. For example, to calculate $\Delta H^\ominus$ for the reaction:

$$NO(g) + \tfrac{1}{2}O_2(g) \rightarrow NO_2(g)$$

we have the cycle shown in Figure 5.16.

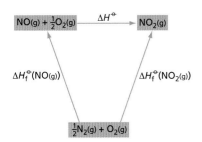

Figure 5.16 Hess's Law cycle for the conversion of nitrogen monoxide to nitrogen dioxide. The enthalpy of formation of the element oxygen is zero, so this term can be ignored.

$$\Delta H_f^\ominus(NO(g)) + \Delta H^\ominus = \Delta H_f^\ominus(NO_2(g))$$
$$+90.4 + \Delta H^\ominus = +33.2$$
$$\Delta H^\ominus = +33.2 - (+90.4) = -57.2 \, kJ \, mol^{-1}$$

5.5 Bond enthalpies

Average bond enthalpies

When two atoms come together, they may form a bond (see section 3.2). To break this bond and separate the two atoms requires energy (ΔH is positive). This is because a force is required to pull the atoms apart, which increases the potential energy of the system. For example, to break the bond in a chlorine molecule requires $242 \, kJ \, mol^{-1}$:

$$Cl_2(g) \rightarrow 2Cl(g) \qquad \Delta H = +242 \, kJ \, mol^{-1}$$

The situation is more complicated if the molecule contains more than two atoms, but fortunately we find that the energy needed to break a particular type of bond between the same atoms is about the same even in different molecules. This value is called the **average bond enthalpy** (or **average bond energy**). If a bond joins an atom of X to an atom of Y, the bond enthalpy is represented by the symbol $E(X\!-\!Y)$.

The **average bond enthalpy**, $E(X\!-\!Y)$, is the enthalpy change when one mole of bonds between atoms of X and atoms of Y are broken in the gas phase:

$$XY(g) \rightarrow X(g) + Y(g) \qquad \Delta H = E(X\!-\!Y)$$

Explaining reaction enthalpies – the formation of hydrogen chloride

The average bond enthalpies of many bonds have been collected together in tables of data. We can use these values of bond enthalpies to explain why some reactions give out so much energy. For example, consider the reaction of hydrogen and chlorine to give hydrogen chloride:

$$H_2(g) \; + \; Cl_2(g) \rightarrow \qquad 2HCl(g)$$

$$H\text{—}H + Cl\text{—}Cl \rightarrow H\text{—}Cl + H\text{—}Cl$$

We can break the reaction down into three steps, as shown in Figure 5.17.

Figure 5.17 The bond-breaking and bond-making steps in the synthesis of hydrogen chloride. The figures are published bond enthalpies, in $kJ\,mol^{-1}$.

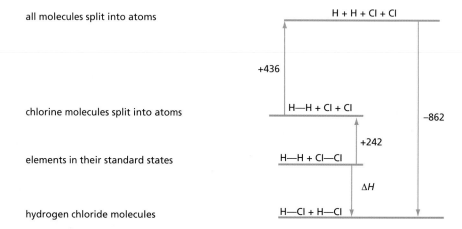

all molecules split into atoms

H + H + Cl + Cl

+436

chlorine molecules split into atoms

H—H + Cl + Cl

−862

+242

elements in their standard states

H—H + Cl—Cl

ΔH

hydrogen chloride molecules

H—Cl + H—Cl

The changes can be represented by equations:

$$H_2(g) \rightarrow 2H(g) \qquad\qquad \Delta H = E(H\text{—}H) = +436\,kJ\,mol^{-1}$$
$$Cl_2(g) \rightarrow 2Cl(g) \qquad\qquad \Delta H = E(Cl\text{—}Cl) = +242\,kJ\,mol^{-1}$$
$$2H(g) + 2Cl(g) \rightarrow 2HCl(g) \qquad \Delta H = -2E(H\text{—}Cl) = -2 \times 431 = -862\,kJ\,mol^{-1}$$

(The first two changes represent bond breaking, an endothermic process, and thus the ΔH values are positive. The third change represents bond making, an exothermic process, and thus ΔH is negative.)

We can use Hess's Law to add these three equations, to obtain:

$$H_2(g) + Cl_2(g) \rightarrow 2HCl(g) \qquad \Delta H = -184\,kJ\,mol^{-1}$$

If we study the values of the bond enthalpies above, we see that while the H—H and H—Cl bonds are strong and have similar values, the Cl—Cl bond enthalpy is much smaller. It is the weakness of the Cl—Cl bond that is the principal reason why the reaction is so exothermic.

Notice that in this reaction we broke two bonds (the H—H and Cl—Cl bonds) and formed two bonds (two H—Cl bonds), and that ΔH equalled the bond enthalpies of the bonds broken minus the bond enthalpies of the bonds formed. For most reactions, the number of bonds broken equals the number of bonds formed.

ΔH for a reaction = the bond enthalpies of the bonds broken minus the bond enthalpies of the bonds formed

Number of bonds broken = number of bonds formed

The combustion of hydrogen

Consider a more complicated example, the combustion of hydrogen, whose measured enthalpy change is as follows:

$$2H_2(g) \quad + \; O_2(g) \; \rightarrow \qquad 2H_2O(g) \qquad\qquad \Delta H = -483\,kJ\,mol^{-1}$$
$$H\text{—}H + H\text{—}H \; + O{=}O \rightarrow H\text{—}O\text{—}H + H\text{—}O\text{—}H$$

Notice that we specify $H_2O(g)$ and not the usual state of water under standard conditions, which is $H_2O(l)$. Liquid water has intermolecular forces between the molecules and these have to be formed as well as the O—H bonds, making the energy of $H_2O(l)$ lower than that of $H_2O(g)$. Bond enthalpies are always quoted for the gas phase so that intermolecular forces do not have to be taken into account.

If we consider forming $H_2O(g)$, we can look up the following bond enthalpies:

bonds broken

$2E(\text{H—H}) = 2 \times 436 \, \text{kJ mol}^{-1}$
$E(\text{O}{=}\text{O}) = 497 \, \text{kJ mol}^{-1}$
total 4 bonds

bonds formed

$4E(\text{O—H}) = 4 \times 460 \, \text{kJ mol}^{-1}$

total 4 bonds

ΔH = enthalpy of bonds broken − enthalpy of bonds formed
$\quad = 1369 - 1840$
$\quad = -471 \, \text{kJ mol}^{-1}$

This calculated value for the enthalpy change is slightly different from the accurate experimental value ($-483 \, \text{kJ mol}^{-1}$), showing that we must not always expect exact agreement between measured energy changes and those calculated from bond enthalpy values. However, this exercise enables us to suggest why the reaction is so exothermic. At first sight all the bonds appear to be of similar strength, but that for oxygen is for two bonds, so that the average for one bond is only $\frac{497}{2} \, \text{kJ mol}^{-1}$. It is the weakness of the O${=}$O bond that is the principal reason why combustion reactions are so exothermic and therefore so important as sources of energy in everyday life.

Worked example 1

a Explain what is meant by the H—I bond enthalpy.
b Write the symbol for the H—I bond enthalpy.
c Write an equation that shows the change brought about in determining the H—I bond enthalpy.

Answer
a This is the energy required to break one mole of H—I bonds in the gas phase.
b $E(\text{H—I})$
c $\text{HI}(g) \rightarrow \text{H}(g) + \text{I}(g)$

Worked example 2

a Write an equation to show the breakdown of methane, $CH_4(g)$, into atoms.
b How is ΔH for this reaction calculated using the C—H bond enthalpy?

Answer
a $\text{CH}_4(g) \rightarrow \text{C}(g) + 4\text{H}(g)$
b $\Delta H = 4 \times E(\text{C—H})$

5.6 Finding bond enthalpies

Enthalpy change of atomisation

One useful piece of data in chemistry is the **standard enthalpy change of atomisation**, $\Delta H_{at}^{\ominus}$. This is the energy required to produce gaseous atoms from an element or compound under standard conditions. It is a quantitative indication of the strength of the bonding in the substance. If a solid breaks up into atoms, it may first melt (accompanied by the **standard enthalpy change of fusion**, $\Delta H_m^{\ominus}$), then evaporate (accompanied by the **standard enthalpy change of vaporisation**, $\Delta H_{vap}^{\ominus}$) and finally, in the gas phase, any remaining bonds break (the sum of all the bond enthalpies). So:

$$\Delta H_{at}^{\ominus} = \Delta H_m^{\ominus} + \Delta H_{vap}^{\ominus} + \text{bond energy terms}$$

For elements, $\Delta H^{\ominus}_{at}$ is the enthalpy change when one mole of atoms is produced. For compounds, it is the enthalpy change when one mole of the substance is broken down completely into atoms.

> The **standard enthalpy change of atomisation**, $\Delta H^{\ominus}_{at}$, for:
> - an element is the enthalpy change that occurs when one mole of gaseous atoms is produced from the element in its standard state under standard conditions
> - compounds is the enthalpy change to convert one mole of the substance in its standard state into gaseous atoms under standard conditions.

Using enthalpy changes of atomisation, we can find bond energies in simple molecules.

Finding the bond enthalpy in simple molecules

The enthalpy change of atomisation of a compound includes the **bond enthalpies** for all bonds in the substance. This enthalpy change can be found, using Hess's Law, from the enthalpy change of formation of the substance and the enthalpy changes of atomisation of the elements it contains. For example to calculate the enthalpy change of atomisation of methane, we can construct the Hess's Law diagram shown in Figure 5.18.

Figure 5.18 Hess's Law cycle to find the average bond enthalpy in methane. All figures are in kJ mol^{-1}.

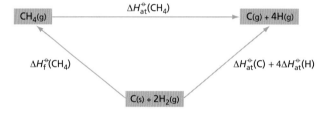

We can see that:

$$\Delta H^{\ominus}_f(CH_4) + \Delta H^{\ominus}_{at}(CH_4) = \Delta H^{\ominus}_{at}(C) + 4\Delta H^{\ominus}_{at}(H)$$
$$-74.8 + \Delta H^{\ominus}_{at} = +714.7 + 4 \times +218$$
$$\Delta H^{\ominus}_{at} = 714.7 + 872 + 74.8$$
$$= +1661.5 \, kJ \, mol^{-1}$$

This enthalpy is the energy needed to break four moles of C—H bonds. Because the four bonds in methane are identical, it is reasonable to allocate $+\frac{1661.5}{4}$ kJ mol^{-1} for one mole of bonds. This gives a value for the average bond enthalpy of the C—H bond in methane, $E(C—H)$, of 415 kJ mol^{-1}.

The same calculation can be carried out for tetrachloromethane, CCl$_4$, to find $E(C—Cl)$. The bonds need to be broken in gaseous tetrachloromethane, rather than the liquid, which is its normal state under standard conditions. This means that we must include the enthalpy change of vaporisation of tetrachloromethane in the calculation. Figure 5.19 shows the Hess's Law cycle used.

Figure 5.19 Hess's Law cycle to find the average bond enthalpy in tetrachloromethane. All figures are in kJ mol^{-1}.

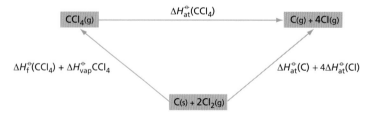

We can see that:

$$\Delta H^{\ominus}_f(CCl_4) + \Delta H^{\ominus}_{vap}(CCl_4) + \Delta H^{\ominus}_{at}(CCl_4) = \Delta H^{\ominus}_{at}(C) + 4\Delta H^{\ominus}_{at}(Cl)$$
$$-135.5 + 12.7 + \Delta H^{\ominus}_{at} = +714.7 + 4 \times 121.1$$
$$\Delta H^{\ominus}_{at} = 714.7 + 484.4 + 135.5 - 12.7$$
$$= +1321.9 \, kJ \, mol^{-1}$$

This gives a value for $E(C—Cl)$ of $\frac{1321.9}{4} = 330 \, kJ \, mol^{-1}$.

Many other covalent substances are liquids under standard conditions, and it is essential to remember to include an enthalpy change of vaporisation term in the calculation of bond enthalpy.

Worked example

a What data are required to calculate the average bond enthalpy of the N—H bond in ammonia, NH_3?

b Calculate the average bond enthalpy, $E(N—H)$.

Answer

a The data required are:

$\Delta H_f^{\ominus}(NH_3(g)) = -46.0\,kJ\,mol^{-1}$

$\Delta H_{at}^{\ominus}(H) = 218.0\,kJ\,mol^{-1}$

$\Delta H_{at}^{\ominus}(N) = 472.7\,kJ\,mol^{-1}$

b Figure 5.20 shows the Hess's Law diagram.

Now try this

Calculate the average bond enthalpies of the following bonds, using the data given below (all in $kJ\,mol^{-1}$).

1 H—Br
2 O—H
3 P—Cl
[$\Delta H_f^{\ominus}(HBr(g)) = -36.2$;
$\Delta H_f^{\ominus}(H_2O(l)) = -285.9$;
$\Delta H_f^{\ominus}(PCl_3(l)) = -272.4$;
$\Delta H_{vap}^{\ominus}(H_2O) = +44.1$; $\Delta H_{vap}^{\ominus}(PCl_3) = +30.7$;
$\Delta H_{at}^{\ominus}(H) = +218.0$; $\Delta H_{at}^{\ominus}(Br) = +111.9$;
$\Delta H_{at}^{\ominus}(O) = +249.2$; $\Delta H_{at}^{\ominus}(P) = +333.9$;
$\Delta H_{at}^{\ominus}(Cl) = +121.3$]

Figure 5.20

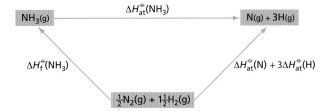

$$\Delta H_f^{\ominus}(NH_3) + \Delta H_{at}^{\ominus}(NH_3) = \Delta H_{at}^{\ominus}(N) + 3\Delta H_{at}^{\ominus}(H)$$
$$\Delta H_{at}^{\ominus}(NH_3) = \Delta H_{at}^{\ominus}(N) + 3\Delta H_{at}^{\ominus}(H) - \Delta H_f^{\ominus}(NH_3)$$
$$= 472.7 + 3 \times 218.0 + 46.0$$
$$= +1172.7\,kJ\,mol^{-1}$$

This gives a value for $E(N—H)$ of $\dfrac{1172.7}{3} = \mathbf{391\,kJ\,mol^{-1}}$.

Finding a bond enthalpy using other known average bond enthalpies

So far we have considered how to find the bond enthalpy in molecules that contain only one type of bond. Many molecules contain several types of bonds. To calculate the bond enthalpies of all of them, we must make the assumption that the bond enthalpy of a bond is the same even though it may be in a different molecule. Bond enthalpies do vary slightly from molecule to molecule so we cannot take their values as being constant to more than a few per cent. However, the universal nature of average bond energies makes them very useful in calculating enthalpies of many different reactions.

Consider the O—H bond. We can calculate its strength from the enthalpy of atomisation of water (see Figure 5.21).

Figure 5.21 Hess's Law cycle to find the enthalpy change of atomisation of water

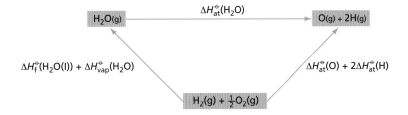

$$\Delta H_f^{\ominus}(H_2O(l)) + \Delta H_{vap}^{\ominus}(H_2O) + \Delta H_{at}^{\ominus}(H_2O) = \Delta H_{at}^{\ominus}(O) + 2\Delta H_{at}^{\ominus}(H)$$
$$-285.9 + 44.1 + \Delta H_{at}^{\ominus}(H_2O) = +249.2 + 2 \times 218.0$$
$$\Delta H_{at}^{\ominus}(H_2O) = 249.2 + 436.0 + 285.9 - 44.1$$
$$= 927\,kJ\,mol^{-1}$$

This gives a value for $E(O{-}H)$ of $\dfrac{927}{2} = 464\,\text{kJ}\,\text{mol}^{-1}$.

We can now use this value to find the strength of the O—O bond in H_2O_2 (see Figure 5.22).

Figure 5.22 Hess's Law cycle to find the enthalpy change of atomisation of H_2O_2

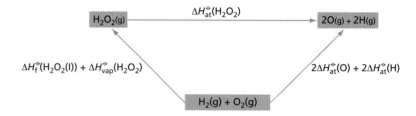

$$\Delta H_f^{\ominus}(H_2O_2(l)) + \Delta H_{vap}^{\ominus}(H_2O_2) + \Delta H_{at}^{\ominus}(H_2O_2) = 2\Delta H_{at}^{\ominus}(O) + 2\Delta H_{at}^{\ominus}(H)$$
$$-187.6 + 52 + \Delta H_{at}^{\ominus}(H_2O_2) = 2 \times 249.2 + 2 \times 218$$
$$\Delta H_{at}^{\ominus} = 498.4 + 436 + 187.6 - 52$$
$$= 1070\,\text{kJ}\,\text{mol}^{-1}$$

H_2O_2 contains two O—H bonds and one O—O bond. If we assume that the value of $E(O{-}H)$ is the same as that in water,

$$2 \times 464 + E(O{-}O) = 1070$$

so $\qquad E(O{-}O) = 142\,\text{kJ}\,\text{mol}^{-1}$

This low value explains why H_2O_2 so readily breaks down into water and oxygen.

Summary

- The **enthalpy change** of a reaction is the difference in enthalpy between the reactants and the products.
- The **enthalpy change of solution**, $\Delta H_{sol}^{\ominus}$, is the enthalpy change when one mole of the substance is completely dissolved in water.
- Enthalpy changes can be found directly by measuring the heat given out or taken in during a reaction, using a **calorimeter**. What is measured is the temperature change; the rest is calculated according to the equation $q = mc\Delta T$, where m is the mass of water in grams (or the volume of solution in cm³), c is the specific heat capacity of water and ΔT is the change in temperature.
- The enthalpy change is given by the equation $\Delta H = -q/n$, where n is the number of moles of reactant taking part in the reaction, or moles of product formed.
- For an **exothermic** reaction ΔH is negative; for an **endothermic** reaction ΔH is positive.
- Most reactions are exothermic, with negative ΔH.
- **Hess's Law** states that ΔH for a reaction is independent of the path taken. If ΔH for a reaction cannot be measured directly, it may often be found using Hess's Law.
- The enthalpy change for a reaction in the gas phase is equal to the bond enthalpies of the bonds broken minus the bond enthalpies of the bonds formed.
- A study of bond enthalpies indicates why ΔH for some reactions is so large.
- Enthalpy changes of atomisation can be used to calculate bond enthalpies.

Some key definitions
- **Standard conditions** are 1.0 bar or 1.0 atm pressure and a specified temperature, usually 298 K.
- The **standard enthalpy change** of a reaction, $\Delta H^{\ominus}$, is the enthalpy change when moles of the reactants as indicated by the equation are completely converted into products under standard conditions.
- The **enthalpy change of neutralisation**, ΔH_{neut}, is the enthalpy change when an acid is neutralised by base to give one mole of water.
- The **standard enthalpy change of combustion**, $\Delta H_c^{\ominus}$, is the enthalpy change when one mole of the substance is completely burnt in oxygen under standard conditions.
- The **standard enthalpy change of solution**, $\Delta H_{sol}^{\ominus}$, is the enthalpy change when one mole of the substance is completely dissolved in water.
- The **standard enthalpy change of formation**, $\Delta H_f^{\ominus}$, is the enthalpy change when one mole of the substance is formed from its elements under standard conditions.
- The **average bond enthalpy**, $E(X{-}Y)$, is the enthalpy change when one mole of bonds between atoms of X and Y are broken in the gas phase:

$$X{-}Y(g) \rightarrow X(g) + Y(g) \qquad \Delta H = E(X{-}Y)$$

- The **standard enthalpy change of atomisation**, $\Delta H_{at}^{\ominus}$, of an element is the enthalpy change when one mole of atoms is formed from the element in its standard state.

Examination practice questions

Please see the data section of the CD for any A_r values you may need.

1 With the prospect that fossil fuels will become increasingly scarce in the future, many compounds are being considered for use in internal combustion engines. One of these is DME or dimethyl ether, CH_3OCH_3. DME is a gas which can be synthesised from methanol.

Methanol can be obtained from biomass, such as plant waste from agriculture.

a Define, with the aid of an equation that includes state symbols, the standard enthalpy change of combustion, $\Delta H_c^{\ominus}$, for DME at 298 K. [3]

b DME may be synthesised from methanol. Relevant enthalpy changes of formation, $\Delta H_f^{\ominus}$, for this reaction are given in the table below.

Compound	$\Delta H_f^{\ominus}$/kJ mol^{-1}
$CH_3OH(l)$	−239
$CH_3OCH_3(g)$	−184
$H_2O(l)$	−286

Use these values to calculate $\Delta H_{reaction}^{\ominus}$ for the synthesis of DME, using the following equation. Include a sign in your answer.

$$2CH_3OH(l) \rightarrow CH_3OCH_3(g) + H_2O(l)$$ [3]

[Cambridge International AS & A Level Chemistry 9701, Paper 23 Q3 a & b June 2012]

2 Carbon disulfide, CS_2, is a volatile, flammable liquid which is produced in small quantities in volcanoes.

a The sequence of atoms in the CS_2 molecule is sulfur to carbon to sulfur.

i Draw a 'dot-and-cross' diagram of the carbon disulfide molecule. Show outer electrons only.

ii Suggest the shape of the molecule and state the bond angle. [3]

b Carbon disulfide is readily combusted to give CO_2 and SO_2.

i Construct a balanced equation for the complete combustion of CS_2.

ii Define the term *standard enthalpy change of combustion*, $\Delta H_c^{\ominus}$. [3]

c Calculate the standard enthalpy change of formation of CS_2 from the following data. Include a sign in your answer.

standard enthalpy change of combustion of $CS_2 = -1110$ kJ mol^{-1}

standard enthalpy change of formation of $CO_2 = -395$ kJ mol^{-1}

standard enthalpy change of formation of $SO_2 = -298$ kJ mol^{-1} [3]

d Carbon disulfide reacts with nitrogen monoxide, NO, in a 1:2 molar ratio. A yellow solid and two colourless gases are produced.

i Construct a balanced equation for the reaction.

ii What is the change in the oxidation number of sulfur in this reaction? [3]

[Cambridge International AS & A Level Chemistry 9701, Paper 23 Q1 June 2013]

3 Many organisms use the aerobic respiration of glucose, $C_6H_{12}O_6$, to release useful energy.

a The overall equation for aerobic respiration is the same as for the complete combustion of $C_6H_{12}O_6$.

i Write the equation for the aerobic respiration of $C_6H_{12}O_6$. [1]

ii Explain, in terms of bond breaking and bond forming, why this reaction is exothermic. [2]

b The table shows some enthalpy changes of combustion, ΔH_c.

Substance	ΔH_c/kJ mol^{-1}
$C(s)$	−394
$H_2(g)$	−286
$C_6H_{12}O_6(s)$	−2801

i What is meant by the term *enthalpy change of combustion*, ΔH_c? [2]

ii The enthalpy change of formation, ΔH_f, of glucose, $C_6H_{12}O_6$, cannot be determined directly. The equation for this enthalpy change is shown below.

$$6C(s) + 6H_2(g) + 3O_2(g) \rightarrow C_6H_{12}O_6(s)$$

Suggest why the enthalpy change of formation of $C_6H_{12}O_6$ **cannot** be determined directly. [1]

iii Use the ΔH_f values in the table to calculate the enthalpy change of formation of $C_6H_{12}O_6$. [3]

[OCR Chemistry A Unit F322 Q1 May 2011]

6 Acids and bases

In this topic the Brønsted–Lowry theory of acids and bases is introduced. This theory describes substances as acids or bases when dissolved in water and also in other solvents. The important quantitative technique of titration is discussed. A detailed explanation of the mole concept in titration calculations is given, including the use of the mole in more complex titrations, which adds emphasis to its importance.

Learning outcomes

By the end of this topic you should be able to:

1.5a) write and construct balanced equations (see also Topic 1)
1.5b) perform calculations, including use of the mole concept, involving volumes and concentrations of solutions, and relate the number of significant figures in your answers to those given or asked for in the question (part, see also Topic 1)
1.5c) deduce stoichiometric relationships from calculations such as those in **1.5b)**
7.2a) show understanding of, and use, the Brønsted–Lowry theory of acids and bases, including the use of the acid-I base-I, acid-II base-II concept
7.2b) explain qualitatively the differences in behaviour between strong and weak acids and bases and the pH values of their aqueous solutions in terms of the extent of dissociation.

6.1 The Arrhenius theory

What is an acid?

Two hundred years ago, substances were called **acids** if they tasted sour (the Latin for sour is *acidus*) or if they changed the colour of some plant extracts (see Figure 6.1). The essential feature of their chemistry was unknown and it was many years before the present-day definition of an acid was recognised. The following is a brief history of how the modern idea of an acid developed. As you can see, it was a very international quest.

Figure 6.1 Lemon juice is a naturally occurring acid. Like other acids, it changes the colour of some plant substances, for example this red cabbage leaf extract.

- 1778 – because most non-metallic oxides dissolved in water to give acidic solutions, the Frenchman Antoine Lavoisier proposed that all acids contain oxygen. This is why he chose the name for the gas oxygen – the word comes from the two Greek words *oxys*, meaning 'acidic, or sharp tasting' and *gonos*, meaning 'generator'.
- 1816 – the British chemist Humphry Davy showed that Lavoisier's view was incorrect when he proved that hydrochloric acid contained hydrogen and chlorine only.
- 1884 – the German Justus von Liebig suggested that acids react with metals to give hydrogen.
- 1884 – as a result of his work on the conductivity of electrolytes, the Swedish chemist Svante Arrhenius defined an acid as a substance that gives hydrogen ions in water.

Arrhenius thought that the hydrogen ion was simply H^+, but we now know that this ion, which is just a proton, only exists by itself in a vacuum. When dissolved in water, a hydrogen ion will be datively bonded to a water molecule to give an **oxonium ion**, H_3O^+ (which is also called the **hydroxonium ion**). This is further hydrated by hydrogen bonding (see Figure 6.2). In chemical equations, it is best to avoid writing just H^+, but instead to write H_3O^+ or $H_3O^+(aq)$ or $H^+(aq)$.

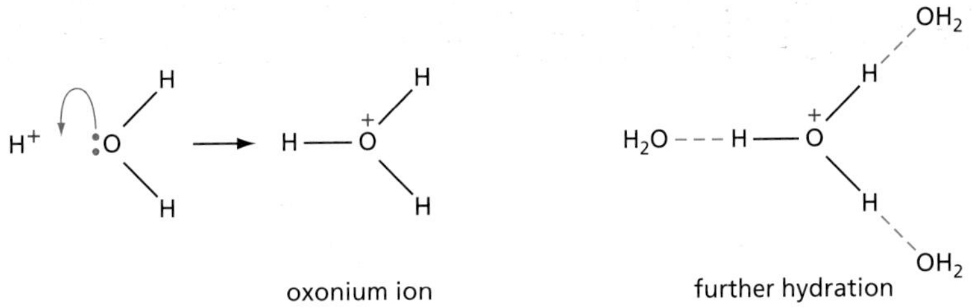

Figure 6.2 The simple hydrogen ion initially forms a dative bond with a water molecule to give the oxonium ion, which is then further hydrated by hydrogen bonding to give mainly $H_5O_2^+$ and $H_9O_4^+$.

oxonium ion   further hydration

Bases and alkalis

The word 'alkali' comes from the Arabic *al qaliy*, which is the calx, or calcined ashes, formed when some minerals or plant or animal material have been strongly heated in air. The ashes would have contained potassium carbonate (pot-ash) and the oxides of various metals. All of these neutralise acids, forming salts. Alkalis were important in the ancient world because soaps were made by heating animal fats with the ashes from plants such as wormwood. (The alkali hydrolysed the glyceryl esters in the fats, to give the sodium or potassium salts of long-chain carboxylic acids, which are the constituents of soap.)

The word 'base' was originally applied to chemistry by the French chemist Guillaume François Rouelle in 1754, for substances that reacted with volatile acids to change them into solid salts. It was then considered to be any substance that reduced the sourness of an acid. Arrhenius limited bases to substances that react with acids (the hydrogen ion) to give the non-acidic product water. This meant that on his definition only metal oxides and hydroxides were bases. We would now consider anything that neutralises an acid to be a base and so would include other substances, such as carbonates.

Today, especially in common usage, the words 'alkali' and 'base' (and the adjectives 'alkaline' and 'basic') are interchangeable, although the following system allows a distinction to be made between them.

- A base is a substance that reacts with an acid.
- An alkali is a base that is soluble in water. Alkalis include the hydroxides of the Group 1 elements (the alkali metals), the soluble hydroxides of the Group 2 elements (the alkaline earth metals), ammonia and the soluble organic amines.

All alkalis produce a pH between 7 and 14 when dissolved in water.

Some salts, such as the carbonates and carboxylates of the alkali metals, also produce solutions in which the pH is greater than 7, because they produce hydroxide ions by reaction with water:

$$CO_3^{2-}(aq) + H_2O(l) \rightleftharpoons HCO_3^-(aq) + OH^-(aq)$$
$$R{-}CO_2^-(aq) + H_2O(l) \rightleftharpoons R{-}CO_2H(aq) + OH^-(aq)$$

They are, therefore, alkalis as well as bases.

Ionic equations

The formation of ionic compounds was discussed in Topic 4. Many ionic compounds dissolve in water to form solutions in which the ions separate from one another and attract water molecules, for example:

$$NaCl(s) + water \rightarrow Na^+(aq) + Cl^-(aq)$$

The '(aq)' indicates that the ions are in aqueous solution and are hydrated, that is, surrounded by water molecules. These solutions conduct electricity.

Many of the reactions undergone by ionic compounds in solution are in fact reactions of the individual ions rather than of the compound as a whole. An example is the common test for the chloride ion, in which a white precipitate is formed when silver nitrate is added to a solution of a chloride, for example sodium chloride:

$$AgNO_3(aq) + NaCl(aq) \rightarrow NaNO_3(aq) + AgCl(s) \tag{1}$$

The white precipitate forms when aqueous silver nitrate is added to a solution of *any* chloride, for example magnesium chloride:

$$2AgNO_3(aq) + MgCl_2(aq) \rightarrow Mg(NO_3)_2(aq) + 2AgCl(s) \tag{2}$$

Moreover, the silver salt does not have to be the nitrate: *any* soluble silver salt will give the same reaction. For example:

$$Ag_2SO_4(aq) + 2KCl(aq) \rightarrow K_2SO_4(aq) + 2AgCl(s) \tag{3}$$

The reaction is therefore seen to be one between the two ions $Ag^+(aq)$ and $Cl^-(aq)$ only: *any* soluble silver salt will react with *any* soluble chloride to give the white precipitate, that is:

$$Ag^+(aq) + Cl^-(aq) \rightarrow AgCl(s)$$

This is the **ionic equation** for the reaction.

Deriving ionic equations from full chemical equations

The procedure consists of the following steps.

- Write the full balanced chemical equation, including state symbols.
- For any ionic compound in aqueous solution, replace the full formula with those of the separate aqueous ions.
- 'Cancel' any ions that appear on both sides of the equation. These do not take part in the reaction, and are described as *spectator ions*.
- What is left is the ionic equation for the reaction.

Worked example

Derive the ionic equation for reaction (1) above.

Answer

There are three aqueous ionic compounds in this equation: $AgNO_3(aq)$ and $NaCl(aq)$ on the left-hand side, and $NaNO_3(aq)$ on the right-hand side. These formulae are replaced by those of their separate ions:

$$Ag^+(aq) + NO_3^-(aq) + Na^+(aq) + Cl^-(aq) \rightarrow Na^+(aq) + NO_3^-(aq) + AgCl(s)$$

(Note that $AgCl(s)$ is not written as its separate ions, since the Ag^+ and Cl^- ions do not exist independently as individual ions in the solid, but are associated with one another in a fixed lattice.)

The 'spectator ions' common to both sides of the equation, $Na^+(aq)$ and $NO_3^-(aq)$, are cancelled:

$$Ag^+(aq) + \cancel{NO_3^-(aq)} + \cancel{Na^+(aq)} + Cl^-(aq) \rightarrow \cancel{Na^+(aq)} + \cancel{NO_3^-(aq)} + AgCl(s)$$

So the resulting ionic equation is:

$$Ag^+(aq) + Cl^-(aq) \rightarrow AgCl(s)$$

The scope of ionic equations

There are three areas in chemistry where ionic equations are useful:

- precipitation reactions (as in the worked example above)
- acid–base reactions
- redox reactions, including acid–metal and water–metal reactions.

The following worked examples cover each of these areas in turn. In each case, the ionic equation is simpler, yet more universal, and shows more clearly than the full chemical equation how reactions are related.

Precipitation reactions

Worked example

Write the ionic equation for the reaction between barium chloride and potassium sulfate.

Answer
- The full balanced equation is:

  $$BaCl_2(aq) + K_2SO_4(aq) \rightarrow 2KCl(aq) + BaSO_4(s)$$

- Separate all aqueous ionic compounds into their separate ions, and cancel the spectator ions that appear on both sides:

  $$Ba^{2+}(aq) + 2Cl^-(aq) + 2K^+(aq) + SO_4{}^{2-}(aq) \rightarrow 2K^+(aq) + 2Cl^-(aq) + BaSO_4(s)$$

- What remains is the ionic equation:

  $$Ba^{2+}(aq) + SO_4{}^{2-}(aq) \rightarrow BaSO_4(s)$$

Now try this

Write the ionic equations for, and list the spectator ions in, the reactions between aqueous solutions of the following pairs of compounds.

1 sodium carbonate and calcium chloride
2 lithium hydroxide and magnesium sulfate

Acid–base reactions

Worked example

Write the ionic equation for the reaction between nitric acid and sodium hydroxide.

Answer
- The full balanced equation is:

  $$HNO_3(aq) + NaOH(aq) \rightarrow NaNO_3(aq) + H_2O(l)$$

- Separate all aqueous ionic compounds into their separate ions, and cancel the spectator ions that appear on both sides:

  $$H^+(aq) + NO_3{}^-(aq) + Na^+(aq) + OH^-(aq) \rightarrow Na^+(aq) + NO_3{}^-(aq) + H_2O(l)$$

- What remains is the ionic equation:

  $$H^+(aq) + OH^-(aq) \rightarrow H_2O(l)$$

  (Note that H_2O is almost completely a covalent compound: only two molecules in 1000 million are ionised as H^+ and OH^-.)

Now try this

Write the ionic equations for, and list the spectator ions in, the reactions between aqueous solutions of the following pairs of compounds.

1 hydrochloric acid and limewater $(Ca(OH)_2(aq))$
2 sulfuric acid and potassium hydroxide

Redox reactions

Worked example 1

Write the ionic equation for the following reaction:

$$\text{chlorine water} + \text{potassium iodide} \rightarrow \text{iodine} + \text{potassium chloride}$$
$$\text{solution} \qquad \text{solution} \qquad \text{solution}$$

Answer
- The full balanced equation is:

  $$Cl_2(aq) + 2KI(aq) \rightarrow 2KCl(aq) + I_2(aq)$$

- Separate all aqueous ionic compounds into their separate ions, and cancel the spectator ions that appear on both sides:

  $$Cl_2(aq) + 2K^+(aq) + 2I^-(aq) \rightarrow 2K^+(aq) + 2Cl^-(aq) + I_2(aq)$$

  (Note that it is only the $K^+(aq)$ ions that are common to both sides – the $I^-(aq)$ ions have been oxidised to $I_2(aq)$.)
- What remains is the ionic equation:

  $$Cl_2(aq) + 2I^-(aq) \rightarrow 2Cl^-(aq) + I_2(aq)$$

Worked example 2

Write the ionic equation for the reaction between magnesium and dilute sulfuric acid.

Answer

- The full balanced equation is:

$$Mg(s) + H_2SO_4(aq) \rightarrow MgSO_4(aq) + H_2(g)$$

- Separate all aqueous ionic compounds into their separate ions, and cancel the spectator ions that appear on both sides:

$$Mg(s) + 2H^+(aq) + \cancel{SO_4^{2-}}(aq) \rightarrow Mg^{2+}(aq) + \cancel{SO_4^{2-}}(aq) + H_2(g)$$

- What remains is the ionic equation:

$$Mg(s) + 2H^+(aq) \rightarrow Mg^{2+}(aq) + H_2(g)$$

Now try this

Write the ionic equation for the following reaction:

zinc metal + hydrochloric acid → zinc chloride solution + hydrogen gas

Ionic equations for redox reactions can also be written using redox half-equations, which are often listed in books of data (see section 7.1 for a description of this method).

Representing acid–base reactions

We now recognise that in an acid–base reaction, as in some other reactions (see pages 120–122) the ions making up the salt are **spectator ions** – they do not take part in a chemical reaction. We can therefore represent these reactions by an ionic equation which omits such spectator ions. For example, in the equation for the neutralisation of hydrochloric acid with sodium hydroxide, the ionic equation is:

$$H^+(aq) + OH^-(aq) \rightarrow H_2O(l)$$

Using simplified ionic equations in this way has the advantages of showing which bonds are being broken and which are formed, and of showing that many apparently different acid–base reactions are, in fact, identical.

According to the Arrhenius theory, acid–base reactions can be divided into three main types:

- between an acid and an alkali:

$$H^+(aq) + OH^-(aq) \rightarrow H_2O(l)$$

- between an acid and an insoluble metal oxide, for example copper(II) oxide or aluminium oxide:

$$2H^+(aq) + CuO(s) \rightarrow H_2O(l) + Cu^{2+}(aq)$$
$$6H^+(aq) + Al_2O_3(s) \rightarrow 3H_2O(l) + 2Al^{3+}(aq)$$

- between an acid and an insoluble metal hydroxide, for example magnesium hydroxide or iron(III) hydroxide:

$$2H^+(aq) + Mg(OH)_2(s) \rightarrow 2H_2O(l) + Mg^{2+}(aq)$$
$$3H^+(aq) + Fe(OH)_3(s) \rightarrow 3H_2O(l) + Fe^{3+}(aq)$$

Worked example

Write the full equation and the ionic equation for the action of nitric acid on magnesium oxide.

Now try this

Write full equations and ionic equations for the following reactions.

1 sulfuric acid with sodium hydroxide
2 hydrochloric acid with iron(III) hydroxide
3 nitric acid with zinc oxide

Answer

The products are magnesium nitrate and water. The full equation is:

$$MgO(s) + 2HNO_3(aq) \rightarrow Mg(NO_3)_2(aq) + H_2O(l)$$

The $NO_3^-(aq)$ ions are unchanged and can be omitted in the ionic equation, but the Mg^{2+} ion changes from the solid to the aqueous state and must be included:

$$MgO(s) + 2H^+(aq) \rightarrow Mg^{2+}(aq) + H_2O(l)$$

⟳ 6.2 The Brønsted–Lowry theory

Donating and accepting protons

The Arrhenius theory can only be applied to reactions in aqueous solution. Many similar reactions take place in solvents other than water or under anhydrous conditions. For example, the reaction:

$$CuO(s) + 2HCl(g) \rightarrow CuCl_2(s) + H_2O(l)$$

is very similar to:

$$CuO(s) + 2HCl(aq) \rightarrow CuCl_2(aq) + H_2O(l)$$

and should therefore be classified as an acid–base reaction.

The Danish chemist Johannes Brønsted and the English chemist Thomas Lowry were working independently but both realised that the important feature of an acid–base reaction was the transfer of a proton from an acid to a base. They therefore defined an **acid** as a substance that could donate a proton and a **base** as a substance that could accept a proton.

According to this theory, we might expect that any substance containing a hydrogen atom could act as an acid. In practice, a substance behaves as an acid only if the hydrogen atom already carries a slight positive charge, $—X^{\delta-}—H^{\delta+}$. This is the case if it is bonded to a highly electronegative atom, such as oxygen, nitrogen, sulfur or a halogen, which are found on the right-hand side of the Periodic Table.

In order to accept a proton, a base must contain a lone pair of electrons that it can use to form a dative bond with the proton (see section 3.7). So, like an acid, a base must contain an atom from the right-hand side of the Periodic Table.

In general, therefore, an acid–base reaction is usually the transfer of a proton from one electronegative atom to another electronegative atom.

> **The Brønsted–Lowry theory of acids and bases**
> - An **acid** is a proton donor.
> - A **base** is a proton acceptor.

Conjugate acid–base pairs

When an acid such as hydrochloric acid loses a proton, a base such as Cl^- is formed (although in this case Cl^- is quite a weak base). A proton could return to this base and re-form HCl. This HCl/Cl^- system is known as a **conjugate acid–base pair**. The changes can be represented by the equations:

$$\underset{\text{acid}}{H\!\!\div\!\!Cl} \rightarrow \underset{\text{proton}}{H^+} + \underset{\text{conjugate base}}{:\!Cl^-} \quad \text{and} \quad \underset{\text{base}}{:\!Cl^-} + \underset{\text{proton}}{H^+} \rightarrow \underset{\text{conjugate acid}}{H\!\!\div\!\!Cl}$$

Hydrochloric acid can donate just one proton – it is called a **monoprotic** acid.

Some acids can donate more than one proton. For example, sulfuric acid is **diprotic** – it can lose two protons in two different steps:

$$H_2SO_4 \rightarrow HSO_4^- + H^+ \quad \text{step 1}$$
$$HSO_4^- \rightarrow SO_4^{2-} + H^+ \quad \text{step 2}$$

Similarly, some bases can accept more than one proton. The carbonate ion is **diprotic** – it can gain two protons in two different steps:

$$CO_3^{2-} + H^+ \rightarrow HCO_3^- \quad \text{step 1}$$
$$HCO_3^- + H^+ \rightarrow H_2CO_3 \quad \text{step 2}$$

> - Acids that can donate a maximum of one, two or three protons are called monoprotic, diprotic or triprotic respectively.
> - Bases that can receive a maximum of one, two or three protons are called monoprotic, diprotic or triprotic respectively.

Water is an exceptional substance as it can behave both as an acid and as a base. Such a substance is called an **ampholyte** and shows **amphoteric** behaviour.

$$H_2O + H^+ \rightarrow H_3O^+ \quad \text{water behaving as a base}$$
$$H_2O \rightarrow H^+ + OH^- \quad \text{water behaving as an acid}$$

Now try this

1 Show the bonding in the conjugate bases of the following acids.
  a HBr
  b HNO_3
  c H_2SO_4
  d HSO_4^-
  e HS^-

2 Show the bonding in the conjugate acids of the following bases.
  a F^-
  b HS^-
  c CO_3^{2-}
  d HSO_4^-
  e H_2O
  f NH_3

Worked example 1

Show the bonding in the conjugate base of hydrogen sulfide, H_2S.

Answer

Hydrogen sulfide behaves as an acid:

$$H \div S \div H \rightarrow H \div S\!:^- + H^+$$

and HS^- is the conjugate base.

Worked example 2

Show the bonding in the conjugate acid of the ethanoate ion, $CH_3CO_2^-$.

Answer

The ethanoate ion behaves as a base:

$$CH_3CO \div O\!:^- + H^+ \rightarrow CH_3CO \div O \div H$$

and ethanoic acid, CH_3CO_2H, is the conjugate acid.

6.3 Brønsted–Lowry acid–base reactions

When an acid gives up its proton to a base, a **proton-transfer reaction** has taken place. This is a Brønsted–Lowry acid–base reaction.

An example is dissolving hydrogen chloride in water. The hydrogen chloride is the acid, and water acts as a base:

$$HCl \rightarrow H^+ + Cl^-$$

acid   proton   conjugate base

$$H_2O\!: + H^+ \rightarrow H_3O^+$$

base   proton   conjugate acid

The change can be shown by 'curly arrows'. Each curly arrow shows the movement of a pair of electrons: the 'tail' shows where the pair of electrons are positioned at the start of the reaction, and the arrow 'head' shows where they finish. These two changes can be represented as follows:

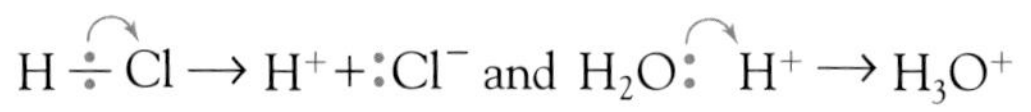

$$H \div Cl \rightarrow H^+ + \!:Cl^- \text{ and } H_2O\!: \; H^+ \rightarrow H_3O^+$$

The H_3O^+ ion is produced by the water molecule forming a dative bond with the proton:

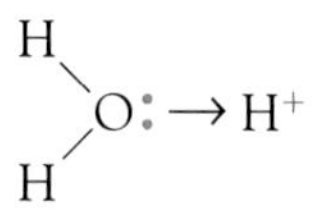

$$\begin{matrix} H \\ \diagdown \\ & O\!: \rightarrow H^+ \\ \diagup \\ H \end{matrix}$$

But, just as all four bonds in the ammonium ion, NH_4^+, are equivalent (see section 3.7), so are the three bonds in the H_3O^+ ion:

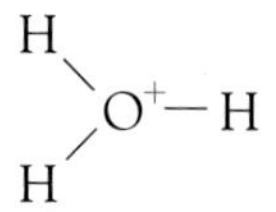

$$\begin{matrix} H \\ \diagdown \\ & O^+\!\!-H \\ \diagup \\ H \end{matrix}$$

So we have two **half-equations** which represent proton movements that happen during the reaction:

$$HCl \rightarrow H^+ + Cl^- \quad \text{and} \quad H_2O + H^+ \rightarrow H_3O^+$$

acid-I   base-I   base-II   acid-II

These can be combined to give the complete equation:

$$\underset{\text{acid-I}}{HCl(g)} + \underset{\text{base-II}}{H_2O(l)} \rightarrow \underset{\text{base-I}}{Cl^-(aq)} + \underset{\text{acid-II}}{H_3O^+(aq)}$$

Notice that there must be four species in the equation: two conjugate acid–base pairs. If the acid or base is diprotic, there are two possible neutralisation reactions. For example, in the neutralisation of sulfuric acid with sodium hydroxide, the following reactions may happen:

$$H_2SO_4 \rightarrow 2H^+ + SO_4^{2-} \qquad \text{reaction 1, when acid:base ratio} \geq 2$$
$$H_2SO_4 \rightarrow H^+ + HSO_4^- \qquad \text{reaction 2, when acid:base ratio} \leq 1$$

This gives two possible neutralisation reactions, depending on the proportions of acid and base present in the reaction.

● With excess acid:

$$\underset{\text{acid-I}}{H_2SO_4(aq)} + \underset{\text{base-II}}{OH^-(aq)} \rightarrow \underset{\text{base-I}}{HSO_4^-(aq)} + \underset{\text{acid-II}}{H_2O(l)}$$

● With excess base:

$$\underset{\text{acid-I}}{H_2SO_4(aq)} + \underset{\text{base-II}}{2OH^-(aq)} \rightarrow \underset{\text{base-I}}{SO_4^{2-}(aq)} + \underset{\text{acid-II}}{2H_2O(l)}$$

The $HSO_4^-(aq)$ ion acts as base in reaction (1) but it can also act as an acid (that is, it is amphoteric):

$$\underset{\text{acid-I}}{HSO_4^-(aq)} + \underset{\text{base-II}}{OH^-(aq)} \rightarrow \underset{\text{base-I}}{SO_4^{2-}(aq)} + \underset{\text{acid-II}}{H_2O(l)}$$

Worked example 1

The carbonate ion is diprotic. Write two overall ionic equations for the reaction of dilute nitric acid with aqueous sodium carbonate, with an excess of either acid or carbonate.

Answer
With an excess of acid:

$$\underset{\text{acid-I}}{2H_3O^+(aq)} + \underset{\text{base-II}}{CO_3^{2-}(aq)} \rightarrow \underset{\text{acid-II}}{H_2CO_3(aq)} + \underset{\text{base-I}}{H_2O(l)}$$

With an excess of carbonate:

$$\underset{\text{acid-I}}{H_3O^+(aq)} + \underset{\text{base-II}}{CO_3^{2-}(aq)} \rightarrow \underset{\text{acid-II}}{HCO_3^-(aq)} + \underset{\text{base-I}}{H_2O(l)}$$

Worked example 2

Write the relevant half-equations and the overall ionic equations for the possible reactions when solid sodium chloride is treated with concentrated sulfuric acid.

Answer
The half-equations are as follows.

● With an excess of acid:

$$H_2SO_4 \rightarrow HSO_4^- + H^+ \quad \text{and} \quad Cl^- + H^+ \rightarrow HCl$$

● With an excess of base:

$$H_2SO_4 \rightarrow SO_4^{2-} + 2H^+ \quad \text{and} \quad Cl^- + H^+ \rightarrow HCl$$

The overall equations are as follows.

● With an excess of acid:

$$\underset{\text{base-II}}{Cl^-(s)} + \underset{\text{acid-I}}{H_2SO_4(l)} \rightarrow \underset{\text{base-I}}{HSO_4^-(s)} + \underset{\text{acid-II}}{HCl(g)}$$

● With an excess of base:

$$\underset{\text{base-II}}{2Cl^-(s)} + \underset{\text{acid-I}}{H_2SO_4(l)} \rightarrow \underset{\text{base-I}}{SO_4^{2-}(s)} + \underset{\text{acid-II}}{2HCl(g)}$$

Now try this

For each of the following equations, decide whether the underlined species is acting as an acid or a base, and state which species is its conjugate base or acid.

1 $Br^- + \underline{HSO_4^-} \rightarrow HBr + SO_4^{2-}$
2 $H_2SO_4 + \underline{HNO_3} \rightarrow HSO_4^- + H_2NO_3^+$
3 $NH_3 + \underline{H_2O} \rightarrow NH_4^+ + OH^-$

6.4 Strong and weak acids

According to the Arrhenius theory, a **strong acid** is one that is completely ionised in aqueous solution, or nearly so, while a **weak acid** is one that is ionised to only a small extent (possibly less than 10%).

According to the Brønsted–Lowry theory, a **strong acid** is one that gives up a proton more easily than the H_3O^+ ion does. This means that when a strong acid such as hydrogen chloride is dissolved in water, the following reaction takes place:

$$HCl(g) + H_2O(l) \rightarrow H_3O^+(aq) + Cl^-(aq)$$

A similar reaction takes place with other strong acids such as sulfuric acid and nitric acid. If the acid is strong, it has a great tendency to give up a proton, and so its conjugate base must be very weak, as it has very little tendency to accept a proton. In the above example, the Cl^- is a very weak base.

A **weak acid** is one that gives up a proton less easily than the H_3O^+ ion does. So if a weak acid such as ethanoic acid, CH_3CO_2H, is dissolved in water, it hardly ionises at all:

$$CH_3CO_2H(l) + H_2O(l) \rightleftharpoons H_3O^+(aq) + CH_3CO_2^-(aq)$$

Since ethanoic acid has little tendency to give up a proton, its conjugate base, the ethanoate ion, must be a fairly strong base, because it has a tendency to accept a proton.

For acids at equal concentrations, the pH values of aqueous solutions of weak acids are higher than those of strong acids, since the concentration of $H^+(aq)$ ions is less.

By observing that stronger acids have a tendency to donate protons to the conjugate bases of weaker acids, an approximate order of acid strength can be found, as shown in Table 6.1. A quantitative basis for this order is discussed in Topic 22.

Table 6.1 Some common acids and bases in order of their strengths

Acid	Strength	Base	Strength
H_2SO_4	very strong (pH < 0)	HSO_4^-	very weak
HCl		Cl^-	
HNO_3		NO_3^-	
H_3O^+	fairly strong (pH = 1)	H_2O	weak
HSO_4^-		SO_4^{2-}	
CH_3CO_2H		$CH_3CO_2^-$	
H_2CO_3	weak (pH = 5)	HCO_3^-	less weak
NH_4^+		NH_3	
HCO_3^-		CO_3^{2-}	
H_2O	very weak (pH = 7)	OH^-	strong

The direction in which an acid–base reaction takes place is governed by energy changes (see Topic 5), just as with any other chemical reaction. The energy change is greatest when a strong acid reacts with a strong base, and these are therefore the ones that react together most readily. If a weak acid reacts with a weak base, the reaction will not go to completion (see Topic 9), and there will be a significant proportion of the original acid and base remaining at equilibrium. For example:

$$NH_3 + CH_3CO_2H \rightleftharpoons NH_4^+ + CH_3CO_2^-$$

6.5 Solvents other than water

The Brønsted–Lowry theory can be applied to acids in solvents other than water. One of the first solvents to be used instead of water was liquid ammonia, with a boiling point of $-33\,°C$. Ammonia, like water, is an ampholyte and can behave as an acid or a base.

As an acid:

$$NH_3 \rightarrow NH_2^- + H^+$$

As a base:

$$NH_3 + H^+ \rightarrow NH_4^+$$

So in liquid ammonia, ammonium chloride, $NH_4^+Cl^-$, behaves as a strong acid, and sodamide, $Na^+NH_2^-$, behaves as a strong base.

> ### Now try this
>
> Write the full balanced equation and the ionic equation for the reaction between ammonium chloride and sodamide in liquid ammonia.

Another solvent that may be used for acid–base reactions is concentrated sulfuric acid. This is used both as an acid and as a solvent in the nitration of benzene (see section 25.3). Very strong acids such as chloric(VII) acid protonate the solvent sulfuric acid:

$$HClO_4 + H_2SO_4 \rightarrow ClO_4^- + H_3SO_4^+$$

whereas 'bases' such as water deprotonate the solvent:

$$H_2O + H_2SO_4 \rightarrow H_3O^+ + HSO_4^-$$

6.6 Acid–base titrations

Calculations involving acid–base reactions

Quantitative information about acid–base reactions can be found by **titration** (see Figure 6.3). This is a technique that is often used in analytical chemistry to measure with a high degree of accuracy the concentration of an acid or base in a sample, by comparing it to the concentration of base or acid in a **standard solution** (a solution in which the concentration of solute in $mol\,dm^{-3}$ is known accurately).

A known volume of a solution of one reactant is measured using a pipette. Into this is run a solution of the other reactant from a burette, until the reaction between the two substances is just complete. For an acid–base reaction, the completion of the reaction is usually found by noting a colour change of an indicator that has been added to the reaction mixture.

Figure 6.3 The stages in carrying out a titration:
a A standard solution is made up of one reactant.
b A measured volume of this standard solution is put into a flask and a solution of the other reactant is run in from a burette.
c The end-point is detected using an indicator, and the volume is read off the burette.

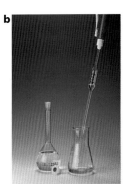

The results of a titration can either be used to find the concentration of one of the solutions or to establish the **stoichiometry** of a chemical reaction, that is, to find how many moles of one substance react with one mole of another substance.

Both of these uses require calculations involving the following two equations, which we met in Topic 1.

The basic equations used in titrations

$$n = \frac{m}{M}$$

n = amount (in mol), m = mass (in g), M = molar mass (in $g\,mol^{-1}$)

$$n = cV$$

c = concentration (in $mol\,dm^{-3}$), V = volume (in dm^3)

Sometimes volumes are measured in cm^3 rather than dm^3. Since $1\,cm^3 = \frac{1}{1000}\,dm^3$, the following equation is used in place of the second equation above:

$$n = c \times \frac{v}{1000}$$

Here, c is the concentration in $mol\,dm^{-3}$ and v is the volume **in cm^3**. This equation is often used in the form:

$$c = \frac{1000}{v} \times n$$

The first step in a titration is usually to make a standard solution. This is done by dissolving an accurately weighed amount of the pure substance in pure water, and making the solution up to a known volume (often $250\,cm^3$) in a calibrated flask. It is essential that the solution is well mixed. Do not do this by shaking but by turning the flask up and down at least five times.

Worked example 1

$0.360\,g$ of sodium hydroxide was made up to $250\,cm^3$ in a calibrated flask. What is the concentration of the solution?

Answer
The molar mass of sodium hydroxide, NaOH, is $(23.0 + 16.0 + 1.0) = 40.0\,g\,mol^{-1}$.

$$n = \frac{m}{M} = \frac{0.360}{40.0} = 9.00 \times 10^{-3}\,mol$$

$$c = \frac{1000}{v} \times n = \frac{1000}{250} \times 9.00 \times 10^{-3}$$

$$= 3.60 \times 10^{-2}\,mol\,dm^{-3} \text{ or } 0.0360\,mol\,dm^{-3}$$

In titration calculations, it is usual to work to 3 significant figures.

Worked example 2

$1.778\,g$ of hydrated ethanedioic acid, $H_2C_2O_4.2H_2O$, were made up to $100\,cm^3$ in a calibrated flask. What is the concentration of the solution?

Answer
The molar mass of $H_2C_2O_4.2H_2O$ is $(2.0 + 24.0 + 64.0 + 2 \times 18.0) = 126.0\,g\,mol^{-1}$

$$n = \frac{m}{M} = \frac{1.778}{126.0} = 1.41 \times 10^{-2}\,mol$$

$$c = \frac{1000}{v} \times n = \frac{1000}{100} \times 1.41 \times 10^{-2}$$

$$= 1.41 \times 10^{-1}\,mol\,dm^{-3} \text{ or } 0.141\,mol\,dm^{-3}$$

Now try this

Calculate the concentration of each of the following solutions.

1 $1.50\,g$ of potassium hydrogensulfate, $KHSO_4$, in $250\,cm^3$
2 $2.45\,g$ of potassium hydrogenphthalate, $C_8H_5O_4K$, in $100\,cm^3$ of solution

Carrying out a titration

Having made a standard solution, the next step is to carry out the titration using this solution and the solution whose concentration we want to find. One of the solutions (usually the acid) is placed in the burette, and a known volume of the other solution is placed in a conical reaction flask using an accurately calibrated pipette. Before their use, the burette and the pipette are thoroughly cleaned and then rinsed out with the solutions they will eventually contain, to ensure that the concentrations remain constant. A few drops of the indicator are added to the flask, and the solution from the burette (called the **titrant**) is slowly added, with shaking, until the first permanent colour change is noted (this is the **end-point**). There are several points to note about the procedure:

- The first and final readings of the burette should be recorded (Figure 6.4).
- Readings should be recorded to the nearest 0.05 cm^3, which is the approximate volume of one drop. So readings are recorded as, for example, 25.00, 25.05 and 26.00 cm^3.
- During the first titration, the end-point is often missed through overshooting, that is, adding slightly more titrant than is necessary *just* to change the indicator colour. However, the reading should still be recorded as a 'rough' titration.
- Further titrations should then be done so that the results of two consecutive titrations differ by no more than 0.10 cm^3.

Figure 6.4 When reading a burette, the volume noted is the level of the bottom of the meniscus. What are the two burette readings, and what volume has been run out of the burette shown in the photographs?

Indicators

The two most common indicators that are used in titrations are methyl orange and phenolphthalein. Their colour changes are shown in Table 6.2 and Figure 6.5.

Table 6.2 The colour changes of two commonly used indicators

Indicator	Colour change when acid is run from the burette	Colour change when alkali is run from the burette
methyl orange	yellow to first tinge of orange	red to first tinge of orange
phenolphthalein	pink to just colourless	colourless to just pink

Finding the concentration of a solution

We are provided with two solutions, A and B, which react together. Solution A, is a standard solution and so its concentration is known. The two volumes used in the titration, one from the pipette and one from the burette, are found by experiment.

The balanced chemical equation shows how many moles of B react with one mole of A.

To find the concentration of solution B, the known values can be set out as shown in Table 6.3. The steps are as follows.

- First work out the amount (in moles) of reactant A:

$$n_A = c_A \times \frac{v_A}{1000}$$

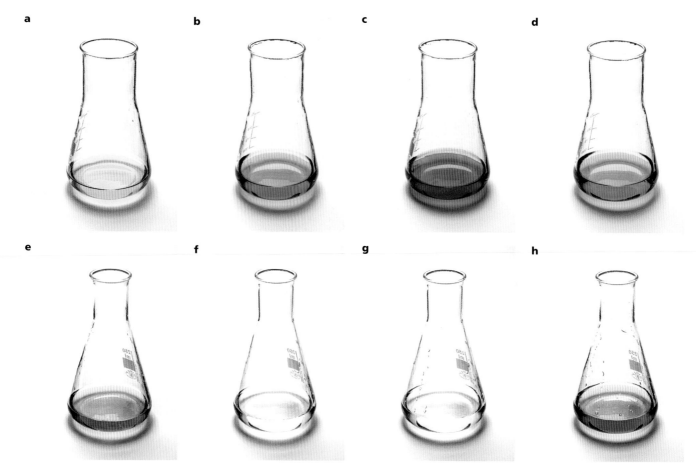

a   b   c   d

e   f   g   h

Figure 6.5 The colour changes of methyl orange (**a** → **b** and **c** → **d**) and phenolphthalein (**e** → **f** and **g** → **h**) at the end-point

Table 6.3 The basis of a titration calculation

Solution A	Solution B
concentration known = c_A	concentration to be found = c_B = ?
volume from titration = v_A	volume from titration = v_B
amount (in moles) = n_A ⟶	amount (in moles) = n_B

- Then use the chemical equation to find the number of moles of B that react with one mole of A.
- In this way you can calculate the amount (in moles) of B used in the titration.
- Finally, work out c_B using the equation:

$$c_B = \frac{1000}{v_B} \times n_B$$

Worked example 1

25.00 cm³ of 0.0500 mol dm⁻³ sodium hydroxide neutralised 20.00 cm³ of hydrochloric acid. What is the concentration of the hydrochloric acid?

Answer
First work out n(NaOH):

$$n = c \times \frac{v}{1000}$$

$$= 0.0500 \times \frac{25.0}{1000} = 1.25 \times 10^{-3}\,mol$$

Next write a balanced equation:

$NaOH(aq) + HCl(aq) \rightarrow NaCl(aq) + H_2O(l)$

This shows that 1 mol of NaOH reacts with 1 mol of HCl. To find $c(HCl)$, use:

$$c = \frac{1000}{V} \times n$$

$$= \frac{1000}{20} \times 1.25 \times 10^{-3}$$

$$= 0.0625\,mol\,dm^{-3}$$

Worked example 2

$23.65\,cm^3$ of $0.0800\,mol\,dm^{-3}$ hydrochloric acid were neutralised by $25.00\,cm^3$ of a solution of sodium carbonate according to the equation:

$2HCl(aq) + Na_2CO_3(aq) \rightarrow 2NaCl(aq) + H_2O(l) + CO_2(g)$

What is the concentration of the sodium carbonate solution?

Answer

$$n(HCl) = c \times \frac{V}{1000} = 0.0800 \times \frac{23.65}{1000} = 1.892 \times 10^{-3}\,mol$$

Since 2 mol of HCl react with 1 mol of Na_2CO_3,

$$n(Na_2CO_3) = \frac{1}{2} \times 1.892 \times 10^{-3} = 9.46 \times 10^{-4}\,mol$$

$$\text{so} \quad c(Na_2CO_3) = \frac{1000}{V} \times n$$

$$= \frac{1000}{25.00} \times 9.46 \times 10^{-4}$$

$$= 0.0378\,mol\,dm^{-3}$$

Finding the purity of a substance, or the amount of water of crystallisation

One use of titrations is to find the purity of a chemical or to find the number of molecules of water of crystallisation in a substance. The following worked examples illustrate these uses.

Worked example 1

$0.982\,g$ of an impure sample of sodium hydroxide was made up to $250\,cm^3$. A $25.00\,cm^3$ pipette-full of this solution was neutralised by $23.50\,cm^3$ of $0.100\,mol\,dm^{-3}$ hydrochloric acid run in from a burette. What is the percentage purity of the sodium hydroxide?

Answer

$$n(HCl) = c \times \frac{V}{1000} = 0.100 \times \frac{23.5}{1000} = 2.35 \times 10^{-3}\,mol$$

Since 1 mol of NaOH reacts with 1 mol of HCl,

$$n(NaOH) \text{ in } 25\,cm^3 = 2.35 \times 10^{-3}\,mol$$

$$n(NaOH) \text{ in } 250\,cm^3 = 2.35 \times 10^{-3} \times \frac{250}{25.0}$$

$$= 2.35 \times 10^{-2}\,mol$$

Since $M(NaOH) = 40.0\,g\,mol^{-1}$,

$$\text{mass of pure NaOH} = 2.35 \times 10^{-2} \times 40.0 = 0.940\,g$$

Hence,

$$\text{percentage purity of NaOH} = \frac{0.940}{0.982} \times 100 = \mathbf{95.7\%}$$

Worked example 2

Washing soda has the formula $Na_2CO_3.xH_2O$. A mass of 1.4280 g of washing soda was made up to 250 cm^3. 25.0 cm^3 of this solution were neutralised by 20.0 cm^3 of 0.0500 mol dm^{-3} hydrochloric acid run in from a burette.

The equation for the reaction is:

$$Na_2CO_3(aq) + 2HCl(aq) \rightarrow 2NaCl(aq) + CO_2(g) + H_2O(l)$$

Find the value of x, and the formula of washing soda.

Answer

$$n(HCl) = c \times \frac{v}{1000} = 0.0500 \times \frac{20.0}{1000} = 1.00 \times 10^{-3}\,mol$$

Since 1 mol of HCl reacts with $\frac{1}{2}$ mol of Na_2CO_3,

$$n(Na_2CO_3) \text{ in the pipette} = \frac{1}{2} \times 1.00 \times 10^{-3} = 5.00 \times 10^{-4}\,mol$$

$$n(Na_2CO_3) \text{ in } 250\,cm^3 = 5.00 \times 10^{-4} \times \frac{250}{25.0} = 5.00 \times 10^{-3}\,mol$$

Since $M(Na_2CO_3) = 106.0\,g\,mol^{-1}$,

mass of anhydrous Na_2CO_3 in 250 cm^3 = $5.00 \times 10^{-3} \times 106.0 = 0.530\,g$

mass of water in the washing soda = (1.4280 − 0.530) = 0.898 g

Since $M(H_2O) = 18.0\,g\,mol^{-1}$,

$$n(H_2O) = \frac{0.898}{18.0} = 4.99 \times 10^{-2}\,mol$$

So 5.00×10^{-3} mol of Na_2CO_3 combine with 4.99×10^{-2} mol of H_2O.

Therefore 1 mol of Na_2CO_3 combines with $\frac{4.99 \times 10^{-2}}{5.00 \times 10^{-3}} = 9.98$ mol of H_2O.

So, to the nearest whole number, 1 mol of Na_2CO_3 combines with 10 mol of H_2O.
Therefore $x = \mathbf{10}$ and the formula of washing soda is $\mathbf{Na_2CO_3.10H_2O}$.

Now try this

1 1.250 g of concentrated sulfuric acid were dissolved in water and made up to 250 cm^3. A 25.0 cm^3 pipette-full of this solution was neutralised by 24.85 cm^3 of 0.102 mol dm^{-3} sodium hydroxide. The equation for the reaction is:

$H_2SO_4(aq) + 2NaOH(aq) \rightarrow Na_2SO_4(aq)$
$+ 2H_2O(l)$

What is the purity of the concentrated sulfuric acid?

2 Ethanedioic acid has the formula $H_2C_2O_4.xH_2O$. 0.900 g of the acid were made up to 250 cm^3. 25.0 cm^3 of this solution were neutralised by 26.75 cm^3 of 0.0532 mol dm^{-3} sodium hydroxide. The equation for the reaction is:

$H_2C_2O_4(aq) + 2NaOH(aq) \rightarrow$
$Na_2C_2O_4(aq) + 2H_2O(l)$

What is the value of x, and hence what is the formula of ethanedioic acid?

Finding the stoichiometry of a reaction

Titration can also be used to work out the stoichiometry of an acid–base reaction. If both the acid and the base are monoprotic, then one mole of the acid must react with one mole of the base. But if one of them is diprotic, there are two possible neutralisation reactions. For example, sulfuric acid and sodium hydroxide could react in two ways:

$$H_2SO_4(aq) + NaOH(aq) \rightarrow NaHSO_4(aq) + H_2O(l)$$
$$H_2SO_4(aq) + 2NaOH(aq) \rightarrow Na_2SO_4(aq) + 2H_2O(l)$$

By making standard solutions of both the acid and the base, the stoichiometry can be found by titration.

Worked example

25.0 cm^3 of 0.105 mol dm^{-3} sodium carbonate neutralised 24.50 cm^3 of 0.213 mol dm^{-3} hydrochloric acid, using methyl orange as an indicator. Construct an equation for the reaction of sodium carbonate with hydrochloric acid under these conditions.

Answer

$$n(Na_2CO_3) = c \times \frac{v}{1000} = 0.105 \times \frac{25.0}{1000} = 2.625 \times 10^{-3}\,mol$$

$$n(HCl) = c \times \frac{v}{1000} = 0.213 \times \frac{24.50}{1000} = 5.2185 \times 10^{-3}\,mol$$

So 1 mol of Na_2CO_3 reacts with $\frac{5.2185 \times 10^{-3}}{2.625 \times 10^{-3}} = 1.988$ mol of HCl.

Therefore 1 mol of Na_2CO_3 reacts with 2 mol of HCl (to the nearest whole number).
The equation is:

$$Na_2CO_3(aq) + 2HCl(aq) \rightarrow 2NaCl(aq) + CO_2(g) + H_2O(l)$$

Now try this

25.0 cm^3 of 0.0765 mol dm^{-3} potassium carbonate, K_2CO_3, neutralised 22.35 dm^3 of 0.0850 mol dm^{-3} hydrochloric acid, using phenolphthalein as an indicator. Construct an equation for the reaction between potassium carbonate and hydrochloric acid under these conditions.

6.7 Back titrations

Sometimes there is no definite end-point to a reaction. One example is when calcium carbonate is being used as the base. Because it is insoluble in water, it reacts very slowly with acid and this means that there is no sharp end-point when an indicator is used.

To overcome this difficulty, a known number of moles of acid, which is calculated to be an excess of that required to react with all the $CaCO_3$, is added to the calcium carbonate. After the reaction has finished, the amount of acid left over can be found by titration against a soluble base such as sodium hydroxide. The acid used up by the calcium carbonate is the total acid added at the start minus the acid left at the end. This method is known as **back titration**.

Back titration

(acid used) = (acid at start)
  − (acid left at end)

Worked example 1

0.765 g of an impure sample of calcium carbonate was dissolved in 25.0 cm^3 of 1.00 mol dm^{-3} hydrochloric acid. The resulting solution was made up to 250 cm^3 in a calibrated flask. 25.0 cm^3 of this solution were neutralised by 24.35 cm^3 of 0.050 mol dm^{-3} sodium hydroxide. What is the purity of the calcium carbonate?

Answer

- First find the amount of acid at the start:

$$n(HCl) \text{ added} = c \times \frac{v}{1000} = 1.00 \times \frac{25.0}{1000} = 2.50 \times 10^{-2} \, mol$$

- Then find the amount of acid left at the end:

$$\text{amount of NaOH used} = c \times \frac{v}{1000} = 0.050 \times \frac{24.35}{1000} = 1.2175 \times 10^{-3} \, mol$$

Since 1 mol of NaOH reacts with 1 mol of HCl,

$$n(HCl) \text{ left over in 250 cm}^3 \text{ flask} = \frac{250}{25} \times 1.2175 \times 10^{-3}$$
$$= 1.2175 \times 10^{-2} \, mol$$

- Now find the amount of acid used up in reaction with calcium carbonate:

$$n(HCl) \text{ used by } CaCO_3 = 2.50 \times 10^{-2} - 1.2175 \times 10^{-2}$$
$$= 1.2825 \times 10^{-2} \, mol$$

- The equation for the reaction is:

$$CaCO_3(s) + 2HCl(aq) \rightarrow CaCl_2(aq) + H_2O(l) + CO_2(g)$$

Since 1 mol of HCl reacts with $\frac{1}{2}$ mol of $CaCO_3$,

$$n(CaCO_3) = \frac{1}{2} \times 1.2825 \times 10^{-2} = 6.413 \times 10^{-3} \, mol$$

- Since $M(CaCO_3) = 40.1 + 12.0 + 48.0 = 100.1 \, g \, mol^{-1}$,

$$\text{mass of pure } CaCO_3 \text{ in sample} = 6.413 \times 10^{-3} \times 100.1 = 0.6419 \, g$$

and $$\text{purity of } CaCO_3 = \frac{0.6419}{0.765} \times 100 = \textbf{83.9\%}$$

Worked example 2

3.920 g of an oxide of formula MO were completely dissolved in 30.0 cm^3 of 2.00 mol dm^{-3} sulfuric acid. The resulting solution was made up to 100 cm^3. 25.0 cm^3 of this solution were neutralised by 27.50 cm^3 of 0.100 mol dm^{-3} sodium hydroxide. What is the relative atomic mass of M? Identify the metal.

Answer

- First find the amount of acid at the start:

$$n(H_2SO_4) \text{ added} = c \times \frac{v}{1000} = 2.00 \times \frac{30.0}{1000} = 6.00 \times 10^{-2} \, mol$$

- Then find the amount of acid left at the end:

$$n(NaOH) = c \times \frac{v}{1000} = 0.100 \times \frac{27.50}{1000} = 2.75 \times 10^{-3} \, mol$$

As the H_2SO_4 is in excess, the equation for the reaction is:

$$H_2SO_4(aq) + 2NaOH(aq) \rightarrow Na_2SO_4(aq) + 2H_2O(l)$$

Since 1 mol of NaOH neutralises $\frac{1}{2}$ mol of H_2SO_4,

$$n(H_2SO_4) \text{ left over in } 100\,cm^3 \text{ flask} = \frac{1}{2} \times 2.75 \times 10^{-3} \times \frac{100}{25}$$
$$= 5.50 \times 10^{-3}\,mol$$

- Now find the amount of acid used up in reaction with MO:

$$n(H_2SO_4) \text{ used by oxide} = 6.00 \times 10^{-2} - 5.50 \times 10^{-3} = 5.45 \times 10^{-2}\,mol$$

- The oxide MO reacts with H_2SO_4 as follows:

$$MO(s) + H_2SO_4(aq) \rightarrow MSO_4(aq) + H_2O(l)$$

We know that 5.45×10^{-2} mol of H_2SO_4 reacted with the oxide. Since one mole of acid reacts with one mole of oxide, the amount of oxide used is 5.45×10^{-2} mol, which has a mass of 3.920 g.

$$M = \frac{mass}{moles} = \frac{3.920}{5.45 \times 10^{-2}} = 71.9\,g\,mol^{-1}$$
$$A_r(M) + A_r(O) = 72 \text{ (to nearest whole number)}$$
$$A_r(M) = 72 - 16 = \mathbf{56}$$

The metal is iron.

Now try this

1 A 0.789 g sample of impure magnesium oxide, MgO, was dissolved in $25.0\,cm^3$ of $1.00\,mol\,dm^{-3}$ sulfuric acid. The resulting solution was made up to $100\,cm^3$ with water. $25.0\,cm^3$ of this final solution were neutralised by $27.80\,cm^3$ of $0.100\,mol\,dm^{-3}$ sodium hydroxide. Calculate the percentage purity of the magnesium oxide.

2 A 0.421 g sample of hydrated lithium hydroxide, $LiOH.xH_2O$, was dissolved in water and made up to $250\,cm^3$. $25.0\,cm^3$ of this solution were neutralised by $20.05\,cm^3$ of $0.0500\,mol\,dm^{-3}$ hydrochloric acid. What is the value of x?

3 (Harder) A 10.00 g sample of a fertiliser which contained ammonium sulfate, $(NH_4)_2SO_4$, was boiled with $25.0\,cm^3$ of $2.00\,mol\,dm^{-3}$ sodium hydroxide. The ammonium sulfate released ammonia as shown by the following equation:

$$(NH_4)_2SO_4(s) + 2NaOH(aq) \rightarrow Na_2SO_4(aq) + 2NH_3(g) + 2H_2O(l)$$

The resulting solution was made up to $250\,cm^3$. $25.0\,cm^3$ of this final solution were neutralised by $24.55\,cm^3$ of $0.0500\,mol\,dm^{-3}$ hydrochloric acid. What is the percentage by mass of ammonium sulfate in the fertiliser?

Summary

- According to the **Brønsted–Lowry theory** of acids and bases, an **acid** is a proton donor and a **base** is a proton acceptor.
- A strong acid has a weak **conjugate base**, and a weak acid has a strong conjugate base.
- Some acids are **diprotic**, and can donate up to two protons (for example, sulfuric acid). Some bases are diprotic, and can accept up to two protons (for example, the carbonate ion).
- The **end-point** at which an acid is exactly neutralised by a base can be determined using an **indicator** such as methyl orange or phenolphthalein.
- A **standard solution** is a solution made up such that its concentration in $mol\,dm^{-3}$ is known.
- In a **titration**, a standard solution takes part in a neutralisation reaction. The volumes of both solutions are measured accurately, so that, for example, the unknown concentration of the other solution may be calculated.
- In a **back titration**, the titration is used to calculate the amount of acid or base left in a solution after reaction with, for example, a solid reactant.

Key reactions and skills

- You need to know, or to be able to work out, equations for the reactions of hydrochloric, nitric and sulfuric acids with:
  - soluble hydroxides, such as sodium hydroxide, NaOH, or ammonium hydroxide, NH_4OH
  - insoluble hydroxides, such as magnesium hydroxide, $Mg(OH)_2$, or iron(III) hydroxide, $Fe(OH)_3$
  - insoluble oxides, such as copper(II) oxide, CuO, or iron(III) oxide, Fe_2O_3
  - soluble carbonates, such as sodium carbonate, Na_2CO_3
  - insoluble carbonates, such as calcium carbonate, $CaCO_3$.
- You need to be able to use the following mathematical formulae in titration calculations:

  - $n = \dfrac{m}{M}$ where n = amount (in mol), m = mass (in g) and M = molar mass (in $g\,mol^{-1}$)

  - $n = cV$ where c = concentration (in $mol\,dm^{-3}$) and V = volume of solution (in dm^3)

  - $n = c \times \dfrac{v}{1000}$ and $c = 1000 \times \dfrac{n}{v}$

    where c = concentration (in $mol\,dm^{-3}$) and v = volume of solution (in cm^3).

Examination practice questions

Please see the data section of the CD for any A_r values you may need.

1 A student carries out experiments using acids, bases and salts.

a Calcium nitrate, $Ca(NO_3)_2$, is an example of a salt. The student prepares a solution of calcium nitrate by reacting dilute nitric acid, HNO_3, with the base calcium hydroxide, $Ca(OH)_2$.

   i Why is calcium nitrate an example of a salt? [1]

   ii Write the equation for the reaction between dilute nitric acid and calcium hydroxide. Include state symbols. [2]

   iii Explain how the hydroxide ion in aqueous calcium hydroxide acts as a base when it neutralises dilute nitric acid. [1]

b A student carries out a titration to find the concentration of some sulfuric acid.

The student finds that $25.00\,cm^3$ of $0.0880\,mol\,dm^{-3}$ aqueous sodium hydroxide, $NaOH$, is neutralised by $17.60\,cm^3$ of dilute sulfuric acid, H_2SO_4.

$$H_2SO_4(aq) + 2NaOH(aq) \rightarrow Na_2SO_4(aq) + 2H_2O(l)$$

   i Calculate the amount, in moles, of $NaOH$ used. [1]

   ii Determine the amount, in moles, of H_2SO_4 used. [1]

   iii Calculate the concentration, in $mol\,dm^{-3}$, of the sulfuric acid. [1]

c After carrying out the titration in b, the student left the resulting solution to crystallise. White crystals were formed, with a formula of $Na_2SO_4.xH_2O$ and a molar mass of $322.1\,g\,mol^{-1}$.

   i What term is given to the '.xH_2O' part of the formula? [1]

   ii Using the molar mass of the crystals, calculate the value of x. [2]

[OCR Chemistry A Unit F321 Q2 (part) January 2010]

2 a i Using the symbol **HZ** to represent a Brønsted-Lowry acid, write equations which show the following substances acting as Brønsted-Lowry bases.

$NH_3 + \rightarrow$

$CH_3OH + \rightarrow$

   ii Using the symbol **B⁻** to represent a Brønsted-Lowry base, write equations which show the following substances acting as Brønsted-Lowry acids.

$NH_3 + \rightarrow$

$CH_3OH + \rightarrow$ [4]

[Cambridge International AS & A level Chemistry, 9701, Paper 41 Q3 a November 2013]

3 a Explain what is meant by the *Bronsted-Lowry* theory of acids and bases. [2]

[Cambridge International AS & A level Chemistry, 9701, Paper 4 Q1 a June 2009]

7 Oxidation and reduction

In Topic 6, acid–base reactions were shown to involve proton transfer. In this topic we show that oxidation–reduction (redox) reactions involve electron transfer. The concept of oxidation number is introduced and used to find the oxidation state of an element. This enables us to write half-equations and, from these, full balanced equations. Volumetric analysis using either potassium manganate(VII) or iodine can be used to find the number of electrons transferred in a reaction.

Learning outcomes

By the end of this topic you should be able to:

4.3b) discuss the finite nature of materials as a resource and the importance of recycling processes

6.1a) calculate oxidation numbers of elements in compounds and ions

6.1b) describe and explain redox processes in terms of electron transfer and changes in oxidation number

6.1c) use changes in oxidation numbers to help balance chemical equations

7.1 What are oxidation and reduction?

Losing and gaining electrons

Oxidation and reduction processes have been in use for about 7000 years in the extraction of metals from their ores, as well as to determine the colours of glazes for earthenware pots. One of the first 'textbooks' about the extraction of metals, *De Re Metallica*, was written by the German scientist Georg Agricola and published in 1556 after his death. As long ago as 1741, the word 'reduce' was used to describe the process of extracting metals from their ores. 'Reduction' referred to the fact that the mass decreased when the ore was converted into its metal. In the eighteenth century, the French nobleman Antoine Lavoisier recognised that these ores were oxides, and by the beginning of the nineteenth century the idea that reactive metals were easily oxidised had become established.

These observations formed the basis of the original definitions of oxidation as the addition of oxygen and reduction as the removal of oxygen. With the use of electrolysis for the extraction of reactive metals such as sodium and aluminium from their ores, this definition became too limited. The reduction of an oxide by heating it with hydrogen or carbon and its reduction in an electrolytic cell are similar chemical processes – both involve the addition of electrons to a positively charged metal ion. The modern definition of **oxidation**, therefore, is the removal of electrons from a substance, while **reduction** is the addition of electrons to a substance. Reduction takes place at the negative electrode (**cathode**) of an electrolytic cell, and oxidation takes place at the positive electrode (**anode**).

- **Oxidation** is the removal of electrons.
- **Reduction** is the addition of electrons.

During electrolysis:

- oxidation takes place at the **anode** – the anode accepts the electrons
- reduction takes place at the **cathode** – the cathode provides the electrons.

Half-equations and redox reactions

The processes of oxidation and reduction happen at the same time in oxidation–reduction reactions (also called **redox reactions**) – one substance is oxidised while the other is reduced. Redox reactions can therefore be broken down into two half-reactions, represented by half-equations that show the electron loss or gain. For example, for the combustion of magnesium to form magnesium oxide:

half-equation 1 (oxidation): $Mg \rightarrow Mg^{2+} + 2e^-$ or $Mg - 2e^- \rightarrow Mg^{2+}$
half-equation 2 (reduction): $O_2 + 4e^- \rightarrow 2O^{2-}$

These half-equations are similar to those used to describe acid–base reactions in Topic 6.

The oxidation half-equation above is written in two ways. The first method shows more clearly that charge is being conserved, while the second method shows more clearly that electron loss is taking place. We will generally use the first method.

To balance the two half-equations together, the number of electrons lost from the magnesium must be the same as the number gained by the oxygen. This can be achieved in two ways:

- add 2 × (half-equation 1) to (half-equation 2) so that 2 mol of magnesium react with 1 mol of oxygen:

$$2Mg(s) + O_2(g) \rightarrow 2MgO(s)$$

- add (half-equation 1) to $\frac{1}{2}$ × (half-equation 2) so that 1 mol of magnesium reacts with $\frac{1}{2}$ mol of oxygen:

$$Mg(s) + \tfrac{1}{2}O_2(g) \rightarrow MgO(s)$$

The complete equation for the combustion of magnesium could easily have been written without using half-equations. With more complicated redox examples, this would be much harder. For example, consider the reduction of aqueous iron(III) sulfate by metallic zinc. The first stage is to identify the oxidised and reduced species. The $Fe^{3+}(aq)$ ions are reduced to $Fe(s)$ while the $Zn(s)$ is oxidised to $Zn^{2+}(aq)$ ions. We can now write the relevant half-equations:

oxidation: $Zn \rightarrow Zn^{2+} + 2e^-$
reduction: $Fe^{3+} + 3e^- \rightarrow Fe$

We must multiply the first equation by 3 and the second equation by 2 in order to balance the electron transfer:

$3Zn \rightarrow 3Zn^{2+} + 6e^-$
$2Fe^{3+} + 6e^- \rightarrow 2Fe$

Adding these together gives:

$$3Zn(s) + 2Fe^{3+}(aq) \rightarrow 3Zn^{2+}(aq) + 2Fe(s)$$

This reaction proceeds because $Zn(s)$ is a more powerful reducing agent than $Fe(s)$. Alternatively, we can say that $Fe^{3+}(aq)$ is a more powerful oxidising agent than $Zn^{2+}(aq)$. This can be found out qualitatively by carrying out the reaction in a test tube: if powdered zinc is added to a yellow-brown solution of $Fe^{3+}(aq)$ ions, the grey zinc dissolves, to be replaced by black specks of iron metal, the brown solution fades to colourless as the $Fe^{3+}(aq)$ ions are replaced by $Zn^{2+}(aq)$ ions, and the tube becomes warm as the exothermic reaction takes place.

To compare strengths of oxidising and reducing agents quantitatively, we need to combine the two half-reactions into an electrochemical cell and measure the voltage produced, as we shall see in Topic 23.

In a redox reaction, the number of electrons lost = the number of electrons gained.

Worked example 1

a Which species has been oxidised and which has been reduced in the following reaction?

$$Zn(s) + CuSO_4(aq) \rightarrow ZnSO_4(aq) + Cu(s)$$

b Write the two relevant half-equations for the reaction in part **a**.

Answer

a $Zn(s)$ is oxidised, and $Cu^{2+}(aq)$ is reduced. (Note that it is not $CuSO_4(aq)$ that is reduced.)
b Oxidation: $Zn \rightarrow Zn^{2+} + 2e^-$; Reduction: $Cu^{2+} + 2e^- \rightarrow Cu$

Worked example 2

In each of the following reactions, identify whether the underlined species has been oxidised, reduced or neither.

a $\underline{Al(s)} + \frac{3}{2}Br_2(l) \rightarrow AlBr_3(s)$
b $2Mg(s) + \underline{TiO_2}(s) \rightarrow 2MgO(s) + Ti(s)$
c $\underline{Ag^+(aq)} + Cl^-(aq) \rightarrow AgCl(s)$
d $\underline{Cl_2(aq)} + SnCl_2(aq) \rightarrow SnCl_4(aq)$

Answer

a Oxidised   b Reduced   c Neither   d Reduced

Now try this

1 Write half-equations and then full equations for the reactions that take place between:
   a aqueous sulfuric acid and magnesium metal
   b liquid bromine and lithium metal
   c oxygen gas and aluminium metal
   d zinc metal and aqueous silver nitrate.
2 In each of the following reactions, identify which species has been oxidised and which has been reduced.
   a $Fe(s) + S(s) \rightarrow FeS(s)$
   b $MnO_2(s) + 4HCl(aq) \rightarrow MnCl_2(aq) + Cl_2(g) + 2H_2O(l)$
   c $2FeCl_2(aq) + Cl_2(g) \rightarrow 2FeCl_3(aq)$
   d $CuO(s) + Cu(s) \rightarrow Cu_2O(s)$
3 In each of the following reactions, identify whether the underlined species has been oxidised, reduced or neither.
   a $\underline{Cu}O(s) + H_2SO_4(aq) \rightarrow CuSO_4(aq) + H_2O(l)$
   b $\underline{Pb}O_2(s) + 4HCl(aq) \rightarrow PbCl_2(aq) + Cl_2(g) + 2H_2O(l)$
   c $4\underline{Fe}(OH)_2(s) + O_2(g) + 2H_2O(l) \rightarrow 4Fe(OH)_3(s)$
   d $\underline{Zn}(s) + 2V^{3+}(aq) \rightarrow Zn^{2+}(aq) + 2V^{2+}(aq)$

7.2 Oxidation numbers

So far we have restricted oxidation and reduction to elements and their ionic compounds. For redox reactions of this type, it is quite clear which species are losing or gaining electrons. However, many compounds contain covalent bonds and for these a simple ionic treatment is inappropriate.

Oxidation number and simple ionic compounds

To get round this difficulty, the concept of **oxidation number** has been developed. This idea applies to both ionic and covalent compounds. For simple ionic compounds, the oxidation number of the element is the same as the charge on the species containing the element, together with its sign. This means that all elements have the oxidation number 0, cations have positive oxidation numbers and anions have negative oxidation numbers. The oxidation number is written in Arabic numbers in brackets, immediately after the species (without a space). So $Cl_2(g)$ contains $Cl(0)$, $FeCl_2$ contains $Fe(+2)$ and MgO contains $O(-2)$.

The convention is helpful when a metal can exist in more than one **oxidation state**. Iron, for example, can have oxidation numbers of +2 and +3, which are distinguished in their formulae by using roman numerals; for example, $FeSO_4$ is iron(II) sulfate and $Fe_2(SO_4)_3$ is iron(III) sulfate.

Oxidation number and covalent molecules

The only covalent substances that contain bonds with no ionic character at all are molecules that contain only a single element, such as hydrogen, H_2, or sulfur, S_8. The atoms of such substances are given an oxidation number of 0. All other covalent bonds have some ionic character, whose magnitude depends on the electronegativity difference between the two bonded atoms (see section 3.10). Oxidation numbers are assigned as though the bond is completely ionic, rather than partially ionic. For example, in carbon dioxide, CO_2, oxygen is more electronegative than carbon. Each oxygen is given the oxidation number −2, just as in an ionic metal oxide. In order to maintain electrical neutrality, the carbon must be assigned the oxidation number +4. In this case it is helpful to include the '+' sign, as in other compounds carbon can have other oxidation numbers, including −4. Of course, we do not suggest that carbon dioxide is actually composed of one C^{4+} and two O^{2-} ions, but the convention indicates that the oxidation of carbon to carbon dioxide is equivalent to a four-electron transfer.

With a few exceptions that will be discussed later, oxygen in its compounds always has an oxidation number of −2. Similarly, in most of its compounds, hydrogen has an oxidation number of +1. This often helps in working out the oxidation numbers of other elements in covalent compounds.

- The oxidation number of atoms in a pure element is 0.
- In a compound, the more electronegative element is given a negative oxidation number (fluorine always has oxidation number −1).
- The sum of all the oxidation numbers in a molecule is 0.
- In most compounds, the oxidation number of oxygen is −2 and that of hydrogen is +1.

Worked example 1

What is the oxidation number of carbon in each of the following compounds?
a CH_4 b CCl_4 c CH_2Cl_2

Answer
a Carbon is more electronegative than hydrogen. Each hydrogen has oxidation number +1, so that of carbon is **−4**.
b Chlorine is more electronegative than carbon. Each chlorine has oxidation number −1. Let the oxidation number of carbon be x. Then $x + 4(-1) = 0$, so the oxidation number of carbon is **+4**.
c If the oxidation number of carbon is x, we have $x + 2(+1) + 2(-1) = 0$, so the oxidation number of carbon is **0**.

Now try this

What is the oxidation number of the underlined element in each of the following compounds?

1 $\underline{Cl}_2O_7$
2 $\underline{P}F_3$
3 $CHCl_3$
4 $\underline{N}_2H_4$
5 $\underline{N}H_2OH$

Worked example 2

What is the oxidation number of the underlined species in each of the following compounds?
a $\underline{S}O_3$ b $\underline{I}Cl$ c $\underline{P}_2O_3$

Answer
In each case let the unknown oxidation number be x.
a $x + 3(-2) = 0$, giving $x = +6$, so S = **+6**.
b $x + (-1) = 0$, giving $x = +1$, so I = **+1**.
c $2x + 3(-2) = 0$, giving $x = +3$, so P = **+3**. (Notice that each atom has its own oxidation number – so *each* P has an oxidation number of +3.)

Oxidation number and 'complex' ions containing covalent bonds

Many ions are not simple, but are made up of several covalently bonded atoms. A familiar example is the sulfate ion, SO_4^{2-}. The oxidation number of each oxygen is -2, but it is not immediately clear what the oxidation number of the sulfur atom is. We can, however, work out the oxidation number of sulfur in potassium sulfate, K_2SO_4, because the sum of all the oxidation numbers is 0. If the oxidation number of sulfur is x, then $2(+1) + x + 4(-2) = 0$, giving $x = +6$. This must also be the oxidation number of sulfur in the SO_4^{2-} ion. It can be found directly by letting the sum of the oxidation numbers equal the charge on the ion: $x + 4(-2) = -2$, giving $x = +6$ as before.

> In an ion, the sum of the oxidation numbers equals the charge on the ion.

Worked example

What is the oxidation number of:
a Cr in $Cr_2O_7^{2-}$ **b** V in VO_2^+?

Answer
Let the unknown oxidation number be x.
a $2x + 7(-2) = -2$, giving $x = \mathbf{+6}$
b $x + 2(-2) = 1$, giving $x = \mathbf{+5}$

Now try this

What is the oxidation number of:

a Al in AlO_2^- **b** P in HPO_3^{2-}
c V in VO^{2+} **d** C in $C_2O_4^{2-}$
e Pb in $PbCl_4^{2-}$ **f** Sn in $Sn(OH)_6^{2-}$?

The oxidation numbers of oxygen and hydrogen

Oxygen is the second most electronegative element, and in most of its compounds it has an oxidation number -2. There are two exceptions.

- As fluorine is more electronegative than oxygen, compounds of oxygen and fluorine are known as oxygen fluorides rather than fluorine oxides. (The names of compounds usually start with the least electronegative element.) For example, in the fluoride OF_2, each fluorine has an oxidation number of -1 and oxygen has an oxidation number of $+2$.
- Peroxides contain the O—O bond. In both the peroxide ion, O_2^{2-}, and in covalent peroxides such as hydrogen peroxide, H_2O_2, each oxygen atom has an oxidation number of -1.

In most of its compounds, hydrogen is joined to a more electronegative element such as carbon or oxygen, so in these compounds the oxidation number of hydrogen is $+1$. But there are also compounds in which hydrogen is combined with a metal that is *less* electronegative than hydrogen (for example, lithium hydride, LiH). In these compounds, the oxidation number of hydrogen is -1.

The usefulness of oxidation numbers

There are two main uses for oxidation numbers.

- In quantitative work, particularly volumetric analysis (as we shall see later in the topic), it is necessary to know the stoichiometry of the reaction being studied, and this is most easily found using oxidation numbers.
- When converting one compound into another, it is necessary to know if the reaction involves oxidation and reduction. This is most easily found by using oxidation numbers.

Oxidation numbers can readily be calculated if the compound is not too complicated. For example, the oxidation number of each carbon in ethane, C_2H_6, is -3. For propane, C_3H_8, the average oxidation number of carbon is $-\frac{8}{3}$, that is, $-2\frac{2}{3}$. This is not very meaningful. If we calculate the oxidation numbers of the individual carbons, it is -3 for the two end carbons and -2 for the middle one, even though they are all considered to be in very similar chemical environments to those in ethane.

We can therefore see that calculating oxidation numbers in organic compounds does not tell us very much. However, *changes* in oxidation numbers in organic compounds can be useful.

In the conversion of ethane into bromoethane:

$$C_2H_6 + Br_2 \rightarrow C_2H_5Br + HBr$$

it is not very helpful to state that the (average) oxidation number of the carbon atoms has increased from -3 to -2. It does show that a two-electron transfer reaction has taken place, but this is more easily established by noting that the two Br(0) atoms have changed their oxidation number to -1. If we calculate the oxidation number of each carbon atom separately, we see that the $-CH_3$ carbon remains unchanged at -3, but the $-CH_2Br$ carbon has increased from -3 to -1 ($x + 2 - 1 = 0$, so $x = -1$).

In more complicated redox reactions, which may not be quantitative, [O] is often used to indicate the addition of oxygen and [H] the addition of hydrogen. In this way, the oxidation of ethanol to ethanoic acid can be represented by:

$$C_2H_5OH + 2[O] \rightarrow CH_3CO_2H + H_2O$$

which shows that a four-electron transfer reaction has taken place ($2[O] + 4e^- \rightarrow 2O(-2)$).

Similarly, the reduction of propene to propane can be represented by:

$$C_3H_6 + 2[H] \rightarrow C_3H_8$$

which shows that a two-electron transfer reaction has taken place ($2[H] \rightarrow 2H(+1) + 2e^-$).

As we saw above with propane, it is best to avoid fractional oxidation numbers. If a simple calculation gives a fractional oxidation number for one type of atom in a formula, this usually means that some atoms have one integer oxidation number and other atoms have another. In Fe_3O_4, for example, the average oxidation number of iron is $+2\frac{2}{3}$, but a better description is to assign one iron atom as Fe(+2) and two as Fe(+3).

7.3 Balancing redox equations

An important use of oxidation numbers is to help establish the stoichiometry of an unknown redox reaction. The overall reaction can be found by combining the half-equations for the oxidation and reduction reactions.

Writing the half-equations

Some half-equations are easy to work out, for example the oxidation of iron(II) ions to iron(III) ions:

$$Fe^{2+} - e^- \rightarrow Fe^{3+} \quad \text{or} \quad Fe^{2+} \rightarrow Fe^{3+} + e^-$$

Others are more difficult, for example the oxidation of nitrite ions to nitrate ions in acid solution. The following stages lead to the relevant half-equation.

- Work out oxidation numbers – nitrite, NO_2^-, is N(+3) and nitrate, NO_3^-, is N(+5).
- Balance for redox – we know that this is a two-electron transfer reaction from the oxidation numbers given (N(+3) $\rightarrow$ N(+5) + 2e$^-$):

$$NO_2^- \rightarrow NO_3^- + 2e^-$$

- Balance for charge – because the reaction is being carried out in acid solution, we add H$^+$ ions to one side of the equation. We need 2H$^+$ on the right-hand side of the equation so that the overall charge is the same on both sides, in this case -1:

$$NO_2^- \rightarrow NO_3^- + 2e^- + 2H^+$$

(If the reaction had been carried out in alkaline solution, we would balance by adding OH$^-$ ions to one side of the equation.)

- Balance for hydrogen – this is done by adding water to one side of the equation. We need to add H_2O on the left-hand side of the equation to balance the 2H$^+$ on the right-hand side:

$$NO_2^- + H_2O \rightarrow NO_3^- + 2e^- + 2H^+$$

- Check for oxygen – the half-equation is now balanced, but it is a good idea to check that it is correct by confirming that the oxygen atoms balance.

To produce a balanced half-equation
- Work out oxidation numbers.
- Balance for redox – add the correct number of e^- to account for the difference between the two oxidation states.
- Balance for charge – add the correct number of H^+ (or OH^-) to one side of the equation so that the total charge is the same on both sides.
- Balance for H – add the correct number of H_2O to one side of the equation so that both sides have the same number of H atoms.
- Check for O – there should now be an equal number of O atoms on both sides of the equation.

Worked example

Produce a balanced half-equation for the reduction of manganate(VII) ions to manganese (II) ions in acid solution.

Answer

Work out oxidation numbers:	MnO_4^- is Mn(+7) and Mn^{2+} is Mn(+2).
Balance for redox:	$MnO_4^- + 5e^- \rightarrow Mn^{2+}$
Balance for charge:	$MnO_4^- + 5e^- + 8H^+ \rightarrow Mn^{2+}$
	(In acid solution, so xH^+ ions are added to one side of the equation to balance for charge: $-1 + (-5) + x(+1) = +2$, giving $x = 8$.)
Balance for H:	$MnO_4^- + 5e^- + 8H^+ \rightarrow Mn^{2+} + 4H_2O$
	(This is done by adding yH_2O to one side of the equation. With $8H^+$ on the left-hand side, we need $4H_2O$ on the right-hand side.)
Check for O:	There are 4O on both sides of the equation.

Now try this

Produce half-equations for the following changes:

1 the oxidation of $Cr(OH)_3$ to CrO_4^{2-} in alkaline solution

2 the reduction of VO_2^+ to V^{2+} in acid solution

3 the oxidation of $H_2C_2O_4$ to CO_2 in acid solution

4 the reduction of $Cr_2O_7^{2-}$ to Cr^{3+} in acid solution

5 the reduction of IO_3^- to I_2 in acid solution

6 the reduction of oxygen in alkaline solution

7 the reduction of H_2O_2 to water in acid solution

8 the oxidation of H_2O_2 to oxygen in alkaline solution

Producing a full equation from two half-equations

A full equation can be written by adding the two half-equations for the oxidised and reduced species together. It is essential that the electrons balance between the two half-equations, and to this end one (or both) of them may have to be multiplied throughout by an appropriate factor.

After adding the half-equations together, the moles of any substance that appears on both sides of the equation (for example, water) can be cancelled.

To produce a full equation
- Write down the two half-equations.
- Multiply each equation so that the number of electrons being given by the reducing agent equals the number of electrons required by the oxidising agent.
- Add the half-equations together.
- Simplify if possible.
- Check that the equation balances for charge.
- Add state symbols.

Produce a balanced equation for the oxidation of iron(II) ions with acidified manganate(VII) ions.

Answer

The two relevant half-equations are:

oxidation: $Fe^{2+} \rightarrow Fe^{3+} + e^-$
reduction: $MnO_4^- + 5e^- + 8H^+ \rightarrow Mn^{2+} + 4H_2O$

We must multiply the first half-equation by 5 so that the electrons will cancel:

$5Fe^{2+} \rightarrow 5Fe^{3+} + 5e^-$

On addition, we then have:

$MnO_4^- + 5Fe^{2+} + 8H^+ \rightarrow Mn^{2+} + 5Fe^{3+} + 4H_2O$

Checking for charge, there are 17 positive charges on both sides of the equation. Finally, we add the state symbols:

$MnO_4^-(aq) + 5Fe^{2+}(aq) + 8H^+(aq) \rightarrow Mn^{2+}(aq) + 5Fe^{3+}(aq) + 4H_2O(l)$

1 Produce balanced equations for the following reactions:
   a the reduction of iodate(V) ions, IO_3^-, to iodine, I_2, by iodide ions in acid solution
   b the oxidation of NO_2^- to NO_3^- with manganate(VII) ions, MnO_4^-, in acid solution
   c the reduction of VO^{2+} to V^{2+} with metallic zinc in acid solution
   d the oxidation of $Fe(OH)_2$ to $Fe(OH)_3$ by oxygen.
2 (Harder) Write balanced equations for the following reactions:
   a the reduction of $Cr_2O_7^{2-}$ to Cr^{3+} with metallic magnesium in acid solution
   b the reduction of nitrate(V) to ammonia with metallic aluminium in alkaline solution
   c the oxidation of $Cr(OH)_3$ to CrO_4^{2-} by H_2O_2 in alkaline solution.

Disproportionation

If an element has three or more oxidation states, then it can act as its own oxidant (oxidising agent) and reductant (reducing agent). Usually, the higher the oxidation state of the element, the more powerful an oxidant it is. If a compound that contains the element in a high oxidation state is mixed with a compound that contains the element in a low oxidation state, the result is that the intermediate oxidation state is formed. For example, if vanadium(+5) is mixed with vanadium(+3), the result is a compound containing vanadium(+4):

$$VO_2^+(aq) + V^{3+}(aq) \rightarrow 2VO^{2+}(aq)$$

If iodine(+5) is mixed with iodine(−1), the product is iodine(0) (see Figure 7.1).

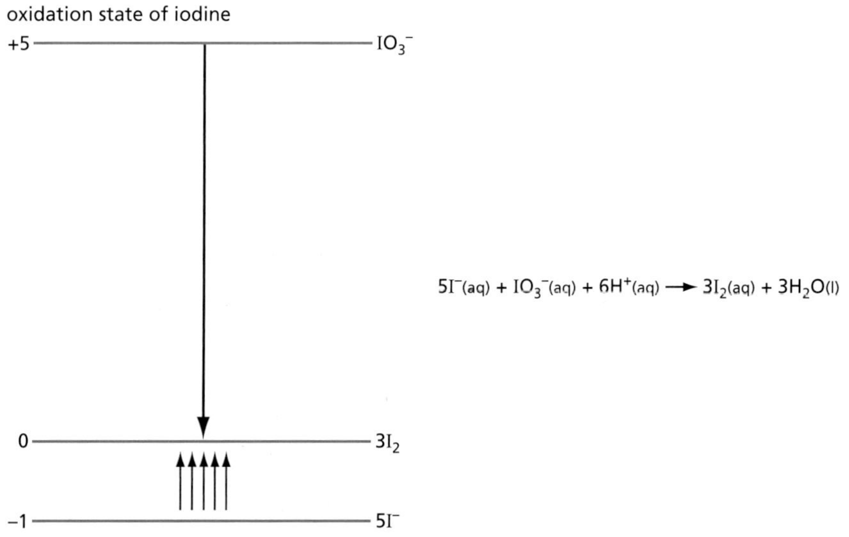

Figure 7.1 Oxidation of iodide, I^-, by iodate(V), IO_3^-. The diagram shows that one iodate ion oxidises five iodide ions.

oxidation state of iodine

+5 ——————————————————— IO_3^-

$5I^-(aq) + IO_3^-(aq) + 6H^+(aq) \longrightarrow 3I_2(aq) + 3H_2O(l)$

0 ——————————————————— $3I_2$

−1 ——————————————————— $5I^-$

Measurements of the voltages of electrochemical cells (see Topic 23) can show why the high oxidation state oxidises the low oxidation state.

Occasionally, the intermediate oxidation state is a more powerful oxidant than the higher oxidation state. This is because the intermediate state has a structure that makes it unstable. The intermediate state then **disproportionates** and breaks down to a mixture of the substances in the higher and lower oxidation states. A familiar example is hydrogen peroxide, H_2O_2, containing $O(-1)$. This spontaneously breaks down to oxygen, $O(0)$, and water, which contains $O(-2)$ (see Figure 7.2).

Figure 7.2 The disproportionation of hydrogen peroxide, H_2O_2. The diagram shows that each molecule of H_2O_2 is converted into one H_2O and $\frac{1}{2}O_2$.

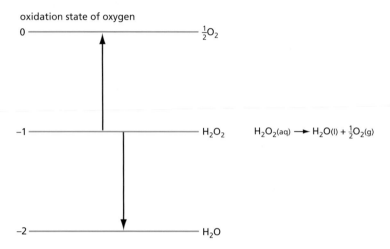

Some more examples of disproportionation are described in section 11.2.

7.4 Redox titrations

Redox titrations involve calculations similar to those described for acid–base titrations (see section 6.6).

The basic equations used in titrations – a reminder

$$n = \frac{m}{M}$$

$$n = c \times \frac{v}{1000} \quad \text{or} \quad c = \frac{1000}{v} \times n$$

n = amount (in mol), m = mass (in g), M = molar mass (in $g\,mol^{-1}$),
c = concentration (in $mol\,dm^{-3}$), v = volume (in cm^3)

Redox titrations are used for two main reasons:

- to find the concentration of a solution
- to determine the stoichiometry of a redox reaction and hence to suggest a likely equation for the reaction.

If the concentration of a solution is unknown and is to be found by a redox titration, a standard solution (one whose concentration is known) that reacts with the unknown solution is required. The balanced equation gives the number of moles that react together. After having carried out the titration, the stages in the calculation are as shown in Table 6.3, page 130. The balanced chemical equation allows us to calculate the number of moles of B from the number of moles of A. This then leads us to the concentration of solution B.

If, instead, the stoichiometry of the reaction is to be found, then the concentrations of both solutions A and B must be known. The volumes from the titration enable us to calculate the numbers of moles that react together.

Although a wide range of oxidising agents can be used in redox titrations, we shall concentrate on two – potassium manganate(VII) and iodine.

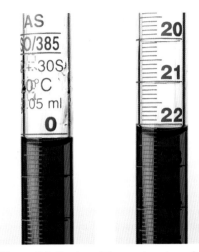

Figure 7.3 The top of the meniscus is read when using potassium manganate(VII) in a titration. Compare the readings with those in Figure 6.4 (page 129).

Figure 7.4 To avoid the solution turning brown, warm the solution or use plenty of acid.

Titrations using potassium manganate(VII) as the oxidant

Usually potassium manganate(VII) titrations are carried out in acid solution. Under these conditions, the following half-equation is relevant:

$$MnO_4^- + 5e^- + 8H^+ \rightarrow Mn^{2+} + 4H_2O$$

You will meet this equation often in redox reactions, and it is a good idea to learn it.

The deep purple manganate(VII) ion is used as its own indicator. This solution is run into the flask from the burette, and in the course of the reaction the colour changes from purple to colourless as nearly colourless manganese(II) ions are produced. At the end-point, the first extra drop of manganate(VII) ions makes the solution turn pink. The following experimental details should be noted.

- Potassium manganate(VII) is not very soluble, and the highest concentration used is $0.02\,mol\,dm^{-3}$.
- The reaction is carried out in the presence of sulfuric acid at approximately $1\,mol\,dm^{-3}$.
- It is usual to read the top, rather than the bottom, of the meniscus of potassium manganate(VII) in the burette because the deep purple colour obscures the reading at the bottom (see Figure 7.3). Because the volume delivered is the difference between the readings, it does not matter whether we read the top or bottom of the meniscus, as long as we do the same for both readings.
- If the titration is carried out too quickly, the solution may turn brown (see Figure 7.4). This is due to the formation of manganese(IV) oxide, MnO_2. This can be avoided either by increasing the acidity of the solution or by warming it.

Worked example 1

In a titration, $25.0\,cm^3$ of a solution of iron(II) sulfate required $22.4\,cm^3$ of $0.0200\,mol\,dm^{-3}$ potassium manganate(VII) solution at the end-point. What is the concentration of the solution of iron(II) sulfate?

Answer
$$n(KMnO_4) = c \times \frac{v}{1000} = 0.0200 \times \frac{22.4}{1000} = 4.48 \times 10^{-4}\,mol$$

oxidation: $Fe^{2+} \rightarrow Fe^{3+} + e^-$
reduction: $MnO_4^- + 5e^- + 8H^+ \rightarrow Mn^{2+} + 4H_2O$

To balance the electrons, 5 mol of Fe^{2+} react with 1 mol of $KMnO_4$. Therefore:

$n(Fe^{2+})$ used is $= 5 \times 4.48 \times 10^{-4} = 2.24 \times 10^{-3}\,mol$

$$c(Fe^{2+}) = \frac{1000}{v} \times n$$

$$= \frac{1000}{25.0} \times 2.24 \times 10^{-3}$$

$$= \mathbf{0.0896\,mol\,dm^{-3}}$$

Worked example 2

25.0 cm^3 of a 0.0200 mol dm^{-3} solution of ethanedioic acid, $H_2C_2O_4$, reacted with 20.0 cm^3 of 0.0100 mol dm^{-3} potassium manganate(VII) solution.

How many moles of ethanedioic acid react with 1 mol of potassium manganate(VII)? Suggest a likely equation for the reaction.

Answer

$$n(H_2C_2O_4) = c \times \frac{V}{1000} \qquad\qquad n(KMnO_4) = c \times \frac{V}{1000}$$

$$= 0.0200 \times \frac{25.0}{1000} = 5.00 \times 10^{-4} \,mol \qquad = 0.0100 \times \frac{20.0}{1000} = 2.00 \times 10^{-4} \,mol$$

Therefore 5 mol of $H_2C_2O_4$ react with 2 mol of $KMnO_4$ (or $2\frac{1}{2}$ mol react with 1 mol). We know from the half-equation for manganate(VII) that 2 mol of $KMnO_4$ receive $2 \times 5 = 10$ mol of electrons, so 5 mol of $H_2C_2O_4$ donate 10 mol of electrons, and 1 mol of $H_2C_2O_4$ donates $\frac{10}{5} = 2$ mol of electrons.

The oxidation number of C in $H_2C_2O_4$ is +3. During the oxidation, 2C lose 2 electrons, so each changes oxidation number from +3 to +4. This suggests that CO_2 may be produced. We shall write the half-equation assuming this:

oxidation: $H_2C_2O_4 \rightarrow 2CO_2 + 2e^- + 2H^+$; reduction: $MnO_4^- + 5e^- + 8H^+ \rightarrow Mn^{2+} + 4H_2O$

We multiply the first half-equation by 5 and the second by 2, and add them together:

$$2MnO_4^- + 5H_2C_2O_4 + 16H^+ \rightarrow 2Mn^{2+} + 8H_2O + 10CO_2 + 10H^+$$

Simplifying and adding state symbols:

$$2MnO_4^-(aq) + 5H_2C_2O_4(aq) + 6H^+(aq) \rightarrow 2Mn^{2+}(aq) + 8H_2O(l) + 10CO_2(g)$$

(Check the charges balance: four positive charges on both sides.)

Now try this

1 A steel nail with a mass of 2.47 g was dissolved in aqueous sulfuric acid and the solution made up to 250 cm^3 in a standard flask. 25.0 cm^3 of this solution reacted with 18.7 cm^3 of 0.0105 mol dm^{-3} potassium manganate(VII) solution. Calculate:
   a the concentration of the iron solution
   b the mass of iron present in the 250 cm^3 flask
   c the percentage by mass of iron in the steel nail.

2 25.0 cm^3 of acidified 0.0370 mol dm^{-3} sodium nitrate(III), $NaNO_2$, solution reacted with 23.9 cm^3 of 0.0155 mol dm^{-3} potassium manganate(VII) solution.
   a Calculate the number of moles of sodium nitrate(III) that react with 1 mol of manganate(VII).
   b Suggest a likely equation for the reaction.

3 A 1.31 g sample of hydrated potassium ethanedioate, $K_2C_2O_4.xH_2O$, was dissolved in acid and made up to 250 cm^3 in a standard flask. 25.0 cm^3 of this solution reacted with 28.5 cm^3 of 0.0100 mol dm^{-3} potassium manganate(VII) solution. Calculate:
   a the amount of potassium manganate(VII) in 25.0 cm^3 of solution
   b the amount of potassium ethanedioate in 250 cm^3 of solution
   c $M_r(K_2C_2O_4.xH_2O)$
   d the value of x.

Titrations using iodine as the oxidant

Iodine is a weak oxidising agent, but it is often used to oxidise the thiosulfate ion, $S_2O_3^{2-}$, to the tetrathionate ion, $S_4O_6^{2-}$:

$$I_2(aq) + 2S_2O_3^{2-}(aq) \rightarrow 2I^-(aq) + S_4O_6^{2-}(aq)$$

This is another reaction that you will meet often, so it is helpful to know the equation well.

The colourless thiosulfate solution is placed in the burette. The iodine solution is initially brown in colour, and as the thiosulfate is added it fades to pale yellow (see Figure 7.5). Finally at the end-point the solution becomes colourless. The end-point is rather indistinct and it is usual to add a few drops of starch solution. This forms an intense blue colour with iodine. The end-point is then a sharp change from blue to colourless, when all the iodine has been used up. The starch solution is not added until the iodine solution is pale yellow – if it is added at the start of the titration, clumps of blue solid may be formed which are difficult to break up.

This titration is not often used just to find the concentration of an iodine solution. There is a large number of oxidising agents that oxidise iodide ions to iodine, and the iodine titration is used mainly to analyse such reactions. If an excess of aqueous potassium iodide is added to the oxidising agent, the amount of iodine liberated is directly related to the amount of oxidising agent used. Table 7.1 lists some of the oxidising agents that may be estimated in this way.

Figure 7.5 The end-point of an iodine titration is made much sharper using starch. A few drops of starch solution are added when the solution is pale yellow; the end-point occurs when the intense blue colour just disappears.

Table 7.1 Some oxidising agents that can be estimated using iodine/thiosulfate. Each mole of iodine reacts with two moles of thiosulfate.

Oxidising agent	Equation	$\dfrac{n(\text{thiosulfate})}{n(\text{oxidising agent})}$
Cl_2	$Cl_2 + 2I^- \rightarrow 2Cl^- + I_2$	2
Br_2	$Br_2 + 2I^- \rightarrow 2Br^- + I_2$	2
IO_3^-	$IO_3^- + 6H^+ + 5I^- \rightarrow 3I_2 + 3H_2O$	6
MnO_4^-	$MnO_4^- + 5I^- + 8H^+ \rightarrow Mn^{2+} + \frac{5}{2}I_2 + 4H_2O$	5
Cu^{2+}	$Cu^{2+} + 2I^- \rightarrow CuI + I_2$	1

Thiosulfate oxidation numbers

The thiosulfate ion, $S_2O_3^{2-}$, is an example of the inappropriate use of oxidation numbers.

The equation:

$$I_2(aq) + 2S_2O_3^{3-}(aq) \rightarrow 2I^-(aq) + S_4O_6^{2-}(aq)$$

shows that the half-equations are as follows:

oxidation: $2S_2O_3^{2-} \rightarrow S_4O_6^{2-} + 2e^-$
reduction: $I_2 + 2e^- \rightarrow 2I^-$

So each thiosulfate ion receives one electron. Some books suggest that the oxidation number of each sulfur atom has changed from +2 to $+2\frac{1}{2}$. Others compare the $S_2O_3^{2-}$ ion with the SO_4^{2-} ion, and assign S(+6) to the central sulfur atom and S(−2) to the other one. On oxidation, the S(−2) atom then becomes S(−1). This may show what is taking place during the reaction more accurately.

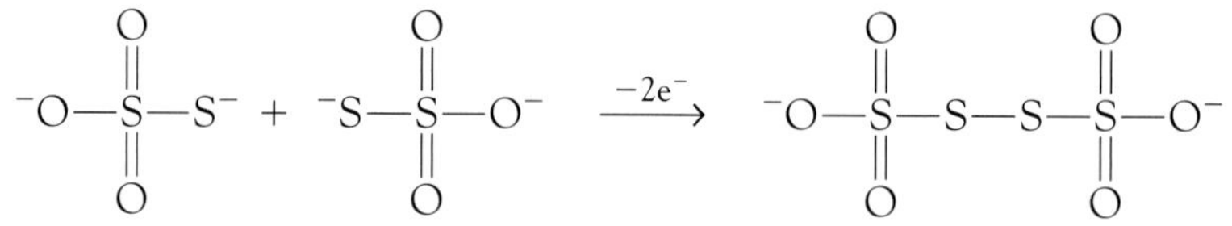

As can be seen, oxidation numbers cease to be useful when applied to the thiosulfate ion. All that needs to be known is that each $S_2O_3^{2-}$ ion loses one electron when it reacts with iodine – it is not necessary to know the oxidation number of each sulfur atom.

Worked example

Although the active ingredient in a commercial bleach is chlorate(I) ions, ClO^-, the concentration is often quoted in terms of free chlorine. This is because the concentration is estimated by liberating iodine from potassium iodide. 10.0 cm³ of the bleach were made up to 250.0 cm³ in a standard flask. 25.0 cm³ of this solution were added to an excess of aqueous potassium iodide.

The iodine liberated reacted with 22.0 cm³ of 0.105 mol dm⁻³ thiosulfate solution. Calculate:

a the concentration of 'chlorine' in the diluted bleach solution
b the concentration of 'chlorine' in g dm⁻³ in the commercial bleach. $A_r(Cl) = 35.5$.

Answer

a First we calculate the amount of thiosulfate used:

$$n(S_2O_3^{2-}) = c \times \frac{v}{1000} = 0.105 \times \frac{22.0}{1000} = 2.31 \times 10^{-3} \, mol$$

The ratio in Table 7.1 (page 147) tells us that $n(Cl_2)$ in 25.0 cm³ of diluted bleach solution is:

$$\frac{1}{2} \times 2.31 \times 10^{-3} = 1.155 \times 10^{-3} \, mol$$

Therefore:

$$c(Cl_2) = \frac{1000}{v} \times n = \frac{1000}{25.0} \times 1.155 \times 10^{-3}$$

$$= \mathbf{0.0462 \, mol \, dm^{-3}}$$

b $M_r(Cl_2) = 2 \times 35.5 \, g \, mol^{-1}$. The original bleach was diluted 25 times. Therefore the concentration of chlorine in the bleach is:

$$25 \times (2 \times 35.5) \times 0.0462 = \mathbf{82.0 \, g \, dm^{-3}}$$

Now try this

1 A $25.0\,cm^3$ sample of $0.0210\,mol\,dm^{-3}$ potassium peroxodisulfate(VI), $K_2S_2O_8$, was treated with an excess of potassium iodide. The iodine liberated reacted with $21.0\,cm^3$ of $0.0500\,mol\,dm^{-3}$ thiosulfate. Calculate:
   a the amount of $S_2O_8^{2-}$, and the amount of I^- used by the peroxodisulfate
   b the amount of I^- that reacts with 1 mol of $S_2O_8^{2-}$.
   c Suggest a likely equation for the reaction between $K_2S_2O_8$ and KI.
2 A $1.00\,g$ sample of brass (an alloy of copper and zinc) was dissolved in nitric acid and made up to $250\,cm^3$ in a standard flask. To a $25.0\,cm^3$ sample of this solution was added an excess of aqueous potassium iodide.

$$Cu^{2+} + 2I^- \rightarrow CuI + \tfrac{1}{2}I_2$$

The iodine liberated reacted with $27.8\,cm^3$ of $0.0425\,mol\,dm^{-3}$ thiosulfate. Calculate:
   a the concentration of Cu^{2+} in the solution
   b the total mass of copper dissolved
   c the percentage of copper in the sample of brass.

7.5 Some uses of electrolysis

The electrolysis of brine

If concentrated aqueous sodium chloride is electrolysed, hydrogen is given off at the cathode and chlorine at the anode.

- At the cathode:

$$H_2O(l) \rightarrow H^+(aq) + OH^-(aq)$$
$$2H^+(aq) + 2e^- \rightarrow H_2(g)$$

Because hydrogen ions are being used up, hydroxide ions are left behind. Adding the equations:

$$2H_2O(l) + 2e^- \rightarrow H_2(g) + 2OH^-(aq)$$

- At the anode:

$$2Cl^-(aq) \rightarrow Cl_2(g) + 2e^-$$

The products, therefore, are hydrogen, chlorine and sodium hydroxide:

$$2NaCl(aq) + 2H_2O(l) \rightarrow 2NaOH(aq) + Cl_2(g) + H_2(g)$$

For most purposes, the chlorine, hydrogen and sodium hydroxide solution need to be collected separately, which can be done using a **diaphragm cell** (see Figure 7.6). This contains an asbestos diaphragm that allows the ions to flow from the anode compartment to the cathode compartment. This flow is established by having the level of the electrolyte higher on the anode side of the diaphragm than on the cathode side.

The incoming solution is saturated brine containing about 25% sodium chloride. The outgoing solution contains about 10% sodium hydroxide and 15% sodium chloride. This solution is evaporated to about one-fifth of its volume, when nearly all the sodium chloride crystallises out, leaving a solution containing 50% of almost pure sodium hydroxide. This can be used as a solution or evaporated to dryness to give the solid product.

Modern plants use a membrane rather than a diaphragm in the cell (see Figure 7.7). The membrane is made of a polymer containing PTFE (see section 28.1), which allows cations but not anions to pass through it. The resulting solution is pure sodium hydroxide, uncontaminated by sodium chloride. The chlorine and hydrogen gases are collected separately.

The principal uses of the three chemicals are listed in Table 7.2.

Figure 7.6 Outline diagram of the diaphragm cell used for the electrolysis of brine

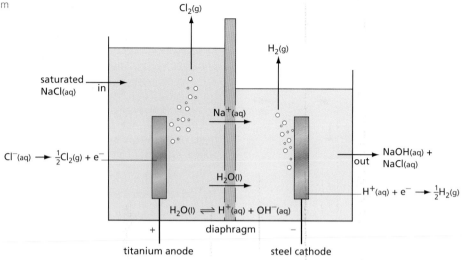

Figure 7.7 Modern plants use a membrane rather than a diaphragm in the cell. This yields very pure sodium hydroxide.

Table 7.2 Principal uses of the products of the electrolysis of brine.

Sodium hydroxide	Hydrogen	Chlorine
making soaps and detergents	in making ammonia	for sterilisation and bleaching
converted into sodium carbonate	for the hydrogenation of oils (see section 14.3)	to make insecticides
used in refining aluminium oxide (see below)	in welding using an oxy–hydrogen flame	to make solvents such as dichloromethane and tetrachloromethane
for making paper		to make CFCs (see section 15.1)
		to make PVC (see section 28.1)

The extraction of aluminium

Most molten aluminium compounds do not conduct electricity. Molten aluminium oxide, Al_2O_3, is a conductor, so its electrolysis yields aluminium. However, it has an extremely high melting point (2060 °C). The oxide is therefore dissolved in molten **cryolite**, sodium hexafluoroaluminate, Na_3AlF_6, which has a melting point lower than 1000 °C.

There are three main stages in the manufacture of aluminium:

- the extraction and purification of the ore, bauxite
- the preparation of cryolite
- electrolysis.

Figure 7.8 Paul Heroult (1863–1914) and Charles Hall (1863–1914), designers of the electrolytic process to produce aluminium

Bauxite is a relatively abundant ore. It is mainly aluminium oxide, but its principal impurity is hydrated iron(III) oxide; it may contain silica and titanium oxide as well. The bauxite is dissolved in 10% aqueous sodium hydroxide under 4 atm pressure at 150 °C. The impurities are largely insoluble and can be filtered off as a sludge. Aluminium hydroxide, $Al(OH)_3$, is then precipitated from the clear solution by cooling it for three days. This precipitation is accelerated by 'seeding' the solution with a crystal of aluminium hydroxide. The hydroxide precipitate is filtered off and heated to convert it into pure aluminium oxide.

$$Al_2O_3(\text{hydrated}) + 2OH^-(aq) + 3H_2O(l) \rightarrow 2Al(OH)_4^-(aq) \quad \text{other impurities are insoluble}$$

$$Al(OH)_4^-(aq) \rightleftharpoons Al(OH)_3(s) + OH^-(aq) \quad \text{precipitation on cooling}$$
$$2Al(OH)_3(s) \rightarrow Al_2O_3(s) + 3H_2O(g) \quad \text{dehydration on heating}$$

The sludge is composed of iron oxide and other impurities. It is washed free of sodium hydroxide before being buried. This treatment is important as untreated sludge is an unattractive brown colour and leaves the soil too alkaline for plants to grow. The sodium hydroxide produced during the precipitation is recycled and used to dissolve more aluminium oxide.

Cryolite is made by dissolving $NaAl(OH)_4$ in hydrofluoric acid and precipitating the product with sodium carbonate:

$$NaAl(OH)_4(aq) + 6HF(aq) + Na_2CO_3(s) \rightarrow Na_3AlF_6(s) + 5H_2O(l) + CO_2(g)$$

The electrolysis is carried out in a steel box whose floor is lined with carbon. This acts as the cathode connection (see Figure 7.9).

Figure 7.9 Outline diagram of the electrolytic cell used to produce aluminium

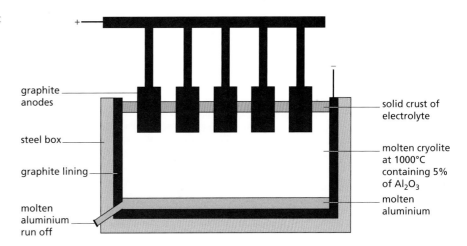

The anodes are blocks of graphite suspended in the molten cryolite. Because aluminium is more dense than cryolite, it sinks to the bottom of the cell as it is formed, creating a molten pool which acts as the cathode. Periodically the aluminium is removed and more aluminium oxide added to maintain a concentration of 5%.

The exact nature of the electrolyte is unknown – it has been suggested that the principal ions present are AlO^+ and AlO_2^-. The reactions taking place at the electrodes may be written as follows, although this is certainly a simplification:

at the cathode: $\quad 2Al^{3+} + 6e^- \rightarrow 2Al(l)$

at the anode: $\quad 3O^{2-} \rightarrow 1\frac{1}{2}O_2(g) + 6e^-$

The oxygen given off at the anode reacts with the carbon, producing carbon dioxide. The graphite blocks therefore have to be renewed regularly.

The process requires enormous amounts of electricity – up to 15 kWh are needed to produce 1 kg of metal. A typical cell operates at 4.5 V and 3×10^5 A, and two-thirds

Figure 7.10 Aluminium is the most abundant metal in the Earth's crust, occurring largely as the ore bauxite. Its high extraction costs mean that recycling is a much more economical way of producing 'new' metal. The photograph shows the effect of aluminium extraction on the landscape.

of the energy goes in heating the cell. Fortunately, aluminium is easy to recycle and this requires only 5% of the energy required to produce it from bauxite. Up to 60% of aluminium used in Europe is from recycled material.

Aluminium is the second most used metal after iron. It has a good strength-to-weight ratio and is resistant to corrosion, especially when anodised (that is, covered with a thin protective oxide film by anodic oxidation). Its principal uses are as follows:

- in lightweight alloys to build cars, ships, aeroplanes, etc.
- for building work, such as for window frames
- in overhead electric cables
- to make packaging, such as cans, foil, etc.

The purification of copper

A major use of copper is for electrical wiring. As the conductivity of copper increases ten-fold when it is more than 99.9% pure, impure copper is purified to this high degree by electrolysis. The impure copper is made the anode of an electrolysis cell, and a small strip of pure copper the cathode. A potential difference of about 0.2–0.4 V is applied. Ions of copper and any more reactive metals go into solution from the anode but the more reactive ions are not discharged and stay in solution as the copper ions gain electrons and are discharged onto the cathode.

Any metals that are less reactive than copper do not dissolve from the anode. These metals drop off the anode as the copper around them dissolves, and fall to the bottom as 'anode sludge'. This contains rare and useful elements, such as silver, gold and selenium. Subsequently this sludge is removed and the rare elements extracted.

Summary

- **Oxidation** is the removal of electrons; **reduction** is the addition of electrons.
- In a **redox reaction**, the number of electrons lost = the number of electrons gained.
- To work out the number of electrons lost or gained, **oxidation numbers** may be used.
- Oxidation numbers can also be used to work out the oxidation and reduction **half-equations** for a redox reaction. The two half-equations, suitably combined together, give the whole oxidation–reduction equation.
- **Titrations** can be carried out based on redox reactions. The most important redox titrations use potassium manganate(VII) or iodine.
- **Electrolysis** is used in many industrial processes, such as the electrolysis of brine, the extraction of aluminium and the purification of copper.

Examination practice questions

Please see the data section of the CD for any A_r values you may need.

1 Radium was discovered in the ore pitchblende by Marie and Pierre Curie in 1898, and the metal was first isolated by them in 1910.

The metal was obtained by first reacting the radium present in the pitchblende to form insoluble radium sulfate which was converted into aqueous radium bromide. This solution was then electrolysed using a mercury cathode and a carbon anode.

a During their electrolysis of aqueous radium bromide, the Curies obtained radium at the cathode and bromine at the anode. Write half-equations for the two electrode reactions that take place during this electrolysis. [2]

[Cambridge International AS & A Level Chemistry 9701, Paper 21 Q2 b November 2009]

2 Chlorine is manufactured by electrolysis from brine, concentrated aqueous sodium chloride.

a i Describe, with the aid of a fully labelled diagram, the industrial electrolysis of brine in a diaphragm cell. State what each electrode is made of and show clearly the inlet for the brine and the outlets for the products.

ii Write a half-equation, with state symbols, for the reaction at **each** electrode.

anode

cathode

iii Name the chemical that is produced in solution in this electrolytic process. [7]

[Cambridge International AS & A level Chemistry, 9701, Paper 22 Q4 June 2011]

8 Rates of reaction

Topic 5 looked at enthalpy changes for reactions. While the sign of ΔH may indicate the direction of chemical change, it gives no information on how fast a reaction occurs. This can only be found out by experiment. The rate of a reaction depends on a number of factors, each of which can be studied separately. It is found that, for a reaction to take place, not only must the molecules of the reactants collide, but they must collide with sufficient energy to get over an activation energy barrier. In a typical chemical reaction, only a tiny fraction of collisions are effective in bringing about chemical change.

Learning outcomes

By the end of this topic you should be able to:

5.2b) construct and interpret a reaction pathway diagram, in terms of the enthalpy change of the reaction and of the activation energy
8.1a) explain and use the term *rate of reaction*
8.1b) explain qualitatively, in terms of collisions, the effect of concentration changes on the rate of a reaction
8.2a) explain and use the term *activation energy*, including reference to the Boltzmann distribution
8.2b) explain qualitatively, in terms both of the Boltzmann distribution and of collision frequency, the effect of temperature change on the rate of a reaction
8.3a) explain and use the term *catalysis*
8.3b) explain that catalysts can be homogeneous or heterogeneous
8.3c) explain that, in the presence of a catalyst, a reaction has a different mechanism, i.e. one of lower activation energy, and interpret this catalytic effect in terms of the Boltzmann distribution
8.3d) describe enzymes as biological catalysts (proteins) which may have specificity.

8.1 Why do reactions take place at different rates?

How quickly a reaction goes, that is, the **rate** of a reaction, is defined and measured in terms of how quickly a reactant is used up, or how quickly a product forms. The rate of reaction varies greatly for different reactions, and under different conditions.

Enthalpy of reaction revisited

In Topic 5, it was seen that the sign of ΔH indicates the likely direction in which a reaction will take place. Most reactions are exothermic, that is, ΔH negative, and very few are endothermic, that is, ΔH positive. We might be tempted to think that if a reaction is highly exothermic, so that the energy of the products is much less than the energy of the reactants, then it will take place very rapidly. This, however, is not necessarily the case. For example, a mixture of methane and oxygen shows no sign of reaction at room temperature and pressure even though $\Delta H^{\ominus}$ for the reaction is $-890\,\mathrm{kJ\,mol^{-1}}$. On the other hand, the reaction between hydrochloric acid and sodium hydroxide takes place very rapidly, even though $\Delta H^{\ominus}$ for this reaction is only $-56\,\mathrm{kJ\,mol^{-1}}$. The study of the rates of chemical reactions is called **kinetics**, while the study of energy changes is called **thermodynamics** (as we saw in Topic 5).

The activation energy barrier

ΔH cannot indicate the rate of reaction because there is an energy barrier between the reactants and products that has to be overcome. This can be thought of as similar to a ridge along the edge of a desk, over which a book must be pushed before it will fall to the floor. Before the book will fall, we first have to push it harder for it to rise over the ridge. In chemical reactions, this ridge is called the **activation energy**, and it is given the symbol E_a. The size of the activation energy (the height of the ridge) does not depend on the size of ΔH (the height of the desk off the floor).

We can draw an **energy diagram** for a reaction, as shown in Figure 8.1. Such a diagram is sometimes called a **reaction profile**. The reactants are higher in energy than the products, but the top of the activation energy peak is higher still.

Figure 8.1 Reaction profiles for the combustion of methane and the neutralisation reaction between sodium hydroxide and hydrochloric acid. The label 'progress of reaction' indicates the change from reactants to products over time.

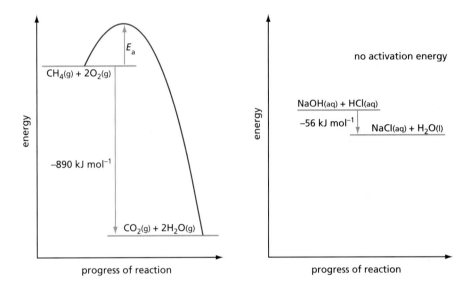

There are therefore two factors that determine whether a reaction takes place or not – the sign of ΔH and the value of E_a.

- If ΔH is positive, the reactants are lower in energy than the products. The reactants are said to be **thermodynamically stable** with respect to the products. Such reactions are unlikely to take place.
- Even if ΔH is negative, the reaction may still not be observed to take place because it has a high activation energy. Because of this, the rate of the reaction is negligible at room temperature. The reactants are said to be **kinetically inert** (or **kinetically stable**).

Table 8.1 summarises the effect of the four combinations of ΔH and E_a on the rate of a reaction.

Table 8.1 The rate of a reaction depends on both the enthalpy change and the activation energy.

ΔH	E_a	Result
+ve	low	usually no reaction, reactants are thermodynamically stable
+ve	high	usually no reaction, reactants are thermodynamically and kinetically stable
−ve	low	reaction likely
−ve	high	usually no reaction, reactants are kinetically stable

 ## 8.2 Measuring rates of reaction

Rates of reaction are studied for two reasons:

- to discover the best conditions to make the reaction go as quickly as possible
- to gather information about the mechanism of a reaction, that is, how the reaction actually takes place.

In this topic, the factors controlling rates are considered, while reaction mechanisms are looked at in more detail in Topic 21. Reaction mechanisms are particularly important when studying organic reactions.

This section deals with how the rates of reactions are measured.

What to measure?

Simple observation tells us that some reactions go faster than others, but to measure the precise rate it is necessary to find out how fast one of the products is being produced or how fast one of the reactants is being used up. This can be achieved using either a physical method or a chemical method of analysis. A physical method can be more convenient than a chemical method, because it does not disturb the reaction being studied. It depends on there being a change in some physical property such as volume, mass or colour during the course of the reaction.

Physical methods of analysis – measuring the volume of a gas given off

The reaction between hydrochloric acid and calcium carbonate may be studied using a physical method of analysis:

$$CaCO_3(s) + 2HCl(aq) \rightarrow CaCl_2(aq) + H_2O(l) + CO_2(g)$$

The reaction gives off a gas, whose volume can be measured. A suitable apparatus is shown in Figure 8.2. If we plot a graph of the volume of carbon dioxide collected against time (see Figure 8.3), the rate of the reaction is given by the slope (or gradient) of the graph. The steeper the gradient, the higher is the rate. The rate is greatest at the start, then decreases and finally becomes zero at the end of the reaction.

Figure 8.2 Apparatus used to study the rate of a reaction that gives off a gas

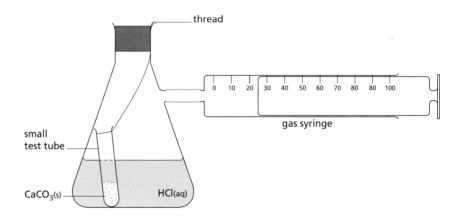

Figure 8.3 A typical graph of volume against time for a reaction that gives off a gas. The gradient gives the rate of reaction at that point. The rate is greatest at the start, then decreases and finally reaches zero at the end of the reaction.

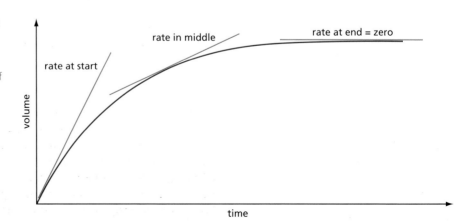

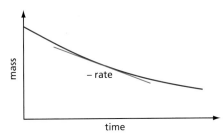

Figure 8.4 A typical graph of mass against time for a reaction that loses a gas to the atmosphere. The rate of the reaction is the negative gradient of the graph.

Physical methods of analysis – measuring the decrease in mass

An alternative method of monitoring the same reaction involves carrying out the reaction in an open flask placed on a top-pan balance. As the gas is given off, the mass of the flask decreases. If we plot a graph of mass against time, the rate of the reaction is the negative gradient of the graph (see Figure 8.4). Note that rates of reaction must always be positive, since a negative rate would mean that the reaction was going backwards. The gradient in Figure 8.4 is negative, so if we take its negative value, we get a positive rate.

The rate of this reaction can be measured under different conditions, for example by changing the temperature, the concentration of acid or the surface area of calcium carbonate, to investigate the conditions under which it will proceed the fastest.

Physical methods of analysis – measuring a change in colour

Sometimes a change in colour of a solution can be used to measure the rate of a reaction. An example of a reaction whose rate can be measured in this way is the oxidation of aqueous potassium iodide by aqueous potassium peroxodisulfate(VI), $K_2S_2O_8$, during which iodine is released:

$$S_2O_8^{2-}(aq) + 2I^-(aq) \rightarrow 2SO_4^{2-}(aq) + I_2(aq)$$

Iodine dissolves in an excess of potassium iodide to give I_3^- ions, which are brown:

$$I_2(aq) + I^-(aq) \rightarrow I_3^-(aq)$$
$$\text{brown}$$

All other chemical species in this reaction are colourless.

The intensity of the brown colour may be measured using an instrument called a **colorimeter** (see Figure 8.5). In this device, light from a light source passes through a filter that selects just one colour of light. This then shines onto a cell containing the reaction mixture, and a detector monitors the amount of light passing through the solution.

Figure 8.5 Outline diagram of a colorimeter

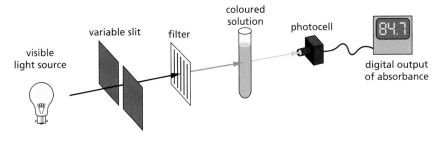

Iodine solution is brown because it absorbs in the blue part of the spectrum and transmits other wavelengths. We therefore select a blue filter for this experiment (see page 412). The colorimeter is first adjusted to zero using water in the cell. The colorimeter provides readings of **absorbance**, which is proportional to the concentration of the iodine. A graph of absorbance against time (see Figure 8.6) is obtained, which is similar in shape to the volume-against-time graph in Figure 8.3.

Chemical methods of analysis – sampling and 'clock' reactions

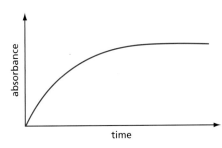

Figure 8.6 A typical graph of absorbance against time. This is equivalent to the volume-against-time graph in Figure 8.3.

Chemical methods of analysis interfere with the reaction being studied. Because of this, either a new reaction mixture needs to be set up for each measurement taken, or small amounts of the reaction mixture, called **samples**, can be extracted and analysed. Unless something is done to stop the reaction, it will continue to take place in the sample while it is being analysed. Therefore, it is necessary to 'freeze' the reaction in the sample as soon as it has been extracted. This can be carried out in one of two ways: either by suddenly cooling the reaction (for example, by running the sample onto ice) or by neutralising a catalyst (for example, hydrogen ions) that may be present. Both methods are time consuming and are avoided if possible.

An alternative is to use a 'clock' method. The rate of the reaction between potassium peroxodisulfate(VI) and potassium iodide described above may be followed in this way. A separate experiment is set up for each determination of rate. The two reactants are mixed in the presence of a known amount of sodium thiosulfate, $Na_2S_2O_3$, and a little starch. The sodium thiosulfate reacts with the iodine produced, converting it back to iodide, in the reaction we met in section 7.4:

$$I_2(aq) + 2S_2O_3^{2-}(aq) \rightarrow 2I^-(aq) + S_4O_6^{2-}(aq)$$

The amount of sodium thiosulfate present is much smaller than the amounts of all the other reagents. When all the sodium thiosulfate has been used up, the iodine (which is still being produced by the reaction) reacts with the starch to give an intense blue colour (see Figure 8.7). For a fixed amount of sodium thiosulfate, the faster the reaction, the shorter the time, t, this takes. The rate of the reaction is then approximately proportional to $1/t$ (see Figure 8.8). A **'clock' reaction** is very convenient but can only be used with a limited number of reactions.

Figure 8.7 Carrying out a 'clock' reaction

Figure 8.8 The graphs show two reactions A and B. They may vary in, for example, concentration or temperature. At y, all the sodium thiosulfate has been used up and the iodine reacts with the starch. The rate of reaction A is approximately equal to y/t_1, and the rate of reaction B is approximately equal to y/t_2. This means that the ratio of the times taken for the 'clock' to stop gives the ratio of the rates: $\dfrac{\text{rate of A}}{\text{rate of B}} = \dfrac{t_2}{t_1}$

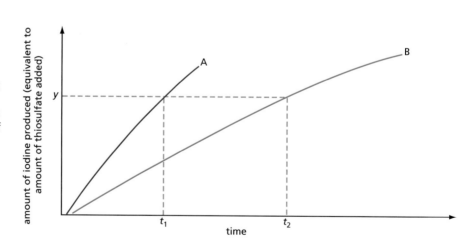

Hydrogen peroxide reacts with potassium manganate(VII) as follows:

$$2MnO_4^- + 5H_2O_2 + 6H^+ \rightarrow 2Mn^{2+} + 5O_2 + 8H_2O$$

Suggest *two* physical methods of measuring the rate of this reaction. For each method, sketch a graph showing how the physical quantity you measure would change over time.

Worked example

Suggest methods of measuring the rates of the following reactions:
a zinc powder with aqueous copper sulfate
b hydrogen peroxide decomposing to give water and oxygen
c $(CH_3)_3CBr + H_2O \rightarrow (CH_3)_3COH + H^+ + Br^-$

Answer
a Allow the zinc to settle, then measure the change in colour of the solution using a colorimeter.
b Measure the volume of gas given off, or measure the decrease in mass.
c Measure the conductivity of the solution.

8.3 Making a reaction go faster – increasing the collision rate

At the molecular level, the rate of reaction depends on two factors:

- how often the reactant molecules hit one another
- what proportion of the collisions have sufficient energy to overcome the activation energy barrier.

The effect of concentration

We can increase the rate at which the reactant particles collide in several ways. The simplest is to increase the **concentration** of the reactants. At a higher concentration, the reactant molecules are closer together and so will collide more frequently. The same is true if we increase the **pressure** for gaseous reactions – the volume gets smaller and so the molecules are closer together. However, changing the pressure has virtually no effect on reactions in the solid or liquid phase, because their volume changes very little when put under pressure, so their particles do not move closer together.

The concentrations of the reactants decrease during the course of a reaction and so the rate also goes down with time (see Figure 8.3). At the end of the reaction, at least one of the reactants has been used up completely and the rate is then zero.

If we repeat the reaction using more concentrated reactants, the initial rate of the reaction will be higher. However, it must not be assumed that if, for example, we double the concentration, the rate will be twice as high. The actual size of the increase in rate is impossible to predict and can only be found by experiment. This study of quantitative kinetics is explored in Topic 21.

Worked example

When 2 g (an excess) of powdered calcium carbonate were added to 50 cm³ of 0.10 mol dm⁻³ hydrochloric acid, the results shown in Table 8.2 were obtained.

Table 8.2

Time/s	10	20	30	40	50	60	70	80
Volume of CO_2 given off/cm³	25	45	60	70	75	78	80	80

a Plot a graph of these results.
b Use the graph to find the rate at the start of the reaction.
c Explain the shape of the graph.

Answer
a

Figure 8.9

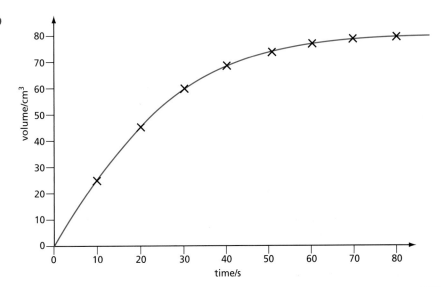

b The gradient of the graph at the beginning shows that 25 cm³ would be given off in 10 s. This is a rate of **2.5 cm³ s⁻¹**.

c The rate decreases (the reaction slows down) as the reaction proceeds, because the acid becomes more dilute. The reaction finally stops when all the acid has been used up.

Now try this

1 The experiment in the worked example above was repeated a using 50 cm³ of 0.20 mol dm⁻³ hydrochloric acid and b using 50 cm³ of 0.050 mol dm⁻³ hydrochloric acid. Draw sketches of the graphs you might expect to be produced and comment on the differences between each graph and that in Figure 8.9.

2 Solutions of potassium iodide and potassium peroxodisulfate(VI) were mixed together and a clock started. A sample of the mixture was placed in the cell of a colorimeter and the absorbance measured at suitable time intervals. The results shown in Table 8.3 were obtained.

Table 8.3

Time/s	30	60	90	120	150	180	210	240	270
Absorbance	0.072	0.13	0.19	0.25	0.29	0.32	0.34	0.35	0.35

a Plot a graph of absorbance against time.

b Find the rate of the reaction i after 60 s and ii after 120 s.

c Explain why these two reaction rates differ.

3 When hydrochloric acid is added to sodium thiosulfate, Na₂S₂O₃(aq), a fine precipitate of sulfur appears:

$$2HCl(aq) + Na_2S_2O_3(aq) \rightarrow 2NaCl(aq) + H_2O(l) + SO_2(aq) + S(s)$$

This gradually makes the solution opaque. The rate of the reaction may be determined by measuring how long it takes for the solution to become so opaque that a cross marked on a piece of paper placed under the beaker containing the reaction mixture just becomes invisible (see Figure 8.10).

The results shown in Table 8.4 were obtained.

Table 8.4

Experiment number	Concentration/mol dm⁻³		Time/s
	HCl	Na₂S₂O₃	
1	0.20	0.40	39
2	0.20	0.30	50
3	0.20	0.20	83
4	0.20	0.10	170

a Plot a graph of the reciprocal of time (1/time) against the concentration of sodium thiosulfate.

b Explain why the rate changes with the concentration of sodium thiosulfate.

c Estimate the concentration of sodium thiosulfate that would give a time of 60 s.

The effect of changing the amount of reactants

If we keep the concentrations of all the reactants the same but, for example, double their amounts, the actual rate of the reaction will be unaffected although the amount of product will be doubled. This means that if we measure the rate from the volume of gas evolved or the mass lost, we will obtain different results for the rate. The actual rate of reaction between calcium carbonate and hydrochloric acid is unaffected if we double the mass of calcium carbonate and double the volume of the acid even though twice the volume of gas is evolved. However, if we are using the apparatus in Figure 8.2 (page 156) the measured rate of reaction will appear to be twice as great because twice as much gas is evolved in the same time. For quantitative work, it is important to define what we mean by the rate of reaction more carefully, and this is considered further in Topic 21.

Figure 8.10 The reaction between hydrochloric acid and sodium thiosulfate produces a precipitate of sulfur. This allows us to study the rate of the reaction – when the cross is no longer visible, the clock is stopped.

The effect of temperature

We can increase the rate of collision by raising the temperature. This causes the particles to move faster (increases their kinetic energy) and so they collide more frequently. The increased rate of collision has only a very small effect on the rate of reaction; the main effect of increased temperature is from the increase in energy, which will be considered later in this topic.

Particle size

Many reactions are **homogeneous**, that is, the reactants are uniformly mixed, either as gases or in solution. Others are **heterogeneous** – there is a boundary between the reactants (for example, between a solid and a solution). In such reactions, the rate of collision will increase if we make the **surface area** of this boundary as large as possible. For example, powdered sugar dissolves faster than lump sugar, water drops evaporate faster than a large puddle, and powdered zinc dissolves in acid faster than lumps of zinc do. The size of the boundary between the reactants in a heterogeneous reaction is described by the **state of division** – for example, a powdered solid is described as 'finely divided'. The efficiency of solid catalysts is also increased if they are finely divided (see page 166).

Worked example

When 2 g (an excess) of lump calcium carbonate were added to 50 cm³ of 0.10 mol dm⁻³ hydrochloric acid, the results shown in Table 8.5 were obtained.

Table 8.5

Time/s	10	20	30	40	50	60	70	80
Volume of CO_2 given off/cm³	5	10	14	18	22	25	28	31

a How does the rate compare with that found in the worked example on page 159?
b Explain why the rate is different.
c What would be the final volume of carbon dioxide? Explain your answer.

Answer
a The rate is lower.
b The surface area of lump calcium carbonate is smaller than that of powdered calcium carbonate, so there are fewer collisions taking place every second.
c The final volume of carbon dioxide would be **80 cm³**. The final amount of carbon dioxide is determined by the amount of acid, because the calcium carbonate is present in excess.

8.4 Making a reaction go faster – overcoming the activation energy barrier

On page 161 it was stated that the energy of the collisions between reacting particles is more important than the rate of collision. Even if a reaction has a small activation energy, only a tiny fraction of the collisions that take place have enough energy to overcome it. These collisions are the **effective collisions**. A typical fraction might be one in ten thousand million (1 in 10^{10}). This is the reason why many reactions are slow at room temperature. There are two ways in which we can increase this fraction:

- give the collisions more energy
- find another route with a lower activation energy.

Giving the collisions more energy

The usual way of increasing the energy of the collisions is to raise the temperature. Although this increases the rate of collision, the main effect is that a higher proportion of the collisions now have enough energy to get over the activation energy barrier – there is a greater number of effective collisions. If we raise the temperature from 25 °C to 125 °C, for example, the increase in the rate of collision goes up by a factor of only 1.2, but the fraction of reactant molecules able to get over the energy barrier could increase by 1000 times. A useful working rule is to say that 'a 10 °C rise in temperature doubles the rate of reaction'. So, for a rise in temperature from 25 °C to 125 °C, the rate might increase by 2^{10} or 1024 times.

Finding another route with a lower activation energy

The second way of increasing the number of collisions that have sufficient energy to overcome the activation energy barrier is to lower the height of the barrier. This is most easily achieved by using a catalyst.

A **catalyst** is a substance that speeds up a chemical reaction without being chemically used up itself.

The catalyst provides an alternative pathway for the reaction, and this pathway has a lower activation energy than that of the uncatalysed reaction (see Figure 8.11).

Figure 8.11 The energy profile for a reaction, with and without a catalyst

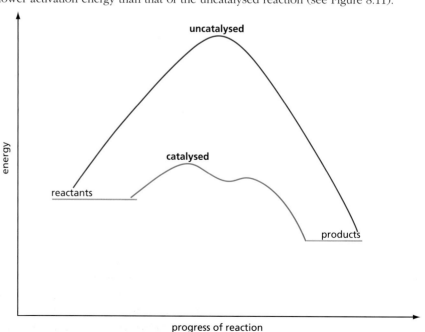

The height of the activation energy barrier may also be changed if a different solvent is used for the reaction. Many inorganic reactions are carried out using water as a solvent, but most organic substances are insoluble in water and so must be dissolved in a different solvent. This has a marked effect on the rate of reaction.

Worked example 1

Return to your graph plotted from Table 8.2 (page 159) for the reaction between hydrochloric acid and calcium carbonate, which was carried out at 25 °C. On the same axes, sketch the graph you might expect if the experiment were repeated at 35 °C rather than at 25 °C.

Answer
The graph should show an initial gradient twice as steep, since the rate will be about twice as fast. The final volume will be the same, because there is the same amount of hydrochloric acid.

Worked example 2

Sketch an energy profile for a slow reaction with a large negative ΔH.

Answer
Since the reaction is slow, there will be a high activation energy. Since ΔH is large and negative, the products are much lower in energy than the reactants.

Now try this

1 a Add to your graph from Worked example 1, above, a sketch of the graph you might expect if the experiment were conducted at 15 °C rather than at 25 °C.
   b Explain the differences and similarities in the graphs.
2 Sketch energy profiles for the following reactions:
   a a fast reaction with ΔH small and negative
   b a fairly fast reaction with ΔH small and positive.

8.5 The collision theory

Rates of collision

As has been discussed, in order for a reaction to take place, the reactant molecules must first collide and secondly have sufficient combined energy to get over the activation energy barrier.

For gaseous reactions, it is possible to calculate how often the molecules collide. This depends on three main factors:

- the size of the molecules – the bigger the molecules, the greater the chance of collision
- the mass of the molecules – the larger the mass, the slower the molecules move at a given temperature and the fewer the collisions
- the temperature – as the temperature is raised, molecules move faster and the rate of collisions increases.

Bigger molecules have larger masses, so the first two factors tend to cancel out. The collision rate therefore depends largely on temperature. So, at a given temperature, most molecules have similar collision rates. At room temperature and pressure, in each cubic decimetre of gas there are about 1×10^{32} collisions every second. If every one of these collisions produced a reaction, all gaseous reactions would be over in a fraction of a second.

For molecules in liquids, there is no simple way to estimate their rate of collision. We would expect the rate to be lower than in gases, because the molecules are moving more slowly and because the solvent molecules get in the way. On the other hand, there are situations when reactant molecules can become trapped in a 'cage' of solvent molecules and this increases their chance of collision. The two effects tend to

Figure 8.12 James Clerk Maxwell, 1831–79, (left) and Ludwig Boltzmann, 1844–1906 (right)

cancel each other out. Studies on reactions both as gases and in solution suggest that the rates are often not much different from one another.

Energy of collision – the energy distribution curve

The molecules in a sample of gas have a range of energies. In order to estimate what fraction of collisions will have sufficient energy to get over the activation energy barrier, we need to know how many molecules in a sample have various different energies – in other words, what their **energy distribution** is. This was first calculated by James Clerk Maxwell and Ludwig Boltzmann (Figure 8.12). The **energy distribution curve** is shown in Figure 8.13. We can see that there are very few molecules with very low energy values. Most molecules have a moderate amount of energy, as shown by the highest point on the graph – this shows the most probable energy. There are a few molecules with very high energy values.

Figure 8.13 The Maxwell–Boltzmann energy distribution curve

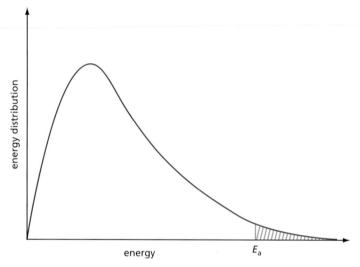

At a given temperature, the shape of the curve is the same for all gases, as it depends only on the kinetic energy of the molecules, which (as we have mentioned) depends only on temperature. The area under the graph gives the total number of molecules. The line E_a is marked, showing a high activation energy for a particular reaction. The shaded area shown is the fraction of molecules with energy greater than this E_a.

If we increase the temperature, the curve has a similar shape but the maximum moves to the right, showing that the most probable energy increases. The energies become more spread out, so there will be fewer molecules with the most probable energy and the maximum is slightly lower. Figure 8.14 shows the effect of raising the temperature by 100 °C. We can see that the shaded area showing the fraction of molecules with energy greater than E_a is much larger at the higher temperature.

Figure 8.14 The effect on the energy distribution curve of increasing the temperature by 100 °C. The curve moves to the right as the temperature rises. The curve is slightly flatter, so that the area under it remains constant.

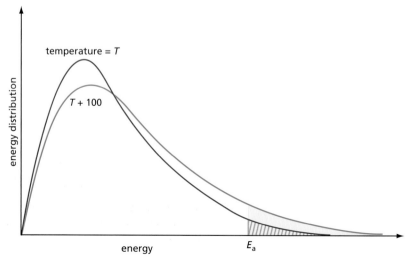

1 a Sketch a graph of the energy distribution of gas molecules at room temperature. Label the graph 'T'. Draw a line to show the activation energy, E_a, of a reaction.
   b On the same axes, sketch the energy distribution at a temperature of about 100 °C below room temperature. Label the graph '$T - 100$'. Shade the areas that show the fractions of molecules with energy greater than E_a at the two temperatures.

2 a Sketch another graph of the energy distribution of gas molecules at room temperature.
   b Draw a line to show the activation energy, E_a, of a reaction.
   c Draw another line to show E_a for a catalysed reaction, in which the activation energy is halved.
   d Explain why the catalysed reaction is so much faster than the uncatalysed reaction.

3 Explain why the collision rate is approximately the same for all gases at room temperature.

8.6 Catalysts

A **catalyst** speeds up a reaction by providing a pathway with a lower activation energy. The effect can be readily shown on the energy distribution graph. On Figure 8.13, if E_a is made smaller, the shaded area is much larger and so the rate is increased.

Catalysts may be **homogeneous**, that is, uniformly mixed with the reactants, or **heterogeneous**, in which case the catalyst and the reactants are in different physical states. There is also an important class of biochemical catalysts called **enzymes**. These three classes will be considered separately.

Homogeneous catalysis

Homogeneous catalysis occurs in both the gas phase and in solution.

● The production of the harmful gas ozone from oxygen near the ground is catalysed by nitrogen dioxide, which is produced in car engines:

$$O_2(g) + NO_2(g) \rightarrow NO(g) + O_3(g)$$

$$NO(g) + \tfrac{1}{2}O_2(g) \rightarrow NO_2(g)$$

● Nitrogen dioxide may also catalyse the oxidation of sulfur dioxide to sulfur trioxide in the atmosphere, and hence speed up the production of acid rain (see section 10.6):

$$NO_2(g) + SO_2(g) \rightarrow SO_3(g) + NO(g)$$

$$NO(g) + \tfrac{1}{2}O_2(g) \rightarrow NO_2(g)$$

● The presence of chlorine atoms from CFCs in the upper atmosphere is responsible for the catalytic destruction of ozone (see section 15.4).
● The rates of many redox reactions are increased by the use of catalysts, usually transition metal compounds. Some of these are considered in section 24.6.
● Many reactions are catalysed by the $H^+(aq)$ ion. Examples are the formation and hydrolysis of esters, for example methyl ethanoate:

$$CH_3CO_2H + CH_3OH \underset{}{\overset{H^+(aq)}{\rightleftharpoons}} CH_3CO_2CH_3 + H_2O$$

The overall mechanism is explained in detail in Topic 18.
● Enzymes are homogeneous catalysts. These are described below and in section 28.4.

Figure 8.15 A catalyst in the form of a gauze. This provides a large surface area, but prevents the catalyst being swept away in the stream of gases.

Heterogeneous catalysis

Two important examples of the use of solid heterogeneous catalysis are in the Contact process and the Haber process (see Topic 10), in which both reagents and products are gaseous. The reaction takes place on the surface of the catalyst, which should ideally be in the form of a fine powder. This is often impracticable because the catalyst would then be lost in the flowing stream of gas. The catalyst is therefore usually made into pellets, a gauze or a fluidised bed.

In the Contact process, the vanadium in the catalyst changes its oxidation state. The mechanism is considered in more detail in section 10.6. The Haber process for the manufacture of ammonia involves a hydrogenation reaction (see section 10.5). Other hydrogenation reactions include those of the alkenes (see Topic 14) and carbonyl compounds (see Topic 26). Many transition metals catalyse hydrogenation reactions, for example platinum and nickel (see section 14.3). This is because they are able to absorb large quantities of hydrogen and form **interstitial hydrides**. In these hydrides, hydrogen atoms are held in the spaces between the metal atoms in the lattice. These hydrogen atoms are able to add on to the multiple bond of a molecule that becomes **adsorbed** on to the metal surface nearby. There are three stages in catalysis involving surface adsorption.

- **Adsorption** – the reactants are first adsorbed onto the surface of the catalyst. The metal catalyst chosen must adsorb the reactants easily but not so strongly that the products do not come off again. Once adsorbed, the bonds in the reacting molecules are weakened.
- **Reaction** – the reactants are held on the surface in such a position that they can readily react together.
- **Desorption** – the products leave the surface, as the bonds between the product and the catalyst are very weak.

The rate of the reaction is controlled by how fast the reactants are adsorbed and how fast the products are desorbed. When the catalyst surface is covered with molecules, there can be no increase in reaction rate even if the pressure of the gaseous reactants is increased.

The efficiency of the catalyst depends on the nature of the catalyst's surface. This is shown by the following effects.

- **Poisoning** – many catalysts are rendered ineffective by trace impurities. For example, hydrogenation catalysts are poisoned by traces of sulfur impurities. This is one reason why nickel is preferred to platinum as a catalyst. Nickel is relatively inexpensive; if a large amount of nickel is used, even if some of it is deactivated by poisoning, enough will remain for it still to be effective. However, platinum is expensive; if a small quantity of platinum is used, all of it might be poisoned.
- **Promotion** – the spacing on the surface of a catalyst is important. For example, it is known that only some surfaces of the iron crystals act as effective catalysts in the Haber process (see section 10.5). The addition of traces of other substances may make the catalyst more efficient by producing **active sites** where the reaction takes place most readily.

The three-way catalytic converter

Ideally, the hydrocarbons in unleaded petrol are converted completely into carbon dioxide and water when burned in a vehicle engine. These are non-polluting products, though the additional carbon dioxide probably has an effect on the Earth's climate (see section 13.4). In practice, however, three other products are formed which are immediately harmful:

- unburnt hydrocarbons
- carbon and carbon monoxide, from incomplete combustion of the fuel
- oxides of nitrogen formed from the nitrogen in the air.

The amount of each impurity produced depends on the type and efficiency of the engine. When starting up and when idling, the proportions of impurities are higher than when driving on the open road. Diesel engines produce higher quantities of carbon particles than petrol engines do. To reduce the amounts of these impurities, engines are fitted with **catalytic converters** (see Figure 8.16). These catalyse the following reactions:

$$CO(g) + NO(g) \rightarrow CO_2(g) + \tfrac{1}{2}N_2(g)$$

$$CO(g) + \tfrac{1}{2}O_2(g) \rightarrow CO_2(g)$$

$$\text{hydrocarbons} + O_2(g) \rightarrow CO_2(g) + H_2O(g)$$

The exhaust gases are passed through the converter, which contains metals such as platinum and rhodium supported on a honeycomb support. Because lead poisons this catalyst, it is essential to use unleaded petrol in a car fitted with a catalytic converter.

Figure 8.16 A catalytic converter

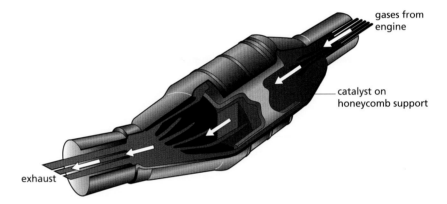

gases from engine

catalyst on honeycomb support

exhaust

Enzymes

Chemical reactions that take place in living cells are very fast, even though they take place at just above room temperature and at low concentrations. This is because they are catalysed by efficient catalysts called **enzymes**. Enzymes are proteins, which are polymers made up of units called amino acids. These units are joined in a chain, like beads in a necklace, to make a polymer. A typical enzyme may contain 500 amino acids and have a relative molecular mass of 50 000.

Enzymes are **globular proteins**. In this type of protein, the amino-acid chain coils into an approximately spherical shape (see Figure 8.17). Globular proteins are soluble in water, but if their globular structure is changed, they become insoluble. An example of this process, which is called **denaturing**, is the setting of egg white when the egg is boiled. Because globular proteins are water soluble, they are homogeneous catalysts, although their mechanism of action resembles heterogeneous catalysis in that the reaction takes place at a specific point on their surface, called the **active site**.

Figure 8.17 Proteins are like a necklace of beads – an enzyme may contain 500 amino acids. The enzyme folds into a globular shape, but it is easily denatured, for example by heat, when it becomes ineffective as a catalyst.

amino acid

unfolded enzyme

in its globular form

denatured

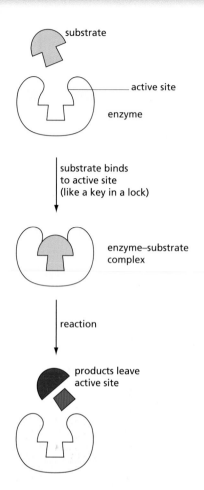

Figure 8.18 Only a molecule of the correct size and shape can attach to the active site and be converted into products.

Most enzymes catalyse only one reaction, or one group of similar reactions. A reactant molecule will only attach to the active site if it has the right size and shape. Once in the active site, the reactant molecule is held in the correct orientation to react and form the products of the reaction. The products leave the active site and the enzyme is then ready to act as a catalyst again. This mechanism of action of an enzyme is described as the **lock-and-key hypothesis**, illustrated in Figure 8.18. The reactant that fits the enzyme's active site is called the **substrate**.

Some examples of enzyme-catalysed reactions are as follows.

- Yeast contains a mixture of enzymes called **zymase**, which converts sucrose into ethanol and carbon dioxide:

$$C_{12}H_{22}O_{11}(aq) + H_2O(l) \rightarrow 4C_2H_5OH(aq) + 4CO_2(g)$$
$$\text{sucrose} \qquad\qquad\qquad \text{ethanol}$$

- Hydrogen peroxide is harmful to cells. It is produced in many metabolic reactions but is rapidly decomposed to water and oxygen in the presence of the catalyst **catalase**:

$$H_2O_2(aq) \rightarrow H_2O(l) + \tfrac{1}{2}O_2(g)$$

- When ethanol is metabolised, it is first converted into ethanal by removing two hydrogen atoms in the presence of an enzyme called a **dehydrogenase**:

$$C_2H_5OH(aq) \rightarrow CH_3CHO(aq) + 2[H]$$
$$\text{ethanol} \qquad\quad \text{ethanal}$$

Every living cell contains at least 1000 different enzymes, each catalysing one of the many different chemical reactions that take place inside the cell. Some enzymes are so efficient that one molecule can catalyse the reaction of 10 000 reactant molecules every second. This means that virtually every collision of the substrate with the active site leads to reaction, showing that enzyme-catalysed reactions have very low activation energies.

In many industrial processes, the enzymes are extracted from the organisms in which they were made (usually a bacteria) This allows specific reactions to be carried out, such as the following:

- conversion of glucose to the sweeter-tasting fructose, using glucose isomerase
- manufacture of new antibiotics, using penicillin G acylase
- manufacture of laundry detergents, using proteases.

Recently, new enzymes have been made using genetic engineering, and these enzymes are now used on the industrial scale. The following compounds are now made industrially with genetically engineered enzymes:

- 1,3-propanediol for making polymers
- sitagliptin, used in the treatment of type 2 diabetes
- paclitaxel (Taxol) for the treatment of breast cancer
- esters used in cosmetics.

Figure 8.19 Formulae of three compounds now made industrially by the use of genetically engineered enzymes

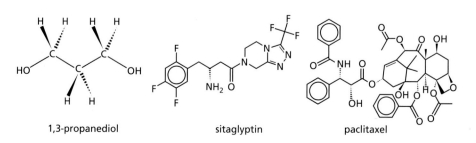

1,3-propanediol   sitaglyptin   paclitaxel

8.7 The effect of light on chemical reactions

A few reactions are affected by light. These are called **photochemical reactions**, and they fall into two main types:

- endothermic reactions in which light provides the energy input
- exothermic reactions that are started off by light.

Most photochemical reactions are of the first type. They use the energy of the light to bring about the reaction. Examples include the breakdown of silver chloride to silver and chlorine, and the bromination of alkanes. The most important photochemical reaction is photosynthesis (see page 210).

Figure 8.20 Photochromic sunglasses which darken on exposure to light

Black-and-white photography

The effect of light on silver chloride was observed by Robert Boyle in the seventeenth century. All the silver halides undergo a photochemical reaction in which the halide slowly decomposes into silver and halogen:

$$AgX(s) \rightarrow Ag(s) + \tfrac{1}{2}X_2(g)$$

A reaction similar to this forms the basis of black-and-white photography. In a black-and-white film, the usual halide is silver bromide, grains of which are suspended in an emulsion of gelatin. On exposure to light, the silver bromide undergoes a photochemical change. It absorbs a photon of light, and an electron is transferred from the bromide ion to the silver ion (see Figure 8.21), leaving the silver bromide in a high-energy state, represented by $AgBr^*$.

When a photograph is taken, the areas of film exposed to light therefore have high-energy silver bromide on them. The film is then 'developed'. The first stage in the development process is reduction of the high-energy silver bromide to silver. This is carried out using a reducing agent, and alkaline benzene-1,4-diol (old name hydroquinone) is the most common:

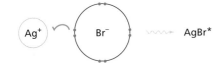

Figure 8.21 Electron transfer in silver bromide

$$2AgBr + HO-\langle\bigcirc\rangle-OH \rightarrow O=\langle\bigcirc\rangle=O + 2Ag + 2HBr$$

Care must be taken over the timing of this stage of the development process. If the reducing agent is left in contact with the film for too long, some of the unexposed silver bromide may be reduced and the film becomes 'fogged'.

The film is then 'fixed' by being treated with a solution of sodium thiosulfate, $Na_2S_2O_3(aq)$, which photographers call 'hypo'. This dissolves any unchanged silver bromide, preventing it from reacting further when the film is exposed to light:

$$AgBr(s) + 2S_2O_3^{2-}(aq) \rightarrow Ag(S_2O_3)_2^{3-}(aq) + Br^-(aq)$$

The film is washed thoroughly at this stage to remove all the silver–thiosulfate complex. What remains is a 'negative' of the image, with light areas of the image showing as dark silver areas.

Chain reactions

A few photochemical reactions are **chain reactions**. These are exothermic reactions that have a high activation energy. The light breaks a bond homolytically in one of the reactants, forming an atom. This free atom is energetic enough to react with another reactant molecule and produce another free atom.

The process then repeats, building up into a **chain reaction**. The light has provided the energy to overcome the activation energy barrier. Examples of chain reactions include the reaction of hydrogen with chlorine, and the chlorination of alkanes (see section 13.5)

Summary

- **Kinetics** is the study of the rates of reactions.
- The **rate** of a reaction is defined as how fast a product appears or how fast a reactant disappears.
- The rate of a reaction is determined by the height of the **activation energy barrier**, and not by the value of ΔH.
- Physical methods of finding the rate of reaction include:
  - measuring the volume of gas given off over time
  - measuring a change in mass over time
  - following a change in colour over time using a colorimeter.
- Chemical methods of finding the rate of reaction include:
  - removing samples at timed intervals, 'freezing' them and then using chemical analysis, for example titration
  - using a 'clock' method.
- The rate of reaction at a given time is shown by the gradient of the graph of amount of product against time (or the negative gradient of the graph of amount of reactant against time).
- The rate increases if the reactants are made more concentrated, because the molecules collide more frequently. For the same reason, increasing the pressure of a gaseous reaction increases the rate.
- The rate increases if the temperature is raised, mainly because there are more collisions with sufficient energy to react.
- A 10 °C rise in temperature often doubles the rate of reaction.
- The area under the **energy distribution curve** shows the fraction of molecules within a certain energy range.
- Only the fraction of molecules with energy greater than the activation energy, E_a, can react. These are shown by the area to the right of E_a on the energy distribution curve.
- A **catalyst** provides an alternative pathway of lower activation energy.
- The rate of a reaction in solution is usually changed if a different solvent is used.
- Some reactions go faster in the presence of light.

Examination practice questions

Please see the data section of the CD for any A_r values you may need.

1 Catalysts speed up the rate of a reaction without being consumed by the overall reaction.

  a Chlorine radicals in the stratosphere act as a catalyst for ozone depletion.

    i Research chemists have proposed possible reaction mechanisms for ozone depletion. The equations below represent part of such a mechanism.
Complete the equations.

$$Cl + O_3 \rightarrow \ldots\ldots\ldots\ldots + \ldots\ldots\ldots\ldots$$
$$ClO + \ldots\ldots\ldots\ldots \rightarrow \ldots\ldots\ldots\ldots + O_2 \quad [2]$$

    ii Write an equation for the overall reaction in **i**.   [1]

  b One of the catalysed reactions that takes place in a catalytic converter is shown below.

$$2CO(g) + 2NO(g) \rightarrow N_2(g) + 2CO_2(g)$$

The catalyst used is platinum/rhodium attached to a ceramic surface.
Outline the stages that take place in a catalytic converter to allow CO to react with NO.   [4]

  c Explain, using an enthalpy profile diagram and a Boltzmann distribution, how the presence of a catalyst increases the rate of reaction.   [7]

  d Explain why many industrial manufacturing processes use catalysts.
Include in your answer ideas about sustainability, economics and pollution control.   [4]

[OCR Chemistry A Unit F322 Q4 (part) January 2011]

2 The diagram below shows, for a given temperature T, a Boltzmann distribution of the kinetic energy of the molecules of a mixture of two gases that will react together, such as nitrogen and hydrogen. The activation energy for the reaction, Ea, is marked.

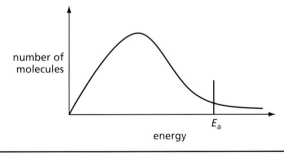

a **i** Draw a new distribution curve, clearly labelled T' for the same mixture of gases at a higher temperature, T';

    ii Mark clearly, as H, the position of the activation energy of the reaction at the higher temperature, T'.   [3]

b Explain the meaning of the term *activation energy*.   [2]

c The reaction between nitrogen and hydrogen to produce ammonia in the Haber process is an example of a large-scale gaseous reaction that is catalysed.

  i State the catalyst used and give the operating temperature and pressure of the Haber process.

  ii On the energy axis of the graph, mark the position, clearly labelled C, of the activation energy of the reaction when a catalyst is used.

  iii Use your answer to **ii** to explain how the use of a catalyst results in reactions occurring at a faster rate.   [3]

d Two reactions involving aqueous NaOH are given below.
$$CH_3CHBrCH_3 + NaOH \rightarrow CH_3CH(OH)CH_3 + NaBr \text{ \textbf{reaction 1}}$$
$$HCl + NaOH \rightarrow NaCl + H_2O \text{ \textbf{reaction 2}}$$
In order for **reaction 1** to occur, the reagents must be heated together for some time.
On the other hand, **reaction 2** is almost instantaneous at room temperature.
Suggest brief explanations why the rates of these two reactions are very different.   [4]

[Cambridge International AS & A level Chemistry, 9701, Paper 21 Q2 June 2010]

3 In many countries, new cars have to comply with regulations which are intended to reduce the pollutants coming from their internal combustion engines. Two pollutants that may be formed in an internal combustion engine are carbon monoxide, CO, and nitrogen monoxide, NO.

  a Outline how **each** of these pollutants may be formed in an internal combustion engine.

  b State the main hazard associated with **each** of these pollutants.   [4]

[Cambridge International AS & A level Chemistry, 9701, Paper 22 Q2 e November 2010]

9 Equilibria

Many chemical reactions do not go to completion. They stop at a point called the equilibrium position, where both reactants and products are present but there seems to be no further reaction. In this topic, we study how a change in reaction conditions brings about a change in the position of equilibrium of a reaction.

Learning outcomes

By the end of this topic you should be able to:

7.1a) explain, in terms of rates of the forward and reverse reactions, what is meant by a *reversible reaction* and *dynamic equilibrium*

7.1b) state Le Chatelier's Principle and apply it to deduce qualitatively (from appropriate information) the effects of changes in temperature, concentration or pressure on a system at equilibrium

7.1c) state whether changes in concentration, pressure or temperature or the presence of a catalyst affect the value of the equilibrium constant for a reaction

7.1d) deduce expressions for equilibrium constants in terms of concentrations, K_c, and partial pressures, K_p

7.1e) calculate the values of equilibrium constants in terms of concentrations or partial pressures from appropriate data

7.1f) calculate the quantities present at equilibrium, given appropriate data

7.1g) describe and explain the conditions used in the Haber process and the Contact process, as examples of the importance of an understanding of chemical equilibrium in the chemical industry.

9.1 Reversible reactions

Many chemical reactions, such as combustion, appear to go completely, converting all the reactants to products. If methane is burnt in an excess of oxygen, no methane is left over. Reactions such as these go to completion because they are highly exothermic, that is, ΔH is large and negative. But there are many reactions, particularly in organic chemistry, for which ΔH is small. These reactions may not go to completion – at the end of the reaction a detectable amount of reactant remains, mixed with the product. Such reactions are called **reversible reactions**.

To decide whether a reaction is reversible or not, the concentrations of reactant and product need to be carefully measured. A common laboratory method of analysis may detect reactants in the final mixture if they are present in as high a proportion as 1%, but more refined techniques can be many times more sensitive.

The reaction of a strong acid and a strong base:

$$H^+(aq) + OH^-(aq) \rightarrow H_2O(l)$$

appears to go to completion but this is not actually the case, as is shown by the fact that pure distilled water has a slight electrical conductivity. This conductivity is due to the ionisation of water:

$$H_2O(l) \rightarrow H^+(aq) + OH^-(aq)$$

indicating that the reaction can go in either direction. Only two molecules in every 10^9 molecules ionise and so the ionisation can usually be ignored, but it becomes important when we study the pH scale of acidity (see Topic 22).

A reaction is considered as having gone to completion if any reactants in the final mixture are present at such low concentrations that they do not affect any tests or reactions that are carried out on the products. In practical terms, this means that if the reaction is more than 99% complete, it can be regarded as having gone to completion.

9.2 The position of equilibrium

Dynamic equilibrium

A reversible reaction appears to stop when no overall change is taking place; the reaction is said to be in **equilibrium**. For example, a reaction may come to be in equilibrium when 75% of the reactants have changed into products. Under these conditions, the composition of the equilibrium mixture is 75% products and 25% reactants, and this composition is known as the **position of equilibrium**.

At equilibrium, there is no overall change from reactants to products and it seems as though the reaction has stopped. This is not actually the case. There is no reason why the **forward reaction**, that is, the reaction between the reactants, should not continue, since there are still reactants present in the mixture. The reason why there is no *overall* change is that the **back reaction**, which turns the products back into reactants, is now going at exactly the same rate. A reversible reaction in which the forward and back reactions are both taking place at the same rate is termed a **dynamic equilibrium**.

For a reversible reaction:

$$\text{reactants} \underset{\text{back reaction}}{\overset{\text{forward reaction}}{\rightleftharpoons}} \text{products}$$

The double-headed arrow $\rightleftharpoons$ indicates a reaction that goes to equilibrium.
At equilibrium, the rate of the forward reaction equals the rate of the back reaction.

Demonstrating dynamic equilibrium

A particularly effective way of showing that an equilibrium is dynamic is to use **isotopic labelling**. In the Haber process, nitrogen and hydrogen gases are allowed to come to equilibrium with ammonia:

$$N_2(g) + 3H_2(g) \rightleftharpoons 2NH_3(g)$$

If deuterium (a heavy isotope of hydrogen) is substituted for some of the hydrogen gas in an equilibrium mixture of the gases, some deuterium becomes incorporated into the ammonia, as can be shown by analysis of the mixture by mass spectroscopy. This shows that the forward reaction is continuing to take place, and this must be balanced by an equal rate of back reaction in order that the same position of equilibrium is maintained (see Figure 9.1).

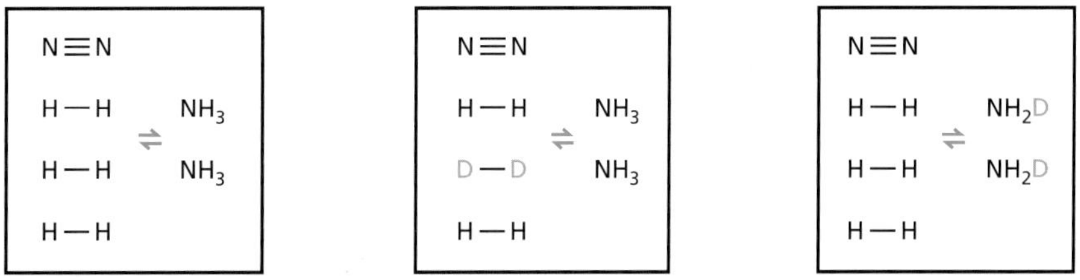

Figure 9.1 If deuterium is substituted for some of the hydrogen gas in the equilibrium mixture, some deuterium becomes bonded to nitrogen in ammonia molecules.

A similar experiment may be carried out with lead chloride, $PbCl_2$, labelled with radioactive lead. Lead chloride is only slightly soluble in cold water, and a saturated solution of the radioactive salt can easily be made. If unlabelled lead chloride is stirred into the saturated solution, no more of the solid can dissolve overall but it is found that some of the radioactive lead still ends up in the solid

Figure 9.2 If solid lead chloride is placed in a saturated solution of lead chloride labelled with radioactive lead ions, Pb^{2+} (shown in red), the solid becomes radioactive.

(see Figure 9.2). This shows that some of the radioactive lead ions in the solution have been precipitated into the solid, and an equal number of lead ions from the solid must have dissolved to keep the solution saturated.

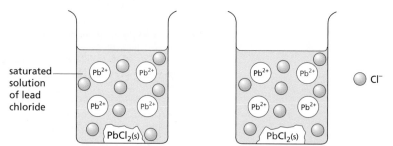

saturated solution of lead chloride

Cl^-

Now try this

1 State, giving a reason, whether ΔH for a reversible reaction is large or small.
2 a What is meant by dynamic equilibrium?
  b Explain why it is reasonable to think that chemical equilibrium is dynamic rather than static.
  c Describe an experiment which shows that a chemical equilibrium is dynamic rather than static.

Worked example

a What is meant by a reversible reaction? Give an example of a reversible reaction.
b How could you show that it is reversible?

Answer

a A reversible reaction is one in which it is possible to detect unchanged reactants in the equilibrium mixture, and in which both the forward and back reactions take place. One example is the neutralisation of a strong acid with a strong base:

$$H^+(aq) + OH^-(aq) \rightleftharpoons H_2O(l)$$

b The reaction in part a must be reversible because pure water conducts electricity, to a small extent.

Moving the position of equilibrium

In Topic 8, the effects of concentration, pressure of gases, temperature, state of division of solids, and catalysts on the rate of reaction were studied. We shall now consider what effect (if any) these factors may have on the position of equilibrium of a reaction.

Some examples of reversible reactions and equilibria

- If acid is added to a yellow solution containing CrO_4^{2-} ions, the solution turns orange owing to the formation of $Cr_2O_7^{2-}$ ions. The reaction can be reversed by the addition of alkali.

$$2CrO_4^{2-}(aq) + 2H^+(aq) \underset{OH^-(aq)}{\overset{H^+(aq)}{\rightleftharpoons}} Cr_2O_7^{2-}(aq) + H_2O(l)$$
yellow   orange

- If alkali is added to a brown solution of iodine, the colour fades owing to the formation of colourless IO^- ions. The colour returns on the addition of acid.

$$I_2(aq) + 2OH^-(aq) \underset{H^+(aq)}{\overset{OH^-(aq)}{\rightleftharpoons}} 2IO^-(aq) + H_2O(l)$$
brown   colourless

- If a mixture of the brown gas NO_2 and the almost colourless gas N_2O_4 at equilibrium is suddenly compressed, the mixture darkens and then the colour slowly fades. The initial increase in colour is due to the increase in concentration of $NO_2(g)$ as the volume decreases, but this decreases as the equilibrium moves in the direction of $N_2O_4(g)$.

$$2NO_2(g) \underset{decreased\ pressure}{\overset{increased\ pressure}{\rightleftharpoons}} N_2O_4(g)$$
brown   colourless

- If a pink solution of cobalt chloride is warmed, the solution goes blue. It changes back to pink on cooling. The forward reaction is endothermic.

$$\underset{\text{pink}}{Co(H_2O)_6^{2+}(aq)} + 4Cl^-(aq) \underset{\underset{\text{cooling}}{\rightleftharpoons}}{\overset{\text{warming}}{\rightleftharpoons}} \underset{\text{blue}}{CoCl_4^{2-}(aq)} + 6H_2O(l)$$

The changes in position of these equilibria can be predicted from **Le Chatelier's Principle**.

Le Chatelier's Principle states that if the conditions of a system in equilibrium are changed (for example, by changing the concentration or pressure or temperature), the position of equilibrium moves so as to reduce that change.

9.3 Changing the concentration

If we increase the concentration of one of the reactants of a reaction in equilibrium, the rate of the forward reaction increases and more products form. The resulting increase in the concentration of the products will then lead to an increased rate of the back reaction until a new position of equilibrium is set up.

The position of equilibrium has moved to the right (see Figure 9.3), with slightly more product being present than at the original position of equilibrium. This is in agreement with Le Chatelier's Principle.

Figure 9.3 An increase in the concentration of one reactant converts more of the other reactant to product and moves the position of equilibrium to the right.

reactant₁ + reactant₂ ⇌ product   system in equilibrium

reactant₁ + reactant₂ ⇌ product   increase in concentration of reactant₁

reactant₁ + reactant₂ ⇌ product   equilibrium moves to the right

In a similar way, if the concentration of one or more of the reactants is reduced (or the concentration of the products is increased), the position of equilibrium moves to the left. The situation may be summarised as shown in Table 9.1.

Table 9.1 The effect on the position of equilibrium of changing the concentration of the reactants or the products

Concentration of reactants	Concentration of products	Direction of movement of the position of equilibrium
increased		to the right
decreased		to the left
	increased	to the left
	decreased	to the right

Worked example 1

What is the effect of increasing the concentration of hydrogen ions on the following reaction?

$$CH_3CO_2^- + H^+ \rightleftharpoons CH_3CO_2H$$

Answer
The equilibrium is pushed over to the right, that is, more ethanoic acid, CH_3CO_2H, is formed.

Worked example 2

Consider the following equilibrium:

$$\underset{\text{acid}}{CH_3CO_2H} + \underset{\text{alcohol}}{C_2H_5OH} \rightleftharpoons \underset{\text{ester}}{CH_3CO_2H_5} + H_2O$$

Explain why removing water leads to more ester being formed in the equilibrium mixture.

Answer
Initially, removing water means that the rate of the back reaction is decreased while that of the forward reaction stays the same. Le Chatelier's Principle predicts that the position of equilibrium will move to oppose the change in the equilibrium mixture. Hence the position of equilibrium moves to the right, producing more ester and water.

Now try this

1 Consider the following equilibrium:
$$H_2(g) + I_2(g) \rightleftharpoons 2HI(g)$$

How will the position of equilibrium alter when each of the following changes is brought about?

a adding more iodine
b adding more hydrogen iodide
c removing hydrogen

9.4 Changing the pressure

In a gaseous reaction, a change in pressure may affect the position of equilibrium in a way similar to changing the concentration in an aqueous reaction. However, for a mixture of gases, an increase in pressure may increase the concentration of both reactants and products to the same extent and may therefore have no effect on the position of equilibrium. This occurs if there are the same number of gas molecules on the left-hand and right-hand sides of the equation. An example is the decomposition of hydrogen iodide:

$$2HI(g) \rightleftharpoons H_2(g) + I_2(g)$$

Here there are two molecules of gas on both sides of the equation, so changing the pressure will not change the position of equilibrium.

At a lower temperature, hydrogen iodide decomposes as follows, producing solid iodine:

$$2HI(g) \rightleftharpoons H_2(g) + I_2(s)$$

Now two gas molecules on the left form only one gas molecule on the right. Because pressure affects only the gaseous components of the mixture, an increase in pressure now will speed up the forward reaction more than the back reaction, so that the equilibrium moves to the right. This again agrees with the prediction from Le Chatelier's Principle, because the increase in pressure can be reduced if the equilibrium moves in the direction in which fewer gas molecules are formed. We can summarise the situation as shown in Table 9.2.

Table 9.2 The effect on the position of equilibrium of increasing the pressure

Number of gaseous reactant molecules	Number of gaseous product molecules	Direction of movement of the position of equilibrium when the pressure is increased
more	fewer	to the right
fewer	more	to the left
same	same	no change

Now try this

What effect (if any) will there be on each of the following equilibria when the pressure is increased?

1. $2HBr(g) \rightleftharpoons H_2(g) + Br_2(g)$
2. $PCl_3(g) + Cl_2(g) \rightleftharpoons PCl_5(g)$
3. $H_2O(l) \rightleftharpoons H_2O(g)$
4. $CaCO_3(s) \rightleftharpoons CaO(s) + CO_2(g)$

Worked example

What is the effect of an increase in pressure on the synthesis of ammonia?

$$N_2(g) + 3H_2(g) \rightleftharpoons 2NH_3(g)$$

Answer
In this reaction, four molecules of gas become two molecules of gas. An increase in pressure favours the products, as these contain fewer gas molecules. So increasing the pressure results in an equilibrium mixture containing more ammonia.

9.5 Changing the temperature

Le Chatelier's Principle can also be used to predict the effect of a temperature change on the position of equilibrium. The most important factor is whether the forward reaction is exothermic (ΔH negative) or endothermic (ΔH positive). If the temperature is raised, the equilibrium will move in the direction which will tend to lower the temperature, that is, the endothermic direction. The effect is summarised in Table 9.3.

Table 9.3 The effect on the position of equilibrium of changing the temperature

Temperature change	Sign of ΔH for forward reaction	Direction of movement of the position of equilibrium
increase	+ve	to the right
decrease	+ve	to the left
increase	−ve	to the left
decrease	−ve	to the right

Careful control of temperature is important in the Contact process and the Haber process, as we shall see later.

Worked example

What is the effect of raising the temperature on the following equilibrium?

$$H_2(g) + I_2(g) \rightleftharpoons 2HI(g) \qquad \Delta H^\ominus = -12\,kJ\,mol^{-1}$$

Answer

The reaction is exothermic (gives out heat). On raising the temperature, the equilibrium will move in the direction that reduces the temperature, that is, in the endothermic direction. The equilibrium will therefore move to the left.

Now try this

What will be the effect of lowering the temperature on each of the following equilibria?

1 $N_2(g) + 3H_2(g) \rightleftharpoons 2NH_3(g)$ $\qquad \Delta H^\ominus = -92\,kJ\,mol^{-1}$
2 $H_2O(l) \rightleftharpoons H^+(aq) + OH^-(aq)$ $\qquad \Delta H^\ominus = +58\,kJ\,mol^{-1}$

9.6 State of division and the addition of catalysts

State of division

In a heterogeneous reaction, a small particle size speeds up a reaction by increasing the surface area. If the reaction is in equilibrium, both the forward and back reactions will be speeded up when the surface area is increased. This increase in speed is the same for both reactions. If, for example, the surface area of a liquid was doubled, the rate at which molecules leave this surface would be doubled, but so too would the rate at which they joined the surface again. The position of equilibrium is therefore unchanged (see Figure 9.4).

Figure 9.4 In **a** and **b** there is the same amount of water, but the surface area in **a** is twice that in **b**. The rates of evaporation and condensation are both twice as fast in **a** as in **b**, but the position of equilibrium is unchanged.

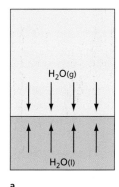
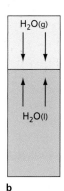

a   b

Catalysts

The effect of adding a catalyst to a mixture in equilibrium is similar to that of increasing the surface area. Both the forward and back reactions are speeded up by the same amount, and the position of equilibrium is unaltered. It may not be immediately obvious why this should be the case, but it can be demonstrated by the following thermodynamic argument. If an exothermic reaction was in equilibrium and then a catalyst was added which made the reaction faster in the exothermic direction only, energy would be obtained. This energy could not have come from the catalyst (which, by definition, is left chemically unchanged at the end). We would have obtained energy from nowhere, which is contrary to the Law of Conservation of Energy (see Figure 9.5).

Figure 9.5 A reaction in equilibrium is placed in contact with a catalyst by opening the trap door. If this changed the position of equilibrium, the piston would move up or down, depending on whether the catalyst shifted the equilibrium in the endothermic or exothermic direction. This movement of the piston could be used to close the trap door, and the process would then be repeated. Perpetual motion would have been created, which is impossible.

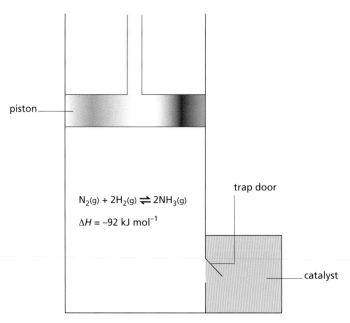

piston

$N_2(g) + 2H_2(g) \rightleftharpoons 2NH_3(g)$

$\Delta H = -92 \text{ kJ mol}^{-1}$

trap door

catalyst

9.7 The equilibrium constant K_c

We have just shown how the position of equilibrium changes when the external conditions are altered. Changing the state of division or adding a catalyst affects the rate of the reaction, but does not affect the position of equilibrium. The factors that *do* affect the position of equilibrium are:

- the concentrations of the reactants and products
- the temperature.

In this section we shall study quantitatively the effect of changes in concentration on the position of equilibrium.

For a reaction in equilibrium, for example,

$$N_2O_4(g) \rightleftharpoons 2NO_2(g)$$

it can be shown that, at a given temperature,

$$K_c = \frac{[NO_2(g)]^2_{eqm}}{[N_2O_4(g)]_{eqm}} \quad \text{which is usually simplified to} \quad \frac{[NO_2]^2}{[N_2O_4]}$$

where K_c is a constant, called the **equilibrium constant**.

The square brackets [ ] indicate molecular concentrations at equilibrium in mol dm^{-3}. The subscript 'c' indicates that this equilibrium constant is calculated using concentrations and can, therefore, be distinguished from another equilibrium constant, K_p, which uses pressures, as we shall see later in the topic.

For the general reaction

$$A + B \rightleftharpoons C + D$$

the **equilibrium constant** is given by the expression

$$K_c = \frac{[\text{products}]}{[\text{reactants}]} = \frac{[C]_{eqm} \, [D]_{eqm}}{[A]_{eqm} \, [B]_{eqm}}$$

Unlike the position of equilibrium, the value of the equilibrium constant does not change with concentration; that is why it is termed a 'constant'. The only factor that affects its value is temperature.

The squared term in the expression

$$K_c = \frac{[NO_2(g)]^2_{eqm}}{[N_2O_4(g)]_{eqm}}$$

arises because there are two NO_2 in the reaction products. We could have written the reaction as:

$$N_2O_4(g) \rightleftharpoons NO_2(g) + NO_2(g)$$

Then the equilibrium constant would have been written as:

$$K_c = \frac{[NO_2(g)]_{eqm} \times [NO_2(g)]_{eqm}}{[N_2O_4(g)]_{eqm}}$$

This is the same result.

There are several ways in which this expression can be derived. The most fundamental is by means of thermodynamics, but it can be approached through experiment or by way of kinetics.

Deriving equilibrium constants

By experiment

The following reaction can be studied:

$$CH_3CO_2H + C_2H_5OH \rightleftharpoons CH_3CO_2C_2H_5 + H_2O$$

A range of mixtures containing different amounts of reactants and products are made up. The mixtures are left to come to equilibrium and are then analysed, for example by determining the amount of ethanoic acid present by titration. For a wide range of mixtures, it is found that the ratio:

$$\frac{[CH_3CO_2C_2H_5][H_2O]}{[CH_3CO_2H][C_2H_5OH]}$$

is constant.

Kinetic approach

If the kinetics of the forward and back reactions are known, the equilibrium constant can be related to the rate constants. For the following reaction:

$$N_2O_4(g) \rightleftharpoons 2NO_2(g)$$

it is found that:

rate of forward reaction at equilibrium $= k_f[N_2O_4]$
rate of back reaction at equilibrium $= k_b[NO_2]^2$

where k_f and k_b are the rate constants for the forward and back reactions respectively (see Topic 21 for an explanation of these rate expressions).

At equilibrium, the two rates are equal, therefore:

$$k_f[N_2O_4] = k_b[NO_2]^2$$

At a given temperature, the two rate constants are fixed, and

$$\frac{[NO_2]^2}{[N_2O_4]} = \frac{k_f}{k_b} = K_c$$

The units of K_c

For the reaction $CH_3CO_2H + C_2H_5OH \rightleftharpoons CH_3CO_2C_2H_5 + H_2O$, the expression for K_c has two concentration terms on the top and two on the bottom:

$$\frac{[CH_3CO_2C_2H_5][H_2O]}{[CH_3CO_2H][C_2H_5OH]}$$

units: $\dfrac{(mol\,dm^{-3})(mol\,dm^{-3})}{(mol\,dm^{-3})(mol\,dm^{-3})} = 1$

The units top and bottom therefore cancel, and the result is that K_c for this reaction has no units.

But for the reaction $N_2O_4(g) \rightleftharpoons 2NO_2(g)$, the expression for K_c has two terms on the top and one term on the bottom:

$$K_c = \frac{[NO_2(g)]^2}{[N_2O_4(g)]}$$

units: $\dfrac{(mol\,dm^{-3})^2}{(mol\,dm^{-3})} = mol\,dm^{-3}$

Worked example

Write the expression for K_c and state its units for each of the following reactions.

a $I_2(aq) + I^-(aq) \rightleftharpoons I_3^-(aq)$

b $N_2(g) + 3H_2(g) \rightleftharpoons 2NH_3(g)$

Answer

a $K_c = \dfrac{[I_3^-]}{[I_2][I^-]}$; units are $mol^{-1}\,dm^3$.

b $K_c = \dfrac{[NH_3]^2}{[N_2][H_2]^3}$; units are $mol^{-2}\,dm^6$.

Now try this

1 Write the expression for K_c and state its units for each of the following reactions.

a $2CrO_4^{2-}(aq) + 2H^+(aq) \rightleftharpoons Cr_2O_7^{2-}(aq) + H_2O(l)$
(Note: H_2O is the solvent; its concentration is constant, so its concentration does not appear in the expression for K_c.)

b $I_2(aq) + 2Fe^{2+}(aq) \rightleftharpoons 2I^-(aq) + 2Fe^{3+}(aq)$

Evaluating K_c for a reaction

In order to determine the value of an equilibrium constant, known amounts of reactants are allowed to come to equilibrium. The amount of one of the substances in the equilibrium mixture is found by experiment. The amounts of the others can then be worked out from the stoichiometric equation, as shown in the following worked example.

Worked example 1

2.0 mol of ethanoic acid and 2.0 mol of ethanol are mixed and allowed to come to equilibrium with the ethyl ethanoate and water they have produced. At equilibrium, the amount of ethanoic acid present is 0.67 mol. Calculate K_c.

Answer
We start by working out the amounts, and hence the concentrations, of the four substances present at equilibrium from the stoichiometric equation:

	CH_3CO_2H	$+$	C_2H_5OH	$\rightleftharpoons$	$CH_3CO_2C_2H_5$	$+$	H_2O
moles at start/mol (given):	2.0		2.0		0		0
moles at equilibrium/mol (given):	0.67						
moles at equilibrium/mol:	0.67		0.67		$(2.0-0.67)$		$(2.0-0.67)$
	0.67 mol		0.67 mol		1.33 mol		1.33 mol
concentrations at equilibrium:	$\dfrac{0.67}{V}$		$\dfrac{0.67}{V}$		$\dfrac{1.33}{V}$		$\dfrac{1.33}{V}$

where V is the total volume of the mixture.

$$K_c = \frac{[CH_3CO_2C_2H_5][H_2O]}{[CH_3CO_2H][C_2H_5OH]} = \frac{1.33/V \times 1.33/V}{0.67/V \times 0.67/V}$$

$$= \mathbf{4.0} \text{ (no units)}$$

K_c has no units because there are the same number of molecules on both sides of the equation. This also means that the volume, V, cancels. The value of K_c, and the position of equilibrium, do not depend on the volume of the reaction mixture.

Worked example 2

0.30 mol of dinitrogen tetraoxide, N_2O_4, is allowed to come to equilibrium with nitrogen dioxide, NO_2. The amount of nitrogen dioxide at equilibrium is 0.28 mol. Calculate:

a the amount of dinitrogen tetraoxide at equilibrium

b the value of K_c at the temperature of the experiment, given that the volume of the containing vessel is 10.0 dm³.

Answer

a

$$N_2O_{4(g)} \rightleftharpoons 2NO_{2(g)}$$

moles at start/mol: 0.30   0

moles at equilibrium/mol: $(0.30 - \frac{1}{2} \times 0.28)$ 0.28

moles of N_2O_4 at equilibrium = **0.16 mol**

b $K_c = \dfrac{[NO_2]^2}{[N_2O_4]} = \dfrac{(0.28/10)^2}{0.16/10} = \textbf{0.049 mol dm}^{-3}$

Now try this

1 When iodic acid is dissolved in water, it ionises slightly according to the following equilibrium:

$$HIO_3(aq) \rightleftharpoons H^+(aq) + IO_3^-(aq)$$

In a 0.50 mol dm⁻³ solution of HIO_3, the concentration of H^+ ions is found to be 0.22 mol dm⁻³.

Construct an expression for K_c, calculate its value and state its units.

2 When aqueous ammonia is added to a solution of a silver salt, the following equilibrium is set up:

$$Ag^+(aq) + 2NH_3(aq) \rightleftharpoons [Ag(NH_3)_2]^+(aq)$$

When equal volumes of 0.1 mol dm⁻³ $Ag^+(aq)$ and 0.2 mol dm⁻³ $NH_3(aq)$ were mixed, the equilibrium concentration of $[Ag(NH_3)_2]^+(aq)$ was found to be 0.035 mol dm⁻³.

Construct an expression for K_c, calculate its value and state its units.

9.8 Using equilibrium constants

Finding the composition of the equilibrium mixture

An equilibrium constant can be used to calculate the composition of an equilibrium mixture. For example, the equilibrium constant for the reaction:

$$CH_3CO_2H + C_2H_5OH \rightleftharpoons CH_3CO_2C_2H_5 + H_2O$$

is 4.0. If we start with 2.0 mol of ethanoic acid and 2.0 mol of ethanol, the composition of the equilibrium mixture can be found by means of some simple algebra as follows.

Let there be x mol of ethyl ethanoate at equilibrium:

	CH_3CO_2H	C_2H_5OH	$CH_3CO_2C_2H_5$	H_2O
moles at start/mol:	2.0	2.0	0	0
moles at equilibrium/mol:	$(2.0-x)$	$(2.0-x)$	x	x
concentrations at equilibrium/mol:	$\dfrac{(2.0-x)}{V}$	$\dfrac{(2.0-x)}{V}$	$\dfrac{x}{V}$	$\dfrac{x}{V}$

$$K_c = \frac{[CH_3CO_2C_2H_5][H_2O]}{[CH_3CO_2H][C_2H_5OH]}$$

$$= \frac{(x/V)^2}{(2.0 - x/V)^2}$$

The volume terms cancel out. This gives us:

$$\frac{x^2}{(2.0 - x)^2} = 4.0$$

Taking square roots of both sides,

$$\frac{x}{(2.0 - x)} = \pm 2.0$$

Either $\quad x = +4.0 - 2x \quad$ so $\quad x = \mathbf{1.33}$

or $\qquad x = -4.0 + 2x \quad$ so $\quad x = \mathbf{4.0}$

The second value must be rejected as it is impossible to produce more than 2.0 mol of product. Therefore, at equilibrium there are 1.3 mol of ethyl ethanoate, 1.3 mol of water, 0.7 mol of ethanoic acid and 0.7 mol of ethanol.

> **Now try this**
>
> At 700 K, the value of K_c for the following reaction is 54.
>
> $$H_2(g) + I_2(g) \rightleftharpoons 2HI(g)$$
>
> 0.5 mol of $H_2(g)$ and 0.5 mol of $I_2(g)$ were mixed at 700 K and allowed to reach equilibrium.
>   What is the composition of the equilibrium mixture?

9.9 The equilibrium constant K_p

For gaseous equilibria, the concentrations of reactants and products can be expressed in terms of the pressures of the different components. If the reaction mixture behaves as an ideal gas, the ideal gas equation $pV = nRT$ (see section 4.13) can be used. The molecular concentration, [X], which is n/V, is then equal to p/RT. This shows that, at a given temperature when T is constant, the pressure of a component is proportional to its molecular concentration.

Thus the equilibrium constant can be written in terms of the pressures of the different components rather than their molecular concentrations. For the equilibrium

$$H_2(g) + I_2(g) \rightleftharpoons 2HI(g)$$

for example, the equilibrium constant can be written

$$K_p = \frac{(p_{HI})^2}{p_{H_2} \times p_{I_2}}$$

where p_{HI}, p_{H_2} and p_{I_2} are the pressures of the different components, called their **partial pressures**. If no other gases are present, the sum of their partial pressures will equal the total pressure P.

If $p_{HI} = 0.50$ atm and $p_{H_2} = p_{I_2} = 0.30$ atm, we have

$$K_p = \frac{(p_{HI})^2}{p_{H_2} \times p_{I_2}} = \frac{0.50^2}{0.30 \times 0.30} = 2.78$$

As there are two partial pressure terms on the top and two on the bottom, K_p has no units (which is also true for K_c as it has two concentration terms on the top and two on the bottom). This means that increasing the total pressure will have no effect on the position of equilibrium. If we had used pascals for the partial pressures, we would have had the same results because we would have multiplied each term by 1.013×10^5 and the four 1.013×10^5 terms would have cancelled out. The situation is more complicated if the number of molecules is different on each side of the equation. For example, for the equilibrium

$$N_2O_4(g) \rightleftharpoons 2NO_2(g)$$

the equilibrium constant is

$$K_p = \frac{(p_{NO_2})^2}{p_{N_2O_4}}$$

If $p_{NO_2} = 0.50$ atm and $p_{N_2O_4} = 0.30$ atm,

$$K_p = \frac{(p_{NO_2})^2}{p_{N_2O_4}} = \frac{0.50^2}{0.30} = 0.83 \text{ atm}$$

K_p would also equal $0.83 \times 1.013 \times 10^5 = 84.4$ kPa.

In this case, changing the total pressure will affect the *position* of equilibrium. It does *not* change the value of K_p, however. If, for example, the total pressure is doubled by compressing the mixture, initially $p_{NO_2}^2$ would increase by 4 times while $p_{N_2O_4}$ would only double. In order to keep K_p constant, p_{NO_2} must then decrease and $p_{N_2O_4}$ must increase. The position of equilibrium has moved to the left, in accordance with Le Chatelier's Principle.

Worked example

The key reaction in the Contact process for manufacturing sulfuric acid is the reaction between SO_2 and O_2:

$$2SO_2(g) + O_2(g) \rightleftharpoons 2SO_3(g) \qquad \Delta H = -195 \text{ kJ mol}^{-1}$$

Write an expression for the equilibrium constant, K_p, for this reaction, and decide what conditions of temperature and pressure would favour the forward reaction. Explain your answer.

Answer

$$K_p = \frac{p_{SO_3}^2}{p_{SO_2}^2 \times p_{O_2}}$$

According to Le Chatelier's Principle, high pressure and low temperature would favour the forward reaction. This is because there are fewer gas molecules on the right-hand side of the equation, and the reaction is exothermic.

Now try this

The Haber process for the industrial manufacture of ammonia uses the following reaction:

$$N_2(g) + 3H_2(g) \rightleftharpoons 2NH_3(g) \qquad \Delta H = -93 \text{ kJ mol}^{-1}$$

Write an expression for the equilibrium constant, K_p, for this reaction, and decide what conditions of temperature and pressure would favour the forward reaction. Explain your answer.

Summary

- Many chemical reactions do not go to completion: they take up a position of **dynamic equilibrium**.
- This **position of equilibrium** may change if the reaction conditions are changed. The direction of change can be predicted using **Le Chatelier's Principle**.
- Increasing the concentration of a reactant moves the position of equilibrium to the right.
- Increasing the pressure in a gaseous reaction moves the equilibrium in the direction that has fewer gas molecules.
- Increasing the temperature moves the position of equilibrium in the endothermic direction.
- Reducing the particle size or increasing the surface area speeds up a reaction, but does not change the position of equilibrium.
- The addition of a catalyst does not affect the position of equilibrium, only how quickly equilibrium is reached.

- For a reaction $A + B \rightleftharpoons C + D$, there is an **equilibrium constant** K_c such that
$$K_c = \frac{[C][D]}{[A][B]}$$
where [ ] are concentrations in mol dm^{-3}. The value of this constant does not change with changes in concentration: it only changes with changes in temperature.
- If the reaction components are all gases, the equilibrium constant can also be expressed in terms of the partial pressures p_A, p_B, p_C and p_D as follows:
$$K_p = \frac{p_C \times p_D}{p_A \times p_B}$$

Examination practice questions

Please see the data section of the CD for any A_r values you may need.

1 When hydrocarbons such as petrol or paraffin wax are burned in an excess of air in a laboratory, carbon dioxide and water are the only products.
When petrol is burned in a car engine, nitrogen monoxide, NO, is also formed.
NO is also formed when nitrosyl chloride, NOCl, dissociates according to the following equation.

$$2NOCl(g) \rightarrow 2NO(g) + Cl_2(g)$$

Different amounts of the three gases were placed in a closed container and allowed to come to equilibrium at 230 °C. The experiment was repeated at 465 °C.
The equilibrium concentrations of the three gases at each temperature are given in the table below.

Temperature/°C	Concentration/mol dm^{-3}		
	NOCl	**NO**	**Cl$_2$**
230	2.33×10^{-3}	1.46×10^{-3}	1.15×10^{-2}
465	3.68×10^{-4}	7.63×10^{-3}	2.14×10^{-4}

a Write the expression for the equilibrium constant, K_c, for this reaction. Give the units.

b Calculate the value of K_c at each of the temperatures given.

[Cambridge International AS & A Level Chemistry 9701, Paper 2 Q3 c June 2008]

2 Ethanoic acid can be reacted with alcohols to form esters, an equilibrium mixture being formed.

$$CH_3CO_2H + ROH \rightleftharpoons CH_3CO_2R + H_2O$$

The reaction is usually carried out in the presence of an acid catalyst.

a Write an expression for the equilibrium constant, K_c, for this reaction, clearly stating the units. [2]

In an experiment to determine K_c a student placed together in a conical flask 0.10 mol of ethanoic acid, 0.10 mol of an alcohol ROH, and 0.005 mol of hydrogen chloride catalyst.
The flask was sealed and kept at 25 °C for seven days.
After this time, the student titrated all of the contents of the flask with 2.00 mol dm^{-3} NaOH using phenolphthalein indicator.
At the end-point, 22.5 cm^3 of NaOH had been used.

b i Calculate the amount, in moles, of NaOH used in the titration.
  ii What amount, in moles, of this NaOH reacted with the hydrogen chloride?
  iii Write a balanced equation for the reaction between ethanoic acid and NaOH.
  iv Hence calculate the amount, in moles, of NaOH that reacted with the ethanoic acid. [4]

c i Use your results from **b** to calculate the amount, in moles, of ethanoic acid present at equilibrium. Hence draw and complete the table below.

	CH$_3$CO$_2$H	ROH	CH$_3$CO$_2$R	H$_2$O
Initial amount/mol	0.10	0.10	0	0
Equilibrium amount/mol				

  ii Use your results to calculate a value for K_c for this reaction. [3]

d Esters are hydrolysed by sodium hydroxide. During the titration, sodium hydroxide reacts with ethanoic acid and the hydrogen chloride, but not with the ester.
Suggest a reason for this. [1]

e What would be the effect, if any, on the amount of ester present if all of the water were removed from the flask and the flask kept for a further week at 25 °C?
Explain your answer. [2]

[Cambridge International AS & A Level Chemistry 9701, Paper 22 Q1 June 2011]

3 Ammonia is one of our most important chemicals, produced in enormous quantities because of its role in the production of fertilisers.
Much of this ammonia is manufactured from nitrogen and hydrogen gases using the Haber process. The equilibrium is shown below.

$$N_2(g) + 3H_2(g) \rightleftharpoons 2NH_3(g) \qquad \Delta H = -92 \text{ kJ mol}^{-1}$$

a i Write an expression for K_c for this equilibrium. [1]
  ii Deduce the units of K_c for this equilibrium. [1]

b A research chemist was investigating methods to improve the synthesis of ammonia from nitrogen and hydrogen at 500 °C. [4]

- The chemist mixed together nitrogen and hydrogen and pressurised the gases so that their total gas volume was 6.0 dm^3.

- The mixture was allowed to reach equilibrium at constant temperature and without changing the total gas volume.

- The equilibrium mixture contained 7.2 mol N_2 and 12.0 mol H_2.
- At 500 °C, the numerical value of K_c for this equilibrium is 8.00×10^{-2}.

  Calculate the amount, in mol, of ammonia present in the equilibrium mixture at 500 °C. [4]

c The research chemist doubled the pressure of the equilibrium mixture whilst keeping all other conditions the same. As expected the equilibrium yield of ammonia increased.

i Explain in terms of Le Chatelier's Principle why the equilibrium yield of ammonia increased. [2]

ii Explain in terms of K_c why the equilibrium yield of ammonia increased. [3]

[OCR Chemistry A Unit F325 Q5 (part) June 2010]

10 Periodicity

A modern form of the Periodic Table is studied in this topic. We shall see how it can be used to show patterns in the properties of some of the chemical elements it contains. The changes in properties in crossing a period (from sodium to argon) or in going down a group (from magnesium to barium) are related to the changes in the elements' atomic structures. Two other common elements, nitrogen and sulfur, are also studied.

Learning outcomes

For the third period (sodium to argon), by the end of this topic you should be able to:

3.3b) understand, in simple terms, the concept of electronegativity and apply it to explain the behaviour of oxides with water (part, see also Topic 3)

9.1a) describe qualitatively (and indicate the periodicity in) the variations in atomic radius, ionic radius, melting point and electrical conductivity of the elements

9.1b) explain qualitatively the variation in atomic radius and ionic radius

9.1c) interpret the variation in melting point and in electrical conductivity in terms of the presence of simple molecular, giant molecular or metallic bonding in the elements

9.1d) explain the variation in first ionisation energy

9.1e) explain the strength, high melting point and electrical insulating properties of ceramics in terms of their giant structure; to include magnesium oxide, aluminium oxide and silicon dioxide

9.2a) describe the reactions, if any, of the elements with oxygen (to give Na_2O, MgO, Al_2O_3, P_4O_{10}, SO_2, SO_3), chlorine (to give NaCl, $MgCl_2$, Al_2Cl_6, $SiCl_4$, PCl_5) and water (Na and Mg only)

9.2b) state and explain the variation in oxidation number of the oxides (sodium to sulfur only) and chlorides (sodium to phosphorus only) in terms of their valence shell electrons

9.2c) describe the reactions of the oxides with water

9.2d) describe and explain the acid/base behaviour of oxides and hydroxides, including, where relevant, amphoteric behaviour in reaction with sodium hydroxide and acids

9.2e) describe and explain the reactions of the chlorides with water

9.2f) interpret the variations and trends in **9.2b), c), d)** and **e)** in terms of bonding and electronegativity

9.2g) suggest the types of chemical bonding present in chlorides and oxides from observations of their chemical and physical properties

9.3a) predict the characteristic properties of an element in a given group by using knowledge of chemical periodicity

9.3b) deduce the nature, possible position in the Periodic Table, and identity of unknown elements from given information about physical and chemical properties.

For Group 2 (magnesium to barium), you should be able to:

10.1a) describe the reactions of the elements with oxygen, water and dilute acids

10.1b) describe the behaviour of the oxides, hydroxides and carbonates with water and with dilute acids

10.1c) describe the thermal decomposition of the nitrates and carbonates

10.1d) interpret, and make predictions from, the trends in physical and chemical properties of the elements and their compounds

10.1e) state the variation in the solubilities of the hydroxides and sulfates

10.2a) describe and explain the use of calcium hydroxide and calcium carbonate (powdered limestone) in agriculture.

For nitrogen and sulfur, you should be able to:

13.1a) explain the lack of reactivity of nitrogen

13.1b) describe and explain the basicity of ammonia, the structure of the ammonium ion and its formation by an acid–base reaction, and the displacement of ammonia from its salts

10.1 Background

150 years ago, when about 50 elements had been discovered, attempts were made to find patterns in their properties. The following three attempts were particularly important.

- In 1829, the German Johann Döbereiner pointed out that some similar elements formed groups of three (which he called 'triads'). The middle element showed properties intermediate between those of the other two. Examples of these triads included lithium, sodium and potassium; calcium, strontium and barium; and chlorine, bromine and iodine.

- In 1864, the Englishman John Newlands showed that if the elements are arranged in order of their relative atomic masses, many show properties similar to the element that is eight places further on. He called this the Law of Octaves, after the eight white notes on a piano which make up an octave. Examples include sodium and potassium; magnesium and calcium; boron and aluminium; oxygen and sulfur; nitrogen and phosphorus; and fluorine and chlorine.

- The first periodic table was independently produced by the German Lothar Meyer and by the Russian Dmitri Mendeleev in 1870. Lothar Meyer was the first to put his ideas down on paper, but publication of his manuscript was delayed for two years, and so Mendeleev is often given sole credit.

Two features of Mendeleev's classification were particularly significant.

- He left gaps in his table for elements which were unknown at that time, and he successfully predicted the existence of the elements scandium, gallium and germanium, all of which were discovered a few years later. He used Sanskrit prefixes to identify his 'missing elements': for example, his name for germanium was eka-silicon, meaning 'one place below silicon' in his table.

- He reversed the order of cobalt and nickel, and of tellurium and iodine, from the order shown by their relative atomic masses. Subsequent work showed that this reversal was correct. He had put the elements in the order of their proton numbers; the fact that their relative atomic masses were in the wrong order was due to the proportions of isotopes in the elements (see section 2.4).

Figure 10.1 Dmitri Mendeleev (1834–1907) collected information about the elements and wrote them down on cards, which he arranged according to relative atomic mass and to the properties of the elements. Along with Lothar Meyer, he formed the Periodic Table we use today.

10.2 A modern form of the Periodic Table

Groups and periods

A modern form of the periodic table is shown in Figure 10.2.

Elements placed under one another in a column are in the same **group** and show many similarities in their physical and chemical properties. Elements are also arranged

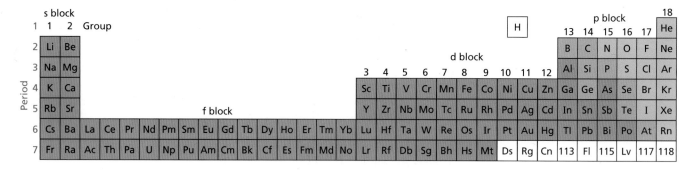

Figure 10.2 A modern form of the Periodic Table. The elements shown in red are good conductors of electricity, and the ones shown in blue are poor conductors. The pink shading indicates that the principal form of the element is a semiconductor. As only a few atoms of the elements at the end of the Periodic Table have been made, these have been left white.

in rows or **periods** across the table. The elements in a period have different physical and chemical properties, but trends become apparent in these properties as we move across a period.

The s, p, d and f blocks

The first two groups form the **s block** (Groups 1 and 2). These have the outer electronic configurations ns^1 and ns^2, respectively, where n is the number of the shell. The last six groups form the **p block** (Groups 13 to 18), in which the p sub-shell is being progressively filled. These elements have the outer electronic configurations ns^2np^1 to ns^2np^6.

In between the s and p blocks is the **d block** (Groups 3–12), in which the d sub-shell is being progressively filled. Our study of the d block is largely restricted to the elements scandium to zinc, and these elements have the outer electronic configurations $4s^23d^1$ to $4s^23d^{10}$, as we shall see in Topic 24.

About a quarter of the known elements belong to the **f block**. The first row, the 4f, includes all the elements from cerium (Ce) to lutetium (Lu) inclusive. In this block, the f sub-shell in the fourth principal shell is being progressively filled. In the past, these elements were called the 'rare earths'. This was a misnomer, as their abundance in the Earth's crust is fairly large – for some it is comparable to that of lead. However, it is true that they are mostly quite thinly spread, so mining them is quite expensive. They are now called the **lanthanoids**, as their properties are similar to those of the element lanthanum (La) that precedes them in the Periodic Table. Although they are not particularly rare, the lanthanoids are difficult to purify from one another because their chemical properties are nearly identical. The elements of the second row of the f block, the 5f, are called the **actinoids**. All the actinoids are radioactive, and most have to be made artificially.

10.3 Periodic trends in the elements of the third period (sodium to argon)

In this section we shall look at the trends in the properties of the elements and their compounds in the third period of the Periodic Table.

Appearance

The elements on the left of the Periodic Table have low values of ionisation energy and electronegativity, and so they show the properties associated with metallic bonding (see section 4.11), for example, they are shiny and conduct electricity. In the middle of the Periodic Table, elements with higher values of ionisation energy and electronegativity are semiconductors: they have a dull shine to them and are poor conductors of electricity (typically 10^{-12} times that of a metal). The elements on the right of the Periodic Table, with the highest values of ionisation energy and electronegativity, are dull in appearance and are such poor conductors that they are used as electrical insulators (their conductivities are only about 10^{-18} times that of a metal).

	Na	Mg	Al	Si	P	S	Cl	Ar
Melting point/°C	98	649	660	1410	44	113	−101	−189
First ionisation energy/kJ mol^{-1}	494	736	577	786	1060	1000	1260	1520
Electronegativity	0.9	1.2	1.5	1.8	2.1	2.5	3.0	—
Electrical conductivity/S cm^{-1}	2.1×10^5	2.3×10^5	3.8×10^5	2.5×10^{-6}	1.0×10^{-11}	5.0×10^{-18}	~0	~0
Atomic radius/nm	0.186	0.160	0.143	0.117	0.110	0.104	0.099	(0.190)

Table 10.1 Some physical properties of the elements in the third period

Interatomic and interionic radii

The term 'atomic radius' does not mean the same thing for every element. Although we can fairly easily measure the distance between two adjacent nuclei in an element (see Figure 10.3), if we are to use those distances to compare one element with another, we must be aware that the interatomic bonding might differ in the two elements. For example, tables of interatomic radii include three values for sodium: its covalent radius is 0.154 nm; its metallic radius is 0.186 nm, and its van der Waals' radius is 0.230 nm. The only radius listed for argon is its van der Waals' radius (0.190 nm), since it does not form compounds. If we are to compare the sizes of the atoms in a sensible way, we need to compare radii of the same type. Comparing the van der Waals' radii of the two elements shows that the sodium atom (0.230 nm) is larger than the argon atom (0.190 nm), as we might expect, since the nuclear attraction for the outer electron shell increases with proton number across the group.

Figure 10.3 The atomic radius is half the internuclear distance.

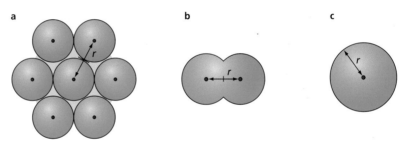

on the left – metallic radius   on the right – covalent radius   for Group 18 – van der Waals' radius

In crossing the third period of the Periodic Table, the proton number increases from 11 to 18. This means that the 3s and 3p outer electrons become more firmly attracted to the nucleus and as a result their distance away from the nucleus becomes less. This is the reason why the atomic radii decrease from sodium to chlorine (see Figure 10.4). As mentioned above, the somewhat larger value of 0.190 nm listed for argon in Table 10.1 is the van der Waals' radius, which is half the distance between the two nuclei when the atoms touch one another in the solid state.

Sodium, magnesium and aluminium are metals and form Na^+, Mg^{2+} and Al^{3+} ions. These ions all have the electronic structure $1s^2 2s^2 2p^6$. As the proton number increases from 11 with sodium to 13 with aluminium, the attraction for the outer electrons increases and they become drawn in closer to the nucleus. This means that the ionic radii get smaller (Na^+ 0.095 nm, Mg^{2+} 0.065 nm, Al^{3+} 0.050 nm).

Melting points

The melting point of an element that has a giant structure is high because many interatomic bonds must be broken for melting to take place. The atomic radii become smaller from sodium to silicon and so the bonding becomes stronger and the melting points become higher, as the bonding electrons are closer to the adjacent nuclei.

Figure 10.4 Atomic radius and proton number

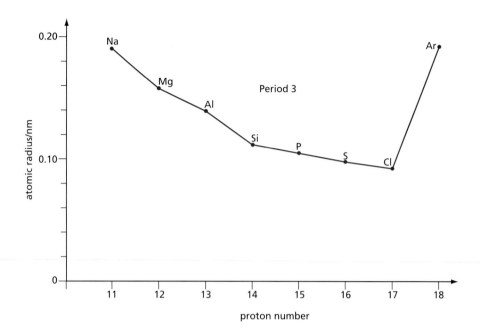

Phosphorus, sulfur and chlorine all form small covalent molecules, P_4, S_8 and Cl_2 respectively. When these substances melt, it is only necessary to break weak intermolecular bonds and not strong interatomic attractions. The melting points decrease in the order sulfur > phosphorus > chlorine (see Figure 10.5), the intermolecular bonds becoming weaker as the molecules become smaller.

The melting point of argon is low, as the attraction between the argon atoms is very small.

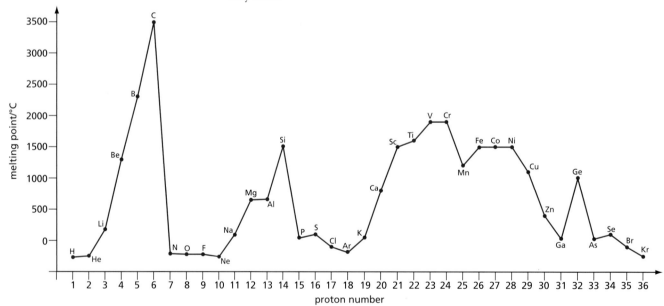

Figure 10.5 Variation of melting point with proton number

Electrical conductivity

Sodium, magnesium and aluminium are metals. They have delocalised electrons that are free to move in the lattice of cations (see section 4.11). Silicon is a semiconductor. The other elements in the third period form covalent bonds with no free electrons, and so are insulators with almost no electrical conductivity.

Ionisation energies

The first ionisation energies generally increase from sodium to argon as the proton number increases. In two instances, this increase does not take place: between magnesium and aluminium, and between phosphorus and sulfur (see Figure 10.6).

Figure 10.6 Variation of first ionisation energy with proton number

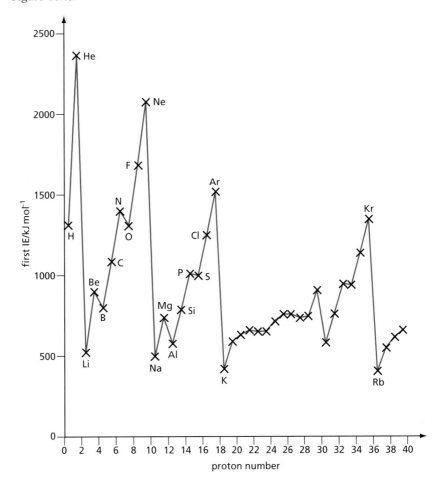

Now try this

1 The two elements silicon and germanium have higher melting points than their neighbours (see Figure 10.5).
  a Suggest the bonding and structure in germanium.
  b Would you expect germanium to be a conductor, a semiconductor or an insulator?
2 Suggest why the melting point of selenium (proton number 34) is higher than that of arsenic (proton number 33).

Worked example

Suggest why:
a the ionisation energy of aluminium is less than that of magnesium
b the ionisation energy of sulfur is less than that of phosphorus.

Answer
a This is because the outer electron in aluminium is in the 3p orbital, and so further from the nucleus and less strongly held than the outer electron in magnesium, which is in the 3s orbital.
b This is because the electron that is being removed from sulfur comes from a doubly-occupied orbital, in which it suffers inter-electron repulsion from the other electron occupying that orbital.

The reactions of the elements with oxygen, chlorine and water

Apart from chlorine itself and argon, all the elements in the third period react directly and exothermically with oxygen and with chlorine, but they often need strong initial heating to overcome the energy barrier to reaction.

with oxygen

$$2Na(l) + \tfrac{1}{2}O_2(g) \rightarrow Na_2O(s)$$

$$Mg(s) + \tfrac{1}{2}O_2(g) \rightarrow MgO(s)$$

$$2Al(s) + 1\tfrac{1}{2}O_2(g) \rightarrow Al_2O_3(s)$$

$$Si(s) + O_2(g) \rightarrow SiO_2(s)$$

$$P_4(l) + 3O_2(g) \xrightarrow{\text{limited oxygen}} P_4O_6(s)$$

$$P_4(l) + 5O_2(g) \rightarrow P_4O_{10}(s)$$
$$S(l) + O_2(g) \rightarrow SO_2(g)$$

$$SO_2(g) + \tfrac{1}{2}O_2(g) \xrightarrow{\text{V}_2\text{O}_5 \text{ catalyst}} SO_3(g)$$

with chlorine

$$Na(l) + \tfrac{1}{2}Cl_2(g) \rightarrow NaCl(s)$$

$$Mg(s) + Cl_2(g) \rightarrow MgCl_2(s)$$

$$2Al(s) + 3Cl_2(g) \rightarrow Al_2Cl_6(s)$$

$$Si(s) + 2Cl_2(g) \rightarrow SiCl_4(l)$$

$$P_4(l) + 6Cl_2(g) \xrightarrow{\text{limited chlorine}} 4PCl_3(l)$$

$$P_4(l) + 10Cl_2(g) \rightarrow 4PCl_5(s)$$

The oxidation numbers of the elements in their oxides and chlorides are equal to the numbers of electrons in the $n = 3$ shell. For the ionic compounds, Na_2O, NaCl, MgO, $MgCl_2$ and Al_2O_3, the metals have formed Na^+, Mg^{2+} and Al^{3+} ions by loss of all the outer electrons. The other oxides and chlorides are covalent. In these covalent compounds, the oxidation number of the element can be found by giving chlorine an oxidation number of −1 and oxygen an oxidation number of −2. With the exceptions of SO_2 and PCl_3, the oxidation number of the element is the same as the number of electrons in its outer shell. This is because all of the electrons are used to make bonds. In SO_2, two of the electrons of the sulfur atom form a lone pair, and only four are used in bonding to oxygen. This then gives sulfur an oxidation number of +4.

Electronegativity increases across the period (see Table 10.1); the difference in electronegativity between the Period 3 element and oxygen or chlorine is high at the left-hand side and gets less and less going across the period. For Na_2O, NaCl, MgO, $MgCl_2$ and Al_2O_3, the electronegativity difference is high and so the salts are ionic; they have high melting points and in the molten state they conduct electricity. The electronegativity differences between Al, Si, P and S and oxygen, and between Si, P and S and chlorine, are low, which suggests covalent bonds. The low melting points of the oxides and chlorides suggest small covalent molecules:

- O_2 is a gas, SO_3 is a low melting point solid and P_4O_{10} sublimes when heated
- l_2Cl_6 readily sublimes, $SiCl_4$ is a liquid and PCl_5 decomposes on melting to give PCl_3, which is a liquid.

Only sodium and magnesium react with water. Sodium reacts with water in the cold, but magnesium needs to be red hot to react with steam:

$$Na(l) + H_2O(l) \rightarrow NaOH(aq) + \tfrac{1}{2}H_2(g)$$

$$Mg(s) + H_2O(g) \rightarrow MgO(s) + H_2(g)$$

Magnesium reacts very slowly with cold water: it takes several hours to collect a test tube of hydrogen.

The reactions of the oxides with water

When an oxide of an element M reacts with water, an M—O—H bond is formed. This bond may ionise in two ways:

$$M\text{—}O\text{—}H \rightarrow M^+ + OH^- \quad \text{or} \quad M\text{—}O\text{—}H \rightarrow MO^- + H^+$$

In the first case the M—O—H bond is behaving as a source of basic hydroxide ion; in the second case, it is behaving as a proton donor – an acid. If the element M has a low electronegativity (<1.5), the first reaction takes place and the hydroxide is basic. If M has a high electronegativity (>2.1), the second reaction takes place and the hydroxide behaves as an acid. If M has an electronegativity between 1.5 and 2.1, both reactions are possible and the hydroxide is amphoteric.

Sodium oxide (electronegativity 0.7) easily dissolves in water to give an alkaline solution:

$$Na_2O(s) + H_2O(l) \rightarrow 2Na^+OH^-(aq)$$

Magnesium oxide (electronegativity 1.2) dissolves slightly in water to give a solution that is just alkaline:

$$MgO(s) + H_2O(l) \rightarrow Mg^{2+}(OH^-)_2(aq)$$

Aluminium oxide does not dissolve in water, but its hydroxide (electronegativity 1.5) is amphoteric: it dissolves in both acids and alkalis.

$$Al(OH)_3(s) + 3H^+(aq) \rightarrow Al^{3+}(aq) + 3H_2O(l)$$
$$Al(OH)_3(s) + OH^-(aq) \rightarrow Al(OH)_4^-(aq)$$

SiO_2 is a weakly acidic oxide. It is the starting material for glasses and ceramics because, when molten, it reacts with bases such as sodium carbonate and calcium carbonate.

P_4O_{10} and SO_2 (electronegativities 2.1 and 2.5) react with water to form the weak acids, H_3PO_4 and H_2SO_3, while SO_3 forms the strong acid, H_2SO_4:

$$P_4O_{10}(s) + 6H_2O(l) \rightarrow 4H_3PO_4(aq)$$

$$SO_2(g) + H_2O(l) \rightarrow H_2SO_3(aq)$$
$$SO_3(g) + H_2O(l) \rightarrow H_2SO_4(aq)$$

10.4 Group 2

Properties of the elements and general trends

The elements of Group 2 show typical metallic behaviour. They form compounds containing M^{2+} ions, and the reactivity of the metals increases down the group.

Table 10.2 Some physical properties of the elements in Group 2

	Mg	Ca	Sr	Ba
Melting point/°C	649	839	769	729
Boiling point/°C	1090	1484	1384	1637
First ionisation energy/kJ mol⁻¹	736	590	548	502
Second ionisation energy/kJ mol⁻¹	1450	1150	1060	966
Ionic radius of M²⁺ ion/nm	0.065	0.099	0.113	0.135

On going from calcium to barium, the first and second ionisation energies decrease. This is because the outer electrons become further away from the nucleus and are less firmly attracted to it. Thus the reactivity of the metals increases from calcium to barium, as shown by their reactions with oxygen and water.

Reactions with oxygen

When heated, the Group 2 metals catch fire and form white oxides. For example, magnesium gives MgO:

$$2Mg(l) + O_2(g) \rightarrow 2MgO(s)$$

Of all the metals, magnesium catches fire the most easily as it has the lowest melting point. It burns with a brilliant white light. Calcium burns with a brick-red flame, strontium burns with a crimson flame and barium with a green flame (see section 19.6).

Reactions with water

As mentioned above, magnesium reacts very slowly with water in the cold to give the oxide and hydrogen. In steam at red heat, the reaction is rapid:

$$Mg(s) + H_2O(g) \rightarrow MgO(s) + H_2(g)$$

It does not give the hydroxide, as magnesium oxide is almost insoluble in water. Because their oxides are soluble in water, calcium, strontium and barium all give the hydroxide and hydrogen. Calcium reacts slowly, strontium more quickly and barium even more quickly:

$$Ba(s) + 2H_2O(l) \rightarrow Ba(OH)_2 + H_2(g)$$

Magnesium oxide is only slightly soluble in water. This means that when it is shaken with water, the pH of the resulting solution is just above 7. However, calcium oxide, CaO, strontium oxide, SrO, and barium oxide, BaO, are all readily soluble in water. When they are shaken with water, they dissolve and the solution has a high pH (>13). For example:

$$CaO(s) + H_2O(l) \rightarrow Ca^{2+}(aq) + 2OH^-(aq)$$

Now try this

Write a balanced equation for each of the following reactions.

1 calcium + oxygen
2 strontium + water

Thermal decomposition of the carbonates and nitrates

On heating, the Group 2 carbonates decompose to give the oxide and carbon dioxide. For example, calcium carbonate decomposes as follows:

$$CaCO_3(s) \rightarrow CaO(s) + CO_2(g)$$

On going down the group, the carbonates become more stable to heat: magnesium carbonate decomposes at 500 °C, and so can be decomposed in the heat of a Bunsen flame, but barium carbonate needs a temperature of over 1000 °C before much carbon dioxide is produced.

The Group 2 nitrates also decompose on heating, to give the oxide, nitrogen dioxide (a brown gas) and oxygen. For example, magnesium nitrate decomposes as follows:

$$2Mg(NO_3)_2(s) \rightarrow 2MgO(s) + 4NO_2(g) + O_2(g)$$

Like the carbonates, a higher temperature is needed to decompose the nitrates of the metals at the bottom of the group compared to those at the top, although all can be decomposed using a Bunsen burner.

Solubility of sulfates and hydroxides

The solubility of the sulfates decreases as the proton number increases, that is they are in the order $MgSO_4 > CaSO_4 > SrSO_4 > BaSO_4$. The solubility of the hydroxides changes in the opposite direction, that is $Mg(OH_2) < Ca(OH_2) < Sr(OH_2) < Ba(OH_2)$.

Uses of Group 2 compounds

Economically, the most important compound of a Group 2 element is calcium carbonate. Large quantities of it are found in the form of limestone, a soft rock which is attractive and is easily carved and decorated. The rock is widely used as a building material although it is readily attacked by acids and dissolves in polluted air, which often contains nitric acid:

$$CaCO_3(s) + 2HNO_3(aq) \rightarrow Ca(NO_3)_2(aq) + H_2O(l)$$

As mentioned above, heating calcium carbonate, in the form of limestone or chalk, breaks it down into calcium oxide (called quicklime) and carbon dioxide, as follows:

$$CaCO_3(s) \rightarrow CaO(s) + CO_2(g)$$

On standing in moist air, the quicklime takes up water and is converted into calcium hydroxide (slaked lime):

$$CaO(s) + H_2O(l) \rightarrow Ca(OH)_2(s)$$

Slaked lime is the main constituent of simple mortar, which has been used since ancient times to bind stones and bricks together in buildings. On standing, calcium hydroxide slowly absorbs carbon dioxide from the air and is converted into calcium carbonate. As calcium carbonate takes up a larger volume than calcium hydroxide, the resulting expansion binds the stones and bricks together.

$$Ca(OH)_2(s) + CO_2(g) \rightarrow CaCO_3(s) + H_2O(l)$$

This reaction forms the basis for the limewater test for carbon dioxide (see section 19.5).

The various forms of lime are used to neutralise acidic soils. The addition of slaked lime rapidly neutralises the soil, but if too much is added the soil may become alkaline. Calcium carbonate, in the form of powdered limestone or chalk, acts more slowly; even an excess will leave the soil approximately neutral.

> **Now try this**
>
> Write an equation for:
>
> **a** the reaction between calcium carbonate and an acid in soil
> **b** the reaction between carbon dioxide and barium hydroxide solution.

10.5 Nitrogen

Unreactivity

Nitrogen makes up 78% by volume of dry air. The element is very unreactive, for two reasons. Firstly, the N≡N bond is very strong ($945\,kJ\,mol^{-1}$) and so its reactions with other elements are usually endothermic. For example, with oxygen:

$$N_2(g) + O_2(g) \rightarrow 2NO(g) \qquad \Delta H = +181\,kJ\,mol^{-1}$$

Secondly, the strong N≡N bond means that reactions involving nitrogen have a very high activation energy, and as a result will be very slow even if they are exothermic; this is the reason why the Haber process to make ammonia requires such extreme conditions.

Ammonia

Nitrogen combines with hydrogen to make ammonia, NH_3:

$$N_2(g) + 3H_2(g) \rightarrow 2NH_3(g) \qquad \Delta H = -92\,kJ\,mol^{-1}$$

The ammonia molecule is a trigonal pyramid in shape. There are four pairs of electrons around the nitrogen atom – three bonding pairs and one lone pair. The presence of a lone pair makes the H—N—H angle slightly less than the tetrahedral angle of 109.5°; the actual value is 107° (see Figure 10.7).

Figure 10.7 Lone pair–bonding pair repulsion makes the bond angle in ammonia 107°.

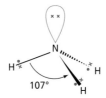

The lone pair easily combines with a proton, making ammonia a weak base. The resulting ammonium ion is exactly tetrahedral. It can be thought of as being formed either by the lone pair of electrons from the nitrogen being donated to the proton (see Figure 10.8a), or by an N^+ ion (isoelectronic with carbon) forming four bonds with hydrogen atoms (see Figure 10.8b).

Figure 10.8 The ammonium ion can either be regarded as being formed **a** by the donation of a pair of electrons from the nitrogen to a proton or **b** by combining four hydrogen atoms with an N^+ ion.

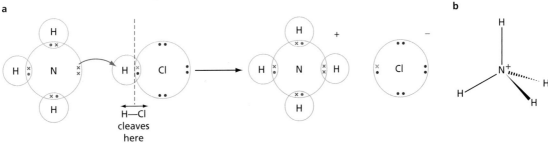

The ammonium ion reacts with strong bases such as NaOH to give ammonia, an alkaline gas that turns moist red litmus paper blue.

$$NH_4^+(s) + OH^-(aq) \rightarrow NH_3(g) + H_2O(l)$$

This is a test for ammonium salts.

The Haber process

Even though the synthesis of ammonia is exothermic, it is a very slow reaction. There are three ways in which the rate can be increased: an increase in pressure, an increase in temperature and the use of a catalyst. All three are used in the Haber process. A typical plant uses a pressure of 250 atmospheres, a temperature of 400 °C and an iron catalyst, promoted with traces of other metals (see section 8.6). A lower temperature would increase the percentage of ammonia in the equilibrium position (remember, the reaction is exothermic) but in practice a low temperature would make the reaction too slow to be economic. This is why a compromise temperature of 400 °C is used. At this temperature, a yield of 15% is obtained in a relatively short time; the ammonia is then removed and the gases continuously recycled to produce further conversion. Although a higher pressure would produce a greater yield of ammonia, and at a faster rate, this high pressure is expensive to produce and maintain (extra-thick pipes and reactor walls would be needed), and make for a less safe process.

The nitrogen is obtained from the air and the hydrogen from the reaction between methane and water over a metal catalyst (usually nickel). The overall reactions may be approximately represented by the following equations:

$$CH_4(g) + 2H_2O(g) \rightarrow CO_2(g) + 4H_2(g)$$
$$2O_2(g) \text{ (plus } N_2(g)) + CH_4(g) \rightarrow CO_2(g) + 2H_2O(g) \text{ (plus } N_2(g))$$

The first reaction is endothermic; the second reaction is exothermic. The heat evolved during the second reaction can be used to heat the catalyst in the first reaction. The two reactions are adjusted so that the resulting gas (after the carbon dioxide has been removed) contains approximately one volume of nitrogen to three volumes of hydrogen.

> ### Now try this
>
> If air is four-fifths nitrogen and one-fifth oxygen, balance the following combined equation to show the production of N_2 and H_2 in a 1:3 ratio:
>
> $$O_2 + 4N_2 + ...CH_4 + ...H_2O \rightarrow$$
> $$...CO_2 + ...H_2 + ...N_2$$

Uses of ammonia

Ammonia forms the basis of nitrogen fertilisers and of explosives.

- In acidic soils, ammonia is injected directly into the ground; in other soils, it is used in the form of ammonium sulfate, ammonium nitrate or urea.
- Ammonia can be catalytically oxidised by air to give NO and NO_2, which are dissolved in water to give nitric acid, HNO_3.
- Nitric acid is used to manufacture the fertiliser ammonium nitrate and to make explosives, such as TNT, dynamite and cordite.

10.6 Sulfuric acid and sulfur dioxide

The principal use of sulfur dioxide is to make sulfuric acid. This process has three stages: the first is burning sulfur in air at 1000 °C to give sulfur dioxide, the second is the oxidation of sulfur dioxide to sulfur trioxide, in a reaction known as the **Contact process**, and the third is dissolving the sulfur trioxide in water to give sulfuric acid:

$$S(s) + O_2(g) \rightarrow SO_2(g) \qquad \Delta H = -308\,\text{kJ mol}^{-1}$$

$$SO_2(g) + \tfrac{1}{2}O_2(g) \rightleftharpoons SO_3(g) \qquad \Delta H = -96\,\text{kJ mol}^{-1}$$

$$SO_3(g) + H_2O(l) \rightarrow H_2SO_4(l) \qquad \Delta H = -130\,\text{kJ mol}^{-1}$$

The second stage is reversible and relatively slow. To speed up the reaction, a catalyst of vanadium(V) oxide, V_2O_5, is used and the equilibrium is displaced to the right by using about three times more air than is needed by the equation. Le Chatelier's Principle also indicates that the yield would be improved by increasing the pressure and using a low temperature. In practice, a temperature of at least 400 °C must be used, as the catalyst becomes ineffective at temperatures lower than this, The pressure is only slightly above atmospheric because the yield is already high, and the extra expense of high pressure would not be economically worthwhile. The sulfur trioxide is not dissolved in water directly, because this reaction is so exothermic that the water would evaporate and a difficult-to-condense mist of concentrated sulfuric acid would be formed. In practice, the sulfur trioxide is dissolved in concentrated sulfuric acid, and then water is added with cooling, to keep the concentration constant:

$$SO_3(g) + H_2SO_4(l) \rightarrow H_2S_2O_7(l) \quad \text{then} \quad H_2S_2O_7(l) + H_2O(l) \rightarrow 2H_2SO_4(l)$$

The main uses of sulfuric acid are in the manufacture of:

- paints and pigments
- detergents and soaps
- dyestuffs.

Acid rain

Sources of sulfur dioxide and nitrogen oxides

Sulfur dioxide is produced whenever sulfur-containing compounds are burnt. Three-quarters of it is produced naturally (see Table 10.3 and Figure 10.9).

Table 10.3 Amounts of sulfur dioxide produced globally from different sources. Dimethyl sulfide is produced from organisms in the sea, and is readily oxidised to sulfur dioxide.

Source	Mass of sulfur dioxide produced/million tonnes per year
volcanoes	20–40
dimethyl sulfide	30
power stations	16
cars	<1

Figure 10.9 As well as the physical destruction caused by a volcanic eruption, huge amounts of sulfur dioxide are also given out into the atmosphere.

Nitrogen combines with oxygen only at high temperatures or in the presence of sparks, such as lightning (Figure 10.10). Initially nitrogen monoxide is formed, but this reacts slowly with oxygen to give nitrogen dioxide. The mixture of nitrogen oxides is often written as NOx to indicate its variable composition. Most of this is produced by human activity (see Table 10.4).

Table 10.4 Amounts of NOx produced globally from different sources

Source	Mass of NOx produced/million tonnes per year
thunderstorms	10
power stations	8
cars	6
industry	6

Pollution and acid rain

The presence of dry sulfur dioxide and nitrogen dioxide gases in the atmosphere is harmful. The principal adverse effects are respiratory problems (particularly in babies, the elderly and bronchitis sufferers) and damage to trees and other vegetation. Nitrogen dioxide also catalyses the formation of ozone, which is a dangerous pollutant at ground level.

In moist air, sulfur dioxide is readily oxidised to give the strong acid sulfuric acid:

$$SO_2(g) + \tfrac{1}{2}O_2(g) + H_2O(l) \rightarrow H_2SO_4(aq)$$

Figure 10.10 Lightning provides the energy to turn nitrogen and oxygen in the atmosphere to nitrogen oxides.

Nitrogen dioxide dissolves in water to give a mixture of nitrous and nitric acids:

$$2NO_2(g) + H_2O(l) \rightarrow HNO_2(aq) + HNO_3(aq)$$

Nitric and sulfuric acids are strongly acidic. When present in rainwater they produce **acid rain**. This is rainfall with a pH lower than 5. Nitrogen dioxide may exacerbate acid-rain formation from sulfur dioxide by catalysing its oxidation to sulfur trioxide:

$$SO_2(g) + NO_2(g) \rightarrow SO_3(g) + NO(g)$$

$$NO(g) + \tfrac{1}{2}O_2(g) \rightarrow NO_2(g)$$

Acid rain causes the following problems.

- It causes aluminium ions in the soil to dissolve. A high concentration of these ions is poisonous to most forms of aquatic life, in particular fish. However, algae flourish under these conditions, particularly if the water also contains high concentrations of phosphate and nitrate ions as a result of excessive use of fertilisers. When the algae die, their decomposition uses up all the oxygen in the water, so that it can no longer support any form of life. The water is said to have undergone **eutrophication.**
- The stonework of buildings is attacked by the acids (see Figure 10.11). Much renovation work has been necessary on old buildings recently because of pollution generated over only the last hundred years.

The harmful effects of acid rain pollution can be minimised by taking the following precautions:

- removing sulfur dioxide from the emissions of power stations as they pass up the chimney stacks
- removing sulfur from diesel fuel and petrol
- using catalytic converters in cars to destroy NOx (see section 8.6).

Figure 10.11 The effects of acid rain on a statue

Summary

- All the Group 2 elements are metals.
- Their atomic (metallic) and ionic radii increase down the group.
- Their ionisation energies and electronegativities decrease down the group.
- Group 2 metals form M^{2+} ions.
- Their nitrates and carbonates become thermally more stable at the bottom of the group.
- Nitrogen is a very unreactive gas and special conditions are required to make ammonia by the **Haber process**.
- Nitrogen compounds are widely used as fertilisers and in the manufacture of explosives.
- Sulfur is used to make sulfuric acid by the **Contact process**.
- Sulfuric acid is used in the manufacture of a wide variety of substances.
- The oxides of nitrogen and sulfur cause **acid rain** which is harmful to life, and attacks buildings.

Key reactions you should know

- Elements with oxygen:

$$2Na(l) + \tfrac{1}{2}O_2(g) \rightarrow Na_2O(s)$$

$$M(s) + \tfrac{1}{2}O_2(g) \rightarrow MO(s) \qquad M = Mg, Ca, Sr, Ba$$

$$2Al(s) + 1\tfrac{1}{2}O_2(g) \rightarrow Al_2O_3(s)$$

$$Si(s) + O_2(g) \rightarrow SiO_2(s)$$

$$P_4(l) + 5O_2(g) \rightarrow P_4O_{10}(s)$$

$$S(l) + O_2(g) \xrightarrow{\text{limited supply of oxygen}} SO_2(g)$$

$$S(l) + 1\tfrac{1}{2}O_2(g) \xrightarrow{\text{excess oxygen}} SO_3(g)$$

$$SO_2(g) + \tfrac{1}{2}O_2(g) \xrightarrow[\text{pressure} > 1\,\text{atm}]{400\,°C,\ V_2O_5\ \text{catalyst}} SO_3(g)$$

- Elements with chlorine:

$$Na(l) + \tfrac{1}{2}Cl_2(g) \rightarrow NaCl(s)$$

$$M(s) + Cl_2(g) \rightarrow MCl_2(s) \qquad M = Mg, Ca, Sr, Ba\ (Cl_2\ \text{could be replaced by}\ Br_2\ \text{or}\ I_2.)$$

$$2Al(s) + 3Cl_2(g) \rightarrow Al_2Cl_6(s)$$

$$Si(s) + 2Cl_2(g) \rightarrow SiCl_4(l)$$

$$P_4(l) + 10Cl_2(g) \rightarrow 4PCl_5(s)$$

- Elements with water:

$$Na(l) + H_2O(l) \rightarrow NaOH(aq) + \tfrac{1}{2}H_2(g)$$

$$M(s) + 2H_2O(l) \rightarrow M(OH)_2(s) + H_2(g) \qquad M = Mg, Ca, Sr, Ba\ (Ba\ gives\ Ba(OH)_2(aq).)$$

$$Cl_2(aq) + H_2O(l) \rightleftharpoons HCl(aq) + HClO(aq)$$

- Oxides and hydroxides with water or acid:

$$Na_2O(s) + H_2O(l) \rightarrow 2NaOH(aq)$$

$$NaOH(aq) + H^+(aq) \rightarrow Na^+(aq) + H_2O(l)$$

$$MO(s) + H_2O(l) \rightarrow M(OH)_2(aq) \qquad M = Ca, Sr, Ba$$

$$M(OH)_2(s) + 2H^+(aq) \rightarrow M^{2+}(aq) + H_2O(l) \qquad M = Mg, Ca, Sr, Ba$$

$$Al(OH)_3(s) + 3H^+(aq) \rightarrow Al^{3+}(aq) + 3H_2O(l)$$
(Also $Al(OH)_3(s) + OH^-(aq) \rightarrow Al(OH)_4^-(aq)$)

$$P_4O_{10}(s) + 6H_2O(l) \rightarrow 4H_3PO_4(aq)$$

$$SO_2(g) + H_2O(l) \rightarrow H_2SO_3(aq)$$

$$SO_3(g) + H_2O(l) \rightarrow H_2SO_4(aq)$$

- Carbon dioxide with Group 2 hydroxides:

$$M(OH)_2(aq) + CO_2(g) \rightarrow MCO_3(s) + H_2O(l) \qquad M = Mg, Ca, Sr, Ba$$

$$MCO_3(s) + CO_2(g) + H_2O(l) \rightarrow M(HCO_3)_2(aq) \qquad M = Mg, Ca\ (in\ hard\ water)$$

- Heat on Group 2 carbonates and nitrates:

$$MCO_3(s) \rightarrow MO(s) + CO_2(g) \qquad M = Mg, Ca, Sr, Ba\ (more\ stable\ down\ the\ group)$$

$$M(NO_3)_2(s) \rightarrow MO(s) + 2NO_2(g) + \tfrac{1}{2}O_2(g) \qquad M = Mg, Ca, Sr, Ba\ (more\ stable\ down\ the\ group)$$

- The Haber process:

$$N_2(g) + 3H_2(g) \xrightarrow[\text{Fe catalyst}]{400\,°C,\ 250\,\text{atm}} 2NH_3(g)$$

- Acid rain on buildings:

$$CaCO_3(s) + 2HNO_3(aq) \rightarrow Ca(NO_3)_2(aq) + CO_2(g) + H_2O(l)$$

Examination practice questions

Please see the data section of the CD for any A_r values you may need.

									H								He
Li	Be											B	C	N	O	F	Ne
Na	Mg											Al	Si	P	S	Cl	Ar
K	Ca	Sc	Ti	V	Cr	Mn	Fe	Co	Ni	Cu	Zn	Ga	Ge	As	Se	Br	Kr

1 This question refers to the elements shown in the portion of the Periodic Table given above.

  a From this table, identify in **each** case **one** element that has the property described. Give the **symbol** of the element in each case.

    i The element that has a molecule which contains exactly eight atoms.

    ii The element that forms the largest cation.

    iii An element that floats on water and reacts with it.

    iv An element that reacts with water to give a solution that can behave as an oxidising agent.

    v An element whose nitrate gives a brown gas on thermal decomposition. [5]

  b i Give the formula of the oxide of the most electronegative element.

    ii Several of these elements form more than one acidic oxide. Give the formulae of **two** such oxides formed by the **same** element. [3]

The formulae and melting points of the fluorides of the elements in Period 3, Na to Cl, are given in the table.

Formula of fluoride	NaF	MgF_2	AlF_3	SiF_4	PF_5	SF_6	ClF_5
m.p./K	1268	990	1017	183	189	223	170

  c i Suggest the formulae of **two** fluorides that could possibly be ionic.

    ii What is the shape of the SF_6 molecule?

    iii In the sequence of fluorides above, the oxidation number of the elements increases from NaF to SF_6 and then falls at ClF_5.

Attempts to make ClF_7 have failed but IF_7 has been prepared.

Suggest an explanation for the existence of IF_7 and for the non-existence of ClF_7. [4]

[Cambridge International AS & A Level Chemistry 9701, Paper 21 Q3 June 2010]

2 Radium was discovered in the ore pitchblende by Marie and Pierre Curie in 1898, and the metal was first isolated by them in 1910.

The metal was obtained by first reacting the radium present in the pitchblende to form insoluble radium sulfate which was converted into aqueous radium bromide. This solution was then electrolysed using a mercury cathode and a carbon anode.

  a Radium has chemical reactions that are typical of Group 2 metals and forms ionic compounds.

    i What is the characteristic feature of the electronic configurations of all Group 2 metals?

    ii Radium sulfate is extremely insoluble. From your knowledge of the simple salts of Group 2 metals, suggest another very insoluble radium salt. [2]

  b i Describe what you would see when magnesium reacts with cold water, and with steam.

    ii Write an equation for the reaction with steam. [5]

  c Radium reacts vigorously when added to water.

    i Write an equation, with state symbols, for this reaction.

    ii State **two** observations that could be made during this reaction.

iii Suggest the approximate pH of the resulting solution.

iv Will the reaction be more or less vigorous than the reaction of barium with water? Explain your answer. [6]

[Cambridge International AS & A Level Chemistry 9701, Paper 21 Q2 a, c & d November 2009]

3 a Chlorine is very reactive and will form compounds by direct combination with many elements.
Describe what you would see when chlorine is passed over separate heated samples of sodium and phosphorus. In **each** case write an equation for the reaction. [4]

b Magnesium chloride, $MgCl_2$, and silicon tetrachloride, $SiCl_4$, each dissolve in or react with water.
Suggest the approximate pH of the solution formed in **each** case.
Explain, with the aid of an equation, the difference between the two values. [5]

[Cambridge International AS & A Level Chemistry 9701, Paper 2 Q3 d & e November 2008]

11 Group 17

In this topic, we study the trends in the properties of the Group 17 elements, the halogens. They are non-metals and show decreasing reactivity with increasing proton number.

Learning outcomes

By the end of this topic you should be able to:

11.1a) describe the colours and the trend in volatility of chlorine, bromine and iodine

11.1b) interpret the volatility of the elements in terms of van der Waals' forces

11.2a) describe the relative reactivity of the elements as oxidising agents (see also Topic 23)

11.2b) describe and explain the reactions of the elements with hydrogen

11.2c) describe and explain the relative thermal stabilities of the hydrides, and interpret these relative stabilities in terms of bond energies

11.3a) describe and explain the reactions of halide ions with aqueous silver ions followed by aqueous ammonia, and with concentrated sulfuric acid

11.4a) describe and interpret, in terms of changes of oxidation number, the reaction of chlorine with cold, and with hot, aqueous sodium hydroxide

11.5a) explain the use of chlorine in water purification

11.5b) state the industrial importance and environmental significance of the halogens and their compounds (e.g. for bleaches, PVC, halogenated hydrocarbons as solvents, refrigerants and in aerosols) (see also Topic 15).

11.1 Physical properties of the elements

Table 11.1 Some physical properties of the elements in Group 17. The importance of the standard electrode potentials is discussed in Topic 23.

Halogen	Chlorine	Bromine	Iodine
Melting point /°C	−101	−7	114
Boiling point /°C	−34	59	184
Appearance at room temperature	yellow-green gas	red-brown liquid	nearly black solid
$E(\text{X—X})/\text{kJ mol}^{-1}$	242	193	151
$E(\text{H—X})/\text{kJ mol}^{-1}$	431	366	299
Standard electrode potential /V	+1.36	+1.07	+0.54

Melting points and boiling points

The halogens exist as diatomic molecules, X_2. The two atoms are bonded together covalently in the solid, liquid and gaseous states. On melting, the van der Waals' forces between the molecules are partially overcome, and on boiling they are almost completely broken. On descending Group 17, the melting points and boiling points increase. This is because the intermolecular forces become stronger as the number of easily polarisable electrons rises (see section 3.17).

Colour

Each halogen absorbs radiation in the near ultraviolet–visible region of the spectrum. The position of the absorption determines the colour of the halogen (see Table 11.2).

Table 11.2 Colours and approximate wavelengths of the absorption bands in the near ultraviolet and visible regions for the halogens

Halogen	Wavelength absorbed/nm	Colour absorbed	Colour of halogen
chlorine	~350	ultraviolet/violet	yellow-green
bromine	~400	blue	red-brown
iodine	~500	green	purple

The colour of iodine in solution varies with the solvent. In non-polar solvents (for example, tetrachloromethane or hexane), it is the same colour as the gas. In more polar solvents (for example, ethanol), it is brown because the polar solvent affects the wavelength at which maximum absorption takes place.

11.2 Some reactions of the elements

Displacement reactions

The oxidising strength of the halogens is in the order $Cl_2 > Br_2 > I_2$. A halogen that is a more powerful oxidising agent displaces one that is less powerful. So chlorine displaces bromine from solutions of bromide ions, for example:

$$Cl_2(aq) + 2Br^-(aq) \rightarrow 2Cl^-(aq) + Br_2(aq)$$

Worked example

State whether or not a reaction would take place if solutions of the following substances were mixed.
a bromine + iodide ions
b iodine + chloride ions
c chlorine + iodide ions
Write an equation for each reaction that takes place.

Answer
a Yes: $Br_2(aq) + 2I^-(aq) \rightarrow 2Br^-(aq) + I_2(aq)$
b No
c Yes: $Cl_2(aq) + 2I^-(aq) \rightarrow 2Cl^-(aq) + I_2(aq)$

Now try this

What observations would you make during each of the experiments described in the worked example?

Reactions with hydrogen

All the halogens can react with hydrogen, but the conditions for the reaction vary going down the group, as shown in Table 11.3.

Table 11.3 Reaction conditions necessary for the halogens' reactions with hydrogen

Halogen	Reaction conditions	Equation
chlorine	explodes when exposed to ultraviolet light	$H_2(g) + Cl_2(g) \rightarrow 2HCl(g)$
bromine	slow reaction when heated	$H_2(g) + Br_2(g) \rightarrow 2HBr(g)$
iodine	reversible reaction when heated	$H_2(g) + I_2(g) \rightleftharpoons 2HI(g)$

The reactivity with hydrogen decreases down the group because the decreasing strength of the H—X bond (see Table 11.1) makes the reaction less exothermic.

Now try this

Carry out the same calculation for the reaction between hydrogen and bromine gases. Does your result fit the trend described above?

Worked example

Use the bond energies given in Table 11.1, and a value of $436\,kJ\,mol^{-1}$ for the H—H bond energy in $H_2(g)$, to calculate the enthalpy change for the reaction between hydrogen and chlorine, and between hydrogen and iodine.

Answer

For $H_2 + Cl_2 \rightarrow 2HCl$, $\Delta H = 436 + 242 - 2 \times 431 = \mathbf{-184\,kJ\,mol^{-1}}$
For $H_2 + I_2 \rightarrow 2HI$, $\Delta H = 436 + 151 - 2 \times 299 = \mathbf{-11\,kJ\,mol^{-1}}$

Thermal stability of the hydrogen halides

The strength of the H—X bonds also determines the thermal stability of the hydrogen halides. This is readily illustrated by the observations that are made when they are heated (a convenient method in the laboratory is to insert a hot metal wire into a test tube of the gas). All the hydrogen halides are colourless gases. Hydrogen chloride does not decompose on heating, so the test tube remains colourless; hydrogen bromide decomposes slightly and the gas turns brown; hydrogen iodide decomposes almost completely and the gas turns purple (see Table 11.4).

Table 11.4 Results of heating the hydrogen halides

Halogen	Reaction (if any)	Comment
chlorine	$HCl(g)$	no decomposition
bromine	$2HBr(g) \rightleftharpoons H_2(g) + Br_2(g)$	some decomposition and gas appears slightly brown
iodine	$2HI(g) \rightarrow H_2(g) + I_2(g)$	almost complete decomposition and gas appears deep purple

Reactions of chlorine with sodium hydroxide

When chlorine gas is passed into cold, dilute aqueous sodium hydroxide, the following reaction takes place and two salts are formed:

$$Cl_2(g) + 2NaOH(aq) \rightarrow NaCl(aq) + H_2O(l) + NaClO(aq)$$

The ionic equation is:

$$Cl_2(g) + 2OH^-(aq) \rightarrow Cl^-(aq) + H_2O(l) + ClO^-(aq)$$

Chlorine itself, being an element, has the oxidation number 0. In the chloride ion, it has an oxidation number of −1. Oxygen has an oxidation number of −2 so that, in the ClO^- ion, the chlorine must have an oxidation number of +1 in order to give an overall charge of −1 on the ion. The ClO^- ion is therefore called the chlorate(I) ion. The '-ate' indicates that the chlorine is combined with oxygen, and the '(I)' shows its oxidation number. NaClO is properly called sodium chlorate(I) but sometimes its common name, 'sodium hypochlorite', is used.

When chlorine is passed into hot, concentrated aqueous sodium hydroxide, a different reaction takes place:

$$3Cl_2(g) + 6NaOH(aq) \rightarrow 5NaCl(aq) + 3H_2O(l) + NaClO_3(aq)$$

The ionic equation is:

$$3Cl_2(g) + 6OH^-(aq) \rightarrow 5Cl^-(aq) + 3H_2O(l) + ClO_3^-(aq)$$

In the ClO_3^- ion, each oxygen has an oxidation number of −2. The three oxygens give a total of −6. The chlorine must have an oxidation number of +5 to give an overall charge of −1 on the ion. The ion is called the chlorate(V) ion and $NaClO_3$ is called sodium chlorate(V); it is sometimes incorrectly called just 'sodium chlorate'.

These reactions of chlorine with sodium hydroxide are examples of **disproportionation**. This term is applied to a reaction in which one element has been both oxidised and reduced, so that its oxidation number goes up and down at the same time.

The halogens also react directly with hydrocarbons (see sections 13.5 and 14.3).

Now try this

1 Heating sodium chlorate(V) above its melting point causes it to undergo another disproportionation reaction, giving a mixture of NaCl and $NaClO_4$. Use oxidation numbers to write a balanced equation for this reaction.

2 Potassium iodide, KI, and potassium iodate(V), KIO_3, react together in acid solution to form iodine, I_2. Write a balanced ionic equation for this reaction.

11.3 Some reactions of the halide ions

Test with aqueous silver nitrate

Most metal halides are soluble in water, but two important metals that form insoluble halides are silver and lead. If a few drops of aqueous silver nitrate are added to a solution of a halide, a silver halide precipitate is formed. The silver halide precipitates are of different colours and have different solubilities in aqueous ammonia, as shown in Table 11.5.

Table 11.5 Identification of the halide ions using aqueous silver nitrate

Halide ion	Appearance and formula of precipitate	Solubility of precipitate in $NH_3(aq)$
Cl^-	white, AgCl	soluble in dilute $NH_3(aq)$
Br^-	cream, AgBr	soluble in concentrated $NH_3(aq)$
I^-	yellow, AgI	insoluble in concentrated $NH_3(aq)$

Figure 11.1 The colours of the silver halides are used in tests for the halide ions.

Worked example

Write fully balanced, and ionic, equations for the reactions between:

a HCl and $AgNO_3$

b KI and $AgNO_3$

Answer

a $HCl + AgNO_3 \rightarrow AgCl + HNO_3$

   $Cl^-(aq) + Ag^+(aq) \rightarrow AgCl(s)$

b $KI + AgNO_3 \rightarrow AgI + KNO_3$

   $I^-(aq) + Ag^+(aq) \rightarrow AgI(s)$

Now try this

Write fully balanced, and ionic, equations for the reaction between aqueous solutions of magnesium bromide, $MgBr_2$, and silver sulfate, Ag_2SO_4.

The solubilities in water of the precipitates AgCl, AgBr and AgI decrease from AgCl to AgI. This explains why they react differently with aqueous ammonia (see Table 11.5).

Aqueous silver ions react with ammonia to form the ion $Ag(NH_3)_2^+$. By Le Chatelier's Principle, in order to dissolve the silver halide precipitate, enough ammonia has to be present to drive the following two equilibria over to the right-hand side:

$$AgX(s) \rightleftharpoons Ag^+(aq) + X^-(aq)$$
$$Ag^+(aq) + 2NH_3(aq) \rightleftharpoons Ag(NH_3)_2^+(aq)$$

Now try this

Explain in words how the addition of $NH_3(aq)$ can cause AgX(s) to dissolve.

If AgX is very insoluble, the first equilibrium lies well over to the left-hand side, and so a large concentration of $NH_3(aq)$ is needed to pull the Ag^+ ions into solution.

Silver iodide is so insoluble that even in concentrated ammonia solution, the concentration of ammonia is not high enough to pull the Ag^+ ions into solution. For

Figure 11.2 Bromine or iodine are the principal products when concentrated sulfuric acid is added to a solid bromide or iodide.

silver chloride, however, which is 1000 times more soluble than silver iodide, even the small concentration in dilute ammonia is sufficient to drive the equilibria over to the right-hand side.

Reaction with concentrated sulfuric acid

When concentrated sulfuric acid is added to a solid metal halide, the hydrogen halide is produced.

With sodium chloride, steamy fumes of hydrogen chloride are evolved and sodium hydrogensulfate is formed:

$$NaCl(s) + H_2SO_4(l) \rightarrow NaHSO_4(s) + HCl(g)$$

(At room temperature, only one of the two H^+ ions available in H_2SO_4 reacts: heating can cause the second one to react, giving Na_2SO_4.)

Concentrated sulfuric acid is a strong oxidising agent. It is unable to oxidise HCl, but it can oxidise HBr and HI. Hence, only a little HBr is formed when concentrated sulfuric acid is added to sodium bromide, the principal products being bromine and sulfur dioxide (see Figure 11.2):

$$NaBr + H_2SO_4(l) \rightarrow NaHSO_4 + HBr(g) \qquad \text{small yield}$$
$$2HBr(g) + H_2SO_4(l) \rightarrow Br_2(l) + 2H_2O(l) + SO_2(g)$$

Hydrogen iodide is so easily oxidised that only a trace of it is found. The sulfuric acid is reduced not only to sulfur dioxide, but further to sulfur, and even to hydrogen sulfide:

$$2HI(g) + H_2SO_4(l) \rightarrow I_2(s) + 2H_2O(l) + SO_2(g)$$
$$HI(g) + H_2SO_4(l) \rightarrow I_2(s) + 2H_2O(l) + S(s) \qquad \text{(not balanced)}$$
$$HI(g) + H_2SO_4(l) \rightarrow I_2(s) + 2H_2O(l) + H_2S(g) \qquad \text{(not balanced)}$$

These reactions demonstrate that the strength of the hydrogen halides as reducing agents is in the order HI > HBr > HCl. This is due to their bond strengths being in the reverse order, H—I < H—Br < H—Cl (see Table 11.1) and shows the ease with which the bonds are broken to give the free halogens.

Now try this

Use oxidation numbers to help you balance the equations for the two reactions of HI with H_2SO_4 shown above, one giving S(s) and the other giving H_2S.

Worked example

State the changes in oxidation numbers of bromine and sulfur in the reaction between Br_2 and H_2SO_4.

Answer

Bromine changes from Br(0) to Br(−1); sulfur changes from S(+6) to S(+4).

11.4 Manufacture and uses of chlorine

The manufacture of chlorine

Chlorine is an important industrial chemical, and is manufactured by the electrolysis of brine. The other products are hydrogen and sodium hydroxide, both of which are important high-value industrial chemicals:

$$2NaCl(aq) + 2H_2O(l) \rightarrow 2NaOH(aq) + Cl_2(g) + H_2(g)$$

This electrolysis process is described in section 7.5.

Water purification

Before being used for drinking, water is first filtered and then harmful micro-organisms are killed by adding a disinfectant. Chlorine is an effective disinfectant at concentrations as low as one part in a million but higher levels are often used to provide a safety margin. If the level exceeds three parts per million, however, the water then has an unpleasant taste and smell.

The added chlorine dissolves and then reacts with the water to produce hydrochloric acid and chloric(I) acid:

$$Cl_2(aq) + H_2O(l) \rightleftharpoons HCl(aq) + HClO(aq)$$

It is thought that the chloric(I) acid produced is the active disinfectant. As acid has been produced by the above reaction, the pH must be raised by adding alkali (often $Ca(OH)_2$). The optimum value is between pH 7 and pH 8.

Other uses of chlorine

As well as being a disinfectant, chlorine is also a powerful bleach, and is often used in the manufacture of paper and fabrics to whiten them. It must be used with care as it attacks most fabrics and can produce poisonous by-products such as dioxin (see section 25.4). The active bleaching agent is most likely the chlorate(I) ion, and so sodium chlorate(I) is often used instead of free chlorine.

A large amount of chlorine is used to make chlorinated organic compounds (see section 13.5). Examples are the plastic PVC (polyvinyl chloride, properly called polychloroethene), used for water pipes and window frames, and solvents such as dichloromethane, used to remove paint and to decaffeinate coffee. Many of these organic compounds have low boiling points and can be used in aerosols or in refrigerators. One of these is chlorodifluoromethane, HCF_2Cl (boiling point 8.9 °C), which is also an important starting material for the manufacture of tetrafluoroethene. Tetrafluoroethene is used to make PTFE (polytetrafluoroethene), which has many uses, including for coating non-stick cookware and for bridge bearings.

A disadvantage of these chlorohydrocarbons is that they destroy the ozone layer if they are allowed to escape into the atmosphere (see section 13.4). Their use is now restricted and other low boiling point compounds, such as propane and butane, are used instead.

A more complicated halogen compound is halothane, $CF_3CHBrCl$, which is widely used as an anaesthetic.

Summary

- The halogens are non-metals that become less reactive with increasing proton number.
- The properties of chlorine, bromine and iodine are similar, and show a steady trend.
- A good test for halide ions is the reaction with silver nitrate solution, followed by the addition of ammonia.
- The ease of oxidation of the hydrogen halides is in the order HCl < HBr < HI, as shown by their reactions with concentrated H_2SO_4 (see below).
- Chlorine has several oxidation states.
- Chlorine is widely used as a disinfectant and bleaching agent.

Key reactions you should know
- Displacement reactions:
  $Cl_2(aq) + 2Br^-(aq) \rightarrow 2Cl^-(aq) + Br_2(aq)$ Cl_2 displaces Br_2, and Br_2 displaces I_2

- Reaction with hydrogen:
  $X_2(g) + H_2(g) \rightarrow 2HX(g)$ X = Cl, Br, I
- Concentrated sulfuric acid on solid halide:
  $X^-(s) + H_2SO_4(l) \rightarrow HX(g) + HSO_4^-(s)$ X = Cl, Br, I
  $2HX(g) + H_2SO_4(l) \rightarrow X_2(g) + SO_2(g) + 2H_2O(l)$ X = Br, I
  $8HX(g) + H_2SO_4(l) \rightarrow 4X_2(g) + H_2S(g) + 4H_2O(l)$ X = I
- Tests with silver nitrate:
  $Ag^+(aq) + X^-(aq) \rightarrow AgX(s)$
- Disproportionation reactions of chlorine:
  $Cl_2(g) + 2NaOH(aq) \rightarrow NaCl(aq) + H_2O(l) + NaClO(aq)$
  $3Cl_2(g) + 6NaOH(aq) \rightarrow 5NaCl(aq) + 3H_2O(l) + NaClO_3(aq)$

Examination practice questions

Please see the data section of the CD for any A_r values you may need.

1 Chlorine and bromine are elements in Group 17 of the Periodic Table.

a Chlorine is used in water treatment.
State **one** advantage and **one** disadvantage of using chlorine in water treatment. [2]

b The electron configuration of bromine contains outermost electrons in the 4th shell.
Using your knowledge of Group 17 elements, complete the electron configuration of bromine.

$$1s^2 2s^2 2p^6 3s^2 3p^6$$ [1]

c Displacement reactions can be used to detect bromide ions in solution.

A student has a solution that contains bromide ions. The student carries out the following experiment.

Step 1
• She bubbles some chlorine gas through a sample of the solution.
• The mixture changes colour.

Step 2
• The student then adds an organic solvent, cyclohexane, to the mixture.
• She shakes the contents and allows the layers to separate.

i Write the **ionic** equation for the reaction that takes place in **step 1**. [1]

ii What colour does the cyclohexane layer turn in **step 2**? [1]

d Chlorine reacts differently with dilute and concentrated aqueous solutions of sodium hydroxide.
• When chlorine reacts with dilute sodium hydroxide, one of the products is sodium chlorate(I). This is the reaction that is used to manufacture bleach.
• When chlorine is reacted with hot concentrated sodium hydroxide, a different reaction takes place.
One of the products is $NaClO_3$, used as a weedkiller.
In each reaction, chlorine has been both oxidised and reduced.

i What term is used to describe a redox reaction in which an element is both oxidised and reduced? [1]

ii Write equations for these two reactions of chlorine with sodium hydroxide. [3]

iii Chlorine forms another chlorate called sodium chlorate(VII), used in the manufacture of matches. Suggest the formula of sodium chlorate(VII). [1]

[OCR Chemistry A Unit F321 Q4 January 2010]

2 Each of the Group VII elements chlorine, bromine and iodine forms a hydride.

a i Outline how the relative thermal stabilities of these hydrides change from HCl to HI.

ii Explain the variation you have outlined in **i**. [3]

b Hydrogen iodide can be made by heating together hydrogen gas and iodine vapour. The reaction is incomplete.
$$H_2(g) + I_2(g) \rightleftharpoons 2HI(g)$$
Write an expression for K_c and state the units. [2]

c For this equilibrium, the numerical value of the equilibrium constant K_c is 140 at 500 K and 59 at 650 K.
Use this information to state and explain the effect of the following changes on the equilibrium position.

i increasing the pressure applied to the equilibrium

ii decreasing the temperature of the equilibrium [4]

d A mixture of 0.02 mol of hydrogen and 0.02 mol of iodine was placed in a 1 dm³ flask and allowed to come to equilibrium at 650 K. Calculate the amount, in moles, of each substance present in the equilibrium mixture at 650 K.

$H_2(g)$	+	$I_2(g)$	$\rightleftharpoons$	$2HI(g)$
initial moles 0.02		0.02		0

[4]

[Cambridge International AS & A level Chemistry, 9701, Paper 21 Q2 November 2012]

12 Introduction to organic chemistry

This topic and those that follow look at the reactions and properties of the vast number of organic compounds – the covalent compounds of carbon. The large variety of organic compounds is described, conventions for naming them and drawing their formulae are explained, and the classification of organic reactions is introduced.

Learning outcomes

By the end of this topic you should be able to:

1.5b) perform calculations, including use of the mole concept, involving reacting masses (part, see also Topic 1)

14.1a) interpret and use the general, structural, displayed and skeletal formulae of the classes of compounds discussed in Topics 13–18

14.1b) understand and use systematic nomenclature of simple aliphatic organic molecules with functional groups mentioned in **14.1a)**, up to six carbon atoms (six plus six for esters and amides, straight chains only)

14.1d) deduce the possible isomers for an organic molecule of known molecular formula

14.1e) deduce the molecular formula of a compound, given its structural, displayed or skeletal formula

14.2a) interpret and use the following terminology associated with organic reactions: functional group, homolytic and heterolytic fission, free radical, nucleophile, electrophile, addition, substitution, elimination, hydrolysis, oxidation and reduction (part, see also Topics 13–18)

14.4a) describe structural isomerism, and its division into chain, positional and functional group isomerism

14.4b) describe stereoisomerism, and its division into geometrical (*cis–trans*) and optical isomerism

14.4c) describe *cis–trans* isomerism in alkenes, and explain its origin in terms of restricted rotation due to the presence of π bonds (see also Topic 14)

14.4d) explain what is meant by a chiral centre and that such a centre normally gives rise to optical isomerism

14.4e) identify chiral centres and *cis–trans* isomerism in a molecule of given structural formula.

12.1 Introduction

At one time there were thought to be two entirely different classes of chemical substances. Those compounds that had been isolated from the living world, from organisms, were called **organic** substances. On the other hand, those elements and compounds occurring in rocks and minerals, or made from them by chemical reactions, were called **inorganic** substances.

It was clear that both types of substances were made from the same elements. For example, the carbon dioxide produced by the fermentation of glucose:

$$C_6H_{12}O_6 \rightarrow 2C_2H_5OH + 2CO_2$$

and that obtained from burning the alcohol that is also produced during fermentation:

$$C_2H_5OH + 3O_2 \rightarrow 2CO_2 + 3H_2O$$

were the same compound as that obtained by heating limestone:

$$CaCO_3 \rightarrow CaO + CO_2$$

Carbon dioxide is an inorganic compound. It was accepted that inorganic compounds could be made from organic compounds, but for many years the reverse process was considered to be impossible. Many ancient doctrines included the concept of 'vitalism', by which it was believed – without much justification – that a vital 'life' force was needed to make organic compounds, so they could only be synthesised within living organisms.

Credit for the shattering of this belief is normally given to the German chemist Friedrich Wöhler, who in 1828 transformed the essentially inorganic salt ammonium cyanate into the organic compound carbamide (urea) (which is used by the body for nitrogen excretion in the urine), by simply heating the crystals gently:

$$NH_4^+ \; OCN^- \xrightarrow{\text{heat}}$$

ammonium cyanate

$$\underset{\text{urea}}{H_2N \overset{\overset{\displaystyle O}{\displaystyle \|}}{\underset{}{C}} NH_2}$$

The two branches of chemistry still remain distinct, however, and are still studied separately at A level and beyond. Organic compounds are compounds of carbon other than its oxides. Today, although several important organic chemicals are still isolated from the living world (penicillin is an example), the majority of the organic compounds we use have been synthesised artificially. Sometimes the starting point is an inorganic chemical, but most often it is a component of crude oil or coal. Both of these are the remains of living organisms, long since dead, decayed and chemically transformed into hydrocarbons. The ultimate origin of all organic chemicals, in nature and in the laboratory, is the inorganic compound carbon dioxide. This is converted into glucose by the transformations of photosynthesis, which require the Sun's energy. Biochemists and organic chemists have only recently unravelled the details of these reactions. The overall process is:

$$6CO_2 + 6H_2O \rightarrow \underset{\text{glucose}}{C_6H_{12}O_6} + 6O_2$$

followed by:

$$\text{glucose} \xrightarrow{\text{biochemical pathways}} \text{polysaccharides, proteins, nucleic acids, etc.}$$

12.2 The stability and variety of organic compounds

Today, organic chemicals (artificially made or naturally occurring) outnumber inorganic ones by 80:1, and number well over 10 million different compounds. Hundreds of thousands of new compounds are being made each year. Why are there so many of them? There are two main reasons:

- carbon atoms form strong bonds with other carbon atoms
- in general, organic compounds are kinetically stable.

We shall look at each of these reasons in turn.

The ability of carbon to form strong bonds to other carbon atoms

Table 12.1 shows how much more stable bonds between carbon atoms are than bonds between the atoms of other elements.

Carbon–carbon bonding allows chains and rings of carbon atoms to be produced by a process called **catenation** (from the Latin word *catena*, meaning 'chain'). What is more, the 4-valent nature of the carbon atom allows branched chains to occur, and the attachment of other atoms, increasing further the number of structures possible.

Table 12.1 The stability of the carbon–carbon bond

Bond	Bond enthalpy/kJ mol^{-1}
C—C	346
N—N	167
O—O	142
Si—Si	222
P—P	201
S—S	226

There are four different ways in which we can arrange four linked carbon atoms (see Figure 12.1).

Figure 12.1 Four ways of linking four carbon atoms

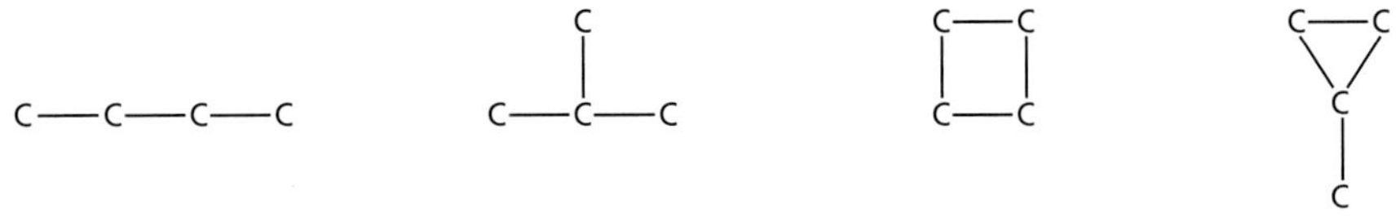

C—C—C—C

C—C—C with C above middle C

square arrangement of four C

triangle of three C with C below

With five carbon atoms there are ten different arrangements, and with six carbon atoms there are no fewer than 17. In a molecule containing 30 carbon atoms, there are more than 4×10^9 different ways of arranging them.

Worked example

Of the ten possible ways of bonding five carbon atoms together, three are open-chain structures (that is, they do not contain a ring of carbon atoms). Draw diagrams of these three.

Answer
One structure will contain a chain of five carbon atoms:

C—C—C—C—C

If we take one carbon atom from the end and join it to the middle of the chain, we produce another structure:

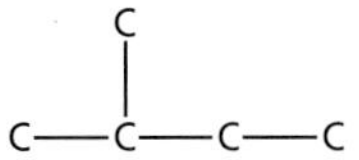

If we do this once more, we arrive at the third of the three open-chain structures:

C—C—C with two C (one above, one below) on second C

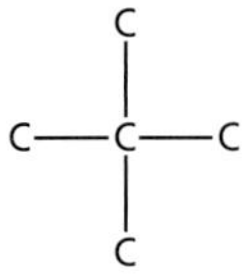

Note that the first structure above represents the *same* chain as any of the following structures:

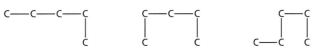

This is because the carbon atoms in a chain are not arranged in a straight line, but are in fact positioned in a zig-zag manner, with each C—C—C angle being 109.5° (see also section 12.6). All of the above are imperfect ways of drawing the five-carbon chain in the two dimensions of a page, but each shows that there is a chain of five carbon atoms.

Now try this

Draw out the five open-chain structures containing six carbon atoms.

The stability of organic compounds

The second reason why there are so many organic compounds is a more subtle one – once they have been formed, organic compounds are remarkably stable. This stability is more kinetic than thermodynamic. For example, although methane is thermodynamically stable with respect to its elements:

$$CH_4(g) \rightarrow C(s) + 2H_2(g) \qquad \Delta H^\ominus = +75 \, kJ \, mol^{-1}$$

it is unstable in the presence of oxygen:

$$CH_4(g) + 2O_2(g) \rightarrow CO_2(g) + 2H_2O(l) \qquad \Delta H_c^\ominus = -890 \, kJ \, mol^{-1}$$

But methane, along with other hydrocarbons (and most other organic compounds), does not react with oxygen unless heated to quite a high temperature – the **ignition point**. This low rate of reaction at room temperature is a result of the high activation energy that many organic reactions possess (see section 8.4). This, in turn, is due to the strong bonds that carbon forms to itself and to other elements. These bonds have to be broken before reaction can occur, and breaking strong bonds requires much energy. Thus methane in oxygen is kinetically stable.

The first stage in the combustion of methane involves the breaking of a C—H bond:

$$CH_3-H \rightarrow CH_3 + H$$

This involves the input of 439 kJ of energy per mole – a large value.

Organic compounds are thermodynamically unstable not only with respect to oxidation, but often also with respect to other transformations. For example, the hydration of ethene to ethanol is exothermic, as is the joining of three molecules of ethyne to form benzene:

$$CH_2=CH_2 + H_2O \longrightarrow C_2H_5OH \qquad \Delta H^\ominus = -43 \text{ kJ mol}^{-1}$$

$$3CH\equiv CH \longrightarrow \bigcirc \qquad \Delta H^\ominus = -635 \text{ kJ mol}^{-1}$$

And yet ethene gas can be collected over water, and ethyne has no tendency to form benzene under normal conditions. If the right conditions and catalysts can be found, however, it is possible to convert one 'stable' organic compound into another 'stable' compound. It is even possible to use the same reagent (hydroxide ions) on the same compound (a bromoalkane) to produce two different products, depending on the conditions of solvent used (see section 15.3 for details):

$$CH_3CHBrCH_3(l) + NaOH(aq) \xrightarrow{\text{warm in water}} NaBr + CH_3CH(OH)CH_3 \qquad (1)$$

$$CH_3CHBrCH_3(l) + NaOH(\text{in ethanol}) \xrightarrow{\text{heat in ethanol}} NaBr + H_2O + CH_2=CHCH_3 \ (2)$$

Figure 12.2 Reaction profiles for 'typical' organic reactions. E_{a1} and E_{a2} are the activation energies for reactions 1 and 2, respectively. The activation energy is the energy that the reactants need in order to undergo the reaction.

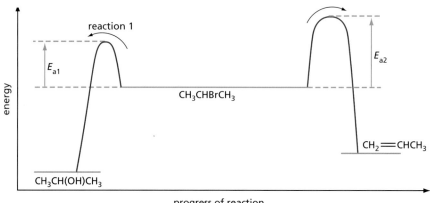

Organic compounds can be considered as occupying stable low-energy 'valleys', with high-energy 'hills' in between them. One catalyst, or set of conditions, can afford a low activation energy 'mountain pass' to a valley next door, whereas a different catalyst can open up a new low-energy route into another valley on the other side.

12.3 The shapes of organic molecules

After Wöhler's conceptual breakthrough in 1828, the next 50 years saw a tremendous increase in activity by organic chemists, making many new compounds and determining their empirical and molecular formulae. However, it was not until after the work of Joseph Le Bel, Jacobus van't Hoff and Friedrich Kekulé in the period 1860–75 that chemists started to think of organic compounds as being made up of separate, three-dimensional molecules. After this period chemists appreciated that the molecular formula was merely a summary of the structure of the molecule, rather than telling the whole story.

We now firmly believe that all organic (and inorganic) molecules possess a three-dimensional structure, and that the various atoms are fixed in their relative positions. The shapes of organic molecules depend ultimately on bond lengths and angles. Bond angles are governed by the mutual repulsion of electron pairs (see section 3.9).

Table 12.2 Bond angles in some organic compounds

Bond angle	Examples	Hybridisation (see section 3.14)
180°	$O=C=O$ $H—C≡N$	sp
120°	H H $C=C$ H H	sp^2
109.5°	H C H H H	sp^3

In each of the molecules carbon dioxide and hydrogen cyanide, in the first row of Table 12.2, the three atoms are collinear (they all lie on the same straight line). In the molecule of ethene, in the second row, all six atoms are coplanar (they all lie in the same plane). Ethene is a 'two-dimensional' molecule. Methane, however, is a three-dimensional molecule, and its shape cannot be easily represented on a two-dimensional page. We use a **stereochemical formula** to show this, as in the last row of Table 12.2. The dashed line represents a bond behind the plane of the paper, whereas the solid wedge represents a bond coming out in front of the plane of the paper. Ordinary lines represent bonds that are in the plane of the paper.

12.4 The different types of formulae

There are six types of formulae that are used to represent organic compounds. We shall illustrate them using the examples of ethanoic acid, the acid occurring in vinegar, and 'iso-octane', one of the constituents of petrol.

The empirical formula

The information that the **empirical formula** gives about a compound is limited to stating which elements are contained in the compound, and their relative ratios. The empirical formula of ethanoic acid is CH_2O, and that of iso-octane is C_4H_9. In section 1.8, we saw how empirical formulae can be calculated from mass or percentage data for compounds.

The molecular formula

The **molecular formula** tells us the actual number of atoms of each element present in a molecule of the compound. This is always a whole-number multiple (×1, ×2, ×3, etc.) of the empirical formula. The molecular formulae of ethanoic acid and iso-octane are $C_2H_4O_2$ and C_8H_{18}, respectively. Although glucose has the same empirical formula as ethanoic acid, CH_2O, its molecular formula, $C_6H_{12}O_6$, is six times its empirical formula.

The structural formula

The **structural formula** shows us how the atoms are joined together in the molecule. The structural formula of ethanoic acid is CH_3CO_2H, or CH_3COOH, and that of iso-octane is $(CH_3)_3CCH_2CH(CH_3)_2$. Structural formulae are used to distinguish between structural isomers (see section 12.6). $C=C$ double bonds are always shown in structural formulae, and sometimes also $C=O$ and $C—C$ bonds, where this makes the structure clearer.

The displayed formula

Similar to the structural formula, the **displayed formula** is more complete. It shows not only which atoms are bonded together, but also the types of bonds (single, double or dative). The displayed formulae for ethanoic acid and iso-octane are shown in Figure 12.3.

Figure 12.3 Displayed formulae show the bonding in the molecule

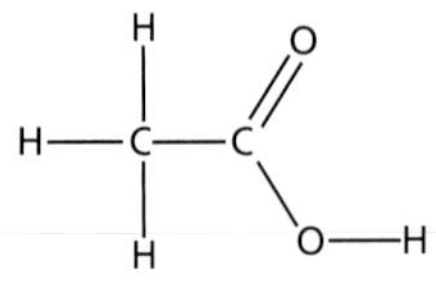

ethanoic acid

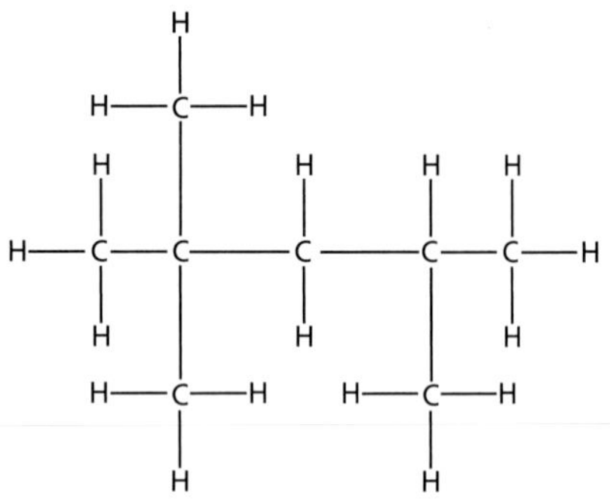

iso-octane

The stereochemical formula

Although the displayed formula tells us much about a molecule, it indicates nothing about its shape. It is difficult to represent the shape of a three-dimensional molecule on a two-dimensional page, so we use the conventions of **stereochemical formulae** illustrated in Figure 12.4.

Figure 12.4 Stereochemical formulae represent the three-dimensional shape of a molecule. The dashed line represents a bond to an atom *behind* the plane of the paper, and the 'wedge-shaped' line represents a bond to an atom *in front of* the plane of the paper.

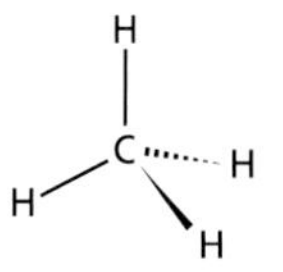

methane, CH_4

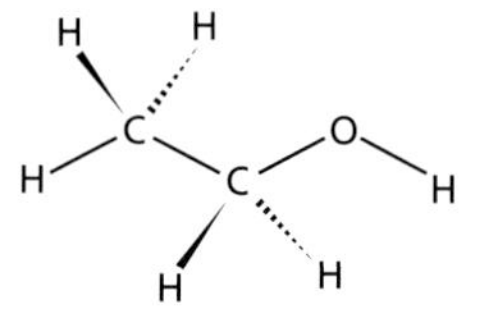

ethanol, C_2H_5OH

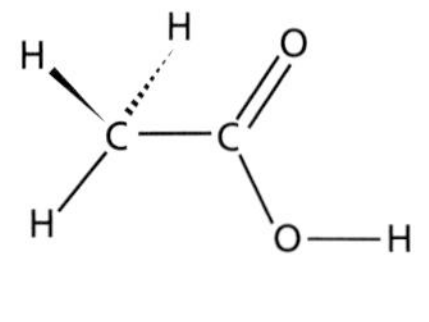

ethanoic acid, CH_3CO_2H

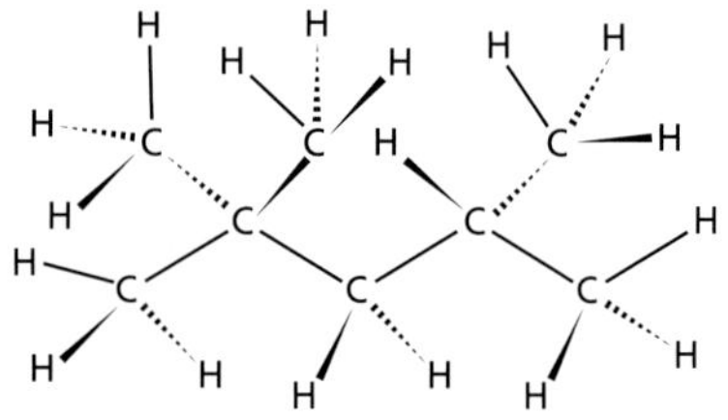

iso-octane, $(CH_3)_3CCH_2CH(CH_3)_2$

The stereochemical formula for iso-octane in Figure 12.4 illustrates how this type of formula can become too complicated and unwieldy for any molecule with more than about five carbon atoms. Quite often, however, you will come across a displayed formula in which just a part has been drawn out to show the stereochemistry around a particular atom and the rest left as a structural formula.

The skeletal formula

The previous five types of formulae have become increasingly more complicated. The **skeletal formula** moves in the other direction. It pares the molecular structure down to its bare minimum, omitting all carbon atoms and all the hydrogen atoms joined to carbon atoms, and therefore shows only the carbon–carbon bonds. It is an excellent way of describing larger and more complicated molecules. The true skeletal formula

Figure 12.5 Skeletal formulae and models of **a** ethanol, **b** ethanoic acid and **c** iso-octane

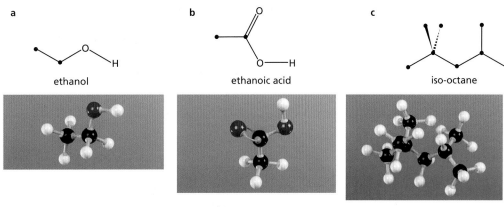

ethanol   ethanoic acid   iso-octane

shows only the carbon–carbon bonds, but, especially before you become familiar with their use, it will be clearer to you if you include a big dot to represent each carbon atom.

Figure 12.5 shows the skeletal formulae of three of the four compounds shown in Figure 12.4, together with pictures of the corresponding molecular models.

The skeletal formula of iso-octane shows more clearly than any of the other formulae how the carbon atoms are joined together. This type of formula will be used in later topics to show the shapes of, and bonding in, the larger molecules we shall come across. Its use for smaller molecules is limited, however, as the skeletal formula for methane is a dot, and that for ethane is a straight line!

Worked example 1

What are the molecular and empirical formulae of the following compounds?
a propane, $CH_3CH_2CH_3$ b ethane-1,2-diol (glycol), $HOCH_2CH_2OH$
c dichlorethanoic acid, $CHCl_2CO_2H$

Answer
a For propane, the molecular formula is C_3H_8, which does not reduce further, so the empirical formula is also C_3H_8.
b For ethane-1,2-diol, the molecular formula of $C_2H_6O_2$ reduces to CH_3O as the empirical formula.
c For dichloroethanoic acid, the molecular formula of $C_2H_2O_2Cl_2$ reduces to the empirical formula $CHOCl$.

Worked example 2

What are the structural and molecular formulae of the following compounds?

a   b   c

Answer
a Structural formula CH_3COCH_3, molecular formula C_3H_6O
b Structural formula $CH_3CH_2CH(CH_3)_2$, molecular formula C_5H_{12}
c Structural formula $CH_3CO_2CH_2CH_3$, molecular formula $C_4H_8O_2$

Now try this

1 Work out the structural, molecular and empirical formulae of the following compounds.

OH

Br   HO   OH

Cl

2 Draw the skeletal formulae of the compounds with the following structural formulae.
   a $(CH_3CH_2)_2CHCH{=}CHCl$ b $(CH_3)_2C(OH)CH_2CH(CH_3)CO_2H$
   c $ClCH_2CH_2CH(CO_2H)_2$

215

The representation of carbon rings in formulae

Rings of atoms are often represented by their skeletal formulae. Common examples are shown in Figure 12.6.

Figure 12.6 Skeletal and structural formulae and models of **a** cyclopropane, **b** cyclopentane, **c** cyclohexane, **d** cyclohexene and **e** benzene

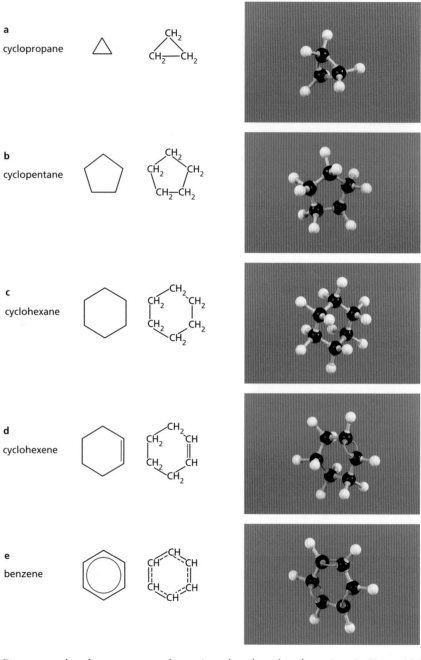

a
cyclopropane

b
cyclopentane

c
cyclohexane

d
cyclohexene

e
benzene

Benzene and cyclopropane are planar rings, but the other three rings in Figure 12.6 are all puckered to some degree. The most common arrangement of cyclohexane is called the 'chair' form, which allows the C—C—C bond angles to be the preferred 109.5° (see Figure 12.7).

Figure 12.7 The 'chair' form of cyclohexane minimises bond strain.

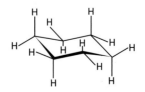

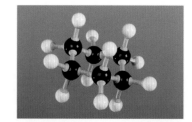

Cyclopropane is a highly strained (and hence reactive) compound. Its bond angles are constrained to be 60° – far removed from the preferred 109.5°. No such problems arise with benzene, however. The internal angles in a (planar) regular hexagon are 120°, which is ideal for carbon atoms surrounded by three electron pairs.

Worked example

What are the molecular formulae of the following compounds?

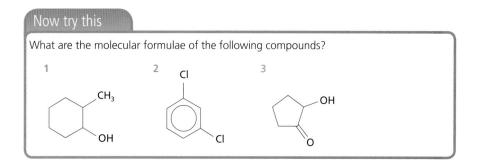

Answer

a Cyclohexane is $(CH_2)_6$, or C_6H_{12}. Two hydrogen atoms have been replaced by chlorines, so the formula is $C_6H_{10}Cl_2$.

b Benzene is C_6H_6. One hydrogen atom has been replaced by CH_3, and another by CO_2H. The molecular formula is $C_8H_8O_2$.

Now try this

What are the molecular formulae of the following compounds?

12.5 Naming organic compounds

As was mentioned earlier, organic compounds were originally extracted from living organisms. The sensible way of naming them incorporated the name (often the Latin version) of the plant or animal concerned. Some common examples are given in Table 12.3.

Table 12.3 The origins of the names of some natural organic compounds

Common name	Structural formula	Isolated from	Latin name
formic acid	HCO_2H	ants	*Formica*
malic acid	CO_2H $|$ $CH(OH)$ $|$ CH_2 $|$ CO_2H	apples	*Malus*
menthol	CH_3 OH $CH(CH_3)_2$	peppermint	*Mentha piperita*
oxalic acid	CO_2H $|$ CO_2H	wood sorrel	*Oxalis*

Table 12.4 Stem names and the numbers of carbon atoms they represent. From 5 upwards, the stem is an abbreviation of the corresponding Greek word for the number.

Stem name	Number of carbon atoms
meth-	1
eth-	2
prop-	3
but-	4
pent-	5
hex-	6
hept-	7
oct-	8

Table 12.5 Stem-suffixes and their meanings

Stem-suffix	Meaning
-an	all C—C single bonds
-en	one C=C double bond
-dien	two C=C double bonds
-trien	three C=C double bonds
-yn	one C≡C triple bond

As more and more organic compounds were discovered during the eighteenth and nineteenth centuries, this naming system became increasingly unsatisfactory for two reasons:

- it could not be applied to those compounds made artificially in the laboratory
- the feat of memory required for a chemist to be able to relate a structure to each name was becoming too great.

So, in the latter part of the nineteenth century, chemists began to devise a systematic, logical form of nomenclature in which it was possible to translate names into the corresponding structural formulae, and vice versa, for previously unknown compounds. The latest version has been adopted by the International Union of Pure and Applied Chemistry (IUPAC), and is (with only a few modifications) completely international. Chemists throughout the world use it, and understand one another, no matter what their first language is. (It has to be said, however, that the *true* language of organic chemistry is the skeletal formula. Sit next to a couple of organic chemists talking 'shop' in a café, and you will see that in no time at all the menu card or paper tablecloth will be scribbled over with the curious hieroglyphics of skeletal formulae.)

All organic compounds contain carbon, and virtually all of them also contain hydrogen. A systematic name for an organic compound is often based on the 'parent' hydrocarbon. Each name consists of the components listed below (not all may be present, however: it depends on the complexity of the compound).

1 A **stem** – this specifies the number of carbon atoms in the longest carbon chain (see Table 12.4). If the carbon atoms are joined in a ring, the prefix 'cyclo' is added.

2 A **stem-suffix** – this indicates the type of carbon–carbon bonds that occur in the compound (see Table 12.5).

So:

CH_3—CH_3 is ethane
CH_3—CH=CH_2 is propene
CH=CH—CH=CH_2 is butadiene.

Arabic numerals may be added before the stem suffix to show the position of any double bond, etc. For example, there are two compounds with the name 'butadiene':

CH_2=CH—CH=CH_2 is buta-1,3-diene
CH_2=C=CH—CH_3 is buta-1,2-diene.

3 A **suffix** – this indicates any oxygen-containing group that might be present (see Table 12.6).

Table 12.6 Suffixes and their meanings

Suffix	Meaning	
-al	an aldehyde group	an aldehyde group structure (C=O with H)
-one	a ketone group	a ketone group structure (C=O)
-ol	an alcohol group	—C—OH
-oic acid	a carboxylic acid group	—C(=O)OH

So:

CH₃—CHO is ethanal
CH₃—CH₂—CO₂H is propanoic acid
CH₃—CO—CH₂—CH₃ is butanone.

4 One or more **prefixes** – these indicate non-oxygen-containing groups, which may be other carbon (alkyl) groups, or halogen atoms which are substituted for hydrogen along the carbon chain (see Table 12.7). The prefixes may also include subsidiary oxygen-containing groups. Prefixes can be combined with one another, and they might in turn be prefixed by 'di' or 'tri' to show multiple substitution.

The position of a substituent along the carbon chain is indicated by an Arabic numeral before the prefix. The carbon atoms in the longest carbon chain are numbered successively from the end that results in the lowest overall numbers in the name. So CH₃—CHCl—CH₂Cl is 1,2-dichloropropane (numbering from the right), and not 2,3-dichloropropane (numbering from the left).

Table 12.7 Prefixes and their meanings

Prefix	Meaning
methyl-	CH₃—
ethyl-	CH₃CH₂—
propyl-	CH₃CH₂CH₂—
…	…
chloro-	Cl—
bromo-	Br—
hydroxy-	HO—

These four simple part-names allow us to describe a multitude of compounds. The following examples will show how the parts are pieced together.

a Cl—CH₂—CH₂—CHO   is   3-chloropropanal

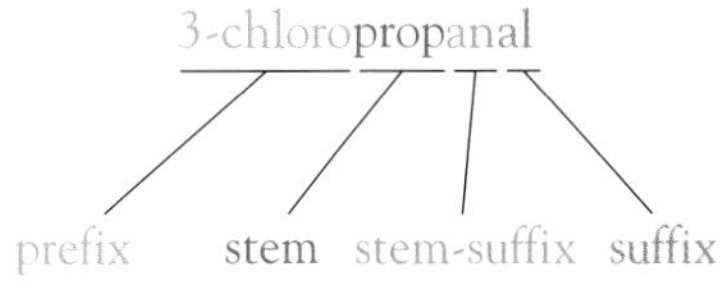

prefix   stem   stem-suffix   suffix

b CH₃—CH(OH)—CH₃   is   propan-2-ol

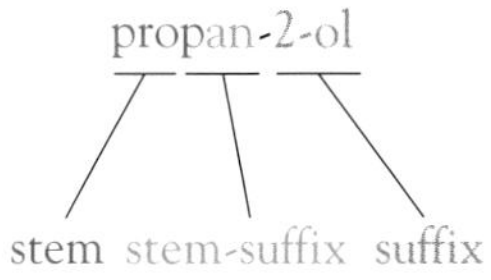

stem   stem-suffix   suffix

c CH₃—CHCl—CH(CH₃)₂   is   2-chloro-3-methylbutane

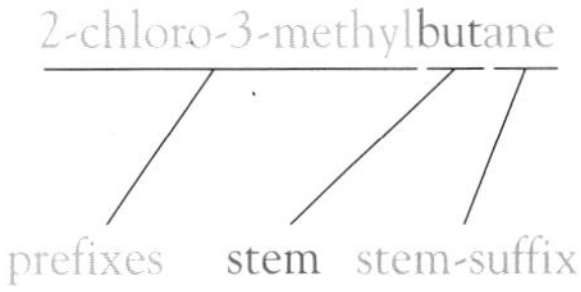

prefixes   stem   stem-suffix

d   CH₃
            \
             C=CH—CH₂OH   is   3-methylbut-2-en-1-ol
            /
          CH₃

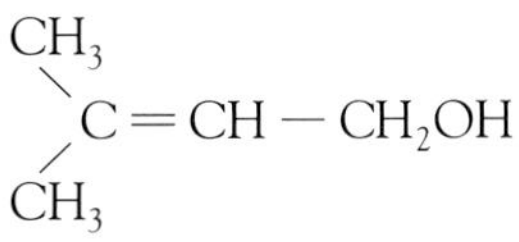

prefix   stem   stem-suffix   suffix

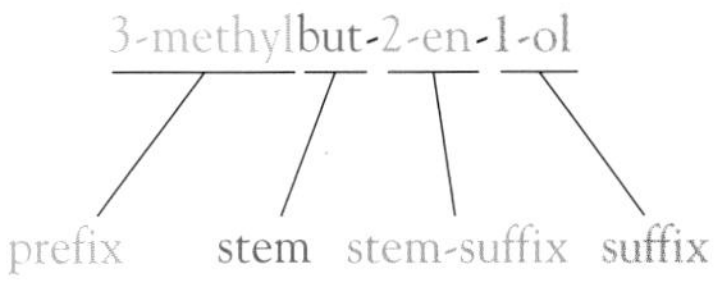

Note that the compound in **a** is not called 1-chloropropanal. If a compound is named as an aldehyde, the aldehyde group is always at carbon atom number 1. Note also that the compound in **c** could equally well have been called 3-chloro-2-methylbutane (that is, numbering from the right). But the name 3-methyl-2-chlorobutane would have been incorrect, because the substituents should be listed alphabetically.

Worked example 1

What are the systematic names of the following compounds?

a $(CH_3)_3OH$
b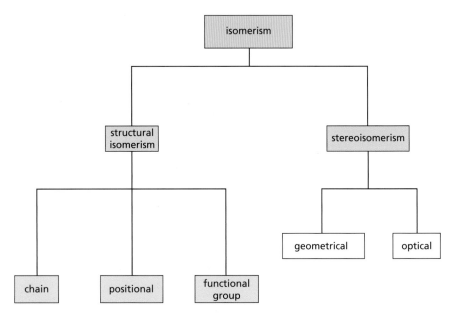
c $H_3C-CH_2-CH(OH)-CH_3$

Answer
a 2-methylpropan-2-ol
b 1,2-dimethylcyclohexane
c butan-2-ol

Worked example 2

Write the structural formulae of the following compounds.
a 3-methylbutan-1-ol
b 1,3-dichloropropane

Answer
a $(CH_3)_2CHCH_2CH_2OH$
b $ClCH_2CH_2CH_2Cl$

Now try this

1 Name the following compounds.
  a $(CH_3)_3CCH(OH)-CH_2CH_3$
  b $CH_3-CH=CH-CH_3$
  c

2 Draw structural formulae for the following compounds.
  a 2-methyl-2,3-dibromopentane
  b 2,2-dimethylpropane

12.6 Isomers

Isomerism is the property of two or more compounds (called **isomers**) that have the same molecular formula but different arrangements of atoms, and hence different structural formulae.

Isomerism in organic compounds can be classified into five different types, according to the scheme shown in Figure 12.8. Some of these are dealt with in detail when they arise in later topics, but all are included here in summary form.

Figure 12.8 The five types of isomerism

```
                        isomerism
                     /            \
         structural                  stereoisomerism
         isomerism                  /            \
       /     |      \         geometrical      optical
   chain positional functional
                     group
```

Structural isomerism

Structural isomers differ as to which atoms in the molecule are bonded to which. There are three types of structural isomerism.

Chain isomerism

Chain isomers have the same number of carbon atoms as one another, but their carbon backbones are different. For example, pentane and 2-methylbutane are isomers with the molecular formula C_5H_{12}:

$$CH_3-CH_2-CH_2-CH_2-CH_3 \qquad CH_3-CH-CH_2-CH_3$$
$$| \atop CH_3$$

Note that

$$CH_3-CH_2-CH_2-CH_2$$
$$| \atop CH_3$$

is not another isomer: this is just another way of writing the formula of pentane. Note also that

$$CH_3 \atop | \atop CH_3-CH_2-CH \atop | \atop CH_3$$

is not another isomer. This is another way of writing the formula of 2-methylbutane. Remember that carbon chains are naturally zig-zag arrangements, and that there is completely free rotation around any C—C single bond; however, structural formulae are often drawn as flat, with 90° bond angles. The formulae in Figure 12.9 all represent different **conformations** of the molecule pentane, and at room temperature a particular molecule will be constantly changing its shape from one conformation to another.

Figure 12.9 Conformations of pentane

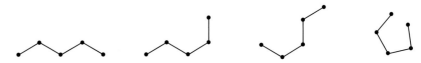

Positional isomerism

This is often included under the general umbrella of structural isomerism. **Positional isomers** have the same carbon backbone, but the positions of their functional group(s) differ. Their systematic names reflect this. For example, there are two alcohols containing three carbon atoms:

$$CH_3-CH_2-CH_2-OH \qquad CH_3-CH-CH_3 \atop | \atop OH$$
$$\text{propan-1-ol} \qquad\qquad \text{propan-2-ol}$$

and there are two compounds with the formula $C_2H_4Cl_2$:

$$Cl-CH_2-CH_2-Cl \qquad CH_3-CHCl_2$$
$$\text{1,2-dichloroethane} \qquad \text{1,1-dichloroethane}$$

(Remember that the carbon chain is numbered from the end that produces the smallest number prefixes in the name, so CH_3-CHCl_2 is not called 2,2-dichloroethane.)

When a carbon chain contains a double bond, the position of the bond can give rise to isomerism. For example, there are two butenes:

$$CH_3-CH_2-CH=CH_2 \qquad CH_3-CH=CH-CH_3$$
$$\text{but-1-ene} \qquad\qquad \text{but-2-ene}$$

Positional isomerism also occurs in ring systems, as shown in Figure 12.10.

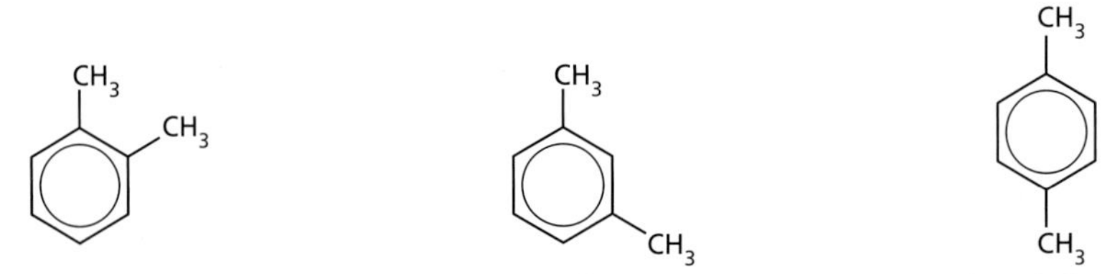

Figure 12.10 Positional isomers in ring molecules. Note that the carbon atoms in the rings are numbered starting at one of the substituted ones, and that in cyclohexanol it is the carbon atom attached to the —OH group that is carbon number 1.

1,2-dimethylbenzene   1,3-dimethylbenzene   1,4-dimethylbenzene

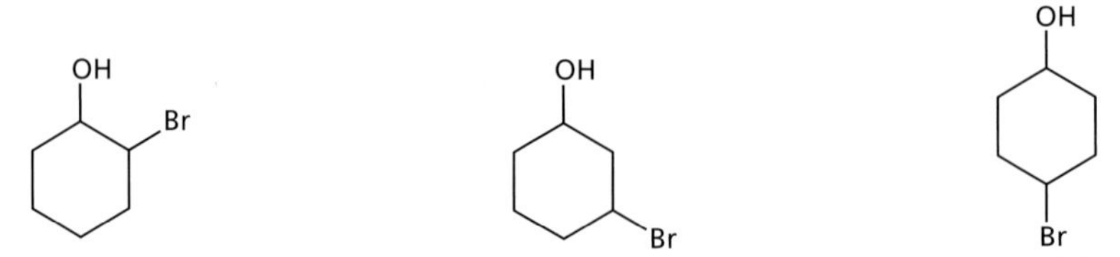

2-bromocyclohexanol   3-bromocyclohexanol   4-bromocyclohexanol

Functional group isomerism

In **functional group isomerism**, the isomers are more different than in the types of isomerism considered so far. Compounds that contain different functional groups often have very different reactions, as we shall see in section 12.7. However, if their molecular formulae are the same, then they are classed as isomers.

For example, the carboxylic acid $CH_3CH_2CO_2H$ (propanoic acid), the ester $CH_3CO_2CH_3$ (methyl ethanoate) and the ketone–alcohol CH_3COCH_2OH (hydroxypropanone) are all isomers, all having the molecular formula $C_3H_6O_2$.

Stereoisomerism

Stereoisomers are very closely related to each other. They contain the same atoms bonded to one another, and the bonding and functional groups are identical. They differ only in the way the atoms are arranged in three-dimensional space ('stereo' is derived from the Greek word *stereos*, meaning 'solid'). There are two types of stereoisomerism – **geometrical** and **optical**.

Geometrical isomerism

This occurs with alkenes that do not have two identical groups on both ends of the C=C double bond. It arises because, unlike the case of a C—C single bond, it is not possible to rotate one end of a C=C double bond with respect to the other (see section 14.1). The following two forms of but-2-ene are therefore distinct and separate isomers:

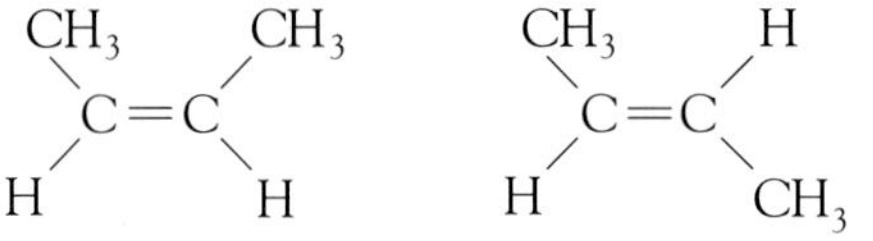

cis-but-2-ene   *trans*-but-2-ene

The isomer with both methyl groups on the same side of the double bond is called the *cis* isomer (from the Latin word *cis*, meaning 'on this side'), and that with the methyl groups on opposite sides is called the *trans* isomer (*trans* meaning 'on the other side'). An alternative name for geometrical isomerism is **cis–trans isomerism**. Note that the carbon atoms on both ends of the double bond must each be bonded to two non-identical groups. The following compounds do *not* exhibit geometrical isomerism because one end, or both ends, of the double bond have two identical atoms or groups attached:

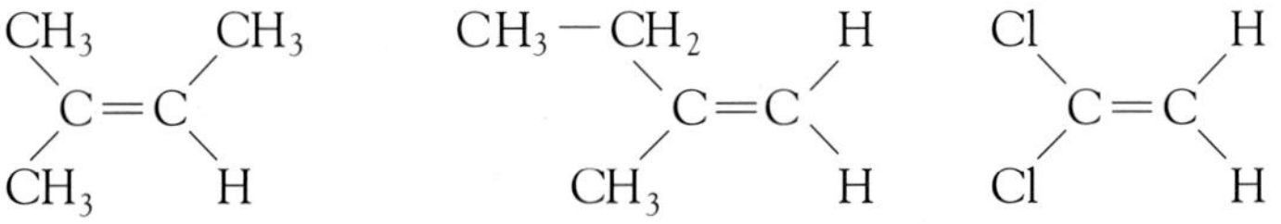

Geometrical isomerism also occurs in ring systems. For example, there are two isomers of 1,3-dimethylcyclobutane, shown in Figure 12.11. (Rotation around a single bond is prevented by linkage in a ring.)

Figure 12.11 Geometrical isomers of 1,3-dimethylcyclobutane

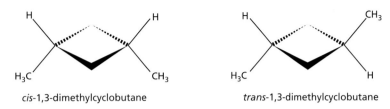

cis-1,3-dimethylcyclobutane   trans-1,3-dimethylcyclobutane

Optical isomerism

Optical isomerism is the most subtle of the five forms of isomerism, as virtually every physical or chemical property of the isomers concerned is identical. The one way in which they can be distinguished is by their effect on the passage of plane-polarised light, hence the name 'optical'. The instrument used to measure this effect is called a polarimeter.

The polarimeter

A **polarimeter** is used to distinguish optical isomers. The device consists of six parts, shown in Figure 12.12. A monochromatic light source (that is, one that produces light of a single wavelength) and a slit produce a thin beam of light, which then passes through a **polariser**. This may be a piece of Polaroid (the material from which Polaroid sunglasses are made); it contains long straight molecules all lined up in the same direction and only allows through photons whose electric field is oscillating in this same direction.

Figure 12.12 Outline diagram of a polarimeter

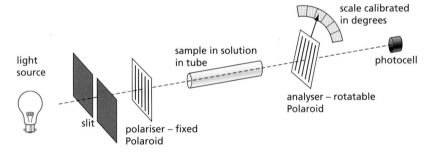

The polarised beam now enters the sample, which is a solution of the compound under investigation, held in a tube 10 cm long. As it comes out of the other end, the beam passes through another piece of Polaroid, the **analyser**, and then into the photocell, which produces an electric current proportional to the intensity of light that falls upon it.

The instrument is calibrated by filling the sample tube with pure solvent, and turning the analyser around its axis until the output from the photocell is a minimum. This occurs when the analyser is at right angles to the polariser. (Electronically, it is more accurate to adjust to a minimum output than to a maximum output.) The angle on the scale is recorded.

The sample tube is now filled with the sample under investigation, and the analyser is turned until the output from the photocell is again a minimum. The new angle on the scale is recorded. The rotation caused by the compound is the difference between the two measured angles.

The actual rotation depends on four factors – the optical path length (the distance through the solution), the concentration of the compound in the solution, the wavelength of light used, and (to a small extent) the temperature. Once these have all been allowed for, the intrinsic rotating power of the compound can be compared with that of others.

The origins of optical isomerism

Optical isomerism arises from the inherent asymmetry of the molecules that make up each isomer. If a molecule has no plane or centre of symmetry, it is called a **chiral** (pronounced 'k-eye-rall', from the Greek word meaning 'handed') molecule. Chiral molecules are not identical to each other, but are related as a mirror image is to its object.

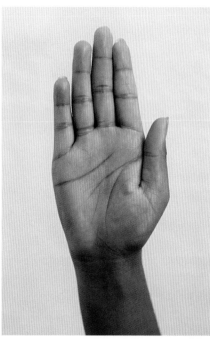

Figure 12.13 Your hands are chiral. They are not superimposable on each other, or on their mirror images, but one is superimposable on the mirror image of the other. Try this experiment. Hold your left hand, outstretched and palm away from you, in front of a mirror. Now hold your right hand, outstretched and palm towards you, next to the mirror. The image of your left hand now looks identical to (is superimposable on) your real right hand!

mirror

isomer I   isomer II

Figure 12.14 The two molecules are optical isomers – mirror images of each other.

The most common cases of chirality occur when a molecule contains a tetrahedral carbon atom surrounded by four different groups or atoms. Such atoms are termed chiral carbon atoms, or chiral centres. There are two ways of arranging the four groups around a chiral tetrahedral atom, and these two arrangements produce molecules that are mirror images of each other (see Figure 12.14).

Optical isomerism most commonly occurs when a carbon atom has four different atoms or groups attached to it.

Two molecules that are non-superimposable mirror images of each other are called **enantiomers**. They are distinguished by the prefixes (+)- or (−)-, or the letters *d*- or *l*-, in front of their names. The letters derive from the Latin words *dextro* and *laevo*, meaning 'right' and 'left', respectively. This relates to the direction in which each isomer rotates the plane of polarised light – either clockwise or anticlockwise (see the panel opposite).

The angles through which the (+) and (−) isomers rotate the plane of polarisation are equal, but they are in opposite directions. A 50:50 mixture of the two isomers therefore has no effect on the passage of plane-polarised light. The clockwise rotation caused by the (+) isomer is exactly cancelled by the anticlockwise rotation caused by the (−) isomer. This equal mixture of the two isomers is called a **racemic mixture** or **racemate**.

Sugars, amino acids and most other biologically important molecules are chiral. Usually only one of the two isomers, (+) or (−), is made in nature (but see the panel opposite for some exceptions). Many pharmaceutical drugs that are designed to interact with living systems also have chiral centres. Enzymes will usually react with only one of the mirror-image isomers. However, when a compound is made in the laboratory, it is usually produced as the racemic mixture of the two mirror-image forms. Not only is this a potential waste of 50% of an expensive product, but it can also cause unforeseen problems if the mixture is administered (see the panel on thalidomide on page 226). Much research effort is therefore put into producing drugs containing molecules of just one optical isomer. Optical purities of greater than 99.5% are the aim.

Some naturally occurring enantiomers

The carvones

You will know the taste of spearmint, whether in the form of chewing gum or toothpaste. You may also know the taste of caraway seeds (Persian cumin or sheema jeerakam). The same compound, carvone, is responsible for both tastes (see Figure 12.15).

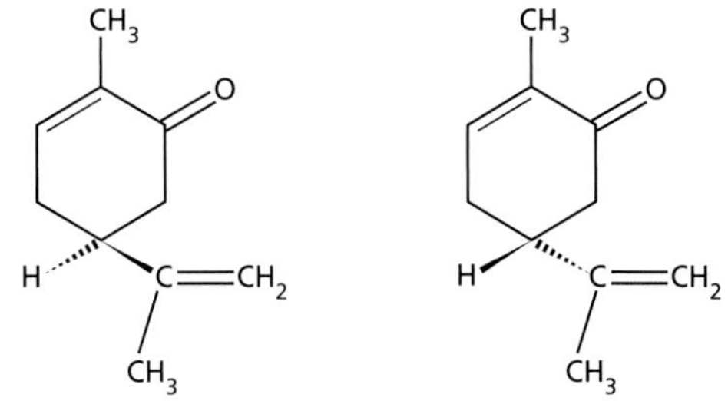

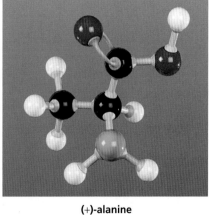

Figure 12.15 Enantiomers of carvone

(–)-carvone
(spearmint)

(+)-carvone
(caraway)

The lactic acids

The common name for 2-hydroxypropanoic acid is lactic acid. It derives its name from the Latin word for 'milk' (*lactis*). It is formed when milk is fermented by certain micro-organisms. Unusually, the acid seems to be produced as a racemic mixture of the two enantiomers. This is in contrast to the lactic acid produced in our muscles when we exercise anaerobically. The product here is pure (+)-lactic acid (see Figure 12.16).

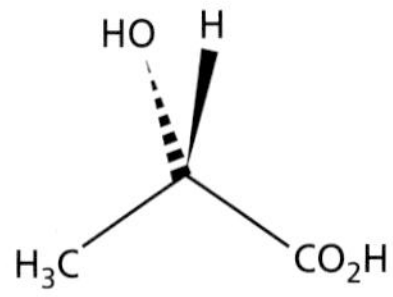

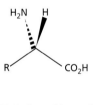

Figure 12.16 (+)-lactic acid is produced in muscle.

The amino acids

Except for glycine, all naturally occurring amino acids are chiral. In almost all cases, the amino acids have the same configuration around their chiral carbon atom (see Figures 12.17 and 12.18).

Figure 12.17 Amino acids are chiral molecules (R = side chain; see also Topic 27).

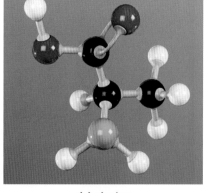

(+)-alanine   (–)-alanine

Figure 12.18 Models of (+)-alanine and (–)-alanine

The cell walls of some bacteria consist of chains of mucopeptides. These are co-polymers of amino acids and sugars. Some of the amino acids used by bacteria, however, are of the opposite configuration to those that normally occur in nature. This has turned out to be their downfall. The antibiotic penicillin interferes with the synthesis of these mucopeptides, by competing with amino acids of the opposite configuration for the active sites of enzymes. The inhibited enzymes can no longer carry out their function of repairing damaged cell walls or synthesising new ones, so the bacteria die.

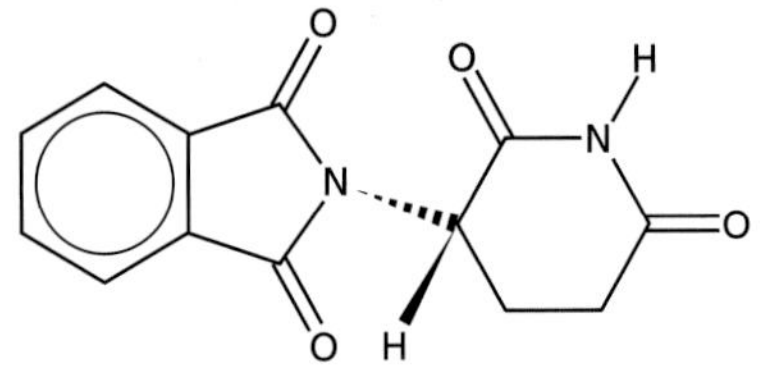

Figure 12.19 Thalidomide

The thalidomide tragedy

Many tranquilisers have been developed whose molecules have been based on the barbituric acid ring system. One of these was thalidomide, shown in Figure 12.19. Originally it seemed to be a very successful, non-toxic sedative with no known side-effects. In the early 1960s, however, it was realised that women who had been prescribed the sedative to treat nausea in early pregnancy ('morning sickness') went on to deliver badly deformed babies. The drug was withdrawn immediately.

Research showed that while the (+) isomer was an effective and safe tranquiliser, the (−) isomer was the culprit in damaging the foetus. Originally it was thought that the problem had been due to contamination by the (−) isomer in some production batches, but subsequent research has shown that racemisation of the pure (+) isomer occurs rapidly as soon as it enters the bloodstream: racemisation is over 50% complete within 10 minutes. Thalidomide will never be marketed again, although analogues that are chirally stable are being investigated as possible anti-inflammatory drugs.

Worked example 1

Draw and name the four alkene isomers with the molecular formula C_4H_8.

Answer

$CH_3—CH_2—CH=CH_2$ but-1-ene

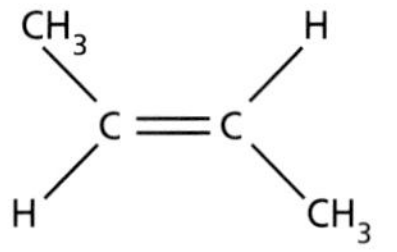

trans-but-2-ene

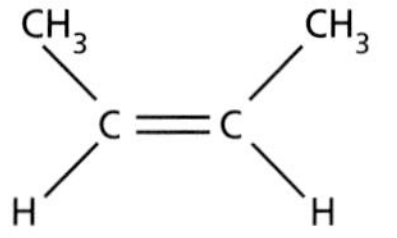

cis-but-2-ene

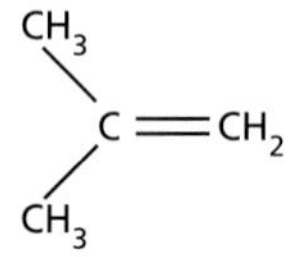

2-methylpropene

Worked example 2

There are four structural or positional isomers of C_4H_9OH. Draw them, name them, and specify which one of them can exist as a pair of optical isomers (that is, which one is chiral).

Answer

$CH_3CH_2CH_2CH_2OH$ butan-1-ol

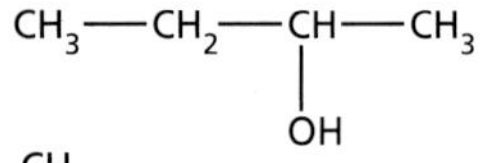

butan-2-ol (this is chiral)

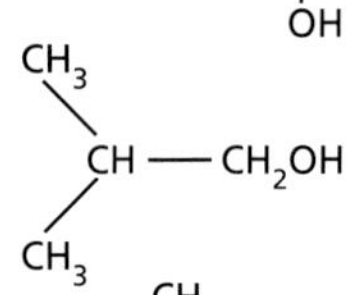

2-methylpropan-1-ol

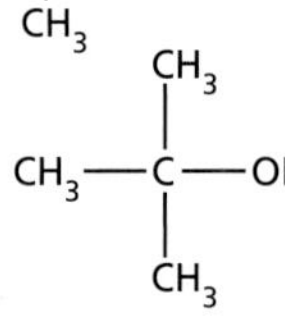

2-methylpropan-2-ol

Now try this

1 There are six alkenes with the molecular formula C_5H_{10}. Draw their skeletal formulae and name them.

2 Draw the skeletal formulae of the four positional isomers of dichloropropane ($C_3H_6Cl_2$). Name them, and specify which contain a chiral centre.

12.7 The different types of organic reactions

Most organic compounds can be represented as:

R—Fg

where Fg represents the functional group and R represents the rest of the molecule. Most organic reactions are those of functional groups. By definition, the rest of the molecule does not change under the conditions of the reaction that changes the functional group. (If it did change, it would have become a functional group itself!) Many functional groups have characteristic reactions, which are covered in detail in Topics 13 to 18 and 26 to 28.

Table 12.8 lists the types of organic reactions we shall be covering.

Table 12.8 Reaction types in organic chemistry

	Reaction type	Functional group	Example
1	radical substitution	alkane	$CH_4 + Cl_2 \rightarrow CH_3Cl + HCl$
2	electrophilic substitution	arene	⬡ $+ HNO_3 \longrightarrow$ ⬡$-NO_2$ $+ H_2O$
3	nucleophilic substitution	halogenoalkane	$OH^- + CH_3Br \rightarrow CH_3OH + Br^-$
		alcohol	$CH_3OH + HBr \rightarrow CH_3Br + H_2O$
4	electrophilic addition	alkene	$CH_2{=}CH_2 + Br_2 \rightarrow BrCH_2CH_2Br + H_2O$
5	nucleophilic substitution	aldehyde or ketone (carbonyl)	$\begin{array}{c} CH_3 \\ {>}C{=}O + HCN \longrightarrow \\ CH_3 \end{array} \begin{array}{c} CH_3 \quad OH \\ C \\ CH_3 \quad CN \end{array}$
6	hydrolysis	ester	$CH_3CO_2CH_3 + H_2O \rightarrow CH_3CO_2H + CH_3OH$
		amide	$CH_3CONH_2 + H_2O \rightarrow CH_3CO_2H + NH_3$
7	condensation	aldehyde or ketone (carbonyl)	$\begin{array}{c} CH_3 \\ {>}C{=}O + H_2N{-}R \longrightarrow \\ CH_3 \end{array} \begin{array}{c} CH_3 \\ C{=}N{-}R + H_2O \\ CH_3 \end{array}$
8	elimination	halogenoalkane	$CH_3CH_2Br + OH^- \rightarrow CH_2{=}CH_2 + Br^- + H_2O$
		alcohol	$CH_3CH_2OH \rightarrow CH_2{=}CH_2 + H_2O$
9	reduction	aldehyde or ketone (carbonyl)	$\begin{array}{c} CH_3 \\ {>}C{=}O + 2[H] \longrightarrow \\ CH_3 \end{array} \begin{array}{c} CH_3 \quad H \\ C \\ CH_3 \quad OH \end{array}$
10	oxidation	alcohol	$CH_3OH + [O] \rightarrow CH_2O + H_2O$
		aldehyde	$CH_2O + [O] \rightarrow HCO_2H$

[O] means an oxygen atom is given by the oxidising agent and [H] means a hydrogen atom is given by the reducing agent.

The meanings of some of the terms used in Table 12.8 are given below.

1 Radical – this is an atom or group of atoms that has an unpaired electron (that is, it has an odd number of electrons). Radicals (sometimes called free radicals) are highly reactive and are intermediates in many of the reactions of alkanes. They have the same number of protons as electrons, so are electrically neutral (unlike cations or anions).

2 Electrophile – this is an atom or group of atoms that reacts with electron-rich centres in other molecules. Electrophiles possess either a positive charge or an empty orbital in their valence shell, or they contain a polar bond producing an atom with a partial charge, shown as δ+.

3 **Nucleophile** – this is an atom or group of atoms that reacts with electron-deficient centres in molecules (such as the $C^{\delta+}$ in $C^{\delta+}$—$Cl^{\delta-}$ compounds). All nucleophiles possess a lone pair of electrons, and many are anions. Nucleophiles are called ligands when they react with transition metals (see section 24.4).

4 **Substitution** – this is a reaction in which one atom or group replaces another atom or group in a molecule.

5 **Addition** – this is a reaction in which an organic molecule (usually containing a double bond) reacts with another molecule to give only one product: A + B → C only.

6 **Hydrolysis** – this is a reaction in which a molecule is split into two by the action of water (often helped by OH^- or H^+ as a catalyst).

7 **Condensation** – this is the opposite of hydrolysis. Two molecules come together to form a bigger molecule, with the elimination of a molecule of water (or other small molecule, such as HCl).

8 **Elimination** – this is a reaction that forms an alkene by the removal of a molecule of H_2O from a molecule of an alcohol, or a molecule of HCl from a molecule of a chloroalkane.

9 **Reduction and oxidation** (**10**) – these have their usual meanings. Reduction is a reaction in which the sum of the oxidation numbers of the atoms in the functional group decreases. Oxidation is the reverse.

Worked example

Classify the following reactions into one (or more!) of the following six categories: substitution, addition, hydrolysis, reduction, condensation, oxidation.

a $CH_3CONH_2 + H_2O \rightarrow CH_3CO_2H + NH_3$
b $(CH_3)_2C{=}O + H{-}NH_2 \rightarrow (CH_3)_2C{=}NH + H_2O$
c $CH_3CHO + 2[H] \rightarrow CH_3CH_2OH$

Answer

a This is the splitting of a molecule into two by the action of water – a hydrolysis reaction. It could also be considered as a substitution reaction, in which —NH_2 is replaced by —OH.
b This reaction is the reverse process – it is a condensation reaction.
c This is both an addition reaction and a reduction.

Now try this

The scheme in Figure 12.20 is an outline of a synthesis of the local anaesthetic benzocaine, starting from 4-nitromethylbenzene. Classify each of the reactions I, II and III into one of the six categories in the worked example.

Figure 12.20

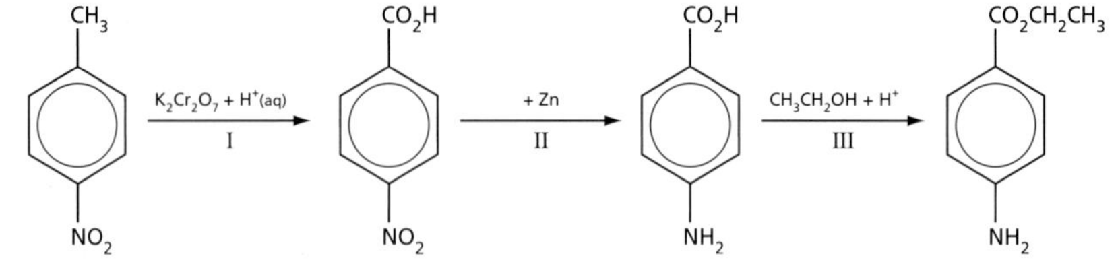

12.8 The different ways of breaking bonds, and how they are represented

Chemical reactions involve the breaking of bonds and the formation of new bonds. In the topics that follow, we shall be looking at the mechanisms of some organic reactions. By **mechanism** we mean a detailed description of which bonds break and which form, and in what order these processes occur. A mechanism also includes a description of how any catalyst that might be involved in a reaction works. There are three different ways in which a covalent bond to a carbon atom may break. Each way produces a different carbon species.

Homolysis

Bond homolysis, often called **homolytic fission**, results in carbon radicals (see Figure 12.21). Homolysis is the splitting of a bond giving an equal share of bonding electrons to each particle (from the Greek word *homo*, meaning 'the same'). In Figure 12.21, the carbon and hydrogen atoms take one electron each. The curly arrows with only half a head ('fish-hook' arrows) represent how each electron in the bond moves when the bond breaks. The carbon atom in the methyl radical $CH_3\bullet$ has only seven electrons around it. Methyl radicals are highly reactive, and readily form bonds with other atoms or molecules to regain their full octet of electrons.

Figure 12.21 Bond homolysis

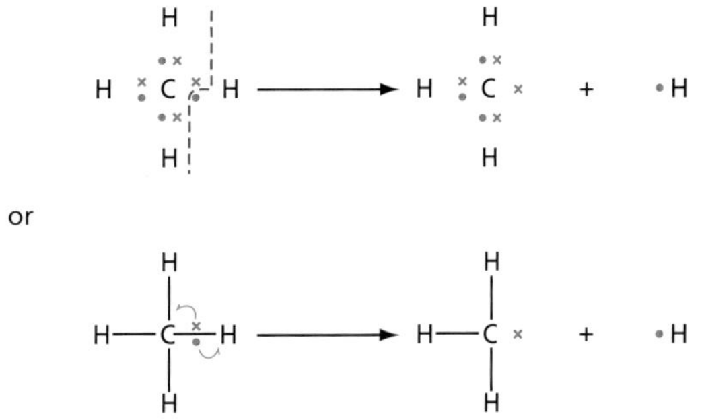

or

Bond heterolysis forming carbocations

Heterolysis involves the splitting of a bond giving an unequal share of bonding electrons to each particle (from the Greek word *hetero* meaning 'other', or 'different'). In Figure 12.22, the chlorine atom has taken both the bonding electrons, leaving the carbon atom with none. The curly arrow describes the movement of the bonded pair of electrons to the chlorine atom. Chlorine, being more electronegative than carbon, has already attracted the bonding electrons partially, forming a bond dipole $^{\delta+}C—Cl^{\delta-}$ (see section 3.11). This movement of the electron pair results in the carbon atom in the methyl **carbocation** having only six electrons in its outer shell, and a single positive charge. It is a strong electrophile.

Figure 12.22 Bond heterolysis forming a carbocation

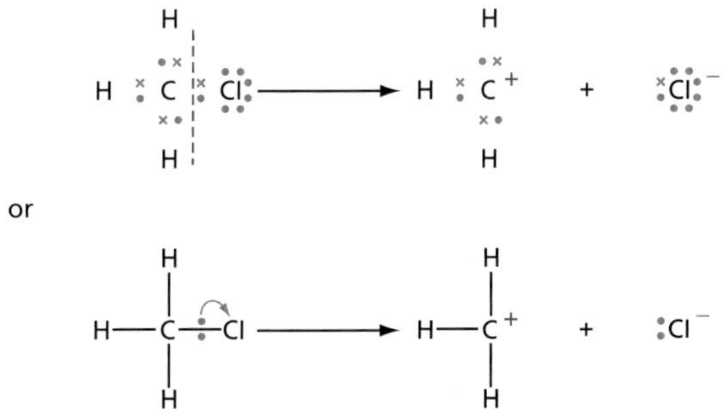

or

Bond heterolysis forming carbanions

This method of splitting a bond to carbon is much less common than that forming carbocations. Carbon is quite electropositive. Only when it is bonded to an even more electropositive element will its bonds split to form **carbanions**. Metals are highly electropositive, and some of them form covalent bonds with carbon. Methyl lithium is the simplest example (see Figure 12.23).

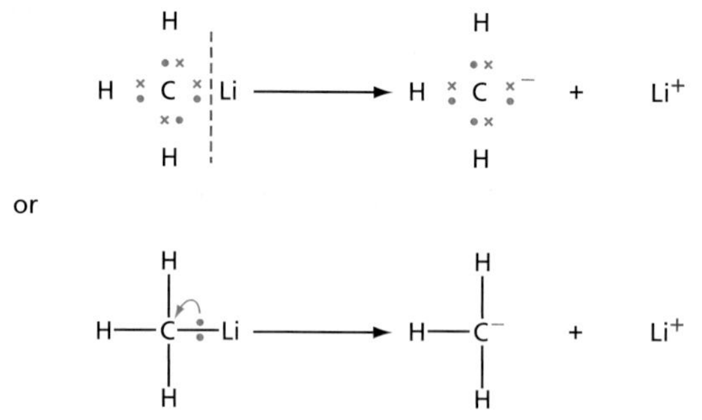

Figure 12.23 Bond heterolysis forming a carbanion

The methyl carbanion formed contains a full octet of electrons in the outer shell of carbon. Its strong negative charge makes it a highly reactive nucleophile. It will react with virtually everything that contains a δ+ atom. The curly arrow describes the movement of the bonded pair of electrons to the carbon atom.

The use of curly arrows

The ⌒ symbol is a useful shorthand which represents the movement of a pair of electrons from the position at the tail of the arrow to the position at the head. Its meaning is very exact, and it should not be used to imply anything more, or less, than this movement of electrons.

An example of the use of curly arrows is the reaction that occurs when hydrogen chloride gas dissolves in water:

$$HCl + H_2O \rightarrow H_3O^+ + Cl^-$$

The movement of electrons in this reaction may be represented as follows:

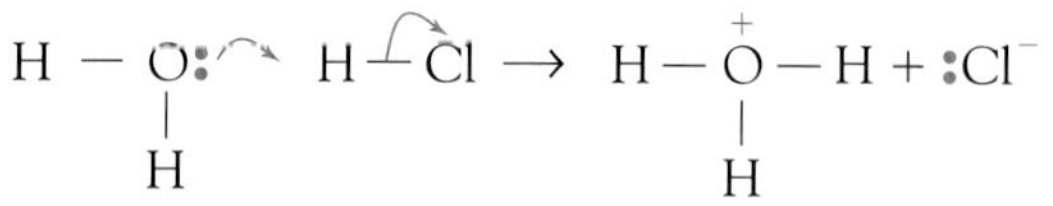

This equation is a highly condensed way of saying:

> A pair of electrons moves from a lone pair of oxygen in water to the space between that oxygen and the hydrogen of the HCl molecule, to form a new O—H bond. At the same time, the two electrons in the H—Cl bond move towards the chlorine atom and become a lone pair on that atom. Since oxygen started off with 'full use' of a lone pair, and has finished up with only a half-share of the electron pair making up the O—H bond, its charge has decreased by one electronic unit (0 to +1). Since the hydrogen being transferred started with a half-share of the electron pair making up the H—Cl bond, and has finished with a half-share of the electron pair making up the new O—H bond, its charge has remained the same (zero). Since chlorine started with a half-share of the electron pair in the H—Cl bond, and has finished with full use of its own lone pair, its charge has increased by one electronic unit (0 to −1).

From the length of the paragraph above, you will appreciate how much shorter the description using curly arrows is! But every time you write out mechanisms using them, try to remember how this paragraph was translated into curly arrows. Their use must be very exact if they are to describe electron movements correctly.

12.9 Homologous series

Compounds containing the same functional group often form **homologous series**. In such a series:

- the molecular formulae of adjacent members of the series differ by a fixed unit (usually CH_2)
- the physical properties, such as boiling point and density, vary regularly from one compound to another
- the molecular formulae of members of the series fit the same general formula.

> **Now try this**
>
> Use curly arrows to describe what happens to the electrons when (ionic) ammonium chloride is made from ammonia and hydrogen chloride:
>
> $NH_3 + HCl \rightarrow NH_4Cl$

Two examples of homologous series are shown in Table 12.9. Their boiling points show a regular trend, as shown in Figure 12.24.

Table 12.9 Some properties of the lower alkanes and alcohols

Alkanes (general formula C_nH_{2n+2})			Alcohols (general formula $C_nH_{2n+1}OH$)		
Name	Formula	Boiling point/°C	Name	Formula	Boiling point/°C
methane	CH_4	−164	methanol	CH_3OH	65
ethane	C_2H_6	−89	ethanol	C_2H_5OH	78
propane	C_3H_8	−42	propanol	C_3H_7OH	97
butane	C_4H_{10}	0	butanol	C_4H_9OH	118
pentane	C_5H_{12}	36	pentanol	$C_5H_{11}OH$	138

Figure 12.24 Boiling points of alkanes and alcohols

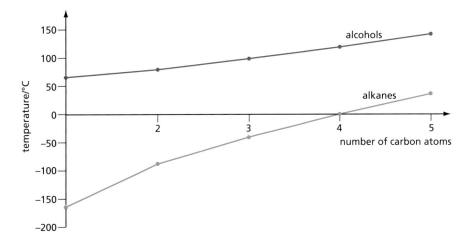

Now try this

For three different homologous series, the formulae of the first two members are as follows:
1 CH_2O, C_2H_4O
2 C_2H_3N, C_3H_5N
3 C_6H_6, C_7H_8
For each series, state:
a the molecular formula of the fifth member
b the general formula.

Worked example

The first and third members of a homologous series have the molecular formulae C_2H_2 and C_4H_6.
a What is the formula of the fourth member of the series?
b What is the general formula of the series?

Answer
a The difference between the two molecular formulae is C_2H_4, which is $2 \times CH_2$. The fourth member will therefore have the formula $C_4H_6 + CH_2$, which is **C_5H_8**.
b The general formula is **C_nH_{2n-2}**.

12.10 Calculating yields in organic reactions

When carried out in the laboratory, organic reactions do not always go to completion. This may be due either to the reaction reaching equilibrium, or to it being so slow that it would take an age to complete. Some organic reactions compete with one another (see pages 212 and 275), and so by-products are formed along with the major product. Even if a reaction goes to completion, that is, every molecule of starting material is converted into a molecule of product, the recovered yield will not be 100%. This is due to material being lost during the purification procedures. For example, if a solid has been purified by recrystallisation, some will inevitably be left in the cold saturated solution after the crystals have been filtered off. If a liquid has been distilled, a small amount of it will remain in the distillation flask, or on the sides of the condenser.

The **percentage yield** is a comparison of the actual yield with the **theoretical yield** – the yield that might have been expected if the reaction took place according to the stoichiometric equation, and there was no loss of product in the purification stages.

$$\text{percentage yield} = \frac{\text{actual yield} \times 100\%}{\text{theoretically possible yield}}$$

For well designed reactions, in the hands of an able experimental chemist, yields of more than 95% are possible. More usually, yields of 60–80% are obtained.

Worked example

The reaction of 10.0 g of benzoyl chloride with concentrated ammonia solution produced 5.63 g of pure recrystallised benzamide. Calculate the percentage yield.

benzoyl chloride   + 2NH₃ →   benzamide   + NH₄Cl

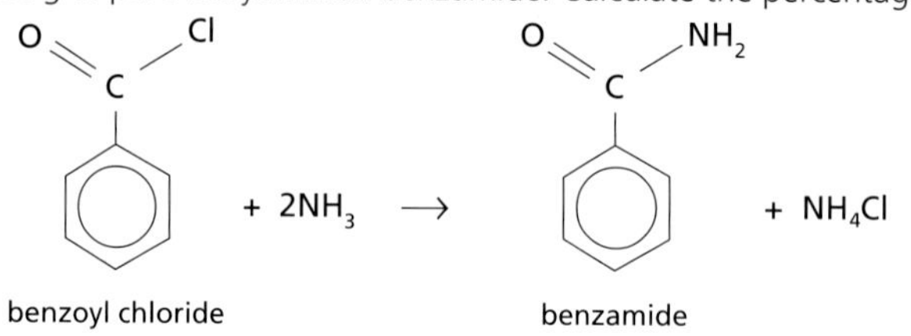

Answer

We first calculate the number of moles of benzoyl chloride used (see section 1.7).

$$M_r(C_6H_5COCl) = 7 \times 12 + 5 \times 1 + 16 + 35.5 = 140.5$$

$$n = \frac{m}{M}$$

$$n(C_6H_5COCl) = \frac{10}{140.5} = 0.071\,17 \text{ mol}$$

From the equation, 1 mol of benzoyl chloride produces 1 mol of benzamide. Therefore:

$$n(C_6H_5CONH_2) = 0.071\,17 \text{ mol}$$

$$m = n \times M_r$$

$$M_r(C_6H_5CONH_2) = 7 \times 12 + 7 \times 1 + 16 + 14 = 121$$

so $m(C_6H_5CONH_2) = 0.071\,17 \times 121 = 8.61 \text{ g}$

and percentage yield $= \dfrac{\text{actual yield} \times 100\%}{\text{theoretical yield}}$

$$= 5.63 \times \frac{100}{8.61}$$

$$= \mathbf{65.4\%}$$

Now try this

Calculate the percentage yield from each of the following reactions.

1 On oxidation, 5.0 g of ethanol produced 4.5 g of ethanoic acid:

$$C_2H_5OH + 2[O] \rightarrow CH_3CO_2H + H_2O$$

2 Reacting 10.0 g of propanol with an excess of hydrogen bromide produced 12.5 g of bromopropane:

$$C_3H_7OH + HBr \rightarrow C_3H_7Br + H_2O$$

Summary

- Carbon forms strong bonds to itself and to many other atoms. This allows organic compounds to be kinetically stable, owing to the high activation barrier to reaction, even though they may be thermodynamically unstable.
- There are six types of formula used to represent organic compounds – **empirical**, **molecular**, **structural**, **displayed**, **stereochemical** and **skeletal**.
- All organic compounds can be named systematically, using a logical system based on the number of carbon atoms in their longest chain.

- There are five types of **isomerism** shown by organic compounds – **chain, positional, functional group, geometrical** and **optical**.
- Most organic reactions can be classified into one of ten general types.
- Many organic reactions do not give a 100% yield of product. The **percentage yield** is a measure of how efficient a reaction is.

Examination practice questions

Please see the data section of the CD for any A_r values you may need.

1 Compound **X** has the molecular formula $C_4H_8O_2$.
   a The molecule of **X** has the following features.
   - The carbon chain is unbranched and the molecule is not cyclic.
   - No oxygen atom is attached to any carbon atom which is involved in π bonding.
   - No carbon atom has more than one oxygen atom joined to it.

   There are five possible isomers of **X** which fit these data. Four of these isomers exist as two pairs of stereoisomers.
   i Draw displayed formulae of **each** of these two pairs.
   ii These four isomers of **X** show two types of stereoisomerism.
     State which type of isomerism each pair shows. [6]
   [Cambridge International AS & A Level Chemistry 9701, Paper 21 Q5 b November 2012]

2 When 0.42 g of a gaseous hydrocarbon **A** is slowly passed over a large quantity of heated copper(II) oxide, CuO, **A** is completely oxidised. The products are collected and it is found that 1.32 g of CO_2 and 0.54 g of H_2O are formed. Copper is the only other product of the reaction.
   a i Calculate the mass of carbon present in 1.32 g of CO_2. Use this value to calculate the amount, in moles, of carbon atoms present in 0.42 g of **A**.
   ii Calculate the mass of hydrogen present in 0.54 g of H_2O. Use this value to calculate the amount, in moles, of hydrogen atoms present in 0.42 g of **A**.
   iii It is thought that **A** is an alkene rather than an alkane. Use your answers to (i) and (ii) to deduce whether this is correct. Explain your answer. [5]
   b Analysis of another organic compound, **B**, gave the following composition by mass: C, 64.86%; H, 13.50%, O, 21.64%.
   i Use these values to calculate the empirical formula of **B**.
   ii The empirical and molecular formulae of **B** are the same. **B** is found to be chiral. Draw displayed formulae of the two optical isomers of this compound, indicating with an asterisk (*) the chiral carbon atom.
   iii There are three other structural isomers of **B** which are not chiral but which contain the same functional group as **B**. Draw these isomers. [7]
   [Cambridge International AS & A Level Chemistry 9701, Paper 23 Q2 November 2011]

3 The structural formulae of six different compounds, **P – U**, are given below.

$CH_3CH=CHCH_2CH_3$
P

$CH_3CH_2CH_2CH_2CH_2OH$
S

$CH_3CH_2COCH_2CH_3$
Q

$HOCH_2CH_2CH(OH)CH_3$
T

$CH_2=CHCH_2CH_2CH_3$
R

$CH_3CH_2CH_2OCH_2CH_3$
U

   a i What is the empirical formula of compound **T**?
   ii Draw the skeletal formula of compound **S**. [2]
   b i Compounds **S** and **U** are isomers.
     What type of isomerism do they show?
   ii Two of the six formulae **P – U** can **each** be drawn in two forms which are known as stereoisomers.
     Which two compounds have formulae that can be drawn in two forms?
     What type of stereoisomerism does each show?
     Identify each compound by its letter. [3]
   c Compound **S** can be converted into compound **R**.
   i What type of reaction is this?
   ii What reagent would you use for this reaction?
   iii Write the structural formula of the compound formed when **T** undergoes the same reaction using an excess of the reagent you have used in **c ii**. [3]
   [Cambridge International AS & A Level Chemistry 9701, Paper 23 Q4 a, b & c November 2011]

13 Alkanes

Functional group:

C—C—H

In this topic the properties and reactions of the simplest of the homologous series, the alkanes, are discussed. Their extraction from crude oil and how they are transformed into fuels and feedstocks for the chemical industry are described, and their reactions with oxygen, chlorine and bromine are studied. Their characteristic reaction of radical substitution is explained in detail.

Learning outcomes

By the end of this topic you should be able to:

1.5b) perform calculations, including use of the mole concept, involving volumes of gases (part, see also Topic 1)

14.1a) interpret and use the general, structural, displayed and skeletal formulae of the alkanes (part, see also Topic 12)

14.2a) interpret and use the following terminology associated with organic reactions: homolytic fission, free radical, initiation, propagation, termination

15.1a) understand the general unreactivity of alkanes, including towards polar reagents

15.1b) describe the chemistry of alkanes as exemplified by the following reactions of ethane: combustion, and substitution by chlorine and by bromine

15.1c) describe the mechanism of free-radical substitution at methyl groups with particular reference to the initiation, propagation and termination reactions

15.1d) explain the use of crude oil as a source of both aliphatic and aromatic hydrocarbons

15.1e) suggest how cracking can be used to obtain more useful alkanes and alkenes of lower M_r from larger hydrocarbon molecules

15.3a) describe and explain how the combustion reactions of alkanes lead to their use as fuels in industry, in the home and in transport

15.3b) recognise the environmental consequences of:
  - carbon monoxide, oxides of nitrogen and unburnt hydrocarbons arising from the internal combustion engine and of their catalytic removal
  - gases that contribute to the enhanced greenhouse effect

15.3c) outline the use of infrared spectrossopy in monitoring air pollution (see also Topic 29).

13.1 Introduction

The **alkanes** are the simplest of the homologous series found in organic chemistry. They contain only two types of bonds, both of which are strong and fairly non-polar (see Table 13.1). (See section 3.10 for an explanation of bond polarity.)

Table 13.1 Carbon–carbon and carbon–hydrogen bonds are strong and non-polar.

Bond	Bond enthalpy/kJ mol^{-1}	Polarity (difference in electronegativity)
C—C	346	0
C—H	413	0.35 ($H^{\delta+}$—$C^{\delta-}$)

Consequently, the alkanes are not particularly reactive. Their old name (the 'paraffins') reflected this 'little (*parum*) **affin**ity'. What few reactions they have, however, are of immense importance to us. They constitute the major part of crude oil, and so are the starting point of virtually every other organic compound made industrially. Apart from their utility as a feedstock for the chemical industry, their major use is as fuels in internal combustion engines, jet engines and power stations. We shall be looking at their role in combustion reactions in detail later in this topic.

The straight-chain alkanes form a homologous series with the general formula C_nH_{2n+2}. As the number of carbon atoms in the molecule increases, the van der Waals' attractions between the molecules become stronger (see section 3.17). This has a clear effect on the boiling points and the viscosities of members of the series (see Figure 13.1 and Table 14.1, page 252).

Figure 13.1 The boiling points and viscosities of the alkanes

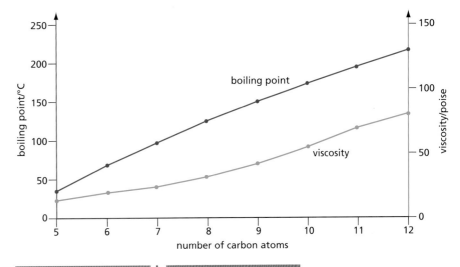

Figure 13.2 Space-filling and ball-and-stick models of ethane, C_2H_6 (**a** and **b**) and butane, C_4H_{10} (**c** to **f**). The space-filling models show the overall shape of the molecule but the ball-and-stick models show the positions of the atoms more clearly. Note that **e** and **f** are identical to **c** and **d**, as they are formed by free rotation of the central C—C bond.

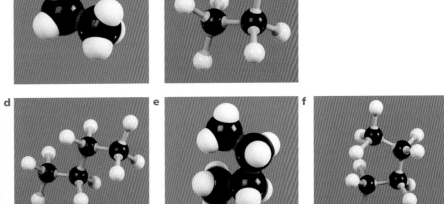

13.2 Isomerism and nomenclature

As was mentioned in section 12.5, the first step in naming an alkane is to identify the longest carbon chain, which gives the stem. Side branches are then identified by prefixes. Each prefix is derived from the stem name denoting the number of carbon atoms the side chain contains, and is preceded by a number denoting the position it occupies on the longest carbon chain. Two examples will make this clear.

Example 1

$$\begin{array}{c} CH_3 \\ | \\ CH \quad\quad CH_3 \\ \diagup \quad\quad \diagup \\ CH_3 \quad\quad CH_2 \end{array}$$

- The longest carbon chain contains four carbon atoms, so the alkane is a derivative of **butane**.
- There is just one side branch, containing one carbon atom (and hence called **methyl**).

235

- The branch is joined to the second carbon atom of the butane chain.
- The name of the alkane is **2-methylbutane**.

Example 2

- The longest carbon chain consists of six carbon atoms (**hexane**).
- There are two one-carbon side branches (two methyl groups, **dimethyl**).
- These are on positions 2 and 4 of the hexane chain. (We count from the right-hand side: counting from the left would put the methyl groups on positions 3 and 5, so we choose to count from the end that gives the *lower* numbers.)
- The name of the alkane is **2,4-dimethylhexane**.

The main type of isomerism shown by the alkanes is chain isomerism, although some of the more complicated structures are capable of existing as optical isomers.

Worked example

What is the alkane of lowest M_r (relative molecular mass) that contains a chiral centre (see section 12.6)? Give its systematic name and molecular formula.

Answer
To be a chiral centre, a carbon atom must have four different groups attached to it. The four lowest M_r groups that are all different to one another are H, CH_3, C_2H_5 and C_3H_7. The alkane is therefore CH_3CH_2—$CH(CH_3)$—$CH_2CH_2CH_3$, whose molecular formula is **C_7H_{16}**. Applying the rules listed in section 12.5, the longest chain in this compound contains six carbon atoms, and the side group (methyl) is on the third carbon atom of the chain. Hence its name is **3-methylhexane**.

Now try this

1 Give the systematic names of the following compounds.
   a $(C_2H_5)_3CH$
   b $(CH_3)_2CH$—$C(CH_3)_2$—$CH(CH_3)_2$
2 Draw the structural and skeletal formulae of the following compounds, and also give their molecular formulae.
   a 2,2,3,3-tetramethylpentane
   b 3,3-diethylhexane

As the number of carbon atoms in an alkane increases, so does the number of possible structural isomers. There are two isomers of C_4H_{10}, three of C_5H_{12}, and five of C_6H_{14} (see Table 13.2). By the time we reach $C_{10}H_{22}$, there are no fewer than 75 isomers, and for $C_{20}H_{42}$ the number of isomers has been calculated to be in excess of 300 000!

Structural (chain) isomers of alkanes undergo the same chemical reactions. They differ slightly in the following two respects, however.

- The more branches an isomer has, the more compact are its molecules, with a smaller surface area of contact between them. This causes the van der Waals' bonding between the molecules to be weaker. The consequence of the weaker intermolecular forces is that the boiling points of branched isomers are *lower* than their straight-chain counterparts (see Table 13.2).
- The isomers that are more branched do not pre-ignite so easily in internal combustion engines. They are therefore much preferred for use in modern high-compression petrol engines.

Table 13.2 The boiling points of some isomers of C_5H_{12} and C_6H_{14}, showing how the boiling point increases with carbon-chain length, but decreases with branching

Number of carbon atoms	Compound	Molecular formula	Skeletal formula	Boiling point/°C
5	pentane	C_5H_{12}		36
	2-methylbutane	C_5H_{12}	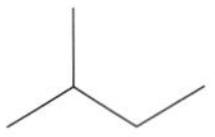	28
	2,2-dimethylpropane	C_5H_{12}	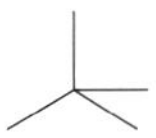	9
6	hexane	C_6H_{14}		69
	3-methylpentane	C_6H_{14}	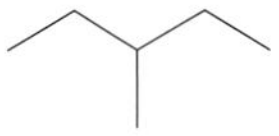	63
	2-methylpentane	C_6H_{14}	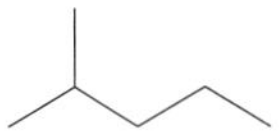	60
	2,3-dimethylbutane	C_6H_{14}	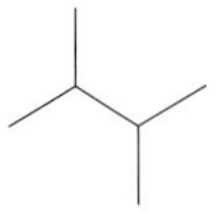	58
	2,2-dimethylbutane	C_6H_{14}	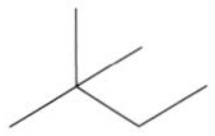	50

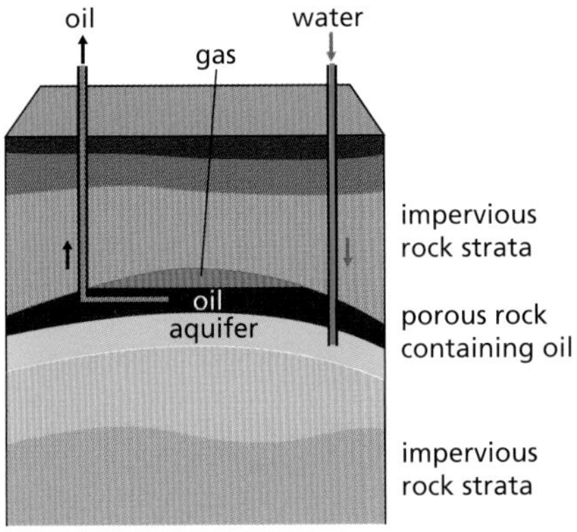

Figure 13.3 'Bottle brush' extraction of oil from rock strata

13.3 The processing of crude oil

Origins

Crude oil, like coal, is a fossil fuel. It was formed millions of years ago when plant (and some animal) remains were crushed and subjected to high temperatures in the absence of air. Under these extreme chemically reducing conditions, oxygen and nitrogen were removed from the carbohydrates and proteins that had made up the organisms, and long carbon chains formed. It is estimated that most of the oil formed in this way, deep underground, has since been transported to the surface by geological movements, and has evaporated. The amount left is (or was) still considerable, however.

Most crude oil is found trapped in a stratum (layer) of porous rock, capped by a dome-shaped layer of impervious rock. It is extracted by drilling through the cap rock, whereupon the natural high pressure of the oil forces it up to the surface. If such pressure is insufficient, water is pumped into the oil reservoir and forces the oil upwards (see Figure 13.3). (Oil is less dense than water.)

Fractional distillation

Crude oil is a mixture of many hundreds of hydrocarbons, ranging in size from one to 40 or so carbon atoms per molecule. The first stage in its processing is a fractional distillation carried out at atmospheric pressure (see Figure 13.4). The boiling point of crude oil ranges continuously from below 20 °C to over 300 °C, and many fractions could be collected, each having a boiling point range of only a few degrees. In practice, the most useful procedure is to collect just four fractions from the primary distillation. Each fraction can then be further purified, or processed, as required. Table 13.3 lists these fractions (and the residue), together with their major uses.

Figure 13.4 The fractional distillation of crude oil

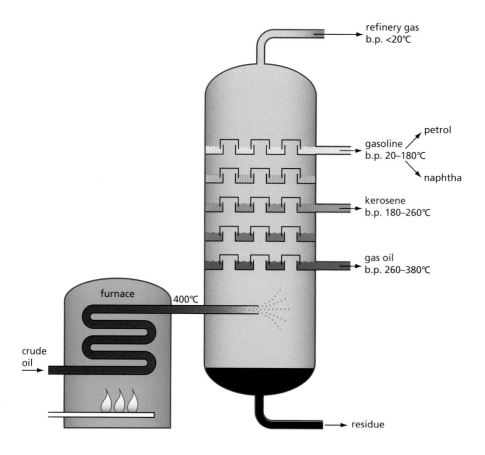

Table 13.3 Fractions from the distillation of crude oil

Fraction	Number of carbon atoms	Boiling point range/°C	Percentage of crude oil	Uses
refinery gas	1–4	<20	2	fuel, petrochemical feedstock
gasoline	5–12	20–180	20	fuel for petrol engines, petrochemical feedstock
kerosene	10–16	180–260	13	aeroplane fuel, paraffin, cracked to give more gasoline
gas oil	12–25	260–380	20	fuel for diesel engines, cracked to give more gasoline
residue	>25	>380	45	lubricating oil, power station fuel, bitumen for roads, cracked to give more gasoline

Cracking – breaking long-chain molecules into shorter ones

Over 90% of crude oil is burnt as fuels of one sort or another. Less than 10% is used as a feedstock to produce the host of organic chemicals (plastics, fibres, dyes, paints, pharmaceuticals) that are derived from crude oil. The relative demand for petrol (and hence for the gasoline fraction) far outstrips the supply from the primary distillation – about twice as much is required as is contained in crude oil. Consequently, methods have been developed to convert the higher boiling, and more abundant, fractions into gasoline. This is done by **cracking** the long-chain molecules into shorter units. Two alternative processes are used – thermal cracking and catalytic cracking. We can illustrate the difference between these using as an example dodecane (boiling point 216 °C), a typical component of the kerosene fraction.

Thermal cracking

Thermal cracking involves heating the alkane mixture to about 800 °C and at moderate pressure, in the absence of air but in the presence of steam. After only a fraction of a second at this temperature, the mixture is rapidly cooled. By this means dodecane might typically be broken into hexane (boiling point 69 °C) and ethene:

$$CH_3(CH_2)_{10}CH_3 \xrightarrow{\text{heat with steam at 800 °C}} CH_3(CH_2)_4CH_3 + 3CH_2{=}CH_2$$

The hexane is required for fuel but the ethene by-product is not wasted, as it is a key feedstock for the plastics, fibres and solvents industries; the conditions for thermal cracking are often chosen so as to optimise ethene production.

The heat energy at 800 °C is sufficient to break the C—C bonds into two carbon free radicals. Long-chain radicals readily split off ethene units, eventually producing shorter-chain alkanes and alkenes.

Very little rearrangement of the chains occurs during thermal cracking.

Catalytic cracking

Catalytic cracking involves heating the alkane mixture to a temperature of about 500 °C and passing it under slight pressure over a catalyst made from a porous mixture of aluminium and silicon oxides (called zeolites). The catalyst causes the carbon chains to undergo internal rearrangements before forming the final products.

A typical set of products from catalytic cracking is shown in the following equation.

$$2CH_3(CH_2)_{10}CH_3 \rightarrow (CH_3)_2CHCH_2CH_2CH_3 + (CH_3)_2CHCH(CH_3)_2 + 6CH_2{=}CH_2$$

The branched-chain alkanes produced by catalytic cracking are useful components of high-octane petrol (see Figure 13.5).

Re-forming

The demand for branched-chain, cyclic and aromatic hydrocarbons as components of high-octane petrol is further satisfied by **re-forming** straight-chain alkanes. The vaporised alkane mixture is passed over a platinum-coated aluminium oxide catalyst at 500 °C and moderately high pressure. For example, heptane can be re-formed as shown in Figure 13.6.

Figure 13.5 In a catalytic cracker (a 'cat cracker'), long-chain molecules are converted into shorter-chain molecules which have more uses.

Figure 13.6 Straight-chain alkanes are re-formed into branched-chain, cyclic and aromatic hydrocarbons. (See Topic 25 for a description of aromatic hydrocarbons.)

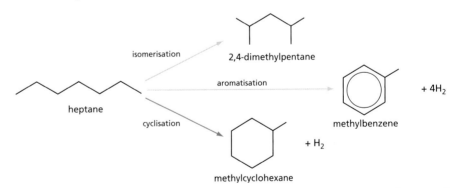

In general, re-forming changes straight-chain alkanes into branched-chain alkanes and cyclic hydrocarbons without the loss of any carbon atoms, but often with the loss of hydrogen atoms.

13.4 Energy sources

Modern society has an almost inexhaustible thirst for energy, in the forms of electricity, space heating or air conditioning for homes and offices, and energy for transport, by road, rail, sea and air.

As we saw in section 5.3, the energy sources available to us can be divided into two categories – renewable and non-renewable. The former category includes the renewable fuel ethanol, obtained from the fermentation of starch-rich plants such as sugar cane, and the alternative energy sources such as wind power, wave power and tidal power. Ethanol from fermentation can be used as a renewable fuel for cars and, conceivably, in power stations, just like the hydrocarbon fossil fuels. In such situations, similar types of pollution could result from its burning. Nevertheless, in practice, it tends to be a 'cleaner' fuel than hydrocarbons.

The alternative energy sources produce no chemical pollution. Their disadvantages are that they are often seasonal and can be unreliable, and they are not readily transportable, and so cannot be used for mobile vehicles. Their energy also cannot be stored for future use. Energy from these alternative sources is usually transformed immediately into electrical energy. Progress is being made to solve these problems, however. One possibility is to use alternative sources of energy to split water into hydrogen and oxygen (see section 23.6). The hydrogen can be transported and stored for future use as a pollution-free fuel.

The non-renewable energy sources include the hydrocarbons natural gas and crude oil, and the carbon-rich coal. These are called **fossil fuels**, having been produced many millions of years ago from trees, giant ferns and marine organisms buried beneath the Earth's surface.

Uranium, the primary fuel for nuclear reactors, was produced an even longer time in the past, when the elements of the Solar System were synthesised in the centres of dying stars billions of years ago. In a sense, it too is a non-renewable energy source. Nuclear reactors produce little chemical pollution, but the risks of radiation pollution from their use and from the reprocessing of their spent fuel have caused them to fall from favour.

Some power stations are now being designed to burn domestic and industrial waste. Although this should decrease the amount of waste going to landfill sites or polluting the environment, care has to be taken to ensure that potentially polluting combustion products are trapped. Another way of recycling combustible waste is to heat it in the absence of oxygen. This produces oil or solid combustible material, which can be used as fuel in a conventional power station.

There are five main pollutants that are formed when fossil fuels are burnt:

- carbon dioxide (this is not strictly a pollutant, since it occurs in nature, but its overproduction can cause problems)
- carbon monoxide
- unburnt hydrocarbons
- nitrogen oxides (see Figure 13.7)
- sulfur dioxide.

Figure 13.7 Nitrogen oxides contribute to low-level ozone and smog. Low-level ozone causes respiratory problems, and peroxyacetyl nitrate (PAN), a major component of smog, is harmful to plants and humans. The reactions are:
$NO_2 \rightarrow NO + O$
$O + O_2 \rightarrow O_3$
hydrocarbons $+ O_2 + NO_2 + $ light $\rightarrow$ PAN

All these pollutant gases absorb infrared radiation at specific and characteristic frequencies, and so their presence and their concentrations in gaseous environments (e.g. automobile exhausts, city streets, the general atmosphere) can be monitored using infrared gas sensors. The principles of infrared spectroscopy are described in section 29.3.

With the exception of carbon dioxide and sulfur dioxide, these can all be removed from vehicle emissions by a catalytic converter (see section 8.6). **Flue-gas desulfurisation (FGD)** is used in power stations to reduce the amount of sulfur dioxide reaching the atmosphere. In this process the flue gases are passed through a slurry of calcium carbonate. Calcium carbonate (limestone) is cheap, and the calcium sulfate produced can be sold as gypsum, a component of plaster.

$$CaCO_3(s) + SO_2(g) \rightarrow CaSO_3(s) + CO_2(g)$$

$$CaSO_3(s) + \frac{1}{2}O_2 \rightarrow CaSO_4(s)$$

The use of the more expensive magnesium carbonate allows the sulfur dioxide to be regenerated and used to make sulfuric acid.

$$MgCO_3(s) \xrightarrow{\text{heat}} MgO(s) + CO_2(g)$$

$$MgO(s) + SO_2(g) \longrightarrow MgSO_3(s)$$

$$MgSO_3(s) \xrightarrow{\text{heat}} MgO(s) + SO_2(g)$$

Magnesium carbonate is used because both it and magnesium sulfite are more easily decomposed by heat than the corresponding calcium salts are (see section 10.4).

The greenhouse effect and carbon dioxide

What is the greenhouse effect?

A greenhouse is a building constructed mainly of glass, which is used to hasten the growth of plants in colder climates, by providing a warmer atmosphere for them. Apart from its obvious role of protecting the contents from the weather, the glass in a greenhouse also warms the inside by means of the **greenhouse effect**. Glass is transparent to visible light, and allows most of the sunlight through. The light is absorbed by objects within the greenhouse (if the objects happen to be plants, useful photosynthesis results – but even a plant-free greenhouse still shows the greenhouse effect). This absorption of light energy causes the objects to warm up. Warm objects emit radiation of a much longer wavelength than the visible spectrum – they give out infrared radiation. Glass is opaque to infrared radiation, so it absorbs it, and warms up in the process. So, much of the energy from the visible light is trapped inside the greenhouse, which becomes warmer and warmer.

Figure 13.8 It is warmer inside a greenhouse than outside.

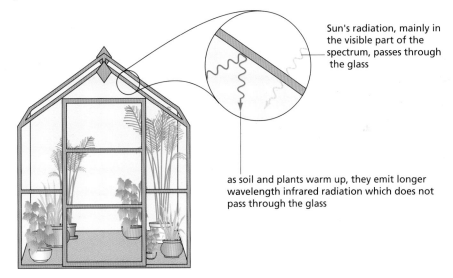

Sun's radiation, mainly in the visible part of the spectrum, passes through the glass

as soil and plants warm up, they emit longer wavelength infrared radiation which does not pass through the glass

The atmosphere around the Earth acts to a certain extent like the glass in a greenhouse. The Sun's rays – visible, ultraviolet and infrared – pass through the atmosphere to the surface of the Earth. Absorption of this energy at the surface causes the land and sea to warm up. They begin to emit infrared radiation, much of which passes straight through the atmosphere and is lost to outer space. But a proportion is absorbed by the gaseous molecules in the atmosphere, and so this part of the energy is not lost, but re-radiated back to the Earth's surface (see Figure 13.9). By this means the surface is kept at a reasonably constant, warm temperature day and night, summer and winter. (Contrast this with the situation on Mars, which has little atmosphere, and where night-time temperatures can be 100°C lower than those during the day.)

Increasing the greenhouse effect of the atmosphere

The greenhouse effect of the atmosphere is essential to provide the right stable environment for life. But there is a fine balance to be struck between not enough of the effect, in which case more of the incident solar energy will be lost, cooling the Earth to another ice age, and too much greenhouse effect, in which case the Earth will become too warm. This could result in melting icecaps, raised sea levels and changes in climatic air circulation.

Carbon dioxide is a natural atmospheric component that is responsible for the greenhouse potential of the atmosphere. Its production from the burning of fossil fuels for heating, transport and electricity generation, causes an increase in that potential. The present concentration of carbon dioxide in the atmosphere is only 0.035%. But 300 years ago it was only 0.028%. This 25% increase has contributed to an average increase in global temperature of 0.5°C.

Figure 13.9 The greenhouse effect warms the Earth.

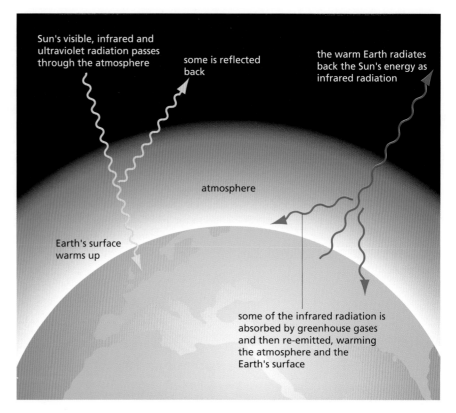

The other major natural greenhouse gas in the atmosphere is water vapour. It has been suggested that a 'runaway' effect might begin to operate – as global warming increases, the warmer atmosphere will be able to absorb more water vapour, which in turn will lead to a greater greenhouse effect, increasing global warming even more. Both CO_2 and H_2O absorb infrared radiation by increasing the vibrational energy of their bonds (see Figure 13.10).

Figure 13.10 Carbon dioxide and water molecules absorb infrared radiation by vibrating their bonds. This is how they increase the greenhouse effect.

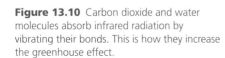

There are many natural ways of reducing atmospheric carbon dioxide.

- Water in the seas dissolves millions of tonnes (but less now than it did, since the average ocean temperature has increased by 0.5 °C in the last 100 years, and gases are less soluble in hot than in cold water).
- Plankton can fix the dissolved carbon dioxide into their body mass by photosynthesis.
- Trees fix more atmospheric carbon dioxide per acre than grass and other vegetation do.
- Blanket peat bogs, which cover ground in temperate climates that have a large annual rainfall, store a large quantity of carbon.

More intensive agriculture, especially cattle-rearing, for example in Australia, also produces greenhouse gases such as methane, which is ten times as effective in this role as carbon dioxide. But methane is produced in smaller quantities than carbon dioxide.

Many of the other pollutant gases that modern society has used have an even greater potential – for example, CFCs are over 5000 times as effective as greenhouse gases as carbon dioxide is but, like methane, they are produced in much smaller quantities. (Their main environmental effect, however, is in destroying the stratospheric ozone layer, as we shall see in section 15.4.)

The case against carbon dioxide is not totally proven, but it seems increasingly firm.

13.5 Reactions of alkanes

Apart from cracking and re-forming, there are two important reactions of alkanes, namely combustion and halogenation. We shall look at each in turn.

1 Combustion

Fuels

As mentioned at the start of the topic, most of the alkanes that are produced from crude oil are burned as fuels. Their complete combustion is a highly exothermic process:

$$CH_4(g) + 2O_2(g) \rightarrow CO_2(g) + 2H_2O(l) \qquad \Delta H_c^\ominus = -890 \, kJ \, mol^{-1}$$
methane
(natural gas)

$$C_{20}H_{42}(l) + 30.5O_2(g) \rightarrow 20CO_2(g) + 21H_2O(l) \qquad \Delta H_c^\ominus = -13\,368 \, kJ \, mol^{-1}$$
eicosane
(fuel for oil-powered power stations)

> ### Worked example
>
> Methane (natural gas) is replacing crude oil and coal as the fuel in some power stations, in an attempt to cut down on greenhouse gas emissions.
>
>   Use the equations above to calculate the percentage reduction in carbon dioxide produced per kJ when natural gas replaces fuel oil in a power station.
>
> #### Answer
> For methane, the equation tell us that 890 kJ of energy are released when 1 mol of carbon dioxide is formed:
> $$\text{carbon dioxide produced per kJ} = \frac{1}{890} = 1.12 \times 10^{-3} \, mol \, kJ^{-1}$$
> For fuel oil, 13 368 kJ of energy are released when 20 mol of carbon dioxide are formed:
> $$\text{carbon dioxide produced per kJ} = \frac{20}{13368} = 1.50 \times 10^{-3} \, mol \, kJ^{-1}$$
> The percentage reduction is therefore given by
> $$100 \times \frac{1.50 - 1.12}{1.50} = \mathbf{25\%}$$

> ### Now try this
>
> The combustion of coal in a coal-fired power station can be represented as follows:
> $$C(s) + O_2(g) \rightarrow CO_2(g)$$
> $$\Delta H_c^\ominus = -394 \, kJ \, mol^{-1}$$
> Assuming that coal is 100% carbon, what would be the percentage reduction in carbon dioxide produced per kJ if natural gas replaces coal in a power station?

Using quantitative combustion data in the determination of empirical and molecular formulae

Under the right conditions, combustion of a hydrocarbon can be carried out quantitatively (that is, in 100% yield). The quantities of the reactants and products can be measured to a high degree of precision, by measuring either their volumes or their masses. The following worked example shows how this can be put to use in finding out the formulae of hydrocarbons.

> ### Worked example
>
> 10 cm³ of a gaseous alkane were mixed with 100 cm³ of oxygen and sparked to cause complete combustion. On cooling to the original temperature (290 K), the volume of gas was found to be 75 cm³. This reduced to 35 cm³ on shaking with concentrated aqueous sodium hydroxide. What is the formula of the alkane?
>
> #### Answer
> At 290 K, all the water produced on combustion will be a liquid. The 75 cm³ of gas must therefore be composed of carbon dioxide, along with an excess of unreacted oxygen. The carbon dioxide will be absorbed by the aqueous sodium hydroxide, so 40 cm³ (75 − 35) of carbon dioxide were produced. Using Avogadro's Law (see section 4.13), if 10 cm³ of alkane produce 40 cm³ of carbon dioxide, then 1 molecule of alkane produces 4 molecules of carbon dioxide. Therefore the alkane contains 4 carbon atoms, so its formula is **C₄H₁₀**.

> ### Now try this
>
> 10 cm³ of a gaseous alkane were mixed with 150 cm³ of oxygen and sparked. On cooling to the original temperature (290 K), the volume of gas was found to be 120 cm³. This reduced to 70 cm³ on shaking with concentrated aqueous sodium hydroxide. What is the formula of the alkane?

Incomplete combustion

If the air supply is limited, the combustion of alkanes is incomplete. The hydrogen atoms in the alkane are always oxidised to water, but the carbon atoms either remain as elemental carbon (producing a black sooty flame), or undergo partial oxidation to carbon monoxide. Both reactions are less exothermic than complete combustion:

$$C_6H_{12} + 9O_2 \rightarrow H_2O(l) + 6CO_2(g) \qquad \Delta H^{\ominus} = -4160 \text{ kJ mol}^{-1}$$
$$C_6H_{12} + 6O_2 \rightarrow 6H_2O(l) + 6CO(g) \qquad \Delta H^{\ominus} = -2470 \text{ kJ mol}^{-1}$$
$$C_6H_{12} + O_2 \rightarrow 6H_2O(l) + 6C(g) \qquad \Delta H^{\ominus} = -1800 \text{ kJ mol}^{-1}$$

2 Halogenation

Alkanes undergo substitution reactions with halogens:

$$CH_4 + X_2 \rightarrow CH_3X + HX$$

Methane reacts vigorously with fluorine, even in the dark and at room temperature. With chlorine and bromine, no reaction occurs unless the reactants are heated or exposed to ultraviolet light. With iodine, no reaction occurs at all. The trend in the calculated ΔH of the reactions matches that of the C—X bond energies (see Table 13.4).

Table 13.4 Bond enthalpies and enthalpies of reaction for halogen substitution reactions with methane

Halogen (X)	C—X bond enthalpy/kJ mol^{-1}	ΔH for the reaction/kJ mol^{-1}
F	484	−475
Cl	338	−114
Br	276	−36
I	238	+27

The mechanism of the halogenation reaction

The halogenation reaction involves radicals (see section 12.8). When initiated (started) by ultraviolet light, experiments have shown that one photon can cause the production of many thousands of molecules of chloromethane from chlorine and methane. Once the photon has started the reaction off, the reaction can continue on its own for a long time. Studies on the effectiveness of various frequencies of ultraviolet light have suggested that the photon initiates the reaction by splitting the chlorine molecule homolytically (see section 12.8):

$$\text{Cl} \;-\; \text{Cl} \longrightarrow \text{Cl}^{\times} + \cdot\text{Cl} \tag{1}$$

We can predict the next step of the reaction by simple bond enthalpy calculations. The chlorine atoms formed by the splitting of the chlorine molecule are highly reactive. They are likely to react with the next molecule they collide with. In the mixture we have just chlorine molecules and methane molecules. Reaction of a chlorine atom with a chlorine molecule:

$$\text{Cl}\cdot \quad \text{Cl}-\text{Cl} \longrightarrow \text{Cl}-\text{Cl} + {}^{\cdot}\text{Cl} \tag{2}$$

might well occur, but it does not lead anywhere. It merely uses up one chlorine atom to produce another.

Reaction with methane could conceivably occur in one of two ways:

$$ \tag{3}$$

$$ \tag{4}$$

In reaction (3), a C—H bond has been broken, and a H—Cl bond has been formed.

$$E(\text{C—H}) = 413\,\text{kJ mol}^{-1}$$
$$E(\text{H—Cl}) = 431\,\text{kJ mol}^{-1}$$

Overall,

$$\Delta H_{(3)} = 413 - 431 = \mathbf{-18\,kJ\,mol^{-1}}$$

In reaction (4), a C—H bond has been broken, and a C—Cl bond has been formed.

$$E(\text{C—Cl}) = 338\,\text{kJ mol}^{-1}$$

Overall,

$$\Delta H_{(4)} = 413 - 338 = \mathbf{+75\,kJ\,mol^{-1}}$$

So reaction (3) is slightly exothermic, whereas reaction (4) is highly endothermic. Reaction (3) is therefore much more likely to take place.

The methyl radical formed in reaction (3) is equally as reactive as a chlorine atom. It is likely to react with the first molecule it collides with. If it collides with a chlorine molecule, the following reaction is possible:

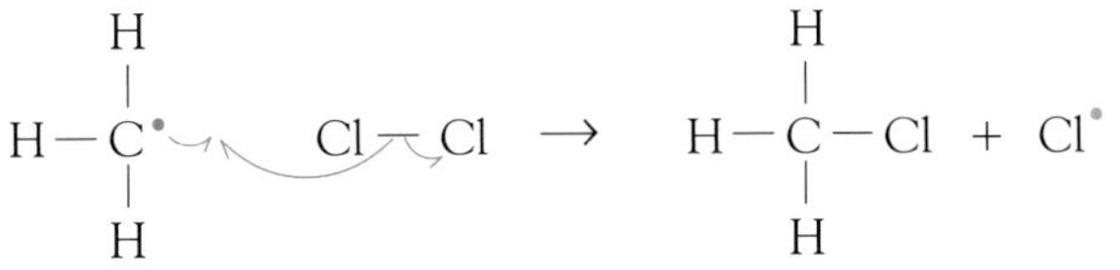

$$\text{(5)}$$

$$\Delta H = E(\text{Cl—Cl}) - E(\text{C—Cl}) = 242 - 338 = \mathbf{-96\,kJ\,mol^{-1}}$$

The reaction is highly exothermic, so it is likely to occur. If the methyl radical collides with a methane molecule, the following 'reaction' could occur:

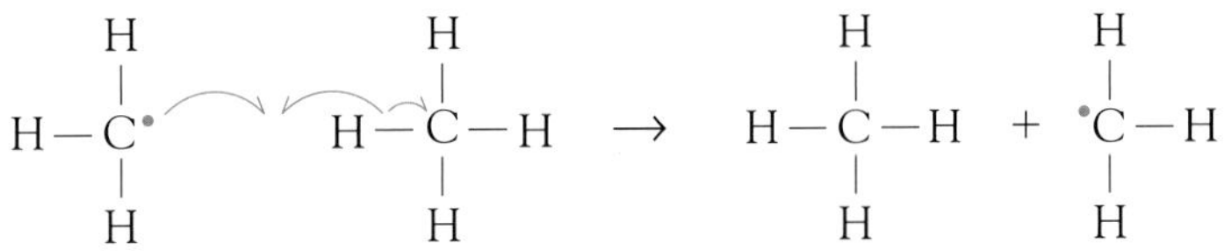

$$\text{(6)}$$

Reaction (6), like reaction (2), does not lead us anywhere. A hydrogen atom has been transferred from a methane molecule to a methyl radical, forming a methane molecule and another methyl radical!

Let us now take stock of the situation. In the initiation reaction (1), we have produced two reactive chlorine atoms. On energetic grounds, the two most likely reactions after that are (3) and (5):

$$\text{Cl}^{\bullet} + \text{CH}_4 \rightarrow \text{CH}_3^{\bullet} + \text{HCl} \qquad \text{(3)}$$
$$\text{CH}_3^{\bullet} + \text{Cl}_2 \rightarrow \text{CH}_3\text{Cl} + \text{Cl}^{\bullet} \qquad \text{(5)}$$

After these two reactions have taken place we have:

• used up one molecule of methane, and one molecule of chlorine
• produced one molecule of chloromethane, and one molecule of hydrogen chloride
• used up a chlorine atom, but regenerated another one.

If we add equations (3) and (5) together, the Cl$^{\bullet}$ radical on the left-hand side of equation (3) 'cancels' with the Cl$^{\bullet}$ radical on the right-hand side of equation (5). The same is true of the CH$_3^{\bullet}$ radical. So the sum of reactions (3) and (5) is:

$$\text{Cl}^{\bullet} + \text{CH}_4 + \text{CH}_3^{\bullet} + \text{Cl}_2 \rightarrow \text{CH}_3^{\bullet} + \text{HCl} + \text{CH}_3\text{Cl} + \text{Cl}^{\bullet}$$

and the overall equation for the reaction is:

$$\text{CH}_4 + \text{Cl}_2 \rightarrow \text{CH}_3\text{Cl} + \text{HCl}$$

Now try this

Use the following bond enthalpies to calculate the ΔH values of reactions (3) and (4) using bromine instead of chlorine. Is reaction (3) still more likely to occur than (4) if chlorine is replaced by bromine?

$$E(\text{C—Br}) = 276\,\text{kJ mol}^{-1}$$
$$E(\text{H—Br}) = 366\,\text{kJ mol}^{-1}$$

The chlorine atom has acted as a homogeneous catalyst (see section 8.6). It has taken part in the reaction, but has not been used up during it.

Reactions (3) and (5) together constitute a never-ending **chain reaction**. They could, in theory, continue until all the methane and chlorine had been converted into chloromethane and hydrogen chloride. This situation does not occur in practice, however. As was mentioned above, one photon initiates the production of many thousands of molecules of chloromethane – but only thousands, not millions. Eventually the chain reaction comprising reactions (3) and (5) stops, because there is a (small) chance that two radicals could collide with each other, to form a stable molecule:

$$Cl^{\bullet} + Cl^{\bullet} \rightarrow Cl_2 \tag{7}$$
$$CH_3^{\bullet} + Cl^{\bullet} \rightarrow CH_3Cl \tag{8}$$
$$CH_3^{\bullet} + CH_3^{\bullet} \rightarrow CH_3{-}CH_3 \tag{9}$$

Any of these reactions would use up the 'catalysts' $Cl^{\bullet}$ and $CH_3^{\bullet}$, and so stop the chain reaction. It would require another photon to split another chlorine molecule in order to restart the chain. Overall, the chlorination of methane is an example of a **radical substitution reaction**. The different stages of the chain reaction are named as follows:

$$Cl_2 \rightarrow 2Cl^{\bullet} \qquad \textbf{initiation}$$

$$\left.\begin{array}{l} Cl^{\bullet} + CH_4 \rightarrow CH_3^{\bullet} + HCl \\ CH_3^{\bullet} + Cl_2 \rightarrow CH_3Cl + Cl^{\bullet} \end{array}\right\} \textbf{propagation}$$

$$\left.\begin{array}{l} Cl^{\bullet} + Cl^{\bullet} \rightarrow Cl_2 \\ CH_3^{\bullet} + Cl^{\bullet} \rightarrow CH_3Cl \\ CH_3^{\bullet} + CH_3^{\bullet} \rightarrow CH_3{-}CH_3 \end{array}\right\} \textbf{termination}$$

By-products of the halogenation reaction

As the reaction takes place, the concentration of methane is being reduced, and the concentration of chloromethane is increasing. Chlorine atoms are therefore increasingly likely to collide with molecules of chloromethane, causing the following reactions to take place:

$$Cl^{\bullet} + CH_3Cl \rightarrow HCl + {}^{\bullet}CH_2Cl$$
$${}^{\bullet}CH_2Cl + Cl_2 \rightarrow CH_2Cl_2 + Cl^{\bullet}$$

Further substitution, giving dichloromethane, trichloromethane and eventually tetrachloromethane, is therefore to be expected. A mixture is in fact formed.

Another interesting set of by-products are chlorinated ethanes. These arise from the ethane produced in termination reaction (9) undergoing a similar set of substitution reactions:

$$Cl^{\bullet} + CH_3CH_3 \rightarrow HCl + CH_3CH_2^{\bullet}$$
$$CH_3CH_2^{\bullet} + Cl_2 \rightarrow CH_3CH_2Cl + Cl^{\bullet}$$

These chloroethanes are produced in only very small amounts, as might be expected, since only a small quantity of ethane is produced in reaction (9). But their presence amongst the reaction products adds weight to the mechanism described here: without the formation of ethane from two methyl radicals, chlorinated ethane could not have been formed.

Commercially, the chlorination of methane is important for the production of degreasing and cleansing solvents. The mixture of products is readily separated by fractional distillation (see Table 13.5). If only one product is required, it can be made to predominate by mixing Cl_2 and CH_4 in suitable proportions.

Table 13.5 The different boiling points of the chloromethanes allow them to be separated by fractional distillation.

Compound	Formula	Boiling point/°C
chloromethane	CH_3Cl	−24
dichloromethane	CH_2Cl_2	40
trichloromethane	$CHCl_3$	62
tetrachloromethane	CCl_4	77

Halogenation of higher alkanes

A useful way of classifying organic halides, alcohols, radicals and carbocations is to place them in one of the three categories **primary**, **secondary** and **tertiary**, depending on the number of alkyl groups joined to the central carbon atom. Table 13.6 describes this classification.

Table 13.6 Primary, secondary and tertiary centres. The methyl group CH_3— is also classified as a primary centre.

Number of alkyl groups	Type of centre	General formula	Examples	
1	primary	R—CH_2—	CH_3CH_2Cl $CH_3CH_2CH_2^•$	chloroethane the prop-1-yl radical
2	secondary	R_2CH—	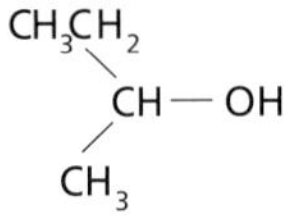 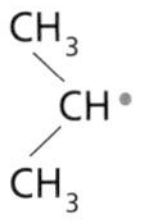 CH_3CH_2 $\quad$ CH—OH CH_3 CH_3 $\quad$ CH$^•$ CH_3	butan-2-ol the prop-2-yl radical
3	tertiary	R_3C—	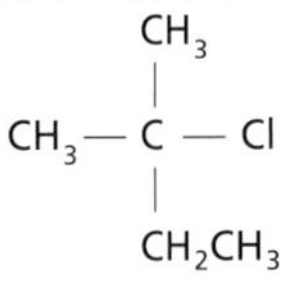 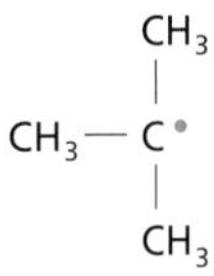 CH_3 CH_3—C—Cl CH_2CH_3 CH_3 CH_3—C$^•$ CH_3	2-chloro-2-methylbutane the 2-methylprop-2-yl radical

There are two monochloropropane isomers: 1-chloropropane, $CH_3CH_2CH_2Cl$, and 2-chloropropane, $CH_3CHClCH_3$. When propane reacts with chlorine, a mixture of the two monochloro- compounds is produced. This is because the chlorine atom can abstract either one of the primary (end-carbon) hydrogens or one of the secondary (middle-carbon) hydrogens:

$$Cl^• + CH_3—CH_2—CH_3 \rightarrow CH_3—CH_2—CH_2^• + H—Cl$$

or

$$Cl^• + CH_3—CH_2—CH_3 \rightarrow CH_3—CH^•—CH_3 + H—Cl$$

Because the ratio of primary hydrogens to secondary hydrogens is $6:2$, or $3:1$, it might be expected on probability grounds that the ratio of 1-chloropropane to 2-chloropropane in the product should also be $3:1$. However, this assumes that the abstraction of a primary hydrogen and of a secondary hydrogen are both equally likely.

In fact, it is found that the ratio of 1-chloropropane to 2-chloropropane is nearly $1:1$, which suggests that the secondary hydrogen atoms are three times more likely to be abstracted, and this cancels out the probability effect.

With bromine, the effect is even more pronounced. Here the ratio of 1-bromopropane to 2-bromopropane is $1:30$, showing that the bromine atom is far more selective as to which hydrogen atom it abstracts. Because of the **inductive effect** of the alkyl groups (see page 248), the secondary prop-2-yl radical is more stable than the primary prop-1-yl radical, and is therefore more likely to form:

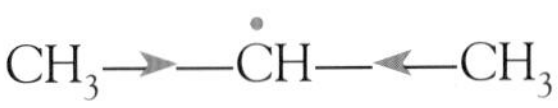

$$CH_3 {\rightarrow}—CH—{\leftarrow}CH_3 \qquad\qquad CH_3—CH_2{\rightarrow}—CH_2^•$$

prop-2-yl radical: two alkyl groups give a double inductive effect $\qquad$ prop-1-yl radical: one alkyl group

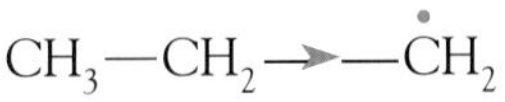

The inductive effect of alkyl groups

C—C bonds are essentially non-polar, owing to the equal electronegativities of the carbon atoms at each end. If one of the carbon atoms is electron deficient in some way (for example, by being surrounded by only seven electrons, as in a radical), the other carbon atom can partially compensate for this deficiency by acting as an electron 'reservoir'. This reservoir effect is not so apparent with C—H bonds, because the bonding electrons are much closer to the hydrogen nucleus.

So the methyl radical:

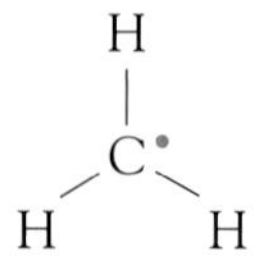

is much less stable than the ethyl radical: $CH_3 \longrightarrow \overset{\bullet}{C}$

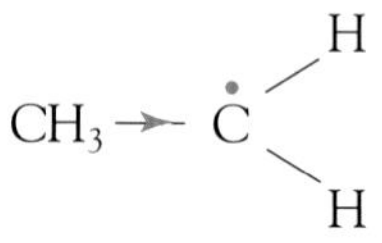

which in turn is less stable than the prop-2-yl radical: $CH_3 \longrightarrow \overset{\bullet}{C}$

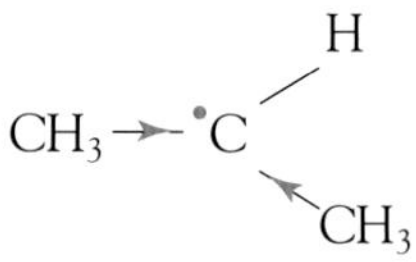

(The arrow on the bond represents the drift of electrons away from the CH_3 group.)

The 2-methylprop-2-yl radical:

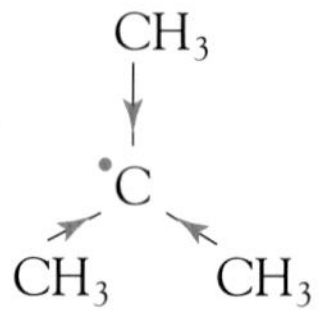

experiences an even greater stabilisation, owing to the inductive effect of *three* alkyl groups.

The same effect is noticed with carbocations. For example,

it is much easier to form the 2-methylprop-2-yl cation: $CH_3 \longrightarrow \overset{+}{C}$

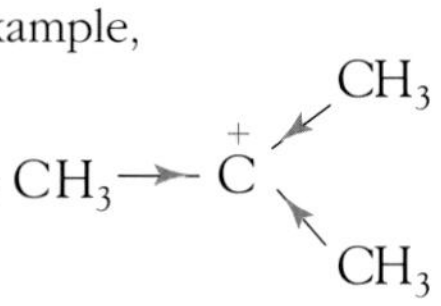

than it is to form the methyl cation:

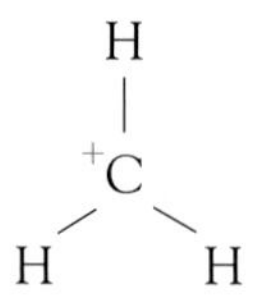

(See Topic 14, page 258 and Topic 15, page 272.)

Worked example

Predict the major product formed during the monobromination of:

a $CH_3CH_2CH_2CH_3$
b $(CH_3)_2CH-CH_3$

Answer

a Of the two possible radicals formed by hydrogen abstraction from butane, the secondary radical $CH_3CH_2CH^{\bullet}CH_3$ is more stable than the primary radical $CH_3CH_2CH_2CH_2^{\bullet}$, so 2-bromobutane is the most likely product.

b Although the ratio of primary hydrogens to tertiary hydrogen is $9:1$ in 2-methylpropane, the tertiary radical $(CH_3)_3C^{\bullet}$ is much more stable than the primary radical $(CH_3)_2CH-CH_2^{\bullet}$, so 2-bromo-2-methylpropane is the most likely product.

Now try this

Draw all the possible monobromoalkanes derived from the reaction between bromine and 2-methylbutane, $(CH_3)_2CH-CH_2CH_3$, and predict, with a reason, which one will be the major product.

Summary

- **Alkanes** form the homologous series of general formula C_nH_{2n+2}.
- They show structural (chain) isomerism and, in some cases, optical isomerism.
- They can be extracted from crude oil by fractional distillation.
- They can be **cracked** and **re-formed** to produce useful fuels and petrochemical feedstocks.
- Their main use is as a source of energy by combustion.
- They react with chlorine or bromine in **free-radical substitution** reactions, giving chloro- or bromoalkanes.
- There are three stages to a radical chain reaction – **initiation**, **propagation** and **termination**.

Key reactions you should know
- Combustion:
$$C_nH_y + (n + \frac{y}{4})O_2 \rightarrow nCO_2 + \frac{y}{2}H_2O$$
- Free-radical substitution:
$$C_nH_{2n+2} + X_2 \rightarrow C_nH_{2n+1}X + HX \qquad (X = Cl \text{ or } Br)$$

Examination practice questions

Please see the data section of the CD for any A_r values you may need.

1 Crude oil is a naturally occurring flammable liquid which consists of a complex mixture of hydrocarbons. In order to separate the hydrocarbons the crude oil is subjected to fractional distillation.

a Explain what is meant by the following terms.
  i *hydrocarbon*
  ii *fractional distillation* [2]

b Undecane, $C_{11}H_{24}$, is a long chain hydrocarbon which is present in crude oil.
  Such long chain hydrocarbons are 'cracked' to produce alkanes and alkenes which have smaller molecules.
  i Give the conditions for **two different** processes by which long chain molecules may be cracked.
  ii Undecane, $C_{11}H_{24}$, can be cracked to form pentane, C_5H_{12}, and an alkene.
    Construct a balanced equation for this reaction. [3]

Pentane, C_5H_{12}, exhibits structural isomerism.

c i Draw the three structural isomers of pentane.
  ii The three isomers of pentane have different boiling points.
    Which of your isomers has the highest boiling point? Suggest an explanation for your answer. [6]

The unsaturated hydrocarbon, **E**, is obtained by cracking hexane and is important in the chemical industry.
The standard enthalpy change of combustion of **E** is $-2059\,kJ\,mol^{-1}$.

d Define the term *standard enthalpy change of combustion*. [2]

When 0.47 g of **E** was completely burnt in air, the heat produced raised the temperature of 200 g of water by 27.5 °C. Assume no heat losses occurred during this experiment.

e i Use relevant data from the data section on the CD to calculate the amount of heat released in this experiment.
  ii Use the data above and your answer to **i** to calculate the relative molecular mass, M_r, of **E**. [4]

f Deduce the molecular formula of **E**. [1]

[Cambridge International AS & A Level Chemistry 9701, Paper 22 Q3 November 2010]

2 A reaction mechanism shows the individual steps that take place during a reaction.

a Methane reacts with bromine in the presence of ultraviolet radiation to form several products.
  Two of these products are bromomethane and hydrogen bromide.

  i Write an equation for the reaction between methane and bromine to make bromomethane and hydrogen bromide. [1]

  ii Name one other bromine-containing organic product which is formed when methane reacts with bromine. [1]

  iii The mechanism for this reaction is called radical substitution.
    Describe the mechanism for the radical substitution of methane by bromine to make bromomethane.

    Use the mechanism to suggest why a small amount of ethane is also formed. [7]

[OCR Chemistry A Unit F322 Q4 (part) January 2010]

14 Alkenes

Functional group:

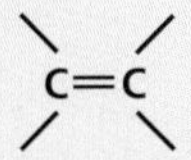

Another class of hydrocarbon, the alkenes, is discussed in this topic. Alkenes contain a carbon–carbon double bond, which is much more reactive than a carbon–carbon single bond. The characteristic reaction of alkenes is electrophilic addition. Many alkenes, especially ethene, are very important industrial chemicals. The addition polymerisation of alkenes and of substituted ethenes results in many useful plastics.

Learning outcomes

By the end of this topic you should be able to:

14.1a) interpret and use the general, structural, displayed and skeletal formulae of the alkenes

14.3a) describe and explain the shapes of, and bond angles in, the ethane and ethene molecules in terms of σ and π bonds, and predict the shapes of, and bond angles in, other related molecules

15.2a) describe the chemistry of alkenes as exemplified, where relevant, by the following reactions of ethene and propene (including the Markovnikov addition of asymmetric electrophiles to alkenes using propene as an example):
  - addition of hydrogen, steam, hydrogen halides and halogens
  - oxidation by cold, dilute, acidified manganate(VII) ions to form the diol
  - oxidation by hot, concentrated, acidified manganate(VII) ions leading to the rupture of the carbon–carbon double bond in order to determine the position of alkene linkages in larger molecules
  - polymerisation (see also Topic 28)

15.2b) describe the mechanism of electrophilic addition in alkenes, using bromine/ethene and hydrogen bromide/propene as examples

15.2c) describe and explain the inductive effects of alkyl groups on the stability of cations formed during electrophilic addition

15.2d) describe the characteristics of addition polymerisation as exemplified by poly(ethene) and PVC

15.2e) deduce the repeat unit of an addition polymer obtained from a given monomer (see also Topic 28)

15.2f) identify the monomer(s) present in a given section of an addition polymer (see also Topic 28)

15.2g) recognise the difficulty of the disposal of poly(alkene)s, i.e. non-biodegradability and harmful combustion products.

14.1 Introduction

The straight-chain **alkenes** form a homologous series with the general formula C_nH_{2n}. They have many reactions in common with the alkanes (alkenes with three or more carbon atoms contain both C—H and C—C bonds, just like alkanes), and their physical properties, such as boiling points, are very similar (see Table 14.1). Reactions of the double bond are very different, however, and these dominate alkene chemistry, the double bond being much more reactive than a C—C single bond.

Table 14.1 Boiling points of some alkenes and alkanes

Number of carbon atoms	Alkane		Alkene	
	Name	Boiling point/°C	Name	Boiling point/°C
1	methane	−162	—	—
2	ethane	−88	ethene	−102
3	propane	−42	propene	−48
4	butane	0	but-1-ene	−6
			trans-but-2-ene	1
			cis-but-2-ene	4
	2-methylpropane	−12	2-methylpropene	−7
5	pentane	36	pent-1-ene	30
6	hexane	69	hex-1-ene	64

Figure 14.1 Two sp² carbon atoms can form a σ bond and a π bond in ethene.

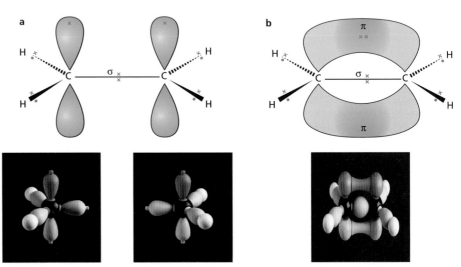

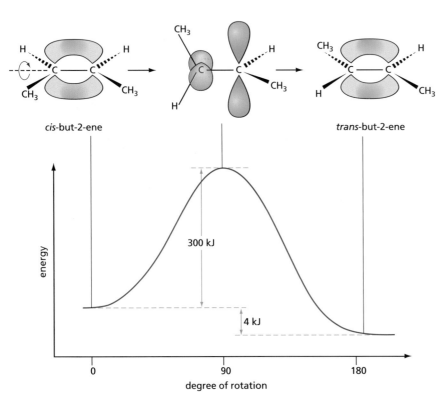

cis-but-2-ene   *trans*-but-2-ene

300 kJ

4 kJ

energy

0   90   180

degree of rotation

Figure 14.2 Hindered rotation around the C=C double bond in the but-2-enes

We saw in section 3.15 that the bonding in ethene consists of a σ-bonded framework of two carbon atoms bonded to each other and to the four hydrogen atoms (see Figure 14.1a). This leaves a p orbital on each carbon atom, which can then overlap sideways, giving the π bond (see Figure 14.1b).

The presence of the π bond confers two special characteristics on the structure and reactivity of alkenes:

- hindered rotation
- reaction with electrophiles.

Hindered rotation

Unlike the C—C σ bond in ethane, which allows complete freedom of rotation at room temperature, the π bond in ethene fixes the two ends of the molecule in their relative orientations. In order to twist one end with respect to the other, it would be necessary to break the π bond and then re-form it. This would require an input of about $300\,\text{kJ}\,\text{mol}^{-1}$, which is far beyond the energy available at normal temperatures. Therefore, no rotation occurs around the C=C double bond (see Figure 14.2). This large energy

252

requirement contrasts with a barrier to rotation of only 12 kJ mol^{-1} in ethane, which is readily available to molecules from thermal energy at room temperature. There is therefore almost completely total freedom of rotation about the C—C bonds in alkanes.

Hindered rotation allows the existence of geometrical (*cis–trans*) isomers (see section 12.6, and also the next section in this topic). Notice from Figure 14.2 that the *trans* isomer of but-2-ene is more stable than the *cis* isomer by 4 kJ mol^{-1}. This difference in stability occurs with most isomers of alkenes, and is significant in some natural alkenes (see the panel below for two examples).

Cis–trans isomerism in nature

Double bonds that occur in particular compounds in nature are usually of a fixed geometry. Whether the double bond is *cis* or *trans* is important to its biological function. Two examples are described here.

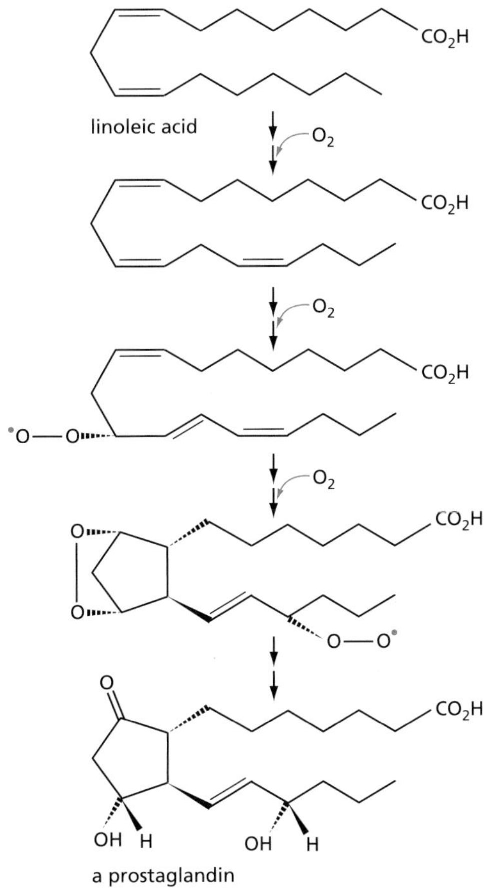

Polyunsaturated oils

The unsaturated fatty acids in certain vegetable oils are considered to be healthier for us than the saturated fatty acids found in animal fats. However, it is not so much the degree of unsaturation that is important, but the type of isomerism that exists around the double bonds. *Cis* double bonds are considered to be more healthy than *trans* ones. Fatty acids containing *trans* double bonds are thought to be no healthier than the fully saturated acids. One of the most healthy is *cis-cis*-linoleic acid, which can be converted in the body into an unsaturated prostaglandin (see Figure 14.3)

The correct balance of the various prostaglandins found in the body is essential for healthy circulation, effective nerve transmission and for maintaining the correct ionic balance within cells.

Retinal

The essential process that converts the arrival of a photon on the retina of the eye into a nerve impulse to the brain involves a molecule called rhodopsin. Part of the molecule is a protein (called opsin) and part is an unsaturated aldehyde called 11-*cis*-retinal. This aldehyde absorbs light strongly at a wavelength of 500 nm, near the middle of the visible spectrum. Within a few picoseconds of the absorption of a photon, the *cis* isomer of retinal isomerises to the *trans* isomer (see Figure 14.4). This causes the aldehyde end of the molecule to move about 0.5 nm away from the hydrophobic hydrocarbon end, which in turn blocks the movement of sodium ions through cell membranes, and triggers a nerve impulse. (A single photon can block the movement of more than a million sodium ions.) Once it has been formed, the *trans*-retinal diffuses away from the opsin protein, and takes several minutes to be re-isomerised to the *cis* isomer.

Figure 14.3 Conversion of *cis-cis*-linoleic acid to an unsaturated prostaglandin

Figure 14.4 Conversion of *cis*-retinal to *trans*-retinal by the absorption of a photon of light

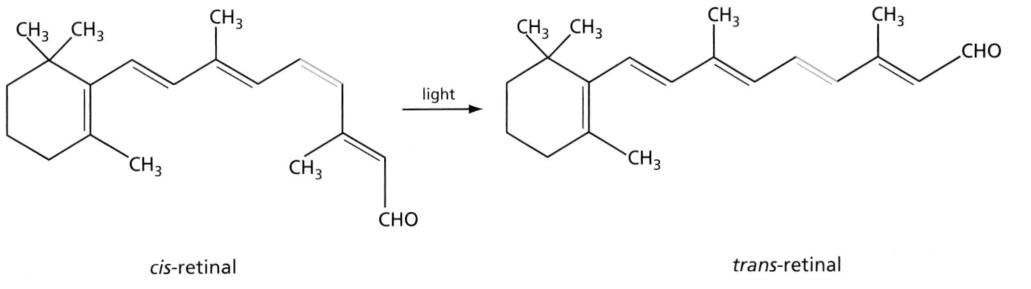

Reaction with electrophiles

The ethene molecule is flat – both carbon atoms and all four hydrogen atoms lie in the same plane. The electrons in the π bond protrude from this plane, both above and below (see Figure 14.1b). This ready availability of electron density attracts a variety of electrophiles (see section 12.7) to an alkene molecule, allowing a much greater range of reactions than occurs with alkanes.

14.2 Isomerism and nomenclature

The principles of nomenclature for alkenes are similar to those used to name alkanes. The longest carbon chain that contains the C=C bond is first identified, and then the side branches are named and positioned along the longest chain. There are two additional pieces of information that the name needs to convey:

- the position of the double bond along the chain
- if appropriate, whether the groups around the double bond are in a *cis* or a *trans* arrangement.

The position of the double bond is designated by including a number before the '-ene' at the end of the name, corresponding to the chain position of the first carbon in the C=C bond.

So we have:

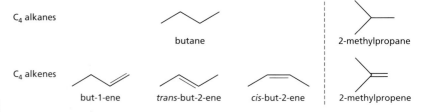

cis-pent-2-ene   and   3-propylhex-1-ene

(Note that although the longest carbon chain contains seven atoms, this chain does not contain the C=C bond, and so is disregarded.)

In addition to the chain (and optical) isomerism shown by the alkanes, alkenes show two other types of isomerism: positional and geometrical. Because of these two additional types of isomerism, there are many more alkene isomers than alkane isomers with a given number of carbon atoms. For example, although there are only two alkanes with the formula C_4H_{10}, there are no fewer than four alkenes with the formula C_4H_8 (see Figure 14.5).

Figure 14.5 Isomers of four-carbon alkanes and alkenes

C_4 alkanes

butane   2-methylpropane

C_4 alkenes

but-1-ene   trans-but-2-ene   cis-but-2-ene   2-methylpropene

Now try this

Apart from 3-methylpent-1-ene, there are 14 other alkene isomers with the molecular formula C_6H_{12}.

1 Three of these have the same carbon skeleton as 3-methylpent-1-ene. Draw their structural formulae, and name them.

2 Of the other 11, five have the same carbon skeleton as hexane, five have the same carbon skeleton as 2-methylpentane, and one has the same carbon skeleton as 2,2-dimethylbutane. Draw out the skeletal formulae of these other 11 isomers.

Worked example

Although the first alkane to show optical isomerism has the molecular formula C_7H_{16} (see the worked example on page 236), the first chiral alkene has the molecular formula C_6H_{12}. Draw its structural formula, and name it.

Answer

The four different groups that are necessary for a carbon atom to be a chiral centre are in this case H, CH_3, C_2H_5 and CH=CH_2. The structure is therefore:

$$CH_3-CH_2-\underset{\underset{H}{|}}{\overset{\overset{CH_3}{|}}{C}}-CH=CH_2$$

The alkene's name is **3-methylpent-1-ene.**

14.3 Reactions of alkenes

The double bond in alkenes is composed of a strong σ bond and a relatively weak π bond. The σ bond has its electron density between the atoms but the π bond electrons are on the surface of the molecule and so they are readily available. Alkenes react by using these π-bond electrons to form new σ bonds to other atoms, leaving the original σ bond intact. Normally, this results in only one molecule of product being produced from two molecules of reactant, that is, **an addition reaction** (see section 12.7):

$$CH_2{=}CH_2 + A{-}B \rightarrow A{-}CH_2{-}CH_2{-}B$$

Addition reactions of alkenes are invariably exothermic. The energy needed to break the π bond and the A—B bond is less than that given out when the C—A and C—B bonds are formed. For example:

$$CH_2{=}CH_2 + H{-}H \rightarrow CH_3{-}CH_3 \qquad \Delta H^{\ominus} = -137\,kJ\,mol^{-1}$$

$$CH_2{=}CH_2 + Br{-}Br \rightarrow BrCH_2{-}CH_2Br \qquad \Delta H^{\ominus} = -90\,kJ\,mol^{-1}$$

Because they are able to 'absorb' (that is, react with) further hydrogen atoms in this way, alkenes are sometimes referred to as **unsaturated hydrocarbons**. Alkanes, on the other hand, which have all the spare valencies of their carbon atoms used up by bonding with hydrogen atoms, are termed **saturated hydrocarbons**.

1 Hydrogenation

Alkenes can be hydrogenated to alkanes by reacting them with hydrogen gas. The reaction does not occur readily without a catalyst, but with platinum it proceeds smoothly at room temperature and pressure. Commercially, alkenes themselves are never hydrogenated to alkanes (the alkenes are valuable feedstocks for further reactions, whereas the alkanes produced have value only as fuels). But the reaction is important in the manufacture of solid or semi-solid fats (margarine) from vegetable oils. In the commercial reaction the less effective, but cheaper, metal nickel is used as a catalyst, under a higher pressure:

$$R{-}CH{=}CH{-}R' + H_2 \xrightarrow{\text{Pt or Ni at 10 atm}} R{-}CH_2{-}CH_2{-}R'$$

Figure 14.6 Partial hydrogenation of sunflower oil produces the semi-solid spread.

2 Addition of bromine

Unlike alkanes, alkenes react with bromine at room temperature, and even in the dark. The reaction takes place either with bromine water, or with a solution of bromine in an organic solvent such as hexane or trichloroethane:

$$CH_2\text{=}CH_2 + Br_2 \xrightarrow{\text{room temperature in hexane}} Br\text{—}CH_2\text{—}CH_2\text{—}Br$$

With bromine water, the major product is 2-bromoethanol, $BrCH_2CH_2OH$. The reaction is very fast – the orange bromine solution is rapidly decolorised. This is an excellent test for the presence of a C=C double bond, since few other compounds decolorise bromine water so rapidly.

Figure 14.7 Alkenes decolorise bromine water

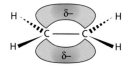

Figure 14.8 The electron-rich π bond induces a dipole in the bromine molecule.

The mechanism of the addition of bromine to an alkene

The reaction is an **electrophilic addition**, like most of the reactions of alkenes. As we have seen, the protruding π bond of ethene is an electron-rich area. This δ− area (see section 3.12) induces a dipole in an approaching bromine molecule, by repelling the electrons of the Br—Br bond (see Figure 14.8).

Eventually, the Br—Br bond breaks completely, heterolytically (see section 12.8), to form a bromide ion. The π electrons rearrange to form a σ bond from one of the carbon atoms to the nearest bromine. The movement of electrons away from the other carbon atom results in the production of a carbocation (see Figure 14.9).

Figure 14.9 The Br—Br bond breaks, and a carbocation is formed.

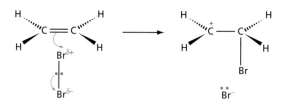

There is a small amount of evidence to suggest that in some cases a lone pair of electrons on the bromine atom can form a dative bond to the cationic carbon atom, spreading out the charge (see Figure 14.10).

Figure 14.10 The lone pair on bromine can form a dative bond.

Finally, the bromide ion acts as a nucleophile and forms a (dative) bond to the carbocation (see Figure 14.11).

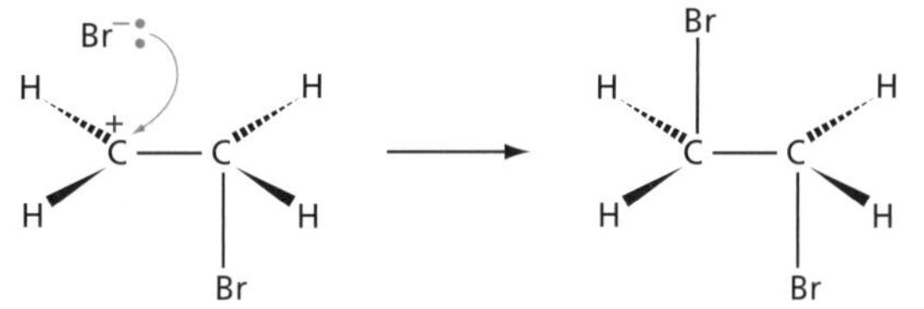

Figure 14.11 The free bromide ion bonds to the carbocation.

Other nucleophiles can also attack the carbocation. In bromine water, for example, the water molecule is present in a much higher concentration than the bromide ion, so is more effective as a nucleophile (see Figure 14.12).

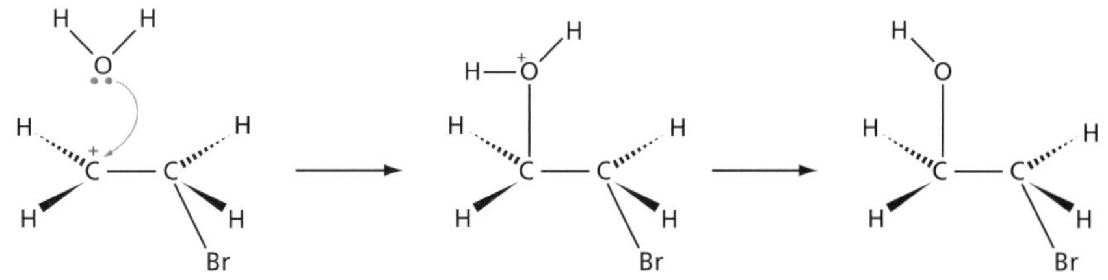

Figure 14.12 In bromine water, the water molecule acts as the nucleophile.

The incorporation of a water molecule, and the consequent production of the bromoalcohol, is good evidence for this suggested mechanism via an intermediate carbocation. Another piece of evidence that supports the two-step mechanism is the incorporation of 'foreign' anions when bromination is carried out in an aqueous solution containing a mixture of various salts:

$$CH_2{=}CH_2 + \begin{cases} Na^+NO_3^- \\ Na^+Cl^- \end{cases} \xrightarrow[\text{with Br}_2]{\text{in water}} \begin{cases} O_2N{-}O{-}CH_2{-}CH_2{-}Br \\ HO{-}CH_2{-}CH_2{-}Br \\ Br{-}CH_2{-}CH_2{-}Br \\ Cl{-}CH_2{-}CH_2{-}Br \end{cases}$$

The ratio of $ClCH_2CH_2Br$ to $O_2NOCH_2CH_2Br$ in the products reflects the $[Cl^-]:[NO_3^-]$ ratio in the solution, showing that the carbocation picks up the first anion it collides with, as would be expected for a cation of such high reactivity.

3 Addition of hydrogen halides

The hydrogen halides HCl, HBr and HI all react with alkenes to give halogenoalkanes. Either the (gaseous) hydrogen halide is bubbled directly through the alkene, or ethanoic acid is used as a solvent. (Aqueous solutions of hydrogen halides such as hydrochloric acid cannot be used for this reaction because they cause the production of alcohols, as we shall see shortly.)

$$CH_2{=}CH_2(g) + HBr(g) \rightarrow CH_3{-}CH_2Br(l)$$
$$\text{bromoethane}$$

The mechanism of this electrophilic addition reaction is similar to that of bromination, but in this case the electrophile already has a permanent dipole. The $H^{\delta+}$ end of the H—Br molecule is attracted to the π bond in the initial step shown in Figure 14.13.

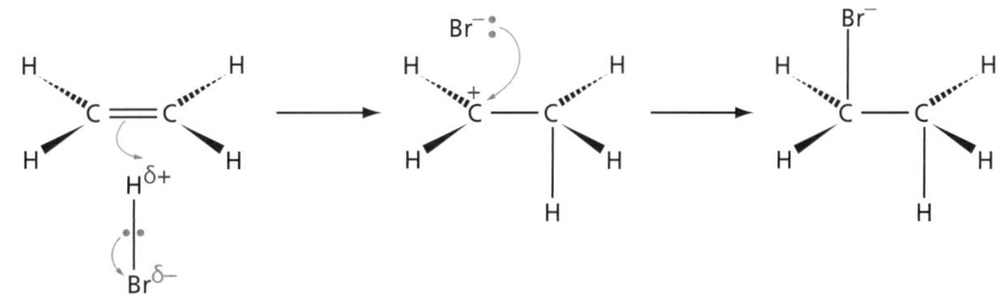

Figure 14.13 The electrophilic addition of hydrogen bromide to ethene

The position of the bromine atom

With the reaction between hydrogen bromide and propene, two different, isomeric, products could form:

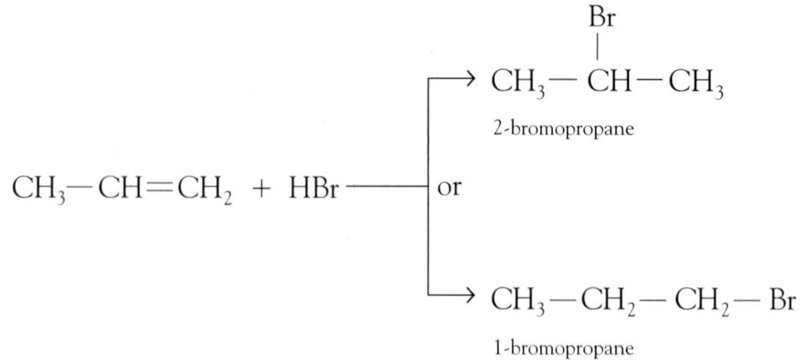

$$CH_3-CH=CH_2 + HBr \longrightarrow \begin{cases} \begin{array}{c} Br \\ | \\ CH_3-CH-CH_3 \\ \text{2-bromopropane} \end{array} \\ \text{or} \\ CH_3-CH_2-CH_2-Br \\ \text{1-bromopropane} \end{cases}$$

When the reaction is carried out, it is found that 2-bromopropane is by far the major product: almost no 1-bromopropane is formed. This can be explained as follows.

The position of the bromine atom along the chain is determined by the position of the positive charge in the intermediate carbocation:

$$CH_3-CH^+-CH_3 + Br^- \longrightarrow CH_3-\overset{\overset{\displaystyle Br}{|}}{CH}-CH_3$$
secondary prop-2-yl
carbocation

$$CH_3-CH_2-CH_2^+ + Br^- \longrightarrow CH_3-CH_2-CH_2-Br$$
primary prop-1-yl
carbocation

This, in turn, is determined by which end of the double bond the $H^{\delta+}$ of the HBr attaches itself to. We saw in the panel on page 248 that the inductive electron-donating effect of methyl groups causes the prop-2-yl radical to be more stable than the prop-1-yl radical. The effect is even more pronounced with carbocations, which have a carbon atom that is surrounded by only six electrons. The secondary prop-2-yl carbocation is *much* more stable than the primary prop-1-yl carbocation, and so the $H^{\delta+}$ adds on to position 1 of the double bond, rather than position 2 (see Figure 14.14).

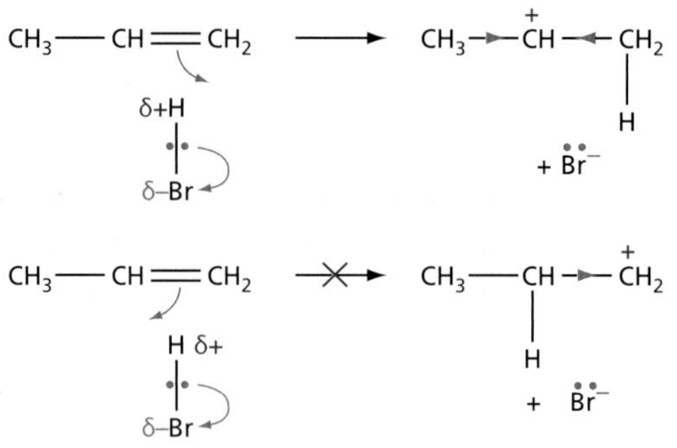

Figure 14.14 When the hydrogen adds on to position 1, the more stable secondary carbocation is formed. The blue arrows represent the inductive effects of the alkyl groups.

Tertiary carbocations such as

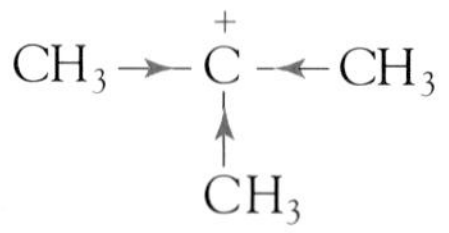

$$CH_3 \rightarrow \overset{+}{\underset{\underset{\displaystyle CH_3}{\uparrow}}{C}} \leftarrow CH_3$$

are even more stable than secondary carbocations. So 2-methylpropene reacts with hydrogen bromide to give 2-bromo-2-methylpropane:

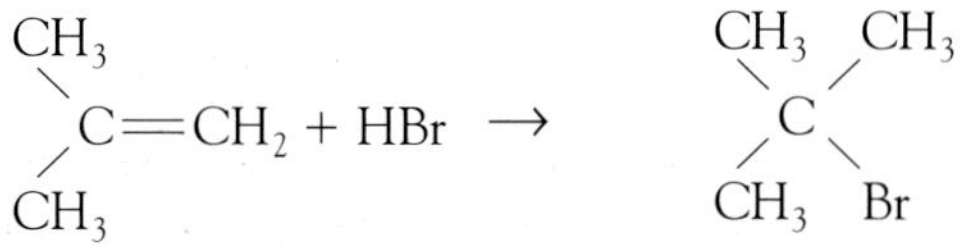

$$\begin{array}{c} CH_3 \\ \diagdown \\ \diagup \\ CH_3 \end{array} C=CH_2 + HBr \rightarrow \begin{array}{c} CH_3 \quad CH_3 \\ \diagdown \quad \diagup \\ C \\ \diagup \quad \diagdown \\ CH_3 \quad Br \end{array}$$

Markovnikov's rule

Markovnikov's rule states that when HX adds to a double bond:

- the hydrogen atom attaches to the carbon that already has the most hydrogens, or
- the electrophile adds in the orientation that produces the most stable intermediate cation.

In 1869 the Russian chemist Vladimir Markovnikov formulated a rule to predict the orientation of the addition of a hydrogen halide to an unsymmetrical alkene (that is, one in which the two ends of the double bond are not the same, with one end having more hydrogen atoms attached to the $C=C$ group than the other). He stated that when HX adds to an unsymmetrical alkene, the hydrogen attaches itself to the least substituted end of the $C=C$ double bond (that is, the end that already has the most hydrogen atoms). A better formulation, based on what we now know of the mechanism, states that an electrophile adds to an unsymmetrical alkene in the orientation that produces the most stable intermediate carbocation.

Worked example

Predict and name the products of the following reactions.

a $CH_2{=}CH{-}CH_2{-}CH_3 + HCl \rightarrow$ b $(CH_3)_2C{=}CH{-}CH_3 + HBr \rightarrow$

Answer

a The reaction could produce either $CH_3{-}C^+H{-}CH_2CH_3$ or $^+CH_2{-}CH_2{-}CH_2CH_3$ as the intermediate cation. The first one is more stable, being a secondary carbocation. This will lead to the formation of $CH_3CHClCH_2CH_3$, 2-chlorobutane, as product.

b The tertiary carbocation $(CH_3)_2C^+{-}CH_2CH_3$ is more stable than the secondary carbocation $(CH_3)_2CH{-}C^+H{-}CH_3$, so the product will be $(CH_3)_2CBr{-}CH_2CH_3$, 2-bromo-2-methylbutane.

Now try this

Predict and name the products of the following reactions.

1

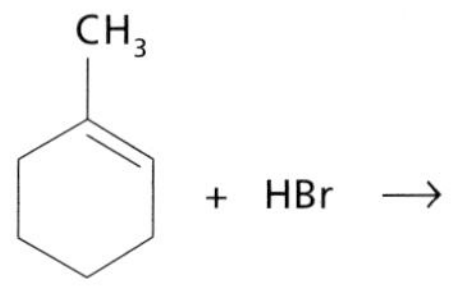

$+ \ HBr \ \longrightarrow$

2 $CH_3CH_2CH_2CH{=}C$

$+ \ HI \ \longrightarrow$

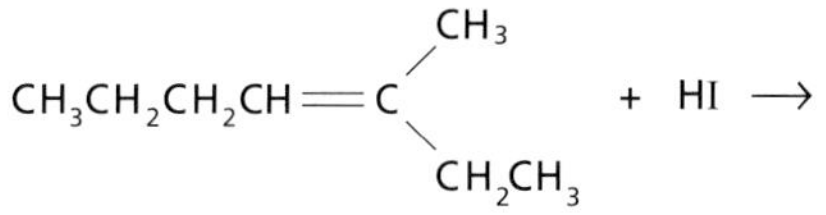

4 Hydration

In the presence of an acid, water undergoes an electrophilic addition reaction with alkenes to produce alcohols. In the laboratory, this is carried out in two stages. Ethene (for example) is absorbed in concentrated sulfuric acid, with cooling, to form ethyl hydrogensulfate:

$$C_2H_4 + H_2SO_4 \rightarrow C_2H_5{-}O{-}SO_3H$$

This is then added to cold water:

$$C_2H_5{-}O{-}SO_3H + H_2O \rightarrow CH_3CH_2OH + H_2SO_4$$

The sulfuric acid therefore acts as a catalyst. The sodium salts of long-chain alkyl hydrogensulfates find a use as anionic detergents.

Industrially, the hydration of alkenes is the major way of manufacturing ethanol, propan-2-ol and butan-2-ol. The alkene and steam are passed over a phosphoric acid catalyst absorbed onto porous pumice (to give it a large surface area) at a pressure of 70 atm and a temperature of 300 °C:

$$CH_2{=}CH_2 + H_2O \rightarrow CH_3{-}CH_2OH$$

Phosphoric acid is used because, like sulfuric acid, it is a non-volatile acid. Although sulfuric acid is the cheaper of the two, it is an oxidising agent, so some by-products are formed with it. The mechanism is an electrophilic addition, with H^+ (from the acid) as the initial electrophile (see Figure 14.15). H^+ is regenerated in the last step, and so can be used again.

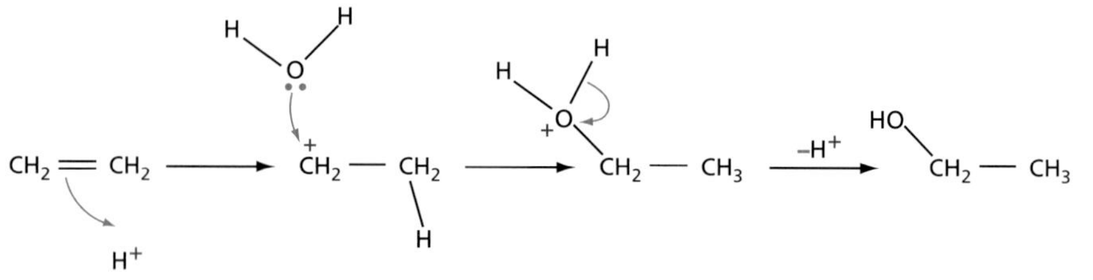

Figure 14.15 The hydration of ethene

Worked example

The hydration reaction follows Markovnikov's rule. What will be the structure of the alcohol formed from the hydration of each of the following?

a propene   b 2-methylbut-2-ene

Answer

a The most stable carbocation formed from $CH_3—CH=CH_2$ and H^+ is $CH_3—C^+H—CH_3$, so propan-2-ol will be produced.

b Likewise, $CH_3CH_2—C^+(CH_3)_2$ is the most stable carbocation formed from 2-methylbut-2-ene, so 2-methylbutan-2-ol will be the product.

Now try this

Predict which alcohols will be formed from the hydration of each of the following alkenes.

1 2-methylbut-1-ene
2 2-methylbut-2-ene

As can be seen from part **a** of the worked example above, the hydration of alk-1-enes always produces alkan-2-ols, rather than alkan-1-ols. The primary alcohols have to be made by a different route. This is reflected in their cost – primary alcohols cost at least twice as much per litre as do the corresponding secondary alcohols. (Alcohols are very useful intermediates from which a variety of organic compounds can be made. Some of their reactions are described in Topic 16.)

5 Oxidative addition with potassium manganate(VII)

Apart from the decolorisation of bromine water, another very good test for the presence of a carbon–carbon double bond is the reaction between alkenes and dilute acidified potassium manganate(VII), $KMnO_4(aq)$. When alkenes are shaken with the reagent, its purple colour disappears. Identification of the product shows that a diol has been formed:

$$CH_2=CH_2 + H_2O + [O] \rightarrow HO—CH_2—CH_2—OH$$

([O] represents an oxygen atom that has been given by the manganate(VII).)

The reaction occurs rapidly at room temperature, whereas potassium manganate(VII) will only oxidise other compounds (such as alcohols and aldehydes) slowly under these conditions. It is therefore a useful test for the presence of an alkene.

Tests for alkenes

- Alkenes rapidly decolorise bromine water (see page 256).
- Alkenes rapidly decolorise dilute potassium manganate(VII).

6 Oxidative cleavage with hot potassium manganate(VII)

When heated with acidified potassium manganate(VII), the $C=C$ double bond splits completely. The products from the oxidation depend on the degree of substitution on the $C=C$ bond (see Table 14.2).

Table 14.2 Products from the cleavage of double bonds by potassium manganate(VII)

If the double bond carbon is:	Then the oxidised product will be:
$CH_2 =$	CO_2
R and H, $C =$	$R - C$ with $=O$ and OH (carboxylic acid)
R and R, $C =$	R and R, $C = O$

Two examples are shown in Figure 14.16.

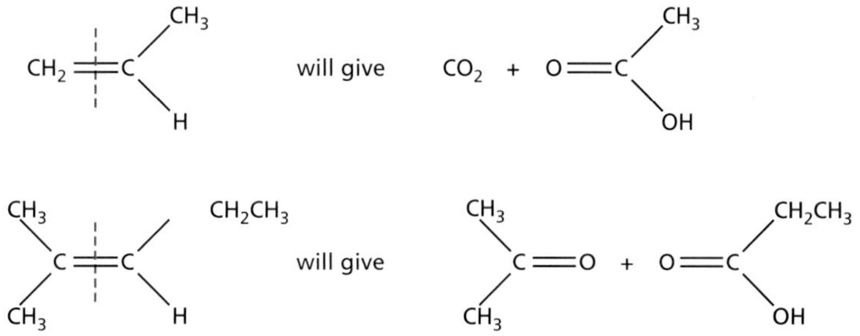

Figure 14.16 Products of oxidative cleavage by potassium manganate(VII)

$CH_2 = C$ with CH_3 and H will give CO_2 + $O = C$ with CH_3 and OH

$C = C$ (with CH_3, CH_3 on left; CH_2CH_3, H on right)   will give $C = O$ (with CH_3, CH_3) + $O = C$ (with CH_2CH_3, OH)

Worked example

Use Table 14.2 to predict the products of the oxidative cleavage by potassium manganate(VII) of the following compounds.

a   (cyclohexane ring) $= CH_2$

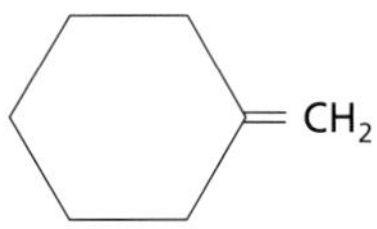

b   (cyclohexene ring with CH_3)

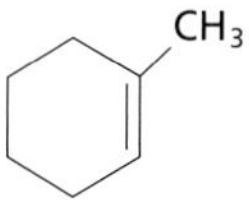

Answer

a Splitting the double bond in the middle gives

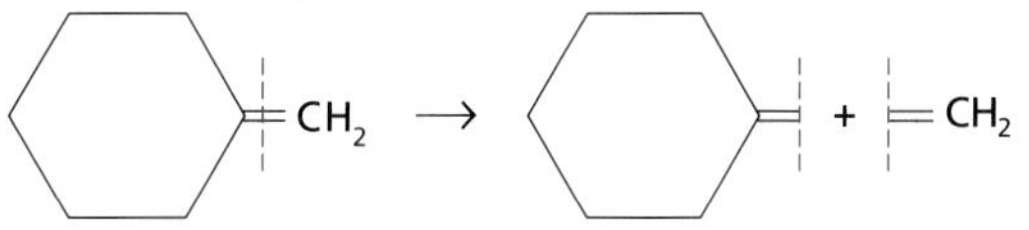

(ring) $= CH_2$ →   (ring) $=$ + $= CH_2$

which, according to Table 14.2, will give carbon dioxide, CO_2, and cyclohexanone:

(cyclohexanone ring) $= O$

b Similar splitting gives

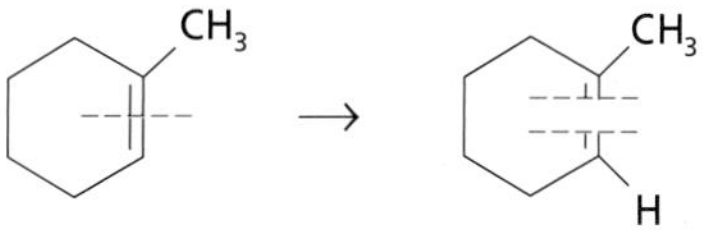

(ring with CH_3)   →   (ring with CH_3 and H)

which eventually will produce

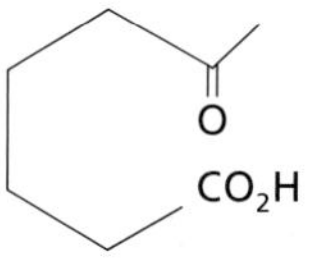

O CO_2H

or $CH_3 - CO - CH_2CH_2CH_2CH_2CO_2H$.

Now try this

Geraniol is an alcohol that occurs in oil of rose and other flower essences. Its structural formula is $(CH_3)_2C = CHCH_2CH_2C(CH_3) = CHCH_2OH$. What are the structures of the three compounds formed by oxidation with hot acidified potassium manganate(VII)?

When monomers join together without the formation of any other products, the resulting polymer is called an **addition polymer**.

14.4 Addition polymerisation

Formation of poly(ethene)

In 1933, chemists working at ICI in Britain discovered by chance that when ethene was subjected to high pressures in the presence of a trace of oxygen, it produced a soft white solid. This solid was poly(ethene), and this was the start of the enormous (multimillion dollar and multimillion tonnage worldwide) industry based on addition polymers.

$$n(CH_2{=}CH_2) \xrightarrow[\text{1000 atm and 200 °C}]{\text{trace of oxygen}} \left[\begin{array}{cc} H & H \\ | & | \\ -C & -C- \\ | & | \\ H & H \end{array}\right]_{n \ (n > 5000)}$$

poly(ethene) (LDPE)

Poly(ethene), commonly called polythene, is termed an **addition polymer** because one molecule is made from many ethene units (**monomers**) adding together, without the co-production of any small molecule (contrast the process of condensation polymerisation – see section 28.3). Just as in all other addition reactions of ethene, the $C{=}C$ double bond is replaced by a $C{-}C$ single bond, and two carbon atoms make two new single bonds to other atoms. The original reaction conditions are still used with little modification to produce **low-density poly(ethene)** or **LDPE**. Sometimes an organic peroxide is used instead of oxygen as an initiator of the radical chain reaction. Using a different catalyst, developed by Karl Ziegler and Giulio Natta in 1953, ethene can be polymerised at much lower pressures to produce **high-density poly(ethene)** or **HDPE**.

$$n(CH_2{=}CH_2) \xrightarrow[\text{in heptane at 50 °C and 2 atm}]{Al(C_2H_5)_3 + TiCl_4} \left[\begin{array}{cc} H & H \\ | & | \\ -C & -C- \\ | & | \\ H & H \end{array}\right]_{n \ (n \cong 100\,000)}$$

poly(ethene) (HDPE)

The major differences between the structures of LDPE and HDPE are that:

- the average chain length of LDPE is much shorter than that of HDPE
- the chains of LDPE are branched, whereas the chains of HDPE are unbranched.

These differences have an effect on the physical properties, and hence the uses, of the two different types of poly(ethene) (see Table 14.3). So, by altering the chain length and the degree of branching, the physical properties of poly(ethene) can be altered to suit the purpose for which it is to be used.

Table 14.3 Properties and uses of LDPE and HDPE

Property or use	Polymer	
	LDPE	HDPE
density	low	high
melting point	approx. 130 °C	160 °C
tensile strength	low	higher
flexibility	very flexible	much more rigid
uses	polythene bags, electrical insulation, dustbin liners	bottles, buckets, crates

Addition polymerisation of other alkenes

The physical properties of a polymer can be changed by altering the structure of the monomer. This is discussed in Topic 28. Here we shall briefly mention just three other common polymers made from substituted ethene molecules, in which one of the hydrogen atoms has been replaced by another atom or group. The three are polypropene, polystyrene (polyphenylethene) and polyvinyl chloride, PVC (polychloroethene). Table 14.4 lists their structures and some of their uses.

Table 14.4 Structures and uses of polypropene, polychloroethene and polyphenylethene

Monomer	Polymer (showing two repeat units)	Uses
propene $CH_2\!=\!CH\!-\!CH_3$	polypropene	packaging, containers, ropes, carpets, thermal underwear and fleeces
chloroethene $CH_2\!=\!CHCl$	polychloroethene (polyvinylchloride, PVC)	guttering, water pipes, window frames, floor coverings
phenylethene	polyphenylethene (polystyrene)	protective packaging, model kits, heat insulation

The disposal of polymers

Not only are plastics and polymers very useful materials for a host of applications, the energy requirements for their manufacture are far less than for equivalent products made from such materials as aluminium, glass or steel. With all these materials, however, we are then left with the problem of how to dispose of them at the end of their useful life. For example, metals from old cars can be recycled and glass can be re-worked, but both of these require a large input of energy. Similarly polymers need a suitable method of disposal, which needs to be safe and environmentally friendly. Methods are now available that fit this description, and which require little or no energy input.

Polyalkenes, being effectively very long-chain alkanes, are chemically inert. They biodegrade only very slowly in the environment. Although there are bacteria that can metabolise straight-chain alkanes once they have been 'functionalised' by oxidation somewhere along the chain, they find branched-chain alkanes more difficult to degrade. Therefore, they can attack HDPE more easily than LDPE. On the other hand, LDPE is more sensitive to being broken down by sunlight. The less compact nature of the chains, and the presence of the weaker tertiary C—H bonds (see Table 14.5), allows photo-oxidation to take place more readily (see Figure 14.17).

Table 14.5 Tertiary C—H bonds are weaker than primary C—H bonds.

Bond	Bond enthalpy/kJ mol⁻¹
$R_3C\!-\!H$	380
$H_3C\!-\!H$	435

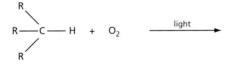

Figure 14.17 Photo-oxidation of tertiary polyalkanes

Poly(chloroethene) is less inert. The C—Cl bond can be attacked by alkalis (see section 15.3) and can undergo rupture in ultraviolet light (C—Cl bond enthalpy = 338 kJ mol⁻¹).

Despite these possible methods of degradation, most addition polymers can potentially cause environmental problems when their useful lives are over. There are four methods that are used, or are being developed, to allow their safe disposal.

1 **Incineration** – the enthalpy change of combustion of all polymers is strongly exothermic. Under the right conditions, incineration can provide useful energy

for space heating or power generation. The process requires a high, carefully controlled temperature if the formation of pollutants such as the poisonous gases carbon monoxide and phosgene, $COCl_2$, is to be avoided. Poly(chloroethene) is a particular problem, as the highly dangerous compound 'dioxin' (see section 25.4) can be formed unless the conditions are carefully controlled. The chlorine atoms in poly(chloroethene) have to end up somewhere, of course. The preferred combustion product is hydrogen chloride gas, which can subsequently be absorbed in water.

2 **Recycling** – many polymers can be melted and re-moulded. For this process to produce materials that industry can use, at a price that is competitive with newly made material, the recycled polymer has to be clean, and of one type (for example, all polyalkene, or all poly(chloroethene)).

3 **Depolymerisation** – heating polymers to high temperatures in the absence of air or oxygen (pyrolysis) can cause their chains to break down into alkene units. The process is similar to the cracking of long-chain hydrocarbons into alkenes (see section 13.3). The monomers produced can be separated by fractional distillation and re-used.

4 **Bacterial fermentation** – under the right conditions, and with an optimal strain of bacteria, some polymers (mainly the polyalkenes and the polyesters, see section 26.4) can be degraded quite rapidly. The solid polymer is usually first fragmented into very small pieces, and partially photolysed. Useful combustible gases are produced during the fermentation, which can be used as an energy source.

14.5 Preparing alkenes

1 Cracking of alkanes

The smaller alkenes ethene, propene and butadiene are made commercially by cracking, as we saw in section 13.3. The method is not of great use for preparing a specific alkene in the laboratory, however, because it gives a random selection of alkenes. For example:

$$CH_3(CH_2)_{10}CH_3 \xrightarrow{\text{heat with steam at } 800\,°C} CH_3(CH_2)_3CH_3 + CH_3CH{=}CH_2 + 2CH_2{=}CH_2$$

Figure 14.18 Alkenes are produced by the cracking of long-chain alkanes (see section 13.3). Crude oil is heated and passed through these catalytic cracking towers.

2 Dehydration of alcohols

This is the reverse of the hydration reaction on page 259. Heating ethanol with concentrated sulfuric acid to 180 °C produces ethene. Concentrated phosphoric acid may also be used. Alcohols also dehydrate when their vapours are passed over strongly heated aluminium oxide:

$$CH_3\text{—}CH_2OH \xrightarrow[\substack{\text{or conc. } H_3PO_4 \text{ at } 200\,°C \\ \text{or pass vapour over strongly heated } Al_2O_3}]{\text{conc. } H_2SO_4 \text{ at } 180\,°C} CH_2\text{=}CH_2 + H_2O$$

3 Elimination of HCl or HBr from chloro- or bromoalkanes

This reaction is described in Topic 15, page 275. For example:

$$KOH + CH_3\text{—}CHCl\text{—}CH_3 \xrightarrow{\text{heat in ethanol}} CH_3\text{—}CH\text{=}CH_2 + KCl + H_2O$$

Sometimes an elimination reaction like the dehydration or elimination of a hydrogen halide described above results in more than one alkene product. For example:

$$CH_3CH_2CHBrCH_3 \xrightarrow[\text{dissolved in ethanol}]{\text{heat with KOH}} \underset{80\%}{CH_3CH\text{=}CHCH_3} + \underset{20\%}{CH_3CH_2CH\text{=}CH_2}$$

Usually, the more substituted alkene (that is, the one with the more alkyl groups attached to the C=C double bond) is more stable than the less substituted alkene. So it is the more substituted alkene that is most likely to be formed, and 80% of the product is but-2-ene.

Summary

- **Alkenes** contain a C=C double bond, and have the general formula C_nH_{2n}.
- In addition to chain isomerism, alkenes show positional isomerism, and also geometrical (*cis–trans*) isomerism due to the restricted rotation around the double bond.
- Their characteristic reaction is **electrophilic addition**. Unsymmetrical electrophiles add in accordance with **Markovnikov's rule**, with the most stable carbocation being formed as intermediate.
- Tests for the presence of a double bond include the decolorisation of bromine water, and the decolorisation of dilute aqueous potassium manganate(VII).
- Ethene and other alkenes form important **addition polymers** such as polythene, polystyrene and PVC.

Key reactions you should know
- Catalytic hydrogenation (addition):

$$CH_2\text{=}CH\text{—}R + H_2 \xrightarrow{Ni} CH_3\text{—}CH_2\text{—}R$$

- Electrophilic additions:

$$CH_2\text{=}CH\text{—}R + Br_2 \xrightarrow{\text{in the dark}} CH_2Br\text{—}CHBr\text{—}R$$
$$CH_2\text{=}CH\text{—}R + HBr \longrightarrow CH_3\text{—}CHBr\text{—}R$$
$$CH_2\text{=}CH\text{—}R + H_2O \xrightarrow{H_3PO_4 \text{ at } 70\,atm \text{ and } 300\,°C} CH_3\text{—}CH(OH)\text{—}R$$

- Oxidations:

$$CH_2\text{=}CH\text{—}R + [O] + H_2O \xrightarrow{\text{cold KMnO}_4} CH_2(OH)\text{—}CH(OH)\text{—}R$$
$$CH_2\text{=}CH\text{—}R + 5[O] \xrightarrow{\text{hot KMnO}_4} CO_2 + HO_2C\text{—}R\ (+ H_2O)$$

- Polymerisation:

$$nCH_2\text{=}CH\text{—}R \xrightarrow[\text{or } AlR_3]{O_2 \text{ at } 1000\,atm} (CH_2\text{—}CHR)_n$$

Examination practice questions

Please see the data section of the CD for any A_r values you may need.

1 Alkenes are a very useful series of hydrocarbons used widely in synthesis. Alkenes are more reactive than alkanes.
  a What is the name of the process used to convert long-chain alkanes into more useful shorter-chain alkenes? [1]
  b Ethene and steam can be converted into ethanol. The equilibrium is shown below.

$$C_2H_4(g) + H_2O(g) \rightleftharpoons C_2H_5OH(g) \qquad \Delta H = -46\,kJ\,mol^{-1}$$

  Le Chatelier's Principle can be used to predict the effect of changing conditions on the position of equilibrium.
  i Name the catalyst used in this reaction. [1]
  ii State Le Chatelier's Principle. [1]
  iii Using Le Chatelier's Principle, predict and explain the conditions that would give the maximum equilibrium yield of ethanol from ethene and steam. [3]
  iv The actual conditions used are 60 atmospheres pressure at 300 °C in the presence of a catalyst. Compare these conditions with your answer to **iii** and comment on why these conditions are used. [3]
  c Alkenes are used to make addition polymers. The repeat unit for an addition polymer is shown below.

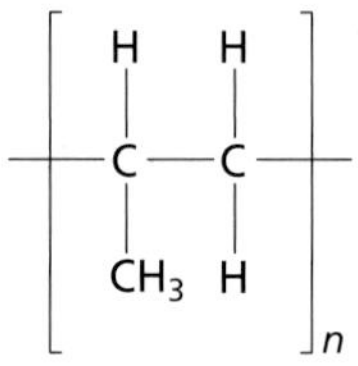

  What is the name of the monomer used to make this polymer? [1]

d Poly(chloroethene) has the repeat unit below.

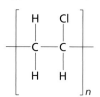

This repeat unit can be written as —CH₂CHCl—.
One way to dispose of poly(chloroethene) is to react it with oxygen at high temperature. This is called incineration.
  i Complete the following equation that shows the reaction taking place during incineration.

$$-CH_2CHCl- + \ldots\, O_2 \rightarrow \ldots\, CO_2 + \ldots\, H_2O + \ldots\, HCl$$
[1]

  ii Research chemists have reduced the environmental impact of incineration by removing the HCl formed from the waste gases.
  Suggest a type of reactant that could be used to remove the HCl. [1]
e The disposal of polymers causes environmental damage. Research chemists are developing polymers that will reduce this environmental damage and increase sustainability.
  Describe two ways in which chemists can reduce this environmental damage. [2]

[OCR Chemistry A Unit F322 Q5 January 2010]

15 Halogenoalkanes

Functional group:

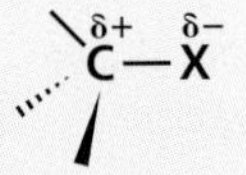

X = F, Cl, Br, I

There are several functional groups that contain atoms (called hetero atoms) other than carbon and hydrogen. In this topic, compounds containing a halogen atom – the halogenoalkanes – are discussed. The halogenoalkanes are a reactive and important class of compound much used in organic synthesis. Their characteristic reaction is nucleophilic substitution.

Learning outcomes

By the end of this topic you should be able to:

14.1a) interpret and use the general, structural, displayed and skeletal formulae of the halogenoalkanes, amines (primary only) and nitriles

16.1a) recall the chemistry of halogenoalkanes as exemplified by:
- the following nucleophilic substitution reactions of bromoethane: hydrolysis, formation of nitriles, formation of primary amines by reaction with ammonia
- the elimination of hydrogen bromide from 2-bromopropane

16.1b) describe the S_N1 and S_N2 mechanisms of nucleophilic substitution in halogenoalkanes including the inductive effects of alkyl groups (see Topic 14)

16.1c) recall that primary halogenoalkanes tend to react via the S_N2 mechanism; tertiary halogenoalkanes via the S_N1 mechanism; and secondary halogenoalkanes by a mixture of the two, depending on structure

16.2a) interpret the different reactivities of halogenoalkanes (with particular reference to hydrolysis and to the relative strengths of the C—Hal bonds)

16.2b) explain the uses of fluoroalkanes and fluorohalogenoalkanes in terms of their relative chemical inertness

16.2c) recognise the concern about the effect of chlorofluoroalkanes on the ozone layer.

15.1 Introduction

There are only a few naturally occurring organic halogen compounds. The most important is thyroxine, the hormone secreted by the thyroid gland. Synthetic organic halogen compounds have found many uses, however. One of the most important uses is as intermediates in organic syntheses: they are readily formed from common materials, but can equally easily be transformed into compounds containing many different functional groups.

Figure 15.1 Thyroxine

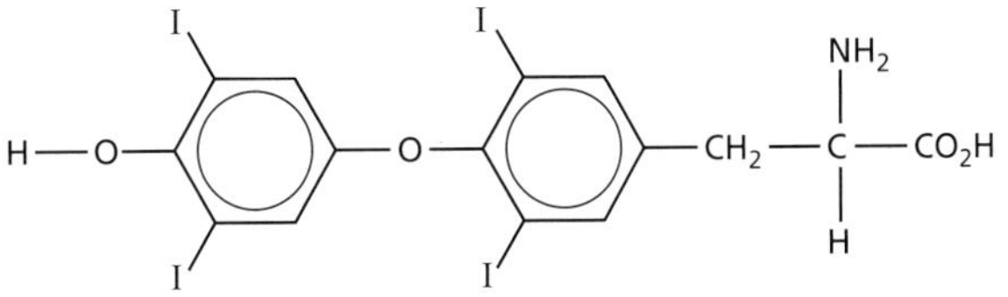

CFCs (chlorofluorocarbons) are inert, volatile liquids that at one time found favour as the circulating fluids in refrigerators and air-conditioning units. They are now being phased out of use because of their destructive effect on the ozone layer, as we shall see in section 15.4. Halothane, $CF_3CHBrCl$, is an important anaesthetic, a successor to chloroform, $CHCl_3$, which was one of the first anaesthetics, introduced in the early nineteenth century. The halons, such as $CBrClF_2$, are useful fire extinguishers. Many polychloroalkanes are used as solvents. The use of certain organochlorine compounds as herbicides and insecticides is described in section 25.4.

Halogenoalkanes are generally colourless liquids, immiscible with and heavier than water, with sweetish smells. Compared with the alkanes, their boiling points are higher, with the boiling point rising from each chloroalkane to the corresponding iodoalkane (see Figure 15.2). The main intermolecular force in halogenoalkanes is the van der Waals' induced dipole force. Therefore their boiling points increase with chain length, just like those of the alkanes.

Figure 15.2 Boiling points of the halogenoalkanes

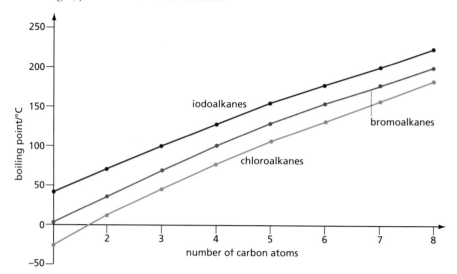

Furthermore, as we would expect, the trend in boiling point with halogen (RCl < RBr < RI) reflects the greater number of electrons available for induced dipole attraction from chlorine to iodine (see section 11.1). In addition, there is a small amount of permanent dipole–dipole attraction between molecules. This will be most important for the chloroalkanes, becoming less so for the bromo- and iodoalkanes (see Figure 15.3).

Figure 15.3 Permanent dipole–dipole attractions in halogenoalkanes

$R^{\delta+}$——$Cl^{\delta-}$ $R^{\delta+}$——$Cl^{\delta-}$ $R^{\delta+}$——$Cl^{\delta-}$ $R^{\delta+}$——$Cl^{\delta-}$

15.2 Isomerism and nomenclature

Halogenoalkanes show chain, positional and optical isomerism. There are, for example, five isomers with the formula C_4H_9Br, as shown in Figure 15.4.

Figure 15.4 Isomers of C_4H_9Br

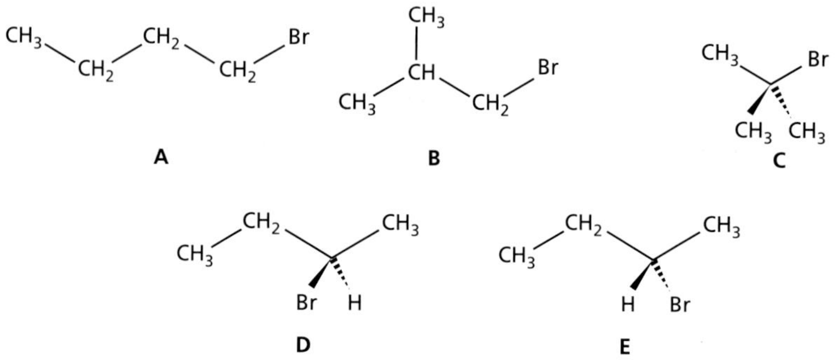

A and the **D–E** pair are positional isomers of one another. **B** and **C** are also positional isomers of each other. **B** and **C** are chain isomers of the other three. **D** and **E** are optical isomers of each other. They form a non-superimposable mirror-image pair (see section 12.6).

A halogenoalkane is classed as **primary**, **secondary** or **tertiary** (see Table 13.6, page 247), depending on how many alkyl groups are attached to the carbon atom joined to the halogen atom:

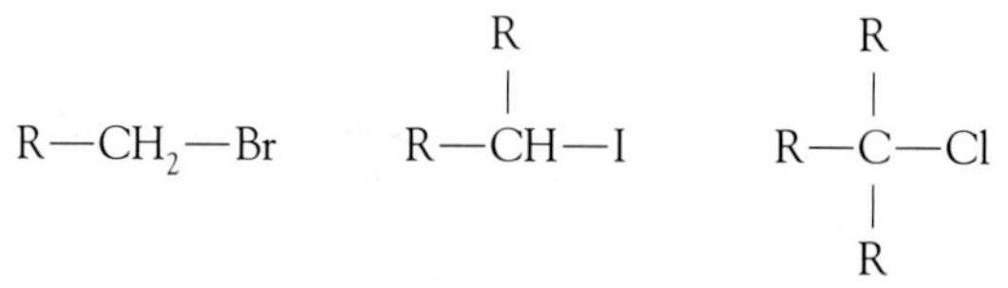

R—CH$_2$—Br   R—CH—I   R—C—Cl

a primary bromoalkane   a secondary iodoalkane   a tertiary chloroalkane

In Figure 15.4, isomers **A** and **B** are primary, **D** and **E** are secondary, and **C** is tertiary.

When a compound contains two halogen atoms, the number of positional isomers increases sharply. There are only two chloropropanes:

$$CH_3—CH_2—CH_2—Cl$$
1-chloropropane

$$CH_3—\overset{\displaystyle Cl}{\overset{|}{CH}}—CH_3$$
2-chloropropane

But there are five dichloropropanes, two of which form mirror-image pairs with each other.

Worked example

Draw skeletal formulae of the five dichloropropanes, and explain which two form a mirror-image pair.

Answer

Figure 15.5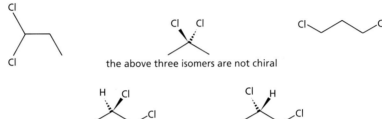

the above three isomers are not chiral

(+) and (–) forms

Now try this

How many positional isomers are there of chlorobromopropane, C_3H_6BrCl? How many of these are capable of forming optical isomers? Draw their structures.

The naming of halogenoalkanes follows the rules given in Topic 12. The longest carbon chain that contains the halogen atom is taken to be the stem. Other side chains are added, and finally the prefix 'fluoro', 'chloro', 'bromo' or 'iodo' is included, together with a numeral showing the position of the halogen atom in the chain:

$CH_3CH_2CHClCH_3$	is 2-chlorobutane
$ClCH_2CH(CH_3)CH_2CH_3$	is 1-chloro-2-methylbutane
$CH_3—CCl_3$	is 1,1,1-trichloroethane.

Now try this

Draw the structural formula of each of the following compounds.

1 1,1,2-trichloropropane
2 2-chloro-2-methylpropane

Worked example

Name the following compounds.

a

b

Answer

a 2,3-dichlorobutane
b 2-bromo-4-chloropentane or 4-bromo-2-chloropentane

Figure 15.6 Ball-and-stick models of the five isomers of C_4H_9Cl. The more spherical the molecule, the lower is the boiling point because of weaker van der Waals' bonding (see section 13.2). What type of isomerism is exhibited by isomers **b** and **c**? What type is exhibited by **d** and **e**?

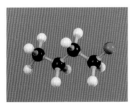
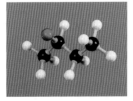
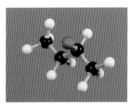
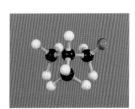
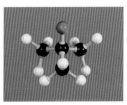

a 1-chlorobutane

b (+)-2-chlorobutane, b.p. 78 °C

c (–)-2-chlorobutane, b.p. 78 °C

d 1-chloro-2-methylpropane, b.p. 69 °C

e 2-chloro-2-methylpropane, b.p. 50 °C

15.3 Reactions of halogenoalkanes

Nucleophilic substitution reactions

Halogens are more electronegative than carbon, and so the C—Hal bond is strongly polarised $C^{\delta+}$—$Hal^{\delta-}$. During the reactions of halogenoalkanes, this bond breaks heterolytically to give the halide ion:

$$-\overset{|}{\underset{|}{C}}{}^{\delta+}\!\!\overset{\frown}{\cdot\!\cdot} X^{\delta-} \rightarrow -\overset{|}{\underset{|}{C}}{}^{+} + :X^{-} \qquad (1a)$$

The C^+ formed is then liable to attack by nucleophiles – anions or neutral molecules that possess a lone pair of electrons:

$$\overset{-}{Nu}\!:\!\overset{\frown}{}\overset{\diagdown\;\diagup}{C^{+}} \rightarrow Nu-\overset{|}{\underset{\diagdown}{C}}{}^{\diagup} \qquad (1b)$$

Often the nucleophile $\overset{..}{Nu}{}^-$ will attack the $C^{\delta+}$ carbon before the halide ion has left:

$$\overset{-}{Nu}\!:\!\overset{\frown}{}\overset{\diagup|}{\underset{\diagdown}{C}}\!\!\overset{\frown}{\cdot\!\cdot} X \rightarrow Nu-\overset{|}{\underset{\diagdown}{C}}{}^{\diagup} + :X^{-} \qquad (2)$$

The dominant reaction of halogenoalkanes is therefore **nucleophilic substitution**. The halide ion is replaced by a nucleophile. Whether mechanism (1a + 1b) or mechanism (2) is followed depends on the structure of the halogenoalkane (primary, secondary or tertiary), the nature of the solvent, and the nature of the nucleophile. The panel below describes the mechanisms of nucleophilic substitution reactions. Table 15.1 illustrates how some common nucleophiles react with a halogenoalkane such as bromoethane.

Table 15.1 Common nucleophiles that react with halogenoalkanes

Nucleophile		Products when reacted with bromoethane
Name	Formula	
water	$H_2\overset{..}{\underset{..}{O}}$	$CH_3CH_2OH + HBr$
hydroxide ion	$H\overset{..}{\underset{..}{O}}{}^{-}$	$CH_3CH_2OH + Br^{-}$
ammonia	$\overset{..}{N}H_3$	$CH_3CH_2NH_2 + HBr$
cyanide ion	$\overset{-}{\underset{..}{C}}{\equiv}N$	$CH_3CH_2CN + Br^{-}$
methoxide ion	$CH_3-\overset{..}{\underset{..}{O}}{}^{-}$	$CH_3CH_2-O-CH_3 + Br^{-}$
methylamine	$\overset{..}{N}H_2-CH_3$	$CH_3CH_2-NH-CH_3 + HBr$

The different mechanisms of nucleophilic substitution reactions

If we carry out kinetics experiments of the kind described in Topic 21 on the hydrolysis of bromoalkanes by sodium hydroxide, we find two different extremes.

The S_N1 reaction

The hydrolysis of 2-bromo-2-methylpropane is a **first-order** reaction (for an explanation of the *order* of a reaction, and the *rate equation*, see Topic 21):

$$CH_3-\overset{\overset{\displaystyle CH_3}{|}}{\underset{\underset{\displaystyle CH_3}{|}}{C}}-Br + OH^{-} \rightarrow CH_3-\overset{\overset{\displaystyle CH_3}{|}}{\underset{\underset{\displaystyle CH_3}{|}}{C}}-OH + Br^{-}$$

$$\text{rate} = k_1[\text{RBr}] \qquad \text{(where RBr is } (CH_3)_3CBr)$$

This means that the rate doubles if we double [RBr], but if we double (or halve) [OH$^-$], the rate does not change at all. The rate depends only on the concentration of the bromoalkane; it is independent of the hydroxide concentration. Hydroxide ions cannot therefore be involved in the **rate-determining step** – that is, the step in the overall reaction that is the slowest; the one that limits the overall rate of reaction. The first of the two mechanisms given on page 270 (equation (1a + 1b)) fits the kinetics equation shown. The mechanism is shown in Figure 15.7.

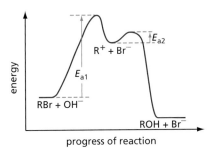

Figure 15.7 The S_N1 hydrolysis of 2-bromo-2-methylpropane

The first step involves only the heterolysis of the C—Br bond, forming the carbocation and a bromide ion. This is the slow step in the reaction, and hydroxide ions do not take part in it. If [OH⁻] were doubled, the rate of the second step (carbocation + OH⁻) might also double. But this second step is already faster than the first one, so the rate of the overall reaction is not affected.

This sequence of events is called the S_N1 mechanism (**S**ubstitution, **N**ucleophilic, unimolecular). The reaction profile of the S_N1 reaction is shown in Figure 15.8.

The activation energy of the first step, E_{a1}, is high, owing to the energy required to break bonds – this is why this is the *slow* step. That of the second step, E_{a2}, is low – oppositely charged ions attract each other strongly.

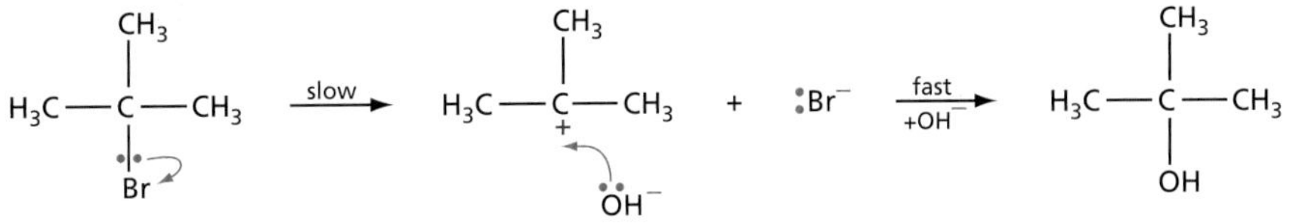

Figure 15.8 Reaction profile for the S_N1 hydrolysis of 2-bromo-2-methylpropane

The S_N2 reaction

The hydrolysis of bromomethane is a **second-order** reaction:

$$CH_3—Br + OH^- \rightarrow CH_3—OH + Br^-$$

$$\text{rate} = k_2[\text{RBr}][\text{OH}^-] \quad \text{(where RBr is } CH_3Br)$$

Doubling either [RBr] or [OH⁻] will double the rate of this reaction. (Doubling both [RBr] and [OH⁻] would increase the rate four-fold.) The hydroxide ion concentration has as much influence on the rate as the bromomethane concentration. We can therefore deduce that both bromomethane and hydroxide are involved in the rate-determining step.

The second mechanism given on page 270 (equation (2)) fits this kinetic relationship. The mechanism is shown in Figure 15.9.

Figure 15.9 The S_N2 hydrolysis of bromomethane

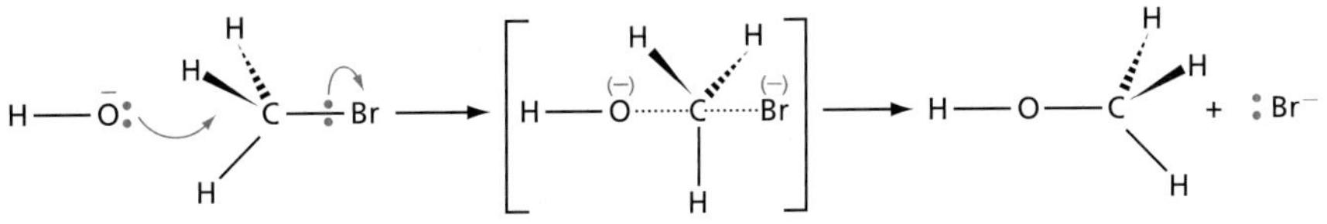

The reaction is a continuous one-step process. The complex shown in square brackets is not an intermediate (unlike the carbocation in the S_N1 reaction) but a **transition state**. It is a half-way stage in the reaction, where the C—Br bond is still getting longer and the O—C bond getting shorter.

This sequence of events is called the S_N2 mechanism (**S**ubstitution, **N**ucleophilic, bimolecular). The energy profile of the S_N2 reaction is shown in Figure 15.10.

Figure 15.10 Reaction profile for the S_N2 hydrolysis of bromomethane

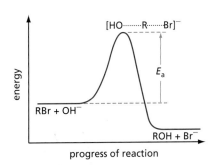

Figure 15.11 Reaction profiles for the S_N1 hydrolysis of primary and tertiary bromoalkanes

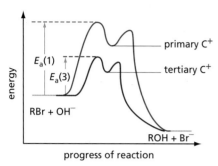

Table 15.2 The S_N1 reaction is fastest with a tertiary bromoalkane.

Compound	Type	Relative rate of S_N1 reaction
CH_3CH_2Br	primary	1
$(CH_3)_2CHBr$	secondary	26
$(CH_3)_3CBr$	tertiary	60 000 000

Figure 15.12 The S_N2 transition state is more crowded with a tertiary bromoalkane.

Table 15.3 The S_N2 reaction is fastest with a primary bromoalkane.

Compound	Type	Relative rate of S_N2 reaction
CH_3CH_2Br	primary	1000
$(CH_3)_2CHBr$	secondary	10
$(CH_3)_3CBr$	tertiary	1

Figure 15.13 Relative rates of S_N1 and S_N2 hydrolysis reactions for primary, secondary and tertiary halogenoalkanes

How the relative rates of S_N1 and S_N2 reactions depend on structure

We saw in Topic 13, page 248, and Topic 14, page 258, that alkyl groups donate electrons to carbocations and radicals by the inductive effect. The stability of carbocations increases in the order primary < secondary < tertiary. As the carbocation becomes more stable, the activation energy for the reaction leading to it also decreases (see Figure 15.11). We therefore expect that the rate of the S_N1 reaction will increase in the order primary < secondary < tertiary. (See Topic 8 for a description of the relationship between the magnitude of the activation energy and the rate of a reaction.) Some observed relative rates are given in Table 15.2.

No carbocations are formed during the S_N2 reaction, and other factors now come into play. The transition state has five groups arranged around the central carbon atom. It is therefore more crowded than either the starting bromoalkane or the alcohol product, each of which has only four groups around the central carbon atom.

Hydrogen atoms are much smaller than alkyl groups. We therefore expect that the more alkyl groups there are around the central carbon atom, the more crowded will be the transition state (see Figure 15.12), and the higher will be the activation energy E_a (see Figure 15.10). This will slow down the reaction (see Table 15.3).

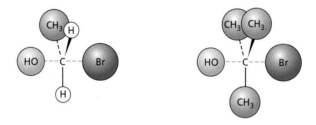

These two effects reinforce each other – the S_N1 reaction is faster with tertiary halogenoalkanes than with primary halogenoalkanes, whereas the S_N2 reaction is faster with primary halogenoalkanes than with tertiary halogenoalkanes. Overall, we expect primary halogenoalkanes to react predominantly by the S_N2 mechanism, tertiary halogenoalkanes to react predominantly by the S_N1 mechanism, and secondary halogenoalkanes to react by a mixture of the two (see Figure 15.13).

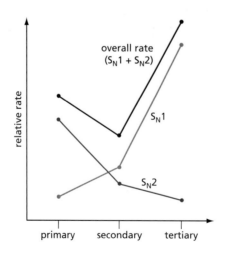

1 Hydrolysis

Some halogenoalkanes are reactive enough to be hydrolysed to form alcohols just by heating with water (in practice, a mixed ethanol–water solvent is used because halogenoalkanes are immiscible with water, but the ethanol is miscible with both):

$$CH_3-\underset{\underset{CH_3}{|}}{\overset{\overset{CH_3}{|}}{C}}-Cl + H_2O \rightarrow CH_3-\underset{\underset{CH_3}{|}}{\overset{\overset{CH_3}{|}}{C}}-OH + HCl$$

The hydrolysis is quicker if it is carried out in hot aqueous sodium hydroxide:

$$CH_3CH_2Br + NaOH(aq) \xrightarrow{\text{boil under reflux}} CH_3CH_2OH + NaBr(aq)$$

The relative rates of hydrolysis of chloro-, bromo- and iodoalkanes can be studied by dissolving the halogenoalkane in ethanol and adding aqueous silver nitrate. As the halogenoalkane slowly hydrolyses, the halide ion is released, and forms a precipitate of silver halide:

$$R-X + H_2O + Ag^+ \rightarrow R-OH + H^+ + AgX(s)$$

Table 15.4 shows the results obtained.

> Aqueous silver nitrate can be used to distinguish between organic chlorides, bromides and iodides.

Table 15.4 Silver halide precipitates formed with halogenoalkanes. The colours of these precipitates are shown in Figure 11.1, page 205.

Compound	Observation on reaction with $AgNO_3 + H_2O$ in ethanol	Colour of precipitate
$(CH_3)_2CH-Cl$	slight cloudiness after 1 hour	white
$(CH_3)_2CH-Br$	cloudiness appears after a few minutes	pale cream
$(CH_3)_2CH-I$	thick precipitate appears within a minute	pale yellow

Table 15.5 Comparing the carbon–halogen bond strengths

Bond	Bond enthalpy/kJ mol^{-1}
C—F	485
C—Cl	338
C—Br	285
C—I	213

Iodoalkanes are therefore seen to be the most reactive halogenoalkanes, and chloroalkanes the least reactive. Fluoroalkanes do not react at all. This Group 17 trend corresponds to the strengths of the C—X bond. The weaker the bond (that is, the smaller the bond enthalpy), the easier it is to break (see Table 15.5).

The bromoalkanes are the halogenoalkanes most often used in preparative organic chemistry. Chloroalkanes react too slowly; iodoalkanes can be too reactive and slowly decompose unless stored in the dark. In addition, iodine is a much more expensive element than bromine. We shall use bromoethane to illustrate subsequent reactions of halogenoalkanes.

2 With ammonia

The ammonia molecule contains a lone pair of electrons. The lower electronegativity of nitrogen compared with oxygen makes ammonia a stronger nucleophile than water. The reaction is as follows:

$$\overset{..}{N}H_3 \quad \underset{CH_3}{\overset{}{C}}H_2 \overset{..}{Br} \rightarrow H_3\overset{+}{N}-\underset{CH_3}{\overset{}{C}}H_2 \quad + \quad :\overset{..}{Br}^-$$

$$H_3\overset{+}{N}-\underset{CH_3}{\overset{}{C}}H_2 \quad + \quad NH_3 \rightleftharpoons H_2N-\underset{CH_3}{\overset{}{C}}H_2 \quad + \quad NH_4^+$$

ethylamine

Because ammonia is a gas, the reactants (in solution in ethanol) need to be heated in a sealed tube, to prevent the ammonia escaping.

The product is called **ethylamine**. Like ammonia itself, ethylamine possesses a lone pair of electrons on its nitrogen atom. The electron-donating ethyl group makes ethylamine even more nucleophilic than ammonia. So, if an excess of bromoethane is used, further reactions can occur:

$$CH_3CH_2\overset{..}{N}H_2 \quad + \quad CH_3CH_2Br \quad \rightarrow \quad \underset{\underset{\overset{|}{\overset{..}{N}H}}{}}{CH_3CH_2 \quad CH_2CH_3} \quad (+\ HBr)$$

diethylamine

$$\underset{\underset{\overset{|}{\overset{..}{N}H}}{}}{CH_3CH_2 \quad CH_2CH_3} \quad + \quad CH_3CH_2Br \quad \rightarrow \quad \underset{\underset{\overset{|}{CH_2CH_3}}{\overset{|}{\overset{.}{N}}}}{CH_3CH_2 \quad CH_2CH_3} \quad (+\ HBr)$$

triethylamine

Finally, even triethylamine possesses a lone pair of electrons on its nitrogen atom. It can react as follows:

$$(CH_3CH_2)_3\overset{..}{N} + CH_3CH_2Br \rightarrow (CH_3CH_2)_4N^+Br^-$$

tetraethylammonium bromide

To avoid these further reactions, an excess of ammonia is used. The properties and reactions of amines are described in Topic 27.

3 With cyanide ions

When a halogenoalkane is heated under reflux with a solution of sodium (or potassium) cyanide in ethanol, a nucleophilic substitution reaction occurs and the halogen is replaced by the cyano group:

$$CH_3CH_2Br + Na^+CN^- \rightarrow CH_3CH_2-C\equiv N + Na^+Br^-$$

The product is called **propanenitrile**. Its name shows that it contains three carbon atoms, although we started with bromoethane, which has just two. The use of cyanide is a good method of increasing the length of a carbon chain by one carbon atom, which can be useful in the synthesis of organic compounds. Nitriles can be hydrolysed to carboxylic acids (see Topic 18) by heating under reflux with dilute sulfuric acid:

$$CH_3CH_2-C\equiv N + 2H_2O + H^+ \rightarrow CH_3CH_2CO_2H + NH_4^+$$

propanoic acid

Nitriles can also be reduced to amines (see Topic 27) by hydrogen with a nickel or platinum catalyst:

$$CH_3CH_2-C\equiv N + 2H_2 \rightarrow CH_3CH_2CH_2NH_2$$

Reflux

Heating under reflux means to heat a flask so that the solvent boils continually. A condenser is placed vertically in the neck of the flask to condense the solvent vapour (see Figure 15.14). This method allows the temperature of the reaction mixture to be kept at the boiling point of the solvent (that is, the highest possible temperature for that particular solvent, so that slow organic reactions can be speeded up as much as possible) without the solvent evaporating away.

Figure 15.14 A reflux apparatus

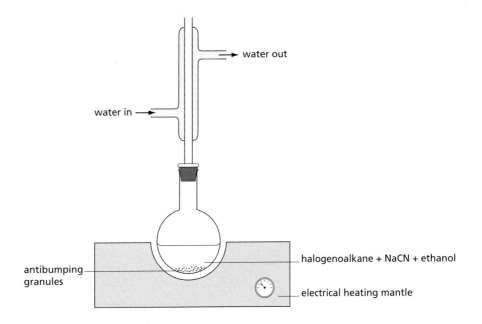

Elimination reactions of halogenoalkanes

Nucleophiles are electron-pair donors, donating to a positive or partially positive carbon atom. Bases are also electron-pair donors, donating a pair of electrons to a positive hydrogen atom.

The hydroxide ion can act as a nucleophile:

$$HO^- : \quad CH_3 \div Br \quad \rightarrow \quad HO—CH_3 \; + \; :Br^-$$

or it can act as a base:

$$HO^- : \quad H \div Cl \rightarrow HO—H \; + \; :Cl^-$$
$$\text{base} \qquad \text{acid} \qquad \text{water}$$

The hydrogen atoms attached to carbon are slightly δ+ (see the table of electronegativities on page 54). They are therefore *very* slightly acidic. When a halogenoalkane is reacted with hydroxide ions under certain conditions, an acid–base reaction can occur at the same time as the carbon–halogen bond is broken:

$$HO^- : \qquad\qquad HO—H$$
$$H \diagdown \; CH_2 \qquad\qquad CH_2$$
$$CH_2 \quad Br \quad \rightarrow \quad CH_2 \quad :Br^-$$

An alkene has been formed by the **elimination** of hydrogen bromide from the bromoalkane. The same reagent (hydroxide ion) can therefore carry out two different types of reaction (substitution or elimination) when reacted with the same bromoalkane. Both reactions do in fact occur at the same time, but the proportion of bromoalkane molecules that undergo elimination can be varied. It depends on three factors:

- the nature of the bromoalkane (primary, secondary or tertiary)
- the strength and physical size of the base
- the solvent used for the reaction.

In general, nucleophilic substitution to form alcohols is favoured by:

- primary bromoalkanes
- bases of weak or medium strength, and small size
- polar solvents such as water
- low temperature.

On the other hand, elimination to form alkenes is favoured by:

- tertiary bromoalkanes
- strong, bulky bases (such as potassium 2-methylprop-2-oxide, $(CH_3)_3C—O^-K^+$)
- less polar solvents such as ethanol
- high temperature.

The most commonly used reagent is potassium hydroxide dissolved in ethanol and heated under reflux. Potassium hydroxide is used because it is more soluble in ethanol than is sodium hydroxide. Table 15.6 shows the results of a study of the effects of solvent and halogenoalkane structure on the elimination : substitution ratio

Table 15.6 Elimination : substitution, (E/S) ratio in bromoalkanes

	Formula of bromoalkane	Type of bromoalkane	Conditions	Ratio, E/S
1	$(CH_3)_2CHBr$	secondary	$2.0\,mol\,dm^{-3}$ OH^- in 60% ethanol*	1.5
2	$(CH_3)_2CHBr$	secondary	$2.0\,mol\,dm^{-3}$ OH^- in 80% ethanol*	2.2
3	$(CH_3)_2CHBr$	secondary	$2.0\,mol\,dm^{-3}$ OH^- in 100% ethanol	3.8
4	$(CH_3)_3CBr$	tertiary	$2.0\,mol\,dm^{-3}$ OH^- in 100% ethanol	13.0

* The remainder is water.

Comparing rows 1 to 3 of Table 15.6, we can see that the elimination : substitution ratio increases as the solvent becomes richer in ethanol. Comparing rows 3 and 4, we can see that the elimination : substitution ratio increases when we replace a secondary bromide by a tertiary one.

Worked example

For each of the following combinations of reagents and conditions, suggest whether substitution or elimination will predominate. State the reasons for your choice.

a heating $CH_3CH_2CH_2Br$ with NaOH(aq)
b heating $(CH_3)_3CBr$ with NaOH in ethanol
c heating $(CH_3)_2CHBr$ with $(CH_3)_3C—O^-K^+$

Answer
a Substitution predominates with a primary bromoalkane and OH^- in a polar solvent.
b Elimination predominates with a tertiary bromoalkane and OH^- in a non-polar solvent.
c Elimination predominates with a secondary bromoalkane and a bulky base.

Now try this

Give the structural formula of the main product of each of the following reactions.

1 $(CH_3)_2CH—CH_2Br$ + NaOH in ethanol
2 $CH_3CH_2CHBrCH_2CH_3$ + NaOH(aq)

15.4 Chlorofluorocarbons and the ozone layer

The **chlorofluorocarbons** (CFCs) have almost ideal properties for use as aerosol propellants and refrigerant heat-transfer fluids. They are chemically and biologically inert (and hence safe to use and handle), and they can be easily liquefied by pressure a little above atmospheric pressure. Although fairly expensive to produce (compared to other refrigerants like ammonia), they rapidly replaced other fluids that had been used. In the early 1970s, however, concern was expressed that their very inertness was a global disadvantage. Once released into the environment, CFCs remain chemically unchanged for years. Being volatile, they can diffuse throughout the atmosphere, and eventually find their way into the stratosphere (about 20 km above the Earth's surface). Here they are exposed to the stronger ultraviolet rays of the Sun. Although the carbon–fluorine bond is very strong, the carbon–chlorine bond is weak enough to be split by ultraviolet light (see Table 15.5). This forms atomic chlorine radicals, which can upset the delicately balanced equilibrium between ozone formation and ozone breakdown.

$$CF_3Cl \xrightarrow{\text{ultraviolet light}} CF_3^{\bullet} + Cl^{\bullet}$$

Some stratospheric chemistry

1 Production of oxygen atoms: $O_2 \rightarrow 2O$
   (by absorption of UV light at 250 nm wavelength)

Once oxygen atoms have formed, they can react with oxygen molecules to produce ozone, which by absorption of ultraviolet light decomposes to re-form oxygen atoms and oxygen molecules.

2 Natural ozone formation: $O + O_2 \rightarrow O_3$
3 Natural ozone depletion: $O_3 \rightarrow O_2 + O$
   (by absorption of UV light at 300 nm wavelength, thus reducing the level of short-wavelength ultraviolet radiation from the Sun's rays at the Earth's surface)

A steady state is set up in which the rate of production of ozone equals the rate of its breakdown. Chlorine atoms disrupt the balance of this steady state by acting as a homogeneous catalyst (see section 8.6) for the destruction of ozone:

4 $Cl^{\bullet} + O_3 \rightarrow ClO^{\bullet} + O_2$
5 $ClO^{\bullet} + O \rightarrow Cl^{\bullet} + O_2$

Reaction 5, in which chlorine atoms are regenerated, involves the destruction of the oxygen atoms needed to make more ozone by the natural formation reaction (2). In this way, chlorine atoms have a doubly depleting effect on ozone concentration in the atmosphere.

It has been estimated that one chlorine atom can destroy over 10^5 ozone molecules before it eventually diffuses back into the lower atmosphere. There it can react with water vapour to produce hydrogen chloride, which can be flushed out by rain as dilute hydrochloric acid.

Once the 'ozone hole' above Antarctica was discovered in 1985, global agreements were signed in Montreal (1989) and London (1990). As a result, the global production of CFCs has been drastically reduced. In many of their applications they can be replaced by hydrocarbons such as propane. It will still take several decades for natural regeneration reactions to allow the ozone concentration to recover, but that recovery is now well underway. Other stratospheric pollutants such as nitric oxide from high-flying aircraft also destroy ozone.

Figure 15.15 The ozone 'hole' is an area over Antarctica where ozone levels are depleted, particularly in spring. A similar depletion is occurring over the Arctic as well.

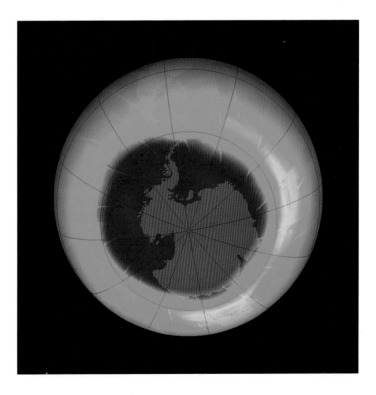

Figure 15.16 Depletion of the ozone layer puts humans at increased risk of skin cancer and damage to the eyes by short-wavelength UV radiation (UV-B).

15.5 Preparing halogenoalkanes

1 From alkanes, by radical substitution

As we saw in section 13.5, this reaction works well with chlorine and bromine. It has limitations in organic synthesis, however, since it is essentially a random process. It gives a mixture of isomers, and multiple substitution often occurs. Nevertheless, it is often used industrially, where fractional distillation can separate the different products. Alkanes that can form only one isomer of a monochloroalkane are the most suitable starting materials.

$$CH_4 + Cl_2 \xrightarrow{\text{light}} CH_3Cl + HCl$$

$$C_2H_6 + Cl_2 \longrightarrow CH_3CH_2Cl + HCl$$

2 From alkenes, by electrophilic addition

We saw in section 14.3 that the addition of chlorine or bromine to alkenes gives 1,2-dihalogenoalkanes:

$$CH_2{=}CH_2 + Br_2 \rightarrow Br{-}CH_2{-}CH_2{-}Br$$

The addition of hydrogen halides to alkenes gives monohalogenoalkanes. The orientation of the halogen follows Markovnikov's rule (page 259). The reaction works with hydrogen iodide as well as with hydrogen chloride and hydrogen bromide:

$$CH_3{-}CH{=}CH_2 + HI \rightarrow CH_3{-}CHI{-}CH_3$$

3 From alcohols, by nucleophilic substitution

These reactions are covered in detail in section 16.3. There are several reagents that can be used:

$$CH_3CH_2OH + HCl \xrightarrow{\text{conc. HCl + ZnCl}_2 + \text{heat}} CH_3CH_2Cl + H_2O$$

$$CH_3CH_2OH + PCl_5 \xrightarrow{\text{warm}} CH_3CH_2Cl + HCl + POCl_3$$

$$CH_3CH_2OH + SOCl_2 \xrightarrow{\text{warm}} CH_3CH_2Cl + HCl + SO_2$$

$$CH_3CH_2OH + HBr \xrightarrow{\text{conc. H}_2SO_4 + \text{NaBr + heat}} CH_3CH_2Br + H_2O$$

$$3CH_3CH_2OH + PI_3 \xrightarrow{\text{heat with P + I}_2} 3CH_3CH_2I + H_3PO_3$$

Summary

- **Halogenoalkanes** contain a $\delta+$ carbon atom.
- Their most common reaction is **nucleophilic substitution**. They also undergo **elimination** reactions to give alkenes.
- Their reactivity increases in the order $C—F < C—Cl < C—Br < C—I$.
- Fluoro- and chloroalkanes are used as solvents, aerosol propellants and refrigerants. CFCs damage the ozone layer by being photolysed to chlorine atoms, which initiate chain reactions destroying ozone.
- Halogenoalkanes can be made from alkanes, alkenes and alcohols.

Key reactions you should know
(R = primary, secondary or tertiary alkyl unless otherwise stated. All reactions also work with Cl instead of Br, but much more slowly.)

- Elimination:

$$R—CH_2—CH_2—Br + OH^- \xrightarrow{\text{heat with NaOH in ethanol}} R—CH\!=\!CH_2 + H_2O + Br^-$$

- Nucleophilic substitutions:

$$R—Br + OH^- \xrightarrow{\text{heat with NaOH in water}} R—OH + Br^-$$

$$R—Br + 2NH_3 \xrightarrow{\text{NH}_3 \text{ in ethanol under pressure}} R—NH_2 + NH_4Br$$

$$R—Br + CN^- \xrightarrow{\text{heat with NaCN in ethanol}} R—C\!\equiv\!N + Br^-$$
$$R—CO_2H \qquad R—CH_2NH_2$$

Examination practice questions

Please see the data section of the CD for any A_r values you may need.

1 Halogenoalkanes have many chemical uses, particularly as intermediates in organic reactions. Three reactions of 1-bromobutane, $CH_3CH_2CH_2CH_2Br$, are shown below.

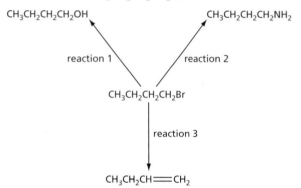

a For **each** reaction, state the reagent and solvent used. [6]

b When 1-iodobutane, $CH_3CH_2CH_2CH_2I$, is reacted under the same conditions as those used in reaction 1, butan-1-ol is formed.
What difference, if any, would there be in the rate of this reaction compared to the reaction of 1-bromobutane? Use appropriate data from the data section on the CD to explain your answer. [3]

Dichlorodifluoromethane, CCl_2F_2, is an example of a chlorofluorocarbon (CFC) that was formerly used as an aerosol propellant. In September 2007, at the Montreal summit, approximately 200 countries agreed to phase out the use of CFCs by 2020.

c State two properties of CFCs that made them suitable as aerosol propellants. [2]

d When CFCs are present in the upper atmosphere, homolytic fission takes place in the presence of ultraviolet light.

i What is meant by the term *homolytic fission*?

ii Suggest an equation for the homolytic fission of CCl_2F_2. [2]

e The most common replacements for CFCs as aerosol propellants are hydrocarbons such as propane and butane.
Suggest **one** disadvantage of these compounds as aerosol propellants. [1]

[Cambridge International AS & A Level Chemistry 9701, Paper 22 Q4 November 2010]

2 This question is about halogenated hydrocarbons.

a Halogenoalkanes undergo nucleophilic substitution reactions with ammonia to form amines. Amines contain the —NH_2 functional group.

For example, 1-bromopropane reacts with ammonia to form propylamine, $CH_3CH_2CH_2NH_2$.

$$CH_3CH_2CH_2Br + 2NH_3 \rightarrow CH_3CH_2CH_2NH_2 + NH_4Br$$

i Iodoethane is reacted with ammonia. Write an equation for this reaction. [2]

ii The first step in the mechanism of the reaction between $CH_3CH_2CH_2Br$ and NH_3 is shown below. It is incomplete.

Complete the mechanism. Include relevant dipoles, lone pairs, curly arrows and the missing product. [3]

b A student investigates the rate of hydrolysis of six halogenoalkanes.
The student mixes $5\,cm^3$ of ethanol with five drops of halogenoalkane. This mixture is warmed to $50\,°C$ in a water bath. The student adds $5\,cm^3$ of aqueous silver nitrate, also heated to $50\,°C$, to the halogenoalkane. The time taken for a precipitate to form is recorded in a results table.
The student repeats the whole experiment at $60\,°C$ instead of $50\,°C$.

Halogenoalkane	Time taken for a precipitate to form/s	
	At 50 °C	At 60 °C
$CH_3CH_2CH_2CH_2Cl$	243	121
$CH_3CH_2CH_2CH_2Br$	121	63
$CH_3CH_2CH_2CH_2I$	40	19
$CH_3CH_2CHBrCH_3$	89	42
$(CH_3)_2CHCH_2Br$	110	55
$(CH_3)_3CBr$	44	21

Describe and explain the factors that affect the rate of hydrolysis of halogenoalkanes. Include ideas about
- the halogen in the halogenoalkanes
- the groups attached to the carbon of the carbon–halogen bond (the type of halogenoalkane)
- the temperature of the hydrolysis.

In your answer you should link the evidence with your explanation. [7]

[OCR Chemistry A Unit F322 Q5 (part) January 2011]

16 Alcohols

Functional group:

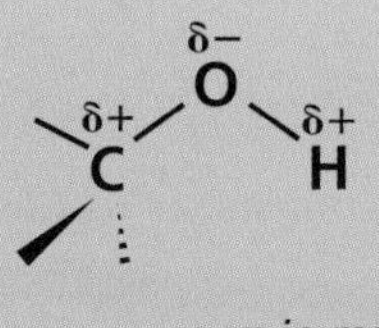

The —OH (hydroxyl) group occurs in many organic compounds. In addition to the alcohols, it forms all or part of the functional group in sugars, phenols and carboxylic acids. In this topic we look at the reactions of the hydroxyl group when it is the only functional group in the molecule, in the alcohols. The hydrogen atom attached to oxygen is slightly acidic, and the δ+ nature of the carbon atom makes it susceptible to nucleophilic substitution reactions. Primary and secondary alcohols also undergo oxidation.

Learning outcomes

By the end of this topic you should be able to:

14.1a) interpret and use the general, structural, displayed and skeletal formulae of the alcohols (including primary, secondary and tertiary)

17.1a) recall the chemistry of alcohols, exemplified by ethanol in the following reactions: combustion, substitution to give halogenoalkanes, reaction with sodium, oxidation to carbonyl compounds and carboxylic acids, dehydration to alkenes, formation of esters by esterification with carboxylic acids (part, see also Topic 26)

17.1b) classify hydroxy compounds into primary, secondary and tertiary alcohols, and suggest characteristic distinguishing reactions, e.g. mild oxidation

17.1c) deduce the presence of a $CH_3CH(OH)$— group in an alcohol from its reaction with alkaline aqueous iodine to form tri-iodomethane.

16.1 Introduction

The **alcohols** form a homologous series with the general formula $C_nH_{2n+1}OH$. They occupy a central position in organic functional group chemistry: they can be readily converted to aldehydes and carboxylic acids by oxidation, and can be formed from them by reduction; and they can be converted to and from halogenoalkanes by nucleophilic substitution.

Alcohols are useful solvents themselves, but they are also key intermediates in the production of esters, which are important solvents for the paints and plastics industries.

The polar —OH group readily forms hydrogen bonds to similar groups in other molecules. This accounts for the following major differences between the alcohols and the corresponding alkanes.

- The lower alcohols (C_1, C_2, C_3 and some isomers of C_4) are totally miscible with water, owing to hydrogen bonding between the alcohol molecules and water molecules (see Figure 16.1). As the length of the alkyl chain increases, van der Waals' attractions predominate between the molecules of the alcohol, and so the miscibility with water decreases.

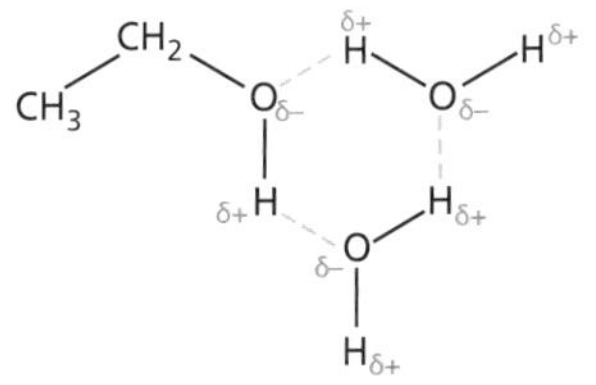

Figure 16.1 Hydrogen bonding between ethanol and water

● The boiling points of the alcohols are all much higher than the corresponding (isoelectronic) alkanes, owing to strong intermolecular hydrogen bonding (see Figure 16.2).

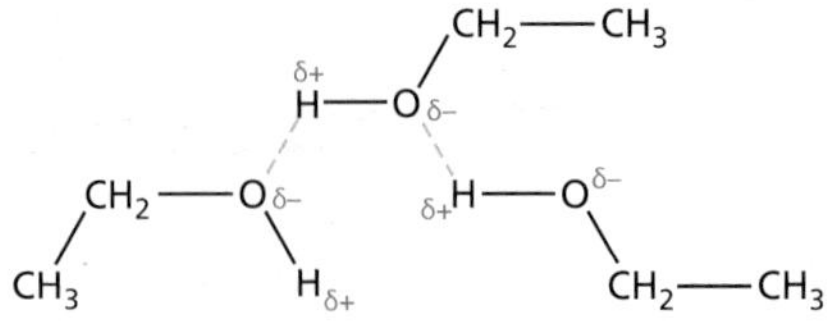

Figure 16.2 Intermolecular hydrogen bonding in ethanol

Table 16.1 and Figure 16.3 show this large difference. As is usual in homologous series, the boiling points increase as the number of carbon atoms increases, owing to increased van der Waals' attractions between the longer alkyl chains. The enhancement of boiling point due to hydrogen bonding also decreases, as the longer chains become more alkane-like.

Table 16.1 Boiling points of some alcohols and alkanes

Number of electrons in the molecule	Alkane		Alcohol		Enhancement of boiling point/°C
	Formula	Boiling point/°C	Formula	Boiling point/°C	
18	C_2H_6	−88	CH_3OH	65	153
26	C_3H_8	−42	C_2H_5OH	78	120
34	C_4H_{10}	0	C_3H_7OH	97	97
42	C_5H_{12}	36	C_4H_9OH	118	82
50	C_6H_{14}	69	$C_5H_{11}OH$	138	69
58	C_7H_{16}	98	$C_6H_{13}OH$	157	59

Figure 16.3 Boiling points of some alcohols and alkanes

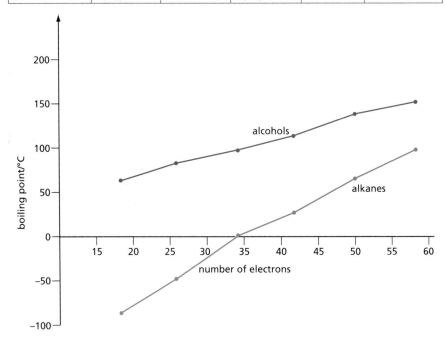

The electron pairs around the oxygen atom in alcohols are arranged in a similar way to those around the oxygen atom in the water molecule, as both have two lone pairs of electrons (see Figure 16.4).

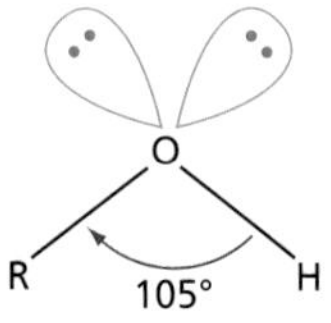

16.2 Isomerism and nomenclature

As was mentioned in section 12.5, the alcohols are named by adding the suffix '-ol' to the alkane stem of the compound. The longest carbon chain that contains the —OH group is chosen as the stem. A numeral precedes the '-ol' to describe the position of the —OH group along the chain.

Figure 16.4 The oxygen atom in an alcohol has two lone pairs of electrons.

Figure 16.5 Molecular models of **a** methanol, **b** ethanol, **c** propan-1-ol, **d** propan-2-ol and **e** butan-1-ol

a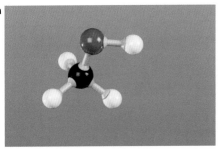

b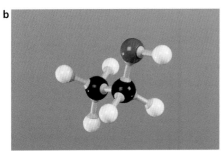

c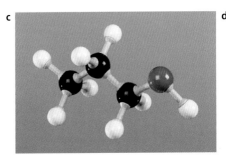

d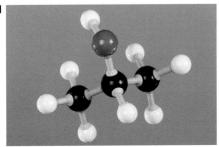

e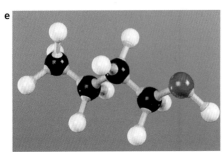

Worked example

Name the following compounds.

a CH_3—$CH(OH)$—CH_2—CH_3
b CH_2=CH—CH_2OH
c CH_3—CH_2—$CH(CH_3)$—CH_2OH

Answer

a The longest (and only) carbon chain contains four carbon atoms. The —OH is on the second atom, so the alcohol is butan-2-ol.
b The three-carbon base is propene. Taking the —OH group to be on position 1, the name is thus prop-2-en-1-ol.
c The longest chain contains four carbon atoms, with a methyl group on position 2 and the —OH group on position 1. The compound is 2-methylbutan-1-ol.

Alcohols show chain, positional and optical isomerism (see section 12.6).

Alcohols are classed as primary, secondary or tertiary, depending on the number of alkyl groups that are attached to the carbon atom joined to the —OH group. Of the five isomers with the formula C_4H_9OH, two (**A** and **B** below) are primary alcohols, two (**C** and **D**) are secondary alcohols, and one (**E**) is a tertiary alcohol.

Now try this

1 Describe the type of isomerism shown by the following pairs of alcohols, whose structures are shown.
  a **A** and **B**   b **B** and **E**
  c **A** and **C**   d **C** and **D**
2 Name the five alcohols whose structures are shown. (Note: alcohols **C** and **D** will have the *same* name.)

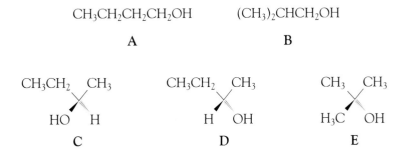

$$CH_3CH_2CH_2CH_2OH \qquad (CH_3)_2CHCH_2OH$$

A   B

The ethers

Ethers are isomers of alcohols, but they do not contain the —OH group and so have none of the reactions of alcohols, and do not form intermolecular hydrogen bonds.

Hence they are fairly inert compounds with boiling points lower than those of the isomeric alcohols. For example:

diethyl ether, $C_4H_{10}O$, b.p. 35°C 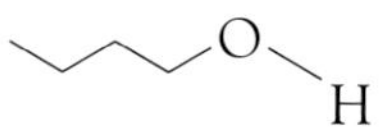   butan-1-ol, $C_4H_{10}O$, b.p. 118°C

They are quite good at dissolving organic compounds, and so are often used as solvents.

16.3 Reactions of alcohols

1 Combustion

All alcohols burn well in air, but only the combustion of ethanol is of everyday importance:

$$C_2H_5OH + 3O_2 \rightarrow 2CO_2 + 3H_2O$$

The use of ethanol as a fuel has been mentioned in section 13.4.

For its uses as a fuel, and as a common solvent, ethanol is marketed as 'methylated' or 'rectified' spirit. This contains about 90% ethanol, together with 5% water, 5% methanol, and pyridine. The poisonous methanol is added to make it undrinkable, and hence exempt from the large excise duty that spirits attract. The pyridine gives it a bitter taste, which makes it unpalatable.

The other reactions of alcohols can be divided into three groups:

- reactions involving the breaking of the O—H bond
- reactions involving the breaking of the C—O bond
- reactions of the $\overset{\diagdown}{\underset{\diagup}{C}}\overset{H}{\underset{OH}{}}$ group (tertiary alcohols cannot show this group of reactions).

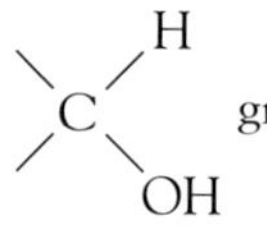

2 Reactions involving the breaking of the O—H bond

With sodium metal

The hydroxyl hydrogen in alcohols is slightly acidic, just like the hydrogen atoms in water, but the alcohols ionise to a lesser extent:

$$H—O—H \rightleftharpoons H—O^- + H^+ \quad K_w = 1.0 \times 10^{-14}\,mol^2\,dm^{-6}$$
$$R—O—H \rightleftharpoons R—O^- + H^+ \quad K = 1.0 \times 10^{-16}\,mol^2\,dm^{-6}$$

Alcohols liberate hydrogen gas when treated with sodium metal:

$$CH_3CH_2OH + Na \rightarrow CH_3CH_2O^-Na^+ + \tfrac{1}{2}H_2(g)$$

The reaction is less vigorous than that between sodium and water. The product is called sodium ethoxide (compare sodium hydroxide, from water). This is a white solid, which is soluble in ethanol and in water (to give a strongly alkaline solution).

All compounds containing —OH groups liberate hydrogen gas when treated with sodium. The fizzing that ensues is a good test for the presence of an —OH group (as long as water is absent).

Esterification

Esters are compounds that contain the $—\overset{O}{\overset{\|}{C}}—O—$ group.

They can be obtained by reacting alcohols with carboxylic acids or with acyl chlorides (see Topics 18 and 26).

alcohol + carboxylic acid → ester + water

For example:

$$CH_3CH_2OH + CH_3\overset{\displaystyle O}{\overset{\displaystyle \|}{C}}{-}OH \xrightarrow[\text{as catalyst}]{\substack{\text{heat with}\\ \text{conc. H}_2\text{SO}_4}} CH_3{-}\overset{\displaystyle O}{\overset{\displaystyle \|}{C}}{-}O{-}CH_2CH_3 + H_2O$$

It has been found that the oxygen in the water comes from the carboxylic acid, and the —O— in the ester comes from the alcohol. The methods of ester preparation will be dealt with in detail in Topics 18 and 26.

Esters are useful solvents, and several are also used, in small quantities, as flavouring agents in fruit drinks and sweets.

3 Reactions involving the breaking of the C—O bond

Although the C—O bond is polarised $C^{\delta+}{-}O^{\delta-}$, it is a strong bond, and does not easily break heterolytically:

$$-\overset{|}{\underset{|}{C}}{-}OH \rightarrow -\overset{|}{\underset{|}{C^+}} + {}^-OH$$

If, however, the oxygen is protonated, or bonded to a sulfur or phosphorus atom, the C—O bond is much more easily broken. As a result, alcohols undergo several **nucleophilic substitution** reactions.

Reaction with hydrochloric acid

Tertiary alcohols react easily with concentrated hydrochloric acid on shaking at room temperature. The reaction proceeds via the S_N1 mechanism (see section 15.3):

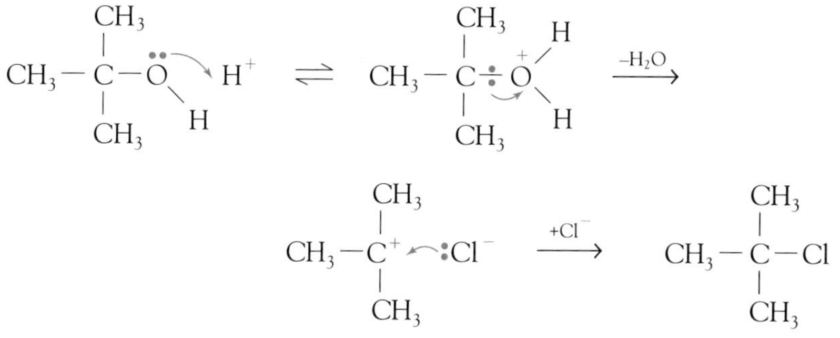

Secondary and primary alcohols also react with concentrated hydrochloric acid, but at a slower rate. Anhydrous zinc chloride needs to be added as a catalyst, and the mixture requires heating:

$$CH_3CH_2OH + HCl \xrightarrow{ZnCl_2,\ heat} CH_3CH_2Cl + H_2O$$

This reaction is the basis of the **Lucas test** to distinguish between primary, secondary and tertiary alcohols (see Table 16.2). It relies on the fact that alcohols are soluble in the reagent (concentrated HCl and $ZnCl_2$) whereas chloroalkanes are not, and therefore produce a cloudiness in the solution.

Table 16.2 The Lucas test

Type of alcohol	Observation on adding conc. HCl + ZnCl₂
R₃COH (tertiary)	immediate cloudiness appears in the solution
R₂CHOH (secondary)	cloudiness apparent within 5 minutes
RCH₂OH (primary)	no cloudiness apparent unless warmed

Reaction with phosphorus(V) chloride or sulfur dichloride oxide

Both of these reagents convert alcohols into chloroalkanes (see section 15.5), and in both cases hydrogen chloride is evolved. This fizzing with PCl_5, and the production of misty fumes, can be used as a test for the presence of an alcohol (as long as water is absent). Both reactions occur on gently warming the reagents.

$$CH_3CH_2OH + PCl_5 \rightarrow CH_3CH_2Cl + POCl_3 + HCl$$
$$CH_3CH_2OH + SOCl_2 \rightarrow CH_3CH_2Cl + SO_2 + HCl$$

Reaction with hydrogen bromide

Alcohols can be converted into bromoalkanes by reaction with either concentrated hydrobromic acid or a mixture of sodium bromide, concentrated sulfuric acid and water (50%). These react to give hydrogen bromide:

$$2NaBr + H_2SO_4 \rightarrow Na_2SO_4 + 2HBr$$

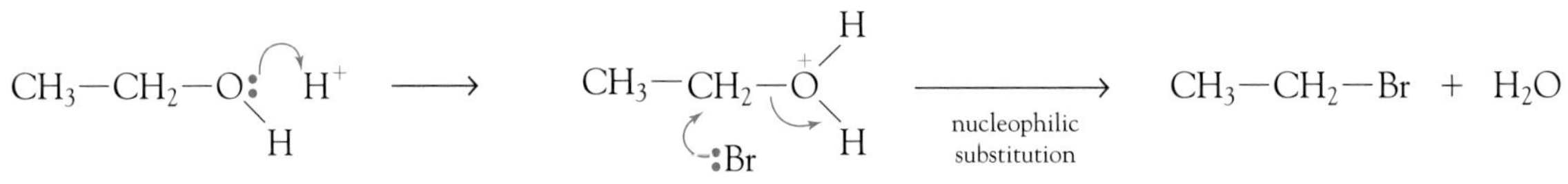

Reaction with red phosphorus and bromine or iodine

These halogens react with phosphorus to give the phosphorus trihalides:

$$2P + 3Br_2 \rightarrow 2PBr_3$$
$$2P + 3I_2 \rightarrow 2PI_3$$

The phosphorus trihalides, for example phosphorus triiodide, then react with alcohols as follows:

$$3CH_3CH_2OH + PI_3 \rightarrow 3CH_3CH_2I + P(OH)_3$$

Reaction with concentrated sulfuric acid

Ethanol and concentrated sulfuric acid react together to give the compound ethyl hydrogensulfate:

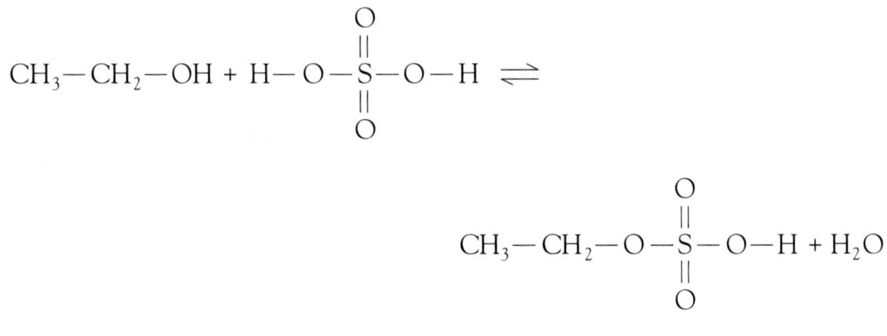

The ethyl hydrogensulfate can then undergo an elimination reaction to form ethene:

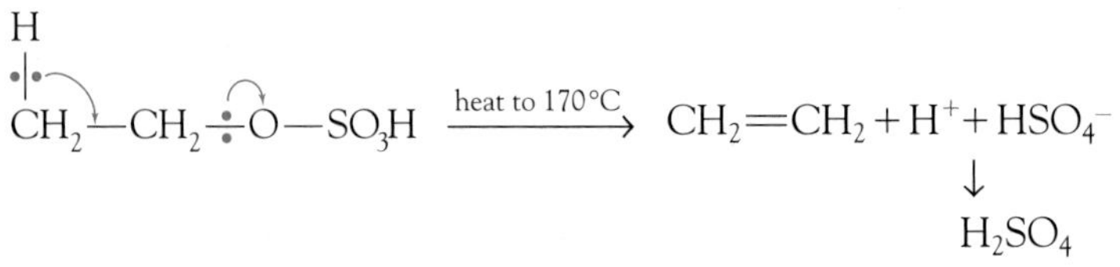

As the sulfuric acid is regenerated on elimination, it is essentially a catalyst for the conversion of the alcohol to the alkene:

$$CH_3CH_2OH \rightarrow CH_2{=}CH_2 + H_2O$$

Concentrated phosphoric acid also acts as a catalyst for the reaction. This is often preferred because, unlike sulfuric acid, it is not also an oxidising agent, and so the formation of by-products is minimised.

$$CH_3CH_2OH \xrightarrow{\text{heat with } H_3PO_4} CH_2{=}CH_2 + H_2O$$

A third method of dehydration is to pass the vapour of the alcohol over strongly heated aluminium oxide:

$$CH_3CH_2OH \xrightarrow{\text{Al}_2\text{O}_3 \text{ at } 350\,^{\circ}\text{C}} CH_2{=}CH_2 + H_2O$$

Tertiary alcohols dehydrate very easily on warming with an acid. The reaction goes via the carbocation, which then loses a proton:

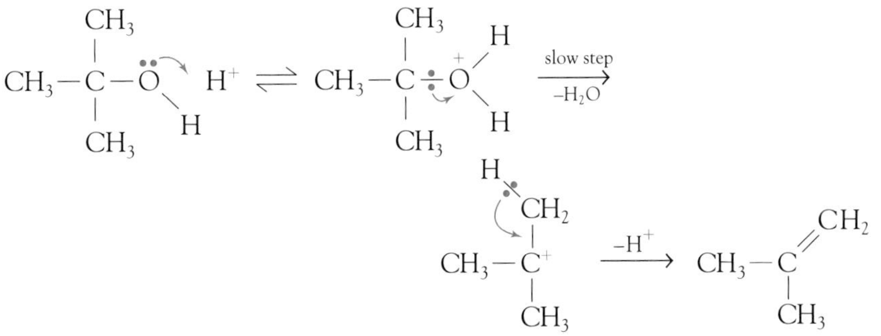

Worked example

Dehydrating longer-chain alcohols can give a mixture of alkenes. Suggest structures for the alkenes possible from the dehydration of the alcohol 1-methylcyclohexanol:

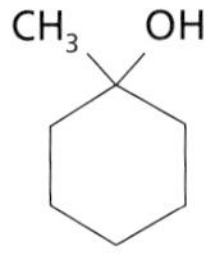

Which alkene will be the most stable?

Answer
The H and the OH of the water which is eliminated come from adjacent carbon atoms, forming the C=C bond between them. There are three possibilities:

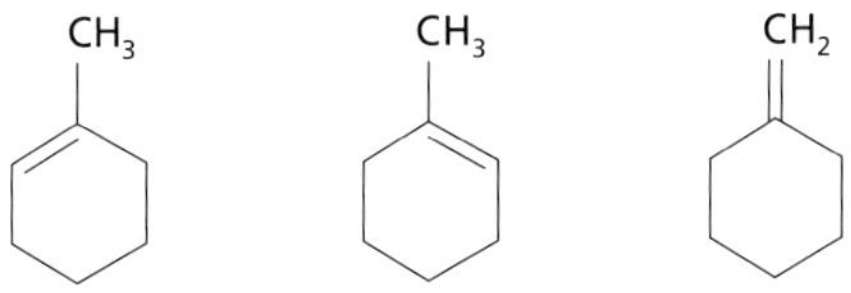

A and B are identical (rotating A by 180° around a vertical axis produces B), so there are *two* possible alkenes. Alkene A/B is the most stable because it has three alkyl groups around the double bond, whereas alkene C has only two (see also section 14.5).

Now try this

Draw all the possible alkenes that could be obtained by the dehydration of 3-methylhexan-3-ol. Which is likely to be the *least* stable of these alkenes?

4 Reactions of the >CH(OH) group

Oxidation

Primary and secondary alcohols are easily oxidised by heating with an acidified solution of potassium dichromate(VI). The orange dichromate(VI) ions are reduced to green chromium(III) ions. Tertiary alcohols, however, are not easily oxidised.

Secondary alcohols are oxidised to ketones (see Topic 17):

$$CH_3\text{-}C(CH_3)(OH)H \xrightarrow[\text{heat}]{K_2Cr_2O_7 + H_2SO_4 \text{ (aq)}} CH_3\text{-}C(CH_3)=O$$

propan-2-ol   propanone

Primary alcohols are oxidised to aldehydes, which, in turn, are even more easily oxidised to carboxylic acids:

$$CH_3CH_2CH_2OH \xrightarrow[\text{heat}]{K_2Cr_2O_7 + H_2SO_4(aq)} CH_3CH_2CH=O \xrightarrow{\text{more oxidant}}$$

propan-1-ol   propanal

$$CH_3CH_2C\begin{smallmatrix}O\\OH\end{smallmatrix}$$

propanoic acid

As soon as any aldehyde is formed it can be oxidised further by the oxidising agent, to the carboxylic acid, and so special techniques are needed to stop the oxidation at the aldehyde stage. One such method makes use of the lower volatility of the alcohol (due to hydrogen bonding) compared with the aldehyde. The reaction mixture is warmed to a temperature that is above the boiling point of the aldehyde, but below that of the alcohol. The aldehyde is allowed to distil out as soon as it is formed, thus avoiding any further contact with the oxidising agent (see Figure 16.6).

Figure 16.6 Apparatus for distilling off the aldehyde as it is formed by the oxidation of an alcohol

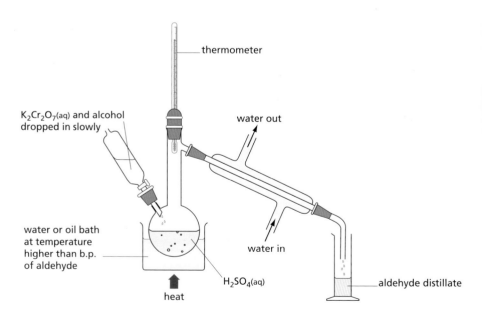

Using an excess of oxidising agent, on the other hand, and heating the reaction under reflux to prevent any escape of the aldehyde before distillation, allows the alcohol to be oxidised all the way to the carboxylic acid.

This oxidation reaction can be used to distinguish between primary, secondary and tertiary alcohols (see Table 16.3).

Table 16.3 The use of potassium dichromate(VI) and sulfuric acid to distinguish between primary, secondary and tertiary alcohols

Type of alcohol	Observation on warming with reagent	Effect of distillate on universal indicator
R_3C—OH (tertiary)	stays orange	neutral (stays green) – only water is produced
R_2CH—OH (secondary)	turns green	neutral (stays green) – ketone is produced
RCH_2—OH (primary)	turns green	acidic (goes red) – carboxylic acid is produced

5 The tri-iodomethane (iodoform) reaction

Alcohols that contain the group $CH_3CH(OH)$—, that is, those that have a methyl group and a hydrogen atom on the same carbon atom that bears the OH group, can be oxidised by alkaline aqueous iodine to the corresponding carbonyl compound $CH_3C(O)$—. This can then undergo the tri-iodomethane reaction:

$$CH_3CH_2-CH\begin{matrix}OH\\\\CH_3\end{matrix} \xrightarrow{I_2 + OH^-(aq)} \left[CH_3CH_2-C\begin{matrix}O\\\\CH_3\end{matrix} \longrightarrow CH_3CH_2-C\begin{matrix}O\\\\CI_3\end{matrix} \right] \xrightarrow{OH^-} CH_3CH_2-C\begin{matrix}O\\\\O^-\end{matrix} + CHI_3$$

Figure 16.7 The tri-iodomethane reaction

The result is shown in Figure 16.7.

Except for ethanol, all the alcohols that undergo this reaction are secondary alcohols, with the OH group on the second carbon atom of the chain, that is, they are alkan-2-ols. The exception, ethanol, is the only primary alcohol to give the **pale yellow precipitate** of tri-iodomethane (iodoform) with alkaline aqueous iodine:

$$CH_3CH_2OH \xrightarrow{I_2 + OH^-(aq)} \left[CH_3CHO \right] \longrightarrow CHI_3(s) + HCO_2^-(aq)$$

The **tri-iodomethane (iodoform) reaction** is thus a very specific test for the $CH_3CH(OH)$— group (or the CH_3C=O group).

Worked example

Which of these alcohols will undergo the iodoform reaction?

A   B   C   D

Answer
Only alcohols **A** and **D** contain the grouping $CH_3CH(OH)$—, so these are the only two to give iodoform. **B** is a tertiary alcohol, whilst **C** is a primary alcohol.

Now try this

P, **Q** and **R** are three isomeric alcohols with the formula $C_5H_{11}OH$. All are oxidised by potassium dichromate(VI) and aqueous sulfuric acid, but only **P** gives an acidic distillate. When treated with alkaline aqueous iodine, **Q** gives a pale yellow precipitate, but **P** and **R** do not react. What are the structures of **P**, **Q** and **R**?

16.4 Preparing alcohols

As we would expect from their central position in organic synthesis, alcohols can be prepared by a variety of different methods.

1 From halogenoalkanes, by nucleophilic substitution

See section 15.3 for details of this reaction.

$$CH_3CH_2Br + OH^-(aq) \rightarrow CH_3CH_2OH + Br^-$$

2 From alkenes, by hydration

See section 14.3 for details. This is the preferred method of making ethanol industrially:

$$CH_2{=}CH_2 + H_2O \xrightarrow[\text{H}_3\text{PO}_4 \text{ at } 300\,°C \text{ and } 70\,atm]{\text{pass vapours over a catalyst of}} CH_3CH_2OH$$

The hydration can also be carried out in the laboratory by absorbing the alkene in concentrated sulfuric acid, and then diluting with water:

$$CH_2{=}CH_2 + H_2SO_4 \rightarrow CH_3CH_2{-}OSO_3H$$
$$CH_3CH_2{-}OSO_3H + H_2O \rightarrow CH_3CH_2OH + H_2SO_4$$

3 From aldehydes or ketones, by reduction

See section 17.3 for details. There are three common methods of reducing carbonyl compounds.

● By hydrogen on a nickel catalyst:

$$CH_3CH_2CHO \xrightarrow{\text{H}_2 + \text{Ni}} CH_3CH_2CH_2OH$$

● By sodium tetrahydridoborate(III) (sodium borohydride) in alkaline methanol:

$$CH_3CH_2COCH_3 \xrightarrow[\text{in methanol}]{\text{NaBH}_4 + \text{OH}^-} CH_3CH_2CH(OH)CH_3$$

● By lithium tetrahydridoaluminate(III) (lithium aluminium hydride) in dry ether:

$$CH_3COCH_3 \xrightarrow[\text{dry ether}]{\text{LiAlH}_4 \text{ in}} CH_3CH(OH)CH_3$$

Lithium tetrahydridoaluminate(III) is a dangerous reagent, producing heat and hydrogen gas (which usually catches fire) with any water, and ether is a hazardous solvent being very volatile and flammable. Hence this method is usually avoided.

4 Preparing ethanol by fermentation

In many parts of the world, the fermentation of sugar or starch is only used to produce flavoured aqueous solutions of ethanol for drinks. But in some countries, for example Brazil, where petroleum is scarce and expensive, ethanol produced by fermentation is used as a fuel for vehicles. The ethanol can be used either alone or as a 25% mixture with petrol. This could become increasingly important for the rest of the world, once oil reserves become depleted.

Yeasts are micro-organisms of the genus *Saccharomyces*. They contain enzymes that not only break glucose down into ethanol and carbon dioxide, but also break down starch or sucrose (from cane or beet sugar) into glucose. They can therefore convert a variety of raw materials into ethanol:

$$\underset{\text{starch}}{(C_6H_{10}O_5)_n} + nH_2O \xrightarrow{\text{yeast}} \underset{\text{glucose}}{nC_6H_{12}O_6}$$

$$C_{12}H_{22}O_{11} + H_2O \rightarrow \underset{\text{glucose}}{C_6H_{12}O_6} + \underset{\text{fructose}}{C_6H_{12}O_6}$$
$$\underset{\text{sucrose}}{}$$

$$\underset{\text{fructose}}{C_6H_{12}O_6} \rightarrow \underset{\text{glucose}}{C_6H_{12}O_6}$$

and finally:

$$C_6H_{12}O_6 \rightarrow 2C_2H_5OH + 2CO_2$$
$$\text{glucose} \qquad \text{ethanol} \qquad \text{carbon dioxide}$$

The conditions required for successful fermentation are:

- yeast
- water
- yeast nutrients (ammonium phosphate is often used)
- warmth (a temperature of 30 °C is ideal)
- absence of air (with oxygen present, the ethanol can be partially oxidised to ethanoic acid, or completely oxidised to carbon dioxide).

The reaction is carried out in aqueous solution. Assuming enough glucose is present, it stops when the ethanol concentration reaches about 15%. Above this concentration, the yeast cells become dehydrated, and the yeast dies. After filtering off the dead yeast cells, the solution is fractionally distilled. The distillate consists of 95% ethanol and 5% water. If required, the remaining 5% water can be removed chemically, by adding either quicklime (calcium oxide):

$$CaO + H_2O \rightarrow Ca(OH)_2$$

or metallic magnesium:

$$Mg + 2H_2O \rightarrow Mg(OH)_2(s) + H_2(g)$$

Figure 16.8 a An electron micrograph of yeast cells. **b** Ethanol can be used on its own or as an additive to petrol to power vehicles. **c** Wine production

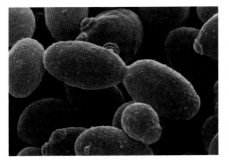  

For most chemical and industrial purposes, however, 95% ethanol is perfectly acceptable.

Apart from the economic advantage of this method of producing alcohol (and hence fuel), the use of ethanol as a fuel has an environmental advantage. As was mentioned in section 13.4, fermentation-ethanol is a renewable fuel. It returns to the atmosphere exactly the same number of molecules of CO_2 as were used in its production by photosynthesis and fermentation:

$$6CO_2 + 6H_2O \xrightarrow{\text{photosynthesis}} C_6H_{12}O_6 + 6O_2 \qquad (1)$$

$$C_6H_{12}O_6 \xrightarrow{\text{fermentation}} 2C_2H_5OH + 2CO_2 \qquad (2)$$

$$2C_2H_5OH + 6O_2 \xrightarrow{\text{combustion}} 6H_2O + 4CO_2 \qquad (3)$$

The carbon dioxide balance is therefore maintained. Indeed, if we add together the three equations (1), (2) and (3) we find no net change in *any* substance. We have obtained the energy we require entirely from the Sun, indirectly through the intermediates glucose and ethanol.

Summary

- **Alcohols** can be prepared from halogenoalkanes, alkenes and carbonyl compounds.
- An important source of ethanol is the fermentation of starch or sucrose by yeast.
- Alcohols are important intermediates in organic synthesis and are useful solvents.
- All alcohols burn to give carbon dioxide and water. The lower members are used as additives in petrol.
- Alcohols can react generally in three different ways: by breaking the O—H bond, by breaking the C—O bond, and by breaking the C—H bond next to the —OH group.
- Two good tests for the —OH group are the effervescence of HCl when reacted with PCl_5, or the effervescence of H_2 when reacted with sodium metal.
- Primary, secondary and tertiary alcohols can be distinguished either by their different oxidation products with $K_2Cr_2O_7$ and H_2SO_4 or by using the **Lucas reagent** (concentrated HCl and $ZnCl_2$).
- The iodoform reaction is a test for the presence of the groups CH_3CO- or $CH_3CH(OH)-$ in a molecule. The products are iodoform and a sodium carboxylate with one fewer carbon atoms than the original alcohol or ketone.

Key reactions you should know
(R, R′ = primary, secondary or tertiary alkyl unless otherwise stated.)

- Combustion:

$$C_2H_5OH + 3O_2 \rightarrow 2CO_2 + 3H_2O$$

- Redox:

$$R{-}OH + Na \rightarrow R{-}O^-Na^+ + \tfrac{1}{2}H_2$$

- Esterification:

$$R{-}OH + HO_2CR' \xrightarrow{\text{heat with conc. } H_2SO_4} R{-}OCOR' + H_2O$$

- Nucleophilic substitutions:

$$R{-}OH + PCl_5 \rightarrow R{-}Cl + POCl_3 + HCl$$

$$R{-}OH + SOCl_2 \rightarrow R{-}Cl + SO_2 + HCl$$

$$R{-}OH + HCl(\text{conc.}) \rightarrow R{-}Cl + H_2O \text{ (best with R = tertiary alkyl)}$$

$$R{-}OH + HBr \xrightarrow{\text{heat with NaBr + } H_2SO_4/H_2O} R{-}Br + H_2O$$

- Eliminations:

$$R{-}CH_2{-}CH_2OH \xrightarrow{Al_2O_3 \text{ at } 350\,°C} R{-}CH{=}CH_2 + H_2O$$

$$R{-}CH_2{-}CH_2OH \xrightarrow{H_2SO_4 \text{ at } 180\,°C} R{-}CH{=}CH_2 + H_2O$$

- Oxidations:

$$RCH_2OH \xrightarrow{\text{heat with } Na_2Cr_2O_7 + H_2SO_4(aq)} R{-}CH{=}O \rightarrow RCO_2H$$

$$R_2CHOH \xrightarrow{\text{heat with } Na_2Cr_2O_7 + H_2SO_4(aq)} R_2C{=}O$$

$$R_3COH \xrightarrow{\text{heat with } Na_2Cr_2O_7 + H_2SO_4(aq)} \text{no reaction}$$

Examination practice questions

Please see the data section of the CD for any A_r values you may need.

1 The structural formulae of six different compounds, **A – F**, are given below. **Each** compound contains four carbon atoms in its molecule.

$$CH_3CH{=}CHCH_3 \qquad CH_3CH_2COCH_3 \qquad CH_2{=}CHCH_2CH_3$$
$$\textbf{A} \qquad\qquad\qquad \textbf{B} \qquad\qquad\qquad \textbf{C}$$

$$CH_3CH_2CH(OH)CH_3 \quad HOCH_2CH_2CH_2CH_2OH \quad CH_3CH_2OCH_2CH_3$$
$$\textbf{D} \qquad\qquad\qquad \textbf{E} \qquad\qquad\qquad \textbf{F}$$

a i What is the empirical formula of compound **E**?
ii Draw the skeletal formula of compound **D**.
iii Structural formulae do not show all of the isomers that may exist for a given molecular formula. Which **two** compounds **each** show **different** types of isomerism and what type of isomerism does each compound show? Identify each compound by its letter. [4]

b Compound **D** may be converted into compound **C**.
i What type of reaction is this?
ii What reagent would you use for this reaction?

iii What is formed when compound E undergoes the same reaction using an excess of the same reagent? [3]

c Compound **A** may be converted into compound **B** in a two-stage reaction.

$$CH_3CH{=}CHCH_3 \xrightarrow{\text{stage I}} \text{intermediate} \xrightarrow{\text{stage II}} CH_3CH_2COCH_3$$

i What is the structural formula of the intermediate compound formed in this sequence?
ii Outline how stage I may be carried out to give this intermediate compound.
iii What reagent would be used for stage II? [4]

d Compounds **D** and **F** are isomers. What type of isomerism do they show? [1]

[Cambridge International AS & A Level Chemistry 9701, Paper 21 Q4 November 2009]

2 The alcohols are an example of an homologous series. The table shows the boiling points for the first four members of straight-chain alcohols.

Alcohol	Structural formula	Boiling point/°C
methanol	CH_3OH	65
ethanol	CH_3CH_2OH	78
propan-1-ol	$CH_3CH_2CH_2OH$	97
butan-1-ol	$CH_3CH_2CH_2CH_2OH$	118

a i What is the general formula of a member of the alcohol homologous series? [1]

ii Deduce the molecular formula of the alcohol that has 13 carbon atoms per molecule. [1]

b Alcohols contain the hydroxyl functional group. What is meant by the term *functional group*? [2]

c i At room temperature and pressure, the first four members of the alcohol homologous series are liquids whereas the first four members of the alkanes homologous series are gases. Explain this difference. [3]

ii Methylpropan-1-ol and butan-1-ol are structural isomers. Methylpropan-1-ol has a lower boiling point than butan-1-ol. Suggest why. [2]

d Alcohols, such as methanol, can be used as fuels.

i Write equations for the complete and incomplete combustion of methanol. [2]

ii Suggest what conditions might lead to incomplete combustion of methanol. [1]

iii In addition to its use as a fuel, methanol can be used as a solvent and as a petrol additive to improve combustion.
State **another** large-scale use of methanol. [1]

e Butan-1-ol can be oxidised by heating under reflux with excess acidified potassium dichromate(VI).
Write an equation for the reaction that takes place.
Use [O] to represent the oxidising agent. [2]

f Butan-1-ol is one of the structural isomers of $C_4H_{10}O$.

i Write the name and draw the structure of the structural isomer of $C_4H_{10}O$ that is a tertiary alcohol. [2]

ii Draw the structure of the structural isomer of $C_4H_{10}O$ that can be oxidised to form butanone. [1]

[OCR Chemistry A Unit F322 Q2 May 2011]

17 Aldehydes and ketones

Functional groups:

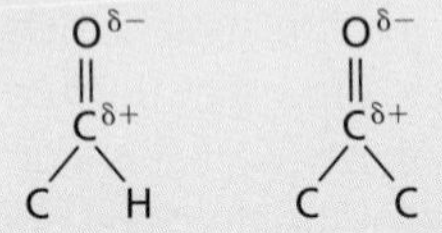

aldehyde   ketone

The carbonyl group, >C=O, contained in aldehydes and ketones is both more polarised, and more polarisable, than the C—O single bond in alcohols and phenols. This topic introduces the last of our four main reaction mechanisms, nucleophilic addition. This is the characteristic way in which carbonyl compounds react and is due to the highly δ+ carbon atom that they contain, combined with their unsaturation. Aldehydes are intermediate in oxidation state between alcohols and carboxylic acids and can undergo both oxidation and reduction reactions, but ketones can only undergo reduction reactions.

Learning outcomes

By the end of this topic you should be able to:

14.1a) interpret and use the general, structural, displayed and skeletal formulae of the aldehydes and ketones

18.1a) describe the formation of aldehydes and ketones from primary and secondary alcohols respectively using $Cr_2O_7^{2-}/H^+$, the reduction of aldehydes and ketones, e.g. using $NaBH_4$ or $LiAlH_4$, and the reaction of aldehydes and ketones with HCN and NaCN

18.1b) describe the mechanism of the nucleophilic addition reactions of hydrogen cyanide with aldehydes and ketones

18.1c) describe the use of 2,4-dinitrophenylhydrazine (2,4-DNPH) reagent to detect the presence of carbonyl compounds

18.1d) deduce the nature (aldehyde or ketone) of an unknown carbonyl compound from the results of simple tests (Fehling's and Tollens' reagents, ease of oxidation)

18.1e) describe the reaction of CH_3CO— compounds with alkaline aqueous iodine to give tri-iodomethane.

17.1 Introduction

Collectively, aldehydes and ketones are known as **carbonyl compounds**. The carbonyl *group*, >C=O, is a subunit of many other functional groups (see Table 17.1), but the term 'carbonyl *compounds*' is reserved for those compounds in which it appears on its own.

Table 17.1 Some functional groups containing the carbonyl group

Group	Class of compound
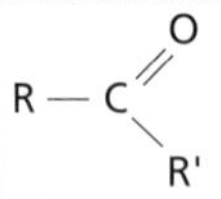	carbonyl compound
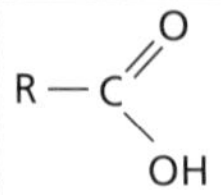	carboxylic acid
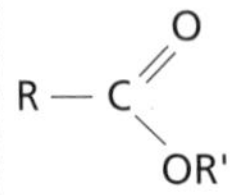	ester
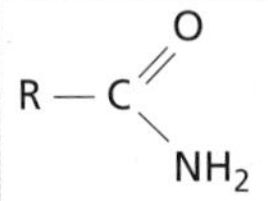	amide
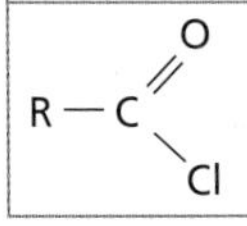	acyl chloride

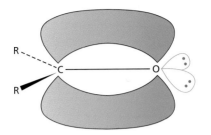

Figure 17.1 Oxygen attracts the bonding electrons away from the carbon atom.

The properties of aldehydes and ketones are very similar to each other: almost all the reactions of ketones are also shown by aldehydes. But aldehydes show additional reactions associated with their lone hydrogen atom.

The bonding of the carbonyl group is similar to that of ethene (see section 14.1). Here, the double bond is formed by the sideways overlap of two adjacent p orbitals, one on carbon and one on oxygen. Because of its higher electronegativity, oxygen attracts the bonding electrons (in both the σ and the π bonds), creating an electron-deficient carbon atom (see Figure 17.1). This unequal distribution of electrons is responsible for the two ways in which carbonyl compounds react.

- The oxygen of the C═O bond can be protonated by strong acids.
- The carbon of the C═O bond, on the other hand, is susceptible to attack by nucleophiles, Nu:

$$\begin{array}{ccc} \diagdown \mathrm{C}{=}\mathrm{O}\ \mathrm{H}^+ & \rightleftharpoons & \diagdown \mathrm{C}^+{-}\mathrm{O}{-}\mathrm{H} \\ {}_{\delta+}\ \ {}_{\delta-} & & \diagup \end{array}$$

$$\mathrm{Nu}{:}\ \ \diagdown \mathrm{C}{=}\mathrm{O} \rightarrow \mathrm{Nu}{-}\overset{|}{\underset{|}{\mathrm{C}}}{-}\mathrm{O}^-$$

The molecules of carbonyl compounds cannot attract one another through intermolecular hydrogen bonds, because they do not contain hydrogen atoms that have a large enough δ+ charge. They can, however, interact with water molecules through hydrogen bonding. The lower members of the series are therefore quite soluble in water (although less so than the corresponding alcohols).

The boiling points of carbonyl compounds are also lower than those of the corresponding alcohols, though significantly higher (by about 40–50 °C) than those of the corresponding alkenes. This is due to the dipole–dipole attractions between the molecules (see Figure 17.2). Table 17.2 lists some of these properties.

Figure 17.2 Dipole–dipole attractions in carbonyl compounds

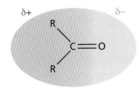 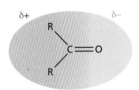

Table 17.2 Properties of some aldehydes and ketones compared with those of alkenes and alcohols

Alkene		Corresponding carbonyl compound			Corresponding alcohol		
Formula	Boiling point/°C	Formula	Boiling point/°C	Solubility in 100 g of water/g	Formula	Boiling point/°C	Solubility in 100 g of water/g
$CH_3CH{=}CH_2$	−48	CH_3CHO	20	∞	CH_3CH_2OH	78	∞
$CH_3CH_2CH{=}CH_2$	−6	CH_3CH_2CHO	49	16	$CH_3CH_2CH_2OH$	97	∞
$CH_3CH_2CH_2CH{=}CH_2$	30	$CH_3CH_2CH_2CHO$	76	7	$CH_3CH_2CH_2CH_2OH$	118	8
$(CH_3)_2C{=}CH_2$	−7	$(CH_3)_2CO$	56	∞	$(CH_3)_2CHOH$	82	∞

Carbonyl compounds have distinctive smells. Ketones smell 'sweeter' than aldehydes – almost like esters. Aldehydes have an astringent smell, but often with a 'fruity' overtone – ethanal smells of apples, for example. The simplest aldehyde, methanal, has an unpleasant, choking smell. Its concentrated aqueous solution, formalin, is used to preserve biological specimens. Aldehydes and ketones of higher molecular mass have important uses as flavouring and perfuming agents. Figure 25.8 (page 419) includes two carbonyl compounds with benzene rings, vanillin and benzaldehyde, that are used as flavouring agents. Some other carbonyl compounds with pleasant smells are shown in Figure 17.3.

Figure 17.3 Some pleasant-smelling carbonyl compounds, and one that occurs naturally in musk (the basis for perfumes)

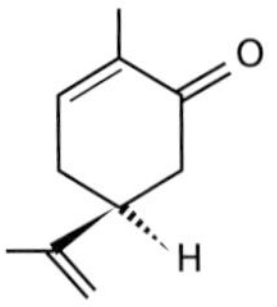

carvone
(spearmint)

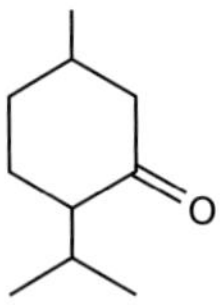

menthone
(peppermint)

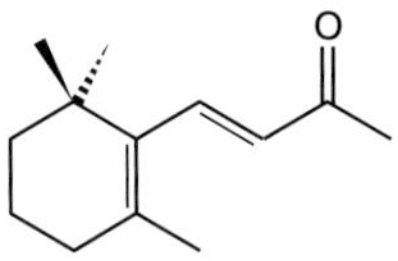

ionone
(violets)

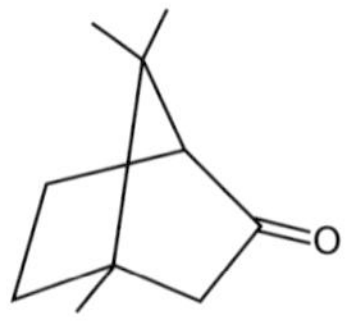

camphor

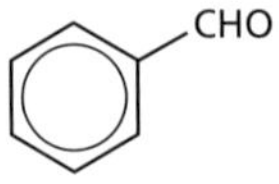

benzaldehyde
(almonds)

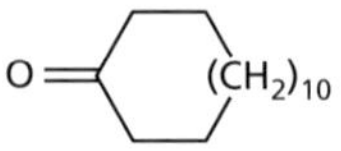

muscone

17.2 Isomerism and nomenclature

In one respect, aldehydes and ketones can be considered as positional isomers of each other: in propanal, CH_3CH_2CHO, the functional group (carbonyl) is on position 1 of the chain, whereas in propanone, CH_3COCH_3, it is on position 2. Because of significant differences in their reactions, however, aldehydes and ketones are named differently, and are considered as two separate classes of compounds.

Aldehydes are named by adding '-al' to the hydrocarbon stem:

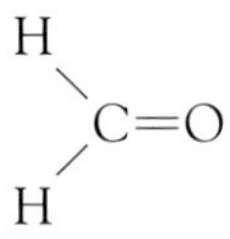

methanal

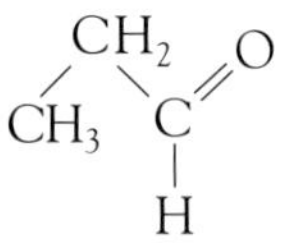

propanal

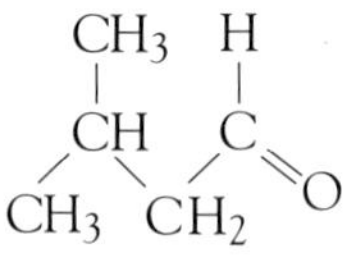

3-methylbutanal

Ketones are named by adding '-one' to the hydrocarbon stem. In higher ketones, the position of the carbonyl group along the chain needs to be specified:

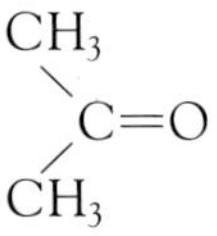

propanone

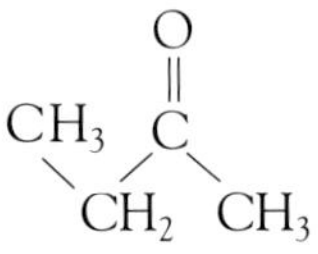

butanone

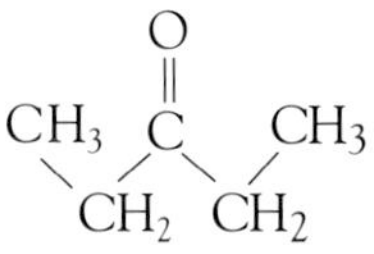

pentan-3-one

Worked example

Draw structures for the following.
a 2-methylpentanal
b hexan-2-one

Answer

a The five-carbon chain is numbered starting at the aldehyde end:

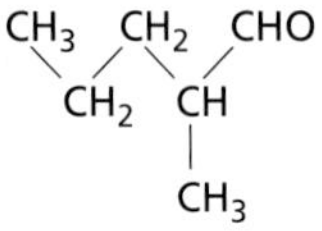

b

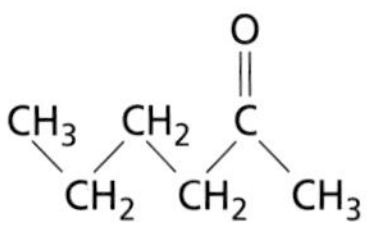

17.3 Reactions of aldehydes and ketones

These reactions can be divided into three main groups:

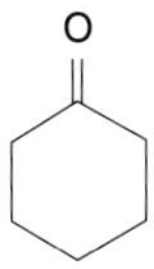

- reactions common to both aldehydes and ketones
- reactions of aldehydes only
- reactions producing tri-iodomethane.

Reactions common to both aldehydes and ketones

Nucleophilic addition of hydrogen cyanide

In the presence of a trace of sodium cyanide as catalyst (or of a base such as sodium hydroxide to produce the salt), hydrogen cyanide adds on to carbonyl compounds:

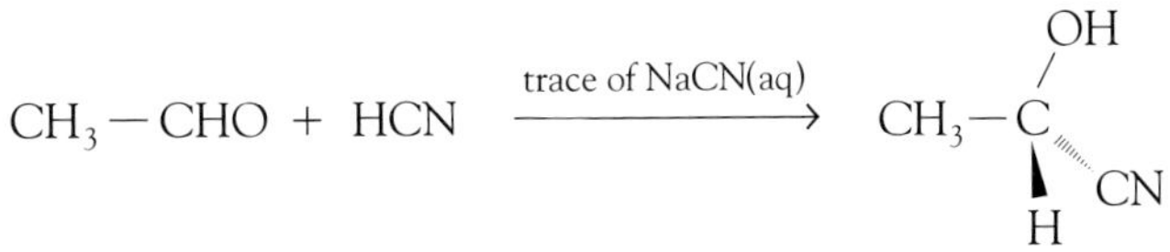

$$CH_3 - CHO + HCN \xrightarrow{\text{trace of NaCN(aq)}} CH_3 - \underset{H}{\overset{OH}{\underset{|}{C}}} CN$$

The product is 2-hydroxypropanenitrile. Like the reaction between sodium cyanide and halogenoalkanes (see section 15.3), the reaction is a useful method of adding a carbon atom to a chain, which can be valuable during organic syntheses. Cyanohydrins can be hydrolysed to 2-hydroxycarboxylic acids (see Topic 18) by heating with dilute sulfuric acid, and can also be reduced to 2-hydroxyamines (see Topic 27):

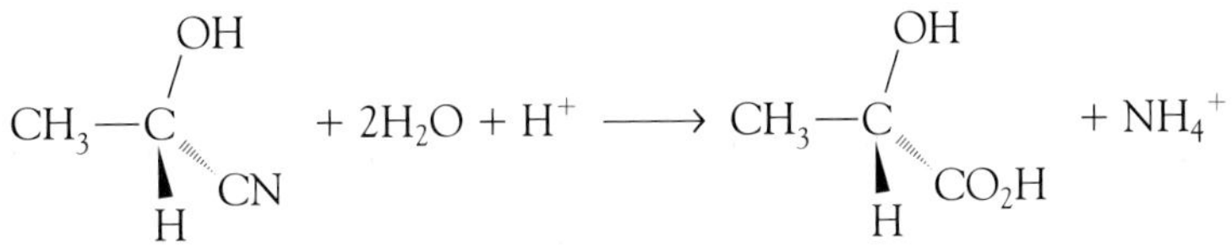

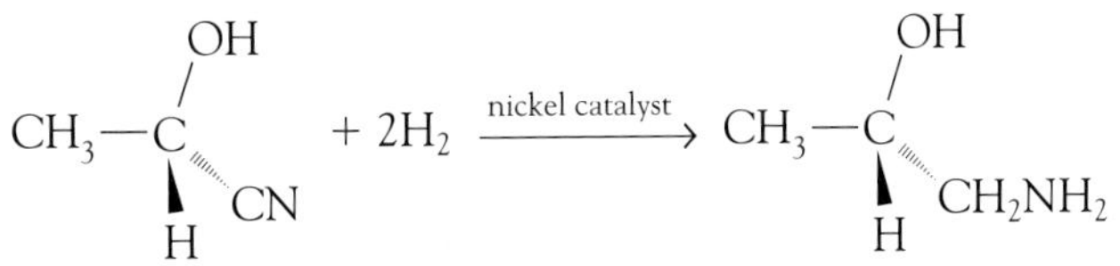

$$CH_3 - \underset{H}{\overset{OH}{\underset{|}{C}}} CN + 2H_2O + H^+ \longrightarrow CH_3 - \underset{H}{\overset{OH}{\underset{|}{C}}} CO_2H + NH_4^+$$

$$CH_3 - \underset{H}{\overset{OH}{\underset{|}{C}}} CN + 2H_2 \xrightarrow{\text{nickel catalyst}} CH_3 - \underset{H}{\overset{OH}{\underset{|}{C}}} CH_2NH_2$$

The mechanism of the addition of hydrogen cyanide to carbonyl compounds

Hydrogen cyanide is neither a nucleophile nor a strong acid. Therefore it cannot react with the carbonyl group by either of the routes shown on page 295. However, as we saw in Topic 15, the cyanide ion is a good nucleophile. It can attack the δ+ carbon of the C=O bond:

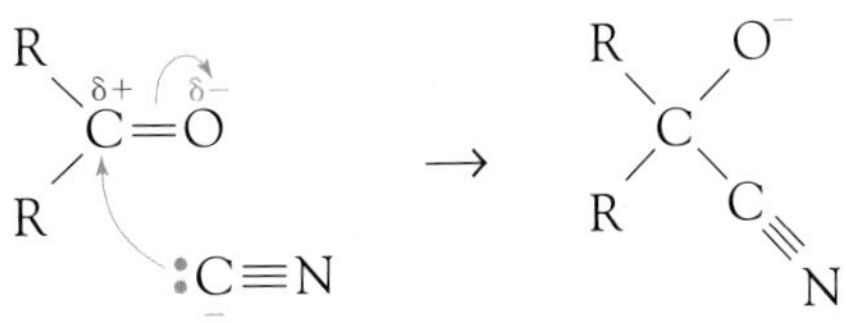

297

The intermediate anion is a strong base, and can abstract a proton from an un-ionised molecule of hydrogen cyanide:

$$R\begin{array}{c}\quad O^-:\\ \diagdown\quad\diagup\\ C\\ \diagup\quad\diagdown\\ R\qquad C\\ \qquad\quad\lVert\lVert\\ \qquad\quad N\end{array}\qquad H\!\div\!C\!\equiv\!N \quad\rightarrow\quad R\begin{array}{c}\quad O-H\\ \diagdown\quad\diagup\\ C\\ \diagup\quad\diagdown\\ R\qquad C\\ \qquad\quad\lVert\lVert\\ \qquad\quad N\end{array}\qquad +:\!\bar{C}\!\equiv\!N$$

So we see that although the cyanide ion is used in the first step, it is regenerated in the second. It is acting as a homogeneous catalyst in this **nucleophilic addition** reaction.

The reduction of carbonyl compounds by complex metal hydrides

Two complex hydrides, sodium borohydride and lithium aluminium hydride, were discovered in the 1940s, and have proved very useful as reducing agents in organic chemistry. They are ionic compounds which contain the tetrahedral anion $[MH_4]^-$. Table 17.3 lists some of their properties and how they are used.

Table 17.3 Some properties of complex inorganic metal hydrides used as reducing agents in organic chemistry

Common name	sodium borohydride	lithium aluminium hydride
Systematic name	sodium tetrahydridoborate(III)	lithium tetrahydridoaluminate(III)
Formula	$NaBH_4$	$LiAlH_4$
Solvent used and reaction conditions	methanol–water mixture, room temperature or warming	totally dry ether, room temperature or at reflux (35 °C)
How the product is separated	adding acid and extracting the product from the aqueous mixture with an organic solvent such as ether	adding water (with care), then acid, and separating the ether solution of the product from the aqueous layer
Hazards during use	relatively safe in alkaline solution, but evolves hydrogen gas with acids	can catch fire with water or when wet, and ether solvents form explosive mixtures with air
Functional groups that can be reduced, and what they are reduced to	carbonyl compounds (to alcohols)	carbonyl compounds (to alcohols) esters (to alcohols) carboxylic acids (to alcohols) amides (to amines) nitriles (to amines)

As the table indicates, the preferred reagent for the reduction of aldehydes and ketones is $NaBH_4$. Although $LiAlH_4$ will carry out the reduction as efficiently, it is both more hazardous and more expensive than $NaBH_4$. However, as the more powerful reductant, it can be used to reduce carboxylic acids and their derivatives, unlike $NaBH_4$.

Each reagent works by transferring a nucleophilic hydrogen atom (with its electrons) to the $\delta+$ carbon of the carbonyl group, forming an intermediate metal alkoxide, which is later decomposed to the alcohol:

$$4\ \begin{array}{c}R\\ \diagdown\\ C=O\\ \diagup\\ R'\end{array} + MH_4^- \rightarrow \left[\left(\begin{array}{c}R\\ R'\diagdown\diagup\\ C-O\\ \diagup\\ H\end{array}\right)_4 M\right]^- \xrightarrow[+H_2O]{+H^+} \begin{array}{c}R\\ R'\diagdown\diagup\\ C-OH\\ \diagup\\ H\end{array} + M(OH)_3$$

Aldehydes give primary alcohols, while ketones give secondary alcohols:

$$CH_3-CHO \xrightarrow[\text{or } LiAlH_4 \text{ in dry ether}]{NaBH_4 \text{ in alkaline aqueous methanol}} CH_3-CH_2OH$$

$$\begin{array}{c}O\\ \lVert\\ C\\ \diagup\ \diagdown\\ CH_3\ \ CH_3\end{array} \xrightarrow[\text{or } NaBH_4 \text{ in alkaline aqueous methanol}]{LiAlH_4 \text{ in dry ether}} \begin{array}{c}HO\quad H\\ \diagdown\ \diagup\\ C\\ \diagup\ \diagdown\\ CH_3\ \ CH_3\end{array}$$

Because these reagents are nucleophilic, needing a $\delta+$ carbon to react with, they do not reduce C=C double bonds, in contrast to catalytic hydrogenation.

Catalytic hydrogenation

Like alkenes, carbonyl compounds can be reduced by hydrogen gas over a nickel or platinum catalyst:

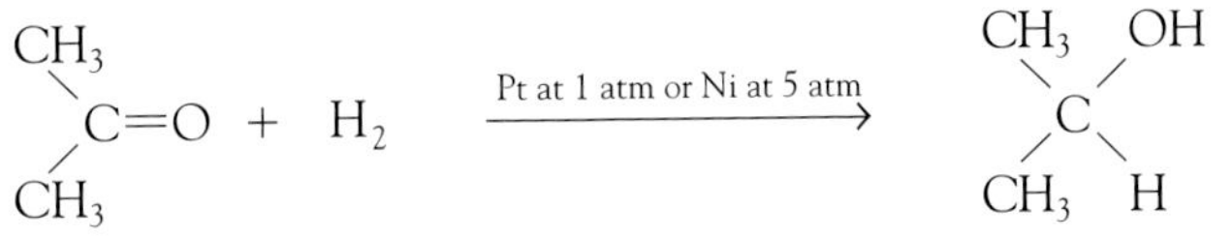

$$CH_3\!\!\diagdown\!\!\underset{CH_3\diagup}{C}\!\!=\!\!O \;+\; H_2 \quad\xrightarrow{\text{Pt at 1 atm or Ni at 5 atm}}\quad CH_3\!\!\diagdown\!\!\underset{CH_3\diagup}{\overset{OH}{\underset{H}{C}}}$$

The mechanism is similar to heterogeneous catalysis (see section 8.6).

Worked example

Suggest the structural formulae of the intermediates and products of the following reactions:

a

$$CH_3CH_2CHO \xrightarrow{\text{HCN + NaCN}} W \xrightarrow{\text{heat with } H_2SO_{4(aq)}} X$$

b

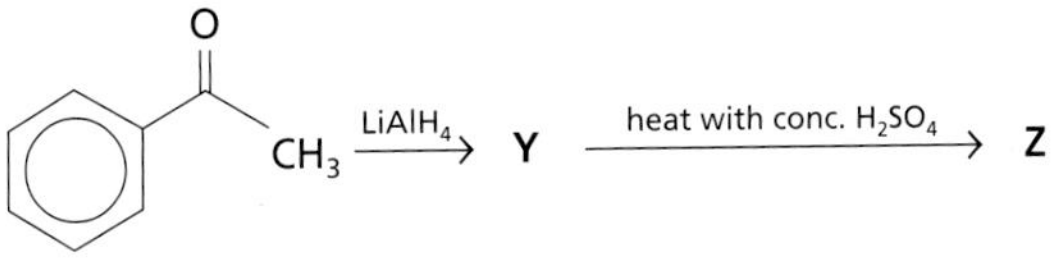

$$\xrightarrow{\text{LiAlH}_4} Y \xrightarrow{\text{heat with conc. } H_2SO_4} Z$$

Answer

a

W is
$$\underset{CH_3CH_2CHCN}{\overset{OH}{\overset{|}{}}}$$

X is
$$\underset{CH_3CH_2CHCO_2H}{\overset{OH}{\overset{|}{}}}$$

b

Y is

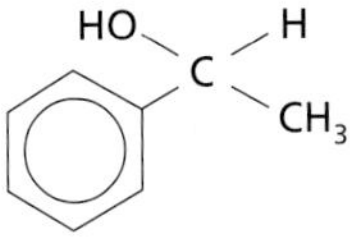

Z is

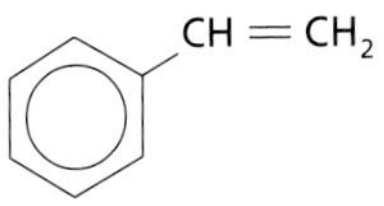

Now try this

1 Suggest reagents for steps **I** and **II** in each of the following transformations, and the structures of the intermediates **A** and **B**:

   a
   $$(CH_3)_2C=O \xrightarrow{\text{I}} A \xrightarrow{\text{II}} CH_3-CH=CH_2$$

   b

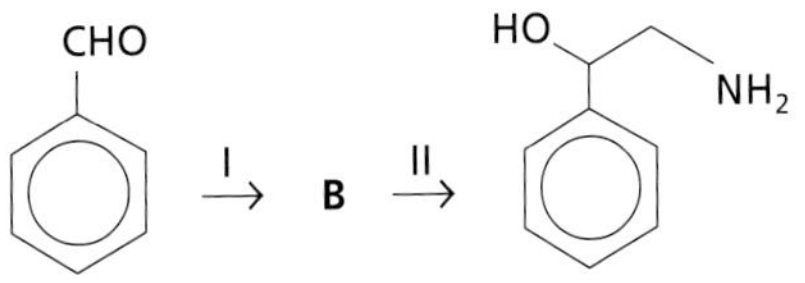

2 Suggest reagents and conditions for the following reactions.
   a $CH_3-CH=CH-CHO \rightarrow CH_3-CH=CH-CH_2OH$
   b $CH_3-CH=CH-CHO \rightarrow CH_3-CH_2-CH_2-CH_2OH$

Condensation reactions

Nitrogen nucleophiles readily add on to carbonyl compounds. But usually, the initially formed addition compounds cannot be isolated. They easily lose water to give stable compounds containing a C=N bond:

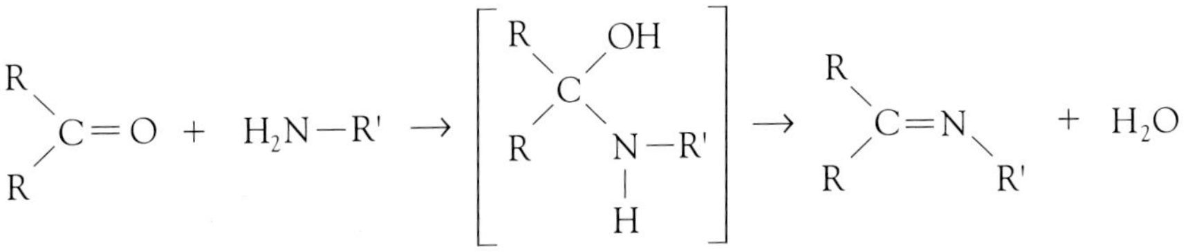

$$\underset{R\diagup}{\overset{R\diagdown}{}}C=O \;+\; H_2N-R' \;\rightarrow\; \left[\underset{R\diagup\quad N-R'\atop |\atop H}{\overset{R\diagdown\quad OH}{C}}\right] \;\rightarrow\; \underset{R\diagup}{\overset{R\diagdown}{}}C=N\diagdown_{R'} \;+\; H_2O$$

Reactions that form water (or any other inorganic small-molecule compound) when two organic compounds react with each other are called **condensation reactions**.

The most important condensation reaction of carbonyl compounds is that with 2,4-dinitrophenylhydrazine (2,4-DNPH). The products are called hydrazones and are crystalline orange solids, which precipitate out of solution rapidly. They can be purified relatively easily by recrystallisation.

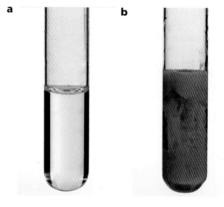

2,4-DNPH   a 2,4-dinitrophenylhydrazone

The formation of an orange precipitate when a solution of 2,4-DNPH is added to an unknown compound is a good test for the presence of a carbonyl compound.

If the melting point of a recrystallised sample is measured and compared with tables of known values, the carbonyl compound can be uniquely identified. This is especially useful if the carbonyl compounds have similar boiling points, and so would otherwise be easily confused (see Table 17.4).

Table 17.4 The melting points of some 2,4-DNPH derivatives of carbonyl compounds

Carbonyl compound	Boiling point/°C	Melting point of 2,4-dinitrophenylhydrazone derivative/°C
$CH_3CH_2CH_2CH_2CHO$	103	108
$CH_3CH_2COCH_2CH_3$	102	156
$CH_3CH_2CH_2COCH_3$	102	144

Figure 17.4 Ethanal reacting with 2,4-DNPH: **a** at the start of the reaction; this forms an orange 2,4-DNPH precipitate, **b**.

Now try this

Two isomers **G** and **H** have the molecular formula C_4H_8O. **G** forms an orange precipitate with 2,4-DNPH whereas **H** does not. **H** decolorises bromine water, but **G** does not. On treatment with hydrogen and nickel, both **G** and **H** give the same compound, butan-1-ol.

Suggest structures for **G** and **H**.

Reactions undergone only by aldehydes

Aldehydes are distinguished from ketones by having a hydrogen atom directly attached to the carbonyl group. This hydrogen atom is not in any way acidic, but the C—H bond is significantly weaker than usual. The —CHO group is readily oxidised to a —COOH group, making aldehydes mild reducing agents. This oxidation, specific to aldehydes, has been used to design the following tests that will distinguish aldehydes from ketones.

Oxidation by acidified potassium dichromate(VI) solution

It was mentioned in section 16.3 that aldehydes are more easily oxidised than alcohols. The reaction takes place on gentle warming, and the colour of the reagent changes from orange to green.

$$CH_3C\overset{O}{\underset{H}{\big\backslash}} + [O] \xrightarrow{K_2Cr_2O_{7(aq)} + H_2SO_{4(aq)}} CH_3C\overset{O}{\underset{OH}{\big\backslash}}$$

Reduction of Fehling's solution

The bright blue **Fehling's solution** is a solution of Cu^{2+} ions, complexed with salts of tartaric acid, in an aqueous alkaline solution. When warmed with an aldehyde, the Cu^{2+} ions are reduced to Cu^+ ions, which in the alkaline solution form a red precipitate of copper(I) oxide. The aldehyde is oxidised to the salt of the corresponding carboxylic acid:

$$CH_3CHO(aq) + 2Cu^{2+}(aq) + 5OH^-(aq) \rightarrow CH_3CO_2^-(aq) + Cu_2O(s) + 3H_2O(l)$$
blue solution   red precipitate

Formation of a silver mirror with Tollens' reagent

Tollens' reagent contains silver ions, complexed with ammonia, in an aqueous alkaline solution. These silver ions are readily reduced to silver metal on gentle warming with an aldehyde. The metal will often silver-plate the inside of the test tube. Once again, the aldehyde is oxidised to the salt of the corresponding carboxylic acid:

$$CH_3CHO + 2Ag^+ + 3OH^- \rightarrow CH_3CO_2^- + 2Ag(s) + 2H_2O$$
silver mirror

A dilute solution of the aldehyde sugar, glucose, was once used to make mirrors for domestic use by means of this reaction.

Figure 17.5 An aldehyde produces a silver mirror with Tollens' reagent.

a   b   c

Aldehydes can be distinguished from ketones by one of the following tests:

- aldehydes produce a red precipitate on warming with Fehling's solution
- aldehydes produce a silver mirror on warming with Tollens' reagent.

The tri-iodomethane (iodoform) reaction

Ketones that contain the group —$COCH_3$ (that is, methyl ketones) undergo the tri-iodomethane (iodoform) reaction on treatment with an aqueous alkaline solution of iodine. Tri-iodomethane, CHI_3, is a pale yellow insoluble solid, with a sweet, antiseptic 'hospital' smell:

$$\begin{matrix} CH_3 \\ \diagdown \\ C=O \\ \diagup \\ CH_3 \end{matrix} + 3I_2 + 4OH^- \longrightarrow CH_3CO_2^- + CHI_3 + 3I^- + 3H_2O$$
pale yellow precipitate

There is one aldehyde, ethanal, that also undergoes the tri-iodomethane reaction:

$$CH_3CHO \xrightarrow{I_2 + OH^-(aq)} CHI_3(s) + HCO_2^-(aq)$$

Figure 16.7 shows the results of the iodoform reaction.

The **tri-iodomethane (iodoform) reaction** is thus a very specific test for the CH_3CO— group (or the $CH_3CH(OH)$— group – see Topic 16).

Worked example 1

Compound **A** has the molecular formula C_4H_8O. It reacts with Fehling's solution. On treatment with sodium tetrahydridoborate(III) it gives **B**, which on warming with concentrated sulfuric acid gives 2-methylpropene.

Identify **A** and **B**.

Answer

A positive Fehling's solution test means that **A** is an aldehyde. Sodium tetrahydridoborate(III) reduces an aldehyde to a primary alcohol. Concentrated sulfuric acid converts this to the alkene. Therefore:

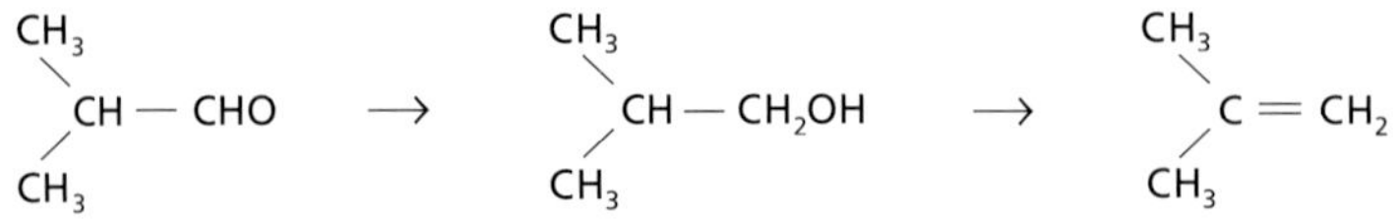

A 2-methylpropanal   B 2-methylpropan-1-ol   2-methylpropene

Now try this

Describe a test (a different one in each case) that you could use to distinguish between the following pairs of compounds.

1 CH_3CH_2CHO and CH_3COCH_3
2 $CH_2{=}CH{-}CH_2OH$ and CH_3CH_2CHO

Worked example 2

Which of these carbonyl compounds will undergo the iodoform reaction?

  A   **B**   **C**   **D**

Answer

Only ketones **A** and **D** contain the grouping $CH_3CO{-}$, so these are the only two to give iodoform. **B** contains the grouping $CH_3CH_2CO{-}$, whilst **C** is an aldehyde.

Now try this

Two carbonyl compounds **E** and **F** have the molecular formula C_4H_8O. **E** gave a yellow precipitate when treated with alkaline aqueous iodine, but **F** does not. Suggest, with a reason, structures for **E** and **F**. Suggest a different test, other than the tri-iodomethane reaction, that could distinguish between these two compounds.

17.4 Preparing carbonyl compounds

From alcohols, by oxidation

See section 16.3 for details of this reaction. For example:

$$CH_3CH_2CH_2OH + [O] \xrightarrow[\text{heat}]{K_2Cr_2O_7 + H_2SO_{4(aq)}} CH_3CH_2CHO + H_2O$$

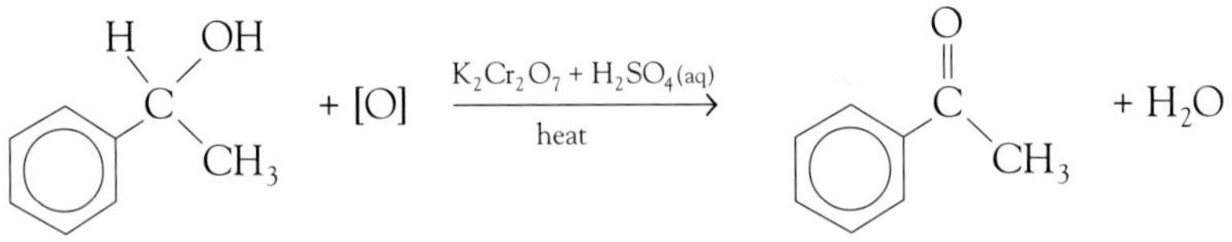

As explained in Topic 16, if an aldehyde is being prepared, the oxidant is added slowly to an excess of the alcohol and the aldehyde is distilled off as it is formed, in order to prevent further oxidation to the acid. (As their molecules are not hydrogen bonded to one another, aldehydes generally have lower boiling points that the corresponding alcohols.) If, however, it is the acid that is required, the reaction mixture is heated under reflux so that any aldehyde that might be formed remains in contact with the oxidising agent, and all is eventually oxidised to the carboxylic acid.

Summary

- **Carbonyl compounds** are those compounds in which the carbonyl group is not associated with other hetero-atoms in the same functional group. There are two types of carbonyl compounds – **ketones** and **aldehydes**.
- Their characteristic reaction type is **nucleophilic addition**.
- They can be reduced to alcohols, and aldehydes can be oxidised to carboxylic acids.
- They undergo **condensation reactions**.
- The condensation reaction with 2,4-dinitrophenylhydrazine, giving an orange precipitate, is a good test for the presence of a carbonyl compound in a sample.
- Aldehydes can be distinguished from ketones by their effects on **Fehling's solution** (deep blue solution → red-brown precipitate) and **Tollens' reagent** (clear colourless solution → silver mirror).
- The iodoform reaction is a test for the presence of the groups CH_3CO- or $CH_3CH(OH)-$ in a molecule. The products are iodoform and a sodium carboxylate with one fewer carbon atoms than the original alcohol or ketone.

Key reactions you should know
(All reactions undergone by R_2CO are also undergone by RCHO.)
- Catalytic hydrogenation:

$$R_2CO + H_2 \xrightarrow{\text{Ni}} R_2CH(OH)$$

- Nucleophilic additions:

$$R_2CO + HCN \xrightarrow{\text{NaCN}} R_2C(OH)CN$$

$$R_2CO \xrightarrow{\text{NaBH}_4} R_2CH(OH)$$

- Condensation reaction:

$$R_2CO + H_2N-R' \longrightarrow R_2C=N-R' + H_2O$$
$$(H_2N-R' = \text{2-DNPH})$$

- Oxidation reactions:

$$RCHO \xrightarrow{Cr_2O_7^{2-} + H_3O^+} RCO_2H$$

$$RCHO \xrightarrow{\text{Fehling's solution (Cu}^{2+})} RCO_2^- + Cu_2O \\ \text{(red ppt)}$$

$$RCHO \xrightarrow{\text{Tollens' reagent (Ag}^+)} RCO_2^- + Ag \\ \text{(silver mirror)}$$

- The tri-iodomethane reaction:

$$RCOCH_3 \xrightarrow{I_2 + {}^-OH(aq)} RCO_2^- + CHI_3$$

Examination practice questions

Please see the data section of the CD for any A_r values you may need.

1 Propanone, CH_3COCH_3, an important industrial solvent, can be converted into another industrially important solvent, MIBK, by the following sequence.

$$2\ CH_3C=O \xrightarrow{\text{step I}} C_6H_{12}O_2 \xrightarrow{\text{step II}} CH_3-C=O$$
(with CH_3 below first carbonyl; **F** under $C_6H_{12}O_2$; $CH=C(CH_3)_2$ below second carbonyl)

$$\mathbf{G}\ (C_6H_{10}O)$$

$$\downarrow \text{step III}$$

$$CH_3C=O$$
(with $CH_2CH(CH_3)_2$ below)

MIBK

a When **F** is formed in step I no other compound is produced. Suggest a structural formula for **F**, which contains one —OH group. [1]

b Compound **G** has two functional groups. Name **one** functional group present in **G** and show how you would identify it. [3]

c **G** is formed from **F** in step II.
Use your answers to **a** and **b** to suggest
  i what type of reaction occurs in step II
  ii a reagent for step II. [2]

d The production of MIBK from **G** in step III involves the hydrogenation of the $>C=C<$ group and is carried out catalytically. A mixture of compounds is formed because the $>C=O$ group is also reduced.
What reagent(s) and solvent are normally used in a laboratory to reduce a $>C=O$ group without reducing a $>C=C<$ group present in the same molecule? [2]

G has a number of structural isomers.

e Draw the displayed formulae of a pair of structural isomers of **G** which contain the CH_3CO- group and which exhibit *cis–trans* isomerism.
Label each structure *cis* or *trans* and give your reasoning. [3]
[Cambridge International AS & A Level Chemistry 9701, Paper 2 Q5 June 2009]

2 Compound **Q**, heptan-2-one, is found in some blue cheeses.

$$CH_3(CH_2)_4COCH_3$$
compound Q

a Compound **Q** may be reduced to **R**. Compound **R** may be dehydrated to give two different products, **S** and **T**.

i Draw the structural formulae of **R**, **S**, and **T**.

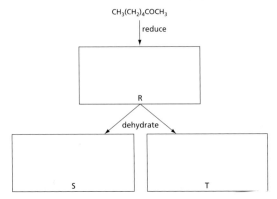

ii State the reagents that would be used for **each** of these reactions in a school or college laboratory.

reduction

dehydration   [5]

b Write the structural formula of the organic compound formed when Q is reacted separately with each reagent under suitable conditions. If you think no reaction occurs, write 'NO REACTION'.

Tollens' reagent		
HCN		
$K_2Cr_2O_7/H^+$		

[5]

c The first stage of cheese making is to produce 2-hydroxypropanoic acid (lactic acid) from milk.

$$CH_3CH(OH)CO_2H$$
lactic acid

Other than the use of a pH indicator, what reagent could you use to confirm the presence of some lactic acid in a sample of heptan-2-one? State what observation you would make.   [2]

[Cambridge International AS & A level Chemistry, 9701, Paper 23 Q4 June 2013]

18 Carboxylic acids and esters

Functional groups:

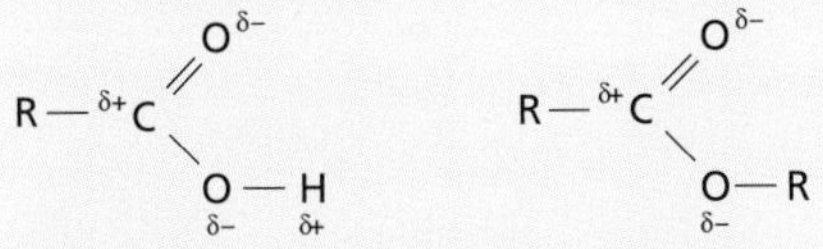

carboxylic acids   esters

In this topic we look at the major class of organic acids, the carboxylic acids, and at some of the compounds derived from them. Apart from acid–base reactions, their chemistry is dominated by nucleophilic substitution reactions, owing to the δ+ carbon atom in the carbonyl group.

Learning outcomes

By the end of this topic you should be able to:

14.1a) interpret and use the general, structural, displayed and skeletal formulae of the carboxylic acids and esters

14.2a) interpret and use the term condensation in relation to organic reactions

19.1a) describe the formation of carboxylic acids from alcohols, aldehydes and nitriles

19.1b) describe the reactions of carboxylic acids in the formation of salts (by the use of reactive metals, alkalis or carbonates), alkyl esters, and alcohols (by the use of LiAlH₄)

19.3a) describe the acid and base hydrolysis of esters

19.3b) state the major commercial uses of esters, e.g. solvents, perfumes, flavourings.

18.1 Introduction

When the carbonyl group is directly joined to an oxygen atom, the **carboxyl group** is formed. This occurs in **carboxylic acids** and **esters**.

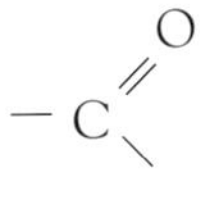

carbonyl group 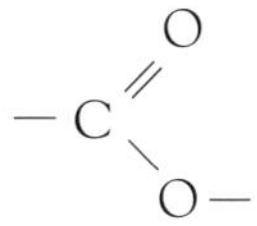   carboxyl group 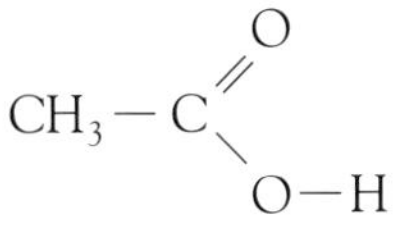   carboxylic acid: (ethanoic acid) 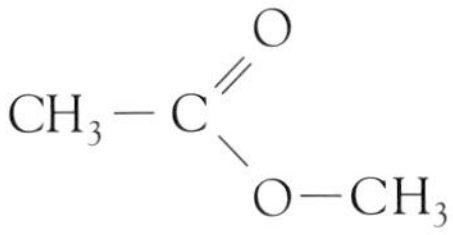   ester: (methyl ethanoate)

The reactions of the carbonyl group are drastically changed by the presence of the electronegative oxygen atom. These compounds have virtually none of the reactions of carbonyl compounds as described in Topic 17. The same is true of two other classes of compounds in which the carbonyl group is directly attached to an electronegative atom, namely the **acyl chlorides** and the **amides**:

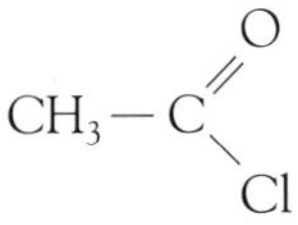

acyl chloride: (ethanoyl chloride) 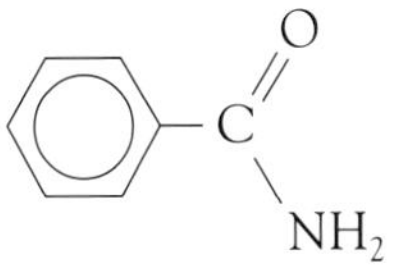   amide: (benzamide)

The reactivity of carboxylic acids is dominated by the tendency of the O—H bond to ionise to give hydrogen ions, hence the incorporation of the word 'acid' in their name. The extent of ionisation is small, however. For example, in a $1.0\,mol\,dm^{-3}$ solution of ethanoic acid, about one molecule in 1000 (0.1%) is ionised:

$$CH_3CO_2H \rightleftharpoons CH_3CO_2^- + H^+$$

Carboxylic acids are therefore classed as **weak acids** (see section 6.4).

The O—H bond breaks heterolytically far more easily in carboxylic acids than in alcohols or in phenols (see section 25.5). The negative charge formed by heterolysis can be delocalised over two electronegative oxygen atoms in the carboxylate ion. This spreading out of charge invariably leads to a stabilisation of the anion (see Figure 18.1 and Figure 3.46, page 62). The effect that changes in the structure of a carboxylic acid have on the stability of its anion, and hence its acidity, is dealt with in section 26.1.

Figure 18.1 The charge is spread in the carboxylate anion, stabilising the ion.

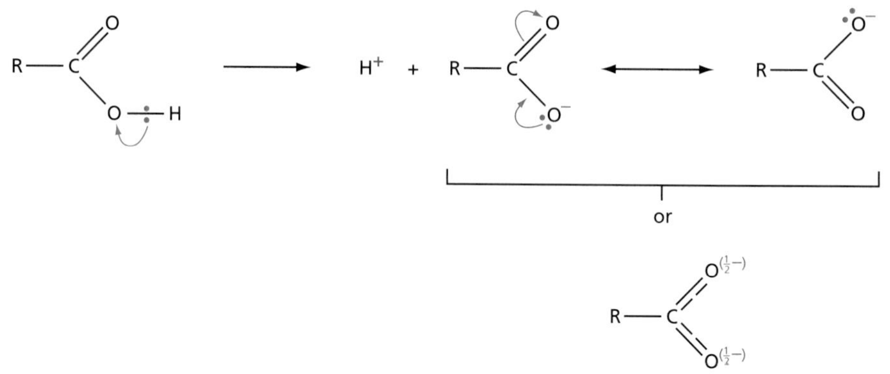

The carboxylic acid group is a strongly hydrogen-bonded one, both to carboxylic acid groups on other carboxylic acid molecules and to solvent molecules such as water. This has effects on two important properties – boiling points and solubilities. Table 18.1 compares these properties for carboxylic acids and the corresponding alcohols.

Table 18.1 Hydrogen bonding increases the boiling points and solubilities in carboxylic acids.

Number of carbon atoms	Alcohol			Carboxylic acid		
	Formula	Boiling point/°C	Solubility in 100g of water/g	Formula	Boiling point/°C	Solubility in 100g of water/g
1	CH_3OH	65	∞	HCO_2H	101	∞
2	CH_3CH_2OH	78	∞	CH_3CO_2H	118	∞
3	$CH_3CH_2CH_2OH$	97	∞	$CH_3CH_2CO_2H$	141	∞
4	$CH_3CH_2CH_2CH_2OH$	118	7.9	$CH_3CH_2CH_2CO_2H$	164	∞
5	$CH_3CH_2CH_2CH_2CH_2OH$	138	2.3	$CH_3CH_2CH_2CH_2CO_2H$	187	3.7

Both in the pure liquid state, and in solution in non-hydrogen-bonding solvents such as benzene, carboxylic acids can form hydrogen-bonded dimers (see Figure 18.2).

Figure 18.2 Carboxylic acids dimerise as two molecules hydrogen bond to each other.

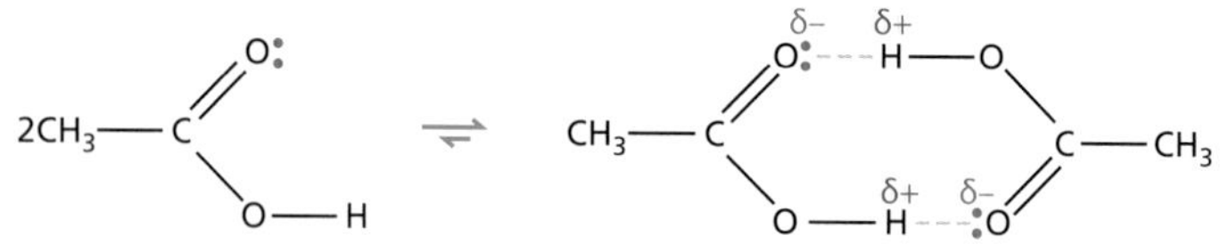

In water, both oxygen atoms and the —OH hydrogen atom of carboxylic acids can hydrogen bond with the solvent (see Figure 18.3).

Figure 18.3 Carboxylic acids hydrogen bond to water molecules.

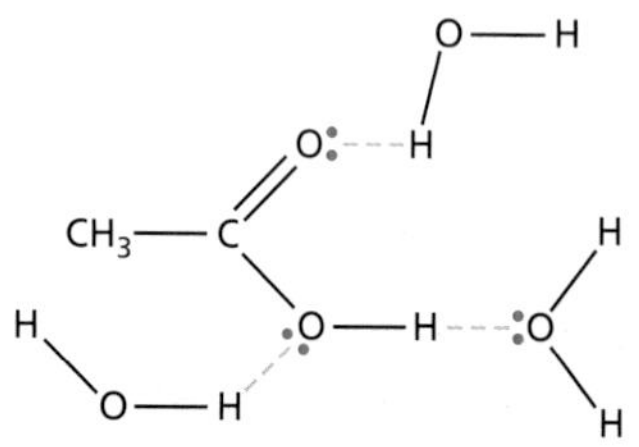

18.2 Isomerism and nomenclature

Carboxylic acids are named by finding the longest carbon chain that contains the acid functional group, and adding the suffix '-oic acid' to the stem. Side groups off the chain are named in the usual way:

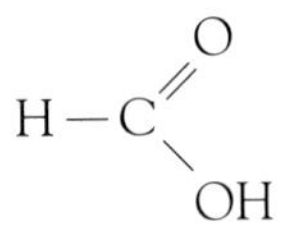

methanoic acid

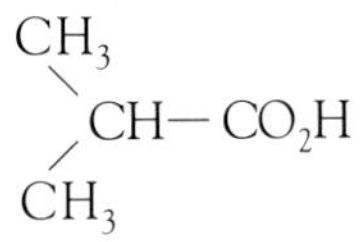

2-methylpropanoic acid

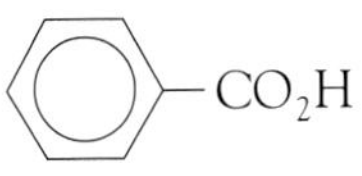

benzoic acid

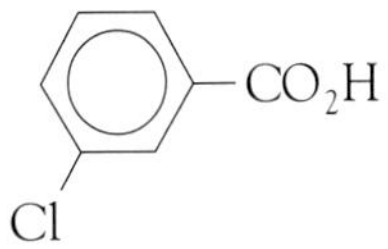

3-chlorobenzoic acid

Esters are named as alkyl derivatives of acids, similar to the naming of salts:

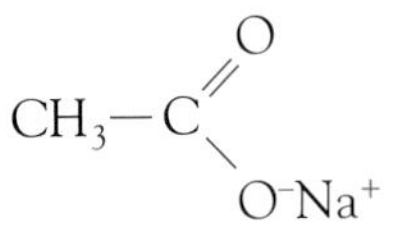

sodium ethanoate

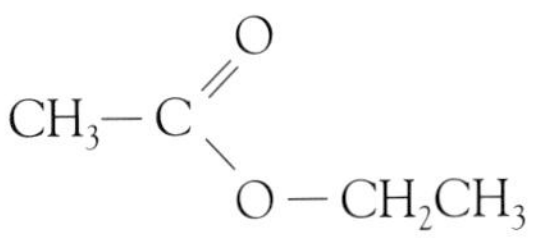

ethyl ethanoate

The isomerism shown by acids is not usually associated with the functional group, but occurs in the carbon chain to which it is attached, as in the following two isomers with the formula $C_4H_8O_2$:

$$CH_3CH_2CH_2CO_2H \qquad (CH_3)_2CHCO_2H$$

butanoic acid   2-methylpropanoic acid

Esters, however, can show a particular form of isomerism, depending on the number of carbon atoms they have in the acid or alcohol part of the molecule. Two isomers of $C_4H_8O_2$ are shown below. Note that although the acid part is usually written first in the formula, by convention, it appears last in the name:

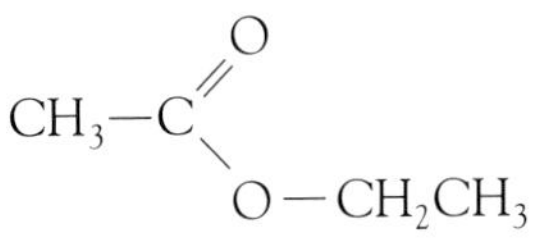

ethyl ethanoate

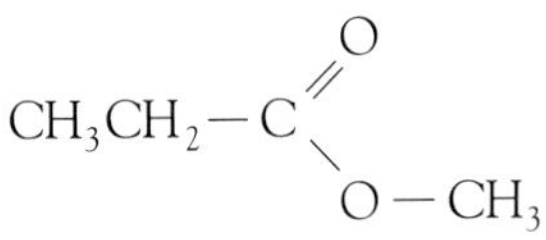

methyl propanoate

Worked example

Draw the structures of the other two isomeric esters with the formula $C_4H_8O_2$, and name them.

Answer

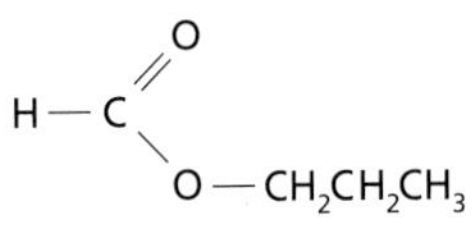

prop-1-yl methanoate

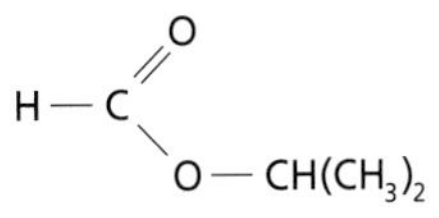

prop-2-yl methanoate

Now try this

Draw and name possible ester isomers with the formula $C_5H_{10}O_2$. How many ester isomers are there with this formula? Indicate in your answer which of the esters are chiral.

18.3 Reactions of carboxylic acids

Reactions of the O—H group

The acidity of carboxylic acids has already been mentioned. They show most of the typical reactions of acids.

With metals

Ethanoic acid reacts with sodium metal, liberating hydrogen gas (compare with alcohols, section 16.3, and phenols, section 25.5):

$$CH_3CO_2H + Na \rightarrow CH_3CO_2^- Na^+ + \tfrac{1}{2}H_2$$

They also react with other reactive metals, such as calcium and magnesium.

With sodium hydroxide

Like phenols, but unlike alcohols, carboxylic acids form salts with alkalis:

$$CH_3CO_2H + NaOH \rightarrow CH_3CO_2^- Na^+ + H_2O$$

Salts are often much more soluble in water than the carboxylic acids from which they are derived. For example, soluble aspirin tablets contain the calcium salt of ethanoylsalicylic acid (see section 25.6):

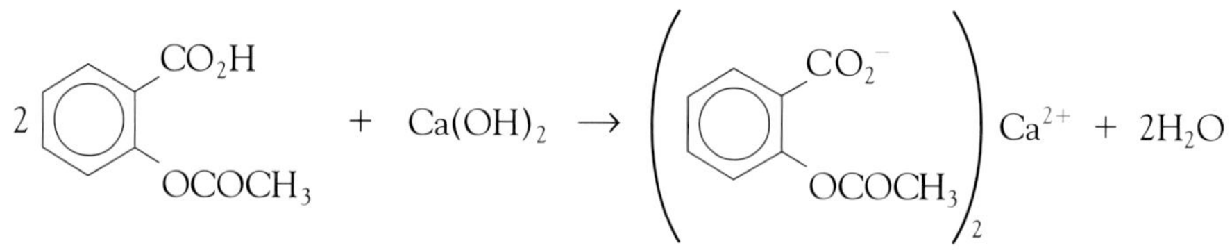

ethanoylsalicylic acid (Aspirin)   calcium ethanoylsalicylate

With sodium carbonate and sodium hydrogencarbonate

Like the usual inorganic acids, but unlike both alcohols and phenols, carboxylic acids react with carbonates, liberating carbon dioxide gas:

$$2CH_3CO_2H + Na_2CO_3 \rightarrow 2CH_3CO_2^- Na^+ + CO_2 + H_2O$$

A similar reaction occurs with sodium hydrogencarbonate:

$$CH_3CO_2H + NaHCO_3 \rightarrow CH_3CO_2^- Na^+ + CO_2 + H_2O$$

These three reactions form the basis of a series of tests by which alcohols, phenols and carboxylic acids can be distinguished from one another, and from other functional groups (see Figure 18.4).

Acids, phenols and alcohols can be distinguished by their reactions with sodium metal, aqueous sodium hydroxide and aqueous sodium carbonate.

Figure 18.4 How to distinguish between alcohols, phenols and carboxylic acids

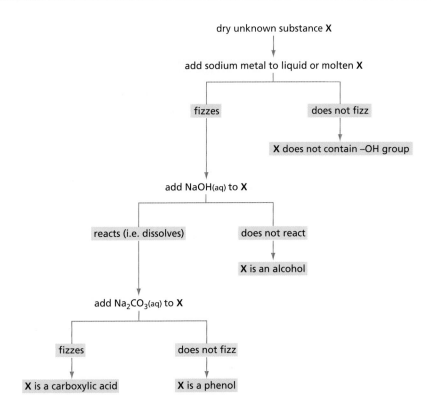

Reactions of the C—OH group

There are two main reactions which replace the —OH group in carboxylic acids with other atoms or groups.

With phosphorus(V) chloride or sulfur dichloride oxide

Carboxylic acids react with these reagents in exactly the same way as do alcohols (see section 16.3). Acyl chlorides are produced:

$$CH_3-C\overset{O}{\underset{OH}{\big\langle}} \ + \ PCl_5 \ \rightarrow \ CH_3-C\overset{O}{\underset{Cl}{\big\langle}} \ + \ POCl_3 \ + \ HCl$$

$$\langle\bigcirc\rangle-C\overset{O}{\underset{OH}{\big\langle}} \ + \ SOCl_2 \ \rightarrow \ \langle\bigcirc\rangle-C\overset{O}{\underset{Cl}{\big\langle}} \ + \ SO_2 \ + \ HCl$$

Acyl chlorides are used to make amide. Further reactions and properties of acyl chlorides are described in Topic 26.

With alcohols in the presence of anhydrous acids

Carboxylic acids react with alcohols (but not with phenols) in an acid-catalysed equilibrium reaction to give esters:

$$CH_3-C\overset{O}{\underset{OH}{\big\langle}} \ + \ CH_3CH_2OH \ \rightleftharpoons \ CH_3-C\overset{O}{\underset{O-CH_2CH_3}{\big\langle}} \ + \ H_2O \ \ (1)$$

The formation of an ester is an example of a *condensation reaction*. This is when two molecules join together with the elimination of a small molecule (often water) (see also sections 26.4 and 28.2).

Typical reaction conditions are:

- heat a mixture of carboxylic acid and alcohol with 0.1 mol equivalent of concentrated sulfuric acid under reflux for 4 hours, or
- heat a mixture of carboxylic acid and alcohol under reflux for 2 hours, while passing dry hydrogen chloride gas through the mixture.

The reaction needs an acid catalyst and, like any equilibrium reaction, it can be made to go in either direction depending on the conditions. To encourage esterification, an excess of one of the reagents (often the alcohol) is used, and the water is removed as it is formed (often by the use of concentrated sulfuric acid):

$$H_2SO_4 + H_2O \rightarrow H_3O^+ + HSO_4^-$$

This drives equilibrium (1) over to the right (see section 9.3). To encourage the reverse reaction (the hydrolysis of the ester), the ester is heated under reflux with a dilute solution of sulfuric acid. This provides an excess of water to drive equilibrium (1) over to the left.

If an ester is made from an alcohol whose —OH group is labelled with an oxygen-18 atom, it is found that the ^{18}O is 100% incorporated into the ester. None of it appears in the water that is also produced. Therefore the water must come from combining the —OH from the carboxylic acid and the —H from the alcohol, rather than vice versa. That is,

The following mechanism explains why this is the case.

The mechanism of esterification/hydrolysis

In a strongly acidic medium, both the alcohol and the carboxylic acid can undergo reversible protonation. However, only the protonated carboxylic acid undergoes further reaction.

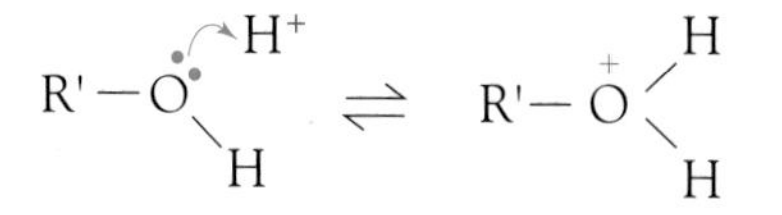

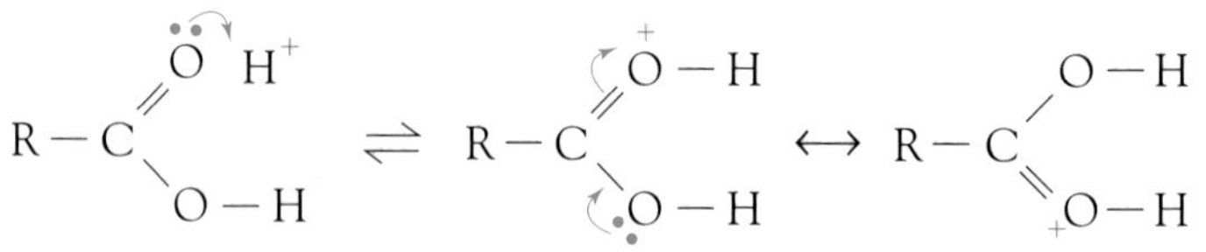

(1)

The protonation of the carboxylic acid in step (1) produces a highly δ+ carbon atom, which can undergo nucleophilic attack by a lone pair of electrons on the oxygen atom of an alcohol molecule:

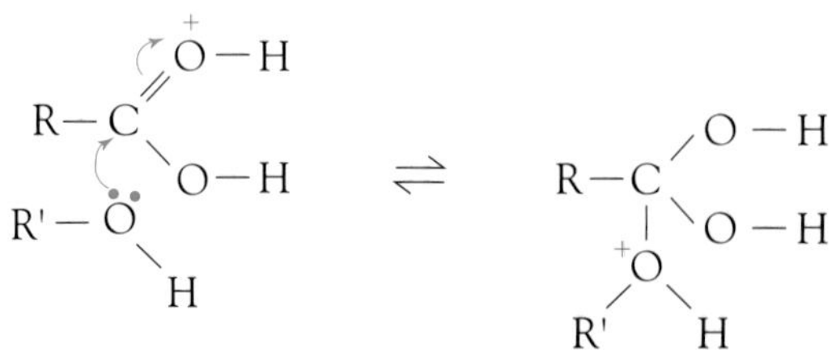

(2)

The three oxygen atoms in this intermediate cation are very similar in basicity. Any one of them could be attached to the proton, and proton transfer from one to another can occur readily:

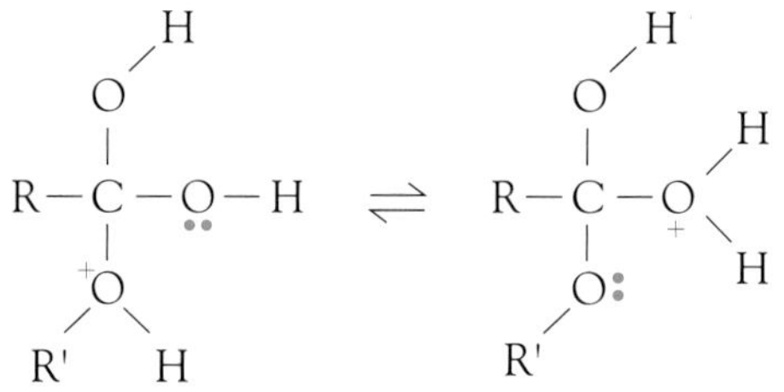

This new cation can undergo a carbon–oxygen bond breaking, which is essentially the reverse of reaction (2), to give a molecule of water, along with the protonated ester:

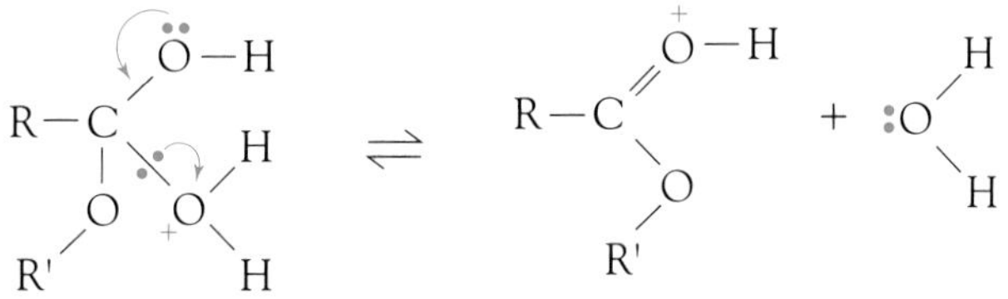

Lastly, the cation loses a proton to regenerate the H^+ catalyst, and form the ester product:

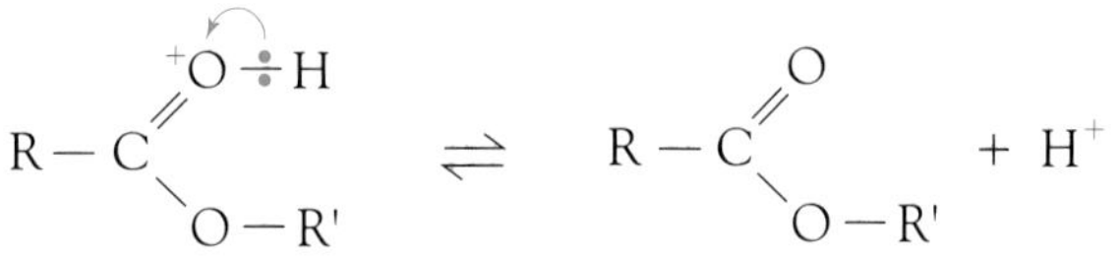

As mentioned above, the reaction is reversible. So is the mechanism. With an excess of water, esters undergo acid-catalysed hydrolysis. The steps in the mechanism for the hydrolysis reaction are the reverse of the above steps for esterification.

Reduction of the –CO$_2$H group

Carboxylic acids can be reduced to alcohols by reacting with lithium tetrahydridoaluminate(III) (lithium aluminium hydride), $LiAlH_4$, in dry ether.

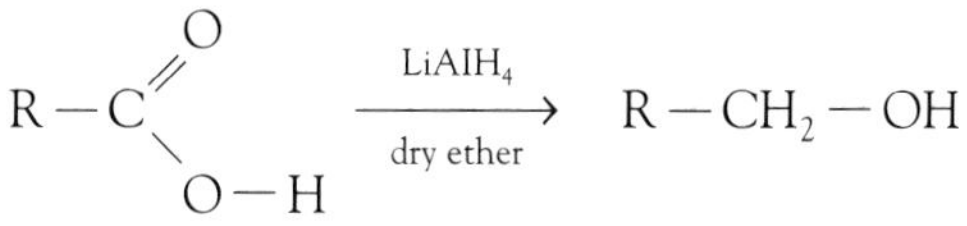

The reaction requires the powerful reducing agent $LiAlH_4$: neither $NaBH_4$ nor $H_2 + Ni$ are strong enough to reduce carboxylic acids. $LiAlH_4$ will also reduce esters (to alcohols) and amides (to amines – see Topic 27).

18.4 Esters

Properties of esters

Esters have the functional group:

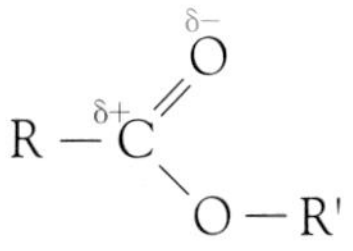

> **Now try this**
>
> Draw the mechanism for the acid-catalysed *hydrolysis* of methyl methanoate.

Many esters are liquids with sweet, 'fruity' smells. They are immiscible with, and usually less dense than, water. Despite containing two oxygen atoms, they do not form strong hydrogen bonds with water molecules. Neither do they form hydrogen bonds with other ester molecules (because they do not contain $\delta+$ hydrogen atoms). Their major intermolecular bonding is van der Waals, supplemented by a small dipole–dipole contribution. Their boiling points are therefore a few degrees higher than those of the alkanes of similar molecular mass. The position of the carboxyl group along the chain has little effect on the strength of intermolecular bonding, and hence boiling point (see Table 18.2).

Table 18.2 Esters have boiling points a little higher than those of corresponding alkanes.

Compound	M_r	Boiling point/°C
hexane 	86	69
propyl methanoate 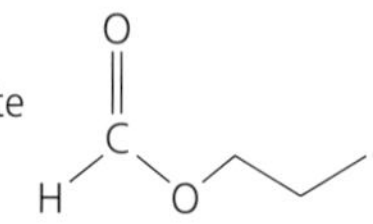	88	81
ethyl ethanoate 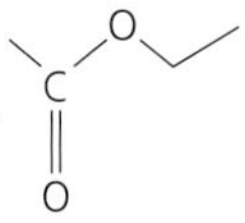	88	77
methyl propanoate 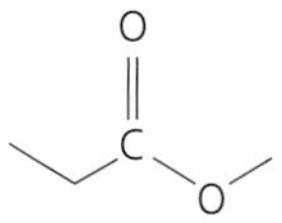	88	79

Reactions of esters

The most common type of reaction that esters undergo is nucleophilic substitution, illustrated here by their hydrolysis.

The hydrolysis of an ester is a slow process, taking several hours of heating under reflux with dilute aqueous acids:

$$CH_3 - C\begin{matrix}O\\\\O-CH_2CH_3\end{matrix} + H_2O \xrightarrow{\text{heat with } H_2SO_4(aq)} CH_3 - C\begin{matrix}O\\\\OH\end{matrix} + C_2H_5OH$$

The mechanism of this acid-catalysed hydrolysis has been discussed in section 18.3: it is the reverse of the acid-catalysed esterification described there.

Ester hydrolysis can also be carried out in alkaline solution. The reaction is quicker than in acid solution: OH⁻ is a stronger nucleophile than water. Additionally, it does not reach equilibrium, but goes to completion. This is because the carboxylic acid produced reacts with an excess of base to form the carboxylate anion:

$$CH_3 - C\begin{matrix}O\\\\O-CH_2CH_3\end{matrix} + Na^+OH^- \xrightarrow{\text{heat with NaOH(aq)}} CH_3 - C\begin{matrix}O\\\\O^-Na^+\end{matrix} + CH_3CH_2OH$$

Now try this

Suggest a mechanism for this reaction, using OH⁻ as the initial nucleophile, and $C_2H_5O^-$ as the base in the last step.

The alkaline hydrolysis of esters has an important application in the manufacture of soap. Fats are glyceryl triesters of long-chain carboxylic acids. Heating fats with strong aqueous sodium hydroxide causes hydrolysis of the triesters. The sodium salts of the long-chain carboxylic acids are precipitated by adding salt (NaCl) to the mixture, and this solid is then washed and compressed into bars of soap (see Figure 18.5). Perfume and colour are also added.

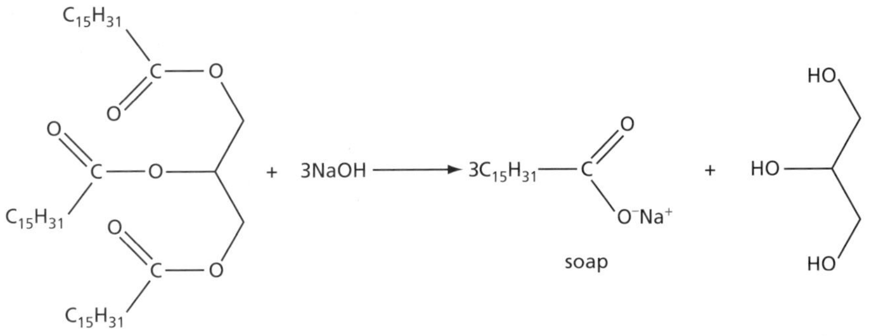

Figure 18.5 Soap is made by the hydrolysis of the glyceryl triesters of fats.

Uses of esters

Esters with small molecular masses find many uses as solvents for paints and varnishes, and also for removing caffeine from coffee and tea to make decaffeinated beverages. Many natural perfumes, flavours and aromas are esters, and many of these are now manufactured commercially to flavour fruit drinks and sweets. It is the ethyl esters formed from alcohol and the various acids in wines that account for much of their 'nose', or aroma, in the glass.

Some of the naturally-occurring esters now made synthetically and used as artificial flavours in foodstuffs are shown in Figure 18.6.

Figure 18.6 Some esters used as flavouring agents

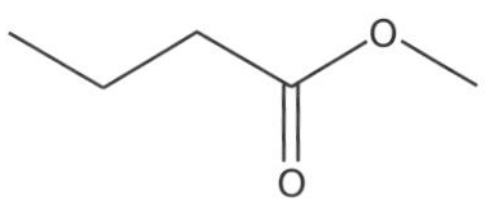

methyl butanoate (apple)

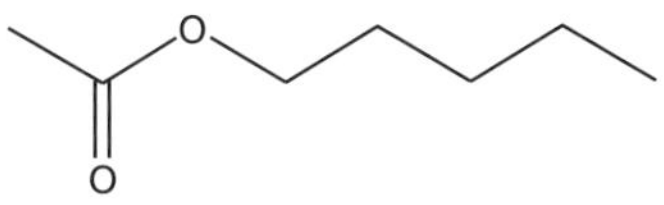

pentyl ethanoate (pear)

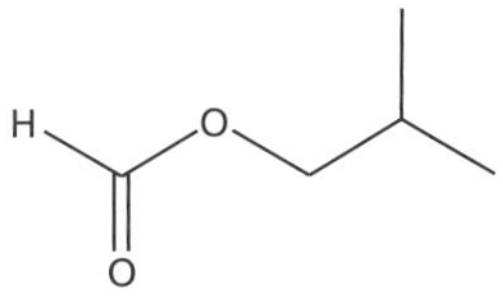

2-methylpropyl methanoate (raspberry)

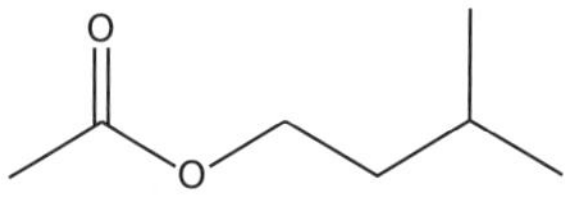

3-methylbutyl ethanoate (banana)

18.5 Preparing carboxylic acids

Carboxylic acids may be prepared by the following reactions.

● By the oxidation of primary alcohols or aldehydes:

$$CH_3CH_2OH + 2[O] \xrightarrow{\text{Na}_2\text{Cr}_2\text{O}_7 + \text{H}_2\text{SO}_4\text{(aq)} + \text{heat}} CH_3CO_2H + H_2O$$

$$CH_3CHO + [O] \xrightarrow{\text{Na}_2\text{Cr}_2\text{O}_7 + \text{H}_2\text{SO}_4\text{(aq)} + \text{heat}} CH_3CO_2H$$

● By the hydrolysis of organic cyanides (nitriles):

$$CH_3 - C \equiv N \xrightarrow{\text{heat with H}_2\text{SO}_4\text{(aq)}} CH_3CO_2H$$

● By the hydrolysis of acid derivatives such as esters, acyl chlorides and amides (see Topics 27 and 28). Although these reactions proceed in high yields, they are not normally useful preparative reactions: acid derivatives are usually made from the carboxylic acids in the first place, so their hydrolysis forms no compounds of further use.

● By the oxidation of aryl side chains. When treated with hot alkaline potassium manganate(VII), aryl hydrocarbons produce benzoic acids by oxidation:

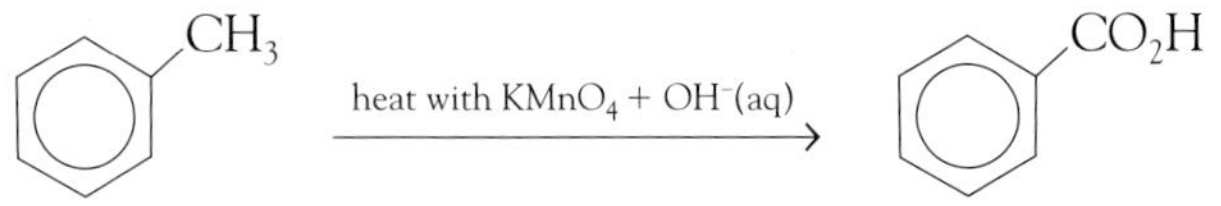

$$\text{C}_6\text{H}_5\text{CH}_3 \xrightarrow{\text{heat with KMnO}_4 + \text{OH}^-\text{(aq)}} \text{C}_6\text{H}_5\text{CO}_2\text{H}$$

Summary

- The **carboxylic acids**, RCO_2H, are weak acids, ionising to the extent of 1% or less in water.
- Carboxylic acids react with alcohols to form **esters**, and with phosphorus(V) chloride to form **acyl chlorides**.
- Esters undergo acid- or base-catalysed **hydrolysis**.

Key reactions you should know
- Carboxylic acids:
  $RCO_2H + Na \rightarrow RCO_2^-Na^+ + \frac{1}{2}H_2$

$RCO_2H + NaOH \rightarrow RCO_2^-Na^+ + H_2O$
$2RCO_2H + Na_2CO_3 \rightarrow 2RCO_2^-Na^+ + CO_2 + H_2O$
$RCO_2H + SOCl_2 \rightarrow RCOCl + SO_2 + HCl$
$RCO_2H + PCl_5 \rightarrow RCOCl + POCl_3 + HCl$

$RCO_2H + R'OH \xrightarrow{\text{heat with conc. H}^+} RCO_2R' + H_2O$

- Esters:

$RCO_2R' \xrightarrow{\text{heat with H}_3O^+ \text{ or OH}^-\text{(aq)}} RCO_2H + R'OH$

Examination practice questions

Please see the data section of the CD for any A_r values you may need.

1 Isomerism occurs in many organic compounds. The two main forms of isomerism are structural isomerism and stereoisomerism. Many organic compounds that occur naturally have molecules that can show stereoisomerism, that is *cis–trans* or optical isomerism.
   a i Explain what is meant by *structural isomerism*.
   ii State **two** different features of molecules that can give rise to **stereoisomerism**. [3]
   Unripe fruit often contains polycarboxylic acids, that is acids with more than one carboxylic acid group in their molecule. One of these acids is commonly known as tartaric acid, $HO_2CCH(OH)CH(OH)CO_2H$.
   b Give the structural formula of the organic compound produced when tartaric acid is reacted with an excess of $NaHCO_3$. [1]
   Another acid present in unripe fruit is citric acid,

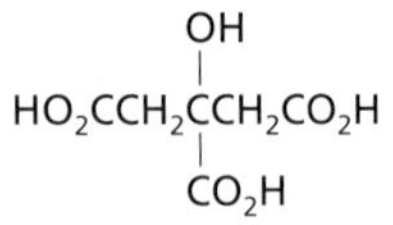

$$\begin{array}{c} \text{OH} \\ | \\ HO_2CCH_2CCH_2CO_2H \\ | \\ CO_2H \end{array}$$

   c Does citric acid show optical isomerism? Explain your answer. [1]
   A third polycarboxylic acid present in unripe fruit is a colourless crystalline solid, **W**, which has the following composition by mass: C, 35.8%; H, 4.5%; O, 59.7%.
   d i Show by calculation that the empirical formula of **W** is $C_4H_6O_5$.
   ii The M_r of **W** is 134. Use this value to determine the molecular formula of **W**. [3]
   A sample of **W** of mass 1.97 g was dissolved in water and the resulting solution titrated with $1.00\,mol\,dm^{-3}$ NaOH. $29.4\,cm^3$ were required for complete neutralisation.

   e i Use these data to deduce the number of carboxylic acid groups present in one molecule of **W**.
   ii Suggest the displayed formula of **W**. [5]
   [Cambridge International AS & A Level Chemistry 9701, Paper 21 Q5 June 2010]

2 a Draw and complete the following reaction scheme which starts with ethanal.
   In **each empty** box, write the **structural formula** of the organic compound that would be formed.

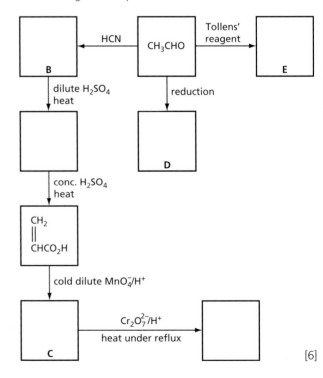

b Write the structural formula for the organic compound formed when, under suitable conditions,

   i compound **C** reacts with compound **D**,

   ii compound **C** reacts with compound **E**. [2]

c Compound **B** is chiral. Draw displayed formulae of the two optical isomers of compound **B**, indicating with an asterisk (*) the chiral carbon atom. [3]

[Cambridge International AS & A Level Chemistry 9701, Paper 2 Q4 June 2009]

19 Practical work

In this topic, we look at some simple practical techniques that are used in the laboratory, such as titrations and the familiar 'wet' tests for finding out which cations and anions are present in an unknown sample in the test tube. Nowadays these methods are very rarely used in analytical laboratories because modern physical methods are much more sensitive and selective, and can readily be automated, making them less labour intensive. The methods do, however, bring together much of the chemistry that has already been met, so they provide a useful alternative way at looking at inorganic and organic chemistry.

Suggested skills

Your practical work will be helped if you are able to:

- take readings with an accuracy determined by the apparatus being used
- produce a numerical result corrected to an appropriate number of significant figures (you are not expected to use statistical methods of analysis or a treatment of errors)
- record clearly and concisely relevant observations (including negative ones)
- draw valid deductions from these observations
- arrive at a likely overall identification of the substance under investigation
- if appropriate, suggest additional experiments (including physical methods of analysis) to confirm your conclusion.

19.1 Thermometric measurements

Measuring enthalpy changes directly

Some reactions, for example neutralisation reactions, take place very quickly. The heat evolved (or absorbed) in these reactions can be measured directly with a thermometer, using an expanded polystyrene cup as a calorimeter.

Examples of such reactions are included in section 5.2.

Measuring enthalpy changes indirectly

Other reactions take place slowly or may require a high temperature to bring them about. The enthalpy change of such a reaction is determined indirectly by carrying out several rapid reactions and then using Hess's Law.

Examples of such reactions are included in section 5.4.

19.2 Titrations

Acid–base titrations

$$H^+(aq) + OH^-(aq) \rightarrow H_2O(l)$$

The most common type of titration is the neutralisation of an acid by a base using an indicator to show the equivalence point.

Examples of such titrations are included in sections 6.6 and 6.7.

Potassium manganate(VII) titrations

For example:

$$MnO_4^-(aq) + 5Fe^{2+}(aq) + 8H^+(aq) \rightarrow Mn^{2+}(aq) + 5Fe^{3+}(aq) + 4H_2O(l)$$

The potassium manganate(VII) solution is placed in the burette. As it is run into the acidified $Fe^{2+}(aq)$ solution in the conical flask, its purple colour disappears. The end-point is when the solution permanently remains pink (that is, pale purple) when the last drop of the manganate(VII) solution is added.

Examples of such titrations are included in section 7.4.

Iodine–thiosulfate titrations

$$I_2(aq) + 2S_2O_3^{2-}(aq) \rightarrow 2I^-(aq) + S_4O_6^{2-}(aq)$$

The sodium thiosulfate solution is added from the burette. The disappearance of the yellow-brown colour of the iodine shows the end-point of these titrations. This end-point is made more obvious by adding a little starch near the end-point; this converts the pale yellow colour into an intense blue colour.

Examples of such titrations are included in section 7.4.

19.3 Appearance

Inorganic

Most inorganic compounds are made up of separate ions, and so they show characteristic reactions associated with each type of ion they contain. Analysing a compound, therefore, involves the separate analysis of cations and of anions.

Since most inorganic compounds are ionic, they are usually solid at room temperature. Much information can be gained from their appearance. Usually, the substance is in the form of a powder. It may be either amorphous, which suggests that it is insoluble in water and has been made by precipitation, or it may be made up of small crystals, which show that it must be soluble, having been prepared by crystallisation from solution. Crystals may absorb moisture from the air – they may be **hygroscopic**. If they are so hygroscopic that they dissolve in the moisture in the air, they are termed **deliquescent**. Some crystals become covered with powder owing to **efflorescence**, that is, the loss of water of crystallisation in a dry atmosphere (see Table 19.1).

Table 19.1 Information that can be gained from observing the appearance of a powder

Type of powder	Inference	Possible type of substance
amorphous	made by precipitation	insoluble, e.g. oxide or carbonate
crystalline	made by crystallisation	soluble, e.g. Group 1 compounds, nitrates, most sulfates and chlorides
hygroscopic	cation of high charge density	Li^+ and most double and triply charged cations
efflorescent	contains much water of crystallisation	$Na_2CO_3.10H_2O$, alums

If the substance is made up of colourless crystals or a white powder, we may infer that we are dealing with a compound from the s block, or possibly from the p or d

Table 19.2 Information that can be gained by observing the colour of a compound

Colour	Possible ions or substance present
colourless	s-block elements; Al^{3+}, Pb^{2+}; d^0 or d^{10} ions, e.g. Cu^+ or Zn^{2+}
blue	Cu^{2+}, VO^{2+}, Co^{2+}
pale green	Fe^{2+}, Ni^{2+}
dark green	Cu^{2+}, Cr^{3+}
purple	Cr^{3+}, MnO_4^-
pale violet	Fe^{3+} (goes yellow-brown in solution)
pink	Mn^{2+}, Co^{2+}
yellow	CrO_4^{2-}, PbO, S
orange	$Cr_2O_7^{2-}$, PbO
red	Pb_3O_4, Cu_2O
brown	PbO_2, Fe^{3+}, Fe_2O_3
black	CuO, MnO_2

block. A coloured compound probably indicates the presence of a transition metal (see section 24.5) or an oxide.

Organic

If the substance is a liquid, it is probably organic. Some organic substances of higher molecular weight are solid, but the most common ones are liquid.

19.4 Solubility in water

When finding out if a substance is soluble in water, only a tiny amount or a drop of the substance should be shaken with $1\,cm^3$ of water. If too much substance is used, it is difficult to see whether any has dissolved.

Inorganic

The solubility of an inorganic substance depends on the relative values of its lattice enthalpy and enthalpy change of hydration (see sections 20.1 and 20.2).

Organic

When a drop of an organic liquid is shaken with water, it may form either a clear solution or a cloudy emulsion. This shows whether the substance is soluble in water or not, and gives an indication of what functional groups are present (see Table 19.3).

Table 19.3 Information that may be deduced about an organic substance from its solubility in water

Solubility in water	Possible functional group present
completely soluble	lower alcohol, carboxylic acid, aldehyde, ketone
slightly soluble	higher alcohol
insoluble – floats on water	hydrocarbon
insoluble – sinks in water	halogenoalkane

19.5 Detection of gases

A common analytical test is to observe whether a gas is given off during a reaction and, if so, to identify it. It is important to consider how the gas was produced and, if possible, to write an equation for the reaction. Table 19.4 lists the most likely gases produced. Some of the tests are illustrated in Figure 19.1.

Table 19.4 The most common gases produced during analysis. Other less common gases include N_2, CO, SO_3, N_2O, NO, Br_2 and I_2.

Gas	Appearance and properties	Test and equation
O_2	colourless, odourless	glowing splint glows more brightly and may be rekindled
H_2	colourless, odourless	when lit, burns at mouth of test tube*: $2H_2(g) + O_2(g) \rightarrow 2H_2O(l)$
CO_2	colourless, almost odourless	limewater turns milky: $Ca(OH)_2(aq) + CO_2(g) \rightarrow CaCO_3(s) + H_2O(l)$
H_2O	condenses on cooler part of test tube	blue cobalt chloride paper turns pink: $CoCl_4^{2-}(aq) + 6H_2O(l) \rightarrow Co(H_2O)_6^{2+}(aq) + 4Cl^-(aq)$
SO_2	colourless, choking, acidic	acidified $K_2Cr_2O_7$ turns green: $Cr_2O_7^{2-}(aq) + 2H^+(aq) + 3SO_2(g) \rightarrow 2Cr^{3+}(aq) + H_2O(l) + 3SO_4^{2-}(aq)$
†HCl	colourless, choking, acidic steamy fumes in moist air	forms white fumes with ammonia gas: $NH_3(g) + HCl(g) \rightarrow NH_4Cl(s)$
Cl_2	pale green, very choking	moist litmus paper is bleached (it may turn red first): $Cl_2(g) + H_2O(l) \rightarrow Cl(aq) + HClO(aq)$
NO_2	brown, choking, acidic	moist litmus paper turns red
NH_3	colourless, choking, pungent	the only common alkaline gas; forms white fumes with HCl: $NH_3(g) + HCl(g) \rightarrow NH_4Cl(s)$

*When pure, hydrogen burns with a quiet, colourless flame. It is nearly always mixed with air and hence burns with a mild explosion. The flame is often tinged with yellow from the sodium ions in the glass of the test tube.
†HBr and HI give the same tests as HCl, but are nearly always given off in the presence of either bromine or iodine, which are easily identified by their colours.

Figure 19.1 Positive tests for **a** carbon dioxide, **b** water vapour, **c** chlorine and **d** ammonia

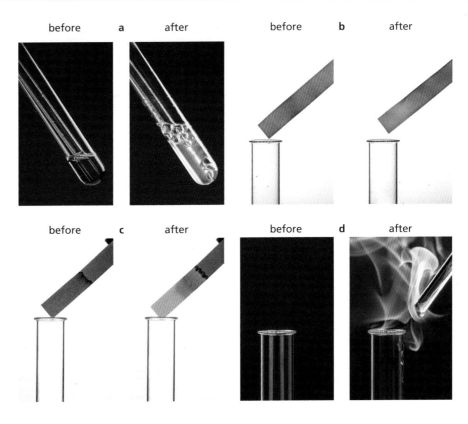

before   **a**   after   before   **b**   after

before   **c**   after   before   **d**   after

The source of the gases is important because it tells us something about the substance under investigation. Table 19.5 lists the most common sources of these gases.

Table 19.5 Possible sources of some common gases

Gas	Likely source	Typical equation
$O_2 + NO_2$ (H_2O if nitrate is hydrated)	heating nitrates	$Ca(NO_3)_2(s) \rightarrow CaO(s) + \frac{1}{2}O_2(g) + 2NO_2(g)$
H_2	HCl(aq) on metal	$Mg(s) + 2HCl(aq) \rightarrow MgCl_2(aq) + H_2(g)$
CO_2	heating carbonates *or* HCl(aq) on carbonates	$CuCO_3(s) \rightarrow CuO(s) + CO_2(g)$ $CaCO_3(s) + 2HCl(aq) \rightarrow CaCl_2(aq) + H_2O(l) + CO_2(g)$
H_2O	heating hydrated salts	$CuSO_4.5H_2O(s) \rightarrow CuSO_4(s) + 5H_2O(g)$
SO_2	HCl(aq) on sulfites	$Na_2SO_3(s) + 2HCl(aq) \rightarrow 2NaCl(aq) + H_2O(l) + SO_2(g)$
HCl	$H_2SO_4(l)$ on chlorides	$KCl(s) + H_2SO_4(l) \rightarrow KHSO_4(s) + HCl(g)$
Cl_2	HCl(aq) with oxidising agents	$4HCl(aq) + MnO_2(s) \rightarrow MnCl_2(aq) + Cl_2(g) + 2H_2O(l)$
NH_3	NH_4^+ and NaOH(aq)	$NH_4Cl(aq) + NaOH(aq) \rightarrow NaCl(aq) + NH_3(g) + H_2O(l)$

19.6 Flame tests

Some cations, particularly those of the s block, give characteristic colours in the flame of a Bunsen burner. A nichrome wire or porcelain rod is first cleaned by dipping it into concentrated hydrochloric acid and heating it in the blue part of the Bunsen burner flame until there is no persistent yellow colour (due to sodium impurities). A speck of the solid to be tested is mixed with a drop of concentrated hydrochloric acid on the end of the wire or rod and heated. The function of the hydrochloric acid is to convert non-volatile substances such as oxides and carbonates into the more volatile chlorides. Typical flame colours are listed in Table 19.6 and shown in Figure 19.2.

Table 19.6 Some typical flame test colours

Colour of flame	Likely ion present
intense scarlet red	Li^+ or Sr^{2+}
persistent, intense yellow	Na^+
lilac	K^+
intermittent brick-red	Ca^{2+}
pale green	Ba^{2+}
blue-green	Cu^{2+}

Figure 19.2 Flame tests

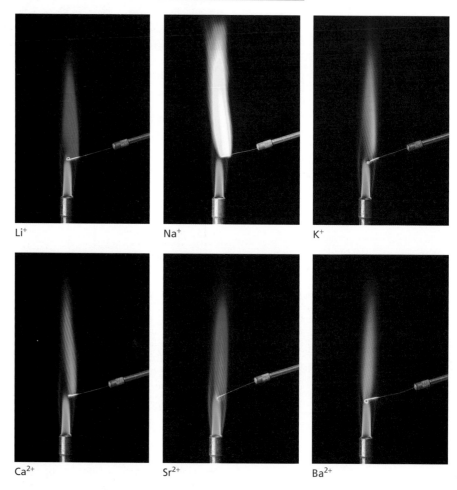

All substances emit light at various visible frequencies, which appear as lines in the spectrum (see section 24.5). Often the lines are distributed across the spectrum and there is no overall predominant colour. Sodium has a single line at 590 nm, which can easily be identified with a direct vision spectroscope. Many substances contain sodium as an impurity and it is difficult to eliminate the yellow colour entirely. Its effect can be partially masked by viewing the colours through blue cobalt glass.

19.7 Precipitation reactions for cations

Hydroxides, except those of Group 1 and ammonium, are insoluble in water and often have characteristic appearances. Some are amphoteric and dissolve in excess sodium hydroxide; others dissolve in excess ammonia to form a complex ion. The addition of aqueous sodium hydroxide (see Table 19.7) or aqueous ammonia (see Table 19.8) to an aqueous solution of the unknown (the tests are useless on the solid) gives valuable information about the cations present.

Reactions with aqueous sodium hydroxide

Table 19.7 A few drops of aqueous sodium hydroxide are added to a solution of the unknown. If a precipitate forms, further sodium hydroxide is added until it is in excess.

Cation		On addition of a few drops of NaOH(aq)	On addition of excess NaOH(aq)
	NH_4^+	no ppt, but smell of NH_3, especially on warming	—
Group 1	Li^{+*}, Na^+, K^+	no ppt	—
Group 2	Mg^{2+}, Ca^{2+}	white ppt	not soluble
	Sr^{2+*}, Ba^{2+*}	no ppt	—
Groups 13 and 14	Al^{3+}, Pb^{2+}	white ppt	soluble
d block	Cr^{3+}	green ppt	soluble
	Mn^{2+}	off-white ppt†	not soluble
	Fe^{2+}	pale green ppt†	not soluble
	Fe^{3+}	brown ppt	not soluble
	Co^{2+}	pink or blue ppt	not soluble
	Ni^{2+}	green ppt	not soluble
	Cu^{2+}	blue ppt	not soluble
	Zn^{2+}	white ppt	soluble
	Ag^+	brown ppt	not soluble

*If the solutions are concentrated, Li^+, Sr^{2+} and Ba^{2+} may give a slight precipitate because their hydroxides are not very soluble (see section 10.4).
†On exposure to air, the off-white precipitate of $Mn(OH)_2$ turns to brown MnO.OH, and the green precipitate of $Fe(OH)_2$ turns to brown FeO.OH.

Figure 19.3 Hydroxide precipitate colours. Can you determine the cation present in each of these solutions?

Summary of equations

- For NH_4^+ ions:

$$NH_4^+(aq) + OH^-(aq) \rightarrow NH_3(g) + H_2O(l)$$

- For Ag^+ ions:

$$2Ag^+(aq) + 2OH^-(aq) \rightarrow Ag_2O(s) + H_2O(l)$$

- For divalent M^{2+} ions:

$$M^{2+}(aq) + 2OH^-(aq) \rightarrow M(OH)_2(s)$$

If the precipitate is soluble in excess:

$$M(OH)_2(s) + 2OH^-(aq) \rightarrow M(OH)_4^{2-}(aq)$$

- For trivalent M^{3+} ions:

$$M^{3+}(aq) + 3OH^-(aq) \rightarrow M(OH)_3(s)$$

If the precipitate is soluble in excess:

$$M(OH)_3(s) + OH^-(aq) \rightarrow M(OH)_4^-(aq)$$

Reactions with aqueous ammonia

Table 19.8 A few drops of aqueous ammonia are added to a solution of the unknown. If a precipitate forms, further ammonia solution is added until it is in excess.

Cation		On addition of a few drops of $NH_3(aq)$	On addition of excess $NH_3(aq)$
Group 1	Li^+, Na^+, K^+	no ppt	—
Group 2	Mg^{2+}	white ppt	not soluble
	Ca^{2+}, Sr^{2+}, Ba^{2+}*	no ppt	—
Groups 13 and 14	Al^{3+}, Pb^{2+}	white ppt	not soluble
d block	Cr^{3+}	green ppt	slightly soluble
	Mn^{2+}	off-white ppt	not soluble
	Fe^{2+}	pale green ppt	not soluble
	Fe^{3+}	brown ppt	not soluble
	Co^{2+}	pink or blue ppt	soluble
	Ni^{2+}	green ppt	soluble
	Cu^{2+}	pale blue ppt	soluble
	Zn^{2+}	white ppt	soluble
	Ag^+	brown ppt	soluble

*Because the OH^- concentration is low, Sr^{2+} and Ba^{2+} ions do not produce a precipitate. A concentrated solution of Ca^{2+} ions may give a faint precipitate.

The ammonia acts in two ways:

- as a weak base, giving a low concentration of OH^- ions, though not sufficient to produce hydroxy complexes
- as a complexing agent, with NH_3 acting as the ligand.

Summary of equations

- For Ag^+ ions:

$$2Ag^+(aq) + 2OH^-(aq) \rightarrow Ag_2O(s) + H_2O(l)$$

With excess aqueous ammonia:

$$Ag_2O(s) + 4NH_3(aq) + H_2O(l) \rightarrow 2Ag(NH_3)_2^+(aq) + 2OH^-(aq)$$

- For divalent M^{2+} ions:

$$M^{2+}(aq) + 2OH^-(aq) \rightarrow M(OH)_2(s)$$

With excess aqueous ammonia:

$$Cu(OH)_2(s) + 4NH_3(aq) + 2H_2O(aq) \rightarrow [Cu(NH_3)_4(H_2O)_2]^{2+}(aq) + 2OH^-(aq)$$
deep blue solution

$$Zn(OH)_2(s) + 4NH_3(aq) \rightarrow Zn(NH_3)_4^{2+}(aq) + 2OH^-(aq)$$
colourless solution

- For trivalent M^{3+} ions:

$$M^{3+}(aq) + 3OH^-(aq) \rightarrow M(OH)_3(s)$$

19.8 Action of acids

Hydrochloric acid

Table 19.9 lists the most common gases given off as a result of treating a solid with aqueous hydrochloric acid.

Table 19.9 Common gases given off when a solid is treated with aqueous hydrochloric acid

Gas evolved	Likely anion present	Typical equation
CO_2	carbonate, CO_3^{2-} or hydrogencarbonate, HCO_3^-	$CuCO_3(s) + 2HCl(aq) \rightarrow CuCl_2(aq) + CO_2(g) + H_2O(l)$ $NaHCO_3(s) + HCl(aq) \rightarrow NaCl(aq) + CO_2(g) + H_2O(l)$
SO_2	sulfite, SO_3^{2-}	$Na_2SO_3(s) + 2HCl(aq) \rightarrow 2NaCl(aq) + SO_2(g) + H_2O(l)$
SO_2 (+ S)	thiosulfate, $S_2O_3^{2-}$	$Na_2S_2O_3(s) + 2HCl(aq) \rightarrow 2NaCl(aq) + SO_2(g) + S(g) + H_2O(l)$
$NO_2 + NO$	nitrite, NO_2^-	$2KNO_2(s) + 2HCl(aq) \rightarrow 2KCl(aq) + NO_2(g) + NO(g) + H_2O(l)$

Hydrogencarbonates can be distinguished from carbonates, as heating a solid hydrogencarbonate gives water as well as carbon dioxide:

$$2NaHCO_3(s) \rightarrow Na_2CO_3(s) + H_2O(l) + CO_2(g)$$

The colourless nitrogen monoxide obtained when nitrite is treated with aqueous hydrochloric acid reacts with oxygen in the air to give more of the brown nitrogen dioxide:

$$NO(g) + \tfrac{1}{2}O_2(g) \rightarrow NO_2(g)$$

If the solid is insoluble in water but dissolves in aqueous hydrochloric acid, it may be a carbonate (if carbon dioxide is given off), a sulfite (if sulfur dioxide is given off) or it may be an insoluble basic oxide. The colour of the resulting solution may help to identify the cation.

Sulfuric acid

If there is no reaction with aqueous hydrochloric acid, a fresh sample of the solid may be treated with a few drops of concentrated sulfuric acid. The reaction must be carried out in a fume cupboard. The mixture may be carefully warmed if there is no reaction in the cold. Table 19.10 shows the likely gases that may be detected.

Table 19.10 Common gases evolved when a solid is treated with concentrated sulfuric acid

Gas	Likely anion present	Typical equation
HCl	chloride, Cl^-	$MgCl_2(s) + 2H_2SO_4(l) \rightarrow Mg(HSO_4)_2(s) + 2HCl(g)$
$HBr + Br_2 + SO_2$	bromide, Br^-	$KBr(s) + H_2SO_4(l) \rightarrow KHSO_4(s) + HBr(g)$ $2HBr(g) + H_2SO_4(l) \rightarrow Br_2(g) + SO_2(g) + 2H_2O(l)$
HI (very little) + $I_2 + SO_2 + H_2S$ (+ S)	iodide, I^-	$NaI(s) + H_2SO_4(l) \rightarrow NaHSO_4(s) + HI(g)$ $2HI(g) + H_2SO_4(s) \rightarrow I_2(g) + SO_2(g) + 2H_2O(l)$ and other reactions
HNO_3	nitrate, NO_3^-	$KNO_3(s) + H_2SO_4(l) \rightarrow KHSO_4(s) + HNO_3(g)$
CO	methanoate, HCO_2^-	$HCO_2H(l) \rightarrow CO(g) + H_2O(l)$
$CO + CO_2$	ethanedioate, $C_2O_4^{2-}$	$H_2C_2O_4(s) \rightarrow CO(g) + CO_2(g) + H_2O(l)$
CH_3CO_2H	ethanoate, $CH_3CO_2^-$	$CH_3CO_2K(s) + H_2SO_4(l) \rightarrow KHSO_4(s) + CH_3CO_2H(g)$

In these reactions, concentrated sulfuric acid acts as a strong non–volatile acid, displacing more volatile acids as gases. With bromides and iodides, it also acts as an oxidising agent. With methanoates and ethanedioates, it also acts as a dehydrating agent.

Bromine gas is red–brown and iodine vapour is purple. Nitric acid vapour is usually pale brown as it decomposes slightly to nitrogen dioxide. It condenses on the cooler part of the test tube as an oily liquid. Carbon monoxide is colourless and burns quietly with an intense blue flame. Ethanoic acid, CH_3CO_2H, can be identified by its smell of vinegar.

19.9 Precipitation reactions for anions

Aqueous barium chloride

Aqueous barium chloride (or barium nitrate) gives a precipitate with a large number of ions, including carbonate, sulfite and sulfate. In the presence of aqueous hydrochloric acid (or aqueous nitric acid), only sulfate ions give a dense white precipitate. So if the addition of aqueous hydrochloric acid and aqueous barium chloride to a solution of the unknown substance gives a dense white precipitate, sulfate ions are present.

$$Ba^{2+}(aq) + SO_4^{2-}(aq) \rightarrow BaSO_4(s)$$

Aqueous silver nitrate

Aqueous silver nitrate gives precipitates with a large number of ions, including carbonate, chromate(VI) (CrO_4^{2-}), hydroxide, chloride, bromide and iodide. However, in the presence of aqueous nitric acid as well, only chlorides, bromides and iodides give precipitation (see Table 19.11). The colours of these precipitates and their solubilities in aqueous ammonia distinguish them from one another.

Table 19.11 The action of silver ions on halide ions

Anion	Precipitate	Effect of NH_3(aq) on precipitate
Cl^-	white	soluble in dilute NH_3(aq)
Br^-	cream or pale yellow	soluble in concentrated NH_3(aq)
I^-	deep yellow	insoluble in concentrated NH_3(aq)

Typical equations

$$Cl^-(aq) + Ag^+(aq) \rightarrow AgCl(s)$$

$$AgCl(s) + 2NH_3(aq) \rightarrow Ag(NH_3)_2^+(aq) + Cl^-(aq)$$

The different colours and solubilities of the silver halides are discussed in section 11.3.

Worked example

Substance **A** is a white amorphous powder. It is insoluble in water. **A** dissolves in aqueous hydrochloric acid, with effervescence, to give a colourless solution **B**. The gas evolved, **C**, turns limewater milky. When aqueous sodium hydroxide was added to solution **B**, a faint white precipitate was formed that did not dissolve in excess. On addition of aqueous ammonia to solution **B**, no precipitate was formed. Substance **A** coloured a Bunsen burner flame green. Explain all these observations, identify **A**, **B** and **C**, and give equations for the reactions that take place.

Answer

The inferences that can be drawn from the observations are summarised in Table 19.12. (It is relatively easy to identify the substance as barium carbonate. However, to answer the question fully, it is important to explain all the observations.)

Table 19.12

Observation	Inference
white, amorphous powder	not a transition metal
insoluble in water	not Group 1; not NO_3^-
dissolves in HCl with effervescence	suggests CO_3^{2-} (or HCO_3^-)
gas turns limewater milky	CO_2 produced from CO_3^{2-} (or HCO_3^-)
solution **B** + NaOH(aq) gives a faint ppt	Sr^{2+} or Ba^{2+} present
solution **B** + NH_3(aq) gives no ppt	Sr^{2+} or Ba^{2+} present
flame test green	Ba^{2+}

A is barium carbonate, $BaCO_3(s)$. (It is not $Ba(HCO_3)_2$, because it is insoluble.)
B is barium chloride solution, $BaCl_2(aq)$.
C is carbon dioxide, $CO_2(g)$.
The equation for the reactions are as follows:

$$BaCO_3(s) + 2H^+(aq) \rightarrow H_2O(l) + CO_2(g) + Ba^{2+}(aq)$$
$$CO_2(g) + Ca(OH)_2(aq) \rightarrow CaCO_3(s) + H_2O(l)$$
$$Ba^{2+}(aq) + 2OH^-(aq) \rightarrow Ba(OH)_2(s)$$
$$\text{(faint ppt)}$$

Now try this

1 **D** is a hygroscopic brown solid. It is readily soluble in water to give a green solution. With aqueous sodium hydroxide, this solution gives a dark green precipitate, **E**, that does not dissolve in an excess of the reagent. **E** turns brown on exposure to air. **D** shows no reaction with aqueous hydrochloric acid, but with concentrated sulfuric acid it frothed and gave off an acidic gas, **F**, as well as a red-brown gas, **G**. When moist air is blown over the mouth of the test tube, **F** gives steamy fumes, and it also gives dense white fumes when a drop of ammonia on a glass rod is brought near. The addition of aqueous nitric acid and silver nitrate to a solution of **D** produced a precipitate that turned green when aqueous ammonia was added.

Explain all these observations, identify **D**, **E**, **F** and **G**, and write equations for the reactions that take place.

2 **H** is a white amorphous powder. It is insoluble in water but dissolves readily in aqueous hydrochloric acid, with no evolution of gas, to give a colourless solution **I**. To a portion of **H**, aqueous sodium hydroxide is added. A white precipitate forms that dissolves in an excess of the reagent. To a separate portion of **H**, aqueous ammonia is added. This also gives a white precipitate that dissolves in an excess of the reagent. A flame test on a sample of **I** is negative.

Explain all these observations, identify **H** and **I**, and write equations for the reactions that take place.

19.10 Organic tests

The solubility of an organic substance in water gives an indication of what group it contains (see section 19.4). Further specific tests can then be used to provide more evidence (see Table 19.13).

Table 19.13 Some common tests used in organic chemistry

Test	Observation	Possible functional group present
shake with bromine water (see page 256)	bromine water decolorised	alkene
add small piece of sodium metal (see pages 284 and 308)	$H_2(g)$ evolved	alcohol, carboxylic acid
add solid PCl_5 (see pages 286 and 309)	HCl evolved	alcohol, carboxylic acid
warm with acidified potassium dichromate (see pages 287–289)	goes from orange to green	primary or secondary alcohol, aldehyde
add sodium carbonate (see page 308)	CO_2 evolved	carboxylic acid
add a few drops of cold acidified $KMnO_4$ (see page 260)	goes from purple to brown to colourless	alkene
add 2,4-DNPH (see page 300)	orange ppt	aldehyde, ketone
warm with Fehling's solution (see page 301)	red ppt	aldehyde
gently warm with Tollens' reagent (see page 301)	silver mirror	aldehyde
boil under reflux with NaOH(aq) and test with HNO_3/$AgNO_3$ (see page 273)	ppt forms	chloride, bromide, iodide
warm with NaOH(aq) and I_2(aq)	yellow ppt forms	methyl ketone or $CH_3CH(OH)^-$

All these functional group tests are described in detail in section 30.5.

Worked example

A colourless liquid **J** dissolves easily in water. Liquid **J** gives steamy fumes when treated with PCl_5. It does not react with sodium carbonate. When warmed with acidified potassium dichromate, the solution turns green. When this green solution is distilled, the distillate **K** gives a precipitate with 2,4-DNPH but does not give a silver mirror with Tollens' reagent.

Explain all these observations, identify **J** and **K**, and write equations for the reactions that take place.

Answer

The inferences that can be drawn from the observations are summarised in Table 19.14.

Table 19.14

Observation	Inference
soluble in water	lower alcohol, carboxylic acid, aldehyde or ketone
gives fumes with PCl_5	alcohol or carboxylic acid
no reaction with Na_2CO_3	not acid, therefore alcohol
potassium dichromate goes green	primary or secondary alcohol
distillate reacts with 2,4-DNPH	aldehyde or ketone
no reaction with Tollens' reagent	not aldehyde

Therefore **J** is a secondary alcohol. It is probably propan-2-ol, $CH_3CHOHCH_3$, as higher secondary alcohols are only slightly soluble in water. **K** would then be propanone CH_3COCH_3.

The equations for the reactions are as follows:

$$CH_3CHOHCH_3 + PCl_5 \rightarrow CH_3CHClCH_3 + POCl_3 + HCl$$
$$CH_3CHOHCH_3 + [O] \rightarrow CH_3COCH_3 + H_2O$$

Now try this

1 A colourless liquid **L** was insoluble in water and had a density greater than 1. It did not decolorise bromine water. When boiled under reflux with sodium hydroxide, some of the liquid dissolved. When the aqueous layer was treated with an excess of nitric acid and silver nitrate, a pale yellow precipitate **M** was formed that was soluble in concentrated ammonia.

   Suggest the group present in **L** and write an equation for the action of concentrated ammonia on **M**.

2 A colourless liquid **O** was soluble in water. It gave fumes when treated with PCl_5. When warmed with acidified potassium dichromate, there was no change in colour. With sodium carbonate it effervesced and gave off a gas.

   Suggest the group present in **O** and write equations for its reactions with PCl_5 and sodium carbonate.

Summary

- Enthalpy changes can be measured with a calorimeter, either directly or indirectly.
- The three most common types of titration are acid–base, redox with potassium manganate(VII) and iodine–thiosulfate.
- Cations can be identified by:
  - the colour of the salt
  - flame tests
  - the action of aqueous sodium hydroxide and aqueous ammonia on a solution.
- Anions can be identified by:
  - the action of heat on the solid
  - the action of $HCl(aq)$ and $H_2SO_4(l)$ on the solid
  - the action of $Ag^+(aq)$ ions or $Ba^{2+}(aq)$ ions on solutions.
- Organic substances can be identified by their solubility in water and by a range of specific tests.

Examination practice questions

Please see the data section of the CD for any A_r values you may need.

1 The carbonates of Group 2 in the Periodic Table decompose on heating forming an oxide and carbon dioxide.
X is any Group 2 cation (e.g. Mg^{2+}).

$$XCO_3 \rightarrow XO + CO_2$$

This decomposition occurs because the positively charged cations polarise (distort) the C—O bond in the carbonate ion causing the ion to break up. The charge density of the Group 2 cations decreases down the group. This affects the decomposition rate.

You are to plan an experiment to investigate how the rate of decomposition of a Group 2 carbonate varies as the group is descended. The rate can be conveniently measured by finding the time taken to produce the same volume of carbon dioxide from each carbonate.

a i Predict how the rate of decomposition of the Group 2 carbonates will change as the group is descended. Explain this prediction in terms of the charge density of the cation as the group is descended.

ii Display your prediction in the form of a sketch graph, clearly labelling the axes. [3]

b In the experiment you are about to plan, identify the following:

i the independent variable

ii the dependent variable. [2]

c Draw a diagram of the apparatus and experimental set up you would use to carry out this experiment. Your apparatus should use only standard items found in a school or college laboratory and show clearly the following:

i the apparatus used to heat the carbonate

ii how the carbon dioxide will be collected. Label each piece of apparatus used, indicating its size or capacity. [2]

d Using the apparatus shown in **c** design a laboratory experiment to test your prediction in **a**. In addition to the standard apparatus present in a laboratory you are provided with the following materials: samples of the carbonates of magnesium, calcium, strontium and barium, a stop-watch/clock with second hand. Give a step-by-step description of how you would carry out the experiment by stating

i the gas volume you would collect from each carbonate

ii how you would calculate the mass of each carbonate to ensure that this volume of carbon dioxide is produced

iii how you would control the factors in the heating so that different carbonates can be compared. [4]

e State a hazard that must be considered when planning the experiment and describe precautions that should be taken to keep risks to a minimum. [2]

f Draw a table with appropriate headings to show the data you would record when carrying out your experiments and the values you would calculate in order to construct a graph to support or reject your prediction in **a**. The headings **must** include the appropriate units. [2]

g This simple experiment is likely to produce only approximate results.
Suggest an improvement to your apparatus or an alternative apparatus that may improve the reliability of the results. [1]

[Cambridge International AS & A Level Chemistry 9701, Paper 51 Q1 June 2011]

2 Hydrated iron(II) sulfate can be represented as $FeSO_4.xH_2O$ where x is the number of molecules of H_2O for each $FeSO_4$. When the compound is heated, it loses the molecules of water leaving anhydrous iron(II) sulfate.
A suggested equation is:

$$FeSO_4.xH_2O(s) \rightarrow FeSO_4(s) + xH_2O(g)$$

An experiment is carried out to attempt to determine the value of x.

- An open crucible is weighed and the mass recorded.
- A sample of hydrated iron(II) sulfate is added to the crucible and the new mass recorded.
- The crucible with hydrated iron(II) sulfate is heated strongly for five minutes and allowed to cool back to room temperature.
- The crucible with the contents is reweighed and the mass recorded.

a Calculate the relative formula masses, M_r, of $FeSO_4$ and H_2O.
[A_r: H, 1.0; O, 16.0; S, 32.1; Fe, 55.8] [1]

b The results of several of these experiments are recorded on the following page. Copy the table and process the results to calculate both the number of moles of anhydrous iron(II) sulfate and the number of moles of water.
Record these values in the additional columns of the table. You may use some or all of the columns. Masses should be recorded to **two decimal places**, while the numbers of moles should be recorded to **three significant figures**. Label the columns you use. For each column you use include units where appropriate and an expression to show how your values are calculated. You may use the column headings A to G for these expressions (e.g. A–B).

A	B	C	D	E	F	G
mass of crucible /g	mass of crucible + FeSO$_4$.xH$_2$O /g	mass of crucible + FeSO$_4$ /g				
15.20	17.03	16.20				
15.10	17.41	16.41				
14.95	17.33	16.25				
15.15	17.70	16.54				
15.05	17.79	16.55				
14.90	17.88	16.53				
14.92	18.18	16.70				
15.30	18.67	17.14				
15.07	18.64	17.02				
15.01	18.80	17.04				

[2]

c Plot a graph to show the relationship between the number of moles of anhydrous iron(II) sulfate, FeSO$_4$ (x-axis), and the number of moles of water (y-axis). Draw the line of best fit. It is recommended that you do not include the origin in your choice of scaling. [3]

d Circle and label on the graph any point(s) you consider to be anomalous. For each anomalous point give a different reason why it is anomalous clearly indicating which point you are describing. [3]

e Determine the slope of the graph. You must mark clearly on the graph any construction lines and show clearly in your calculation how the intercepts were used in the calculation of the slope. [3]

f Comment on the reliability of the data provided in b. [1]

g i Use the value of the slope of your graph calculated in (e) to suggest the correct formula for hydrated iron(II) sulfate.

ii Explain your answer to i. [2]

[Cambridge International AS & A Level Chemistry 9701, Paper 51 Q2 June 2013]

20 Further energy changes

In Topic 5, we showed how we could measure the standard enthalpy change, $\Delta H^\ominus$, of a reaction, and at how this value is determined by the relative bond strengths in the reactants and in the products. In this topic, we look how the $\Delta H^\ominus$ values for ionic reactions are determined by the forces of attraction between the ions. We then show how energy considerations can explain some trends in the properties of Group 2 compounds. The criteria that ΔH is negative cannot be used for endothermic reactions. The concept of entropy as a measure of disorder is introduced and the fact that the total entropy change must be positive for a spontaneous process is discussed. For chemical changes, the concept of Gibbs free energy provides a simpler method of determining the feasibility of a reaction.

Learning outcomes

By the end of this topic you should be able to:

2.3g) explain and use the term *electron affinity* (part, see also Topic 2)

5.1b) explain and use the term *lattice energy* (ΔH negative, i.e. gaseous ions to solid lattice) (part, see also Topic 5)

5.1d) explain, in qualitative terms, the effect of ionic charge and of ionic radius on the numerical magnitude of a lattice energy

5.2a) apply Hess's Law to construct simple energy cycles, and carry out calculations involving such cycles and relevant energy terms, with particular reference to the formation of a simple ionic solid and of its aqueous solution, and Born–Haber cycles (including ionisation energy and electron affinity) (part, see also Topic 5)

5.3a) explain that entropy is a measure of the 'disorder' of a system, and that a system becomes more stable when its energy is spread out in a more disordered state

5.3b) explain the entropy changes that occur during a change in state, e.g. (s) → (l); (l) → (g); (s) → (aq), during a temperature change, and during a reaction in which there is a change in the number of gaseous molecules

5.3c) predict whether the entropy change for a given process is positive or negative

5.3d) calculate the entropy change for a reaction, $\Delta S^\ominus$, given the standard entropies, $S^\ominus$, of the reactants and products

5.4a) define standard Gibbs free energy change of reaction by means of the equation $\Delta G^\ominus = \Delta H^\ominus - T\Delta S^\ominus$

5.4b) calculate $\Delta G^\ominus$ for a reaction using the equation $\Delta G^\ominus = \Delta H^\ominus - T\Delta S^\ominus$

5.4c) state whether a reaction or process will be spontaneous by using the sign of $\Delta G^\ominus$

5.4d) predict the effect of temperature change on the spontaneity of a reaction, given standard enthalpy and entropy changes

10.1f) interpret and explain qualitatively the trend in the thermal stability of the Group 2 nitrates and carbonates in terms of the charge density of the cation and the polarisability of the large anion

10.1g) interpret and explain qualitatively the variation in solubility of the Group 2 hydroxides and sulfates in terms of relative magnitudes of the enthalpy change of hydration and the corresponding lattice energy.

20.1 The Born–Haber cycle – finding the strengths of ionic bonds

Lattice enthalpy

In Topic 5 we saw that the following enthalpy changes are measures of the strengths of different types of bonding:

● intermolecular forces:

$$M(s) \rightarrow M(l) \qquad \text{enthalpy change of fusion, } \Delta H^\ominus_m$$
$$M(l) \rightarrow M(g) \qquad \text{enthalpy change of vaporisation, } \Delta H^\ominus_{vap}$$

- metallic:

$$M(s) \rightarrow M(g) \qquad \text{enthalpy change of atomisation, } \Delta H^{\ominus}_{at}$$

- covalent:

$$X{-}Y(g) \rightarrow X(g) + Y(g) \qquad \text{bond enthalpy, } E(X{-}Y)$$

In the same way, we could use the enthalpy change for the following process:

$$M^+X^-(s) \rightarrow M^+(g) + X^-(g)$$

as a measure of the strength of ionic bonding. It does not measure the strength of a *single* ionic bond between two ions, because there are many ionic bonds of different strengths in all directions in an ionic lattice (in NaCl, for example, each Na^+ ion is surrounded by six Cl^- ions, and it is attracted to each of them), but nevertheless it allows useful comparisons to be made. The convention usually used is to quote the enthalpy change for the reverse process, that is, the enthalpy change when one mole of ionic solid is formed from its isolated gaseous ions. This quantity is known as the **lattice enthalpy (LE)** (or lattice energy). It has a large negative value because the formation of an ionic compound is a strongly exothermic process. For example, for sodium chloride:

$$Na^+(g) + Cl^-(g) \rightarrow NaCl(s) \qquad LE = -787\,kJ\,mol^{-1}$$

The **lattice enthalpy** is the enthalpy change when one mole of the solid is formed from its isolated ions in the gas phase.

Lattice enthalpies are calculated in a similar way to bond enthalpies, except that the additional factor of electron transfer must be taken into account. For sodium chloride, the energy associated with this process is the sum of the following two energies:

- the ionisation energy (IE) of sodium, for the process $Na(g) \rightarrow Na^+(g) + e^-$
- the electron affinity (EA) of chlorine, for the process $Cl(g) + e^- \rightarrow Cl^-(g)$.

We met ionisation energies in section 2.13. The definition of electron affinity is as follows.

The **electron affinity** (EA) of an atom (or ion) is defined as the energy change that occurs when one mole of electrons combines with one mole of gaseous atoms (or ions). That is, the electron affinity is the energy change for the following process:

$$X(g) + e^- \rightarrow X^-(g)$$

If there is a space in an atom's outer shell for an extra electron, the effective nuclear charge usually attracts an electron sufficiently strongly to make its electron affinity exothermic. However, a second electron affinity is always positive because the negative charge on the ion repels the second electron. For example, for oxygen:

$$O(g) + e^- \rightarrow O^-(g) \qquad \Delta H = -141\,kJ\,mol^{-1}$$
$$O^-(g) + e^- \rightarrow O^{2-}(g) \qquad \Delta H = +790\,kJ\,mol^{-1}$$

Adding these two gives:

$$O(g) + 2e^- \rightarrow O^{2-}(g) \qquad \Delta H = +649\,kJ\,mol^{-1}$$

Figure 20.1 shows how the lattice enthalpy of sodium chloride may be worked out. Such a cycle constructed to find a lattice enthalpy is called a **Born–Haber cycle**.

The values of all the terms in this Born–Haber cycle can be measured, with the exception of the lattice enthalpy. (It is not physically possible to produce a mole of isolated gaseous ions – the electrostatic forces would be too great – let alone

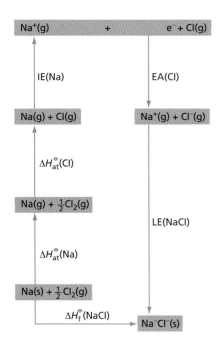

Figure 20.1 The Born–Haber cycle for sodium chloride

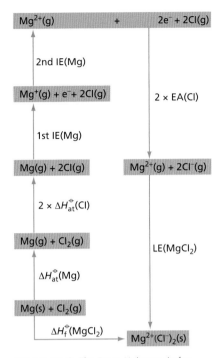

Figure 20.2 The Born–Haber cycle for magnesium chloride

condense them into a mole of solid lattice.) But the lattice enthalpy can be calculated as follows:

$$\Delta H_{at}^{\ominus}(Na) + \Delta H_{at}^{\ominus}(Cl) + IE(Na) + EA(Cl) + LE(NaCl) = \Delta H_f^{\ominus}(NaCl)$$
$$+107.3 + 121.3 + 495.8 + (-348.6) + LE(NaCl) = -411.2$$
$$LE(NaCl) = -411.2 - (+375.8)$$
$$= -787 \, kJ \, mol^{-1}$$

The situation is slightly more complicated if the ions carry more than one charge. For example, for magnesium chloride, the energy change associated with the process:

$$Mg(g) \rightarrow Mg^{2+}(g) + 2e^-$$

is the sum of the first and second ionisation energies of magnesium (it is *not* just the second ionisation energy of magnesium). In addition, as the formula is $MgCl_2$, we need to produce two moles of chloride ions. The Born–Haber cycle is shown in Figure 20.2.

$$\Delta H_{at}^{\ominus}(Mg) + 2 \times \Delta H_{at}^{\ominus}(Cl) + 1st \, IE(Mg) + 2nd \, IE(Mg) + 2 \times EA(Cl) + LE(MgCl_2)$$
$$= \Delta H_f^{\ominus}(MgCl_2)$$
$$+147.1 + 2 \times 121.3 + 737.7 + 1450.7 + (2 \times -348.6) + LE(MgCl_2)$$
$$= -641.3$$
$$LE(MgCl_2) = -641.3 - (+1880.9)$$
$$= -2522 \, kJ \, mol^{-1}$$

Now try this

Construct a Born–Haber cycle for each of the following compounds, and use it to calculate the lattice energy. Use the data given below in your calculations (all in $kJ \, mol^{-1}$).
1 Na_2O
2 MgO

[$IE(Na) = +494$; $IE(Mg) = +736$; $IE(Mg^+) = +1450$; $EA(O) = -141$; $EA(O^-) = +790$; $\Delta H_f^{\ominus}(Na_2O) = -414$; $\Delta H_f^{\ominus}(MgO) = -602$; $\Delta H_{at}^{\ominus}(Na) = +107$; $\Delta H_{at}^{\ominus}(Mg) = +147$; $E(O{=}O) = +496$]

The magnitude of the lattice enthalpy

The value of the lattice enthalpy of magnesium chloride is much larger than that for sodium chloride. There are two reasons for this. The first is that there are more cation-to-anion attractions because there are twice as many chloride ions. The second reason is that each of these attractions is much stronger, because the magnesium ion carries twice the charge of the sodium ion.

There are other factors that determine the size of the lattice enthalpy. The most important of these is how closely the ions approach each other (the sum of their ionic radii). The effects on the lattice enthalpy of the distance between the ions and the charges on the ions are illustrated in Table 20.1.

Table 20.1 For ionic compounds of the same type, the shorter the inter-ionic distance, the greater the value of the lattice enthalpy. Compounds containing doubly or triply charged ions have much larger values of lattice energy.

Compound	Inter-ionic distance/nm	Charges on ions	Lattice energy/kJ mol⁻¹
LiF	0.211	1, 1	−1036
NaCl	0.279	1, 1	−787
CsI	0.385	1, 1	−604
BeF₂	0.167	2, 1	−3505
MgCl₂	0.259	2, 1	−2522
BaI₂	0.363	2, 1	−1877
Li₂O	0.210	1, 2	−2814
MgO	0.210	2, 2	−3791
Al₂O₃	0.189	3, 2	−16470

Lattice enthalpies are determined by the magnitudes of:
- the charges on the ions
- the inter-ionic distance
- the type of lattice.

Worked example

a Write an equation that represents the chemical change associated with the lattice enthalpy of calcium fluoride.

b Which two elements, one from Group 1 and the other from Group 17, form a compound with the most negative value of lattice enthalpy?

Answer

a $Ca^{2+}(g) + 2F^-(g) \rightarrow CaF_2(s)$

b Lithium and fluorine, because they have the smallest ionic radii.

Now try this

1 a Draw a Born–Haber cycle for barium sulfide, BaS.

  b Use the following values (all in $kJ\,mol^{-1}$) to calculate the lattice enthalpy of barium sulfide:

  $\Delta H_{at}(Ba) = +180$; $\Delta H_{at}(S) = +249.2$; 1st IE(Ba) = +502; 2nd IE(Ba) = 966; 1st EA(S) = −141.4; 2nd EA(S) = +790.8; $\Delta H_f(BaS) = −595.8$

2 Explain why the following substances have different values of lattice enthalpy from that of sodium chloride:

  a potassium chloride   b magnesium sulfide   c calcium chloride.

Ceramics

Ceramics are substances with giant structures that contain either strong covalent bonds with some ionic character (e.g. SiO_2) or strong ionic bonds with some covalent character (e.g. MgO and Al_2O_3). They are hard but brittle. They are hard because of the strong interatomic forces that hold the crystal together, and brittle because any deformation brings similarly charged ions closer together and leads to increased repulsion. If the ions have multiple charges, the lattice energy, melting point and Young modulus (a measure of stiffness) all increase, as Table 20.2 shows.

Table 20.2 The properties of some ionic substances

Substance	Charges on ions	Lattice energy/ $kJ\,mol^{-1}$	Melting point/°C	Young modulus/ $GN\,m^{-2}$
NaCl	1, 1	−787	801	44
MgO	2, 2	−3791	2852	245
Al_2O_3	3, 2	−16470	2072	525

The high melting points mean that ceramics are used for furnace linings. The linings for blast furnaces are made of bricks with a high content of aluminium oxide, and those for steel converters are mainly magnesium oxide, made by heating magnesium carbonate (see section 10.4).

As the charges in the ionic substance increase, so does the covalent character of the bonds. By the time the charge reaches +4, as in Si^{4+}, the bond can be considered to be almost totally covalent. Ceramics such as silicon(IV) oxide, SiO_2, silicon(IV) nitride, Si_3N_4, silicon(IV) carbide, SiC, and tungsten carbide, WC, have high Young moduli, and are useful in adding stiffness to plastics or metals. For example:

- glass fibres embedded in epoxy resins ('glass-reinforced plastic', GRP) are used in boat construction
- the tiles covering the space shuttle contain silicon(IV) oxide fibres
- the blades in some jet engines are stiffened with silicon(IV) carbide.

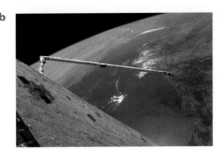

Figure 20.3 Ceramics provide many useful properties.
a Fine china is hard but brittle.
b The tiles covering the space shuttle have an extremely low thermal conductivity, and so insulate the inside of the shuttle from the heat at re-entry.
c GRP is tough and does not corrode, and so is used extensively in boat-building.

Some ceramics can lose their electrical resistance completely when cooled and act as **superconductors** while still at relatively high temperatures compared to those near to absolute zero that are needed to cause superconductivity in metals. These ceramics are mixtures of metal oxides, usually including copper oxide. One of them, with the approximate formula $YBa_2Cu_3O_7$, is a superconductor at $90\,K$. In the future, these superconducting ceramics could be used to transmit electricity over large distances without any loss in energy, or to make highly efficient electrical devices.

20.2 Enthalpy changes of solution and hydration

Enthalpy change of solution

The solubilities of salts in water show wide variations, with no obvious pattern. One of the factors determining solubility is whether the **enthalpy change of solution**, ΔH_{sol}, of the salt is positive or negative. This enthalpy change is the energy associated with the process:

$$M^+X^-(s) \rightarrow M^+(aq) + X^-(aq)$$

We must be careful to specify the dilution of the final solution when quoting this enthalpy change. On dilution, the ions in a solution become more extensively hydrated (an exothermic process) and also move further apart (an endothermic process). The relative importance of these two effects changes with dilution, affecting the values of ΔH_{sol} in a complicated way. A quoted single value of ΔH_{sol} refers to an 'infinitely dilute' solution. This value cannot be determined directly by experiment and must be found by a process of extrapolation. In practice, there comes a point when further dilution has no measurable effect on the value of ΔH_{sol}, and this may be taken as infinite dilution.

> The **enthalpy change of solution**, ΔH_{sol}, is the enthalpy change when one mole of solute dissolves in an infinite volume of water.

If ΔH_{sol} is positive, the salt is likely to be insoluble or of very low solubility. If ΔH_{sol} is approximately zero or negative, then the salt is likely to be soluble or very soluble.

Enthalpy change of hydration

The value of ΔH_{sol} depends on two energy changes. The first is the energy that holds the lattice together, that is, the lattice energy. The second is the energy given out when gaseous ions are hydrated. This is called the **enthalpy change of hydration**, ΔH_{hyd}.

> The **enthalpy change of hydration**, ΔH_{hyd}, is the enthalpy change when one mole of gaseous ions dissolves in an infinite volume of water.

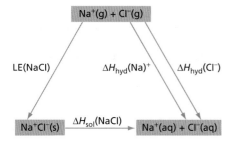

Figure 20.4 Hess's Law cycle for the dissolving in water of sodium chloride

Table 20.3 Absolute enthalpy changes of hydration for some common ions

Ion	ΔH_{hyd}/kJmol^{-1}
H$^+$	−1120
Li$^+$	−544
Na$^+$	−435
Mg^{2+}	−1980
Ca^{2+}	−1650
Al^{3+}	−4750
F$^-$	−473
Cl$^-$	−339

For sodium chloride, we can construct the cycle shown in Figure 20.4. The value of the lattice enthalpy is found using the Born–Haber cycle (see Figure 20.1); it is −787 kJ mol^{-1}. ΔH_{sol} can be found experimentally: its value is +3.9 kJ mol^{-1}. Therefore:

$$\Sigma(\Delta H_{hyd}) = LE(NaCl) + \Delta H_{sol}(NaCl)$$
$$= -787 + 3.9$$
$$= -783.1 \, kJ \, mol^{-1}$$

where $\Sigma(\Delta H_{hyd}) = \Delta H_{hyd}(Na^+) + \Delta H_{hyd}(Cl^-)$.

We can now see why similar substances can show wide variations in solubility. Both the lattice enthalpy and the enthalpy change of hydration are large negative quantities. A small percentage change in either of these has a very large effect on the difference between the two terms: a 0.5% change in LE would alter ΔH_{sol} by 100%, for example. These variations can be used to explain the differences in solubility of some compounds of Group 2 (see section 20.6).

The Hess's Law cycle does not enable us to find ΔH_{hyd} for the individual ions, because it provides only the *sum* of the enthalpy changes of hydration for the cation and anion. In order to find their individual values, we must assign a value to one of the ions. The value for the H$^+$ ion is generally agreed to be −1120 kJ mol^{-1} and, by using this value, we can calculate the **absolute enthalpy change of hydration** of ions, as shown in Table 20.3.

The values in Table 20.3 show that the absolute enthalpy change of hydration depends on the ability of the ion to attract water molecules. Small, highly charged ions have the most negative values of ΔH_{hyd} (see section 4.9).

Enthalpy changes of hydration are more negative:
- if the ion is small
- if the ion has two or three units of charge on it.

Worked example

a Draw an energy cycle to show the connection between solid magnesium chloride, its gaseous ions and its aqueous ions.

b Use the following values to calculate the value of $\Delta H_{hyd}(MgCl_2)$:

   LE(MgCl$_2$) = −2375 kJ mol^{-1}; $\Delta H_{sol}(MgCl_2)$ = −155.1 kJ mol^{-1}

c Compare this value with the value obtained from the separate absolute enthalpy changes of hydration for Mg^{2+} and Cl$^-$ in Table 20.3.

Answer

a See Figure 20.5.

b From Figure 20.5:

$$\Delta H_{hyd}(MgCl_2) = \Delta H_{hyd}(Mg^+) + 2\Delta H_{hyd}(Cl^-)$$
$$= LE + \Delta H_{sol}$$
$$= -2375 - 155.1$$
$$= \mathbf{-2530 \, kJ \, mol^{-1}}$$

c From Table 20.3:
$$\Delta H_{hyd}(Mg^+) + 2\Delta H_{hyd}(Cl^-) = -1980 + 2 \times (-339)$$
$$= \mathbf{-2658 \, kJ \, mol^{-1}}$$

This value agrees with that in **b** to within 5%: a fair agreement

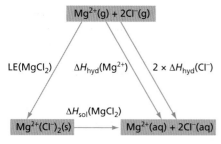

Figure 20.5

> ### Now try this
>
> 1 Draw a Born–Haber cycle for aluminium fluoride, $AlF_3(s)$.
> 2 Use the following values (all in $kJ\,mol^{-1}$) to calculate the lattice energy of aluminium fluoride:
>
> $\Delta H_f^{\ominus}(AlF_3) = -1504.1$; $\Delta H_f^{\ominus}(Al) = 324.3$; $\Delta H_f^{\ominus}(F) = 79.0$; 1st IE(Al) = 577; 2nd IE(Al) = 1820; 3rd IE(Al) = 2740; EA(Cl) = −328
>
> 3 Draw an energy diagram to show the connection between $AlF_3(s)$, its gaseous ions and its aqueous ions.
> 4 Use the figures in Table 20.3 and the lattice enthalpy value obtained in **2** to calculate $\Delta H_{sol}(AlF_3)$.

20.3 The concept of entropy

Spontaneous and reversible processes

A **spontaneous change** is one that, in the absence of any barrier, such as activation energy, takes place naturally in the direction stated. Not all chemical changes that occur spontaneously are exothermic, and so the principle that ΔH is negative cannot always be used as the criterion of feasibility. In order to find a more general principle, which will enable us to predict if a reaction is spontaneous or not, we need to consider what happens to the energy that is released or absorbed in a chemical reaction.

Spontaneous processes often take place because the potential energy of the system decreases – for example, a weight falls to the ground or an electric current flows through a resistor from a point of high potential to one of lower potential. In both these examples, the reduction in potential energy is accompanied by an increase in heat, which is random kinetic energy. Potential energy is much more ordered than heat energy. In potential energy the energy mostly moves in a definite direction: in heat energy the energy is random and moves in every direction. Random motion is much more probable than directed motion and that is why systems tend to become more disordered.

The measure of this disorder is called **entropy**, and is given the symbol S. Formally a change of entropy is defined using the equation $\Delta S = q/T$, where q is the amount of heat transferred at an absolute temperature T. We can see why the temperature is important. An amount of random heat energy q added to a relatively ordered system at a low temperature T_1 causes more disorder than if the same amount of heat was added to a much less ordered system at a higher temperature T_2. An analogy that has been used is that a sneeze or cough causes more disturbance in a quiet library than a sneeze or cough in a noisy, busy street.

The general principle for a spontaneous change is that the **total entropy change ΔS_{tot} is positive**. This is known as the Second Law of Thermodynamics (the Law of Conservation of Energy being the first).We know that heat spontaneously flows from a hot body at temperature T_2 to a cold body at temperature T_1. The removal of an amount of energy at temperature T_2 is accompanied by an entropy decrease $-q/T_2$. The addition of this energy to a cold body at temperature T_2 is accompanied by an entropy increase of $+q/T_2$. The total entropy change is $-q/T_2 + q/T_1$ (see Figure 20.6). This is positive if $T_2 > T_1$, which agrees with the observation that heat flows spontaneously form a hot to a cold body. If $T_2 = T_1$, the system is in equilibrium but any heat transfer is so infinitely slow that no useful work can be obtained from it. This special condition when $\Delta S_{tot} = 0$ is known as a **reversible change**.

Figure 20.6 When a quantity of heat q flows from a hot body at temperature T_2 to a cold body at temperature T_1, $\Delta S_{tot} = -q/T_2 + q/T_1$. This is positive if $T_2 > T_1$.

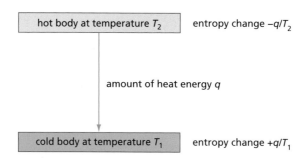

hot body at temperature T_2 — entropy change $-q/T_2$

amount of heat energy q

cold body at temperature T_1 — entropy change $+q/T_1$

Entropy values

Unlike H, S can be given an absolute value. Using the Third Law of Thermodynamics, which states that **at absolute zero the entropy of a pure, crystalline substance is zero**, we have a base from which we can measure absolute entropy values. This law fits in with our idea that under these conditions a system is at the most ordered that it can be; it also has experimental support from measurements at very low temperatures. Values for some elements are listed in Table 20.4. As entropy change is q/T, its units are $\mathrm{J\,mol^{-1}\,K^{-1}}$.

Table 20.4 Values of absolute entropies $S^{\ominus}$ for some elements. These are at 298 K and 1.0 bar.

Element	He	Ar	C(diamond)	C(graphite)	Na	K	Mg	Fe
$S^{\ominus}/\mathrm{J\,mol^{-1}\,K^{-1}}$	126.0	154.7	2.4	5.7	51.2	64.2	32.7	27.3

If we look at Table 20.4, we can find some trends about absolute entropies. These may be summarised as follows. $S^{\ominus}$ values:

- increase as A_r increases
- increase from solids to liquids to gases
- decrease if the element is soft.

We find similar trends in the entropies of the compounds listed in Table 20.5.

Table 20.5 Values of absolute entropies for some compounds

Compound	NaCl(s)	Na_2CO_3(s)	KCl(s)	H_2O(l)	C_2H_5OH(l)	CO(g)	CO_2(g)
$S^{\ominus}/\mathrm{J\,mol^{-1}\,K^{-1}}$	72.1	135.0	82.6	69.9	160.7	197.6	213.6

Some entropy changes

An entropy change is defined as q/T. We can use this to calculate the increase in entropy when 10 kJ of heat flows from a hot body at 500 K to a cold body at 400 K. We have:

$$\text{entropy change of hot body} = \frac{-10\,000}{500} = -200\,\mathrm{J\,mol^{-1}\,K^{-1}}$$

$$\text{entropy change of cold body} = \frac{+10\,000}{400} = +250\,\mathrm{J\,mol^{-1}\,K^{-1}}$$

$$\text{total entropy change} = -200 + 250 = +50\,\mathrm{J\,mol^{-1}\,K^{-1}}$$

Another example when it is easy to calculate the entropy change is the melting of a solid at its melting point or the evaporation of a liquid at its boiling point. Under these conditions the change is reversible and so ΔS_{tot} is zero. This means that $\Delta S = q/T$, where q is the molar latent heat. At 273 K, the latent heat of fusion of ice is 6.01 kJ mol^{-1}. This gives the entropy change as $6010/273 = 22.0\,\mathrm{J\,mol^{-1}\,K^{-1}}$.

Entropy change of a chemical reaction

The entropy change of chemical reaction can be calculated in a similar way to that used for calculating enthalpy changes. But we must remember that although the standard enthalpy change of formation of elements is zero, their standard entropies are not.

> **Now try this**
>
> What is the entropy change when 30 kJ of heat flows from a hot body at 100 °C to a cold body at 27 °C?

> **Now try this**
>
> 1 The latent heat of vaporisation of water is 2260 J g^{-1}. What is its entropy change of evaporation at 100 °C in J mol^{-1} K^{-1}?
> 2 The latent heat of fusion of lead is 23.0 J g^{-1} at 327 °C. What is its entropy of fusion?

Figure 20.7 Calculating the entropy change for the formation of sodium chloride. At 0 K, $\Delta S = 0$ as ΔS for all the substances = 0.

Worked example

What is $\Delta S^\ominus$ for the reaction $2Na(s) + Cl_2(g) \rightarrow 2NaCl(s)$?

Answer

Figure 20.7 shows the entropy changes for the reaction.

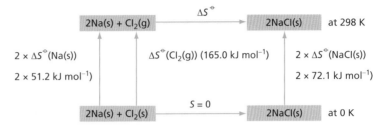

From Figure 20.7:

$$2 \times 51.2 + 165.0 + \Delta S^\ominus = 2 \times 72.1$$
$$\Delta S^\ominus = 142.2 - 267.4 = -125.2$$
$$= \mathbf{-125.2\, J\, mol^{-1}\, K^{-1}}$$

It is not surprising that the reaction is accompanied by an entropy decrease as a gas is being used up. This entropy decrease is only possible because the reaction is so exothermic that the entropy of the heat given off more than compensates for this decrease.

Now try this

What is $\Delta S^\ominus$ for the following reactions?

1 $CaCO_3(s) \rightarrow CaO(s) + CO_2(g)$
2 $C_2H_4(g) + H_2O(l) \rightarrow C_2H_5OH(l)$
3 $N_2(g) + 3H_2(g) \rightarrow 2NH_3(g)$
[$\Delta S^\ominus$/J mol^{-1} K^{-1}: CaCO$_3$(s) 92.9; CaO(s) 39.7; H$_2$O(l) 69.9; C$_2$H$_5$OH(l) 160.7; H$_2$O(g) 130.6; H$_2$(g) 130.6; N$_2$(g) 191.6; NH$_3$(g) 192.3; C$_2$H$_4$(g) 219.5]

Some other examples of entropy changes for different types of reactions are given in Table 20.6.

Table 20.6 Some examples of entropy changes

Example	$\Delta S^\ominus$/J K^{-1} mol^{-1}	Type of change
mixing two gases (one mole of each)	22	mixing
$C_6H_6(s) \rightarrow C_6H_6(l)$	3.6	melting
$C_6H_{14}(l) \rightarrow C_6H_{14}(g)$	88	evaporation
$K(s) + I_2(s) \rightarrow KI(s)$	−17	combination
$MgCO_3(s) \rightarrow MgO(s) + CO_2(g)$	175	decomposition
$NaCl(s) \rightarrow Na^+(aq) + Cl^-(aq)$	43	solution
$Ca^{2+}(aq) + CO_3^{2-}(aq) \rightarrow CaCO_3(s)$	−204	precipitation

20.4 Gibbs free energy

In a chemical reaction, we have to consider the entropy changes that take place both in the system (that is the reaction that is happening) and in the surroundings (which absorbs or gives out heat). Many chemical reactions are exothermic and the entropy increase in the surroundings more than compensates for any entropy decrease that may be taking place in the system. We have:

$$\Delta S^\ominus_{total} = \Delta S^\ominus_{system} + \Delta S^\ominus_{surroundings}$$

$$\Delta S^\ominus_{surroundings} = \frac{q}{T} = \frac{-\Delta H}{T}$$

It is always possible to calculate the total entropy change from these two terms but it is more convenient if they are combined together in a single function. This function is called the Gibbs free energy function after Josiah Willard Gibbs (see Figure 20.8) is given the symbol ΔG.

Figure 20.8 Josiah Willard Gibbs (1839–1903), the father of chemical thermodynamics

It is a measure of the total entropy increase associated with the reaction and is a measure of the driving force of that reaction. If the reaction has a decrease in free energy, it is feasible. We have:

$$\Delta S^{\ominus}_{total} = \Delta S^{\ominus}_{system} + \Delta S^{\ominus}_{surroundings}$$

$$\frac{-\Delta G^{\ominus}}{T} = \Delta S^{\ominus}_{system} - \frac{\Delta H^{\ominus}}{T}$$

This equation can be written in the more familiar form:

$$\Delta G^{\ominus} = \Delta H^{\ominus} - T\Delta S^{\ominus}$$

We can use this equation as follows:

- if $\Delta G^{\ominus}$ is negative, the reaction is feasible
- if $\Delta G^{\ominus} = 0$, the reaction is in equilibrium
- if $\Delta G^{\ominus}$ is positive, the reaction will not go spontaneously to completion.

The entropy increase associated with a decrease in $\Delta G^{\ominus}$ can be given another interpretation. Whereas $-\Delta H^{\ominus}$ is a measure of the total heat energy q lost from the system, $-\Delta G^{\ominus}$ is the *maximum* amount of that total energy that is available to do work. Usually $\Delta G^{\ominus}$ is less negative than $\Delta H^{\ominus}$ but occasionally the reverse is true.

Calculating $\Delta G^{\ominus}$

We can calculate $\Delta G^{\ominus}$ for a reaction in a similar way to that we calculated $\Delta H^{\ominus}$, using free energies of formation rather than enthalpies of formation.

Worked example

Calculate $\Delta G^{\ominus}$ for the reaction $C_2H_6(g) + 4\frac{1}{2}O_2(g) \rightarrow 2CO_2(g) + 3H_2O(l)$

Answer

Figure 20.9 shows the free energies of formation for the reaction.

Figure 20.9 Calculating $\Delta G^{\ominus}$ from $\Delta G^{\ominus}_f$ values

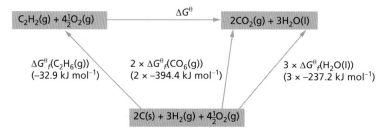

From Figure 20.9:

$$-32.9 + \Delta G^{\ominus} = -788.8 - 711.6$$
$$\Delta G^{\ominus} = -1500.4 - (-32.9)$$
$$= \mathbf{-1467.5\,kJ\,mol^{-1}}$$

(We should not be surprised that $\Delta G^{\ominus}$ is so large and negative. The reaction is highly exothermic and this more than compensates for $\Delta S^{\ominus}$ being negative because of the reduction in the number of gas molecules.)

Now try this

Calculate $\Delta G^{\ominus}$ for

1 the combustion of propane
   [$\Delta G^{\ominus}_f(C_3H_8(g)) = -23.4\,kJ\,mol^{-1}$]
2 the chlorination of ethane to give ethyl chloride.
   [$\Delta G^{\ominus}_f/kJ\,mol^{-1}$: $C_2H_6(g)$ −32.9; $C_2H_5Cl(g)$ −52.9; $HCl(g)$ −95.2]

Variation of $\Delta G^{\ominus}$ with temperature

The values of $\Delta H^{\ominus}$ and $\Delta S^{\ominus}$ for a reaction vary slightly with changes in temperature, but the effect is very small compared to the effect that changes in temperature have on $\Delta G^{\ominus}$. Increasing temperature makes the $T\Delta S^{\ominus}$ term increasingly important. For example, over the temperature range 298 K to 1500 K, $\Delta H^{\ominus}$ and $\Delta S^{\ominus}$ for the decomposition of calcium carbonate change by only a few per cent but $\Delta G^{\ominus}$ changes from $+131\,kJ\,mol^{-1}$ at 298, to $+16\,kJ\,mol^{-1}$ at 1000 K and $-64\,kJ\,mol^{-1}$ at 1500 K. To a good approximation, we can regard $\Delta H^{\ominus}$ and $\Delta S^{\ominus}$ as constant and calculate how $\Delta G^{\ominus}$ changes with temperature using the equation $\Delta G^{\ominus} = \Delta H^{\ominus} - T\Delta S^{\ominus}$.

Worked example

For the reaction $C_2H_4(g) + HCl(g) \rightarrow C_2H_5Cl(g)$, at 298 K,
$\Delta H^\ominus = -96.7\,kJ\,mol^{-1}$ and $\Delta G^\ominus = -25.9\,kJ\,mol^{-1}$.
Calculate the value of $\Delta G^\ominus$ at 1000 K.

Answer

$$\Delta G^\ominus = \Delta H^\ominus - T\Delta S^\ominus$$
At 298 K, $-T\Delta S^\ominus = -25.9 - (-96.7) = +70.8\ kJ\,mol^{-1}$
$$\Delta S^\ominus = \frac{-70800}{298} = -239\,J\,mol^{-1}\,K^{-1}$$
At 1000 K, $\Delta G^\ominus = -96.7 - \left(1000 \times \dfrac{-239}{1000}\right)$
$$= \mathbf{+142\,kJ\,mol^{-1}}$$

This shows that as the temperature increases, the addition reaction becomes less feasible.

Now try this

1. Calculate the temperature at which $\Delta G^\ominus$ for the addition of HCl to ethene becomes zero.
2. When KNO_3 dissolves in water at 298 K, $\Delta H^\ominus = +34.8\,kJ\,mol^{-1}$ and $\Delta G^\ominus = +0.3\,kJ\,mol^{-1}$.
   a. Calculate $\Delta S^\ominus$.
   b. Calculate $\Delta G^\ominus$ at 320 K.
   c. Comment on the effect of temperature on the solubility of KNO_3.
3. Heptane is converted into methylbenzene industrially:

   $$C_7H_{16}(l) \rightarrow C_7H_8(l) + 4H_2(g)$$

   At 300 K, $\Delta H^\ominus = +211\,kJ\,mol^{-1}$ and $\Delta G^\ominus = +110\,kJ\,mol^{-1}$.

   Estimate $\Delta G^\ominus$ at:
   a. 600 K
   b. 900 K
   c. Explain why your estimated values will be unreliable.

Applying free energy changes

The free energy of a reaction tells us whether the reaction proceeds spontaneously and this depends on the signs and magnitudes of both $\Delta H^\ominus$ and $\Delta S^\ominus$ for the reaction. As $\Delta H^\ominus$ and $\Delta S^\ominus$ can each be positive or negative, there are four possibilities.

$\Delta H^\ominus$ and $\Delta S^\ominus$ both positive

These reactions are said to be entropy driven. These endothermic reactions, which may not be feasible at room temperature, become feasible if the temperature is raised (Le Chatelier's Principle). The following are some examples:

- melting and boiling
- decomposition reactions
- electrolysis
- dissolving (in some cases).

It is easy to see why both $\Delta H^\ominus$ and $\Delta S^\ominus$ are positive for melting and boiling (see Figure 20.10). The change is endothermic because intermolecular bonds are being broken. There is an increase in the entropy because disorder increases from solid to liquid to gas.

Most decomposition reactions (for example, the cracking of alkanes and the thermal decomposition of calcium carbonate) are endothermic because the total bond enthalpy in the products is less than that in the reactants. The energy required to break relatively strong bonds is not recovered by the formation of fewer or weaker bonds. Decomposition reactions are accompanied by an increase in entropy because the change in number of molecules, Δn, is positive.

Figure 20.10 ΔH and ΔS both positive: **a** melting ice and **b** boiling water

a

b

Decomposition may be brought about by electrolysis rather than by heating. The reaction can then take place at a temperature at which $\Delta G^{\ominus}$ is positive, because it is being driven by the passage of an electric current through a potential difference. For the electrolysis of water, the minimum voltage needed to bring about decomposition is 1.23 V. This point is discussed in more detail in Topic 23.

Liquids of similar polarity (for example, hexane and heptane) mix together in all proportions. If the polarity of the liquids is different, however, $\Delta H^{\ominus}$ becomes so large and positive that they may be only partially miscible (for example, butan-1-ol and water). The solubility then increases if the temperature is raised. This is usually the situation when a covalent solid dissolves in a liquid (for example, benzoic acid in water).

The dissolving of ionic solids in water is an extremely complex process. Because ions hydrate to a greater extent in dilute solution, dissolving an ionic solid in water may be exothermic when the solution is dilute, but endothermic when it becomes saturated. Two entropy terms operate:

- an entropy increase because the ions in the solid are free to move in solution
- an entropy decrease because water molecules that were originally free to move become restricted by hydration of the ions.

This complex interplay between enthalpy and entropy factors makes it very difficult to explain why some ionic compounds are very soluble in water while others are highly insoluble. For ions with a single charge, the overall entropy change is usually positive and this means that compounds of Group 1 are water soluble, even when the enthalpy of solution is positive. For ions with a multiple charge, the overall entropy change is usually negative. Many compounds in Group 2 have a positive enthalpy of solution and combining this with an adverse entropy term makes them insoluble (see page 345).

$\Delta H^{\ominus}$ and $\Delta S^{\ominus}$ both negative

These reactions are said to be enthalpy driven. These exothermic reactions are feasible at low temperatures. Here are some common examples:

- condensation and freezing (see Figure 20.11)
- addition and combination reactions
- electrochemical cells
- precipitation.

These reactions are the reverse of the reactions in which both $\Delta H^{\ominus}$ and $\Delta S^{\ominus}$ are positive. Although addition and combination reactions are feasible at room temperature, it may be that the rate of reaction is then so slow that a catalyst has to be used. An example of this is the catalytic hydrogenation of an alkene.

a

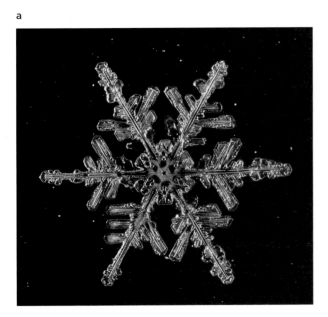

b

Figure 20.11 ΔH and ΔS both negative: **a** formation of a snow crystal and **b** rain from clouds

$\Delta H^{\ominus}$ negative and $\Delta S^{\ominus}$ positive

These reactions are feasible at all temperatures. The reactants are said to be **metastable** under all conditions because they exist only because the activation energy of the reaction is so high. They are thermodynamically unstable, but kinetically inert. The following are some examples:

- a few decomposition reactions
- organic combustion reactions
- combustion of explosives (see Figure 20.12 and the panel on page 343).

Figure 20.12 ΔH negative and ΔS positive: **a** fireworks and **b** explosive decomposition of ammonium nitrate

a

b

Some substances decompose exothermically because the total bond enthalpy in the products is higher than that in the reactants. Examples include the decomposition of ozone and dinitrogen oxide:

	$\Delta H^{\ominus}$/kJ mol^{-1}	$\Delta S^{\ominus}$/J K^{-1} mol^{-1}
$O_3(g) \rightarrow 1\frac{1}{2}O_2(g)$	−142.7	+68.7
$N_2O(g) \rightarrow N_2(g) + \frac{1}{2}O_2(g)$	−82.0	+74.4

This means that ozone and dinitrogen oxide cannot be synthesised directly by the reverse of the reactions above. Ozone is made by the combination of oxygen atoms with oxygen molecules, and dinitrogen oxide is made by the thermal decomposition of ammonium nitrate:

$$NH_4NO_3(s) \rightarrow N_2O(g) + 2H_2O(g)$$

If the reaction rapidly gives off large quantities of gas at high temperature, an explosion may result. An example is the decomposition of propane-1,2,3-trinitrate (commonly called nitroglycerine):

	$\Delta H^{\ominus}$/kJ mol^{-1}	$\Delta S^{\ominus}$/J K^{-1} mol^{-1}
$2C_3H_5N_3O_9 \rightarrow 3N_2 + 5H_2O + 6CO_2 + \frac{1}{2}O_2$	−3617	+1840

$\Delta H^{\ominus}$ positive and $\Delta S^{\ominus}$ negative

Reactions of this type are not spontaneously feasible and have to be driven. An example is photosynthesis (see Figure 20.13), which must be continuously supplied with energy from sunlight.

	$\Delta H^{\ominus}$/kJ mol^{-1}	$\Delta S^{\ominus}$/J K^{-1} mol^{-1}
$6CO_2(g) + 6H_2O(l) \rightarrow C_6H_{12}O_6(s) + 6O_2(g)$	+2803	−225

Figure 20.13 ΔH positive and ΔS negative: photosynthesis in chloroplasts

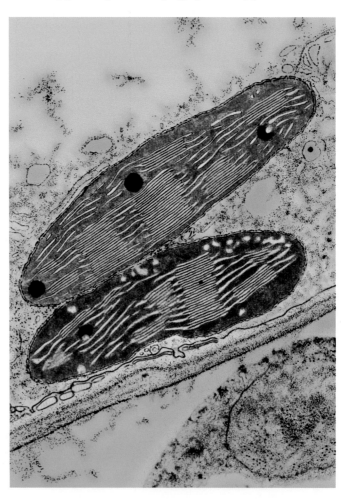

Worked example

For the following processes, explain why $\Delta H^{\ominus}$ and $\Delta S^{\ominus}$ have the signs (positive or negative) shown.

a $Mg(s) + O_2(g) \rightarrow MgO(s)$ $\Delta H^{\ominus}$ –ve, $\Delta S^{\ominus}$ –ve
b $C_2H_5OH(l) \rightarrow C_2H_4(g) + H_2O(l)$ $\Delta H^{\ominus}$ +ve, $\Delta S^{\ominus}$ +ve

Answer

a $\Delta H^{\ominus}$ is negative because relatively weak bonds in magnesium and oxygen are converted into strong ionic bonds in magnesium oxide. $\Delta S^{\ominus}$ is negative because $\Delta n = -1$.
b $\Delta H^{\ominus}$ is positive because a relatively weak π bond is being formed. $\Delta S^{\ominus}$ is positive because $\Delta n = +1$.

Now try this

For the following processes, explain why $\Delta H^{\ominus}$ and $\Delta S^{\ominus}$ have the signs (positive or negative) shown.

1 $C_2H_4(g) + H_2(g) \rightarrow C_2H_6(g)$ $\Delta H^{\ominus}$ –ve, $\Delta S^{\ominus}$ –ve
2 $H_2O(s) \rightarrow H_2O(g)$ $\Delta H^{\ominus}$ +ve, $\Delta S^{\ominus}$ +ve
3 $H_2O_2(l) \rightarrow H_2O(l) + O_2(g)$ $\Delta H^{\ominus}$ –ve, $\Delta H^{\ominus}$ +ve
4 $NH_4NO_3(s) + (aq) \rightarrow NH_4^+(aq) + NO_3^-(aq)$ $\Delta H^{\ominus}$ +ve, $\Delta S^{\ominus}$ –ve

Explosives

An explosion is a chemical reaction that produces sound, as well as a great deal of heat and light. Sound is produced if the speed of the ejected gases exceeds the speed of sound, $330\,m\,s^{-1}$ in air, resulting in the propagation of a shock wave.

Some explosions are produced by **propellants**, fuels used to drive rockets (Figure 20.14a) or to set in motion a shell or bullet in the barrel of a gun. These reactions must take place quickly and smoothly. This is achieved by using a source of heat near one small part of the propellant to initiate the reaction and then the heat produced by the reaction here sets off the propellant in contact with it. The speed of propagation is comparatively slow, probably only a few metres per second, and lasts all the time the bullet or shell is in the barrel of the gun. Propagation of the explosion by heat is a characteristic of **low explosives**. Common examples of low explosives include gunpowder (used in firework rockets) and cellulose trinitrate (used as cordite in the cartridges of shells or bullets). Low explosives may be set off using a match and fuse, or by being hit with a percussion cap. The simplified equations for the decompositions are as follows:

$$3C(s) + S(s) + 2KNO_3(s) \rightarrow K_2S(s) + 3CO_2(g) + N_2(g)$$
gunpowder

$$C_6H_7N_3O_{11}(s) \rightarrow 4\tfrac{1}{2}CO(g) + 1\tfrac{1}{2}CO_2(g) + 3\tfrac{1}{2}H_2O(g) + 1\tfrac{1}{2}N_2(g)$$
cellulose trinitrate

Cellulose trinitrate contains enough oxygen for all the products to be gaseous. The carbon monoxide produced burns to form carbon dioxide in the air around the explosion.

Other types of explosive are set off by a shock wave. This process is called **detonation** and is characteristic of **high explosives** (see Figure 20.14b). The whole explosion is nearly instantaneous as the shock wave travels at the speed of sound, which in a solid can be as high as $1000\,m\,s^{-1}$. The explosion produced is much more intense than that from a low explosive. High explosives are used in mining, for the demolition of buildings and in the warheads of shells. A high explosive would be disastrous if used in the cartridge of a rifle – the explosion would be so violent that the gun would explode instead of the bullet being sent out of the barrel.

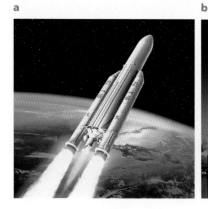

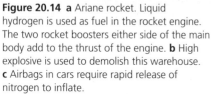

Figure 20.14 a Ariane rocket. Liquid hydrogen is used as fuel in the rocket engine. The two rocket boosters either side of the main body add to the thrust of the engine. **b** High explosive is used to demolish this warehouse. **c** Airbags in cars require rapid release of nitrogen to inflate.

Common examples of high explosives include TNT (see section 25.3) and propane-1,2,3-trinitrate. The latter, by itself, is highly dangerous and sensitive to shock. Alfred Nobel made his fame and fortune by showing that when it was absorbed in clay, a much more stable explosive is produced, called **dynamite**. Propane-1,2,3-trinitrate contains enough oxygen for complete combustion, but TNT needs additional oxygen, often supplied by ammonium nitrate. A simplified equation representing the decompositions of TNT and ammonium nitrate is as follows:

$$C_7H_5N_3O_6(s) + 3\tfrac{1}{2}NH_4NO_3(s) \rightarrow 7CO(g) + 5N_2(g) + 9\tfrac{1}{2}H_2O(g)$$

High explosives must be set off with a detonator. This contains a small quantity of a high explosive, such as mercury fulminate or lead azide, that is very sensitive to shock. Detonators can be used to set off either high explosives or low explosives, because the shock wave produced is sufficiently intense to set off a high explosive and enough heat is given out to set off a low explosive.

$$Hg(CNO)_2(s) \rightarrow Hg(l) + 2CO(g) + N_2(g)$$
mercury fulminate

$$Pb(N_3)_2(s) \rightarrow Pb(s) + 3N_2(g)$$
lead azide

This rapid release of nitrogen is used to inflate airbags in cars, using sodium azide, NaN_3 (see Figure 20.14c).

The manufacture of explosives is very dangerous and these chemicals have caused many fatal accidents. On no account should their preparation be attempted in the laboratory.

20.5 Trends in Group 2 compounds

Thermal stability of the carbonates and nitrates

We have already pointed out that the carbonates and nitrates of Group 2 require increasingly higher temperatures before they break down (see section 10.4). If we look at the general reactions

$$MCO_3(s) \rightarrow MO(s) + CO_2(g)$$

and

$$M(NO_3)_2(s) \rightarrow MO(s) + 2NO_2(g) + \tfrac{1}{2}O_2(g)$$

we can see that the entropy changes for both reactions will be positive because one mole of each of the carbonates gives off one mole of gas and one mole of each of the nitrates gives off 2½ moles of gas. Because the number of moles of gas evolved is the same for all the carbonates, their entropy change of decomposition should be similar; a similar argument applies to the entropy change of decomposition for all the nitrates. This means that the ease of decomposition of the carbonates and nitrates is largely determined by their enthalpy changes. This is shown by the data in Tables 20.7 and 20.8.

Table 20.7 Values of ΔH, ΔG and ΔS for the decomposition of the Group 2 carbonates

	Mg	Ca	Sr	Ba
$\Delta H/\text{kJ mol}^{-1}$	+100.3	+178.3	+234.6	+269.3
$\Delta G/\text{kJ mol}^{-1}$	+48.3	+130.4	+184.1	+218.1
$\Delta S/\text{J mol}^{-1}\text{K}^{-1}$	+174	+161	+168	+172

Table 20.8 Values of ΔH, ΔG and ΔS for the decomposition of anhydrous Group 2 nitrates. Group 2 nitrates are usually obtained as hydrated salts, which will make the actual values different, but the general trend should remain the same

	Mg	Ca	Sr	Ba
$\Delta H/\text{kJ mol}^{-1}$	+255.4	+369.7	+452.6	+505.0
$\Delta G/\text{kJ mol}^{-1}$	+122.7	+241.8	+320.8	+374.2
$\Delta S/\text{J mol}^{-1}\text{K}^{-1}$	+445	+429	+442	+439

At room temperature ΔG for all the reactions is positive. If we assume that the values of ΔH and ΔS do not change much with temperature, ΔG will approach zero when $T\Delta S \approx \Delta H$. The temperature at which this occurs will be lowest when ΔH has the smallest positive value, that is at the top of the group. This agrees with the observation that the carbonates and nitrates at the top of the group decompose at a lower temperature than those at the bottom.

The reason why ΔH becomes more positive as the proton number increases is because the size of the cation increases. During decomposition, the large carbonate and nitrate ions become converted into oxide ions at the same time as carbon dioxide or nitrogen dioxide and oxygen are being given off. The ease with which this happens is affected by the cations that are next to the carbonate or nitrate ions. The two charges on the small magnesium cation create a much stronger electrostatic field than the two charges on the much larger barium cation. We say that the magnesium ion has a higher *charge density*. This high charge density changes the shape of the anion making it more like the shapes of the products (Figure 20.15). This change in shape of the anion is most marked when the anion is large and its electrons are not so firmly attracted to the nucleus. We say that the carbonate and nitrate ions are easily *polarised*.

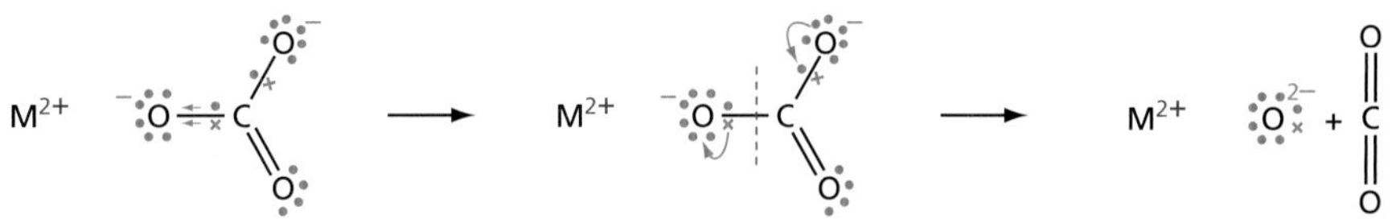

Figure 20.15 The small magnesium ion has a high charge density and changes the shape of the carbonate ion to a greater degree than does the much larger barium ion. A similar change in shape happens with the nitrate ion.

> **Now try this**
>
> Assuming that ΔH and ΔS remain constant, use the data in Table 20.7 to calculate the temperatures at which ΔG for the decomposition of magnesium carbonate and barium carbonate become zero.

Solubility of Group 2 sulfates and hydroxides

The solubility of the sulfates in Group 2 is in the order $MgSO_4 > CaSO_4 > SrSO_4 > BaSO_4$ (see section 10.4). These are determined by the free energy of solution. The entropy changes for the dissolving will be very similar as in all cases $MSO_4(s)$ is being converted into M^{2+} and SO_4^{2-} ions. The solubility changes are, therefore, largely determined by the enthalpy changes of solution. These can be calculated using the thermochemical cycle shown in Figure 20.16. This shows us that:

$$\Delta H_{sol} = +\Delta H_{hyd}(M^{2+}) + \Delta H_{hyd}(SO_4^{2-}) - LE(MSO_4)$$

For the sulfates, the values of lattice energies are dominated by the large sulfate ion, so there is a comparatively small change from magnesium to barium: they vary by only $500\,\text{kJ mol}^{-1}$ from magnesium sulfate to barium sulfate (see Table 20.9). The enthalpy changes of hydration change to a greater extent (by $617\,\text{kJ mol}^{-1}$) from the small magnesium ion to the much larger barium ion. The value for $\Delta H^{\ominus}_{hyd}(SO_4^{2-})$

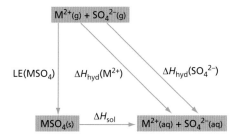

Figure 20.16 Thermochemical cycle for the solubility of Group 2 sulfates

is constant, $-1160 \, \text{kJ mol}^{-1}$. So the general trend is for $\Delta H^{\ominus}_{\text{sol}}$ to become more positive (by $117 \, \text{kJ mol}^{-1}$) going down the group. This is the principal reason for the decreasing solubility.

Table 20.9 Thermochemical data for the solubilities of the Group 2 sulfates

M	$-\text{LE(MSO}_4)/$ kJ mol^{-1}	$\Delta H_{\text{hyd}}(M^{2+})/$ kJ mol^{-1}	$\Delta H_{\text{hyd}}(SO_4^{2-})/$ kJ mol^{-1}	$\Delta H_{\text{sol}}/$ kJ mol^{-1}	Solubility/mol in 100 g of water
Mg	2959	-1890	-1160	-91	2.2×10^{-1}
Ca	2704	-1562	-1160	-18	1.5×10^{-3}
Sr	2572	-1414	-1160	-2	7.1×10^{-4}
Ba	2459	-1273	-1160	$+26$	1.1×10^{-6}

The solubility of the hydroxides is in the opposite order, that is Mg < Ca < Sr < Ba. If we apply a similar Born–Haber cycle to the hydroxides as we did for the sulfates, we have:

$$\Delta H_{\text{sol}} = \Delta H_{\text{hyd}}(M^{2+}) + 2\Delta H_{\text{hyd}}(OH^-) - \text{LE(MOH)}$$

$2 \times \Delta H_{\text{hyd}}(OH^-)$ has the constant value $-1100 \, \text{kJ mol}^{-1}$. The lattice enthalpies for the hydroxides change more down the group than do the lattice enthalpies for the sulfates, and this is the reason for the reversed trend in solubility (see Table 20.10).

Table 20.10 Thermochemical data for the solubilities of the Group 2 hydroxides

M	$-\text{LE(M(OH)}_2)/$ kJ mol^{-1}	$\Delta H_{\text{hyd}}(M^{2+})/$ kJ mol^{-1}	$2 - \Delta H_{\text{hyd}}(OH^-)/$ kJ mol^{-1}	$\Delta H_{\text{sol}}/$ kJ mol^{-1}	Solubility/mol in 100 g of water
Mg	2993	-1890	-1100	3	1.6×10^{-5}
Ca	2644	-1562	-1100	-18	2.5×10^{-3}
Sr	2467	-1414	-1100	-47	3.4×10^{-3}
Ba	2320	-1273	-1100	-53	4.1×10^{-2}

Because the hydroxide ion is small, the lattice enthalpies of the Group 2 hydroxides are sensitive to changes in size of the cation and vary in value by $673 \, \text{kJ mol}^{-1}$ from magnesium hydroxide to barium hydroxide. The changes in $\Delta H^+_{\text{hyd}}(M^{2+})$ are less than this ($617 \, \text{kJ mol}^{-1}$). So in contrast to the sulfates, ΔH^+_{sol} becomes more negative, by $56 \, \text{kJ mol}^{-1}$, on descending the group and the solubility increases.

Summary

- Experimentally-derived lattice enthalpies can be found using a **Born–Haber cycle**.
- The **enthalpy change of solution** is determined by the lattice enthalpy and the **enthalpy change of hydration**.
- Polarisation of the large anion by the doubly charged cation determines the stability of the Group 2 nitrates and carbonates.
- There is an entropy change ΔS when a quantity of heat q is passed at an absolute temperature T.
- There is an entropy increase when a system becomes more disordered, for example melting, boiling, and increase in the number of gas molecules.
- The total entropy of the system plus surroundings must increase for a spontaneous reaction.
- This total entropy increase is related to the Gibbs free energy $\Delta G^{\ominus}$. The more negative $\Delta G^{\ominus}$ is, the greater the driving force of the reaction.
- The solubility of Group 2 sulfates decrease with increasing proton number. This is because the enthalpy change of solution becomes increasingly unfavourable. The reverse effect is shown by the Group 2 hydroxides.

Some key definitions

- The **lattice enthalpy**, LE, is the enthalpy change when one mole of the solid is formed from its isolated ions in the gas phase.
- The **ionisation energy**, IE, is the minimum energy change required to remove one mole of electrons from one mole of atoms in the gas phase.
- The **electron affinity**, EA, is the energy change when one mole of electrons is added to one mole of atoms in the gas phase.
- The **enthalpy change of solution**, $\Delta H^{\ominus}_{\text{sol}}$, is the enthalpy change when one mole of solute is dissolved in an infinite volume of water.
- The **enthalpy change of hydration**, $\Delta H^{\ominus}_{\text{hyd}}$, is the enthalpy change when one mole of isolated ions in the gas phase is dissolved in an infinite volume of water.

Examination practice questions

Please see the data section of the CD for any A_r values you may need.

1 Calcium chloride, $CaCl_2$, is an important industrial chemical used in refrigeration plants, for de-icing roads and for giving greater strength to concrete.

a Show by means of an equation what is meant by the lattice energy of calcium chloride. [1]

b Suggest, with an explanation, how the lattice energies of the following salts might compare in magnitude with that of calcium chloride.
i calcium fluoride, CaF_2 ii calcium sulfide, CaS [3]

c Use the following data, together with additional data from the data section on the CD, to calculate the lattice energy of $CaCl_2$.

standard enthalpy change of formation of $CaCl_2$	$-796\,kJ\,mol^{-1}$
standard enthalpy change of atomisation of $Ca(s)$	$+178\,kJ\,mol^{-1}$
electron affinity per mole of chlorine atoms	$-349\,kJ\,mol^{-1}$

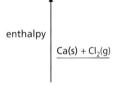

enthalpy

$\underline{Ca(s) + Cl_2(g)}$

[3]

d When a solution of $CaCl_2$ is added to a solution of the dicarboxylic acid, malonic acid, the salt calcium malonate is precipitated as a white solid. The solid has the following composition by mass: Ca, 28.2%; C, 25.2%; H, 1.4%; O, 45.2%.
i Calculate the empirical formula of calcium malonate from these data.
ii Suggest the structural formula of malonic acid. [3]
[Cambridge International AS & A Level Chemistry 9701, Paper 41 Q2 November 2009]

2 a i What is meant by the term *enthalpy change of hydration*, $\Delta H^{\ominus}_{hyd}$?
ii Write an equation that represents the $\Delta H^{\ominus}_{hyd}$ of the Mg^{2+} ion.
iii Suggest a reason why $\Delta H^{\ominus}_{hyd}$ of the Mg^{2+} ion is greater than $\Delta H^{\ominus}_{hyd}$ of the Ca^{2+} ion.
iv Suggest why it is impossible to determine the enthalpy change of hydration of the oxide ion, O^{2-}. [5]

b The enthalpy change of solution for $MgCl_2$, $\Delta H^{\ominus}_{sol}$ ($MgCl_2(s)$), is represented by the following equation.
$$MgCl_2(s) + aq \rightarrow Mg^{2+}(aq) + 2Cl^- -(aq)$$
Describe the simple apparatus you could use, and the measurements you would make, in order to determine a value for $\Delta H^{\ominus}_{sol}$ ($MgCl_2(s)$) in the laboratory. [4]

c The table below lists data relevant to the formation of $MgCl_2(aq)$.

enthalpy change	value/$kJ\,mol^{-1}$
$\Delta H^{\ominus}_f$ ($MgCl_2(s)$)	-641
$\Delta H^{\ominus}_f$ ($MgCl_2(aq)$)	-801
lattice energy of $MgCl_2(s)$	-2526
$\Delta H^{\ominus}_{hyd}$ ($Mg^{2+}(g)$)	-1890

By constructing relevant thermochemical cycles, use the above data to calculate a value for
i $\Delta H^{\ominus}_{sol}$ ($MgCl_2$ (s)),
ii $\Delta H^{\ominus}_{hyd}$ ($Cl^-(g)$). [3]

d Describe and explain how the solubility of magnesium sulfate compares to that of barium sulfate. [4]
[Cambridge International AS & A Level Chemistry 9701, Paper 42 Q1 June 2012]

3 a Describe and explain how the solubilities of the sulfates of the Group 2 elements vary down the group. [3]

b The following table lists some enthalpy changes for magnesium and strontium compounds.

Enthalpy change	Value for magnesium/ $kJ\,mol^{-1}$	Value for strontium/ $kJ\,mol^{-1}$
lattice enthalpy of $M(OH)_2$	-2993	-2467
enthalpy change of hydration of $M^{2+}(g)$	-1890	-1414
enthalpy change of hydration of $OH^-(g)$	-550	-550

i Use the above data to calculate values of $\Delta H^{\ominus}_{sol}$ for $Mg(OH)_2$ and $Sr(OH)_2$.

ii Use your results in **i** to suggest whether $Sr(OH)_2$ is more or less soluble in water than is $Mg(OH)_2$. State any assumptions you make.

iii Suggest whether $Sr(OH)_2$ would be more or less soluble in hot water than in cold. Explain your reasoning.   [5]

c Calcium hydroxide, $Ca(OH)_2$, is slightly soluble in water.

  i Write an expression for K_{sp} for calcium hydroxide, and state its units.

  ii $25.0\,cm^3$ of a saturated solution of $Ca(OH)_2$ required $21.0\,cm^3$ of $0.0500\,mol\,dm^{-3}$ HCl for complete neutralisation. Calculate the $[OH^-(aq)]$ and the $[Ca^{2+}(aq)]$ in the saturated solution, and hence calculate a value for K_{sp}.

iii How would the solubility of $Ca(OH)_2$ in $0.1\,mol\,dm^{-3}$ NaOH compare with that in water? Explain your answer.

[6]

[Cambridge International AS & A Level Chemistry 9701, Paper 42 Q2 June 2010]

21 Quantitative kinetics

Topic 8 gave an overview of how to measure rates of reaction, and what factors increase the number of collisions with enough energy to overcome the activation energy barrier. In this topic, we look at rates of reaction quantitatively. The analysis of the collision theory, the simplest model to explain kinetic data, is developed. Analysing the effects of concentration changes on the reaction rate enables us to write a rate equation, and from this we can obtain information about the mechanism of the reaction.

Learning outcomes

By the end of this topic you should be able to:

8.1c) explain and use the terms *rate equation, order of reaction, rate constant, half-life of a reaction, rate-determining step* (see also Topic 8)

8.1d) construct and use rate equations of the form rate = $k[A]^m[B]^n$ (for which m and n are 0, 1 or 2), including:
- deducing the order of a reaction, or the rate equation for a reaction, from concentration–time graphs or from experimental data relating to the initial rates method and half-life method
- interpreting experimental data in graphical form, including concentration–time and rate–concentration graphs
- calculating an initial rate using concentration data

8.1e) show understanding that the half-life of a first-order reaction is independent of concentration and use the half-life of a first-order reaction in calculations

8.1f) calculate the numerical value of a rate constant, for example by using the initial rates or half-life method

8.1g) for a multi-step reaction, suggest a reaction mechanism that is consistent with the rate equation and the equation for the overall reaction and predict the order that would result from a given reaction mechanism (and vice versa)

8.1h) devise a suitable experimental technique for studying the rate of a reaction, from given information

8.2c) explain qualitatively the effect of temperature change on a rate constant and hence the rate of a reaction

8.3e) outline the different characteristics and modes of action of homogeneous, heterogeneous and enzyme catalysts, including the Haber process, the catalytic removal of oxides of nitrogen from the exhaust gases of car engines (see also Topic 13), the catalytic role of atmospheric oxides of nitrogen in the oxidation of atmospheric sulfur dioxide (see also Topic 10), the catalytic role of Fe^{2+} or Fe^{3+} in the $I^-/S_2O_8^{2-}$ reaction, and the catalytic role of enzymes (including the explanation of specificity using a simple lock and key model but excluding inhibition).

21.1 The rate of a reaction

An expression for the rate of a reaction

In Topic 8, several different ways to measure the rate of a reaction were seen. Examples included the gradient of a graph of volume of gas produced against time (see page 156) and the gradient of a graph of absorbance against time (see page 157). In order to study rates of reaction quantitatively, a precise definition of **rate** is needed. This may be expressed either as how the concentration of a product P increases with time, or how the concentration of a reactant R decreases with time.

$$\text{rate} = \frac{\text{change in concentration of a product}}{\text{time}}$$

or $\quad \text{rate} = \dfrac{-\text{change in concentration of a reactant}}{\text{time}}$

The negative sign in the second expression reflects the fact that the concentration of the reactant is decreasing and therefore produces a positive value for the rate, as it should. The expressions above may be written:

$$\text{rate} = \frac{\Delta[P]}{\Delta t} \quad \text{or} \quad \text{rate} = \frac{-\Delta[R]}{\Delta t}$$

Using calculus notation, this becomes:

$$\text{rate} = \frac{d[P]}{dt} \quad \text{or} \quad \text{rate} = \frac{-d[R]}{dt}$$

Rate is measured in units of concentration per unit time, $\text{mol}\,\text{dm}^{-3}\,\text{s}^{-1}$.

Calculus notation

Although we are not going to use calculus in this topic, its notation is useful in representing the slope of a line at a particular point.

The expression $\dfrac{\Delta[P]}{\Delta t}$ means that during the time interval Δt (which may be of any length, for example 1 second, 20 minutes, etc.), between the two times t_1 and t_2, the concentration of P, written as $[P]$, has changed from $[P]_1$ to $[P]_2$:

$$\frac{\Delta[P]}{\Delta t} = \frac{([P]_2 - [P]_1)}{(t_2 - t_1)}$$

$\dfrac{\Delta[P]}{\Delta t}$ therefore measures the *average* rate of reaction over this time interval. As we decrease the time interval Δt to smaller and smaller values, the value of $\Delta[P]$ will become smaller and smaller too, and the ratio $\dfrac{\Delta[P]}{\Delta t}$ will approximate more and more to the exact slope of the concentration–time graph at a particular point. In the limit, when Δt is (virtually) zero, the ratio becomes the *exact* rate at a particular point. The expression is then written as $\dfrac{d[P]}{dt}$.

Measuring rates by physical and chemical analysis

Before investigating the rate of a particular reaction, we need to know the overall (stoichiometric) equation so that we can decide what method can be used to follow it. For example, if we were studying the following esterification reaction:

$$\underset{\text{ethanoic acid}}{CH_3CO_2H} + \underset{\text{ethanol}}{C_2H_5OH} \rightarrow \underset{\text{ethyl ethanoate}}{CH_3CO_2C_2H_5} + H_2O$$

the decrease in the amount of ethanoic acid, as found by titration, could be followed.

The rate at a given time t is then:

$$\frac{-\Delta[CH_3CO_2H]}{\Delta t} \quad \text{or} \quad \frac{-d[CH_3CO_2H]}{dt}$$

We could also express the rate as any of the following:

$$\frac{-d[C_2H_5OH]}{dt} \quad \text{or} \quad \frac{+d[CH_3CO_2C_2H_5]}{dt} \quad \text{or} \quad \frac{+d[H_2O]}{dt}$$

All of these would give the same numerical value of the rate.

For the reaction:

$$\underset{\text{dinitrogen tetraoxide}}{N_2O_4(g)} \quad \rightarrow \quad \underset{\text{nitrogen dioxide}}{2NO_2(g)}$$

however,

$$\frac{-d[N_2O_4]}{dt} \neq \frac{+d[NO_2]}{dt}$$

because *two* moles of nitrogen dioxide are produced for each *one* mole of dinitrogen tetraoxide used up. When there are different coefficients in the equation like this, it is usual to define the rate in terms of the substance with a coefficient of 1.

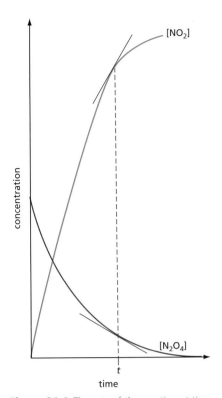

Figure 21.1 The rate of the reaction at time t is the gradient of the concentration–time graph at that point. The rate is either $-d[N_2O_4]/dt$ or $+\frac{1}{2}d[NO_2]/dt$.

So in this case:

$$\text{rate} = \frac{-d[N_2O_4]}{dt} = \frac{1}{2}\frac{d[NO_2]}{dt}$$

Figure 21.1 illustrates this point.

As we mentioned in Topic 8, although chemical analysis can be used to follow how the amount of one substance changes in a reaction, physical methods are usually preferred. For the reaction above, for example, we could follow the change in volume if we were carrying out the reaction in the gas phase, because the number of moles of gas is changing. Alternatively, if we carried out the reaction in solution, we could follow the change in colour (dinitrogen tetraoxide is pale yellow, while nitrogen dioxide is dark brown).

The esterification reaction is difficult to follow using a physical method because there is no obvious change in a physical property as the reaction proceeds. Here we would have to use a chemical method of analysis. Samples of the reaction mixture are extracted at measured time intervals and the amount of ethanoic acid remaining is found by titration with alkali. One problem with this method is that the reaction continues to take place in the sample until the titration is complete, so that the concentration of ethanoic acid at time t is difficult to measure accurately. This reaction is carried out using an acid catalyst, and the reaction in the samples can be effectively stopped by immediately adding the exact amount of alkali needed to neutralise the acid catalyst which was added to the reaction mixture at the start of the reaction. Other reactions can be stopped by appropriate methods, such as rapid cooling of the samples.

Factors affecting the rate of a reaction – a quantitative approach

In Topic 8, we looked at the various factors that determine the rate of a reaction. In this topic, we examine quantitatively how some of these factors affect the rate. This quantitative data often gives information about how the reaction takes place, that is, about the **mechanism** of the reaction. This allows us to study chemical reactions at the most fundamental level.

We have already discussed qualitatively how the following factors affect the rate of a reaction:

- concentration (or pressure for gas reactions)
- temperature
- catalysts
- state of division
- nature of the solvent
- light.

For convenience, when we carry out kinetics experiments in the laboratory, we usually study homogeneous reactions in aqueous solution. The principal factors that we can study under these conditions are the effects of concentration, temperature and homogeneous catalysts. These will now be considered in some detail.

21.2 The effect of concentration

In this section we shall study a single reaction – the iodination of propanone – but the principles established can be applied to many other reactions.

In acid solution, iodine reacts with propanone as follows:

$$CH_3COCH_3(aq) + I_2(aq) \rightarrow CH_2ICOCH_3(aq) + H^+(aq) + I^-(aq)$$

It would be tempting to predict that the rate of the reaction depends on the concentration of both the iodine and the propanone but, as we shall see, this is not the case. Information about the kinetics of a reaction can be found only by experiment, and does not always agree with what might be expected from the stoichiometric equation.

- The stoichiometric equation *cannot* be used to predict how concentration affects the rate of a reaction.
- Kinetic data must be found by experiment.

The stoichiometric equation is used to decide which substance we are going to follow during the reaction in order to measure the rate. In this case, the concentration of iodine is most easily measured, either using a colorimeter (see section 8.2) or by titration with sodium thiosulfate (see section 7.4). It is much more difficult to follow the change in concentration of either propanone or the products of the reaction.

Because we are investigating the effect of concentration on the rate of the reaction, all other factors that might affect the rate should be kept constant. Accurate control of temperature is essential. The reaction should be carried out in a water bath whose temperature is controlled to within 1 K. For very accurate work, thermostatic water baths whose temperature varies by less than 0.1 K are available (see Figure 21.2). Once the reaction has started, the reactants are shaken well in order to mix them thoroughly.

Figure 21.2 A thermostatic water bath keeps the temperature of the reaction constant, removing one factor that might affect the rate.

Initial investigations show that the reaction is not affected by light, but that it is affected by $H^+(aq)$ ions, which act as a catalyst. As this is a homogeneous catalyst, we shall treat $H^+(aq)$ ions in the same way as we treat the other two reactants. We now have three substances whose concentration may affect the rate: $I_2(aq)$, $CH_3COCH_3(aq)$ and $H^+(aq)$. In order to find out how each of them affects the rate of the reaction, we must vary each in turn while keeping the other two constant. There are two main ways of designing experiments to achieve this:

- keeping all reactants in excess except the one being studied
- the initial-rates method.

Keeping all reactants in excess except one

If the concentrations of propanone and hydrogen ions are chosen to be much higher than that of iodine, then during the course of the reaction, their concentrations vary so little that they may be taken as constant. Table 21.1 illustrates why this is the case.

Table 21.1 A typical set of concentrations used to study the iodination of propanone. The concentration of hydrogen ions increases because as well as being a catalyst, it is also being produced by the reaction.

Reactant	Concentration at start/ $mol\,dm^{-3}$	Concentration at end/ $mol\,dm^{-3}$
I_2	0.01	0.0
CH_3COCH_3	1.00	0.99
H^+	0.50	0.51

A mixture of propanone and sulfuric acid is placed in a thermostatically controlled water bath. The solution of iodine is also placed in the bath in a separate container. At the start of the experiment, the two solutions are mixed, shaken well and a stop-clock started. At regular times, the concentration of iodine is found, and a graph is plotted of iodine concentration against time (see Figure 21.3).

The graph is a straight line. This shows that the rate, given by the negative gradient of the graph, is constant. The rate does not change as the iodine concentration falls – the rate is independent of the iodine concentration.

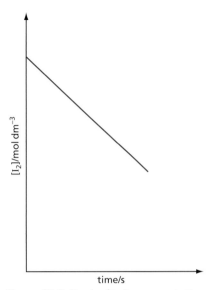

Figure 21.3 Graph of iodine concentration against time for the iodination of propanone

Figure 21.4 The three most common types of concentration–time graphs: **a** zero order, **b** first order, **c** second order. $[R]_0$ is the concentration at time $t = 0$.

The **half-life** of a reaction is the time taken for the concentration of a reactant to decrease to half its initial value.

Figure 21.5 Straight-line graphs are obtained for the following plots:
a zero order: $[R]$ against time
b first order: $\log_{10}[R]$ against time
c second order: $1/[R]$ against time.

Analysing the results – the order of the reaction with respect to each reactant

The **order of reaction** with respect to a reactant, X, tells us to what extent the concentration of that reactant has an effect on the reaction rate. In this case, since the rate of the reaction is independent of the iodine concentration, we can write:

$$\text{rate} = \frac{-d[I_2]}{dt} = k, \text{ a constant}$$

We can indicate the fact that the rate is not affected by the iodine concentration by writing:

$$\text{rate} = k[I_2]^0$$

Any term raised to the power of zero is equal to 1, so the term could be omitted altogether. We include it to emphasise that the effect of iodine concentration on the rate has been studied and found to be **zero order**. This result may seem surprising, and emphasises the point that we cannot predict which reactants determine the reaction rate by looking at the stoichiometric equation.

Of course, the rate is not independent of *all* the reactants, and the concentration of at least one of them must affect the rate.

The order with respect to each reactant can be found by studying concentration–time graphs like the one obtained for iodine. The three common types of rate dependence are shown in Figure 21.4.

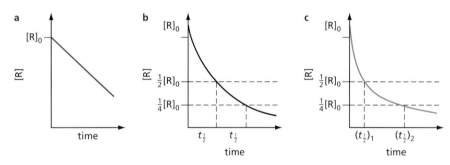

The first example, Figure 21.4a, is like Figure 21.3 for iodine. The graph is a straight line and the rate is constant. This shows that the rate is zero order with respect to that reactant, so that rate $= k[R]^0$. (Since $[R]^0 = 1$, rate $= k$.)

In the second example, Figure 21.4b, the curve is an exponential, and has a constant half-life, $t_{1/2}$. By 'constant half-life' we mean that the successive half-lives are the same. That is, the time taken for the concentration of the reactant to decrease from $[R]_0$ to $\frac{1}{2}[R]_0$ is the same as the time taken for the concentration to decrease from $\frac{1}{2}[R]_0$ to $\frac{1}{4}[R]_0$, and so on.

In the third example, Figure 21.4c, successive half-lives become longer. This indicates an order greater than 1. Since orders higher than 2 are most unusual, the rate is likely to be second order, that is, rate $= k[R]^2$.

In order to confirm that the reaction is of the order suggested by the concentration–time graph, the results can be used to plot a graph that is a straight line. For a zero-order reaction, the graph of $[R]$ against time is a straight line, a first-order reaction gives a straight line for a graph of $\log_{10}[R]$ against time, and a second-order reaction gives a straight-line graph when $1/[R]$ is plotted against time (see Figure 21.5).

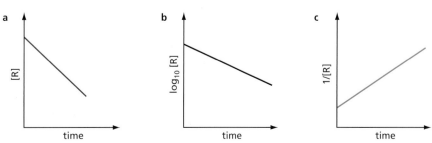

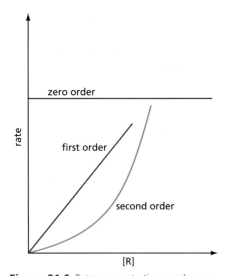

Figure 21.6 Rate–concentration graphs can be used to distinguish between zero-order, first-order and second-order reactions. The zero-order graph is a straight line parallel to the x-axis. The first-order graph is a straight line passing through the origin. The second-order graph is a parabola starting at the origin.

Another, less reliable, way to check the order is to measure the gradient of the concentration–time graph at various times in order to find the rates. These rates can then be plotted against concentration: Figure 21.6 shows how this distinguishes between the three possible orders.

- If it is found by experiment that

  $$rate = k[X]^1$$

  then the order with respect to X is 1.
- If it is found that

  $$rate = k[X]^2$$

  then the order with respect to X is 2.
- There are some reactions where changing the concentration of a reactant has *no* effect on the rate. In this case we can write

  $$rate = k[X]^0$$

  and we say that the order with respect to X is zero. (Note that $n^0 = 1$, a constant, no matter what the value of n.)

The initial-rates method

Analysing the effect of concentration of a reactant on the rate of a reaction by the method of keeping all reactants except one in excess has the advantage that a large amount of data can be collected from each experiment. The results may be plotted so that a straight-line graph is obtained, possibly up to the point when the reaction is 90% completed. If this is the case, we can be confident that the order suggested by the results is the one that should be taken as correct.

Sometimes, however, it is not possible for all the reactants to be in excess. In the iodination of propanone experiment, for example, the iodine concentration must always be small, as this is the reactant whose concentration we are following during the course of the reaction. Another method is therefore needed to find the order of reaction with respect to the propanone and with respect to the acid.

This is done by carrying out a series of experiments in which the initial concentration of all the reactants is kept the same except for the one under investigation, whose concentration is varied. At the start of each experiment, the **initial rate** is found from the gradient of the concentration–time graph (see Figure 21.7). By studying how the initial rate changes when the concentration of this one reactant is varied, we can find the order with respect to that reactant.

The **initial rate** of a reaction is the slope (tangent) of the concentration–time graph at the start of the reaction, when $t = 0$.

Figure 21.7 The initial rate is found from the gradient at the start of the reaction, that is, when $t = 0$.

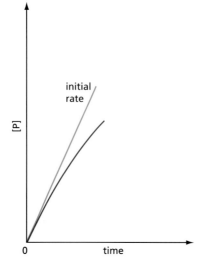
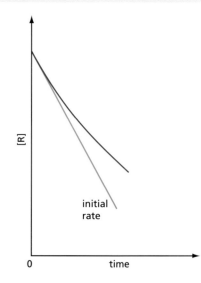

Table 21.2 shows a typical set of results for the iodination of propanone, in which the concentrations of first hydrogen ions and then propanone were varied. From experiments 1, 2 and 3, as $[H^+]$ is reduced in the ratio $5:3:1$, so the initial rate decreases in the ratio $5:3:1$. Therefore, the reaction is first order with respect to $[H^+]$.

Table 21.2 A typical set of results for the iodination of propanone. Initial rates were calculated from the iodine concentration–time graph. Because the order of the reaction with respect to iodine has already been found, it is not necessary to vary $[I_2]$.
$[I_2]$ is kept smaller than the concentrations of the other reactants because the change in $[I_2]$ is being used to monitor the reaction rate. Note that at the start of the reaction, when the initial rate is measured, $[I_2]$ is constant at $0.002\,mol\,dm^{-3}$.

Experiment	$[H^+]/$ $mol\,dm^{-3}$	$[I_2]/$ $mol\,dm^{-3}$	$[CH_3COCH_3]/$ $mol\,dm^{-3}$	Initial rate/ $mol\,dm^{-3}\,s^{-1}$
1	0.50	0.002	0.50	2.0×10^{-5}
2	0.30	0.002	0.50	1.2×10^{-5}
3	0.10	0.002	0.50	4.0×10^{-6}
4	0.50	0.002	0.30	1.2×10^{-5}
5	0.50	0.002	0.10	4.0×10^{-6}

From experiments 1, 4 and 5, as $[CH_3COCH_3]$ is reduced in the ratio $5:3:1$, so the rate again decreases in the ratio $5:3:1$. Therefore, the reaction is also first order with respect to $[CH_3COCH_3]$.

In the case of the iodination of propanone experiment, it is easy to obtain an accurate value for the initial rate because the graph of $[I_2]$ against t is a straight line. This is not usually the case, and then the determination of the initial rate, found from the gradient when $t = 0$, is much more difficult.

Sometimes the initial rate may be found by a 'clock' method. The initial rate of the reaction between peroxodisulfate(VI) ions and iodide ions:

$$S_2O_8^{2-}(aq) + 2I^-(aq) \rightarrow 2SO_4^{2-}(aq) + I_2(aq)$$

can be found by adding a known small amount of thiosulfate ions and a little starch. Initially the iodine produced reacts with the thiosulfate ions:

$$I_2(aq) + 2S_2O_3^{2-}(aq) \rightarrow 2I^-(aq) + S_4O_6^{2-}(aq)$$

but, as we saw in section 8.2, after the thiosulfate ions have been used up, the iodine reacts with the starch to give a blue colour. The time taken, t, for this colour to appear may be used as an approximate measure of the initial rate. If, for example, $[S_2O_3^{2-}] = 0.0030\,mol\,dm^{-3}$, the equation shows that this will be used up when $[I_2] = 0.0015\,mol\,dm^{-3}$. Suppose that the time taken, t, is $60\,s$. Then:

$$\text{initial rate of reaction} = \frac{-\Delta[I_2]}{\Delta t} = \frac{0.0015}{60}$$
$$= 2.5 \times 10^{-5}\,mol\,dm^{-3}\,s^{-1}$$

This is only an approximate value because it assumes that the concentration–time graph is a straight line not only near the origin, but up to the point when the blue colour appeared (see Figure 21.8). The approximation is reasonable provided the reaction is only a small way towards completion.

Figure 21.8 A 'clock' method gives a value for the initial rate that is lower than the true value, because it assumes that the concentration–time graph is a straight line until the 'clock' stops.

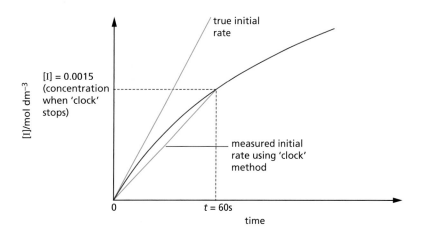

21.3 The rate equation

Arriving at the rate equation

When the order of a reaction with respect to each reactant has been found, the results are combined together in the form of a **rate equation**. For the iodination of propanone, we have the following results:

$$\text{rate} \propto [I_2]^0 \qquad \text{rate} \propto [H^+]^1 \qquad \text{rate} \propto [CH_3COCH_3]^1$$

These may be combined to give the following equation:

$$\text{rate} = k[I_2]^0[H^+]^1[CH_3COCH_3]^1$$

The constant k is known as the **rate constant**. Strictly speaking, it is not necessary to include the $[I_2]^0$ term (which is equal to 1), and the exponent '1' after $[H^+]$ and $[CH_3COCH_3]$ could be omitted. It is, however, sometimes useful to include these terms in the rate equation as they emphasise the effect of varying the concentration of each reactant.

The reaction is now described as follows. It is:

- zero order with respect to $[I_2]$
- first order with respect to $[H^+]$
- first order with respect to $[CH_3COCH_3]$
- of total order 2, or **second order overall**.

The total order is the sum of the exponents from the rate equation. In this case, the total order is:

$$0 + 1 + 1 = 2$$

Finding the rate constant

In order to work out a value for the rate constant, k, we can use a set of readings such as those in Table 21.2. For example, if we use the figures in experiment 1, including the units:

$$\text{rate} = k[I_2]^0[H^+]^1[CH_3COCH_3]^1$$

$$2.0 \times 10^{-5}\,\text{mol}\,\text{dm}^{-3}\,\text{s}^{-1} = k \times 1\ (\text{no units}) \times 0.50\,\text{mol}\,\text{dm}^{-3} \times 0.50\,\text{mol}\,\text{dm}^{-3}$$

$$k = \frac{2.0 \times 10^{-5}\,\text{mol}\,\text{dm}^{-3}\,\text{s}^{-1}}{1 \times 0.50 \times 0.50\,\text{mol}^2\,\text{dm}^{-6}}$$

$$= 8.0 \times 10^{-5}\,\text{mol}^{-1}\,\text{dm}^3\,\text{s}^{-1}$$

The units for the rate constant depend on the total order of the reaction.

Units of k
- overall first-order reaction: s^{-1}
- overall second-order reaction: $\text{mol}^{-1}\,\text{dm}^3\,\text{s}^{-1}$
- overall third-order reaction: $\text{mol}^{-2}\,\text{dm}^6\,\text{s}^{-1}$

The first-order case is interesting because k has no units of concentration. This means that it can be determined directly from a change in a physical property such as volume or colour. As long as it is known that this property is proportional to the concentration, it is not necessary to use the actual concentration. For example, the rate of decomposition of hydrogen peroxide may be studied by measuring the volume of oxygen evolved, and the rate of bromination of methanoic acid by measuring the decrease in absorbance of the bromine. One key reading must be determined as accurately as possible; this is the final reading, when the reaction is complete. This enables us to construct a graph of how the concentration of a reactant (or the volume of gas, or the absorbance) decreases with time, which finishes at zero. Such a graph can be used to evaluate the rate constant for a first-order reaction (see Figures 21.9 and 21.10).

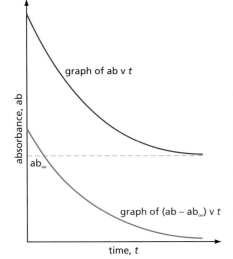

Figure 21.9 Graph showing how a property of a reactant, such as the absorbance, changes with time for a first-order reaction. The final reading ab$_\infty$ may not be zero, so this value has to be subtracted from all the readings. An example is using a colorimeter to measure the decrease in absorbance of bromine as it reacts with methanoic acid.

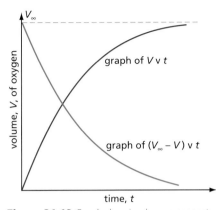

Figure 21.10 Graph showing how a property of a product, such as the volume of a gas, changes with time for a first-order reaction. The volume readings are each subtracted from the final volume to produce the required exponential curve. An example is the volume of oxygen produced during the decomposition of hydrogen peroxide.

These exponential 'concentration'–time graphs for first-order reactions can be used to find the half-life of the reaction, which is related to the rate constant by the following equation:

$$k = \frac{2.30 \log_{10} 2}{t\frac{1}{2}} = \frac{0.693}{t\frac{1}{2}}$$

The rate equation shows how the rate changes with the concentration of the reactants (and the concentration of a homogeneous catalyst, if one is present). The rate constant is unaffected by changes in these concentrations. That is why it is called a rate *constant*. If we exclude heterogeneous catalysts and light, the only factor that changes the value of a rate constant is temperature.

- The **rate constant** is the constant of proportionality in a rate equation.
- It is unaffected by changes in concentration.
- It changes with temperature.

Worked example

In tetrachloromethane at 45 °C, dinitrogen pentaoxide, N_2O_5, decomposes as follows:

$$N_2O_5 \rightarrow 2NO_2 + \frac{1}{2}O_2$$

The rate of the reaction was measured at different times. The results are shown in Table 21.3.

Table 21.3 Rates and reactant concentrations at different times. $1\,\mu mol = 10^{-6}\,mol$.

$[N_2O_5]/mol\,dm^{-3}$	2.21	2.00	1.79	1.51	1.23	0.82
Rate/$\mu mol\,dm^{-3}\,s^{-1}$	22.7	21.0	19.3	15.7	13.0	8.3

a Suggest a method of following the reaction.
b Plot a graph of rate against $[N_2O_5]$.
c Use your graph to find the order of the reaction with respect to N_2O_5.
d Calculate the rate constant, giving its units.
e What will be the shape of the graph of $[N_2O_5]$ against time?

Answer
a Either measure the volume of oxygen given off, or measure the absorbance, since nitrogen dioxide is coloured.
b See Figure 21.11.

Figure 21.11

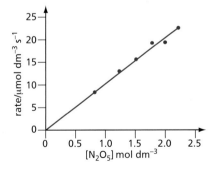

c The graph is a straight line passing through the origin, so the reaction is first order.

d Rate = $k[N_2O_5]$ so $k = \dfrac{\text{rate}}{[N_2O_5]}$ = gradient of graph = **$1.0 \times 10^{-5}\,s^{-1}$**

e The graph will show an exponential decay with a constant half-life.

Now try this

In alkaline solution, iodide ions react with chlorate(I) ions as follows:

$$I^-(aq) + ClO^-(aq) \rightarrow Cl^-(aq) + IO^-(aq)$$

The reaction can be followed by measuring the absorbance of $IO^-(aq)$ ions at 400 nm in a colorimeter. Table 21.4 shows a series of measured initial rates.

Table 21.4

Experiment number	$[I^-(aq)]/mol\,dm^{-3}$	$[ClO^-(aq)]/mol\,dm^{-3}$	Initial rate/ $mol\,dm^{-3}\,s^{-1}$
1	0.0010	0.00073	4.5×10^{-5}
2	0.0010	0.0010	6.2×10^{-5}
3	0.0010	0.0014	8.7×10^{-5}
4	0.00073	0.0010	4.6×10^{-5}
5	0.0014	0.0010	8.6×10^{-5}

1 Calculate the order of the reaction with respect to $I^-(aq)$ ions and $ClO^-(aq)$ ions. Explain your answer.
2 Write a rate equation for the reaction.
3 Calculate the rate constant, stating the units.

21.4 Reaction mechanisms

Proposing a mechanism

The rate equation is often used as a basis for suggesting a likely mechanism for the reaction. Most reactions can be broken down into a number of steps, one of which has a high activation energy that determines the overall rate of the reaction. This step is called the **rate-determining step**. An analogy is a group of people buying a paper at the local newsagent. If one person arrives, it takes them 1 second to pick up the paper, 10 seconds to pay for it and 1 second to leave the shop. If ten people arrive at the same time, it takes 10 seconds for them all to pick up their paper and 100 seconds for them all to pay, but they still take only 1 second each to leave the shop. There will be a queue to pick up the paper and at the checkout, but not on the way out of the shop. Any step that takes place after the rate-determining step has no effect on the overall rate. In the rate equation for the iodination of propanone,

$$\text{rate} = k[I_2]^0[H^+]^1[CH_3COCH_3]^1$$

iodine does not appear in the rate equation, so any step involving iodine must come after the rate-determining step. It is also reasonable conclude that the first step is the reaction of a proton with propanone, as both H^+ and CH_3COCH_3 appear as first-order terms in the rate equation. This reaction is an acid–base reaction. Such reactions are usually fast and reversible (H^+ exists as H_3O^+ in aqueous solution). So we write:

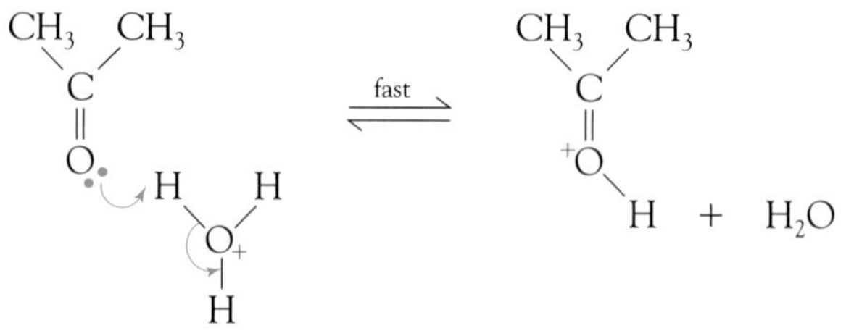

The second step, involving the breaking of a C—H bond, probably controls the rate of the reaction: C—H bond breaking is known to be much slower than O—H bond breaking. In this step, a water molecule could act as a base, taking a proton off the protonated propanone and re-forming H_3O^+. A possible second step is therefore:

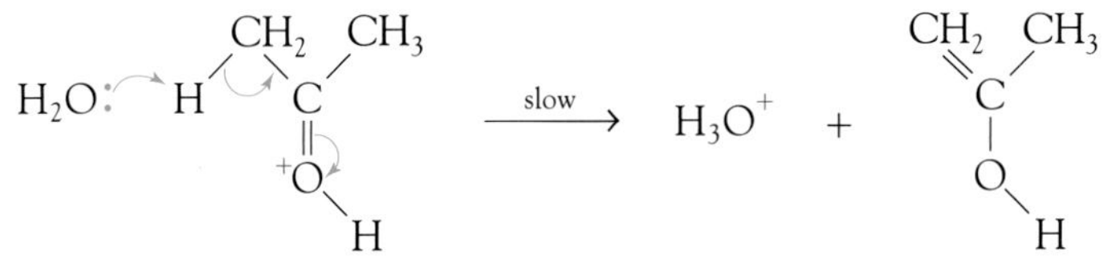

Iodine reacts rapidly with compounds containing the C=C group, so a fast step follows:

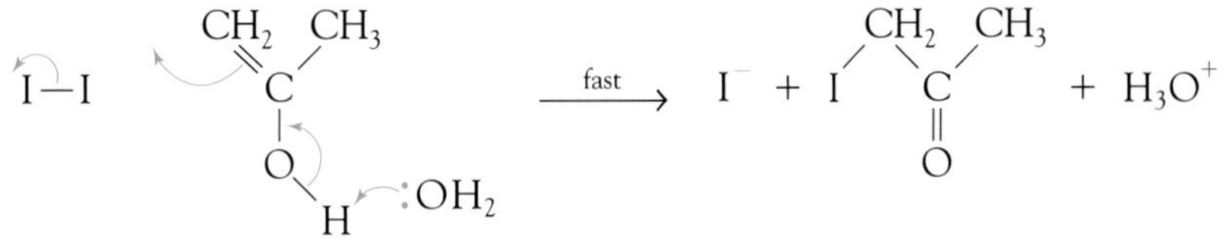

Adding up all the individual steps in the mechanism, we arrive at the overall stoichiometric equation:

$$I_2(aq) + CH_3COCH_3(aq) \rightarrow CH_2ICOCH_3(aq) + H^+(aq) + I^-(aq)$$

Mechanistically we can define a homogeneous catalyst as a substance that appears in the rate equation but not in the stoichiometric equation. If we consider just the first two steps, of which the second is the rate-determining step, H_3O^+ is a **homogeneous catalyst**.

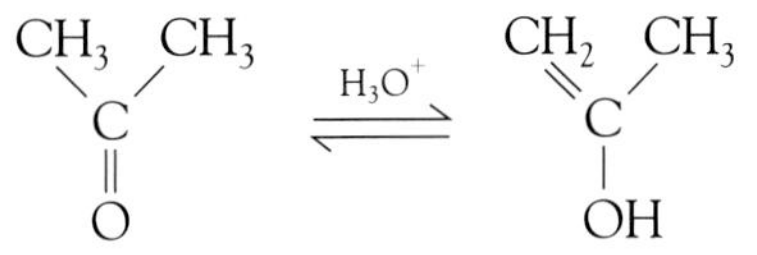

Note, however, that the last of the three steps described above produces $H^+(aq)$, which, as we have seen, is the catalyst for the first two steps. A reaction which produces its own catalyst like this is called **autocatalytic reaction**.

Substances such as:

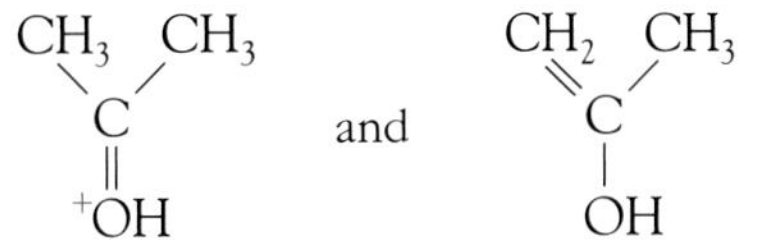

are known as **intermediates**, because they are produced during the reaction but are not part of the final products.

Testing the mechanism

We have now put forward a *possible* mechanism for the reaction, which is consistent with the kinetic data. We do not yet know if this mechanism is the *most likely* one and so we need to carry out further experiments to confirm the mechanism. Some of the techniques used are listed below.

● **Use a wider range of concentrations** – the experimentally determined rate equation may hold over only a limited range of concentrations. For example, the proposed mechanism for the iodination of propanone predicts that at very low concentrations of iodine, the zero-order dependence of the iodine changes to first-order dependence. This is because the rate of the last reaction will be given by the expression:

rate = $k'[I_2][CH_2{=}C(OH)CH_3]$

where k' is the rate constant for the third step shown above, and so this rate will decrease as $[I_2]$ decreases. At very low concentrations of iodine, the third step could become so slow that it effectively becomes the rate-determining step. This has been shown to be the case, which fits in with the mechanism we have suggested above.

- **Use sophisticated analytic techniques** – these may be able to detect the presence of the suggested intermediates. The use of nuclear magnetic resonance (see Topic 29) shows that acidified propanone contains about one molecule in 10^6 of the enol form, $CH_2\!\!=\!\!C(OH)CH_3$, an intermediate proposed by the mechanism.
- **Do experiments on the intermediates** – some intermediates are stable enough for experiments to be carried out on them. For example, some organic halides form tertiary carbocations (see section 14.3) that can be isolated.
- **Use isotopic labelling** – if an atom is labelled with an isotope (not necessarily a radioactive one), the label may indicate which bond has been broken in a reaction. For example, when some esters labelled with ^{18}O are hydrolysed, the ^{18}O appears in the alcohol and not in the acid group. This shows that the acyl oxygen bond, and not the alkyl oxygen bond, is the one that is broken:

ester   acid   alcohol

- **Kinetic isotope effect** – deuterium behaves slightly differently from hydrogen; for example, the C—D bond is slightly harder to break than the C—H bond. This means that if this bond is broken during the rate-determining step, a compound containing C—D bonds reacts between 5 and 10 times as slowly as one with C—H bonds. This is the case with the iodination of propanone: CD_3COCD_3 reacts considerably more slowly than CH_3COCH_3.

    On the other hand, the fact that C_6H_6 and C_6D_6 nitrate at the same rate shows that the rate-determining step is the initial attack by the NO_2^+ ion (see section 25.3), and not the elimination of the proton.

- **Change the solvent** – the rate of ionic reactions changes with the polarity of the solvent. For example, the rate of hydrolysis of 2-bromo-2-methylpropane is raised by the addition of sodium chloride. The sodium chloride increases the polarity of the solvent and increases the ionisation of the bromide:

$$(CH_3)_3CBr \rightleftharpoons (CH_3)_3C^+ + Br^-$$

On the other hand, the addition of sodium bromide reduces the overall rate. This is because the ionisation of $(CH_3)_3CBr$ is suppressed by the high concentration of Br^- ions (Le Chatelier's Principle), and this can be larger than the positive effect caused by the increased polarity of the solvent.

No amount of experimental work can ever prove that the proposed mechanism is the correct one. In particular, the role of the solvent is always uncertain, as there is no way in which its concentration can be varied without changing the overall polarity.

Order and molecularity

Some complex reactions have orders that are negative or fractional. Other reactions have an order that changes with concentration, and only a detailed mathematical treatment shows why this is so.

While the order of reaction may be non-integral, the molecularity of a reaction is integral. The **molecularity** is the number of species in the rate-determining step. This must be integral, probably 1 or 2. For many reactions, the order and molecularity are the same, and this can cause confusion between them.

1 For each of the following reactions, suggest a mechanism that is compatible with the rate equation.

a $H_2O_2(aq) + 3I^-(aq) + 2H^+(aq)$
$$\rightarrow 2H_2O(l) + I_3^-(aq)$$
rate = $k[H_2O_2(aq)][I^-(aq)]$

b $ClO^-(aq) + I^-(aq)$
$$\rightarrow IO^-(aq) + Cl^-(aq)$$
rate = $k[ClO^-(aq)][I^-(aq)][OH^-(aq)]^{-1}$
(Hint: $[OH^-][H^+]$ is a constant.)

c $BrO_3^-(aq) + 5Br^-(aq) + 6H^+(aq)$
$$\rightarrow 3H_2O(l) + 3Br_2(aq)$$
rate = $k[H^+(aq)][Br^-(aq)][BrO_3^-(aq)]$

2 Suggest why the following reactions have two terms in their rate equations.

a $C_2H_5CO_2CH_3 + H_2O$
$$\rightarrow C_2H_5CO_2H + CH_3OH$$
rate = $k_1[C_2H_5CO_2CH_3][H^+] + k_2[C_2H_5CO_2CH_3][OH^-]$

b $H_2(g) + I_2(g) \rightleftharpoons 2HI(g)$
rate = $k_1[H_2][I_2] - k_2[HI]^2$
(Hint: what *two* reactions are taking place in each case?)

Hydrogen cyanide, HCN, adds on to ethanal, CH_3CHO, to give $CH_3CHOHCN$. Two mechanisms have been proposed for this reaction:

$CH_3CHO + H^+ \rightarrow [CH_3CHOH]^+$ and $[CH_3CHOH]^+ + CN^- \rightarrow CH_3CHOHCN$ (1)

or

$CH_3CHO + CN^- \rightarrow [CH_3CHOCN]^-$ and $[CH_3CHOCN]^- + H^+ \rightarrow CH_3CHOHCN$ (2)

The rate equation is rate = $k[CN^-][CH_3CHO][H^+]^0$.

a Which mechanism is consistent with the rate equation? Explain your answer.
b Which step in this mechanism is the rate-determining step?

Answer

a Mechanism (1) is excluded because $[H^+]$ is in the first equation and so would appear in the rate equation. Mechanism (2) is consistent with the rate equation because the step involving H^+ appears after the rate-determining step.
b The first step is the rate-determining step.

21.5   Reaction profiles

Intermediates and transition states

If we draw the reaction profile for a multi-step reaction such as that outlined in section 21.4, it is important to distinguish an **intermediate** (symbol **I**), which has an energy minimum, from a **transition state** (symbol **TS‡**), at the top of the energy curve, which has an energy maximum. An intermediate is a definite chemical species that exists for a finite length of time. A transition state has no permanent lifetime of its own – it exists for just a few femtoseconds (10^{-15} s) when the molecules are in contact with each other. Even a reactive intermediate, with a lifetime of only a microsecond, has a long lifetime in comparison with the time that colliding molecules are in contact.

A simple one-step reaction, for example the S_N2 hydrolysis of a primary halogenoalkane (see section 15.2), has a single energy maximum (see Figure 21.12).

Figure 21.12 The reaction profile for a one-step reaction. There is a single transition state at the energy maximum.

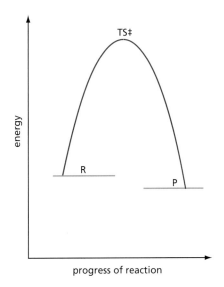

progress of reaction

Figure 21.13 The reaction profile for a two-step reaction. There is an intermediate I and also two transition states. In **a**, the first step is rate determining; in **b**, the second step is rate determining.

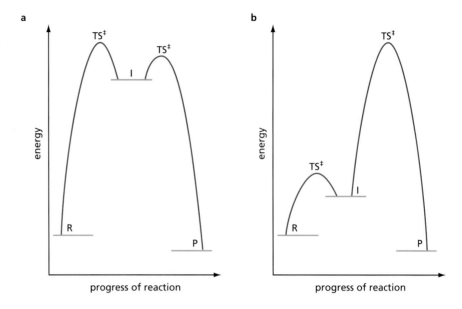

If the reaction has two steps, on the other hand, there is an intermediate, I, and two transition states. An example is the hydrolysis of 2-bromo-2-methylpropane, which is an S_N1 reaction (see section 15.2). The first step is rate determining and so has the higher activation energy (see Figure 21.13a).

The structure of the transition state

It is possible to guess the structure of the transition state using **Hammond's postulate**. This states that the structure of the transition state will resemble that of the intermediate nearest to it in energy. In Figure 21.13a, the structures of the two transition states resemble that of the intermediate, I. In Figure 21.13b, the structure of the first transition state resembles that of the intermediate, while that of the second will be somewhere between that of the intermediate and the product.

While it is not possible to find the exact energy of the transition state except by experiment, some common-sense rules can be followed. For example, if the stage involves bond breaking, the activation energy is high, but if the stage is a reaction between ions of opposite charge, the activation energy is low.

Steps with high activation energy:
- between two neutral molecules
- between ions of similar charge
- if a bond breaks to form free radicals.

Steps with low activation energy:
- between two free radicals
- between ions of opposite charge
- acid–base reactions.

21.6 Some examples of catalysis

The Haber process

The Haber process produces ammonia by combining hydrogen gas with nitrogen gas. The temperature and pressure required to bring about the reaction are discussed in section 10.5, but it is also essential to use a catalyst. The catalyst is made of iron that has traces of potassium oxide and aluminium oxide in it. These impurities improve the efficiency of the catalyst and are called **promoters**. Because the iron is a solid and is in a different phase from the reactants, which are gases, it is a **heterogeneous catalyst**. Because the reaction takes place on the surface of the catalyst, it is essential that the iron is finely divided so as to make the surface area as large as possible. This can be achieved by reducing powdered iron oxide with hydrogen to make the iron catalyst.

The way the catalyst works is not completely understood. It is thought that the hydrogen and nitrogen are adsorbed onto the surface of the iron, where the interatomic bonds are weakened sufficiently for the hydrogen atoms to add onto the nitrogen atoms in three steps. Breaking the nitrogen–nitrogen triple bond needs a lot of energy and is thought to be the rate-determining step.

The three-way catalytic converter

This has been covered in section 8.6, page 166.

Oxidation of sulfur dioxide: the formation of 'acid rain'

A mixture of sulfur dioxide and air is very slowly converted into sulfur trioxide. In the presence of oxides of nitrogen, the reaction proceeds rapidly, the nitrogen monoxide acting as a homogeneous catalyst by the following mechanism:

$$NO(g) + \tfrac{1}{2}O_2(g) \rightarrow NO_2(g)$$

$$NO_2(g) + SO_2(g) \rightarrow SO_3(g) + NO(g)$$

Sulfur dioxide dissolves in water to give sulfurous acid, which is a weak acid, but sulfur trioxide forms sulfuric acid, which is a strong acid:

$$SO_2(g) + H_2O(l) \rightarrow H_2SO_3(aq)$$
$$SO_3(g) + H_2O(l) \rightarrow H_2SO_4(aq)$$

Being a strong acid, sulfuric acid attacks buildings much more rapidly than sulfurous acid. It is also much more damaging to plants because it makes the soil very acidic.

The iodide–peroxydisulfate(VI) reaction

Iodide reacts slowly with aqueous peroxydisulfate(VI) ions to produce iodine and sulfate ions:

$$2I^-(aq) + 2S_2O_8{}^{2-}(aq) \rightarrow I_2(aq) + 2SO_4{}^{2-}(aq)$$

The reaction is catalysed by the addition of some transition metal cations, for example $Fe^{3+}(aq)$:

$$2Fe^{3+}(aq) + 2I^-(aq) \rightarrow 2Fe^{2+}(aq) + I_2(aq)$$
$$2Fe^{2+}(aq) + S_2O_8{}^{2-}(aq) \rightarrow Fe^{3+}(aq) + 2SO_4{}^{2-}(aq)$$

> ### Now try this
>
> Considering ionic attractions and repulsions, explain why Fe^{3+} or Fe^{2+} speed up the reaction between I^- and $S_2O_8{}^{2-}$, whereas $VO_3{}^-$, ClO^- and $NO_3{}^-$ are not effective catalysts.

Summary

- The **rate** of a reaction is given by the following expressions:
  $$\text{rate} = \frac{\Delta[P]}{\Delta t} \quad \text{or} \quad \text{rate} = \frac{-\Delta[R]}{\Delta t}$$
- The units of rate are $mol\,dm^{-3}\,s^{-1}$.
- The **order** of the reaction with respect to a reactant shows how the concentration of that reactant affects the rate of the reaction. The order for each reactant is found by experiment, and these orders are combined together in a **rate equation**.
- The proportionality constant in the rate equation is called the **rate constant**. The rate constant does not vary with concentration but it does vary with temperature.
- A **homogeneous catalyst** does not appear in the overall stoichiometric equation, but its concentration does appear in the rate equation.
- If the reactants and their coefficients in the rate equation are the same as those in the stoichiometric equation, the reaction may take place in a single step.
- If the reactants and their coefficients in the rate equation differ from those in the stoichiometric equation, the reaction takes place in more than one step.
- The step with the highest activation energy is the **rate-determining step**. The number of species that take part in the rate-determining step is known as the **molecularity** of the reaction.
- Reactants whose concentrations appear in the rate equation react before or at the rate-determining step. Reactants whose concentrations do not appear in the rate equation but do appear in the stoichiometric equation react after the rate-determining step.
- **Transition states** are at the maxima in the energy profile of a reaction.
- **Intermediates** are at the minima in the energy profile of a reaction.
- A **heterogeneous catalyst** is one that is in a different phase from the reactants. Heterogeneous catalysts are important in many well-known industrial reactions.

Examination practice questions

Please see the data section of the CD for any A_r values you may need.

1 In the late 19th century the two pioneers of the study of reaction kinetics, Vernon Harcourt and William Esson, studied the rate of the reaction between hydrogen peroxide and iodide ions in acidic solution.

$$H_2O_2 + 2I^- + 2H^+ \rightarrow 2H_2O + I_2$$

This reaction is considered to go by the following steps.

step 1 $H_2O_2 + I^- \rightarrow IO^- + H_2O$
step 2 $IO^- + H^+ \rightarrow HOI$
step 3 $HOI + H^+ + I^- \rightarrow I_2 + H_2O$

The general form of the rate equation is as follows.

$$\text{rate} = k[H_2O_2]^a[I^-]^b[H^+]^c$$

a Suggest how the appearance of the solution might change as the reaction takes place. [1]
b Suggest values for the orders a, b and c in the rate equation for each of the following cases.

Case	Numerical value		
	a	b	c
step 1 is the slowest overall			
step 2 is the slowest overall			
step 3 is the slowest overall			

[3]

A study was carried out in which both $[H_2O_2]$ and $[H^+]$ were kept constant at $0.05\,mol\,dm^{-3}$, and $[I^-]$ was plotted against time. The following curve was obtained.

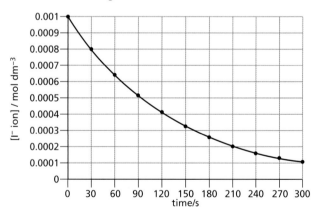

If you are aiming for a top-level grade you will need to draw relevant construction lines on the graph to show your working. Draw them using a pencil and ruler.

c Calculate the initial rate of this reaction and state its units. [2]
d Use half-life data calculated from the graph to show that the reaction is first order with respect to $[I^-]$. [2]

e Use the following data to deduce the orders with respect to $[H_2O_2]$ and $[H^+]$, explaining your reasoning. [2]

$[H_2O_2]/mol\,dm^{-3}$	$[H^+]/mol\,dm^{-3}$	Relative rate
0.05	0.05	1.0
0.07	0.05	1.4
0.09	0.07	1.8

f From your results, deduce which of the three steps is the slowest (rate-determining) step. [1]

[Cambridge International AS & A Level Chemistry 9701, Paper 4 Q2 November 2008]

2 a The reaction between iodide ions and persulfate ions, $S_2O_8^{2-}$, is slow.

$$2I^- + S_2O_8^{2-} \rightarrow I_2 + 2SO_4^{2-} \qquad \textbf{1}$$

The reaction can be speeded up by adding a small amount of Fe^{2+} or Fe^{3+} ions. The following two reactions then take place.

$$2I^- + 2Fe^{3+} \rightarrow I_2 + 2Fe^{2+} \qquad \textbf{2}$$
$$2Fe^{2+} + 2S_2O_8^{2-} \rightarrow 2Fe^{3+} + 2SO_4^{2-} \qquad \textbf{3}$$

i What type of catalysis is occurring here?

ii The rates of reactions **2** and **3** are both faster than that of reaction **1**. By considering the species involved in these reactions, suggest a reason for this.

iii The following reaction pathway diagram shows the enthalpy profile of reaction **1**.

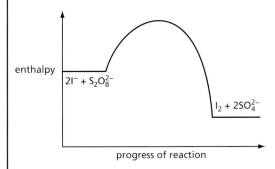

Use the same axes to draw the enthalpy profiles of reaction **2** followed by reaction **3** starting reaction **2** at the same enthalpy level as reaction **1**. [4]

b The oxidation of SO_2 to SO_3 in the atmosphere is speeded up by the presence of nitrogen oxides.

i Describe the environmental significance of this reaction.

ii Describe a major source of SO_2 in the atmosphere.

iii By means of suitable equations, show how nitrogen oxides speed up this reaction. [4]

[Cambridge International AS & A Level Chemistry 9701, Paper 4 Q4 June 2009]

3 Acetals are compounds formed when aldehydes are reacted with an alcohol and an acid catalyst. The reaction between ethanal and methanol was studied in the inert solvent dioxan.

$$CH_3CHO + 2CH_3OH \overset{H^+}{\rightleftharpoons} CH_3CH(OCH_3)_2 + H_2O$$
$$\text{ethanal} \qquad \text{methanol} \qquad\qquad \text{acetal A}$$

a When the initial rate of this reaction was measured at various starting concentrations of the three reactants, the following results were obtained.

Experiment number	$[CH_3CHO]/mol\,dm^{-3}$	$[CH_3OH]/mol\,dm^{-3}$	$[H^+]/mol\,dm^{-3}$	Relative rate
1	0.20	0.10	0.05	1.00
2	0.25	0.10	0.05	1.25
3	0.25	0.16	0.05	2.00
4	0.20	0.16	0.10	3.20

i Use the data in the table to determine the order with respect to each reactant.

ii Use your results from part **i** to write the rate equation for the reaction.

iii State the units of the rate constant in the rate equation.

iv Calculate the relative rate of reaction for a mixture in which the starting concentrations of all three reactants are 0.20 mol dm^{-3}. [6]

b The concentration of the acetal product was measured when experiment number 1 was allowed to reach equilibrium. The result is included in the following table.

	$[CH_3CHO]/mol\,dm^{-3}$	$[CH_3OH]/mol\,dm^{-3}$	$[H^+]/mol\,dm^{-3}$	$[acetal\ A]/mol\,dm^{-3}$	$[H_2O]/mol\,dm^{-3}$
at start	0.20	0.10	0.05	0.00	0.00
at equilibrium	$(0.20 - x)$			x	
at equilibrium				0.025	

i Copy and complete the second row of the table in terms of x, the concentration of acetal **A** at equilibrium. You may wish to consult the chemical equation above.

ii Using the [acetal **A**] as given, 0.025 mol dm^{-3}, calculate the equilibrium concentrations of the other reactants and products and write them in the third row of the table.

iii Write the expression for the equilibrium constant for this reaction, K_c, stating its units.

iv Use your values in the third row of the table to calculate the value of K_c. [9]

[Cambridge International AS & A Level Chemistry 9701, Paper 41 Q2 November 2011]

4 a Catalysts can be described as homogeneous or heterogeneous.

i What is meant by the terms *homogeneous* and *heterogeneous*?

ii By using iron and its compounds as examples, outline the different modes of action of homogeneous and heterogeneous catalysis.

Choose **one** example of each type, and for **each** example you should
- state what the catalyst is, and whether it is acting as a homogeneous or a heterogeneous catalyst,
- write a balanced equation for the reaction,
- outline how the catalyst you have chosen works to decrease the activation energy. [8]

b The reaction between SO_2, NO_2 and O_2 occurs in two steps.

$$NO_2 + SO_2 \rightarrow NO + SO_3 \qquad \Delta H_1^{\ominus} = -88 \text{ KJ mol}^{-1}$$

$$NO + \tfrac{1}{2}O_2 \rightarrow NO_2 \qquad \Delta H_2^{\ominus} = -57 \text{ KJ mol}^{-1}$$

The activation energy of the first reaction, E_{a_1}, is higher than that of the second reaction, E_{a_2}.
Construct a fully-labelled reaction pathway diagram for this reaction, labelling E_{a_1}, E_{a_2}, $\Delta H_1^{\ominus}$ and $\Delta H_2^{\ominus}$.

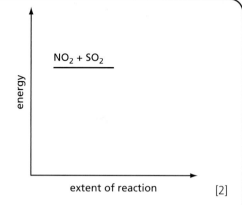

[2]

[Cambridge International AS & A Level Chemistry 9701, Paper 41 Q3 November 2012]

22 Quantitative equilibria

In this topic we use equilibrium constants to provide a quantitative measure of the acidity of acids and bases. The relationship between the pH scale and the hydrogen ion concentration in a solution is described. This scale covers the whole range of acidity found in the laboratory. We also look at the concepts of solubility product and partition coefficients.

Learning outcomes

By the end of this topic you should be able to:

7.2c) explain the terms pH, K_a, pK_a, K_w and use them in calculations

7.2d) calculate [H^+(aq)] and pH values for strong and weak acids and strong bases

7.2e) explain the choice of suitable indicators for acid–base titrations, given appropriate data

7.2f) describe the changes in pH during acid–base titrations and explain these changes in terms of the strengths of the acids and bases

7.2g) explain how buffer solutions control pH, and describe and explain their uses, including the role of HCO_3^- in controlling pH in blood

7.2h) calculate the pH of buffer solutions, given appropriate data

7.2i) show understanding of, and use, the concept of solubility product, K_{sp}

7.2j) calculate K_{sp} from concentrations and vice versa

7.2k) show understanding of the common ion effect

7.3a) state what is meant by partition coefficient; calculate and use a partition coefficient for a system in which the solute is in the same molecular state in the two solvents.

22.1 The dissociation of weak acids

An acid, HA, dissociates in water as follows:

$$HA(aq) + H_2O(l) \rightleftharpoons H_3O^+(aq) + A^-(aq)$$

HA and H_3O^+ are Brønsted–Lowry acids (see section 6.2) – they can donate a proton. H_2O and A^- are their conjugate bases – they can accept a proton. If the equilibrium lies over to the left, H_3O^+ is a stronger acid than HA, and HA is said to be a **weak acid**.

We may write the equilibrium expression (see section 9.7) for this reaction as follows:

$$K_c = \frac{[H_3O^+(aq)][A^-(aq)]}{[H_2O(l)][HA(aq)]}$$

Pure water has a molar mass of $18.0\,g\,mol^{-1}$ and a density of $1.00\,g\,cm^{-3}$. Because $1\,dm^3$ of water has a mass of $1000\,g$,

$$[H_2O(l)] = \frac{1000}{18.0 \times 1} = 55.5\,mol\,dm^{-3}$$

In any dilute aqueous solution, [H_2O(l)] is very little different from this value and so may be taken as a constant. The value of $55.5\,mol\,dm^{-3}$ can be combined with the equilibrium constant, K_c, and a new constant can be defined, called the **acid dissociation constant**, K_a, as follows:

$$K_a = \frac{[H^+(aq)][A^-(aq)]}{[HA(aq)]}$$

Table 22.1 lists the K_a values for some common weak acids.

Table 22.1 Acid dissociation constants for some weak acids

Name of acid	Equilibrium	$K_a/\text{mol dm}^{-3}$
sulfurous acid	$H_2SO_3 \rightleftharpoons HSO_3^- + H^+$	1.3×10^{-2}
nitrous acid	$HNO_2 \rightleftharpoons NO_2^- + H^+$	5.6×10^{-4}
carbonic acid	$H_2CO_3 \rightleftharpoons HCO_3^- + H^+$	2.0×10^{-4}
methanoic acid	$HCO_2H \rightleftharpoons HCO_2^- + H^+$	1.6×10^{-4}
ethanoic acid	$CH_3CO_2H \rightleftharpoons CH_3CO_2^- + H^+$	1.7×10^{-5}
benzoic acid	⬡—CO_2H $\rightleftharpoons$ ⬡—$CO_2^- + H^+$	6.3×10^{-5}
hydrocyanic acid	$HCN \rightleftharpoons CN^- + H^+$	4.9×10^{-10}

K_a values can be used to calculate the concentration of hydrogen ions in solutions of weak acids. For example, if a solution of $0.020\,\text{mol dm}^{-3}$ ethanoic acid is made up, some of it will dissociate to produce $x\,\text{mol dm}^{-3}$ of both $H^+(aq)$ and $A^-(aq)$:

	$[HA(aq)]$	$[H^+(aq)]$	$[A^-(aq)]$
initial/mol dm^{-3}	0.020	0	0
equilibrium/mol dm^{-3}	$(0.020 - x)$	x	x

$$K_a = \frac{x^2}{(0.020 - x)} = 1.7 \times 10^{-5}\,\text{mol dm}^{-3}$$

To avoid having to solve a quadratic equation, the solution to this equation can be simplified by assuming that x is small, so that $(0.020 - x) \approx 0.020$. This is usually justified because the degree of ionisation of a weak acid is very small. Hence:

$$x^2 = 0.020 \times 1.7 \times 10^{-5}$$
$$x = \sqrt{(0.020 \times 1.7 \times 10^{-5})} = 5.8 \times 10^{-4}\,\text{mol dm}^{-3}$$

(Without the approximation, the answer would have been $5.75 \times 10^{-4}\,\text{mol dm}^{-3}$. The difference is insignificant because it is within the limits of the accuracy to which the data is given.)

For an acid whose dissociation constant is K_a and whose initial concentration is $c\,\text{mol dm}^{-3}$, we may use the formula $[H^+] = \sqrt{(K_a \times c)}$. Notice that this equation can only be used if there are no added $H^+(aq)$ ions or $A^-(aq)$ ions: it *cannot* be used for calculations involving buffer solutions (see section 22.3).

For a weak acid:

$[H^+] = \sqrt{(K_a \times c)}$

Now try this

Using K_a values from Table 22.1, calculate the hydrogen ion concentration in each of the following aqueous solutions.

1 $0.0036\,\text{mol dm}^{-3}$ methanoic acid
2 $1.3\,\text{mol dm}^{-3}$ hydrocyanic acid

Worked example

Calculate the hydrogen ion concentration in aqueous $0.0050\,\text{mol dm}^3$ hydrofluoric acid, for which $K_a = 5.6 \times 10^{-4}\,\text{mol dm}^{-3}$.

Answer
$[H^+] = \sqrt{(K_a \times c)} = \sqrt{(5.6 \times 10^{-4} \times 5.0 \times 10^{-3})} = \mathbf{1.7 \times 10^{-3}\,\text{mol dm}^{-3}}$

22.2 The ionic product of water (K_w) and the pH scale

The ionic product of water

Pure water conducts electricity slightly, so it must be ionised to a small extent:

$$2H_2O(l) \rightleftharpoons H_3O^+(aq) + OH^-(aq)$$

For this equilibrium:

$$K_c = \frac{[H_3O^+(aq)][OH^-(aq)]}{[H_2O(l)]^2}$$

Because $[H_2O(l)]$ is a constant, we can define a new constant called the **ionic product of water**, K_w, such that $K_w = [H_3O^+(aq)][OH^-(aq)]$. Because $[H_3O^+] = [H^+(aq)]$ (hydrogen ions in water are all hydrated), this expression is often simplified to $K_w = [H^+][OH^-]$.

Experimentally, K_w is found to have a value of $1.0 \times 10^{-14}\,mol^2\,dm^{-6}$ at $25\,°C$.

The ionic product of water, K_w:

$$K_w = [H^+][OH^-] = 1.0 \times 10^{-14}\,mol^2\,dm^{-6}\ \text{at } 25\,°C$$

This expression shows that in an acidic solution there will be a few hydroxide ions, and that in an alkaline solution there will be a few hydrogen ions. Even in a concentrated solution of sodium hydroxide, there are still some hydrogen ions – the equilibrium of the reaction:

$$H_3O^+(aq) + OH^-(aq) \rightleftharpoons 2H_2O(l)$$

is displaced to the right by the hydroxide ions added, but a few $H_3O^+(aq)$ ions remain.

Worked example

Calculate the hydrogen ion concentration, $[H_3O^+]$, in a $2.0\,mol\,dm^{-3}$ solution of sodium hydroxide.

Answer

$$[H_3O^+] = \frac{K_w}{[OH^-]} = \frac{1.0 \times 10^{-14}}{2.0} = \mathbf{5 \times 10^{-15}\,mol\,dm^{-3}}$$

In a **neutral solution**, the concentrations of H_3O^+ and OH^- are equal:

$$[H_3O^+] = [OH^-]$$

Therefore:

$$K_w = [H_3O^+][OH^-] = [H_3O^+]^2$$

and

$$[H_3O^+]^2 = 1.0 \times 10^{-14}\,mol^2\,dm^{-6}$$
$$[H_3O^+] = \sqrt{(1.0 \times 10^{-14})} = 1.0 \times 10^{-7}\,mol\,dm^{-3}$$

The pH scale

The hydrogen ion concentration is a measure of the acidity of a solution. In different solutions, the hydrogen ion concentration, $[H^+]$, can have a wide range of values (from about $1\,mol\,dm^{-3}$ to $1.0 \times 10^{-14}\,mol\,dm^{-3}$). To make the numbers representing acidity easier to deal with, the **pH scale** was introduced. This scale is defined by the expression:

$$pH = -\log_{10}[H^+]$$

(or, more precisely, as $pH = -\log_{10}([H^+]/mol\,dm^{-3})$, because logarithms can only be taken of a quantity without units).

So in a **neutral solution**, in which $[H^+] = 1.0 \times 10^{-7}\,mol\,dm^{-3}$,

$$\log_{10}[H^+] = -7.0$$
$$pH = 7.0$$

We can calculate the pH of an **acidic solution**, for example one in which $[H^+] = 3.0 \times 10^{-2}\,mol\,dm^{-3}$, as follows:

$$[H^+] = 3.0 \times 10^{-2}\,mol\,dm^{-3}$$

$$\log_{10}[H^+] = -1.52 \qquad \text{(taking } \log_{10})$$

$$pH = 1.52 \qquad \text{(changing sign)}$$

Now try this

Calculate the pH of each of the following solutions:

1 a $0.002\,mol\,dm^{-3}$ solution of HCl(aq)
2 concentrated HCl(aq), in which $[HCl] = 18\,mol\,dm^{-3}$.

To find the pH of an **alkaline solution**, for example one in which $[OH^-] = 2.5 \times 10^{-3}\,mol\,dm^{-3}$, we must first work out $[H^+]$ using K_w:

$$[H^+][OH^-] = 1.0 \times 10^{-14}\,mol^2\,dm^{-6}$$

$$[H^+] = \frac{1.0 \times 10^{-14}}{2.5 \times 10^{-3}} = 4.0 \times 10^{-12}\,mol\,dm^{-3}$$

$$\log_{10}[H^+] = -11.40$$

$$pH = 11.40$$

The pH of strong acids and bases

We can calculate the pH of a solution of a strong acid or a strong base if we assume it to be completely ionised. For a monoprotic acid, that is, one that releases only one H^+ ion per molecule of acid, $[H^+] = [acid]$. For a monoprotic base, that is, one that reacts with only one H^+ ion, $[OH^-] = [base]$.

For a strong diprotic acid such as sulfuric acid, the situation is complicated by the fact that although the first ionisation is that of a strong acid, the second is not:

$$H_2SO_4 \rightleftharpoons H^+ + HSO_4^- \qquad K_a \text{ very large}$$
$$HSO_4^- \rightleftharpoons H^+ + SO_4^{2-} \qquad K_a = 1.0 \times 10^{-2}\,mol\,dm^{-3}$$

As an approximation, we can ignore the second ionisation and treat sulfuric acid as a strong monoprotic acid.

For strong diprotic bases such as barium hydroxide, $Ba(OH)_2$, we may assume complete ionisation and so $[OH^-] = 2 \times [base]$.

pH and pK_a

The symbol 'p' means '$-\log_{10}$'. It can also be used with K_a and other similar equilibrium constants:

$$pK_a = -\log_{10} K_a \qquad\qquad \text{(or, more strictly, } pK_a = -\log 10(K_a/mol\,dm^{-3}))$$

For ethanoic acid, $K_a = 1.7 \times 10^{-5}\,mol\,dm^{-3}$ and $pK_a = 4.76$. Sometimes 'pOH' is used to represent '$-\log_{10}[OH^-]$'. Notice that $pH + pOH = 14$.

Worked example

Calculate the pH of each of the following solutions.
a a $0.050\,mol\,dm^{-3}$ solution of hydriodic acid, HI (a strong acid)
b a $0.30\,mol\,dm^{-3}$ solution of hydrofluoric acid, HF ($K_a = 5.6 \times 10^{-4}\,mol\,dm^{-3}$)
c a $0.40\,mol\,dm^{-3}$ solution of sodium hydroxide

Answer
a Because HI is a strong acid,
   $[HI] = [H^+] = 0.050\,mol\,dm^{-3}$
   $\log_{10}[H^+] = -1.3$ and pH = **1.3**
b $[H^+] = \sqrt{(K_a \times c)} = \sqrt{(5.6 \times 10^{-4} \times 0.30)} = 1.30 \times 10^{-2}\,mol\,dm^{-3}$
   $\log_{10}[H^+] = -1.9$ and pH = **1.9**
c Because $[H^+][OH^-] = 1.0 \times 10^{-14}\,mol^2\,dm^{-6}$,

$$[H^+] = \frac{1.0 \times 10^{-14}}{0.4} = 2.5 \times 10^{-14}\,mol\,dm^{-3}$$

$\log_{10}[H^+] = -13.6$ and pH = **13.6**

Now try this

Calculate the pH of each of the following solutions.

1 $0.5\,mol\,dm^{-3}$ HCl
2 $2.5\,mol\,dm^{-3}$ KOH(aq)
3 $0.5\,mol\,dm^{-3}$ benzoic acid (Use the K_a value from Table 22.1.)

22.3 Buffer solutions

What is a buffer?

Many experiments, particularly in biochemistry, need to be carried out in solutions of constant pH. Although it is impossible to make a solution whose pH is totally unaffected by the addition of small quantities of acid or alkali, it is possible to make a solution, called a **buffer solution**, whose pH remains *almost* unchanged. A moderately concentrated solution of a strong acid or alkali behaves in this way, and can be used to provide solutions of nearly constant pH in the 0–2 or 12–14 ranges. Dilute solutions of strong acids and bases are useless as buffers between pH 2 and pH 12, because the addition of small quantities of acid or alkali changes their pH considerably. This intermediate range of pH is very useful, particularly for biochemical experiments, and so other means of making buffer solutions in this range have been devised.

A weak acid, by itself, acts as a poor buffer solution because its pH drops sharply when a small quantity of acid is added. A mixture of a weak acid and one of its salts, however, behaves as a good buffer solution. In such a mixture, there is a high concentration of $A^-(aq)$ from the salt, as well as the undissociated acid. So the equilibrium:

$$HA(aq) \rightleftharpoons H^+(aq) + A^-(aq)$$

reservoir of acid   reservoir of base

is well over to the left, which increases the pH of the solution. When a small amount of strong acid is added, most of the extra $H^+(aq)$ ions react with the reservoir of $A^-(aq)$, and so the equilibrium moves to the left to remove the added H^+. This tends to minimise the decrease in pH. When a small amount of strong base is added, most of the extra $OH^-(aq)$ ions react with the reservoir of $HA(aq)$, and this tends to minimise the increase in pH.

Worked example

Write equations to show how the buffer solution described above removes:
a added H^+ ions   b added OH^- ions.

Answer
a $A^-(aq) + H^+(aq) \rightarrow HA(aq)$ b $HA(aq) + OH^-(aq) \rightarrow A^-(aq) + H_2O(l)$

Finding the pH of a buffer solution

To calculate the pH of a buffer solution of known composition, we start with the equation:

$$K_a = \frac{[H^+(aq)][A^-(aq)]_{eq}}{[HA(aq)]_{eq}}$$

Because the presence of excess $A^-(aq)$ suppresses the ionisation of HA, the equilibrium concentrations of $A^-(aq)$ and $HA(aq)$, $[A^-(aq)]_{eq}$ and $[HA(aq)]_{eq}$, are almost identical to their initial concentrations, $[A^-(aq)]_{in}$ and $[HA(aq)]_{in}$, which were added to make the buffer solution. Therefore:

$$K_a = \frac{[H^+(aq)][A^-(aq)]_{in}}{[HA(aq)]_{in}}$$

and

$$[H^+(aq)] = \frac{K_a[HA(aq)]_{in}}{[A^-(aq)]_{in}}$$

This is the equation used for buffer solution calculations: notice that the equation $[H^+] = \sqrt{(K_a \times c)}$ *must not* be used in this case, because $[H^+] \neq [A^-]$.

> A **buffer solution** is one whose pH remains nearly constant on the addition of small quantities of acid or base. For a buffer solution:
>
> $$[H^+(aq)] = \frac{K_a[HA(aq)]_{in}}{[A^-(aq)]_{in}}$$
>
> For a buffer solution, note that:
> $$[H^+(aq)] \neq \sqrt{(K_a \times c)}$$

Calculate the pH of a solution made by mixing $100\,cm^3$ of $0.10\,mol\,dm^{-3}$ ethanoic acid ($K_a = 1.76 \times 10^{-5}\,mol\,dm^{-3}$) with $100\,cm^3$ of $0.20\,mol\,dm^{-3}$ sodium ethanoate.

Answer

After mixing, the volume is $200\,cm^3$ and so the solution contains $0.050\,mol\,dm^{-3}$ ethanoic acid and $0.10\,mol\,dm^{-3}$ sodium ethanoate.

$$[H^+(aq)] = \frac{K_a[HA(aq)]_{in}}{[A^-(aq)]_{in}} = \frac{1.76 \times 10^{-5} \times 0.050}{0.10} = 8.5 \times 10^{-6}\,mol\,dm^{-3}$$

pH = **5.1**

1 Calculate the pH of the buffer solution formed by adding $0.003\,mol$ of NaOH to $100\,cm^3$ of $0.1\,mol\,dm^{-3}$ methanoic acid ($K_a = 1.6 \times 10^{-4}\,mol\,dm^{-3}$).

2 A buffer solution is made by mixing $100\,cm^3$ of $0.20\,mol\,dm^{-3}$ Na_2HPO_4 with $500\,cm^3$ of $0.30\,mol\,dm^{-3}$ NaH_2PO_4. For the equilibrium:

$$H_2PO_4^-(aq) \rightleftharpoons H^+(aq) + HPO_4^{2-}(aq)$$

$K_a = 6.3 \times 10^{-7}\,mol\,dm^{-3}$. Calculate the pH of the solution.

3 (Harder) Calculate the change in pH when $0.050\,cm^3$ of $1.00\,mol\,dm^{-3}$ hydrochloric acid is added to:

a $100\,cm^3$ of a solution of $0.000\,10\,mol\,dm^{-3}$ hydrochloric acid

b $100\,cm^3$ of a solution that contains $0.10\,mol\,dm^{-3}$ ethanoic acid and $0.10\,mol\,dm^{-3}$ sodium ethanoate.

K_a for ethanoic acid is $1.7 \times 10^{-5}\,mol\,dm^{-3}$.

(Hint: assume that all the hydrochloric acid reacts with ethanoate ions.)

Using buffer solutions

There are many situations where it is essential to control the pH. In the following examples, just the addition of an acid or an alkali is sufficient to keep the pH in the required range without buffering.

- In swimming pools, sterilisation is usually carried out using chlorine. This makes the water acidic (see section 11.4). Solid calcium hydroxide is added to bring the pH to just above 7.
- Acidic soils are treated with calcium carbonate or calcium hydroxide.
- For acid-loving plants, ammonium sulfate may be added to the soil. This acts as a weak acid and can bring the pH of the soil to pH4, which is suitable for many azaleas and rhododendrons.

In other cases, more accurate control of pH is essential. In the human body, the production of carbon dioxide by respiration lowers the pH of the blood from about pH7.5 to pH7.3. The pH of blood does not fall below this value because blood acts as an efficient buffer. There are several ways in which this buffering is accomplished, of which the following are the most important.

- Proteins are made up of amino acids (see Topic 27). These contain both acidic and basic groups and can therefore act as buffers, removing H_3O^+ or OH^-:

$$R{-}NH_2 + H^+ \rightarrow R{-}NH_3^+$$
$$R{-}CO_2H + OH^- \rightarrow R{-}CO_2^- + H_2O$$

- The acid H_2CO_3, derived from dissolved carbon dioxide, is buffered by the presence of the hydrogencarbonate ion, HCO_3^-.

Many other processes, including the use of shampoos and other hair treatments, developing photographs, medical injections and fermentation, all require strict pH control. This is achieved by the use of an appropriate buffer solution.

Write equations to show how the H_2CO_3/HCO_3^- buffer system reacts with:

a added H^+ ions

b added OH^- ions.

22.4 Titration curves and indicators

Titration curves for different acids and alkalis

During a titration, acid is usually put into the burette and run into alkali in the presence of an indicator. This indicator suddenly changes colour at the **end-point**, which is the point at which the number of moles of acid is exactly balanced by the same amount of alkali. As we shall see, this does not always mean that the solution at this point has a pH of 7.

Figure 22.1 In this experimental set-up to record titration curves, the reaction mixture is stirred magnetically and the change in pH is recorded on the meter.

Figure 22.2 The graphs show the pH changes as $1.0\,mol\,dm^{-3}$ acid is run into $100\,cm^3$ of $0.10\,mol\,dm^{-3}$ alkali. The end-point comes after $10\,cm^3$ of acid have been added. The purpose of making the acid ten times as concentrated as the alkali is to ensure there is very little pH change due to dilution.

Either the acid or the alkali, or both, may be strong or weak, so that there are four possible combinations. If an acid is run into an alkali, the pH changes as shown in Figure 22.2.

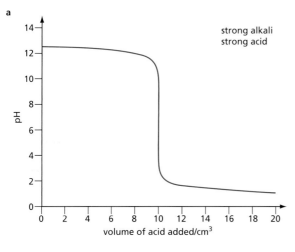
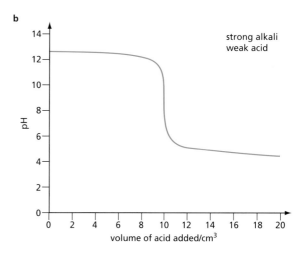
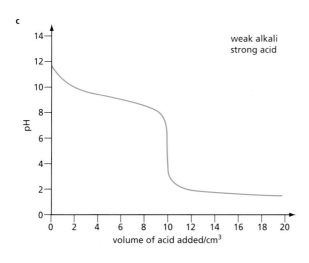
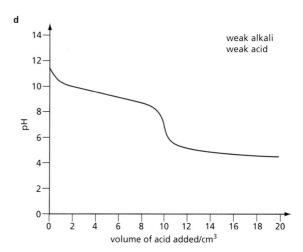

With a strong alkali, such as sodium hydroxide, the pH starts at 13 and then decreases slowly until 10 cm³ of acid have been added. At this point, the **equivalence point**, there is a sharp drop in pH – to about pH 3 with a strong acid or to about pH 5 with a weak acid.

With a weak alkali, such as ammonia, the pH starts at about 11 and falls with an S–shaped curve to about pH 7 until 10 cm³ of acid have been added. At this point there is a drop – sharply with a strong acid to about pH 2 and slowly with a weak acid to about pH 5.

To find the equivalence point accurately, an indicator must be chosen which changes colour when the curve is steepest. The colour will then change when only one further drop of acid is added. Two common indicators are methyl orange and phenolphthalein. Methyl orange changes colour over the pH range 3.2 to 4.4 and phenolphthalein over the pH range 8.2 to 10.0. The titration curves show us that methyl orange is suitable for titrations between a strong or weak alkali and a strong acid, and phenolphthalein is suitable for titrations between a strong alkali and a strong or weak acid. Neither indicator gives a clear end-point with a distinct colour change in a titration between a weak alkali and a weak acid. This is summarised in Table 22.2.

Figure 22.3 shows the titration curves when alkali is added to acid, rather than the other way around. The pH ranges over which methyl orange and phenolphthalein change colour are also shown.

The steep portions of Figures 22.2 and Figure 22.3 are the parts where the pH changes most rapidly with added alkali or acid. It is at these points that an indicator will change colour most rapidly. The pH range over which an indicator changes colour is usually about 2 pH units, so it is important to choose an indicator whose colour-change range is covered by the steep portion of the titration curve.

Conversely, the flat portions of the graphs show where the pH changes most slowly on addition of acid or alkali. These flat portions are, therefore, where the best buffering action occurs.

Table 22.2 The suitability of methyl orange and phenolphthalein as indicators for titrations between strong and weak acids and alkalis

Alkali	Acid	Indicator
strong	strong	methyl orange or phenolphthalein
strong	weak	phenolphthalein
weak	strong	methyl orange
weak	weak	neither

Figure 22.3 Titration curves starting with 100 cm³ of 0.10 mol dm⁻³ acid and adding 1.0 mol dm⁻³ alkali. The curves are a reflection of those in Figure 22.2. The ranges of the indicators are shown beside the steep portions of the curves.

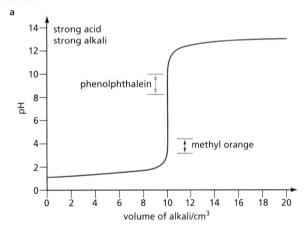

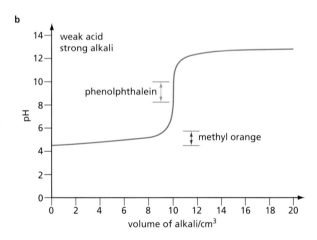

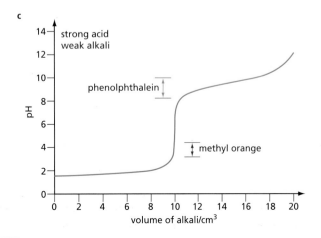

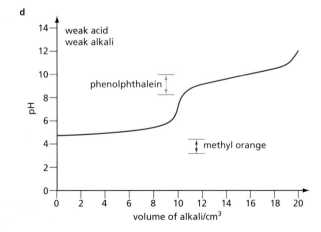

> ### Worked example
>
> Some $0.10\,mol\,dm^{-3}$ sodium hydroxide is added to $25.0\,cm^3$ of $0.10\,mol\,dm^{-3}$ hydrochloric acid. Calculate the pH of the resulting solution after the addition of:
>
> a $24.9\,cm^3$ of alkali   b $50.0\,cm^3$ of alkali.
>
> #### Answer
>
> a Amount of HCl remaining $= c = \dfrac{v}{1000} = 0.10 \times \dfrac{(25.0 - 24.9)}{1000} = 1.0 \times 10^{-5}\,mol$
>
> Volume of solution $= 25.0 + 24.9 = 49.9\,cm^3$
>
> $[H^+] = \dfrac{n}{v} = \dfrac{1.0 \times 10^{-5}}{49.9 \times 10^{-3}} = 2.0 \times 10^{-4}\,mol\,dm^{-3}$ and pH $= $ **3.7**
>
> b The acid has now all reacted, so we need to calculate the pH from the concentration of hydroxide ions in the solution.
>
> Amount of NaOH remaining $= c \times \dfrac{v}{1000} = 0.10 \times \dfrac{(50.0 - 25.0)}{1000} = 2.5 \times 10^{-3}\,mol$
>
> Volume of solution $= 25.0 + 50.0 = 75.0\,cm^3$
>
> $[OH^-] = \dfrac{n}{v} = \dfrac{2.5 \times 10^{-3}}{75.0 \times 10^{-3}} = 0.033\,mol\,dm^{-3}$
>
> $[H^+] = \dfrac{K_w}{[OH^-]} = \dfrac{10^{-14}}{0.033} = 3.0 \times 10^{-13}\,mol\,dm^{-3}$ and pH $= $ **12.5**

22.5 Solubility of salts

An important example of heterogeneous equilibrium is dissolving an ionic substance in water. Take, for example, solid silver chloride in contact with a saturated solution:

$$Ag^+Cl^-(s) \rightleftharpoons Ag^+(aq) + Cl^-(aq)$$

We can write:

$$K_c = \frac{[Ag^+(aq)][Cl^-(aq)]}{[Ag^+Cl^-(s)]}$$

At a given temperature, the $[Ag^+Cl^-(s)]$ term is a constant, irrespective of the amount of silver chloride present. We can absorb this $[Ag^+Cl^-(s)]$ term into the equilibrium constant and define a new equilibrium constant:

$$K_{sp} = [Ag^+(aq)][Cl^-(aq)]$$

where K_{sp} is called the **solubility product**. For AgCl(s), the value of K_{sp} is $1 \times 10^{-10}\,mol^2\,dm^{-6}$. In a saturated solution of silver chloride in water, $[Ag^+(aq)] = [Cl^-(aq)]$; since their product is $1 \times 10^{-10}\,mol^2\,dm^{-6}$, each of them must equal $\sqrt{(1 \times 10^{-10})} = 1 \times 10^{-5}\,mol\,dm^{-3}$.

> ### Worked example
>
> Calculate the solubility of $SrSO_4$, for which $K_{sp} = 5.0 \times 10^{-7}\,mol^2\,dm^{-6}$.
>
> #### Answer
>
> $K_{sp} = [Sr^{2+}][SO_4^{2-}] = 5.0 \times 10^{-7}\,mol^2\,dm^{-6}$
>
> So in a saturated solution
>
> $[Sr^{2+}] = [SO_4^{2-}] = \sqrt{(5.0 \times 10^{-7})}$
> $= 7.1 \times 10^{-4}\,mol\,dm^{-3}$
>
> One mole of $SrSO_4$ dissolves to give one mole of Sr^{2+} ions, so the solubility of $SrSO_4$ is **$7.1 \times 10^{-7}\,mol\,dm^{-3}$**.

The situation is more complex if the salt has ions with different charges, for example calcium hydroxide. The solubility of $Ca(OH)_2$ is $1.5 \times 10^{-2}\,mol\,dm^{-3}$. But here one mole of $Ca(OH)_2$ dissolves to give one mole of Ca^{2+} ions and two moles of OH^- ions.

Let $[Ca(OH)_2]$ in the saturated solution be x. Then $[Ca^{2+}] = x$ and $[OH^-] = 2x$. The solubility product is given by:

$$K_{sp} = [Ca^{2+}][OH^-]^2$$
$$= x \times (2x)^2$$
$$= 4x^2$$
$$= 4 \times (1.5 \times 10^{-2})^3 = 1.35 \times 10^{-5}\,mol^3\,dm^{-9}$$

The connection between solubility and solubility product is summarised in Table 22.3.

Table 22.3 The relationship between the solubility ($x\,mol\,dm^{-3}$) and solubility product for salts with different ionic types

Charge on cation	Charge on anion	Example	Expression for K_{sp}
1	1	AgBr	$x^2\,mol^2\,dm^{-6}$
2	2	$MgCO_3$	$x^2\,mol^2\,dm^{-6}$
1	2	Ag_2CrO_4	$4x^3\,mol^3\,dm^{-9}$
2	1	PbI_2	$4x^3\,mol^3\,dm^{-9}$
3	1	$Fe(OH)_3$	$27x^4\,mol^4\,dm^{-12}$

Now try this

1 The solubility of nickel hydroxide is $1.0 \times 10^{-4}\,mol\,dm^{-3}$. What is its solubility product?
2 The solubility product of silver sulfate is $1.6 \times 10^{-5}\,mol^3\,dm^{-9}$. What is its solubility?

These connections between solubility and solubility product can only be used if no other ions have been added which would disturb the equilibrium. If, for example, we have a saturated solution of silver chloride, its solubility will be decreased if we increase either the $[Ag^+(aq)]$ or the $[Cl^-(aq)]$ term, for example, by adding silver nitrate or sodium chloride. This would cause silver chloride to precipitate out of the solution, by what is known as the **common ion effect**.

Now try this

Aqueous sodium hydroxide is added to aqueous calcium chloride to precipitate calcium hydroxide. The final concentration of hydroxide ions is $0.50\,mol\,dm^{-3}$. What is the solubility of the calcium hydroxide remaining in solution? [K_{sp} $Ca(OH)_2 = 1.4 \times 10^{-5}\,mol^3\,dm^{-9}$]

Worked example

If sodium chloride is added to a saturated solution of silver chloride, so that $[Cl^-(aq)] = 0.10\,mol\,dm^{-3}$, what is the maximum value for $[Ag^+(aq)]$ in the solution? [$K_{sp}(AgCl) = 1.0 \times 10^{-10}\,mol^2\,dm^{-6}$]

Answer

$$K_{sp} = [Ag^+][Cl^-] \quad so \quad [Ag^+] = \frac{K_{sp}}{[Cl^-]} = \frac{1.0 \times 10^{-10}}{0.10} = \mathbf{1.0 \times 10^{-9}\,mol\,dm^{-3}}$$

22.6 The partition coefficient

A special case of equilibrium is how a solute divides between two immiscible liquid phases. If we take a solute such as iodine, which is soluble in a variety of solvents, it is not surprising to find that it is not equally soluble in all the solvents. Taking the two immiscible solvents water and hexane, we find that iodine is much more soluble in hexane than it is in water. Methanol, on the other hand, is much more soluble in water than it is in hexane.

Now try this

Considering the various intermolecular forces that might operate, explain why iodine is more soluble in hexane than it is in water, whereas methanol is more soluble in water.

Figure 22.4 Iodine partitioned between water and hexane

When some iodine crystals are shaken with a mixture of hexane and water until no further change takes place, and the two layers allowed to separate (see Figure 22.4), we find that the ratio of the concentrations of iodine in each layer is a constant. This is true if we use only a small amount of iodine, or a much larger amount. This constant is the equilibrium constant for the change:

$$I_2(aq) \rightleftharpoons I_2(hexane)$$

$$K = \frac{[I_2(hexane)]}{[I_2(aq)]}$$

K is the **partition coefficient** of iodine between hexane and water. An alternative phrase is **distribution coefficient**. Like all equilibrium constants, the value of a partition coefficient changes with temperature.

It is important always to write an equation alongside a value of a partition coefficient. Following the usual equilibrium constant expression:

$$K = \frac{[products]}{[reactants]}$$

we need to know which solution is the 'product' and which is the 'reactant'. There is no universally accepted convention, but usually the organic solution is the 'product' on the right-hand side.

The equilibrium is a dynamic one: iodine molecules are constantly crossing the interface between the two solvents. Equilibrium is established when the rates of these two processes are equal:

rate of leaving aqueous layer = $k_1[I_2(aq)]$
rate of leaving hexane = $k_2[I_2(hexane)]$

Figure 22.5 At equilibrium, the rates at which I_2 leaves each layer are equal.

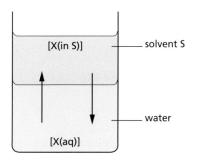

(k_1 and k_2 are rate constants).
At equilibrium,

$$k_1[I_2(aq)] = k_2[I_2(hexane)]$$

$$\frac{[I_2(hexane)]}{[I_2(aq)]} = \frac{k_1}{k_2} = K$$

Notice that partition coefficients do not usually have units.

The technique of **solvent extraction** depends on partition. This is used when an organic compound is extracted from an aqueous solution by an organic solvent, which is immiscible with water. Examples would be if the aqueous solution contained a perfume component or a pharmaceutical extracted from a plant, or contained the product formed during a synthetic reaction carried out in a laboratory. If we know the value of the partition coefficient for the compound, we can calculate how much would be extracted into the organic layer.

Industrially, octan-1-ol is often used as the organic solvent.

Worked example

The partition coefficient for the distribution of cyclohexanone between water and octanol is 24.

(aq) $\rightleftharpoons$ (in octanol) $K = 24$

Calculate the mass of cyclohexanone that would be extracted if $20\,cm^3$ of octanol was shaken with $100\,cm^3$ of an aqueous solution containing $2.5\,g$ of cyclohexanone.

Answer

Let [X] be the concentration of cyclohexanone expressed in $g\,cm^{-3}$; let the mass of cyclohexanone extracted be $x\,g$; the mass remaining in the aqueous layer will therefore be $(2.5 - x)\,g$

$$K = \frac{[X(\text{octanol})]}{[X(\text{aq})]}$$

$$[X(\text{octanol})] = [X(\text{aq})] \times K$$

$$\frac{x}{20} = \left(\frac{2.5 - x}{100}\right) \times 24$$

$$5x = (2.5 - x) \times 24$$

$$5x = 60 - 24x$$

$$\boldsymbol{x = 2.07\,g}$$

This process has extracted about 83% of the 2.5 g of cyclohexanone that was in the water.

If we wanted to extract more cyclohexanone, we could repeat the process. But this would require another $20\,cm^3$ of octanol. However, if we were to split the original $20\,cm^3$ of octanol into two $10\,cm^3$ portions, and use these for successive extractions, our yield of extracted material would increase. The following calculations will make this clear.

1 Let the mass of cyclohexanone extracted by the first $10\,cm^3$ portion of octanol be $y\,g$. The mass remaining in the water will therefore be $(2.5 - y)\,g$

$$\frac{y}{10} = \left(\frac{2.5 - y}{100}\right) \times 24$$

$$10y = (2.5 - y) \times 24$$

$$10y = 60 - 24y$$

$$\boldsymbol{y = 1.765\,g}$$

This first extraction results in $2.5 - 1.765 = 0.735\,g$ of cyclohexanone remaining in the aqueous layer.

2 We can now calculate how much of this remaining cyclohexanone can be extracted by using the second $10\,cm^3$ portion of octanol.

Let the mass of cyclohexanone extracted by the second $10\,cm^3$ portion of octanol be $z\,g$. The mass remaining in the water will be $(0.735 - z)\,g$.

$$\frac{z}{10} = \left(\frac{0.735 - z}{100}\right) \times 24$$

$$10z = 17.64 - 24z$$

$$\boldsymbol{z = 0.519\,g}$$

The total mass of cyclohexanone extracted is $y + z = 1.765 + 0.519 = \boldsymbol{2.28\,g}$

The use of two successive extractions has now raised the percentage extracted from

$$83\% \text{ to } 100 \times \left(\frac{2.28}{2.5}\right) = \boldsymbol{91\%}$$

If a higher degree of extraction is required (when dealing with expensive perfumery components, for example), the process can be automated into a continuous extraction apparatus.

Now try this

When $100\,cm^3$ of an aqueous solution containing 4.0 g of ketone **Y** was shaken with $25\,cm^3$ of hexane, 3.0 g of **Y** was extracted into the hexane.

a Calculate the partition coefficient of **Y** between hexane and water.
b What volume of hexane would be needed to extract 90% of **Y** from the $100\,cm^3$ of its aqueous solution, using only one extraction?

Summary

- In dilute aqueous solution, $[H_2O]$ is a constant, $55.5\,mol\,dm^{-3}$.
- For a weak acid, HA, its degree of ionisation may be calculated using the formula:

  $$K_a = \frac{[H^+][A^-]}{[HA]}$$

- In aqueous solution, the product $[H^+][OH^-]$ is a constant, the **ionic product of water**, K_w, whose value is $1.0 \times 10^{-14}\,mol^2\,dm^{-6}$ at room temperature.
- $pH = -\log_{10}[H^+]$
- For a strong acid or base, $[H^+]$ or $[OH^-] = c$, where c is the concentration of the acid or base.
- For a weak acid, $[H^+] \approx \sqrt{(K_a \times c)}$
- A **buffer solution** is one that resists changes in pH on the addition of small quantities of acids or bases.
- A buffer solution is usually a mixture of a weak acid and the salt of that weak acid. The pH of the buffer solution can be found using the formula:

  $$[H^+] = \frac{K_a[\text{acid}]_{initial}}{[\text{salt}]_{initial}}$$

- When an acid is titrated with base, there is a rapid change in pH at the equivalence point. An appropriate **indicator** can be used to find this equivalence point.
- When a solid salt M^+X^- is in equilibrium with its saturated solution, the product $[M^+][X^-]$ is a constant, the **solubility product**, K_{sp}.
- When an acid is titrated with base, there is a rapid change in pH at the equivalence point. An appropriate **indicator** can be used to find this equivalence point.
- If either excess M^+ ions or X^- ions are added, the solubility of the salt decreases; this is called the **common ion effect**.
- When a substance X dissolves in two immiscible liquids A and B, the ratio:

  $$\frac{\text{solubility of X in solvent A}}{\text{solubility of X in solvent B}}$$

  is a constant; this constant is called the **partition coefficient**.

Examination practice questions

Please see the data section of the CD for any A_r values you may need.

1 a The K_a values for some organic acids are listed below.

Acid	K_a/mol dm^{-3}
CH_3CO_2H	1.7×10^{-5}
$ClCH_2CO_2H$	1.3×10^{-3}
Cl_2CHCO_2H	5.0×10^{-2}

  i Explain the trend in K_a values in terms of the structures of these acids.

  ii Calculate the pH of a 0.10 mol dm^{-3} solution of $ClCH_2CO_2H$.

  iii Use a copy of the following axes to sketch the titration curve you would obtain when 20 cm^3 of 0.10 mol dm^{-3} NaOH is added gradually to 10 cm^3 of 0.10 mol dm^{-3} $ClCH_2CO_2H$.

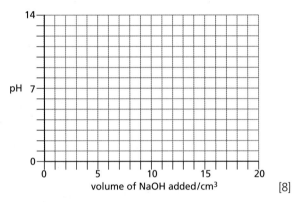

volume of NaOH added/cm^3 [8]

b i Write suitable equations to show how a mixture of ethanoic acid, CH_3CO_2H, and sodium ethanoate acts as a buffer solution to control the pH when either an acid or an alkali is added.

  ii Calculate the pH of a buffer solution containing 0.10 mol dm^{-3} ethanoic acid and 0.20 mol dm^{-3} sodium ethanoate.   [4]

[Cambridge International AS & A Level Chemistry 9701, Paper 4 Q1 b & c June 2009]

2 a i With the aid of a fully-labelled diagram, describe the standard hydrogen electrode.

  ii Use the *Data Booklet* on the CD to calculate the standard cell potential for the reaction between Cr^{2+} ions and $Cr_2O_7^{2-}$ ions in acid solution, and construct a balanced equation for the reaction.

  iii Describe what you would see if a blue solution of Cr^{2+} ions was added to an acidified solution of $Cr_2O_7^{2-}$ ions until reaction was complete.

b A buffer solution is to be made using 1.00 mol dm^{-3} ethanoic acid, CH_3CO_2H, and 1.00 mol dm^{-3} sodium ethanoate, CH_3CO_2Na.

Calculate to the nearest 1 cm^3 the volumes of each solution that would be required to make 100 cm^3 of a buffer solution with pH 5.50. Clearly show all steps in your working.

K_a (CH_3CO_2H) $= 1.79 \times 10^{-5}$ mol dm^{-3} [4]

c Write an equation to show the reaction of this buffer solution with each of the following.

  i added HCl

  ii added NaOH   [2]

d Choose **one** reaction in organic chemistry that is catalysed by an acid, and write the structural formulae of the reactants and products.   [3]

[Cambridge International AS & A Level Chemistry 9701, Paper 42 Q2 June 2013]

23 Electrochemistry

In this topic, we study how a chemical reaction can produce electricity. The standard electrode potential, $E^{\ominus}$, of an electrode system is a measure of the ease of reduction of an atom or ion. The difference between two electrode potentials gives rise to a cell potential, $E^{\ominus}_{cell}$, the size and sign of which indicates the feasibility of a reaction. In contrast, many reactions that are not otherwise feasible can be carried out by electrolysis – by passing an electric current through the reaction mixture. By choosing appropriate conditions, a wide variety of useful chemicals can be manufactured.

Learning outcomes

By the end of this topic you should be able to:

6.2a) state and apply the relationship $F = Le$ between the Faraday constant, the Avogadro constant and the charge on the electron

6.2b) predict the identity of the substance liberated during electrolysis from the state of electrolyte (molten or aqueous), position in the redox series (electrode potential) and concentration

6.2c) calculate the quantity of charge passed during electrolysis, and the mass and/or volume of substance liberated during electrolysis, including those in the electrolysis of $H_2SO_4(aq)$ and of $Na_2SO_4(aq)$

6.2d) describe the determination of a value of the Avogadro constant by an electrolytic method

6.3a) define the terms *standard electrode (redox) potential* and *standard cell potential*

6.3b) describe the standard hydrogen electrode

6.3c) describe methods used to measure the standard electrode potentials of metals or non-metals in contact with their ions in aqueous solution, and ions of the same element in different oxidation states

6.3d) calculate a standard cell potential by combining two standard electrode potentials

6.3e) use standard cell potentials to explain/deduce the direction of electron flow in a simple cell, and predict the feasibility of a reaction

6.3f) deduce from $E^{\ominus}$ values the relative reactivity of elements of Group 17 (the halogens) as oxidising agents

6.3g) construct redox equations using the relevant half-equations (see also Topic 24)

6.3h) predict qualitatively how the value of an electrode potential varies with the concentrations of the aqueous ions

6.3i) use the Nernst equation, e.g. $E = E^{\ominus} + (0.059/z) \log \frac{[\text{oxidised species}]}{[\text{reduced species}]}$ to predict quantitatively how the value of an electrode potential varies with the concentrations of the aqueous ions; examples include $Cu(s) + 2e^- \rightleftharpoons Cu^{2+}(aq)$, $Fe^{3+}(aq) + e^- \rightleftharpoons Fe^{2+}(aq)$, $Cl_2(g) + 2e^- \rightleftharpoons 2Cl^-(aq)$

6.4a) state the possible advantages of developing other types of cell, e.g. the H_2/O_2 fuel cell and the nickel-metal hydride and lithium-ion rechargeable batteries.

23.1 The electrochemical cell

Measuring electrode potentials

If a metal, for example zinc, is placed in a solution containing its ions, an equilibrium is set up:

$$Zn(s) \rightleftharpoons Zn^{2+}(aq) + 2e^-$$

Initially the metal begins to dissolve in the solution as metal ions, leaving the electrons on the undissolved metal. This leaves the metal negatively charged. Soon an equilibrium is established in which the rate of dissolving is balanced by ions recovering electrons and re-forming the metal (see Figure 23.1). The more reactive the metal, the further to the right is the position of equilibrium and the larger is the negative charge on the metal. Therefore measuring this charge gives us a measure of the reactivity of the metal.

Figure 23.1 The equilibrium established by zinc metal in contact with its ions. A double layer of charges is formed at the surface of the metal.

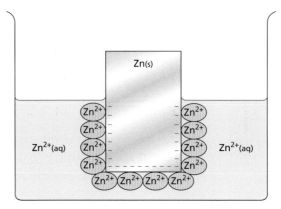

This charge cannot be measured by connecting one terminal of a voltmeter to the metal because a voltmeter measures **potential difference** (p.d.) and needs its other terminal to be connected to a second electrode at a different electrical potential. It would be convenient if this second electrode could be made by putting an inert metal, such as platinum, into the solution. Unfortunately this does not help, as platinum sets up its own potential which varies from experiment to experiment, depending on the solution it is immersed in, so inconsistent results are obtained. Another approach must be used.

The solution to the problem is to regard the zinc/zinc ion system as a **half-cell** and connect it to another half-cell. This will allow us to measure the potential difference between the two half-cells accurately and consistently. This means that we can only *compare* reactivities, rather than measuring them absolutely. The other half-cell could be, for example, a copper/copper ion system (see Figure 23.2).

Figure 23.2 Two half-cells are combined to make an electrochemical cell.

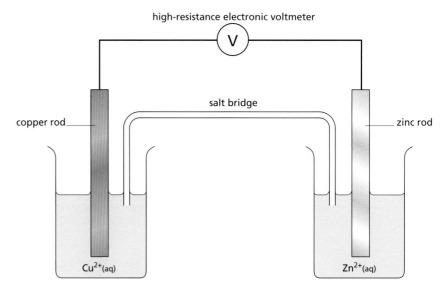

The voltage produced by the two half-cells depends on the conditions. As long as a high-resistance voltmeter is used, so that only a very small current is taken from the cell, and if the conditions are standard (298 K, 1.00 bar or 1.00 atm, solutions of $1.00\,mol\,dm^{-3}$), the voltage is the **standard cell e.m.f.**, $E^{\ominus}_{cell}$. (If an appreciable current is drawn from the cell, the measured voltage would be smaller than this.)

The need for a salt bridge

As mentioned above, the voltage is measured using a high-resistance electronic voltmeter. Even so, some electrons must flow and, because zinc is more reactive than copper, the following half-reactions take place to a small extent:

at the zinc electrode: $Zn(s) \rightarrow Zn^{2+}(aq) + 2e^-$
at the copper electrode: $Cu^{2+}(aq) + 2e^- \rightarrow Cu(s)$

This means that the zinc *solution* becomes positively charged, with an excess of $Zn^{2+}(aq)$ ions, and the copper *solution* becomes negatively charged, with an excess of $SO_4^{2-}(aq)$ ions, for example, if copper sulfate has been used. Because of this, the current would cease unless the circuit were completed by electrically connecting the two solutions. This cannot be done with a piece of wire, which passes only electrons, because we need to move positive *ions* one way and negative *ions* the other. (It does not matter which ions actually move, because very few of them are transferred compared with those already in the solutions.) The circuit is completed by a **salt bridge** dipping into the two solutions. Salt bridges are made from either a strip of filter paper soaked in an electrolyte (see Figure 23.3) or a bent tube packed with an electrolyte jelly, with porous plugs at the ends that allow ions to flow while minimising the mixing of the electrolytes by diffusion. The electrolyte in the salt bridge is usually potassium nitrate. This is used because all potassium compounds and all nitrates are soluble, and so no precipitate will form with any ions in contact with it.

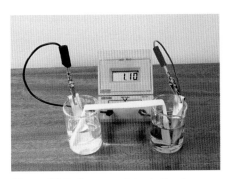

Figure 23.3 Copper and zinc half-cells connected together

The standard hydrogen electrode

The convention for determining the sign of the e.m.f of a cell is

$$E^\ominus_{cell} = E^\ominus_{\text{right-hand electrode}} - E^\ominus_{\text{left-hand electrode}}$$

For the copper/zinc cell shown in Figure 23.2, $E^\ominus_{cell} = -1.075\,V$, the negative sign showing that the zinc electrode is more negative than the copper electrode. This fits in with what we know of the relative reactivities of copper and zinc: zinc ions are much less readily reduced to the metal than copper ions. This value tells us the voltage of a standard zinc half-cell compared with a standard copper half-cell. Conventionally, it has been agreed to compare all half-cells against an electrode called the **standard hydrogen electrode (SHE)**. By international agreement, this has been assigned $E^\ominus = 0.00\,V$. The reaction is:

$$H^+(aq) + e^- \rightleftharpoons \tfrac{1}{2}H_2(g)$$

> The **standard hydrogen electrode** is the half-cell represented by the equation
>
> $H^+(aq) + e^- \rightleftharpoons \tfrac{1}{2}H_2(g)$, Pt under
>
> standard conditions.
>
> Standard conditions are:
> $p = 1.00\,atm$, $T = 298\,K$ and all concentrations = $1.00\,mol\,dm^{-3}$.

This reaction is very slow to reach equilibrium, and to speed it up the hydrogen is bubbled over a platinum catalyst. The platinum surface is **platinised** – it has a spongy layer of platinum electrolysed onto it, which increases its effective surface area. The platinum is also an inert electrode, and transfers electrons to and from the circuit without taking part in any chemical reaction.

Figure 23.4 shows the standard hydrogen electrode. Its half-cell reaction is represented as:

$$H^+(aq) + e^- \rightleftharpoons \tfrac{1}{2}H_2(g),\ Pt$$

Figure 23.4 The standard hydrogen electrode

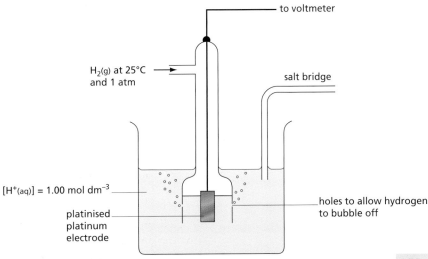

Standard electrode potential, $E^\ominus$

All half-cell reactions are written in the following format, with the electrons on the left:

$$\text{oxidised form} + ne^- \rightleftharpoons \text{reduced form}$$

If the voltage produced by a half-cell is measured against the standard hydrogen electrode, the cell voltage is

$$E^\ominus_{cell} = E^\ominus_{\text{right-hand electrode}} - E^\ominus_{\text{left-hand electrode}}$$

But since $E^\ominus_{\text{left-hand electrode}} = 0.00\,\text{V}$ for the SHE,

$$E^\ominus_{cell} = E^\ominus_{\text{right-hand electrode}}$$

and the cell voltage is the **standard electrode potential**, $E^\ominus$ of the half-cell. The reduced form of the substance is not necessarily the free element – it may be the element in a lower oxidation state. For example, in a solution containing $1.00\,\text{mol dm}^{-3}$ of $Fe^{2+}(aq)$ ions and $1.00\,\text{mol dm}^{-3}$ of $Fe^{3+}(aq)$ ions, electrons are transferred from the $Fe^{2+}(aq)$ ions to the $Fe^{3+}(aq)$ ions. In order to measure the voltage for this half-cell, these electrons can be carried away on a platinum wire placed in the solution. A shiny platinum wire is used for this electrode, because it is not acting as a catalyst. The half-cell is written:

$$Fe^{3+}(aq) + e^- \rightleftharpoons Fe^{2+}(aq), Pt$$

and a diagram of it is shown in Figure 23.5.

A common error is to draw this cell with two electrodes, one dipping into $Fe^{2+}(aq)$ ions and the other dipping into $Fe^{3+}(aq)$ ions. However, both ions need to be in the same solution so that the electrode potential of the *mixture* can be measured relative to the SHE.

Many standard electrode potentials have been measured, and a few of the more important ones are listed in Table 23.1.

These standard electrode potentials are sometimes called **redox potentials**. A large negative value of $E^\ominus$ indicates a highly reactive metal that is easily oxidised. A large positive value of $E^\ominus$ indicates a highly reactive non-metal that is easily reduced. The order of these $E^\ominus$ values matches the observed chemical reactivity of the substances concerned, though the large negative value for lithium is surprising, since chemically it is the least reactive of the Group 1 elements.

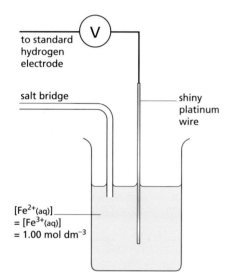

to standard hydrogen electrode

salt bridge

shiny platinum wire

[Fe²⁺(aq)] = [Fe³⁺(aq)] = 1.00 mol dm⁻³

Figure 23.5 The half-cell used to measure $E^\ominus$ for the Fe^{3+}/Fe^{2+} system. Here, $[Fe^{3+}]$ and $[Fe^{2+}]$ are shown as $1.00\,\text{mol dm}^{-3}$. This is not essential – the important feature is that they are equal.

Cell diagrams

Instead of sketching the two half-cells, a **cell diagram** can be drawn. Cell diagrams are not really diagrams at all: they are an indication of all the chemical species involved in the whole cell make-up (including state symbols and inert electrodes, where used), written out on a single line.

Figure 23.2 (page 382) is a picture of a cell in which a zinc half-cell is compared with a copper half-cell. The cell diagram for these two half-cells is:

$$Cu(s) \,|\, Cu^{2+}(aq) \,\vdots\, Zn^{2+}(aq) \,|\, Zn(s)$$

The solid vertical lines indicate a phase boundary, with the reduced forms on the outside, and the dotted line represents the salt bridge. (Sometimes a double dotted line is used for the salt bridge.) The concentrations of the solutions are assumed to be $1.00\,\text{mol dm}^{-3}$ unless a different concentration is shown underneath the ions. The e.m.f. of this cell is the potential differences between the zinc rod (right-hand half-cell) and that of the copper rod (left-hand half-cell).

Table 23.1 Values of $E^{\ominus}$ for some common half-reactions. In water, in which $[H^+(aq)] = 10^{-7}\,mol\,dm^{-3}$ (that is, the conditions are non-standard), $E(H^+/\frac{1}{2}H_2) = -0.41\,V$ and $E(\frac{1}{2}O_2/H_2O) = +0.82\,V$.

Oxidised form	Reduced form	$E^{\ominus}/V$
Weakest oxidising agents	strongest reducing agents	
$Li^+(aq)$	$Li(s)$	−3.03
$K^+(aq)$	$K(s)$	−2.92
$Na^+(aq)$	$Na(s)$	−2.71
$Mg^{2+}(aq)$	$Mg(s)$	−2.37
$Al^{3+}(aq)$	$Al(s)$	−1.66
$Zn^{2+}(aq)$	$Zn(s)$	−0.76
$Fe^{2+}(aq)$	$Fe(s)$	−0.44
$H^+(aq)$	$\frac{1}{2}H_2(g), Pt$	0.00
$Cu^{2+}(aq)$	$Cu(s)$	+0.34
$\frac{1}{2}I_2(aq)$	$I^-(aq), Pt$	+0.54
$Fe^{3+}(aq)$	$Fe^{2+}(aq), Pt$	+0.77
$Ag^+(aq)$	$Ag(s)$	+0.80
$\frac{1}{2}Br_2(aq)$	$Br^-(aq), Pt$	+1.09
$\frac{1}{2}O_2(g), Pt + 2H^+(aq)$	$H_2O(l)$	+1.23
$\frac{1}{2}Cl_2(aq)$	$Cl^-(aq), Pt$	+1.36
$MnO_4^-(aq) + 8H^+(aq)$	$Mn^{2+}(aq), Pt$	+1.51
$\frac{1}{2}F_2(aq)$	$F^-(aq), Pt$	+2.87
strongest oxidising agents	weakest reducing agents	

Worked example

Write a cell diagram for:

a a magnesium electrode compared with the standard hydrogen electrode
b a $Fe^{3+}(aq)/Fe^{2+}(aq)$ half-cell compared with a zinc half-cell.

Answer

a $Pt, \frac{1}{2}H_2(g)\,|\,H^+(aq)\,\vdots\,Mg^{2+}(aq)\,|\,Mg(s)$

(Note: some textbooks use 'Pt $[\frac{1}{2}H_2(g)]$' rather than 'Pt, $\frac{1}{2}H_2(g)$'.)

b $Zn(s)\,|\,Zn^{2+}(aq)\,\vdots\,Fe^{3+}(aq), Fe^{2+}(aq)\,|\,Pt$

Here, $Fe^{3+}(aq)$ and $Fe^{2+}(aq)$ are separated by a comma because they are in the same phase.

Now try this

Write cell diagrams for:

1 a $Cr^{3+}(aq)/Cr^{2+}(aq)$ half-cell compared with a copper half-cell
2 an $Al^{3+}(aq)/Al(s)$ half-cell compared with a chlorine half-cell
3 a $[MnO_4^-(aq) + 8H^+(aq)]/Mn^{2+}(aq)$ half-cell compared with a bromine half-cell (include $H^+(aq)$ ions and $H_2O(l)$)
4 an $[O_2(g) + 2H^+(aq]$ half-cell compared with the SHE half-cell.

Calculating the standard e.m.f. of a cell, $E^{\ominus}_{cell}$

The values of $E^{\ominus}$ shown in Table 23.1 can be used to work out the standard e.m.f. of a cell, $E^{\ominus}_{cell}$. Conventionally this is the voltage of the right-hand half-cell measured against the left-hand half-cell. So:

The standard e.m.f. of a cell

$E^{\ominus}_{cell} = E^{\ominus}_{\text{right-hand half-cell}} - E^{\ominus}_{\text{left-hand half-cell}}$

Worked example

a Calculate $E^{\ominus}_{cell}$ for a copper electrode compared with a zinc electrode.

b Calculate $E^{\ominus}_{cell}$ for the following cell:

$Pt \mid Fe^{2+}(aq), Fe^{3+}(aq) \vdots Ag^+(aq) \mid Ag(s)$

Answer

a $E^{\ominus}_{cell} = +0.34 - (-0.76) = \mathbf{+1.10\,V}$
(It is a good idea always to include the sign.)

b $E^{\ominus}_{cell} = +0.80 - (+0.77) = \mathbf{+0.03\,V}$

Now try this

Draw the experimental set-up, write the cell diagram and calculate $E^{\ominus}_{cell}$ for each of the following cells.

1 a magnesium electrode compared with a zinc electrode
2 a chlorine electrode compared with a bromine electrode
3 an oxygen electrode compared with a copper electrode

23.2 Using $E^{\ominus}_{cell}$ values to measure the feasibility of reactions

When the two electrodes of a cell are connected to each other, electrons flow from the negative electrode to the positive electrode. Chemical changes take place at each electrode that reduce the voltage produced by the cell (the terminal p.d.), and finally the reaction stops because the two electrodes are at the same electrical potential. If $E^{\ominus}_{cell}$ is positive, the reaction corresponding to the cell diagram takes place from left to right. For example, the cell diagram:

$$Zn(s) \mid Zn^{2+}(aq) \vdots Cu^{2+}(aq) \mid Cu(s) \qquad E^{\ominus}_{cell} = +1.10\,V$$

corresponds to the reaction:

$$Zn(s) + Cu^{2+}(aq) \rightarrow Zn^{2+}(aq) + Cu(s)$$

and, as $E^{\ominus}_{cell}$ is positive, the reaction is feasible and should take place if zinc is added to copper sulfate solution. In this case, experiment confirms that the prediction is correct. In other cases, this does not happen. There are two reasons why this may be so, and we shall look at each in turn.

Non-standard conditions

When determining $E^{\ominus}_{cell}$, all solutions have a concentration of $1.00\,mol\,dm^{-3}$. This is unlikely to be the case if the reaction is carried out in a test tube. For the zinc/copper sulfate reaction, the voltage is so large and positive that changing the concentrations has no effect. However, the situation is different if $E^{\ominus}_{cell}$ is less than $+0.2\,V$ – under these conditions, a fairly small change in concentration could make the voltage become negative and the reaction is then not feasible. Non-standard conditions are discussed further in section 23.3.

High activation energy

Many reactions, particularly those involving gases and solutions, have a high activation energy. Although the $E^{\ominus}_{cell}$ value may indicate that the reaction is feasible, it is often so slow that no change is observed if it is carried out in the test tube. For example, $E^{\ominus}_{cell}$ for Cu^{2+}/Cu is $+0.34\,V$. This suggests that hydrogen gas should react with copper sulfate solution, but no reaction is observed when hydrogen is bubbled through a solution of copper sulfate. The slowness of the conversion of gases into ions is the reason why platinised platinum is used in the standard hydrogen electrode.

Just as the conversion of $H_2(g)$ to $H^+(aq)$ ions has a high activation energy, so too does the reverse process. This means that the production of hydrogen (and oxygen as well) by electrolysis often requires a greater voltage than is predicted by the $E^{\ominus}_{cell}$ values. This means that hydrogen and oxygen often need an **overvoltage** before they are discharged.

Worked example

Use $E^{\ominus}$ values from Table 23.1 to calculate $E^{\ominus}_{cell}$ for the following reactions, and hence predict whether they are feasible:

a $Zn(s) + Mg^{2+}(aq) \rightarrow Zn^{2+}(aq) + Mg(s)$

b $Cl_2(aq) + H_2O(l) \rightarrow \frac{1}{2}O_2(g) + 2Cl^-(aq) + 2H^+(aq)$

Answer

a $E^{\ominus}_{cell} = E^{\ominus}(Mg^{2+}/Mg) - E^{\ominus}(Zn^{2+}/Zn) = -2.37 - (-0.76) = \mathbf{-1.61\,V}$
  The reaction is not feasible.

b $E^{\ominus}_{cell} = E^{\ominus}(\frac{1}{2}Cl_2/Cl^-) - E^{\ominus}(\frac{1}{2}O_2/H_2O) = +1.36 - (+1.23) = \mathbf{+0.13\,V}$

  The reaction is feasible. Cl^- and H_2O are the reduced forms.

Now try this

1 Use $E^{\ominus}$ values from Table 23.1 to calculate $E^{\ominus}_{cell}$ and hence predict whether the following reactions are feasible.
  (Hint: you may find section 7.1 helpful in writing the relevant half-equations.)

  a $\frac{1}{2}H_2(g) + Ag^+(aq) \rightarrow H^+(aq) + Ag(s)$
  b $MnO_4^-(aq) + 5Cl^-(aq) + 8H^+(aq) \rightarrow Mn^{2+}(aq) + 4H_2O(l) + 2\frac{1}{2}Cl_2(aq)$
  c $Fe(s) + 2Fe^{3+}(aq) \rightarrow 3Fe^{2+}(aq)$
  d $Ag^+(aq) + Fe^{2+}(aq) \rightarrow Fe^{3+}(aq) + Ag(s)$

2 Iodide reacts slowly with aqueous peroxydisulfate(VI) ions to produce iodine and sulfate ions:

  $$2I^-(aq) + 2S_2O_8^{2-}(aq) \rightarrow I_2(aq) + 2SO_4^{2-}(aq)$$

  It has been found that the best catalysts for this reaction are transition metal cation-pairs whose $E^{\ominus}$ values lie between those of I_2/I^- ($+0.54\,V$) and $S_2O_8^{2-}/SO_4^{2-}$ ($+2.01\,V$). Thus Fe^{3+}/Fe^{2+} ($E^{\ominus} = +0.77\,V$), Ce^{4+}/Ce^{3+} ($E^{\ominus} = +1.70\,V$) and Co^{3+}/Co^{2+} ($E^{\ominus} = +1.81\,V$) are all effective. Anion-pairs such as ClO^-/Cl^- ($E^{\ominus} = +0.89\,V$) are not effective, however.
  a Suggest why the reaction can be catalysed by cations, but not by anions.
  b Use the $E^{\ominus}$ data given above to calculate $E^{\ominus}$ for each of the two redox reactions, and hence show that each reaction is feasible.

Figure 23.6 An untreated iron gate rusts when exposed to air and water.

Corrosion

All reactive metals **corrode** – they react with oxygen and water in the air. Some, for example aluminium and chromium, form a thin layer of oxide that protects them from further attack. Iron readily **rusts** to form hydrated iron(III) oxide, $Fe_2O_3.H_2O$ or FeO.OH. Unlike the oxides of aluminium and chromium, this does not stick well to the metal surface. It easily flakes off and exposes more of the surface to further corrosion.

Rusting is a complex electrochemical process that takes place most readily under the following conditions:

- some of the iron is in contact with air while other regions are not
- the iron is in contact with water containing salt or other ionic substances.

In contact with air, the following reaction takes place:

$$1\tfrac{1}{2}O_2 + 3H_2O + 6e^- \rightarrow 6OH^- \qquad E^\ominus = +0.81\,V \text{ at pH 7}$$

In the air-free region, iron dissolves:

$$2Fe \rightarrow 2Fe^{3+} + 6e^- \qquad\qquad E^\ominus = -0.04\,V$$

The overall equations are:

$$2Fe(s) + 1\tfrac{1}{2}O_2(g) + 3H_2O(l) \rightarrow 2Fe^{3+} + 6OH^-$$

$$2Fe^{3+} + 6OH^- \rightarrow 2FeO.OH(s) + 2H_2O$$

The electrons flow in the metal from the air-rich to the air-free region. This takes place more quickly if the resistance of the solution that completes the circuit is lowered by the addition of salt.

Corrosion can be minimised in several ways. These include:

- painting or covering the iron in plastic to exclude air
- coating iron with another metal, for example zinc, tin or chromium
- alloying the iron, for example with chromium
- fixing a more reactive metal, such as magnesium, to the iron. The reactive metal preferentially dissolves and the process is therefore called **sacrificial protection** or **anodic protection**. Magnesium dissolves, having a more negative $E^\ominus$ value ($-2.37\,V$) than iron, keeping the iron negative and discouraging the reaction:

$$Fe \rightarrow Fe^{3+} + 3e^-$$

23.3 The Nernst equation

Non-standard conditions

In Figure 23.2, we showed a zinc half-cell connected to a copper half-cell. Under standard conditions, $E^\ominus_{cell} = +1.10\,V$ and, when the cell passes current, electrons flow from the zinc to the copper rod. If we now reduce the concentration of the zinc ions in the zinc half-cell, the equilibrium

$$Zn^{2+}(aq) + 2e^- \rightleftharpoons Zn(s)$$

is displaced to the left and the negative charge on the zinc rod becomes bigger. Qualitatively this can be predicted from Le Chatelier's Principle (see section 9.2), and it always happens when the concentration of the oxidised form is reduced below the standard value of $1.0\,mol\,dm^{-3}$.

This change in the value of $E^\ominus_{cell}$ from non-standard conditions can be calculated using the Nernst equation, named after Walther Nernst (Figure 23.7).

Figure 23.7 Walther Nernst

For a half-cell against a standard hydrogen electrode at room temperature, this equation can be written as:

$$E = E^\ominus + (0.059/z) \log \frac{[\text{oxidised species}]}{[\text{reduced species}]}$$

where z is the number of electrons added to the oxidised species to form the reduced species.

A ten-fold change in concentration only affects the $E^\ominus$ value by 0.059 V for a single electron transfer and 0.030 V for the transfer of two electrons. These are very small changes and this is the reason why $E^\ominus$ values are such a good guide to the feasibility of the reaction, even under non-standard conditions.

Worked example 1

What is E for a copper electrode dipping into a solution of $0.00010\,\text{mol}\,\text{dm}^{-3}$ $Cu^{2+}(aq)$ ions?
$[Cu^{2+} + 2e^- \rightleftharpoons Cu \quad E^\ominus = +0.34\,V]$

Answer

$$E = E^\ominus + (0.059/z) \log \frac{[\text{oxidised species}]}{[\text{reduced species}]}$$

The reduced species is metallic copper which is also the reduced species in the standard electrode potential. This means that the value of [reduced species] has been incorporated into $E^\ominus$ and can be ignored.

$$E = +0.34 + \frac{0.059}{2} \log 1.0 \times 10^{-4}$$

$$= +0.34 + \frac{(0.059 \times -4)}{2}$$

$$= +0.34 - 0.118 = +0.22$$

$$= \mathbf{+0.22\,V}$$

However if the concentration changes are very large, the change in $E^\ominus_{\text{cell}}$ becomes significant. We will look at the following three examples:

- for a sparingly soluble salt
- for changes in pH
- for the formation of complex ions. (This will be considered in section 24.4.)

Sparingly soluble salt

Worked example 2

A silver electrode dips into a saturated solution of silver bromide which has a solubility of $7.1 \times 10^{-7}\,\text{mol}\,\text{dm}^{-3}$. What is E for this half-cell?
$[Ag^+ + e^- \rightleftharpoons Ag \quad E^\ominus = 0.80\,V]$

Answer

$$E = E^\ominus + (0.059/z) \log \frac{[\text{oxidised species}]}{[\text{reduced species}]}$$

The value of [reduced species] is incorporated in $E^\ominus$ and is ignored.

$$[Ag^+] = 7.1 \times 10^{-7}\,\text{mol}\,\text{dm}^{-3}$$

$$E = 0.80 + \frac{0.059}{1} \times (-6.15)$$

$$= \mathbf{+0.44\,V}$$

Changes in pH

A typical pH meter consists of a glass electrode (which allows the passage of H^+ ions) attached to a reference electrode by a salt bridge. Frequently the H^+ concentration varies between $1.0\,\text{mol}\,\text{dm}^{-3}$ (e.g. $1.0\,\text{mol}\,\text{dm}^{-3}$ HCl) and $10^{-14}\,\text{mol}\,\text{dm}^{-3}$ (e.g. $1.0\,\text{mol}\,\text{dm}^{-3}$ NaOH). This has a profound effect on the E_{cell} values and forms the basis

Now try this

1 K_{sp} for AgCl is $2.0 \times 10^{-10}\,\text{mol}^2\,\text{dm}^{-6}$. What is E for a silver electrode when it is placed in
  a a saturated solution of silver chloride
  b in a solution of silver chloride in $0.10\,\text{mol}\,\text{dm}^{-3}$ KCl?
2 (Harder) Brine is saturated sodium chloride solution, whose concentration is $0.50\,\text{mol}\,\text{dm}^{-3}$. When this solution is electrolysed, chlorine gas is formed until a saturated solution is formed. Because chlorine reacts with water, the concentration of free chlorine in a saturated solution is only $0.030\,\text{mol}\,\text{dm}^{-3}$. Calculate E for the reaction $Cl_2(aq) + 2e^- \rightleftharpoons 2Cl^-(aq)$ under these conditions.

for the accurate measurement of pH. Because pH = −log[H⁺], we can write the Nernst equation in the form

$$E_{cell} = E_{cell}^{\ominus} - 0.059 \times pH$$

Worked example 3

The reference electrode in a pH meter is a silver/silver chloride electrode with

$E_{cell}^{\ominus} = +0.23\,V$

What is the E_{cell} value when it and a glass electrode dip into a solution of pH 3?

Answer

$E_{cell} = E_{cell}^{\ominus} - 0.059 \times pH$
$E_{cell} = +0.23 - 3 \times 0.059$
$\quad\quad = \textbf{+0.053\,V}$

23.4 The Faraday constant

When current flows from the copper/zinc cell, the following changes take place:

at the negative electrode: $Zn(s) \rightarrow Zn^{2+}(aq) + 2e^{-}$
at the positive electrode: $Cu^{2+}(aq) + 2e^{-} \rightarrow Cu(s)$

During electrolysis, however, current is forced through the cell in the other direction and the reverse processes take place:

at the cathode: $Zn^{2+}(aq) + 2e^{-} \rightarrow Zn(s)$
at the anode: $Cu(s) \rightarrow Cu^{2+}(aq) + 2e^{-}$

In order to deposit one mole of zinc and dissolve one mole of copper, two moles of electrons need to be passed through the cell. One mole of electrons carries a charge of $-96\,500\,C$. This is the same charge as the charge on one electron, $1.603 \times 10^{-19}\,C$, multiplied by the Avogadro constant, $6.022 \times 10^{23}\,mol^{-1}$. Numerically, the same amount of charge is carried by one mole of protons; this quantity is called the **Faraday constant**, F, and has a value of $96\,500\,C\,mol^{-1}$.

The **coulomb** (symbol C) is the unit of electrical charge. One coulomb is the amount of charge that is passed when one ampere flows for one second. Hence the amount of charge passed in time t is:

$$I \times t$$

Figure 23.8 Michael Faraday (1791–1867). One of his many achievements was to establish a quantitative basis for electrochemistry.

The amount of charge on one mole of $z+$ ions is zF, and so the number of moles deposited is given by:

$$amount = \frac{charge\ passed}{charge\ per\ mole\ of\ ions} = \frac{I \times t}{zF}\ mol$$

where z is the charge on the ion, I the current in amperes and t the time in seconds.

The amount liberated may be converted to a mass by multiplying by the molar mass. Alternatively, if the products are gaseous, the amount may be multiplied by $24\,dm^3$ to obtain the volume of gas produced at room temperature and pressure.

Electrolysis may also be used to find the value of the Avogadro constant. The value of the Faraday constant is first determined from an electrolysis experiment, for example by depositing copper from a solution containing $Cu^{2+}(aq)$ ions. A steady current, I, is passed for a measured time, t (in seconds), and the mass, m, of copper deposited is measured. We have:

$$\frac{m}{63.5} = \frac{I \times t}{2F}$$

The electrolysis experiment gives a value for the Faraday constant of $96\,500\,C\,mol^{-1}$, that is, $96\,500\,C$ is one mole of electrons. Experiments in physics have shown that the charge on each electron is $1.60 \times 10^{-19}\,C$.

Now try this

1 Calculate the mass of aluminium produced when a current of 1000 A is passed through a cell containing Al^{3+} ions for 1 hour.

2 a Show that 0.0016 mol of oxygen is produced when a current of 1.0 A is passed through sulfuric acid solution for 10 minutes.

  b Calculate the volume of oxygen that should be liberated.

  c Suggest why the actual volume collected may be less than this.

3 Explain why:

  a hydrogen and oxygen are evolved when aqueous sodium hydroxide is electrolysed

  b when aqueous copper sulfate is electrolysed with platinum electrodes, the cathode becomes plated with copper and oxygen is evolved at the anode

  c when aqueous copper sulfate is electrolysed with copper electrodes, the anode dissolves.

Figure 23.9 Voltage–current graph for a copper/zinc cell. At 1.10 V, no current flows. Above 1.10 V, electrolysis takes place. The reaction runs in reverse and it is $Zn^{2+}(aq)$ that is reduced.

Thus:

$$L \times 1.603 \times 10^{-19} = 96\,500$$

$$L = \frac{96\,500}{1.603 \times 10^{-19}}$$

$$= 6.03 \times 10^{23}\,\text{mol}^{-1}$$

Worked example

Calculate the mass of copper that dissolves when a current of 0.75 A is passed through a zinc/copper cell for 45 minutes. [$A_r(Cu) = 63.5$]

Answer

$$\text{amount} = \frac{I \times t}{zF} = \frac{0.75 \times 45 \times 60}{2 \times 96\,500} = 0.0105\,\text{mol}$$

mass of copper $= 0.0105 \times 63.5 = \textbf{0.67\,g}$

23.5 Electrolysis

Pushing the cell reaction in the opposite direction

If the electrodes of a cell are connected together, electrons flow from the negative to the positive terminal. If, instead, the terminals are connected to an external voltage supply and the applied voltage is gradually increased, electrons still flow as before until the external voltage equals E_{cell}. At this voltage no current flows. If the external voltage is increased still further, current flows in the opposite direction and **electrolysis** takes place (see Figure 23.9).

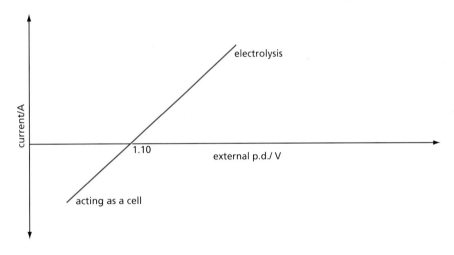

During electrolysis, positively charged ions, called **cations**, gain electrons at the negatively charged electrode (the **cathode**). At the same time negatively charged ions, called **anions**, give up electrons at the positively charged electrode (the **anode**).

    E_{cell} therefore is the minimum voltage required to bring about electrolysis. In practice, the voltage used for electrolysis is always greater than this minimum. There are two reasons for this.

- The cell has resistance, so an additional voltage is needed to drive current through the cell.
- The cell discharge often requires an 'overvoltage' to overcome a high activation energy associated with the discharge. This is particularly important when hydrogen or oxygen is being produced by electrolysis of aqueous solutions.

Molten salts

When solid, a salt will not conduct electricity because the ions are not free to move. When molten, however, the ions can move and the salt starts to conduct electricity. The conductivity often increases as the temperature is raised because there is less viscous drag on the ions to limit how fast they can move.

The electrolysis of molten salts is used to produce reactive elements, such as the alkali metals, that cannot be formed by electrolysing aqueous solutions, because they react with water. For example:

$$2NaCl(l) \xrightarrow{\text{electrolyse molten NaCl}} 2Na(l) + Cl_2(g)$$

Selective discharge

In an electrolysis cell, the electrolyte may contain several cations or anions. Under these circumstances, the one that is discharged is the one that requires the least energy. This is called **selective discharge**. For example, if a mixture of copper chloride and zinc chloride is electrolysed, copper rather than zinc is deposited at the cathode. This is because $E^{\ominus}$ for the Cu^{2+}/Cu half-cell is closer in value to $E^{\ominus}$ for the Cl_2/Cl^- half-cell than $E^{\ominus}$ for the Zn^{2+}/Zn half-cell is. This shows that copper requires a smaller potential difference to be discharged than zinc does (see Table 23.2).

Table 23.2 In a solution containing copper chloride and zinc chloride, copper rather than zinc is deposited at the cathode because a smaller voltage is needed to bring about the electrolysis.

Cation	Anion	$E^{\ominus}_{cell}$
$E^{\ominus}(Zn^{2+}/Zn) = -0.76\,V$	$E^{\ominus}(\frac{1}{2}Cl_2/Cl^-) = +1.36\,V$	2.12 V
$E^{\ominus}(Cu^{2+}/Cu) = +0.34\,V$	$E^{\ominus}(\frac{1}{2}Cl_2/Cl^-) = +1.36\,V$	1.02 V

In aqueous solutions, $H_3O^+(aq)$ and $OH^-(aq)$ ions are present and so selective discharge takes place. If dilute sulfuric acid is electrolysed, for example, hydrogen is liberated at the cathode because the only cation present is the $H_3O^+(aq)$ ion. There are, however, both $OH^-(aq)$ ions and $SO_4^{2-}(aq)$ ions in solution. A study of $E^{\ominus}$ values (see Table 23.3) predicts that oxygen should be liberated at the anode:

$$2OH^-(aq) \rightarrow H_2O(l) + O_2(g) + 2e^-$$

Table 23.3 The $E^{\ominus}_{cell}$ values predict that hydrogen and oxygen are evolved in the electrolysis of dilute sulfuric acid.

Cathode	Anode	$E^{\ominus}_{cell}$
$E^{\ominus}(H^+/\frac{1}{2}H_2) = 0.0\,V$	$\frac{1}{2}O_2/OH^- = +1.23\,V$ $[OH^-] = 10^{-14}\,mol\,dm^{-3}$	1.23 V
$E^{\ominus}(H^+/\frac{1}{2}H_2) = 0.0\,V$	$\frac{1}{2}S_2O_8^{2-}/SO_4^{2-} = 2.01\,V$	2.01 V

Because both hydrogen and oxygen have a high overvoltage, they are often not discharged even though $E^{\ominus}$ values suggest that they should be. For example, the electrolysis of brine usually yields chlorine rather than oxygen and it is possible in aqueous solution to electroplate objects with metals (for example, nickel) that have negative values of $E^{\ominus}$ for their half-cells.

23.6 Other types of cells

Fuel cells

When the gases heated by burning a fossil fuel are used to drive a piston or to turn the blades of a turbine, the overall efficiency is always low, no matter how well the plant is designed. The typical efficiency of a motor car is 20%, and that of a power station is 40%. This is a thermodynamic limitation: the only way to raise the efficiency is to use a higher operating temperature. This in turn produces fresh problems.

In order to achieve higher efficiencies, efforts have been made to convert the fuel directly into electrical energy by means of a **fuel cell**. The obvious pollution-free fuel is hydrogen, which can be burnt in air to produce water. The reactions are the reverse of the electrolysis of water. In acidic solution, we have:

at the negative plate: $\qquad H_2(g) \rightarrow 2H^+(aq) + 2e^-$

at the positive plate: $\qquad \frac{1}{2}O_2 + 2H^+ + 2e^- \rightarrow H_2O(l)$

In alkaline conditions:

at the negative plate: $\qquad H_2(g) + 2OH^-(aq) \rightarrow 2H_2O(l) + 2e^-$

at the positive plate: $\qquad \frac{1}{2}O_2 + H_2O(l) + 2e^- \rightarrow 2OH^-(aq)$

The problem is that the reactions at the electrodes are slow, even when the electrodes are made of platinum, which acts as a catalyst. There is a further complication for cars in that the hydrogen must be transported, either in heavy cylinders or at very low temperatures as a liquid. Efforts are being made to carry the hydrogen in the form of transition metal hydrides, but these are costly and can absorb only a limited amount of hydrogen.

As an alternative to transporting hydrogen to fuel cars, methanol is a fuel that is easy to carry and can be broken down to produce carbon dioxide and hydrogen. This introduces an additional stage in the process, and there is also the danger that some poisonous carbon monoxide may be produced in addition to the carbon dioxide. Attempts are therefore being made to produce fuel cells that use methanol directly, but these cells are only partially successful as they have to be operated above room temperature.

In the final analysis, fuel cells are really only another form of storage battery. In the first place, the hydrogen must have been produced chemically, possibly by electrolysis, and this process is not pollution-free. At present, a promising line of development is to use a hybrid system, running the car on electricity in towns where pollution is the main problem, and then, on the open road, using a traditional fuel to recharge the batteries.

Rechargeable batteries

Theoretically, most cells can be recharged when they run down, but in practice there are difficulties if the recharging is carried out quickly. The problem may be that metals are not re-deposited in an even layer, or that gaseous hydrogen and oxygen are given off.

The familiar lead–acid accumulator (the traditional 'car battery') has lead plates dipping into moderately concentrated sulfuric acid. After the first charging, the positive plate becomes covered with a layer of lead(IV) oxide. During discharge, the following reactions take place:

at the negative plate: $\qquad Pb(s) + H_2SO_4(aq) \rightarrow PbSO_4(s) + 2H^+(aq) + 2e^-$

at the positive plate: $\quad PbO_2(s) + H_2SO_4(aq) + 2H^+(aq) + 2e^- \rightarrow PbSO_4(s) + 2H_2O(l)$

During recharging, the reactions are reversed. Because sulfuric acid is a good conductor of electricity, the internal resistance of the lead–acid cell is very low. This means that it can produce very large currents, making it suitable for powering the starter motor in a car. Its disadvantage is its low power-to-weight ratio and the fact that the battery must be kept upright.

Because lead–acid cells are so heavy, other rechargeable cells have been developed which are more expensive but which give a better power-to-weight ratio. A common one is the nickel-hydride or NiMH cell. NiMH cells have largely taken over from the older nickel-cadmium cells because they have about three times the storage capacity. Like the nickel-cadmium cell, the positive electrode is nickel hydroxide and the electrolyte is potassium hydroxide. The main difference is that instead of cadmium

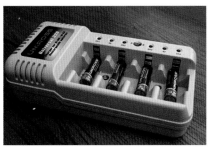

Figure 23.10 Rechargeable batteries. The chemical reactions that take place at each electrode must be reversible. Side reactions limit the number of recharging cycles that can be achieved – for most cells this is about 1000.

(which is highly toxic), the negative electrode is a metal M that can absorb hydrogen to make a hydride. When charging, the following changes take place:

at the positive electrode: $Ni(OH)_2 + OH^- \rightarrow NiO(OH) + H_2O + e^-$

at the negative electrode: $H_2O + M + e^- \rightarrow OH^- + MH$

The metal M is either a mixture of rare earth metals or a mixture of transition metals. A big advantage of NiMH cells is that when they are fully charged any oxygen given off combines with the metal hydride to form water. A cell can, therefore, be sealed and used in any position. A disadvantage is that the charge leaks away and after a fortnight a cell will have lost about half of its charge.

Other modern rechargeable cells are based on lithium. They have the following advantages over the lead–acid cell: they are light, they produce a large voltage and they can be used in a sealed container. Their disadvantages are that they are expensive and must be treated carefully as they contain very reactive chemicals. In these cells, lithium ions move from the positive to the negative electrode (not electrons as in the lead–acid cell). The negative electrode is a mixed lithium-transition metal oxide such as $LiCoO_2$. The positive electrode is graphite and the electrolyte is a lithium salt dissolved in an organic solvent.

When charging, the following reactions take place:

at the positive electrode: $LiCoO_2 \rightarrow CoO_2 + Li^+ + e^-$

at the negative electrode: $Li^+ + e^- + graphite \rightarrow Li(graphite)$

At the positive electrode, some of the cobalt changes from Co(III) to Co(IV), releasing electrons (which go to the positive terminal of the charger) and lithium ions (which migrate through the electrolyte to the negative electrode). At the negative electrode, lithium ions are attracted by the free electrons in the layers of graphite and form lithium atoms sandwiched between the layers of graphite. The electrons are supplied by the negative terminal of the charger.

During discharge, the reverse of the charging reactions takes place. At the negative electrode each lithium atom gives up its outer electron to form lithium ions. These electrons flow through external circuit, giving up their energy. At the same time, lithium ions move through the electrolyte and combine with CoO_2 to give $LiCoO_2$. This process requires the addition of the electrons which have moved through the external circuit. Other transition metals, such as iron, manganese and nickel, are also used instead of cobalt in the positive electrode.

In a discussion about the merits of each type of rechargeable cell, the following are some of the factors that need to be considered.

- Cost: lead–acid is the cheapest.
- Power-to-weight ratio: lithium is the best.
- Safety: NiMH can be used in any position and the chemicals they contain are relatively safe.

Summary

- A metal dipping into a solution of its ions forms a **half-cell**. When this is connected to another half-cell by means of salt bridge, an **electrochemical cell** is set up. Under standard conditions, the voltage set up by this cell is the **standard cell e.m.f.**, $E^{\ominus}_{cell}$, for the cell.

  $$E^{\ominus}_{cell} = E^{\ominus}_{\text{right-hand half-cell}} - E^{\ominus}_{\text{left-hand half-cell}}$$

- If the left-hand half-cell is the **standard hydrogen electrode**, the standard cell e.m.f. is the **standard electrode potential**, $E^{\ominus}$, for the right-hand half-cell.
- The value of $E^{\ominus}$ measures the oxidising/reducing power of the half-cell system, and the feasibility of redox reactions can be predicted from $E^{\ominus}_{cell}$ values.
- A redox reaction that is predicted to be feasible from $E^{\ominus}_{cell}$ values might not take place in practice, either because the conditions are non-standard or because the activation energy is very high.
- If the conditions are non-standard, the **Nernst equation** in the form $E = E1 + (0.059/z) \log \frac{[\text{oxidised species}]}{[\text{reduced species}]}$ can be used to calculate the cell voltage.

- **Electrolysis** is the process of driving a reaction that is not thermodynamically feasible by passing an electric current through it.
- The ions that are discharged in electrolysis are the ones that require the least energy.
- The amount of substance in moles dissolved or deposited during electrolysis is given by:

  $$\text{amount} = I \times \frac{t}{zF}$$

  where I is the current, t the time in seconds, z the charge on the ion and F the Faraday constant.
- If e is the charge on an electron, then $Le = F$.

Examination practice questions

Please see the data section of the CD for any A_r values you may need.

1 The electrolytic purification of copper can be carried out in an apparatus similar to the one shown below.

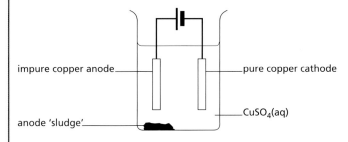

impure copper anode

pure copper cathode

anode 'sludge'

$CuSO_4(aq)$

The impure copper anode contains small quantities of metallic nickel, zinc and silver, together with inert oxides and carbon resulting from the initial reduction of the copper ore with coke. The copper goes into solution at the anode, but the silver remains as the metal and falls to the bottom as part of the anode 'sludge'. The zinc also dissolves.

a i Write a half-equation including state symbols for the reaction of copper at the anode.

ii Use data from the data section on the CD to explain why silver remains as the metal.

iii Use data from the data section on the CD to predict what happens to the nickel at the anode.

iv Write a half-equation including state symbols for the main reaction at the cathode.

v Use data from the data section on the CD to explain why zinc is not deposited on the cathode.

vi Suggest why the blue colour of the electrolyte slowly fades as the electrolysis proceeds. [7]

b Most of the current passed through the cell is used to dissolve the copper at the anode and precipitate pure copper onto the cathode. However, a small proportion of it is 'wasted' in dissolving the impurities at the anode which then remain in solution. When a current of 20.0 A was passed through the cell for 10.0 hours, it was found that 225 g of pure copper was deposited on the cathode.

i Calculate the following, using appropriate data from the data section on the CD.
- number of moles of copper produced at the cathode
- number of moles of electrons needed to produce this copper
- number of moles of electrons that passed through the cell

ii Hence calculate the percentage of the current through the cell that has been 'wasted' in dissolving the impurities at the anode. [4]

c Nickel often occurs in ores along with iron. After the initial reduction of the ore with coke, a nickel–iron alloy is formed. Use data from the data section on the CD to explain why nickel can be purified by a similar electrolysis technique to that used for copper, using an impure nickel anode, a pure nickel cathode, and nickel sulfate as the electrolyte. Explain what would happen to the iron during this process. [2]

[Cambridge International AS & A Level Chemistry 9701, Paper 41 Q3 November 2010]

2 Chlorine gas and iron(II) ions react together in aqueous solution as follows.

$$Cl_2 + 2Fe^{2+} \rightarrow 2Cl^- + 2Fe^{3+}$$

a The following diagram shows the apparatus needed to measure the $E^{\ominus}_{cell}$ for the above reaction.

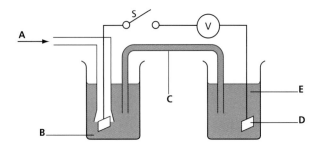

i Identify what the five letters **A – E** in the above diagram represent.

ii Use the data section on the CD to calculate the $E^{\ominus}_{cell}$ for this reaction, and hence decide which direction (left to right, or right to left) electrons would flow through the voltmeter V when switch S is closed. [7]

b Iron(III) chloride readily dissolves in water.

$$FeCl_3(s) \rightarrow Fe^{3+}(aq) + 3Cl^-(aq)$$

i Use the following data to calculate the standard enthalpy change for this process.

Species	$\Delta H^{\ominus}_f$/kJ mol^{-1}
$FeCl_3(s)$	−399.5
$Fe^{3+}(aq)$	−48.5
$Cl^-(aq)$	−167.2

ii A solution of iron(III) chloride is used to dissolve unwanted copper from printed circuit boards. When a copper-coated printed circuit board is immersed in $FeCl_3(aq)$, the solution turns pale blue.

Suggest an equation for the reaction between copper and iron(III) chloride and use the data section on the CD to calculate the $E^{\ominus}$ for the reaction. [4]

[Cambridge International AS & A Level Chemistry 9701, Paper 4 Q1 June 2008]

3 a State the relationship between the Faraday constant, F, the charge on the electron, e, and the Avogadro number, L. [1]

b If the charge on the electron, the A_r and the valency of copper are known, the value of the Avogadro number can be determined experimentally. This is done by passing a known current for a known time through a copper electrolysis cell, and weighing the mass of copper deposited onto the cathode.

i Draw a diagram of suitable apparatus for carrying out this experiment. Label the following: power supply (with + and − terminals); anode; cathode; and ammeter.

State the composition of the electrolyte.

The following are the results obtained from one such experiment.

current passed through the cell = 0.500 A
time current was passed through cell = 30.0 min
initial mass of copper cathode = 52.243 g
final mass of copper cathode = 52.542 g

ii Use these data and relevant information from the data section on the CD to calculate a value of L to **3 significant figures**. [9]

c Use relevant information from the data section on the CD to identify the substances formed at the anode and at the cathode when aqueous solutions of the following compounds are electrolysed. [5]

Compound	Product at anode	Product at cathode
AgF		
FeSO$_4$		
MgBr$_2$		

[Cambridge International AS & A Level Chemistry 9701, Paper 42 Q3 June 2011]

24 The 3d block

In this topic we study the 3d block, which includes many familiar metals such as iron and copper. The chemical properties of these metals and their compounds contrast with those of the metals in the s and p blocks. For example, the 3d-block elements can exist in several oxidation states and their ions are coloured, whereas the s- and p-block elements usually show only one oxidation state and have colourless ions. The complicated chemistry of the 3d block can be explained in terms of their electronic configurations and in particular by the closeness in energy of the 3d, 4s and 4p orbitals.

Learning outcomes

By the end of this topic you should be able to:

12.1a) explain what is meant by a *transition element*, in terms of d-block elements forming one or more stable ions with incomplete d orbitals

12.1b) sketch the shape of a d orbital

12.1c) state the electronic configuration of each of the first row transition elements and of their ions

12.1d) contrast, qualitatively, the melting points and densities of the transition elements with those of calcium as a typical s-block element

12.1e) describe the tendency of transition elements to have variable oxidation states

12.1f) predict from a given electronic configuration, the likely oxidation states of a transition element

12.2a) describe and explain the reactions of transition elements with ligands to form complexes, including the complexes of copper(II) and cobalt(II) ions with water and ammonia molecules and hydroxide and chloride ions

12.2b) define the term *ligand* as a species that contains a lone pair of electrons that forms a dative bond to a central metal atom/ion including monodentate, bidentate and polydentate ligands, define the term *complex* as a molecule or ion formed by a central metal atom/ion surrounded by one or more ligands, describe transition metal complexes as linear, octahedral, tetrahedral or square planar, and state what is meant by co-ordination number and predict the formula and charge of a complex ion, given the metal ion, its charge, the ligand and its co-ordination number

12.2c) explain qualitatively that ligand exchange may occur, including the complexes of copper(II) ions with water and ammonia molecules and hydroxide and chloride ions

12.2d) describe and explain the use of Fe^{3+}/Fe^{2+}, MnO_4^-/Mn^{2+} and $Cr_2O_7^{2-}/Cr^{3+}$ as examples of redox systems (see also Topic 23)

12.2e) predict, using $E^{\ominus}$ values, the likelihood of redox reactions

12.3a) describe the splitting of degenerate d orbitals into two energy levels in octahedral and tetrahedral complexes

12.3b) explain the origin of colour in transition element complexes resulting from the absorption of light energy as an electron moves between two non-degenerate d orbitals

12.3c) describe, in qualitative terms, the effects of different ligands on absorption, and hence colour, using the complexes of copper(II) ions with water and ammonia molecules and hydroxide and chloride ions as examples

12.3d) apply the above ideas of ligands and complexes to other metals, given information

12.4a) describe the types of stereoisomerism shown by complexes, including those associated with bidentate ligands: *cis-trans* isomerism, e.g. *cis-* and *trans*-platin $Pt(NH_3)_2Cl_2$, and optical isomerism, e.g. $[Ni(NH_2CH_2CH_2NH_2)_3]^{2+}$

12.4b) describe the use of cisplatin as an anticancer drug and its action by binding to DNA in cancer cells, preventing cell division

12.5a) describe and explain ligand exchanges in terms of competing equilibria (also see Topic 22)

12.5b) state that the stability constant, K_{stab}, of a complex ion is the equilibrium constant for the formation of the complex ion in a solvent from its constituent ions or molecules

12.5c) deduce expressions for the stability constant of a ligand substitution

12.5d) explain ligand exchange in terms of stability constants, K_{stab}, and understand that a large K_{stab} is due to the formation of a stable complex ion.

24.1 Introduction

The elements from scandium to zinc inclusive comprise the **3d block**. The 3d subshell contains five orbitals, each able to accommodate two electrons (see Topic 2), and so this block contains ten elements. The electronic configuration of scandium is $[Ar]3d^1 4s^2$ and that of zinc is $[Ar]3d^{10} 4s^2$. A 3d-block element is sometimes defined as one of the elements in which the 3d subshell is being progressively filled. This is not strictly speaking correct, because copper has the electronic configuration $[Ar]\ 3d^{10}4s^1$. It is more accurate to state that the 3d block contains elements with electronic configurations from $[Ar]\ 3d^1 4s^2$ to $[Ar]3d^{10} 4s^2$ inclusive.

Originally this block was known as the 'transition metals', because some of their properties show a gradual change between those of the reactive metal calcium in Group 2 to the much less reactive metal gallium in Group 13. The term **transition metal** is now reserved for those metals in the block that show properties characteristically different from those in the s and p blocks: they can exist in more than one oxidation state, and eir ions are often coloured. We exclude both scandium and zinc from the class of transition metals, for the following reasons.

- Scandium forms only the colourless Sc^{3+} ion, which is isoelectronic with the Ca^{2+} ion and has no electrons in the 3d subshell.
- Zinc forms only the colourless Zn^{2+} ion, which is isoelectronic with the Ga^{3+} ion and has 10 electrons in the 3d subshell.

It is the elements that form ions with *some* electrons in the 3d subshell that exhibit the special properties of transition metals.

The 3d elements and the transition metals

- The **3d block** contains elements in which the 3d subshell is being progressively filled.
- The 3d block includes all the elements with the electronic configurations $[Ar]3d^1 4s^2$ to $[Ar]3d^{10} 4s^2$ inclusive.
- The **transition metals** form some compounds containing ions with an incomplete d subshell.

The 4s subshell is filled before the 3d subshell because it is lower in energy (see section 2.12), even though the 4s subshell is part of a shell whose average distance is further away from the nucleus. The difference in energy between 4s and 3d is very small, however, which means that both s and d electrons may be involved in bonding.

The order of energy levels 4s < 3d holds only as far as calcium. With the increasing nuclear charge from 21 in scandium to 30 in zinc, the energy levels of both the 4s and 3d orbitals decrease, but the 3d level decreases faster than the 4s (see Figure 24.1 and Figure 2.25, page 34). At scandium their energy levels are almost the same, but subsequently the order changes so that 3d is slightly lower than 4s. When an ion is formed, the 4s electrons are removed before the 3d electrons. The reason for this reversal in energy levels is discussed in the next section.

Figure 24.1 The 3d and 4s energy levels on crossing the 3d block. The graph is not to scale.

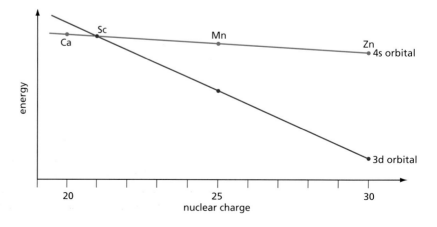

24.2 Properties of the metals

Table 24.1 The main properties of the elements in the 3d block

	Sc	Ti	V	Cr	Mn	Fe	Co	Ni	Cu	Zn
Electronic configuration [Ar]	$3d^1 4s^2$	$3d^2 4s^2$	$3d^3 4s^2$	$3d^5 4s^1$	$3d^5 4s^2$	$3d^6 4s^2$	$3d^7 4s^2$	$3d^8 4s^2$	$3d^{10} 4s^1$	$3d^{10} 4s^2$
Melting point/°C	1541	1660	1887	1857	1244	1535	1495	1453	1083	420
Boiling point/°C	2831	3287	3377	2672	1962	2750	2870	2732	2567	907
Density/g cm^{-3}	2.99	4.54	6.11	7.19	7.44	7.87	8.90	8.90	8.96	7.13
Metallic radius/nm	0.164	0.146	0.135	0.129	0.132	0.126	0.125	0.124	0.128	0.135
Conductivity/mS m^{-1}	1.6	1.2	4.0	7.9	0.54	10.2	16.0	14.6	59.9	16.9
M^{2+} radius/nm	—	0.090	0.079	0.073	0.067	0.061	0.078	0.070	0.073	0.075
M^{3+} radius/nm	0.083	0.067	0.064	0.062	0.062	0.055	0.053	0.056	—	—
ΔH_{at}/kJ mol^{-1}	378	470	514	397	281	416	425	430	338	131
First IE/kJ mol^{-1}	632	661	648	653	716	762	757	736	745	908
Second IE/kJ mol^{-1}	1240	1310	1370	1590	1510	1560	1640	1750	1960	1730
Third IE/kJ mol^{-1}	2390	2720	2870	2990	3250	2960	3230	3390	3350	3828
$E^{\ominus}$(M^{2+}/M)/V	—	−1.63	−1.20	−0.91	−1.18	−0.44	−0.28	−0.25	+0.34	−0.76
$E^{\ominus}$(M^{3+}/M^{2+})/V	—	−0.37	−0.26	−0.41	+1.49	+0.77	+1.82	—	—	—
Electronegativity	1.36	1.54	1.63	1.66	1.55	1.83	1.88	1.91	1.90	1.65

Electronic configuration

As the proton number increases by one unit, an extra electron is usually added to the 3d subshell. There are two exceptions to this general trend:

- chromium is $[Ar] 3d^5 4s^1$, and not $[Ar] 3d^4 4s^2$
- copper is $[Ar] 3d^{10} 4s^1$, and not $[Ar] 3d^9 4s^2$.

This suggests that $3d^5$ (with a half-filled subshell) and $3d^{10}$ (with a full subshell) are energetically preferred configurations, avoiding the inter-electron repulsion in the 4s orbital that occurs with the $4s^2$ configuration. Both these configurations have a symmetrical 3d cloud of electrons that screens the nucleus more effectively than other configurations.

Melting points, boiling points and conductivities

The melting and boiling points of the 3d-block metals are generally much higher than those of the s- and p-block metals. This indicates that not only the 4s electrons but also the 3d electrons are involved in the metallic bonding. When melting and boiling points are plotted against proton number, both graphs have two maxima, with an intermediate minimum at manganese (see Figure 24.2). These minima suggest that the half-filled subshell of 3d electrons is not readily available for metallic bonding. We might expect the same effect in chromium, but it occurs to a lesser degree because the nuclear charge is lower.

The conductivities of the 3d-block metals are similar to those of calcium (29.2 mS m^{-1}); the conductivity of copper is particularly high.

Metallic radii, ionic radii and densities

The metallic radii tend to decrease across the block as the increasing nuclear charge attracts the outer electrons more strongly. There are minor variations to this trend that can be rationalised in terms of the different strengths of the metallic bonding – for example, the metallic radii of both copper and zinc are larger than those of the preceding metals, and their metallic bonding is weaker, as shown by their comparatively low melting points. This decrease in metallic radius means that the densities of the transition metals are higher than that of calcium (1.55 g cm^{-3}); those at

Figure 24.2 Melting and boiling points of the 3d-block metals

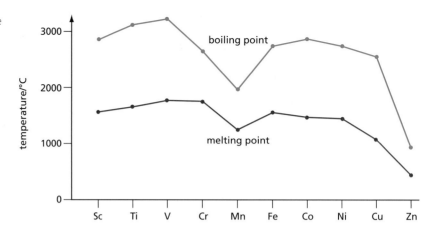

the end of the block have the highest densities, as they have the smallest atomic radii and the highest relative atomic masses.

There is no obvious trend in the M^{2+} ionic radii, but the M^{3+} ionic radii tend to decrease across the block as the nuclear charge increases and attracts the electrons more strongly. This decrease in ionic radius is accompanied by an increase in charge density of the ion, leading to an increased stability of the complexes formed towards the right of the block.

Ionisation energies

A graph of ionisation energies against proton number (see Figure 24.3) shows three main features.

Figure 24.3 Ionisation energies of the 3d-block metals

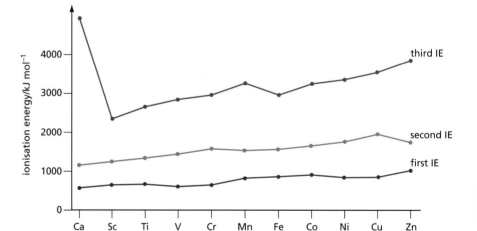

- The first and second ionisation energies increase only slightly across the block.
- The second ionisation energies are only slightly higher than the first.
- The third ionisation energies are significantly higher, and show a characteristic d-subshell pattern.

The first ionisation energies involve the removal of a 4s electron. This is outside the 3d subshell and is partially screened by it. As the nuclear charge increases across the block, the additional d electrons shield the effect of the increasing nuclear charge so that the 4s electron experiences only a small extra attraction. The same effect is shown by the second ionisation energies, with the exception of those for chromium and copper – these have rather higher second ionisation energies because the second electron being removed is a 3d electron, which does experience the increased nuclear charge.

The third ionisation energies involve the removal of a 3d electron. The pattern of five values steadily rising, followed by a drop and then five more steadily rising, mirrors the ionisation energies involving the removal of 2p electrons on crossing the second period (see Figure 2.33, page 41). The drop from manganese to iron is a reflection of the fact that two electrons in the same orbital repel each other (see Figure 24.4).

Figure 24.4 The electronic configurations of manganese and iron. The formation of the Fe^{3+} ion involves the removal of an electron from a doubly occupied 3d orbital; this electron is more easily removed because it is repelled by the other electron in the orbital.

3d and 4s energy levels

The relative energies of the 3d and 4s orbitals change on crossing the block (see Figure 24.1). Up to calcium, the energy of the 4s orbital is lower than that of the 3d, even though, on average, the electron is further away from the nucleus. This is because the 4s electron spends some time very near to the nucleus, where it experiences the full attraction of the unscreened nuclear charge. A 4s electron is said to be more **penetrating** than a 3d electron.

After calcium, the 4s electrons are screened from the effect of the increasing nuclear charge by the addition of the 3d electrons, which are nearer to the nucleus. The energy of the 4s electrons therefore decreases only slightly on crossing the 3d block. This small decrease in energy explains the small increase in first and second ionisation energies on crossing the 3d block from scandium to zinc. The 3d electrons, however, are not screened to the same extent and so experience a greater effective nuclear charge, becoming progressively more tightly held by the nucleus. The energy of the 3d electrons therefore decreases significantly on crossing the 3d block, as is shown by the sharper rise in the third ionisation energies (see Figure 24.3). (From scandium to zinc, the second ionisation energies increase by 40%, whereas the third ionisation energies increase by 60%.)

Worked example 1

Using [Ar] to represent the argon core, give the electronic configurations of the following species.
a Cr   b Cu^{2+} c V^{2+}

Answer
a $[Ar]3d^5 4s^1$ b $[Ar]3d^9$ c $[Ar]3d^3$

Now try this

1 Using [Ar] to represent the argon core, give the electronic configurations of the following species.
   a Cu   b Co^{2+} c Ti^{3+}
2 Briefly explain the following.
   a The second ionisation energy of copper is higher than the second ionisation energy of zinc.
   b It is difficult to oxidise $Mn^{2+}(aq)$ to $Mn^{3+}(aq)$.
   c On passing from scandium to titanium, the increase in the third ionisation energy is much larger than the increase in the first ionisation energy.

Briefly explain the following observations.

a The first ionisation energy of cobalt is only slightly larger than the first ionisation energy of iron.
b The third ionisation energy of iron is much lower than the third ionisation energy of manganese.
c The metallic radius of vanadium is smaller than that of titanium.

Answer

a The increase in nuclear charge from 26 to 27 is screened by the addition of an extra 3d electron. The effective nuclear charge, and hence the attraction of the 4s electrons, therefore increases only slightly.
b The third electron is being removed from a d^6 configuration, which has two electrons in one d orbital. These two electrons repel each other, making each one easier to remove.
c The increase in nuclear charge from 22 to 23 makes the inner clouds of electrons contract, and the radius of the atom decreases.

24.3 Variable oxidation states

Table 24.2 shows the more familiar oxidation states of some of the 3d elements. Many other oxidation states are known which can be stabilised under special conditions.

Table 24.2 The more familiar oxidation states of some of the 3d elements are marked ●. Other oxidation states are known but are formed only under special conditions. All the elements have an oxidation state of 0 in the metal.

Oxidation state	Cr	Mn	Fe	Cu	Zn
VII		●			
VI	●				
V					
IV					
III	●		●		
II		●	●	●	●
I				●	

Table 24.2 shows the following main features.

● Chromium and manganese show their highest oxidation state in CrO_4^{2-} and MnO_4^{-}. In these, their oxidation numbers are equal to the sum of the numbers of 3d and 4s electrons; this shows that all of these electrons are used in the bonding.
● Manganese, iron, copper and zinc all form M^{2+} ions by loss of the $4s^2$ electrons.
● Chromium and iron form M^{3+} ions as their third ionisation energies are relatively low.
● Copper forms Cu(I) compounds as it has a single 4s electron.

The presence of a high and a low oxidation state in chromium, manganese and iron means that these elements have important uses in redox chemistry, as is shown by the values of $E^\ominus$ in Table 24.3.

Table 24.3 Values of $E^\ominus$ for three common redox systems. The use of manganate(VII) and acid dichromate(VI) is discussed in Topic 16. Titrations using MnO_4^{-} are discussed in section 7.4.

Redox system	$E^\ominus$/V
$Cr_2O_7^{2-}/Cr^{3+}$	+1.33
MnO_4^{-}/Mn^{2+}	+1.52
Fe^{3+}/Fe^{2+}	+0.77

In titrations, dichromate(VI) and manganate(VII) are always used in acidic solution.

1 Write the half-equation (i.e. including electrons) showing the reduction of each of these oxidants in acidic solution.
2 Write the balanced ionic equation for the reaction between each of these oxidants with Fe^{2+} ions, and calculate the $E^\ominus_{cell}$ value for each reaction.

24.4 Complex formation

Ligands

The 3d metal ions are relatively small and have a high charge density. They therefore attract groups containing lone pairs of electrons. These groups are known as **ligands**. Ligands are bases, and also nucleophiles (see Table 15.1, page 270), and their ability to be attracted to metal ions follows a similar pattern to the nucleophilic strength found in organic chemistry. For common ligands an approximate order of attraction is as follows:

$$H_2O < \text{halide ions} < NH_3 < CN^-$$

The ligand bonds to the metal ion to form a **complex ion**. Six, four or two ligands combine with one metal ion, so the formation of a complex ion may be represented as the donation of six or four or two pairs of electrons to the metal ion. A dative covalent bond (co-ordinate bond) is formed between each ligand and the metal ion.

Ligands contain atoms of p-block elements that are more electronegative than the 3d-block atom. This means that the ligand attracts the electrons of the bond away from the metal. The result is that the metal atom is approximately neutral, and the charge is spread to the outside of the ion (see Figure 24.5).

Figure 24.5 The formation of a complex ion. Water is shown as the ligand in this example. The metal ion could be M^{2+} rather than M^{3+}.

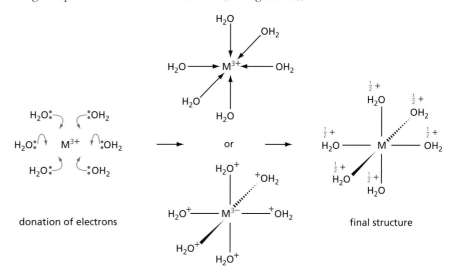

In the example in Figure 24.5, the oxidation number of the metal is simply the same as the charge on the complex ion. If the ligand is charged, for example a Cl^- ion, the oxidation number of the metal is found in the same way as in any other ion (see section 7.2). If the oxidation number of copper in the $CuCl_4^{2-}$ ion is x, then:

$$x + 4 \times (-1) = -2 \quad \text{and} \quad x = +2$$

So the oxidation number of copper in this complex ion is +2. If a simple formula does not make it clear which groups are attached to the metal ion, it is common practice to enclose the complex ion in square brackets. An example of this is $Cr(H_2O)_6Cl_3$, which does not make it clear which ligands are attached to the chromium ion. It could, for example, be $[Cr(H_2O)_6]Cl_3$ (which has no chlorine ligands but has three free chloride ions) or $[Cr(H_2O)_5Cl]Cl_2.H_2O$ (which has one chlorine ligand and two free chloride ions).

A **complex ion** is formed when **ligands** donate a pair of electrons to a metal ion. The 3d-block metals form stable complex ions for the following reasons.

- Their ions are small and have a high charge density.
- They have 3d orbitals of low energy that can accommodate electrons donated by the ligands.

The shapes of complex ions

The number of atoms surrounding a central atom is called the **co-ordination number**. Thus a complex such as $[Cr(H_2O)_6]^{3+}$ has a co-ordination number of 6. A complex of formula ML_6^{3+} has six pairs of electrons and its shape using VSEPR arguments (see section 3.9) is a regular octahedron (see Figure 24.6).

Figure 24.6 The shapes of complex ions with six ligands, with four ligands and with two ligands

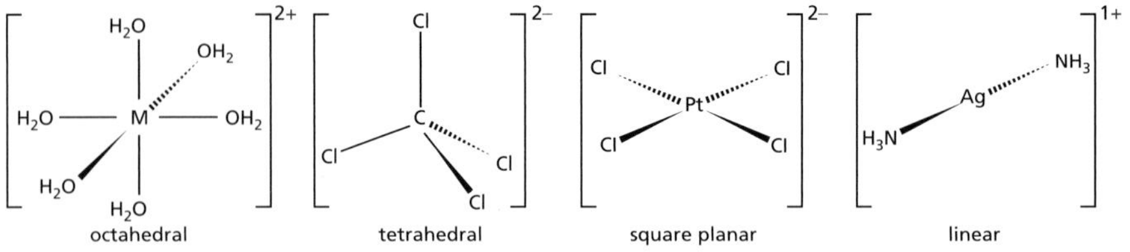

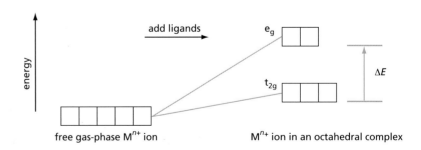

If the complex contains two ligands (co-ordination number 2), it is linear; examples are the $CuCl_2^-$ and $[Ag(NH_3)_2]^+$ ions. With four ligands, VSEPR arguments predict that the complex is tetrahedral and this is usually the case; examples are the $CoCl_4^{2-}$ and $[Zn(NH_3)_4]^{2+}$ ions. However, there are complexes with four ligands that are square planar, which shows that factors other than electron repulsion must be taken into account. An example of a square planar complex is cisplatin, which is discussed on page 410.

Bonding in complex ions

In aqueous solution, Cu^{2+} ions hydrate to give the $Cu(H_2O)_6^{2+}$ ion. Each water molecule is bonded to the central Cu^{2+} ion by a dative covalent bond. In order to form this ion, there need to be six empty orbitals to receive the six lone pairs from the six water molecules. There are also nine 3d electrons in the free Cu^{2+} ion that need five orbitals to accommodate them. Thus at least 11 orbitals are required in all. These are obtained by hybridising the six ligand lone-pair orbitals, together with the one 3s, the three 3p and the five 3d orbitals of the metal ion.

The energies and geometries of all of the resulting orbitals need not concern us, except for one important outcome: under the influence of the ligand lone pairs, the five 3d orbitals split into two groups with different energies. In the simple Cu^{2+} ion, the five 3d orbitals all have the same energy (they are said to be **degenerate**), but in the presence of six ligands spaced octahedrally, the orbitals divide into two groups – a group of two, called the e_g, and a group of three, called the t_{2g} (see Figure 24.7). The lobes of the three t_{2g} orbitals are directed in between the six ligands (see Figure 24.8), and their energy is only slightly increased when the complex is formed. The two e_g orbitals, however, have lobes which point in the direction of the six ligands; this means that any electrons in these orbitals will experience inter-electron repulsion with the ligand lone pairs when the complex is formed, and so their energy is higher than that of electrons in the t_{2g} orbitals.

Figure 24.7 The splitting of the 3d orbitals in an octahedral field

Figure 24.8 The shapes of the three t_{2g} and the two e_g orbitals. The t_{2g} orbitals point in between the ligands; the e_g orbitals point towards the ligands.

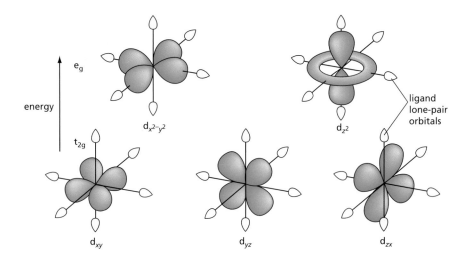

In tetrahedral complexes, the changes in the energy levels of the e_g and t_{2g} orbitals is reversed. The t_{2g} orbitals now point towards the ligands. This means that they can be used to form strong bonds, but it also means that any electrons that are already in the orbitals will be strongly repelled. The e_g orbitals now lie between the ligands and any electrons in them are relatively unaffected by the formation of a complex.

Ligand-exchange reactions

When ammonia is added to a solution containing aqueous Cu^{2+} ions, four of the water molecules are substituted by ammonia, producing the cuprammonium ion. This is an example of a **ligand-exchange reaction**, the substitution of one ligand by another:

$$Cu(H_2O)_6^{2+}(aq) + 4NH_3(aq) \rightarrow [Cu(NH_3)_4(H_2O)_2]^{2+}(aq) + 4H_2O(l)$$

This is a reversible reaction and if the purple complex $[Cu(NH_3)_4(H_2O)_2]^{2+}(aq)]$ is diluted, the pale blue colour of the $[Cu(H_2O)_6]^{2+}(aq)]$ ion is restored. There is competition between the H_2O and NH_3 molecules to attach themselves to the copper ion. We have an equilibrium with a constant given by the expression:

$$K = \frac{[[Cu(NH_3)_4(H_2O)_2]^{2+}][H_2O]^4}{[Cu(H_2O)_6]^{2+}[NH_3]^4}$$

As $[H_2O]$ is in large excess and virtually a constant, we can include it in the equilibrium constant and write:

$$K_{stab} = \frac{[[Cu(NH_3)_4(H_2O)_2]^{2+}]}{[Cu(H_2O)_6][NH_3]^4}$$

where K_{stab} is known as the **stability constant**. Strictly speaking, there are stability constants for each of the equilibria as H_2O molecules are progressively substituted by NH_3 molecules, but here we will only consider the overall constant. The value of K_{stab} depends on how firmly the ligands bind to the metal atom. In general, atoms with high electronegativity bond weakly and those with lower electronegativity bind more strongly. Thus water (which binds via the very electronegative oxygen atom) forms weak bonds while the CN^- ion (which binds via the carbon atom that has much lower electronegativity) forms strong bonds and will have a large value of K_{stab} (see Table 24.4). A high charge on the metal ion also makes K_{stab} larger; compare $\frac{[[Fe(CN)_6]^{3-}]}{[Fe^{3+}][CN^-]^6}$ and $\frac{[[Fe(CN)_6]^{4-}]}{[Fe^{2+}][CN^-]^6}$.

Table 24.4 Values of K_{stab} for some complexes. The values are usually so large that they are often given in the form $\log K_{stab}$.

Complex	K_{stab}	$\log K_{stab}$
$[CuCl_4]^{2-}$	4.2×10^5	5.62
$[Cu(NH_3)_4]^{2+}$	1.3×10^{13}	13.1
$[Fe(CN)_6]^{3-}$	10×10^{31}	31
$[Fe(CN)_6]^{4-}$	1.0×10^{24}	24
$[Ag(NH_3)_2]^+$	1.7×10^7	7.23

Now try this

1 A solution of copper sulfate had ammonia added until $[NH_3] = 0.050\,mol\,dm^{-3}$. Calculate the ratio of $[Cu(NH_3)_4(H_2O)_2]^{2+}$ to $[Cu(H_2O)_6]$ under these conditions.

2 (Harder – use the Nernst equation on page 389.)

$E^{\ominus}_{cell}$ for $Fe^{3+} + e^- \rightleftharpoons Fe^{2+}$ is +0.77 V

Calculate $E^{\ominus}_{cell}$ for $[Fe(CN)_6]^{3-} + e^- \rightleftharpoons [Fe(CN_6)]^{4-}$.

A more complicated ligand-exchange reaction occurs when concentrated hydrochloric acid is added to aqueous Cu^{2+} ions. Here, four H_2O ligands are replaced by Cl^- ions, but the remaining two water molecules are expelled, leaving the tetrahedral tetrachlorocopper(II) ion (see Figure 24.9):

$$Cu(H_2O)_6^{2+}(aq) + 4Cl^-(aq) \rightarrow [CuCl_4]^{2-}(aq) + 2H_2O(l)$$

Figure 24.9 Solutions containing **a** the $Cu(H_2O)_6^{2+}$ ion and **b** the $[CuCl_4]^{2-}$ ion

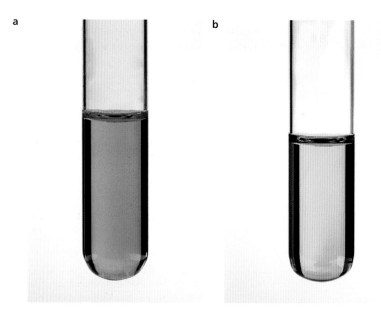

a

b

Deprotonation reactions

The water molecules in the hydrated complex ions play an important role. When sodium hydroxide solution is added to a solution containing $Fe^{3+}(aq)$ ions, a precipitate of iron(III) hydroxide, $Fe(OH)_3(s)$, is formed. At first this looks like a simple precipitation reaction:

$$Fe^{3+}(aq) + 3OH^-(aq) \rightarrow Fe(OH)_3(s)$$

This equation suggests that the reaction is similar to the precipitation of, for example, barium sulfate when $SO_4^{2-}(aq)$ ions and $Ba^{2+}(aq)$ ions are mixed together:

$$Ba^{2+}(aq) + SO_4^{2-}(aq) \rightarrow BaSO_4(s)$$

However, this ignores the role of the water molecules. The positive charge of the $Fe(H_2O)_6^{3+}$ ion is partially spread over the surface of the complex. This enables the $Fe(H_2O)_6^{3+}$ ion to act as a weak acid ($K_a = 10^{-5}\,mol\,dm^{-3}$) and to undergo an acid–base reaction with water molecules:

$$Fe(H_2O)_6^{3+}(aq) + H_2O(l) \rightarrow [Fe(H_2O)_5OH]^{2+}(aq) + H_3O^+(aq)$$

If the H_3O^+ ions are removed by the addition of a strong base such as sodium hydroxide, further acid–base reactions can take place:

$$[Fe(H_2O)_5OH]^{2+}(aq) + H_2O(l) \rightarrow [Fe(H_2O)_4(OH)_2]^+(aq) + H_3O^+(aq)$$
$$[Fe(H_2O)_4(OH)_2]^+(aq) \rightarrow Fe(OH)_3(s) + H_3O^+(aq) + 2H_2O(l)$$

The overall reaction with hydroxide ions is:

$$Fe(H_2O)_6^{3+}(aq) + 3OH^-(aq) \rightarrow Fe(OH)_3(s) + 6H_2O(l)$$

The result is that the $Fe(H_2O)_6^{3+}$ ion has been **deprotonated** by a series of acid–base reactions and water ligands have been released.

Worked example 1

a State the number of d electrons in the Fe^{2+} ion.
b Show, using curly arrows, the formation of the $Fe(H_2O)_6^{2+}$ ion.
c Write an equation for the conversion of $Fe(H_2O)_6^{2+}$ into the $[Fe(CN)_6]^{4-}$ ion on the addition of aqueous KCN.
d State the type of reaction represented by this change.

Answer
a 6
b See Figure 24.10.

Figure 24.10

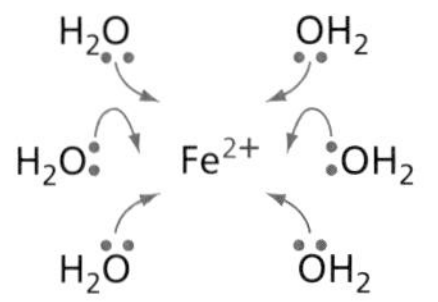

c $Fe(H_2O)_6^{2+}(aq) + 6CN^-(aq) \rightarrow [Fe(CN)_6]^{4-}(aq) + 6H_2O(l)$
d This is a ligand-exchange reaction.

Worked example 2

a Write an equation for the action of aqueous sodium hydroxide on the $Cu(H_2O)_6^{2+}(aq)$ ion.
b State the type of reaction that is represented by this change.

Answer
a $Cu(H_2O)_6^{2+}(aq) + 2OH^-(aq) \rightarrow Cu(OH)_2(s) + 6H_2O(l)$
b This is a deprotonation or acid–base reaction.

Complex ions nomenclature

When naming a complex ion, the following rules apply.

- A cation has the usual metal name, for example, copper.
- An anion has the metal name with an 'ate' ending, for example, chromate.
- The oxidation state is indicated in the usual way, for example, iron(III).
- The ligands are given specific names, for example, chloro (Cl^-), aqua (H_2O), hydroxo (OH^-), ammine (NH_3), cyano (CN^-).
- The number of ligands is indicated by the prefixes di-, tri-, tetra-, penta-, hexa-.

So the complex ion $Fe(H_2O)_6^{3+}$ is the hexaaquairon(III) ion, and $Cu(NH_3)_4^{2+}$ is tetraamminecopper(II).

Name the following complex ions.

a $Fe(CN)_6^{3-}$
b $[Cu(NH_3)_4(H_2O)_2]^{2+}$
c $[Fe(H_2O)_5OH]^{2+}$

Answer

a hexacyanoferrate(III) ion
b tetraamminediaquacopper(II) ion
c pentaaquahydroxoiron(III) ion

Now try this

Name the following complex ions.

a $Cr(NH_3)_6^{3+}$
b $CuCl_4^{2-}$
c $Zn(OH)_4^{2-}$
d $[CrCl_2(H_2O)_4]^+$

Chelates

Bidentate ligands

Ligands such as H_2O and CN^- are attached by one coordinate bond to the metal ion. If the ligand contains two groups that have a lone pair of electrons, it may form two bonds to the metal atom, forming a ring. Such a ligand is called a **chelate**, a name derived from the Greek word for a crab's claw. Stable complexes result if five- or six-membered rings are produced by the chelate and the metal ion.

Two ligands that readily form chelates are 1,2-diaminoethane, $H_2NCH_2CH_2NH_2$, and the ethanedioate ion, $^-O_2CCO_2^-$. These form five-membered rings (see Figure 24.11) and are called **bidentate** ligands because they join by two bonds.

Chelates form particularly stable complex ions, partly because they form strong bonds to the metal ion, but also because there is an additional entropy effect that adds to their stability. For example, a chelate is formed in which three ethanedioate ions bond to Fe^{3+}. Four species become seven after the reaction, so the formation of this chelate is accompanied by an increase in entropy:

$$Fe(H_2O)_6^{3+} + 3C_2O_4^{2-} \rightarrow Fe(C_2O_4)_3^{3-} + 6H_2O$$

Another way of visualising the increase in entropy is to consider the effect after one end of the chelate has become bonded to the metal ion. Once this end is secured, it becomes much more likely that the other end will be in the right position to bond too.

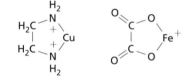

Figure 24.11 Five-membered rings formed between 1,2-diaminoethane and the Cu^{2+} ion, and between the ethanedioate ion and the Fe^{3+} ion.

Multidentate ligands – edta

Some chelates form more than two bonds with the metal ion. A particularly important one is **e**thylene **d**iamine **t**etraacetic **a**cid, abbreviated as edta, which is used in the form of its disodium salt, containing $(edtaH_2)^{2-}$.

This has six pairs of electrons able to bond to a metal ion, and so forms a hexadentate chelate. edta is used to remove Ca^{2+} and Mg^{2+} ions from hard water.

$$Ca(H_2O)_6^{2+}(aq) + (edtaH_2)^{2-}(aq) \rightarrow Ca(edta)^{2-}(aq) + 2H+(aq) + 6H_2O(l)$$

This trapping of metal ions is called **sequestering**. It alters the chemical properties of the metal ions, and can be used to counteract the effect of poisoning by heavy metal ions such as lead.

Analysing tap water

The formation of stable complexes between edta and Mg^{2+} and Ca^{2+} ions is used to estimate the hardness of water by titration with edta. A sample of the water being tested is mixed with a buffer solution at pH 10. A few drops of an indicator, solochrome black, are added and a solution of edta is run in from the burette.

Solochrome black forms a red complex with magnesium ions. If solochrome black is represented as $HIn^{2-}(aq)$, the reaction is

$$MgIn^-(aq) + (edtaH_2)^{2-}(aq) \rightarrow Mg(edta)^{2-}(aq) + H^+(aq) + HIn^{2-}(aq)$$

red complex   **blue solution**

During the titration, the free calcium and magnesium ions first react with the edta:

$$Mg(H_2O)_6^{2+}(aq) + (edtaH_2)^{2-}(aq) \rightarrow Mg(edta)^{2-}(aq) + 2H^+(aq) + 6H_2O(l)$$

When all the free calcium and magnesium ions have reacted with edta, the colour changes from red to blue as the $MgIn^-(aq)$ complex breaks down.

Worked example

A few drops of solochrome black indicator were added to $50.0 \, cm^3$ of tap water. A $0.0100 \, mol \, dm^{-3}$ solution of edta in a buffer at pH 10 was added from a burette and the indicator changed from red to blue after the addition of $7.55 \, cm^3$.

a Suggest two substances that could be used to make the buffer solution.
b Calculate the total concentration of calcium and magnesium ions in the sample of hard water.

Answer

a Ammonia and an ammonium salt, for example, ammonium chloride (see section 22.3).
b $n(edta) = c \times \dfrac{v}{1000}$

$$= 0.0100 \times \frac{7.55}{1000}$$

$$= 7.55 \times 10^{-5} \, mol$$

This amount of edta reacted with $50.0 \, cm^3$ of tap water. Since one mole of calcium or magnesium ions reacts with one mole of edta, the concentration of calcium and magnesium ions in the tap water is given by:

$$c(Ca^{2+} + Mg^{2+}) = \frac{1000}{v} \times n$$

$$= \frac{1000}{50} \times 7.55 \times 10^{-5}$$

$$= 1.51 \times 10^{-3} \, mol \, dm^{-3}$$

Isomerism in complex ions

Some of the types of isomerism found in organic chemistry (see section 12.6) are also found in complex ions. An example of structural isomerism occurs in compounds of formula $CrCl_3.6H_2O$. Three such compounds are known with different properties (see Table 24.5). In particular, they have different numbers of free Cl^- ions (as shown by their reactions with Ag^+ ions) and free water molecules (as shown by the number of water molecules that are easily removed by dehydration).

Table 24.5 The properties of the three isomers of $CrCl_3.6H_2O$

Colour	Number of free Cl^- ions	Number of free H_2O molecules	Structure
purple	3	0	$[Cr(H_2O)_6{}^{3+}]3Cl^-$
blue-green	2	1	$[Cr(H_2O)_5Cl]^{2+}2Cl^-.H_2O$
green	1	2	$[Cr(H_2O)_4Cl_2]^+Cl^-.2H_2O$

Complexes can also show stereochemistry. A good example of this is the complex ion $[Ni(en)_2(NH_3)_2]^{2+}$, where 'en' is used as an abbreviation for 1,2-diaminoethane, $NH_2CH_2CH_2NH_2$. This complex can exist in *cis* and *trans* forms. The *cis* form (but not the *trans*) has a chiral centre (the nickel atom) and can be resolved into optical isomers (see Figure 24.12).

Figure 24.12 The stereochemistry of $[Ni(en)_2(NH_3)_2]^{2+}$

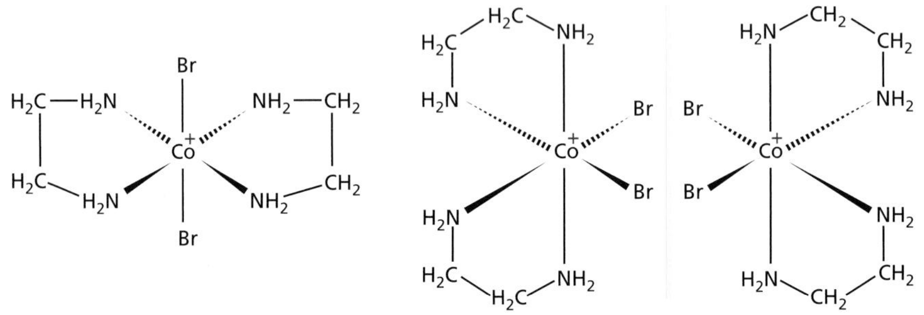

trans form   *cis* form

Ni^{2+} also forms a complex with three 1,2-diaminoethane ligands. This has only the *cis* form as the distance is too large for the ligand to stretch across the *trans* positions. The *cis* form can also be resolved into optical isomers (see Figure 24.13).

Figure 24.13 Optical isomerism in the $[Ni(NH_2CH_2CH_2NH_2)_3]^{2+}$ ion

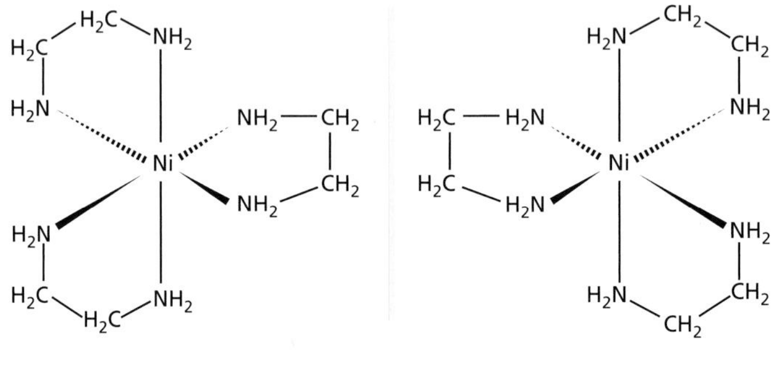

Cis- and *trans*-platin

An important example of *cis/trans* isomerism is the drug cisplatin. This has a Pt atom joined to two chlorine atoms and two ammonia molecules. It is a square planar complex and can, therefore, exist in both *cis* and *trans* forms.

cis-platin   *trans*-platin

Cis-platin (but not *trans*-platin) is very effective in treating some forms of cancer. One of the chlorine atoms in *cis*-platin is easily hydrolysed to give the $[(NH_3)_2Cl(H_2O)]^+$ ion. This ion binds to one of the four bases in DNA, usually guanine (see Topic 27, page 489), to give the $[PtCl(guanine-DNA)(NH_3)_2]^+$ ion. This ion can cross-link with another DNA chain by displacement of the chlorine from the complex. The cross-linking inhibits

DNA replication, particularly in cancer cells that are undergoing rapid cell division, and the cell dies. There is no obvious reason why *trans*-platin is ineffective; one theory is that it becomes deactivated before it can attach itself to DNA.

24.5 Colour in the d block

The origin of colour

All atoms and molecules absorb in the ultraviolet region of the spectrum because this radiation has enough energy to excite their outer electrons. The ultraviolet region is outside the visible range of the spectrum, so absorption in the ultraviolet leaves a substance colourless. Some substances, however, also have outer electron levels that are sufficiently close together for visible radiation to have enough energy to bring about electronic excitation. Under these circumstances, the substance appears coloured. The resulting colour seen is white light minus the colour being absorbed. The colour we see is therefore the **complementary** colour to the colour being absorbed (see Figure 24.14).

To absorb in the visible region, the substance must have two energy levels that are very close together. Close energy levels come about in two ways, namely **charge transfer** and **d-to-d transitions**.

Figure 24.14 The colours observed when absorption takes place in the visible region of the spectrum

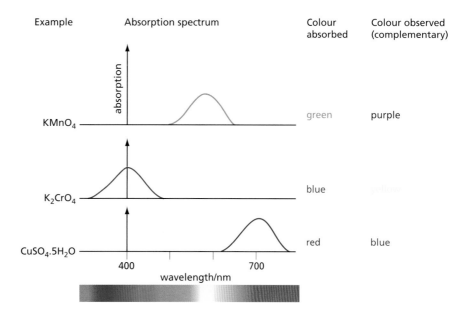

Charge transfer

If a substance contains bonds which are on the borderline between ionic and covalent, it may change from one bonding type to another by absorbing visible light. This effect is shown by some solids which are coloured (for example PbO, which is orange, and AgI, which is yellow) even though the ions they contain are colourless. In the d block, another example is CuO, which is black even though the Cu^{2+} ion is blue.

d-to-d transitions

In the d block, colour usually arises because of d-to-d electronic transitions. The t_{2g} and the e_g orbitals are close in energy, and electrons in the t_{2g} orbitals may be excited into the e_g orbitals by the absorption of a photon of visible light. Using the Planck equation,

$$\Delta E = hf$$

the frequency associated with the energy difference between the orbitals (ΔE in Figure 24.7) corresponds to the frequencies of visible light.

Ions that have no t_{2g} electrons, for example $Sc(H_2O)_6^{3+}$, are colourless because there are no t_{2g} electrons to promote. Similarly d^{10} ions, for example $Zn(H_2O)_6^{2+}$, are also colourless because there is no empty space in the e_g level to receive an extra electron (see Figure 24.15).

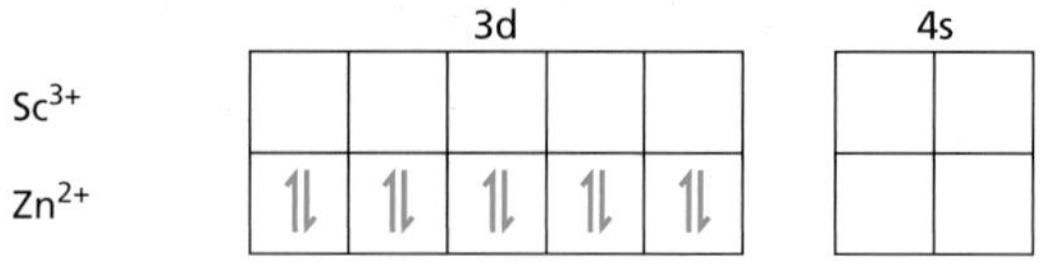

Figure 24.15 The electronic structures of the $Sc(H_2O)_6^{3+}$ and $Zn(H_2O)_6^{2+}$ ions

	3d					4s	
Sc^{3+}							
Zn^{2+}	⇅	⇅	⇅	⇅	⇅		

It is fairly easy to explain the colours of the Cu^{2+} complexes. This is because there is only a single space in the e_g orbitals to receive an excited electron, and as a result there is only one absorption band. The absorption is a band rather than a sharp line (as happens with atomic spectra in the gas phase) because the two energy levels are spread out under the influence of the adjacent ligands. The colour of the complex then depends on where the maximum of this absorption is in the visible spectrum.

- For the simple Cu^{2+} ion (anhydrous $CuSO_4$), the ligands create such a very weak field that the two energy levels are nearly the same; absorption is then in the infrared and this means that the substance is colourless.
- For the $[Cu(H_2O)_6]^{2+}$ ion, absorption is partially in the red end of the spectrum and the complex appears pale blue.
- For the $[Cu(NH_3)_4(H_2O)_2]^{2+}$ ion, the absorption is in the green part of the spectrum and the colour is a more intense purple.

This change in the position of the absorption band is caused by the increase in the electrostatic field created by the ligands, which results in a larger splitting in the energy levels (ΔE in Figure 24.7) and a higher frequency of absorption (see Figure 24.16).

Figure 24.16 Colours of some Cu^{2+} compounds. As the ligand field becomes stronger, the splitting between the d orbitals increases and the absorption moves from the infrared into the visible region of the spectrum.

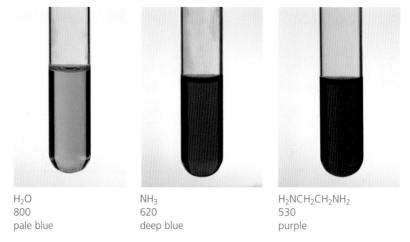

ligand	H_2O	NH_3	$H_2NCH_2CH_2NH_2$
absorption maximum/nm	800	620	530
colour	pale blue	deep blue	purple

24.6 Catalytic properties

The transition metals often act as catalysts. In the Haber process (see section 10.5) and the Contact process (see section 10.6), the catalysts are in the solid state. Transition metals can also act as catalysts in solution. A good example is the oxidation of iodide ions by peroxodisulfate(VI) ions:

$$2I^-(aq) + S_2O_8^{2-}(aq) \rightarrow 2SO_4^{2-}(aq) + I_2(aq)$$

This reaction is normally quite slow, most likely because the negatively charged ions repel each other. However, it is catalysed by the addition of a number of d-block metal ions, for example $Fe^{2+}(aq)$. A possible mechanism is:

$$2Fe^{2+}(aq) + S_2O_8^{2-}(aq) \rightarrow 2Fe^{3+}(aq) + 2SO_4^{2-}(aq)$$
$$2Fe^{3+}(aq) + 2I^-(aq) \rightarrow 2Fe^{2+}(aq) + I_2(aq)$$

Table 24.6 Relevant values of $E^\ominus$ for catalysis by Fe^{2+} ions

System	$E^\ominus$/V
$S_2O_8^{2-}/2SO_4^{2-}$	+2.01
Fe^{3+}/Fe^{2+}	+0.77
$I_2/2I^-$	+0.54

Each of the two parts of this mechanism involves a reaction between two ions of opposite charge, which is a favourable situation.

Both theory and experiment can be used to support this mechanism.

- The relevant values of standard electrode potentials are given in Table 24.6. These show that the postulated mechanism is thermodynamically feasible.

$$2Fe^{2+}(aq) + S_2O_8^{2-}(aq) \rightarrow 2Fe^{3+}(aq) + S_2O_4^{2-}(aq) \qquad E^\ominus_{cell} = +1.34\,V$$
$$2Fe^{3+}(aq) + 2I^-(aq) \rightarrow 2Fe^{2+}(aq) + I_2(aq) \qquad E^\ominus_{cell} = +0.23\,V$$

- The two reactions can be tested experimentally. If a solution of $S_2O_8^{2-}$ ions is added to Fe^{2+} ions, Fe^{3+} ions are produced. If Fe^{3+} ions are added to I^- ions, iodine is liberated.

Summary

- The **3d block** includes the elements from scandium to zinc inclusive.
- The elements titanium to copper show features associated with **transition metals**, that is, variable oxidation states and coloured ions.
- The elements of the 3d block have high melting points, boiling points and densities.
- The first and second ionisation energies increase only slightly across the block from scandium to zinc, as 4s electrons are being removed which are shielded from the nuclear attraction by the inner 3d electrons. The third ionisation energies increase more rapidly as a 3d electron is being removed.
- Most of the elements form M^{2+} ions by loss of the 4s electrons, and some form M^{3+} ions as well.

- The elements chromium and manganese have a maximum oxidation number equal to the sum of the numbers of 3d and 4s electrons.
- The ions of all the 3d-block elements form **complex ions** by receiving electrons from two, four or six **ligands**.
- Complex ions may exist in different forms (isomerism) and in different shapes (stereoisomerism).
- The stability of the complex is measured by the stability constant, K_{stab}.
- Some ligands join onto the metal ions by more than one pair of electrons. These form **chelates**, which are especially stable.
- d^0 and d^{10} ions are colourless. Other d-block ions are coloured because of **d-to-d transitions** in the visible region of the spectrum.

Examination practice questions

Please see the data section of the CD for any A_r values you may need.

1 a Explain what is meant by the term *transition element*. [1]
   b Complete the electronic configuration of
      i   the vanadium atom, $1s^2\,2s^2\,2p^6$ ...
      ii the Cu^{2+} ion. $1s^2\,2s^2\,2p^6$ ... [2]
   c List the **four** most likely oxidation states of vanadium. [1]
   d Describe what you would see, and explain what happens, when dilute aqueous ammonia is added to a solution containing Cu^{2+} ions, until the ammonia is in an excess. [5]
   e Copper powder dissolves in an acidified solution of sodium vanadate(V), $NaVO_3$, to produce a blue solution containing VO^{2+} and Cu^{2+} ions.
      By using suitable half-equations from the data section on the CD, construct a balanced equation for this reaction. [2]
      [Cambridge International AS & A Level Chemistry 9701, Paper 4 Q3 June 2009]

2 One major difference between the properties of compounds of the transition elements and those of other compounds is that the compounds of the transition elements are often coloured.
   a Explain in detail why many transition element compounds are coloured. [3]

b The following graph shows the absorption spectrum of two complexes containing copper.

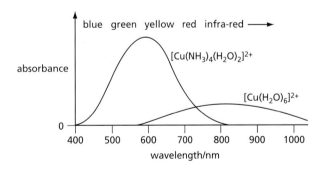

i State the colours of the following complex ions.
   - $[Cu(H_2O)_6]^{2+}$
   - $[Cu(NH_3)_4(H_2O)_2]^{2+}$
ii Using the spectra above give **two** reasons why the colour of the $[Cu(NH_3)_4(H_2O)_2]^{2+}$ ion is deeper (more intense) than that of the $[Cu(H_2O)_6]^{2+}$ ion.

iii Predict the absorption spectrum of the complex $[Cu(NH_3)_2(H_2O)_4]^{2+}$, and sketch this spectrum on a copy of the graph. [6]

c Copper forms a complex with chlorine according to the following equilibrium.

$$Cu^{2+}(aq) + 4Cl^-(aq) \rightleftharpoons [CuCl_4]^{2-}(aq)$$

i Write an expression for the equilibrium constant, K_c, for this reaction, stating its units.

ii The numerical value of K_c is 4.2×10^5. Calculate the $[[CuCl_4]^{2-}]/[Cu^{2+}]$ ratio when $[Cl^-] = 0.20\,mol\,dm^{-3}$. [3]

[Cambridge International AS & A Level Chemistry 9701, Paper 41 Q3 November 2009]

3 a Complete the electronic structures of the Cr^{3+} and Mn^{2+} ions.

Cr^{3+} $1s^2\,2s^2\,2p^6$ ...
Mn^{2+} $1s^2\,2s^2\,2p^6$ ...   [2]

b i Describe what observations you would make when dilute $KMnO_4(aq)$ is added slowly and with shaking to an acidified solution of $FeSO_4(aq)$ until the $KMnO_4$ is in a large excess.

ii Construct an ionic equation for the reaction that occurs. [4]

c By selecting relevant $E^\ominus$ data from the data section on the CD explain why acidified solutions of $Fe^{2+}(aq)$ are relatively stable to oxidation by air, whereas a freshly prepared precipitate of $Fe(OH)_2$ is readily oxidised to $Fe(OH)_3$ under alkaline conditions. [4]

d Predict the organic products of the following reactions and draw their structures. You may use structural or skeletal formulae as you wish.

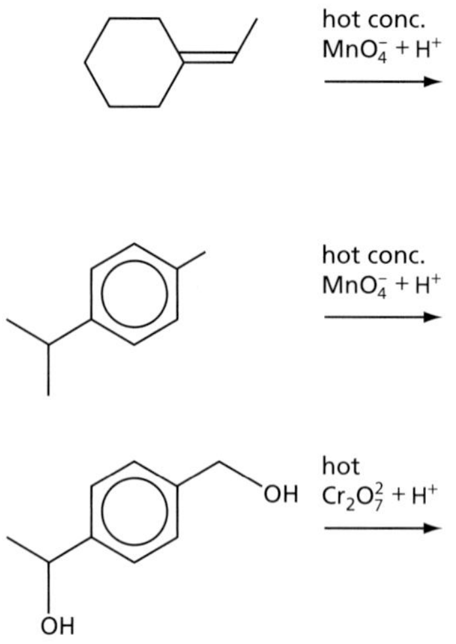

hot conc.
$MnO_4^- + H^+$

hot conc.
$MnO_4^- + H^+$

hot
$Cr_2O_7^{2-} + H^+$

[4]

e $KMnO_4$ and $K_2Cr_2O_7$ are the reagents that can be used to carry out the following transformation.

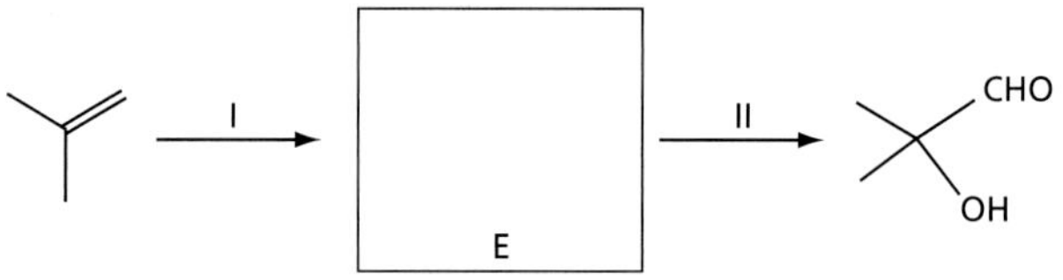

E

i Draw the structure of intermediate **E**.

ii Suggest reagents and conditions for reaction **I**, and for reaction **II**. [3]

[Cambridge International AS & A Level Chemistry 9701, Paper 42 Q4 June 2010]

25 Arenes and phenols

Functional groups:

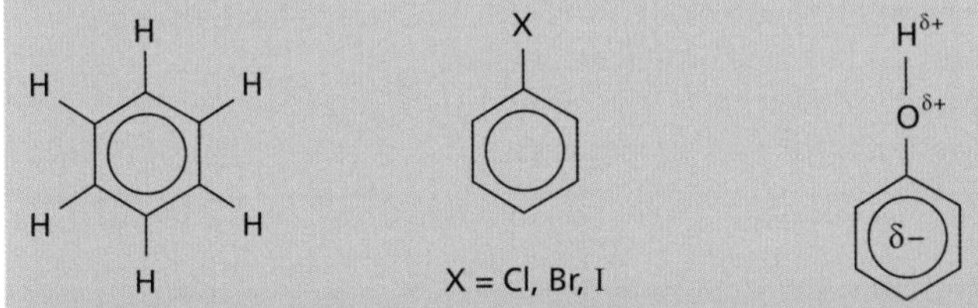

arenes   halogenoarenes   phenols

X = Cl, Br, I

The arenes are a group of hydrocarbons whose molecules contain the benzene ring. Although they are unsaturated, they do not show the typical reactions of alkenes. Their characteristic reaction is electrophilic substitution. Phenols contain a hydroxy group (—OH) joined directly to a benzene ring. The presence of the ring modifies the typical alcohol-like reactions of the —OH group. Likewise, the presence of the —OH group modifies the typical arene-like reactions of the benzene ring. The presence of the benzene ring severely reduces the reactivity of the carbon–halogen bond in halogenoarenes.

Learning outcomes

By the end of this topic you should be able to:

14.1a) interpret and use the general, structural, displayed and skeletal formulae of the following classes of compound: arenes, halogenoarenes and phenols (part, see also Topic 12)

14.1c) understand and use systematic nomenclature of simple aromatic molecules with one benzene ring and one or more simple substituents

14.3a) describe and explain the shape of, and bond angles in, benzene molecules in terms of σ and π bonds

15.4a) describe the chemistry of arenes as exemplified by the following reactions of benzene and methylbenzene: substitution reactions with chlorine and with bromine, nitration, Friedel–Crafts alkylation and acylation, complete oxidation of the side-chain to give a benzoic acid, and hydrogenation of the benzene ring to form a cyclohexane ring

15.4b) describe the mechanism of electrophilic substitution in arenes, as exemplified by the formation of nitrobenzene and bromobenzene, suggest the mechanism of other electrophilic substitution reactions, given data, and describe the effect of the delocalisation of electrons in arenes in such reactions

15.4c) interpret the difference in reactivity between halogenoalkanes and chlorobenzene

15.4d) predict whether halogenation will occur in the side-chain or in the aromatic ring in arenes depending on reaction conditions

15.4e) apply knowledge relating to position of substitution in the electrophilic substitution of arenes

17.2a) recall the chemistry of phenol, as exemplified by the following reactions: with bases, with sodium, with diazonium salts (see also Topic 27), and nitration of, and bromination of, the aromatic ring

17.2b) describe and explain the relative acidities of water, phenol and ethanol.

25.1 Introduction

The structure of the benzene ring

Arenes are hydrocarbons that contain one or more benzene rings. Benzene was first isolated and identified by the English chemist Michael Faraday in 1825. Its structure posed a problem for nineteenth-century chemists. Comparing its molecular formula,

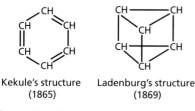

Kekule's structure (1865)   Ladenburg's structure (1869)

Figure 25.1 Two structures that were proposed for benzene in the nineteenth century

Figure 25.2 The German chemist Friedrich August Kekulé von Stradonitz (1829–96) was the first to suggest that benzene had a ring structure.

C_6H_6, to that of the alkane hexane, C_6H_{14}, it can be seen that benzene is highly deficient in hydrogen. Other compounds with very low hydrogen-to-carbon ratios were all known to be highly unsaturated, containing many $C=C$ double bonds: they readily decolorise $KMnO_4(aq)$ and $Br_2(aq)$, and react with HBr, and with H_2O in the presence of acids. Benzene showed none of these reactions. Furthermore, it was apparent that all six hydrogen atoms in benzene were equivalent to each other, in that only one isomer of chlorobenzene, C_6H_5Cl, could be made. There were three isomers of dichlorobenzene, $C_6H_4Cl_2$, however. No open-chain structure would show these properties, but several structures containing rings were suggested in the 1860s. Those proposed by the German chemists Albert Ladenburg and August Kekulé are shown in Figure 25.1.

> ### Now try this
>
> 1 Draw the structural formulae of all possible isomers of $C_6H_4Cl_2$ using the Ladenburg structure, and thus show that there are only three of them.
> 2 Draw the structural formulae of all possible isomers of $C_6H_4Cl_2$ using the Kekulé structure, and thus show that there are four of them.

If two hydrogens are replaced by chlorines, to form dichlorobenzene, it is found that three (and only three) isomers of $C_6H_4Cl_2$ can be made (see section 25.2). Although this fitted with Ladenburg's structure, Kekulé's structure did not quite account for this observation; his proposed structure implied that there should be two isomers of 1,2-dichlorobenzene:

and

In fact, there is only one 1,2-dichlorobenzene. To overcome this disagreement with his structure, Kekulé proposed that the bonding in the benzene ring alternated between two equivalent structures:

and

Now that we accept that π electrons can become delocalised in many-atom π orbitals, formed by the overlapping of p orbitals on adjacent atoms, we have no need to postulate this alternation between the two Kekulé forms. The true bonding in benzene has sometimes been described as being in between the two extremes represented by the two Kekulé forms. We represent this 'in-between' state by a *double-headed arrow* joining the two formulae:

$\longleftrightarrow$

It is important to understand the meaning of the double-headed arrow, $\longleftrightarrow$. This states that there is *only one* structure, which is in between the two 'classical' structures drawn either side of the arrow. The existence of a structure which cannot be represented by a single 'classical' structure, but which is intermediate between several of them, is known as **mesomerism** (from the Latin/Greek work *meso*, meaning 'middle'). The classical structures are called mesomers.

An early representation of the mesomeric structure of benzene was proposed by the German chemist Johannes Thiele in 1899 (see Figure 25.3); this is very similar to the delocalised structure we draw today.

Figure 25.3 The representation of the benzene molecule proposed by Thiele in 1899

The structure of benzene was described in detail in section 3.16. It consists of six carbon atoms arranged in a regular hexagon, each joined to a hydrogen atom and to its neighbours by σ bonds. There are six spare p orbitals, one on each carbon atom, all parallel to each other and perpendicular to the plane of the ring. Each p orbital overlaps equally with both its neighbours, forming a delocalised six-centre molecular π orbital (see Figure 25.4).

Figure 25.4 The delocalised π bond in benzene

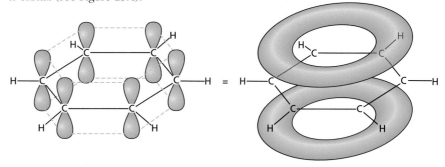

All the bond angles in benzene are 120°. All the C—C bonds have the same length, 0.139 nm. This is intermediate between the length of the C —C bond in an alkane (0.154 nm) and the C=C double bond in an alkene (0.134 nm).

The usual representation used for benzene is either the symmetrical skeletal formula (derived from the Thiele structure) or the Kekulé structure, either skeletal or structural. Figure 25.5 shows these representations. The benzene structure is normally represented by the skeletal formula, as in Figure 25.5a. When we want to show in detail how the π electrons move during reactions, however, we may sometimes use a Kekulé structure, as in Figure 25.5b or 25.5c. Figure 25.6 shows models of the Kekulé and Thiele structures.

a   b   c

Figure 25.5 Three representations of the benzene molecule: **a** the Thiele skeletal formula, **b** the Kekulé skeletal formula, **c** the structural formula for benzene

Figure 25.6 a Ball-and-stick and **b** electron-density molecular models of benzene

a
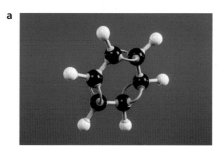

b
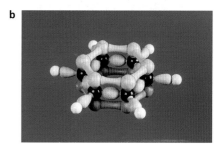

In the skeletal formula, each corner of the hexagon is assumed to contain a carbon atom, to which is attached a hydrogen atom, unless otherwise stated. For example, chlorobenzene can be written as follows:

The Kekulé structure fitted in with the chemical bonding ideas of the 1860s, but was clearly incorrect in one important respect. It predicts that benzene is highly unsaturated – the formula suggests that the molecule contains three double bonds. It ought to undergo addition reactions readily, just like an alkene. In fact, benzene is inert to most reagents that readily add on to alkenes (see Table 25.1).

Table 25.1 A comparison of some reactions of benzene and cyclohexene

Reagent	Benzene	Cyclohexene
shaking with KMnO$_4$(aq)	no reaction	immediate decolorisation
shaking with Br$_2$(aq)	no reaction	immediate decolorisation
H$_2$(g) in the presence of nickel	very slow reaction at 100 °C and 100 atm	rapid reaction at 20 °C and 1 atm

417

Under more severe conditions, benzene can be made to react with one of these reagents, bromine. But now the reaction follows a different course – a substitution reaction occurs, rather than addition, and the necessary reaction conditions are much more extreme.

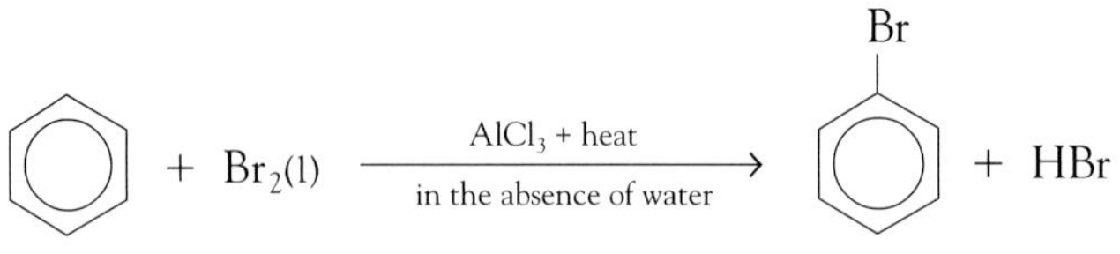

The stability of benzene

The six-centre delocalised π bond is responsible for the following physical and chemical properties of benzene.

- It causes all C—C bond lengths to be equal, creating a planar, regular hexagonal shape.
- It prevents benzene undergoing any of the normal addition reactions that alkenes show.

The π bond also makes benzene more stable than expected. Not only is benzene much less reactive than cyclohexene, it is also thermodynamically more stable. We can demonstrate this by comparing the actual and the 'calculated' enthalpy changes of hydrogenation. The hydrogenation of alkenes is an exothermic process:

$$CH_2{=}CH_2 + H_2 \rightarrow CH_3{-}CH_3 \qquad \Delta H^{\ominus} = -137\,\text{kJ}\,\text{mol}^{-1}$$

For many higher alkenes, the enthalpy changes of hydrogenation are very similar to each other, and average about $118\,\text{kJ}\,\text{mol}^{-1}$. Furthermore, for dienes and trienes (containing two and three double bonds, respectively), the enthalpy changes of hydrogenation are simple multiples of this value (see Table 25.2). So we might expect benzene to follow the same trend.

Table 25.2 Enthalpy changes of hydrogenation of some alkenes

Alkene	Formula	$\Delta H^{\ominus}_{\text{hydrogenation}}/$ $\text{kJ}\,\text{mol}^{-1}$	$\Delta H^{\ominus}$ per C=C
cis-but-2-ene	$CH_3{-}CH{=}CH{-}CH_3$	−119	−119
cyclohexene	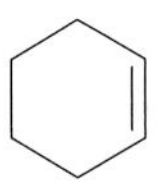	−118	−118
butadiene	$CH_2{=}CH{-}CH{=}CH_2$	−236	−118
cyclohexadiene	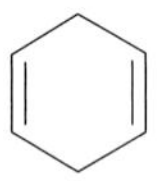	−232	−116
'cyclohexatriene'	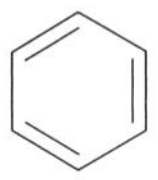	[−354] (predicted)	[−118] (assumed)

If we measure the experimental enthalpy change of hydrogenation of benzene, we find that it is far less exothermic than that predicted for 'cyclohexatriene' in Table 25.2:

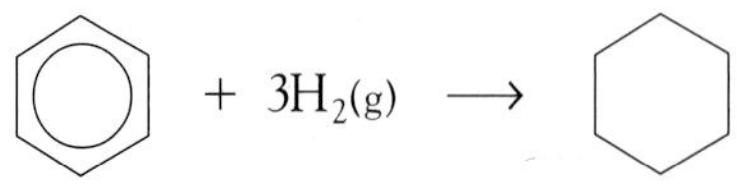

$$\Delta H^{\ominus}_{\text{hydrogenation}} = -205\ \text{kJ mol}^{-1}$$

Figure 25.7 Enthalpy changes of hydrogenation of benzene and the cyclohexenes

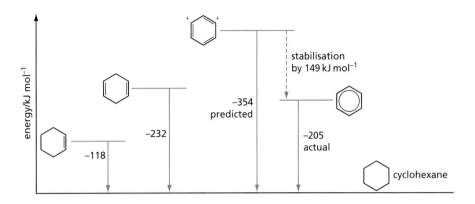

So benzene is more stable than 'cyclohexatriene' by $354 - 205 = 149\,\mathrm{kJ\,mol^{-1}}$ (see Figure 25.7).

The stabilisation of $149\,\mathrm{kJ\,mol^{-1}}$ is known variously as the **stabilisation energy**, the **delocalisation energy** or the **resonance energy**.

Other arenes

Compounds that contain rings of delocalised electrons are called **aromatic compounds**. The name was originally applied to certain natural products that had strong, pleasant aromas, such as vanilla-bean oil, clove oil, almond oil, thyme oil and oil of wintergreen. All of these oils contained compounds whose structures were found to include a benzene ring. The term 'aromatic' eventually became associated with the presence of the ring itself, whether or not the compound had a pleasant aroma (see Figures 25.8 and 25.9).

Figure 25.8 Some pleasant-smelling naturally occurring aromatic compounds

vanillin
(oil of vanilla bean)

benzaldehyde
(oil of almonds)

thymol
(oil of thyme)

methyl salicylate
(oil of wintergreen)

eugenol
(oil of cloves)

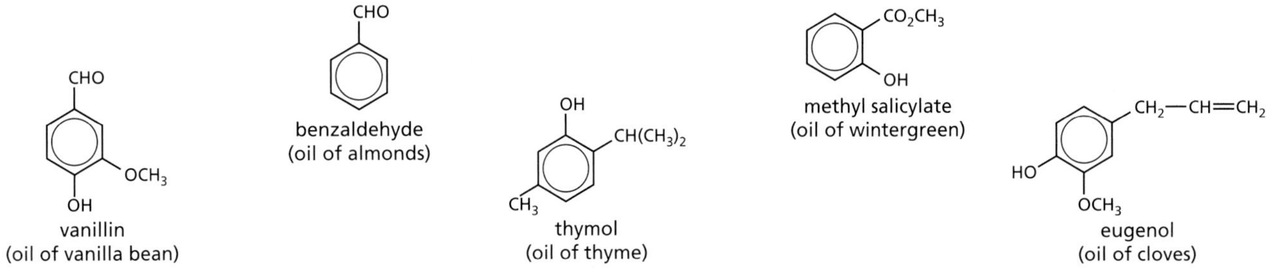

Figure 25.9 Some not-so-pleasant-smelling aromatic compounds made in the laboratory

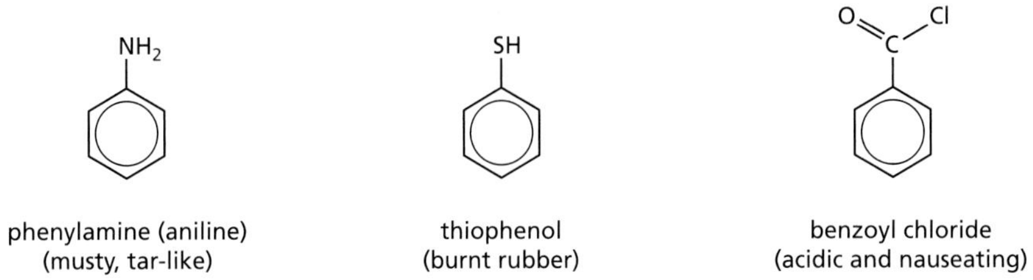

phenylamine (aniline)
(musty, tar-like)

thiophenol
(burnt rubber)

benzoyl chloride
(acidic and nauseating)

There are two ways in which benzene rings can join together. Two rings could be joined by a single bond, as in biphenyl:

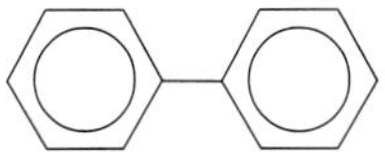

Or the two rings could share two carbon atoms in common (with their π electrons), as in naphthalene:

More rings can fuse together, giving such compounds as anthracene and pyrene:

anthracene   pyrene

Notice that with each successive ring fused together, the hydrogen-to-carbon ratio decreases, from 1:1 in benzene to 5:8 in pyrene. Eventually, as many more rings fuse together, a sheet of the graphite lattice (graphene) would result (see Topic 4).

Many of the multiple-ring arenes, such as pyrene, are strongly carcinogenic (cancer producing). Even benzene itself is a highly dangerous substance, causing anaemia and cancer on prolonged exposure to its vapour. In the past it was used as a laboratory solvent, but its use is now severely restricted, although it is still added to some brands of unleaded petrol to increase their anti-knock rating.

Physical properties of arenes

Benzene and most alkylbenzenes are strongly oily-smelling colourless liquids, immiscible with, and less dense than, water. They are non-polar, and the only intermolecular bonding is due to the induced dipoles of van der Waals' forces. Their boiling points are similar to those of the equivalent cycloalkanes (see Table 25.3), and increase steadily with relative molecular mass as expected.

Table 25.3 Boiling points of some cyclohexanes and arenes

Compound	Formula	Boiling point/°C	Compound	Formula	Boiling point/°C
cyclohexane	C_6H_{12}	81	benzene	C_6H_6	80
methylcyclohexane	C_7H_{14}	100	methylbenzene	C_7H_8	111
			ethylbenzene	C_8H_{10}	136
			propylbenzene	C_9H_{12}	159

25.2 Isomerism and nomenclature

Aromatic compounds with more than one substituent on the benzene ring can exist as positional isomers. There are three dichlorobenzenes:

1,2-dichlorobenzene   1,3-dichlorobenzene   1,4-dichlorobenzene

The terms *ortho-*, *meta-* and *para-* are sometimes used as prefixes to represent the relative orientations of the groups (see Table 25.4).

Table 25.4 Orientations of substituents on the benzene ring

Orientation	Prefix	Abbreviation	Example
1,2-	ortho-	o-	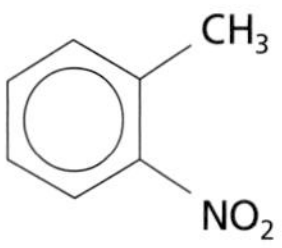
1,3-	meta-	m-	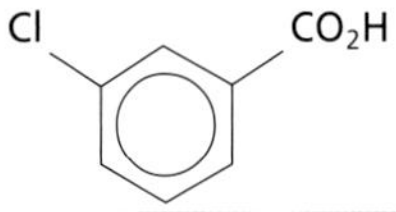
1,4-	para-	p-	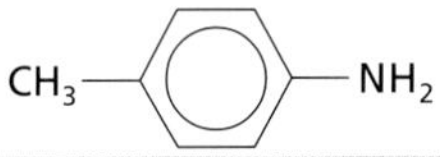

If the two substituents are different, one of them is defined as the 'root' group, in the following order of precedence

$$-CO_2H > -OH > -CH_3 > -halogen > -NO_2 \quad (NO_2 \text{ is called the nitro group})$$

and the ring carbons numbered from the carbon atom holding that group. For example:

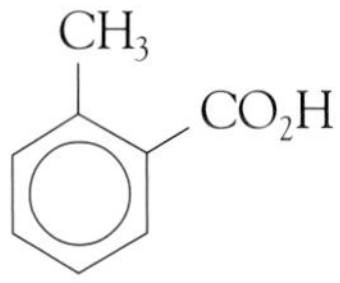

is 2-methylbenzoic acid

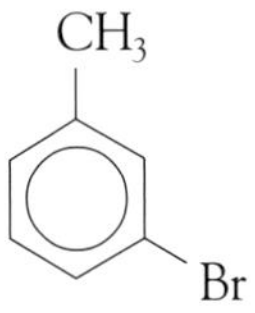

is 3-bromomethylbenzene

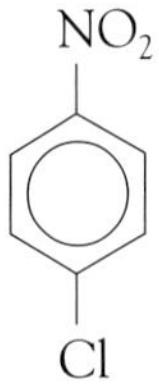

is 4-nitrochlorobenzene.

Worked example

Draw out all possible positional isomers of $C_6H_3Br_2OH$ and name them.

Answer

There are six isomers. Their names and formulae are as follows:

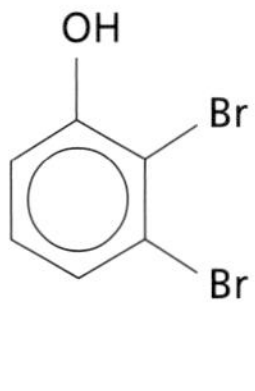

2,3-dibromophenol 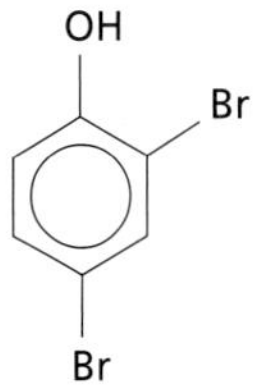   2,4-dibromophenol 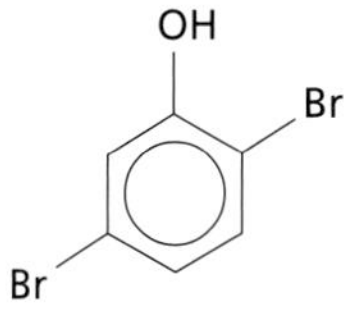   2,5-dibromophenol

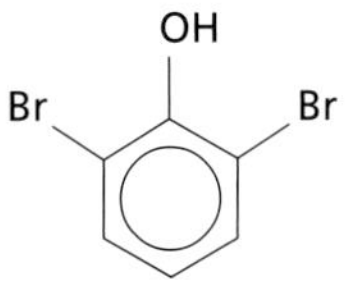

2,6-dibromophenol 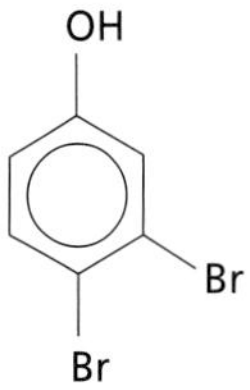   3,4-dibromophenol 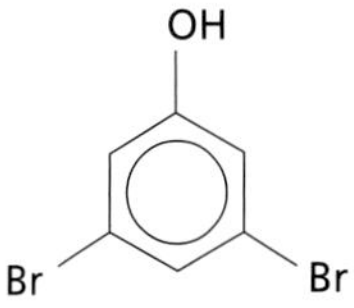   3,5-dibromophenol

Now try this

How many isomers are there of trichloromethylbenzene, $C_6H_2Cl_3(CH_3)$? What are their names?

If the benzene ring is a 'substituent' on an alkyl or alkenyl chain, it is given the name **phenyl**:

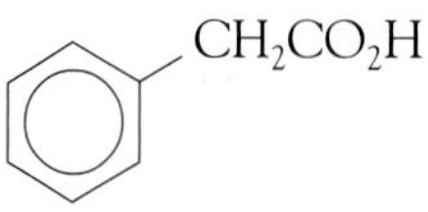

phenylethanoic acid 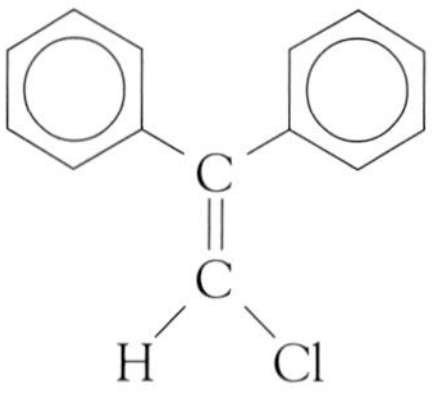   2,2-diphenylchloroethene

25.3 Reactions of arenes

Combustion

Benzene and methylbenzene (whose old name, toluene, is still in use) are components of many brands of unleaded petrol. In sufficient oxygen, they burn completely to carbon dioxide and steam:

$$C_6H_6 + 7\tfrac{1}{2}O_2 \rightarrow 6CO_2 + 3H_2O$$

If liquid arenes are set alight in the laboratory, they burn with very smoky flames. Much soot is produced because there is insufficient oxygen for complete combustion. A smoky flame is an indication of a compound with a high C:H ratio.

$$C_6H_6 + 1\tfrac{1}{2}O_2 \rightarrow 6C\ (s) + 3H_2O$$

Just as with hydrogenation, the enthalpy of combustion of benzene is less exothermic than expected, because of the stability due to the six delocalised π electrons.

Electrophilic substitution in benzene

In a similar way to the π bond in alkenes, the delocalised π bond in benzene is an area of high electron density, above and below the six-membered ring. Benzene therefore reacts with electrophiles. Because of the extra stability of the delocalised electrons, however, the species that react with benzene have to be much more powerful electrophiles than those that react with ethene. Bromine water and aqueous acids have no effect on benzene.

The electrophiles that react with benzene are all positively charged, with a strong electron-attracting tendency. The other major difference between benzene and alkenes is what happens after the electrophile has attacked the π bond. In alkenes, an anion 'adds on' to the carbocation intermediate. In benzene, on the other hand, the carbocation intermediate loses a proton, so as to re-form the ring of π electrons. This demonstrates how stable the delocalised system is.

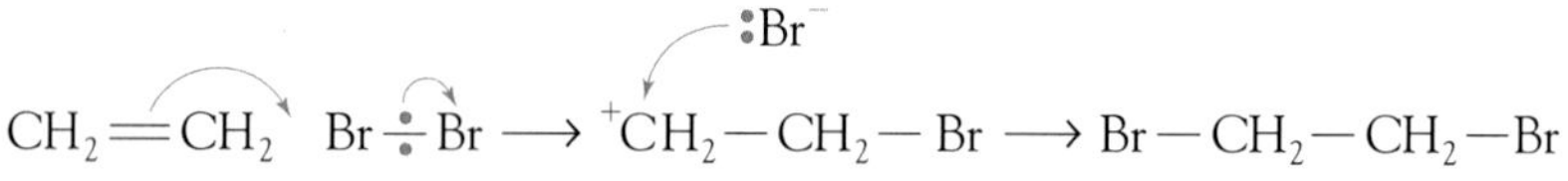

carbocation intermediate

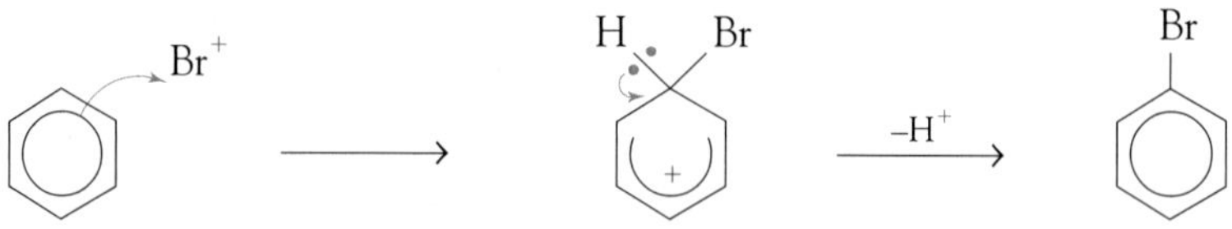

carbocation intermediate

Alkenes react by **electrophilic addition**.
Arenes react by **electrophilic substitution**.

Bromination

Benzene will react with non-aqueous bromine on warming in the presence of anhydrous aluminium chloride or aluminium bromide, or iron(III) chloride, or even just iron metal. In the latter case, an initial reaction between iron and bromine provides the iron(III) bromide catalyst:

$$2Fe(s) + 3Br_2(l) \rightarrow 2FeBr_3(s)$$

Anhydrous aluminium or iron(III) halides contain electron-deficient atoms. They can react with the bromine molecule by accepting one of the lone pairs of electrons on bromine:

$$\text{Br}-\text{Br}: \quad \text{Al}-\text{Br} \longrightarrow \text{Br}-\overset{+}{\text{Br}}-\overset{-}{\text{Al}}-\text{Br} \qquad (1a)$$

This causes strong polarisation of the Br—Br bond, weakening it, and eventually leading to its heterolytic breaking:

$$\text{Br} \overset{+}{\cdot} \text{Br}-\text{Al}-\text{Br} \longrightarrow \text{Br}^+ + \text{Br}-\overset{-}{\text{Al}}-\text{Br} \qquad (1b)$$

The bromine cation that is formed is a powerful electrophile. It becomes attracted to the π bond of benzene. It eventually breaks the ring of electrons and forms a σ bond to one of the carbon atoms of the ring:

$$\text{(benzene ring)} \quad \text{Br}^+ \longrightarrow \text{(ring)} \overset{\text{H}}{\underset{\text{Br}}{}} \qquad (2)$$

Two of the six π electrons are used to form the (dative) bond to the bromine atom. The other four π electrons are spread over the remaining five carbon atoms of the ring, in a five-centre delocalised orbital (see Figure 25.10).

Figure 25.10 The four π electrons are delocalised over five carbons

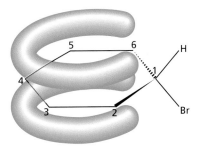

The distribution of the four π electrons is not even. They are associated more with carbon atoms 3 and 5 than with atoms 2, 4 and 6. The positive charge is therefore distributed over atoms 2, 4 and 6. This is best represented by the following classical mesomers:

We shall return to this feature of the intermediate carbocation when we look at orientation effects on page 427.

The intermediate carbocation then loses a proton, to re-form the sextet of π electrons:

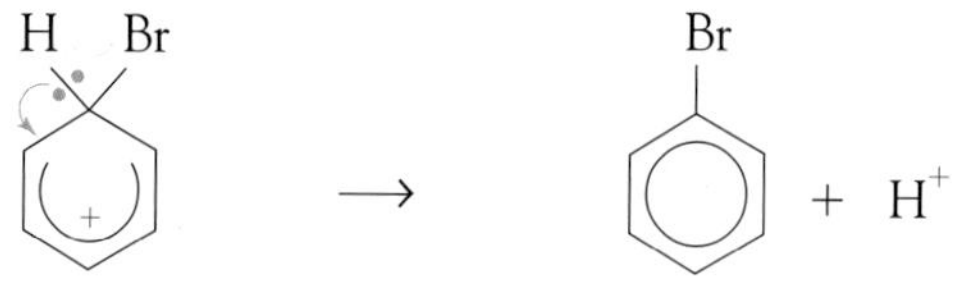

$$\tag{3}$$

The final stage regenerates the catalyst, by the reaction between the proton formed in step (3) with the $[AlBr_4]^-$ formed in step (1b):

$$H^+ + AlBr_4^- \rightarrow HBr + AlBr_3$$

Further substitution can occur with an excess of bromine, to form dibromobenzene, and even tribromobenzene. For the reactions of bromobenzene, see section 25.4.

Chlorination

Just as with bromine, chlorine in the presence of an electron-acceptor such as aluminium chloride will substitute into the benzene ring:

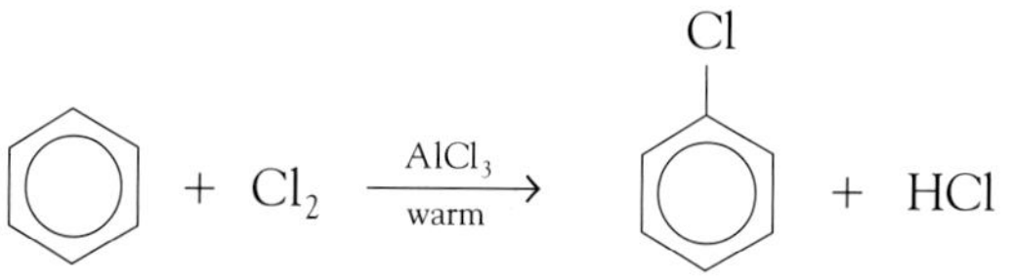

The properties and reactions of chlorobenzene are discussed in section 25.4.

> **Now try this**
>
> Suggest a mechanism for the chlorination of benzene.

Nitration

Benzene does not react with nitric acid, even when the acid is concentrated. But a mixture of concentrated nitric and concentrated sulfuric acids produces nitrobenzene:

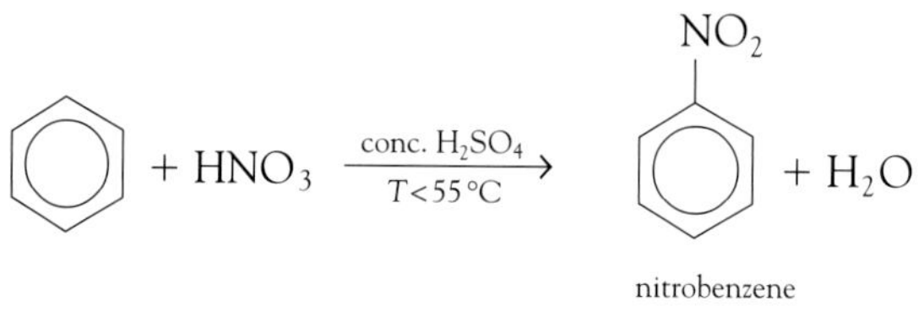

nitrobenzene

Further nitration to dinitrobenzene (and also some oxidation) can occur unless the temperature is controlled to be below about 55 °C.

The role of the concentrated sulfuric acid in this reaction is to produce the powerful electrophile needed to nitrate the benzene ring. By analogy with bromination:

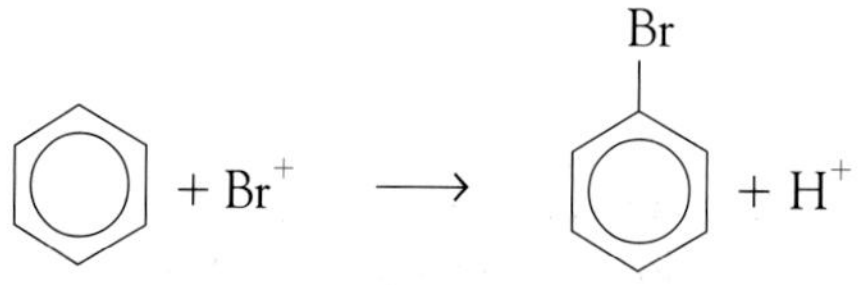

we might expect the electrophile for nitration to be NO_2^+:

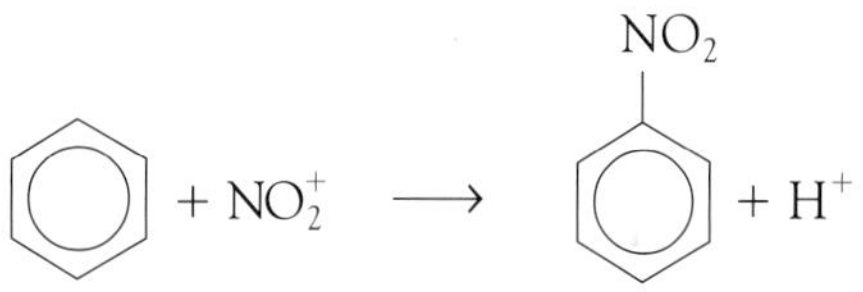

and this indeed seems to be the case.

Over the years, various pieces of evidence have been collected to substantiate the claim of NO_2^+ (known as the nitryl cation or nitronium ion) to be the electrophile.

- Stable salts containing the nitryl cation exist, for example nitryl chlorate(VII), $NO_2^+ClO_4^-$, and nitryl tetrafluoroborate(III), $NO_2^+BF_4^-$. Each of these, when dissolved in an inert solvent, nitrates benzene smoothly and in high yield to give nitrobenzene.
- The infrared spectrum of a compound tells us about the types of bonds in the compound, and its shape. The infrared spectrum of a solution of nitric acid in sulfuric acid shows a strong absorbance at $2350\,cm^{-1}$. This absorbance is also present in the spectra of nitryl salts, but is not present in that of nitric acid in the absence of sulfuric acid. It is almost identical to a similar absorbance in the spectrum of carbon dioxide. CO_2 and NO_2^+ are isoelectronic linear molecules (see section 3.13), and so would be expected to have similar infrared spectra.
- Various physico-chemical data (for example, the depression of freezing point) show that when one molecule of HNO_3 is dissolved in concentrated sulfuric acid, *four* particles are formed. (The freezing point of a liquid is lowered if substances are dissolved in it, and the amount of the depression depends upon the mole ratio of solute to solvent.)

These three pieces of evidence suggest that the following reaction occurs when the two acids are mixed:

$$2H_2SO_4 + HNO_3 \rightarrow H_3O^+ + NO_2^+ + 2HSO_4^-$$

The mechanism of the nitration of benzene

Sulfuric acid is a stronger acid than nitric acid. It is so strong that it donates a proton to nitric acid. In this reaction, nitric acid is acting as a base!

$$H_2SO_4 + HNO_3 \rightarrow HSO_4^- + H_2NO_3^+$$

The protonated nitric acid then loses a water molecule, and this is then protonated by another molecule of sulfuric acid:

$$H_2NO_3^+ \rightarrow H_2O + NO_2^+$$
$$H_2SO_4 + H_2O \rightarrow HSO_4^- + H_3O^+$$

The nitronium ion then attacks the benzene ring in the usual way, forming the carbocation intermediate, which subsequently loses a proton:

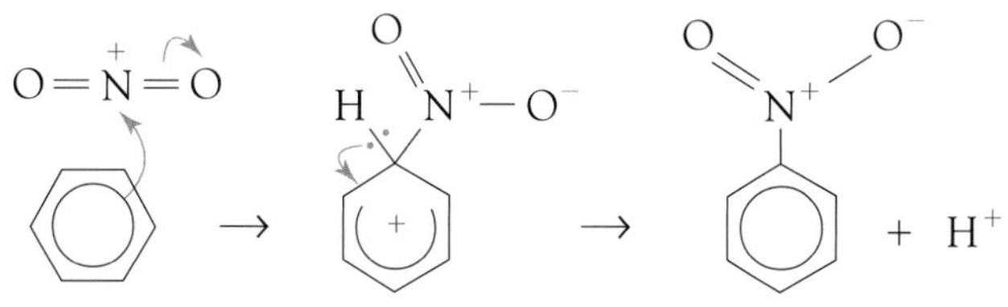

The nitration of benzene and other arenes is an important reaction in the production of explosives (see Figure 25.11)

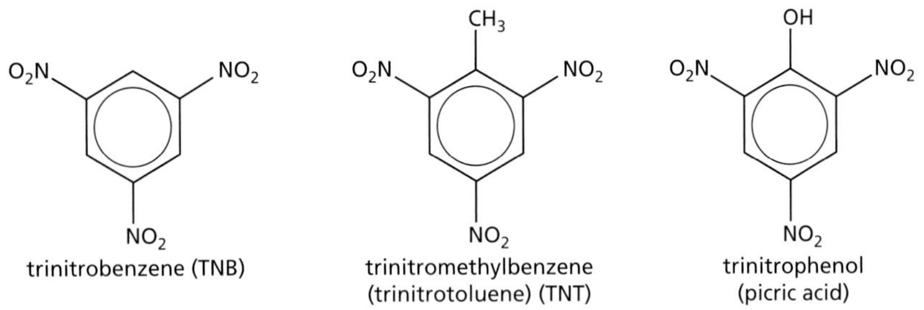

Figure 25.11 Three powerful explosives that are poly-nitro compounds

trinitrobenzene (TNB)

trinitromethylbenzene (trinitrotoluene) (TNT)

trinitrophenol (picric acid)

Also, via subsequent reduction, nitration is an important route to aromatic amines, which are used to make a variety of dyes (see Topic 27).

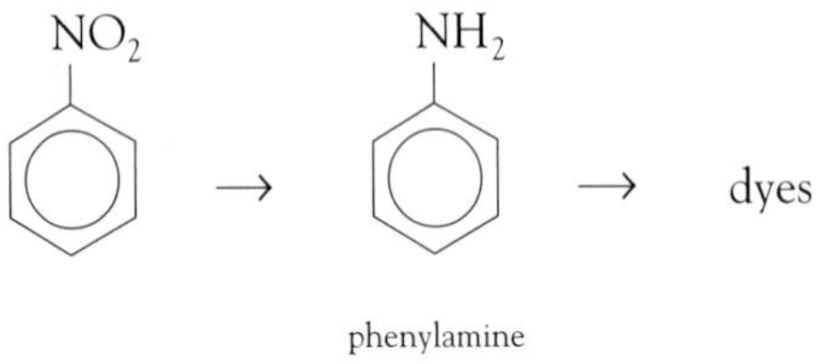

phenylamine

Alkylation (the Friedel–Crafts reaction)

In 1877 the French chemist Charles Friedel and his co-worker, American chemist James Crafts, discovered that when benzene is heated with a chloroalkane in the presence of aluminium chloride, the alkyl group attaches to the benzene ring.

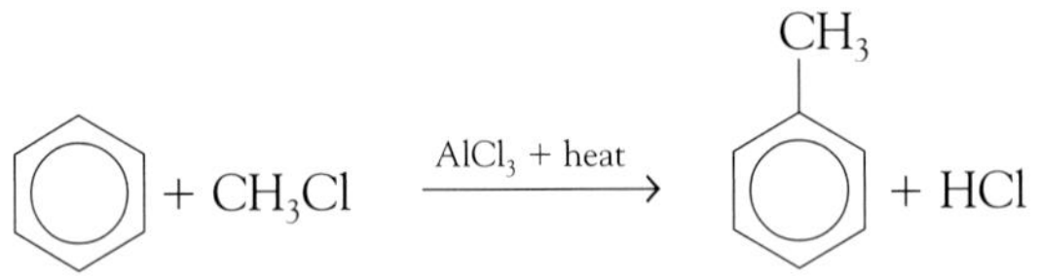

The reaction goes via the formation of an intermediate carbocation.

The carbocation is the electrophile.

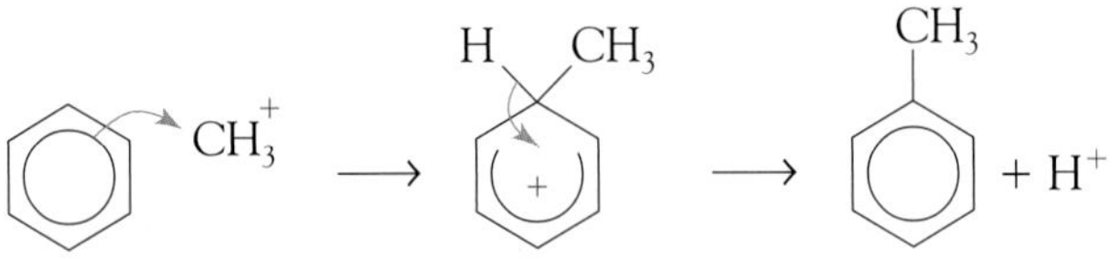

Acylation (Friedel–Crafts acylation)

If an acyl chloride (see section 26.3, page 442) is used instead of a chloroalkane in the Friedel–Crafts reaction, a phenylketone is produced.

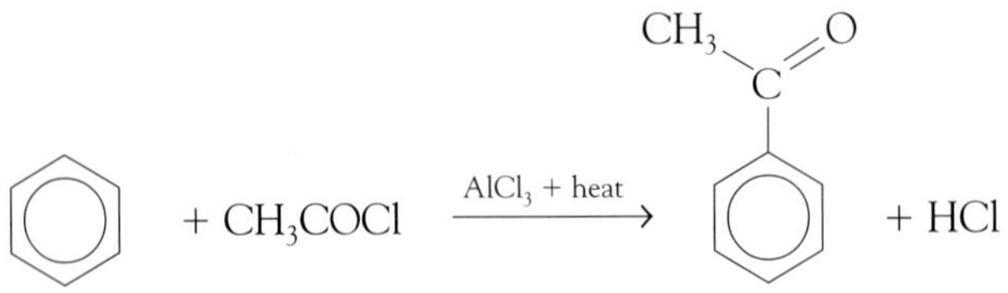

The reaction between an acyl chloride and aluminium chloride produces the acylium ion. The intermediate is formed from the attack of the acylium ion on the benzene ring.

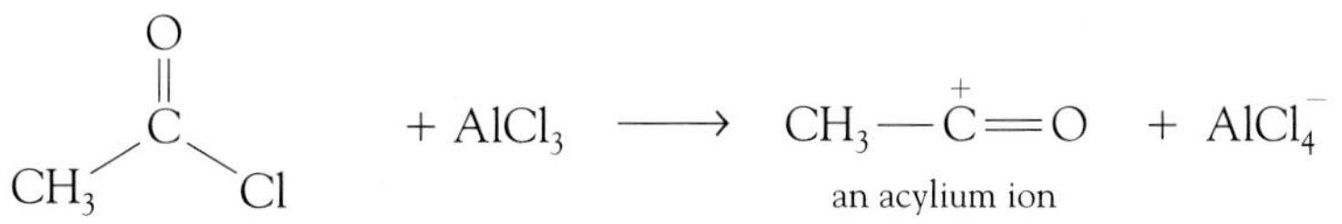

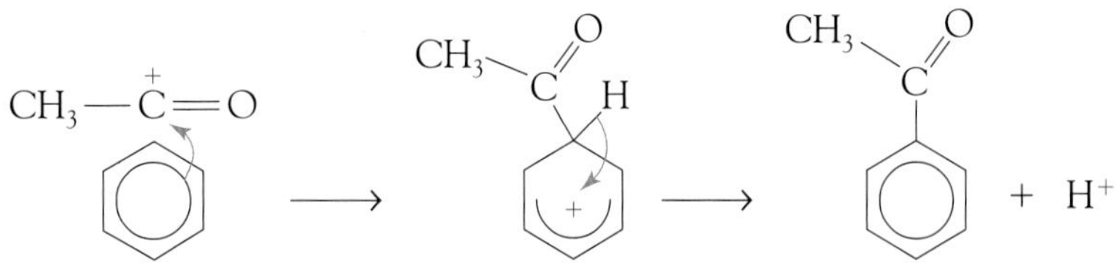

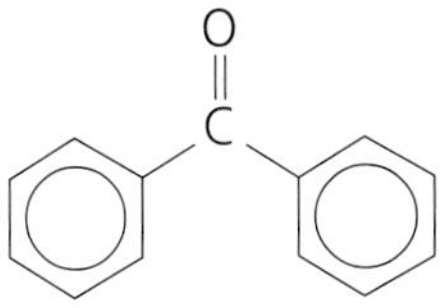

Worked example

Draw the structural formulae of the products you would expect from the reaction of benzene and aluminium chloride with

a CH_3CH_2Cl
b $(CH_3)_2CH—COCl$.

Answer

a

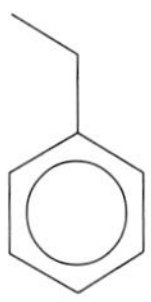

b

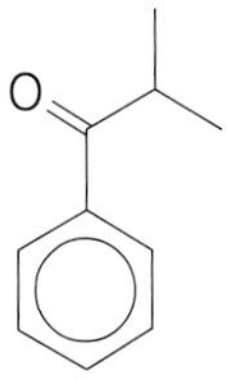

Now try this

What organochlorine compounds are needed to synthesise the following compounds from benzene?

1 $C(CH_3)_3$

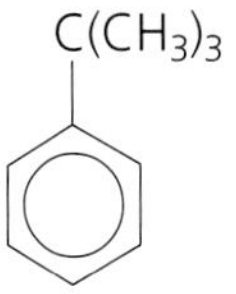

2

Electrophilic substitution in substituted arenes – the orientation of the incoming group

When methylbenzene is treated with nitric and sulfuric acids, the three possible mono-nitro compounds are formed in the following ratios:

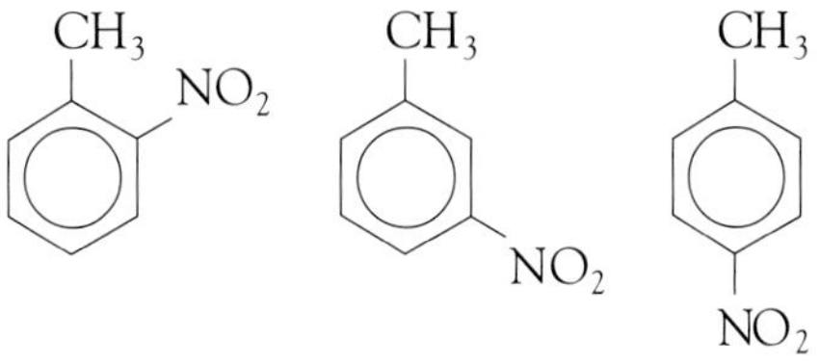

58%   4%   38%

If the NO_2^+ electrophile had attacked the ring in a purely random way, the distribution should have been 40:40:20 (there being two 2-positions and two 3-positions, but only one 4-position). This non-random attack is seen in other reactions too (see Table 25.5).

Table 25.5 Orientation in some electrophilic substitution reactions

Reaction	Percentage of product			Ratio (ortho + para) / meta
	ortho (1,2)	meta (1,3)	para (1,4)	
CH₃—⟨⟩ + Br₂ → CH₃—⟨⟩—Br	33	1	66	99 : 1
OH—⟨⟩ + HNO₃ → OH—⟨⟩—NO₂	50	1	49	99 : 1
NO₂—⟨⟩ + HNO₃ → NO₂—⟨⟩—NO₂	6	93.5	0.5	1 : 14
CO₂H—⟨⟩ + HNO₃ → CO₂H—⟨⟩—NO₂	19	80	1	1 : 4

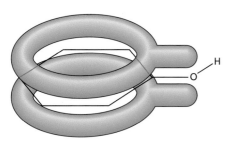

Figure 25.12 Delocalisation of the lone pair in 2,4-directing substituents

The data in Table 25.5 can be interpreted as follows.

- The orientation of the incoming group (NO₂ or Br) depends on the substituent *already in the ring*, and not on the electrophile.
- Some substituents favour both 2- and 4-substitution, whereas other substituents favour 3-substitution, at the expense of both 2- and 4-substitution.

If we look closely at the types of substituents that are 2,4-directing, we find that either they are capable of donating electrons to the ring by the inductive effect, or they have a lone pair of electrons on the atom joined to the ring. This lone pair can be incorporated into the π system by sideways overlap of p orbitals (see Figure 25.12).

On the other hand, all those substituents that favour 3-substitution have a δ+ atom joined directly to the ring (see Table 25.6).

Table 25.6 Substituents and their effects on the benzene ring

2- and 4-directing substituents	3-directing substituents
CH₃ ⇀ Ar	O=N⁺(—O⁻)—Ar
H—Ö—Ar	$^{δ-}$O=C$^{δ+}$(H)—Ar
H₂N—Ar	$^{δ-}$N≡C$^{δ+}$—Ar
	$^{δ-}$O=C$^{δ+}$(R)—Ar

The orientation effects of different substituents

To explain why electron-donating substituents are 2,4-directing, and why electron-withdrawing groups are 3-directing, we need to return to the electron distribution in the carbocation intermediate, mentioned on page 423.

In the bromination of benzene, the intermediate can be represented as follows.

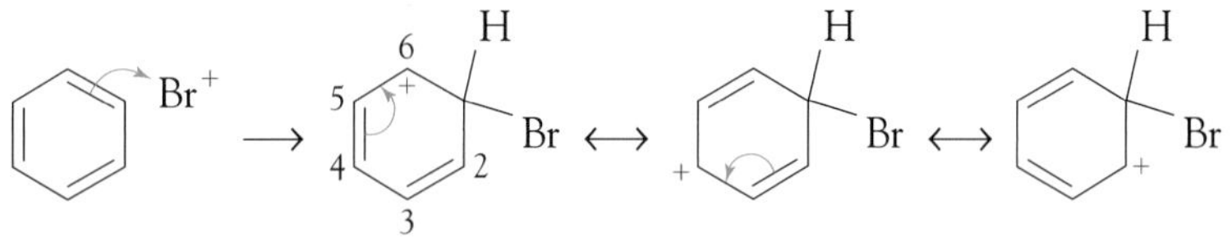

The + charge (that is, the electron deficiency) is due to only four π electrons being spread over five carbon atoms. But the + charge is not evenly spaced. As can be seen above, it is mainly on carbon atoms 2, 4 and 6.

Electron-donating groups at these positions will therefore be more effective at stabilising the intermediate, by spreading out its charge, than if the electron-donating group were in position 3 or 5, one carbon removed from the + charge:

If the methyl group is situated in the 2- or 4-position relative to the bromine atom, the intermediate is *much* more stable compared to the intermediate in the substitution of benzene.

If the methyl group is situated in the 3-position relative to the bromine atom, the intermediate is only a *little* more stable compared to the intermediate in the substitution of benzene.

Now try this

The bromination of nitrobenzene produces mainly 3-bromonitrobenzene. Draw out the mesomers of the intermediate carbocation and use a similar argument to the one above to explain why production of 3-bromonitrobenzene is favoured over that of 2- or 4-bromonitrobenzene.

Addition reactions of the benzene ring

When it reacts with electrophiles, benzene always undergoes substitution. However, under some conditions it can be forced to undergo addition reactions. One of these is the addition of hydrogen.

Ethene adds on hydrogen over a nickel catalyst readily at room temperature and slight pressure. Benzene, being more stable, requires an elevated temperature and pressure:

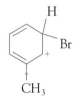

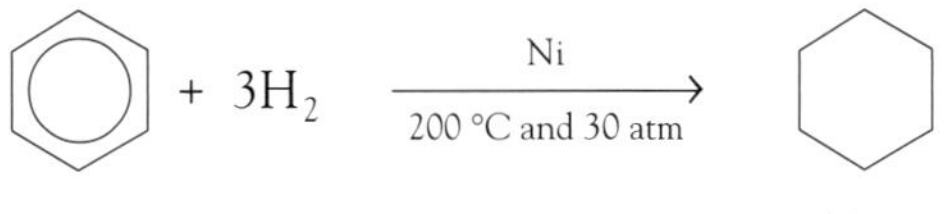

$$+ \; 3H_2 \quad \xrightarrow[\text{200 °C and 30 atm}]{\text{Ni}}$$

cyclohexane

Substitution in the side chain

Methylbenzene and other alkylbenzenes very readily undergo radical substitution with chlorine or bromine in the presence of ultraviolet light (just like alkanes – see section 13.5), or by boiling in the absence of ultraviolet light.

$$CH_3 \quad + \; Cl_2 \quad \xrightarrow[\text{or boiling}]{\text{UV light}} \quad CH_2Cl \quad + \; HCl$$

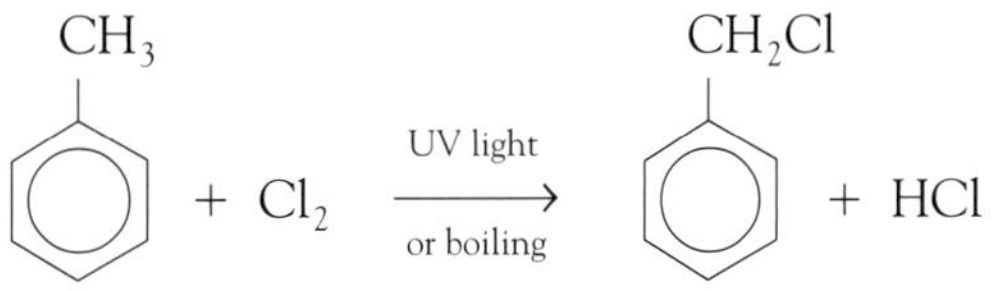

Just as with alkanes, more than one chlorine atom can be substituted. But substitution occurs only in the side chain during this reaction, not in the ring.

In ethylbenzene, the hydrogen atoms on the carbon atom adjacent to the benzene ring are substituted much more readily than the three hydrogens on the other carbon, so the following reaction gives a good yield of the product:

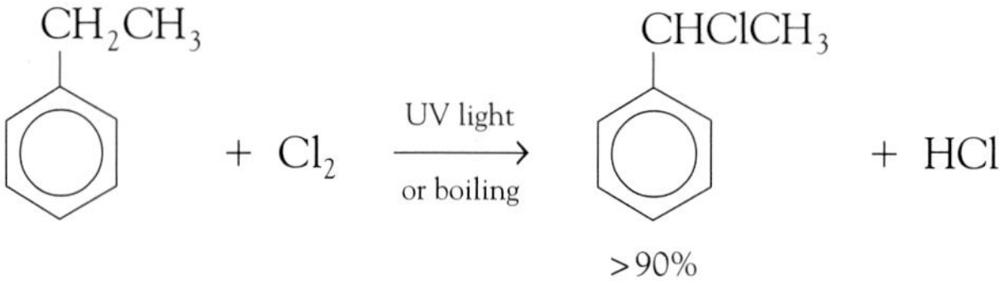

Oxidation of the side chain

When alkylbenzenes are treated with hot alkaline potassium manganate(VII), oxidation of the whole side chain occurs, leaving the carbon atom closest to the ring as a carboxylate or carboxylic acid group:

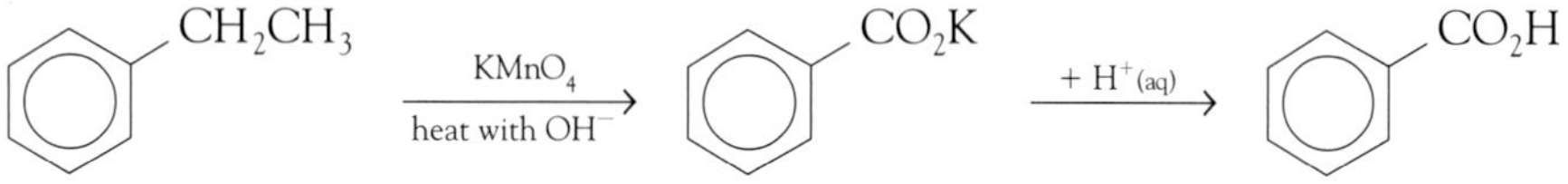

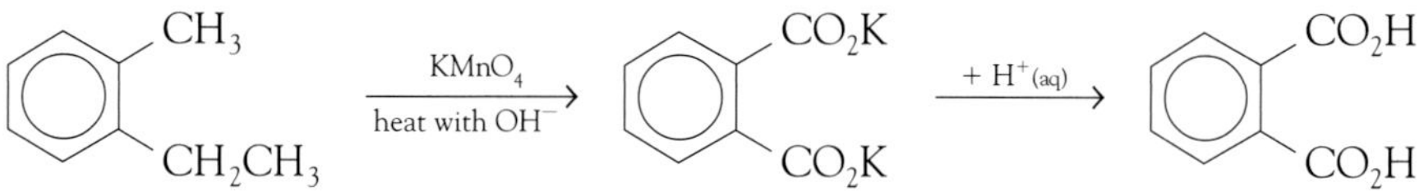

Worked example

Three hydrocarbons **A**, **B** and **C** with the formula C_9H_{12} were oxidised by hot potassium manganate(VII).

- Hydrocarbon **A** gave benzoic acid, $C_6H_5CO_2H$.
- Hydrocarbon **B** gave benzene-1,2-dioic acid:

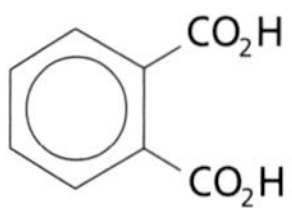

- Hydrocarbon **C** gave benzene-1,2,4-trioic acid:

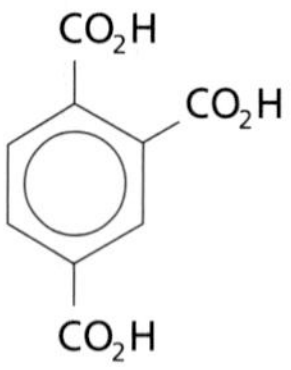

Suggest the structures of **A**, **B** and **C**.

Answer

Since **A** gave benzoic acid, all three 'extra' carbon atoms must be in the same side chain. So **A** is:

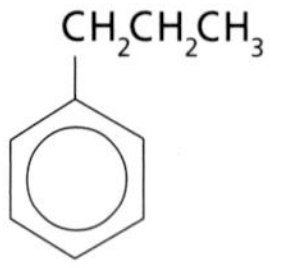

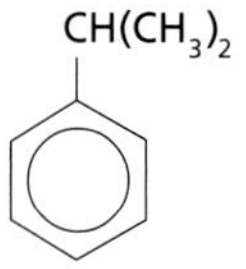

Compound **B** must contain two side chains, since two carboxylic acid groups are left after oxidation. What is more, the chains must be on adjacent carbons in the ring, as a 1,2-dicarboxylic acid is formed. So **B** is:

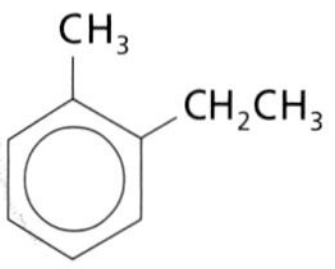

By similar reasoning, **C** must be:

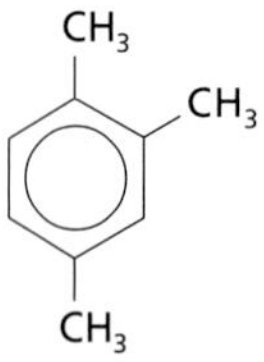

Now try this

Suggest structures for the aromatic carboxylic acids which will be produced when the following compounds are oxidised by hot potassium manganate(VII). (All these compounds are isomers with the molecular formula $C_{11}H_{14}$.)

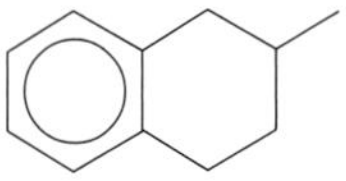

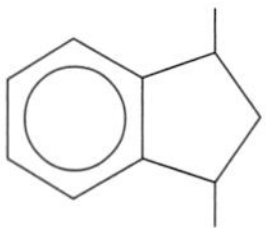

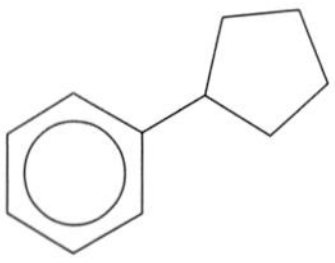

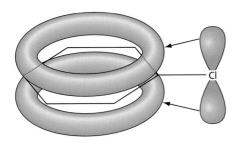

Figure 25.13 Delocalisation of the lone pair in chlorobenzene

25.4 Halogenoarenes

We saw on pages 423–424 how bromobenzene and chlorobenzene can be made from benzene. The reactions of the ring in halogenobenzenes are similar to those of benzene.

Halogenoarenes undergo electrophilic substitution, and can be nitrated:

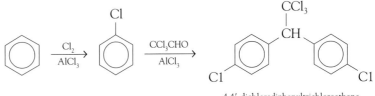

However, unlike halogenoalkanes, halogenoarenes cannot be hydrolysed, even by boiling in aqueous sodium hydroxide. The carbon–halogen bond is stronger in halogenoarenes than it is in halogenoalkanes, possibly due to an overlap of p electrons similar to that in phenol (see Figure 25.13, and compare it with Figure 25.12).

In addition to this, the carbon attached to the halogen is not accessible to the usual nucleophilic reagents that attack halogenoalkanes, since its δ+ charge is shielded by the negative π cloud of the ring. This means that halogenoarenes are inert to all nucleophiles.

Certain halogenoarenes find important uses as insecticides and herbicides (see the panel below).

Organochlorine insecticides and herbicides

Insecticides

Chlorobenzene used to be made in large quantities as an intermediate in the production of the insecticide DDT:

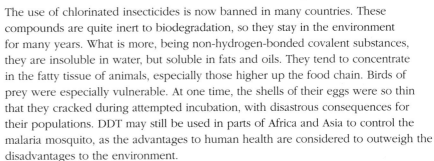

4,4′-dichlorodiphenyltrichloroethane
(DDT)

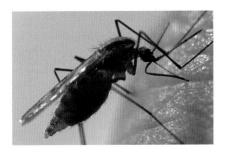

Figure 25.14 Malaria is caused by a microscopic parasite that is spread by the bite of the *Anopheles* mosquito. Malaria affects over 300 million people worldwide and kills around 3 million a year. DDT treatment of the *Anopheles* mosquito saves millions of human lives, so the environmental price is considered worth paying.

The use of chlorinated insecticides is now banned in many countries. These compounds are quite inert to biodegradation, so they stay in the environment for many years. What is more, being non-hydrogen-bonded covalent substances, they are insoluble in water, but soluble in fats and oils. They tend to concentrate in the fatty tissue of animals, especially those higher up the food chain. Birds of prey were especially vulnerable. At one time, the shells of their eggs were so thin that they cracked during attempted incubation, with disastrous consequences for their populations. DDT may still be used in parts of Africa and Asia to control the malaria mosquito, as the advantages to human health are considered to outweigh the disadvantages to the environment.

Herbicides

The compounds 2,4-dichlorophenoxyethanoic acid (2,4-D) and 2,4,5-trichlorophenoxyethanoic acid (2,4,5-T) have been successfully used for many years as selective weedkillers for broad-leaved weeds. They have little effect on narrow-leaved plants such as grass and cereals, and so can be used as lawn or field weedkillers. Their action mimics that of the natural plant growth hormone indole ethanoic acid (indole-acetic acid, IAA). They are members of a group of compounds called hormone weedkillers. Their structures are shown in Figure 25.15.

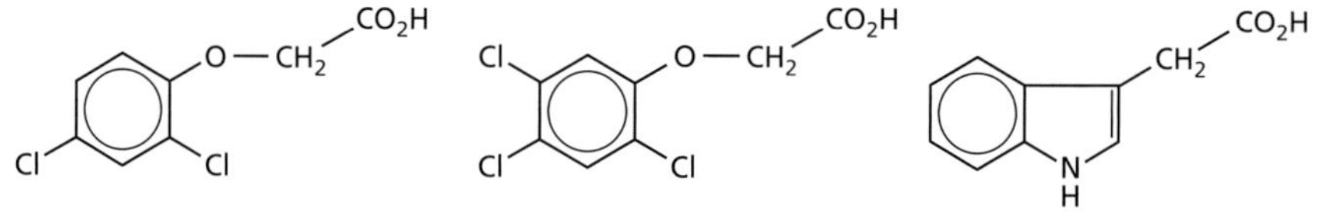

Figure 25.15 The structures of some hormone weedkillers (herbicides)

2,4-D   2,4,5-T   IAA

2,4-D and 2,4,5-T were components of 'Agent Orange', sprayed as a jungle defoliant by the US army during the Vietnam war. It was during this large-scale use that an unforeseen problem came to light. During the manufacture of the intermediate chlorinated phenol, a small quantity of a highly toxic impurity, tetrachlorodibenzodioxin (or 'dioxin') was also produced (see Figure 25.16). The temperature of the reaction has to be controlled very carefully if the amount of dioxin is to be kept at a low level.

Unlike 2,4-D or 2,4,5-T, dioxin is not soluble in water. If ingested, or absorbed through the skin or lungs, it becomes concentrated in the fatty tissues of the body. It has now been discovered that dioxin is one of the most poisonous chemicals known – it is strongly carcinogenic (cancer producing), teratogenic (produces malformation of the foetus) and mutagenic (causes mutations). It causes skin burns and ulcers that heal only very slowly. Improved methods of production of 2,4-D and 2,4,5-T are now in place, but the risk of accidentally producing large quantities of dioxin is still present unless effective controls are in place. Other non-chlorine-containing hormone weedkillers are preferred – they are equally effective, but more expensive.

The use of chlorinated herbicides is now severely restricted, just as is the use of chlorinated insecticides.

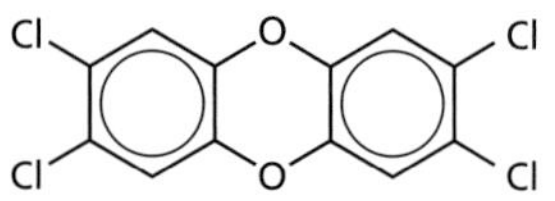

tetrachlorodibenzo-1,4-dioxin
(TCDD, 'dioxin')

Figure 25.16 Dioxin – a highly toxic molecule

25.5 Phenols

Phenols contain two functional groups – the —OH group of the alcohols, and the phenyl ring of the arenes. Their reactions are for the most part the sum of the two sets, but with significant modifications.

Nomenclature

Many of the compounds illustrated in Figure 25.8 (page 419) are phenols. If the only other groups on the benzene ring are halogen atoms, nitro, amino or alkyl groups, the compounds are named as derivatives of phenol itself:

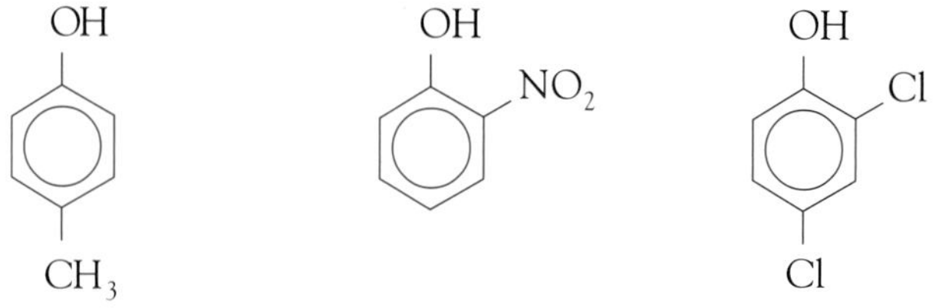

4-methylphenol
(p-methylphenol)

2-nitrophenol
(o-nitrophenol)

2,4-dichlorophenol

If, however, the other group is an aldehyde, ketone or carboxylic acid group, the phenolic —OH becomes a 'hydroxy' substituent:

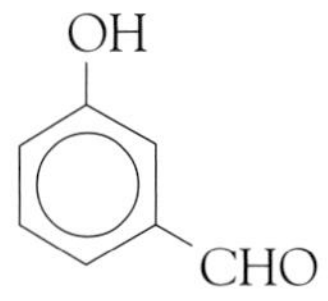

3-hydroxybenzaldehyde
(also known as
3-hydroxybenzenecarbaldehyde)

4-hydroxybenzoic acid
(also known as
4-hydroxybenzenecarboxylic acid)

Worked example

Name the following compounds:

a

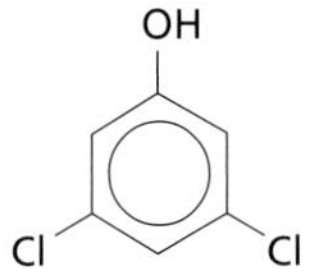

b

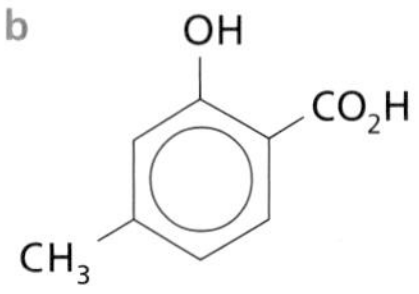

Answer

a 3,5-dichlorophenol   b 2-hydroxy-4-methylbenzoic acid

Now try this

Draw the structural formulae of the following compounds.

1 2,4,6-trimethylphenol
2 3,4,5-trihydroxybenzoic acid

Reactions of the —OH group

The C—O bond in phenol is very strong, as a result of the delocalisation of the lone pair of electrons on oxygen over the arene ring. There are no reactions in which it breaks, unlike the situation with the alcohols (see section 16.3).

Reactions of the O—H bond

Acidity

Phenols are more acidic than alcohols:

$$R{-}O{-}H \; \rightleftharpoons \; RO^- + H^+ \qquad\qquad K_a = 1.0 \times 10^{-16}\,\text{mol dm}^{-3}$$

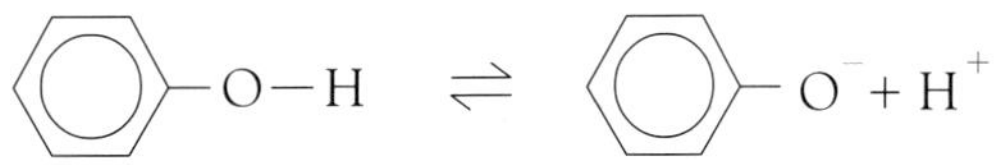

$K_a = 1.3 \times 10^{-10}\,\text{mol dm}^{-3}$

The negative charge of the anion can be delocalised over the benzene ring. Figure 25.17 shows various ways in which this can be represented.

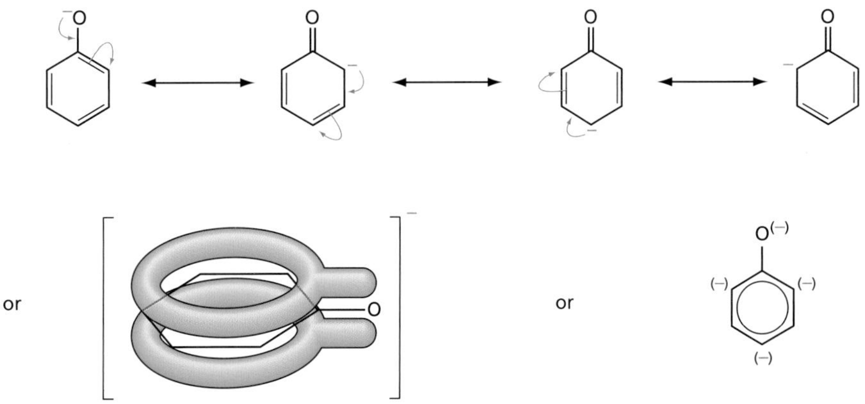

Figure 25.17 Representations of the phenol anion

Consequently, phenol not only reacts with sodium metal, giving off hydrogen gas:

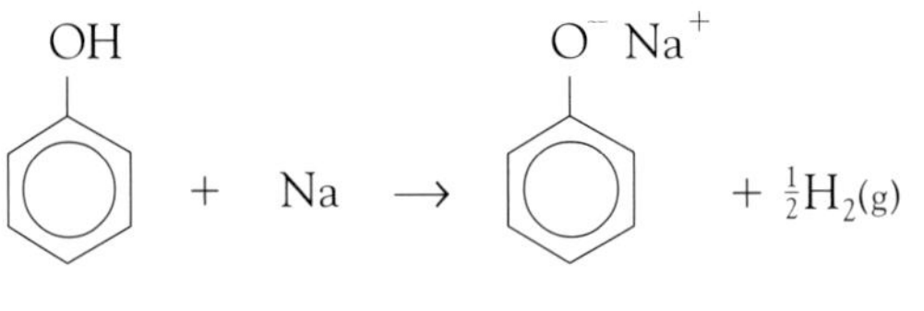

sodium phenoxide
(a white solid)

but, unlike alcohols, it also dissolves in aqueous sodium hydroxide:

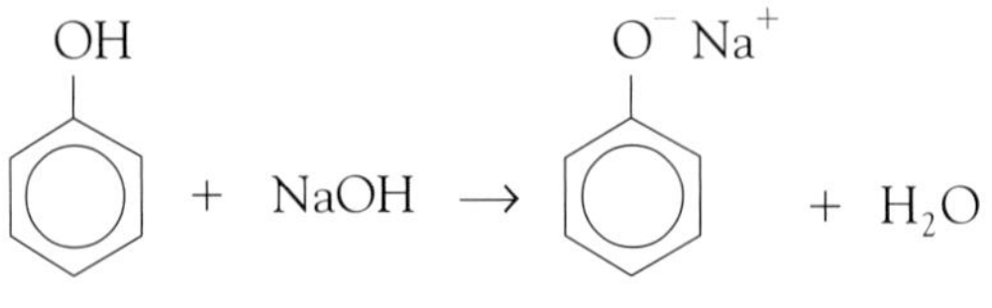

Phenol is visibly acidic: the pH of a $0.1\,mol\,dm^{-3}$ solution in water is 5.4, so it will turn universal indicator solution yellow. An old name for phenol is carbolic acid (see section 25.6).

Esterification

Because of the delocalisation over the ring of the lone pair on the oxygen atom, phenol is not nucleophilic enough to undergo esterification in the usual way, that is, by heating with a carboxylic acid and a trace of concentrated sulfuric acid (for the mechanism of esterification, see section 18.3).

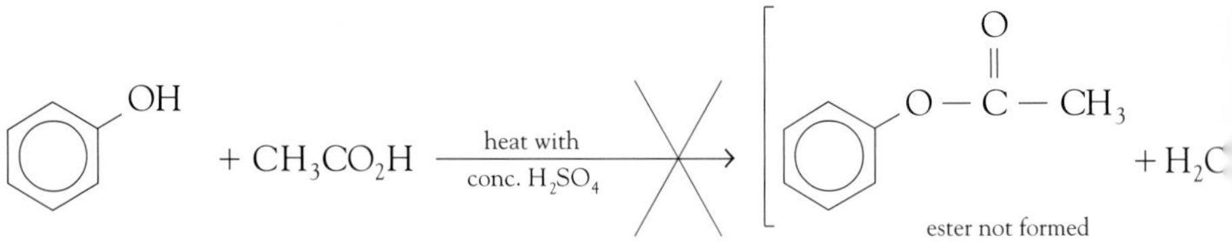

ester not formed

Phenol can, however, be esterified by adding an acyl chloride:

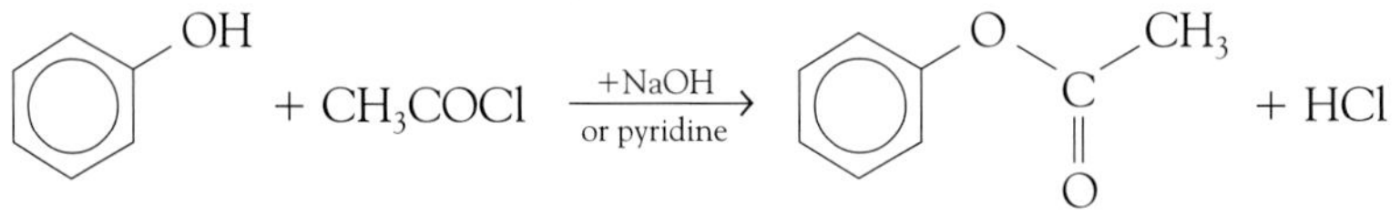

phenyl ethanoate

See section 26.3 for a fuller description of the conditions used for this reaction.

Substitution reactions of the benzene ring

As we mentioned on page 428, phenols are more susceptible to electrophilic attack than benzene, owing to the delocalisation of the lone pair of electrons on oxygen. This allows phenol to react with reagents that are more dilute, and also to undergo multiple substitution with ease.

Nitration

When treated with *dilute* aqueous nitric acid (no sulfuric acid is needed) phenol gives a mixture of 2- and 4-nitrophenols:

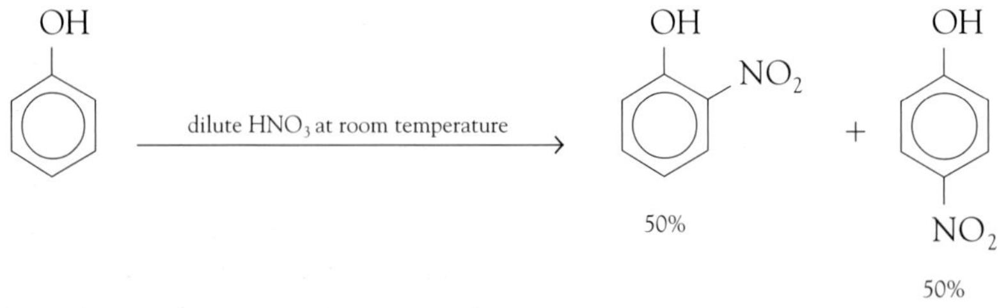

Bromination

Phenol decolorises a dilute solution of bromine in water at room temperature, giving a white precipitate of 2,4,6-tribromophenol. No aluminium bromide is needed. Contrast this with the conditions needed for the bromination of benzene on page 423.

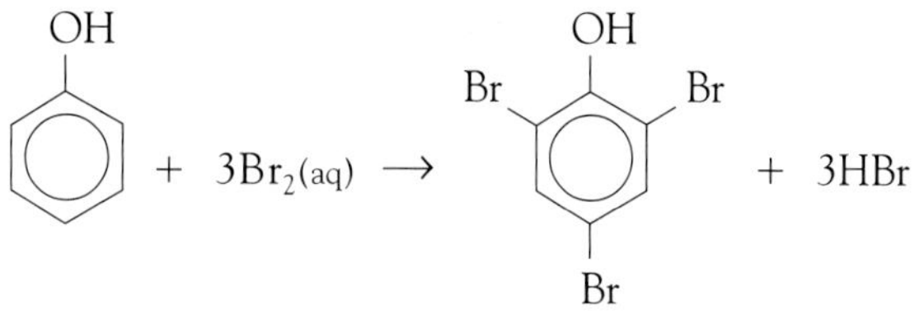

A similar product, formed by the action of chlorine water on phenol, is used in dilute solution as the antiseptic TCP (see page 436).

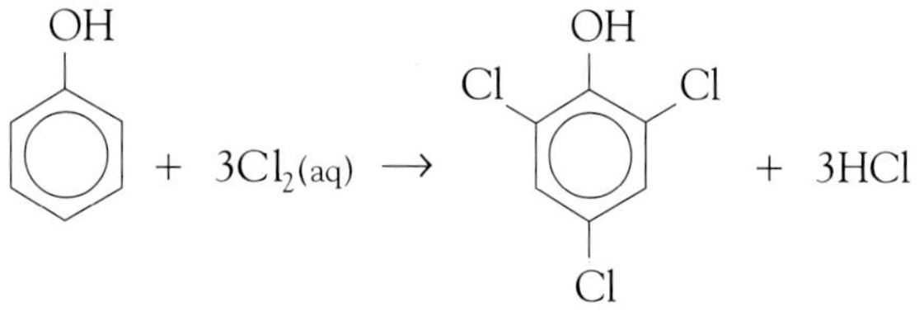

2,4,6-trichlorophenol
(TCP)

Coupling

Phenols couple with diazonium salts (see section 27.3) to form azo dyes:

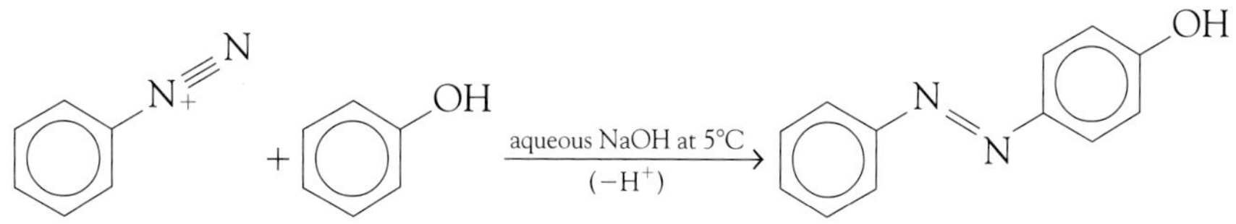

phenyldiazonium ion   an azo dye

A specific test for phenols

When a dilute solution of iron(III) chloride is added to a dilute solution of a phenol, a coloured complex is formed. The colour depends on the substituents on the ring. With phenol itself, a purple coloration is observed:

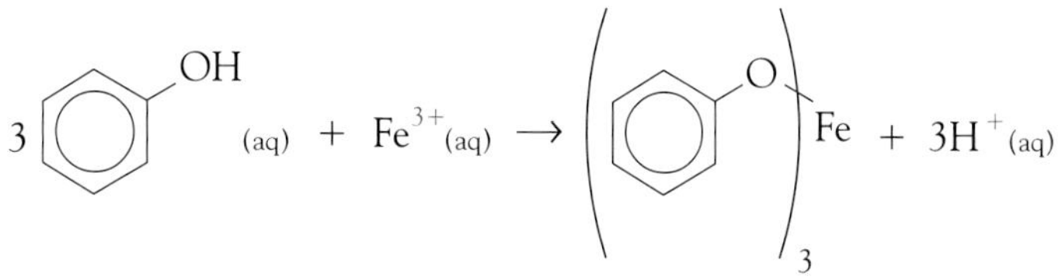

25.6 Some important phenols

Antiseptics

A dilute solution of phenol in water (known as carbolic acid) was one of the first disinfectants to be used in medicine, by Joseph Lister in Glasgow, Scotland, in 1867. Phenol itself is unfortunately too corrosive to be of general use as an antiseptic. Many chloro derivatives have been found to be more potent antiseptics than phenol itself. They can be used in much lower concentrations, which reduces their corrosive effect. Two of the most common are trichlorophenol (TCP) and chloroxylenol (see Figure 25.19). Thymol occurs in oil of thyme, and is an excellent non-toxic antiseptic, as well as being a fungicide.

Figure 25.18 The Scottish doctor Joseph Lister (1827–1912) was the first to use phenol as an antiseptic during surgery. Deaths from infections following operations were much reduced as a result, though the corrosive nature of phenol did not help the skin to heal.

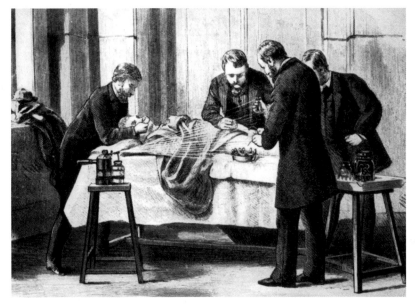

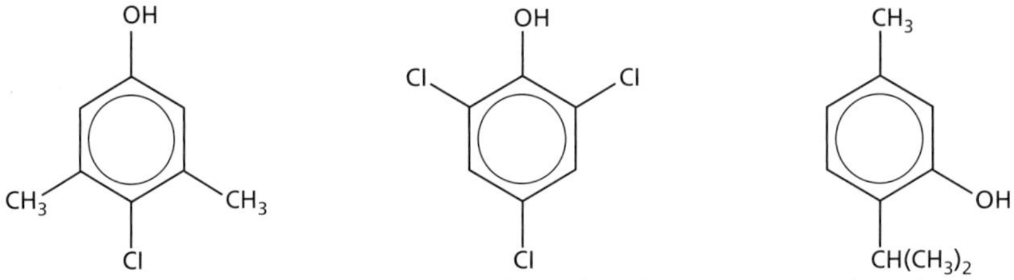

Figure 25.19 Three antiseptics that are derivatives of phenol

chloroxylenol ('Dettol')   2,4,6-trichlorophenol   thymol

Analgesics

A significant number of pharmaceutical drugs contain phenolic groups, or are derived from phenols.

The painkilling and fever-reducing properties of an extract of willow bark have been known in many countries since at least the sixteenth century. In the nineteenth century, the active ingredient, salicylic acid (2-hydroxybenzoic acid), was isolated and purified (the name 'salicylic' derives from the Latin name for willow, *Salix*). The therapeutic use of salicylic acid was limited, however, because it caused vomiting and bleeding of the stomach. In 1893 an ester derived from salicylic acid and ethanoic acid was found to have far fewer side-effects:

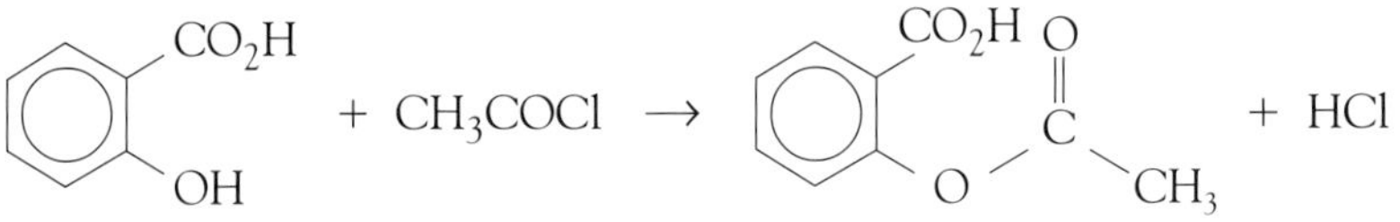

$$+ \ CH_3COCl \ \rightarrow \qquad + \ HCl$$

2-hydroxybenzoic acid   ethanoyl 2-hydroxybenzoic acid
(salicylic acid)   (aspirin)

Aspirin is the most widely used of all analgesics. However, it still retains some of the stomach-irritating effects of salicylic acid. A less problematic painkiller is paracetamol, which is another phenol.

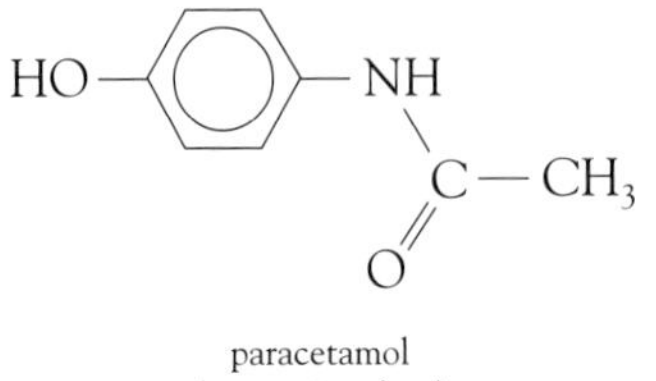

paracetamol
(acetaminophen)

25.7 Preparing arenes and phenols

- Benzene and methylbenzene are formed during the cracking and re-forming of fractions from the distillation of crude oil (see section 13.3).
- The —OH group of phenols can be introduced onto an arene ring by the following sequence (see section 27.3 for details):

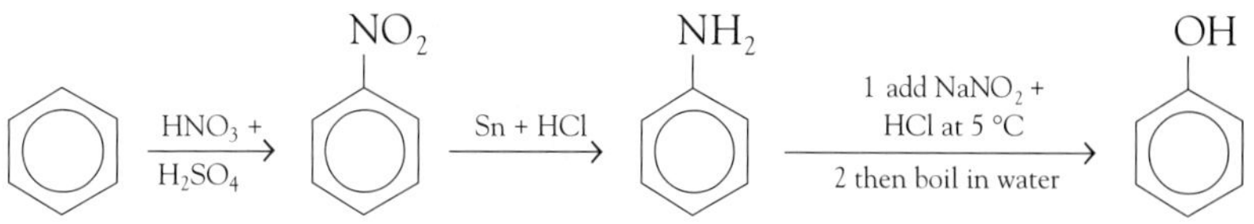

Summary

- **Arenes** are hydrocarbons that contain one or more benzene rings.
- The benzene ring contains a delocalised group of six π electrons. This confers great stability on the system.
- Despite their unsaturation, arenes do not undergo the usual addition reactions associated with alkenes. Their preferred reaction type is **electrophilic substitution**.
- **Aromatic** compounds can be halogenated in either the ring or the side chain, depending on the conditions used.
- Aromatic compounds with alkyl side chains can be oxidised by potassium manganate(VII) to benzenecarboxylic acids.
- Under forcing conditions with hydrogen, the benzene ring can undergo addition rather than substitution.
- **Phenols** are more acidic than alcohols, and can only be esterified by using acyl chlorides.
- The C—O bond in phenols is very strong, and no reactions occur in which it breaks.
- The benzene ring in phenol is much more susceptible to electrophilic attack than is the ring in benzene itself.
- Like alcohols and carboxylic acids, phenol reacts with sodium. Unlike alcohols, it reacts with sodium hydroxide; unlike carboxylic acids, it does *not* react with carbonates.

Key reactions you should know
- Electrophilic substitutions:

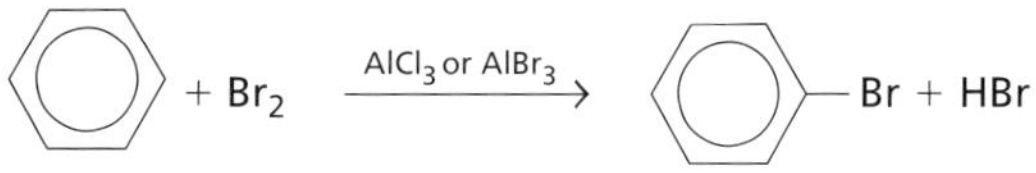

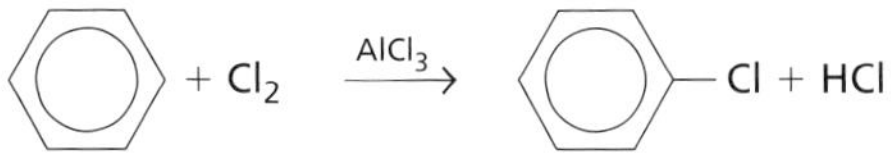

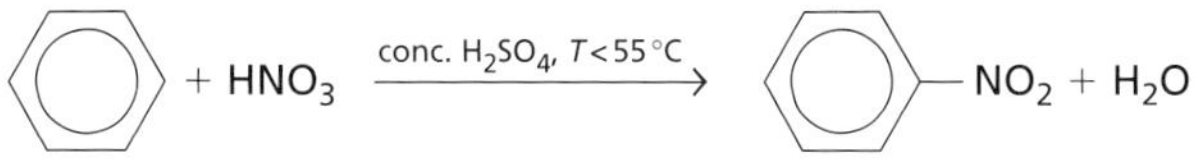

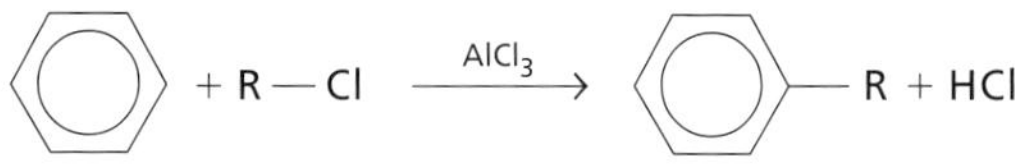

- Side-chain reactions:

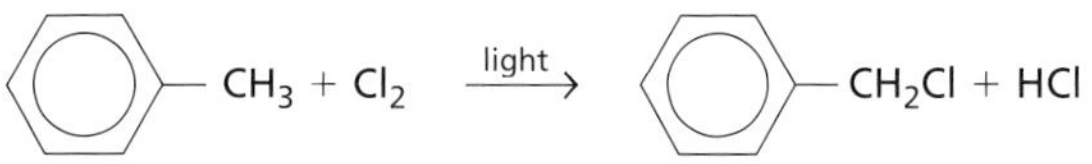

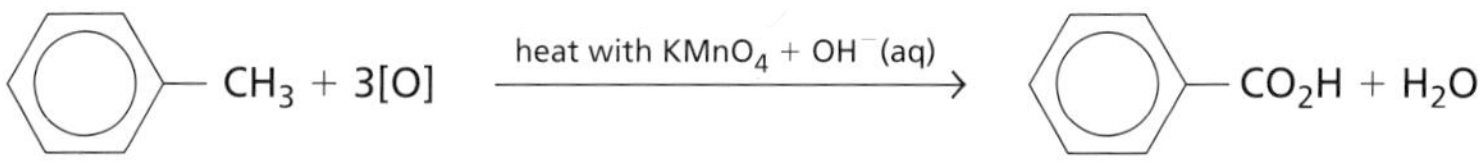

- Reactions of phenols:

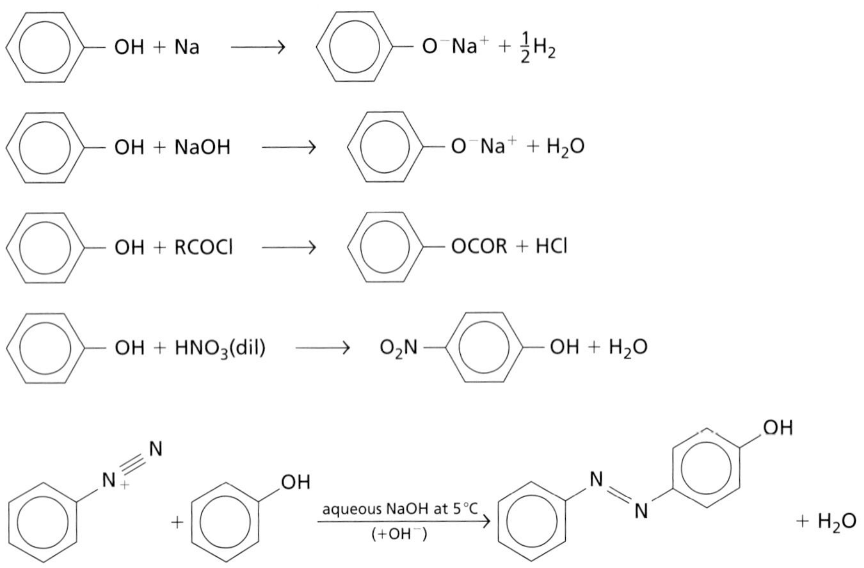

$$\text{C}_6\text{H}_5\text{—OH} + \text{Na} \longrightarrow \text{C}_6\text{H}_5\text{—O}^-\text{Na}^+ + \tfrac{1}{2}\text{H}_2$$

$$\text{C}_6\text{H}_5\text{—OH} + \text{NaOH} \longrightarrow \text{C}_6\text{H}_5\text{—O}^-\text{Na}^+ + \text{H}_2\text{O}$$

$$\text{C}_6\text{H}_5\text{—OH} + \text{RCOCl} \longrightarrow \text{C}_6\text{H}_5\text{—OCOR} + \text{HCl}$$

$$\text{C}_6\text{H}_5\text{—OH} + \text{HNO}_3(\text{dil}) \longrightarrow \text{O}_2\text{N—C}_6\text{H}_4\text{—OH} + \text{H}_2\text{O}$$

aqueous NaOH at 5 °C
(+OH⁻)

+ H₂O

Examination practice questions

Please see the data section of the CD for any A_r values you may need.

1 The substituted benzene compound

can be further substituted.
If **Y** is an electron-withdrawing group, the next substitution will be in position 3.
If **Y** is an electron-releasing group, the next substitution will be mostly in position 4.

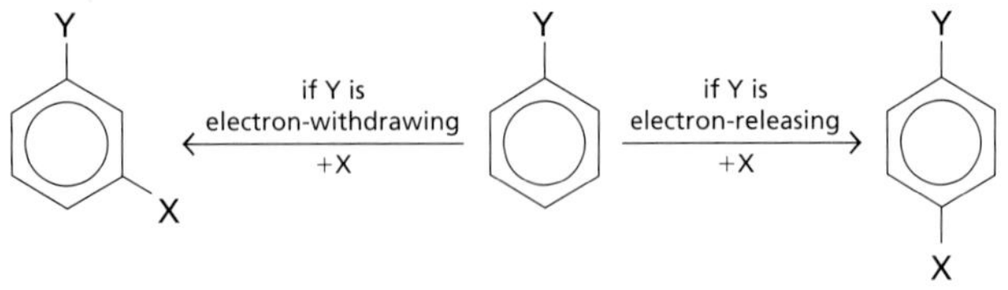

if Y is
electron-withdrawing
+X

if Y is
electron-releasing
+X

The following table lists some electron-withdrawing and electron-releasing substituents.

Electron-withdrawing groups	Electron-releasing groups
—NO₂	—CH₃
—COCH₃	—CH₂Br
—CO₂H	—NH₂

Use the above information to draw relevant structural formulae. [5]

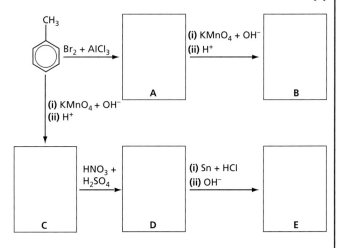

[Cambridge International AS & A Level Chemistry 9701, Paper 4 Q6 June 2008]

2 Cyclohexanol and phenol are both solids with low melting points that are fairly soluble in water.

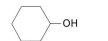

cyclohexanol   phenol

a Explain why these compounds are more soluble in water than their parent hydrocarbons cyclohexane and benzene. [2]

b Explain why phenol is more acidic than cyclohexanol. [2]

c For **each** of the following reagents, draw the structural formula of the product obtained for **each** of the two compounds. If no reaction occurs write **no reaction**. [7]

Reagent
Na(s)
NaOH(aq)
Br_2(aq)
I_2(aq) + OH$^-$(aq)
an excess of acidified $Cr_2O_7^{2-}$(aq)

d Choose **one** of the five reagents that could be used to distinguish between cyclohexanol and phenol. Describe the observations you would make with each compound. [2]

[Cambridge International AS & A Level Chemistry 9701, Paper 41 Q4 November 2009]

26 Carboxylic acids – derivatives and further reactions

Functional groups:

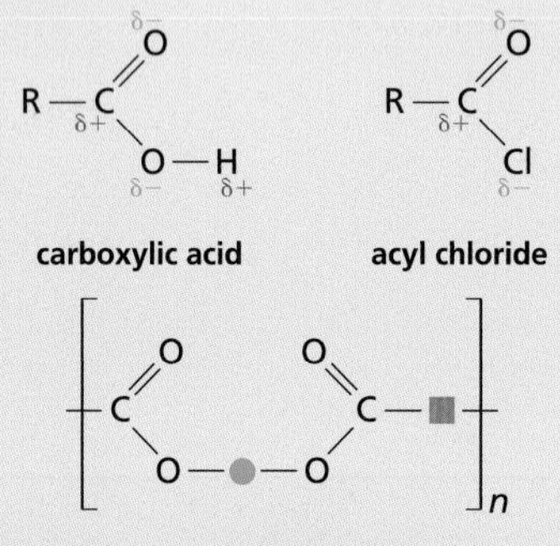

carboxylic acid   acyl chloride

polyester

In this topic we look at the effect of structure on the acidity of carboxylic acids, and at further reactions of their derivatives.

Learning outcomes

By the end of this topic you should be able to:

14.1a) interpret and use the general, structural, displayed and skeletal formulae of the acyl chlorides

17.1a) recall the chemistry of alcohols, exemplified by ethanol, in the formation of esters by acylation with acyl chlorides using ethyl ethanoate and phenyl benzoate as examples

19.1b) describe the reactions of carboxylic acids in the formation of acyl chlorides

19.1c) recognise that some carboxylic acids can be further oxidised: the oxidation of methanoic acid, HCO_2H, with Fehling's and Tollens' reagents, and the oxidation of ethanedioic acid, HO_2CCO_2H, with warm acidified manganate(VII)

19.1d) explain the relative acidities of carboxylic acids, phenols and alcohols

19.1e) use the concept of electronegativity to explain the acidities of chlorine-substituted ethanoic acids

19.2a) describe the hydrolysis of acyl chlorides

19.2b) describe the reactions of acyl chlorides with alcohols, phenols, ammonia and primary amines

19.2c) explain the relative ease of hydrolysis of acyl chlorides, alkyl chlorides and aryl chlorides including the condensation (addition-elimination) mechanism for the hydrolysis of acyl chlorides.

26.1 The effect of structure on acid strength

As we saw in Topic 18, carboxylic acids are weak acids, and dissociate to a small extent when in aqueous solution:

$$CH_3CO_2H \rightleftharpoons CH_3CO_2^- + H^+$$

They are more acidic than alcohols because the negative charge on the anion can be delocalised over two electronegative oxygen atoms. (see Figure 26.1).

The increasing acidities of alcohols, phenols and carboxylic acids are explained by the increasing ability of the molecular structures to delocalise the negative charge in the alkoxide, phenoxide and carboxylate anions. Their relative acidities are demonstrated by their different reactions with sodium, sodium hydroxide and sodium carbonate (see Table 26.1).

Table 26.1 The use of Na, NaOH and Na_2CO_3 to illustrate the relative acidities of alcohols, phenol and carboxylic acids.

Reagent	Observation with		
	Hexanol	Phenol	Hexanoic acid
Na(s)	$H_2(g)$ evolved	$H_2(g)$ evolved	$H_2(g)$ evolved
NaOH(aq)	no reaction	dissolves	dissolves
Na_2CO_3(aq)	no reaction	no reaction	$CO_2(g)$ evolved

Atoms or groups that draw electrons away from the $-CO_2^-$ group will help the anion to form, and this causes the acid to be more dissociated (that is, to become a stronger acid). On the other hand, groups that donate electrons to the $-CO_2^-$ group will cause the acid to become weaker (see Table 26.2).

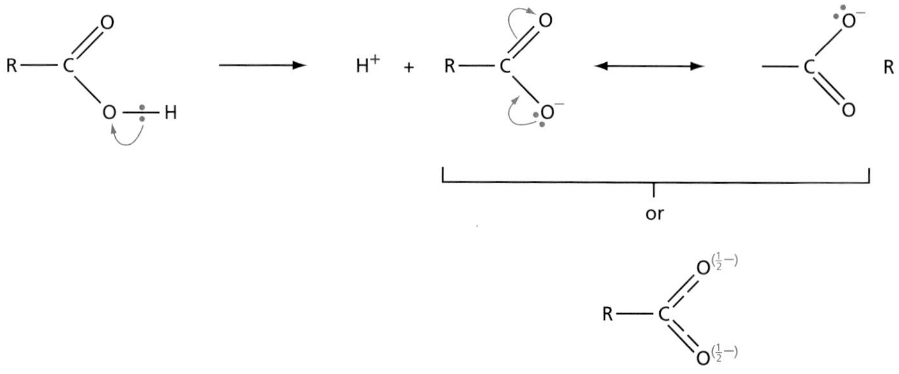

Figure 26.1 In the carboxylate anion, the charge is spread, increasing the stability of the ion.

Electron-donating groups decrease the acid strength of carboxylic acids, whereas electron-withdrawing groups increase their acid strength.

Table 26.2 The acidity of some carboxylic acids. Remember from Topic 22 that $pK_a = -\log_{10} K_a$, where $K_a = \dfrac{[H^+(aq)][A^-(aq)]}{[HA(aq)]}$ for acids undergoing the dissociation $HA \rightleftharpoons H^+ + A^-$.

Formula of acid	pK_a	Percentage dissociation in $1.0\,mol\,dm^{-3}$ aqueous solution
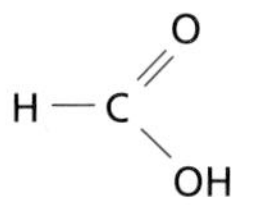 H—C(=O)OH	3.75	1.3%
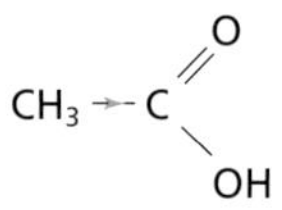 CH_3—C(=O)OH	4.76	0.42%
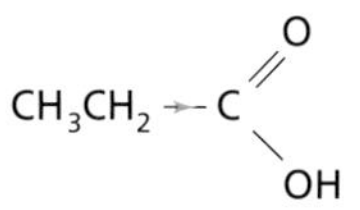 CH_3CH_2—C(=O)OH	4.87	0.36%
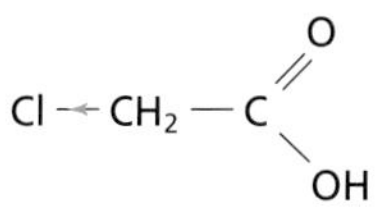 Cl—CH_2—C(=O)OH	2.87	3.7%
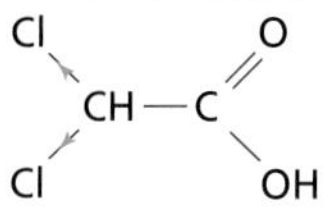 Cl_2CH—C(=O)OH	1.26	21%
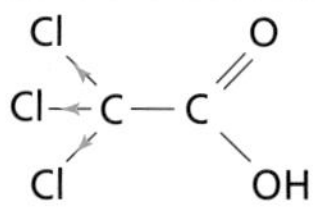 Cl_3C—C(=O)OH	0.66	59%

Worked example

The pK_a of fluoroethanoic acid is 2.57.

a Does this mean it is a stronger or a weaker acid than chloroethanoic acid? Explain your answer.

b What pK_a would you expect for difluoroethanoic acid?

1 Use the data in Table 26.2 to predict the pK_a values of:
   a $(CH_3)_3C—CO_2H$
   b $Cl—CH_2—CH_2—CO_2H$
   Explain your reasoning.
2 Explain why the pK_a values of 2-chlorobutanoic acid ($pK_a = 2.86$) and 4-chlorobutanoic acid ($pK_a = 4.53$) differ so much. Predict the pK_a value for 3-chlorobutanoic acid.

Answer

a The lower the pK_a, the larger is K_a. This means that the acid is more dissociated, and therefore stronger. Hence fluoroethanoic acid is a stronger acid than chloroethanoic acid. This is due to the greater electron-withdrawing ability of the highly electronegative fluorine atom.
b Difluoroethanoic acid would be expected to have a lower pK_a than dichloroethanoic acid – about **1.0**.

The effect of electron-donating or electron-withdrawing groups is seen even when such groups are situated on the opposite side of a benzene ring to the —CO_2H group (see Table 26.3).

Formula of acid	pK_a	Percentage dissociation in $1.0\,mol\,dm^{-3}$ aqueous solution
⬡—CO₂H	4.20	0.80%
CH₃—⬡—CO₂H	4.37	0.73%
Cl—⬡—CO₂H	3.99	1.0%

Table 26.3 The effect on pK_a for benzoic acid of electron-donating and electron-withdrawing groups

26.2 The further oxidation of some carboxylic acids

Unlike other aliphatic carboxylic acids, **methanoic acid** does not contain an alkyl chain attached to the —CO_2H group. Instead, the H—CO group, consisting of a hydrogen atom attached to a carbonyl group, has some reactions in common with aldehydes. In particular, it undergoes oxidation with common oxidants such as acidified dichromate(VI).

$$3H—CO_2H + Cr_2O_7^{2-}(aq) + 8H^+(aq) \rightarrow 2Cr^{3+}(aq) + 3CO_2(g) + 7H_2O(l)$$

It also reacts with both Fehling's solution and Tollens' reagent (see page 301).

$$H—CO_2H(aq) + 2Cu^{2+}(aq) + 6OH^-(aq) \rightarrow CO_3^{2-}(aq) + Cu_2O(s) + 4H_2O(l)$$

Ethanedoic acid contains two adjacent carbonyl groups. The proximity of two δ+ carbon atoms weakens the C–C bond sufficiently for the molecule to be readily oxidised by warm acidified manganate(VII).

$$5HO_2C—CO_2H(aq) + 2MnO_4^-(aq) + 6H^+(aq) \rightarrow 10CO_2(g) + 2Mn^{2+}(aq) + 8H_2O(l)$$

This forms the basis of an accurate volumetric method of analysing solutions containing ethanedioic acid or its salts.

In a similar type of reaction, **α-ketoacids** are oxidised by alkaline H_2O_2 with the loss of a carbon atom.

$$R—CO—CO_2H + H_2O_2 + 3OH^- \rightarrow R—CO_2^- + CO_3^{2-} + 3H_2O$$

26.3 Acyl chlorides

In section 18.3 we saw how acyl chlorides can be prepared from carboxylic acids. Most acid derivatives are prepared from carboxylic acids by the reactions described in this topic. Figure 26.2 shows a chart summarising the interrelationships between the various derivatives.

Figure 26.2 Interconversions between carboxylic acids and their derivatives

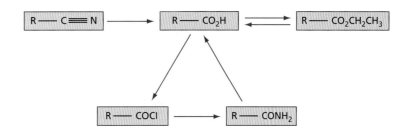

Now try this

Copy the chart in Figure 26.2 and write on each arrow the correct reagents and conditions for each reaction.

Properties of acyl chlorides

Acyl chlorides are usually liquids which fume in moist air. They are immiscible with water, but react slowly with it and eventually dissolve (see below). They are not hydrogen bonded: their main intermolecular attractions are a combination of van der Waals' forces and dipole–dipole forces (see Figure 26.3).

Figure 26.3 Dipole–dipole attractions in acyl chlorides

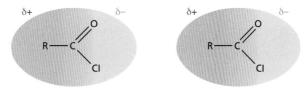

The extra electron-withdrawing effect of the carbonyl group, resulting in extra dipole–dipole attractions, has the effect of increasing the boiling point by about 10–15 °C compared with the boiling point of the halogenoalkane with similar shape (see Table 26.4).

Table 26.4 Strong dipole–dipole attractions increase the boiling points in acyl chlorides.

Chloroalkane	M_r	Boiling point /°C	Acyl chloride	M_r	Boiling point /°C	Difference /°C
![structure] Cl	78.5	37	![structure] O, Cl	78.5	52	15
![structure] Cl	92.5	68	![structure] O, Cl	92.5	80	12

Reactions of acyl chlorides

The electronegativity of the oxygen, and the easily polarised C=O double bond, have a dramatic effect on the reactivity of acyl chlorides compared with that of chloroalkanes.

With water

Acyl chlorides react readily with water:

$$CH_3{-}COCl + H_2O \xrightarrow{\text{room temperature, complete within minutes}} CH_3{-}COOH + HCl$$

Compare this with the conditions for the hydrolysis of a halogenoalkane:

$$CH_3{-}CH_2Cl + H_2O \xrightarrow{\text{heat under pressure at 100 °C for 14 days}} CH_3{-}CH_2OH + HCl$$

As a result of the speedy reaction with water acyl chlorides produce steamy fumes of HCl(g) when a few drops of water are added to them. This is a good test for them.

Now try this

Gather together all the information given in sections 15.3, 25.4 and this section to write an explanation of why the rates of hydrolysis of these three chloro compounds differ in this way. Would you expect the same relative rates with NH_3 or NaOH as the reagent, instead of H_2O?

Together with what we have already seen in sections 15.3 and 25.4, we note that in the reaction:

$$R—Cl + H_2O \rightarrow R—OH + HCl$$

the relative rates of hydrolysis are as follows:

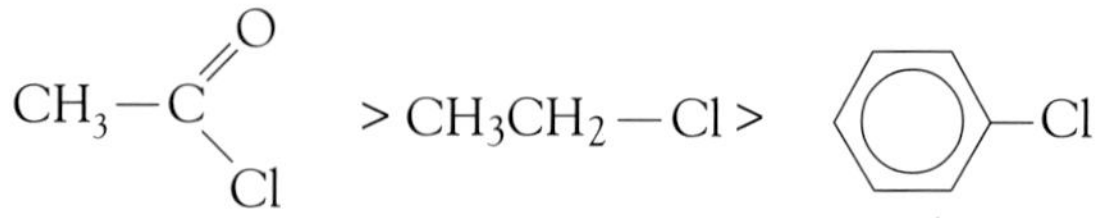

With alcohols or phenols

Acyl chlorides react readily with alcohols or phenols, forming esters:

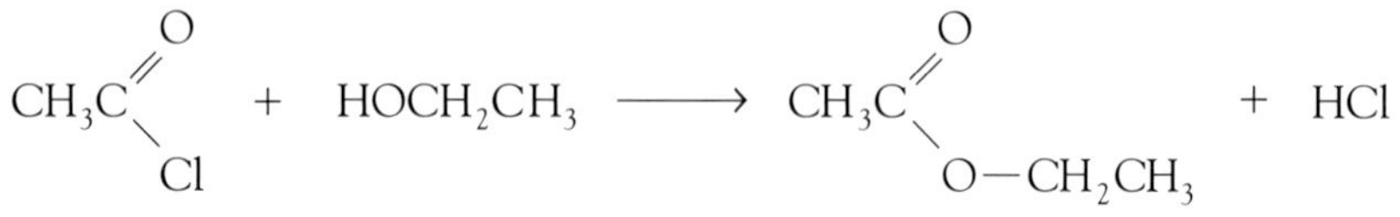

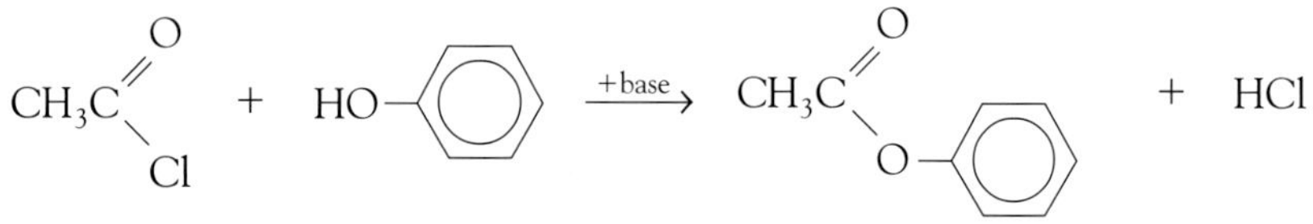

As mentioned in section 25.5, phenols are not so nucleophilic as alcohols, because the lone pair on the oxygen atom is delocalised over the ring. The acylation of phenols is therefore usually carried out under basic conditions (NaOH or pyridine have been used), when the more nucleophilic phenoxide ion is formed as an intermediate:

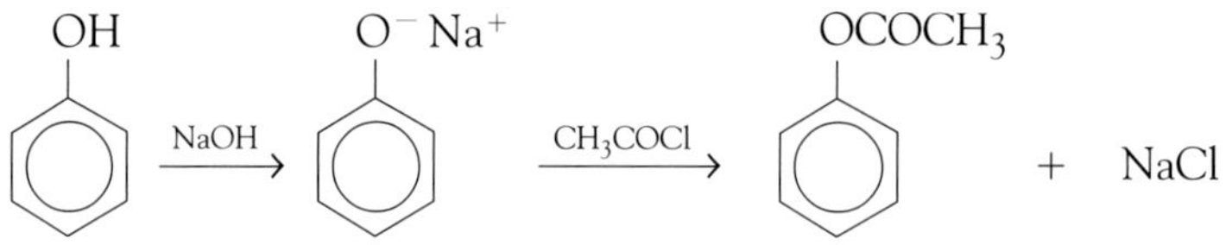

phenyl ethanoate

Because phenols do not react directly with carboxylic acids, this is the only method for preparing phenyl esters (see Topic 25, page 434).

With ammonia and amines

Acyl chlorides react with ammonia, forming amides. The reaction is vigorous, and white smoky fumes of $NH_4Cl(s)$ are often seen if an excess of ammonia is used.

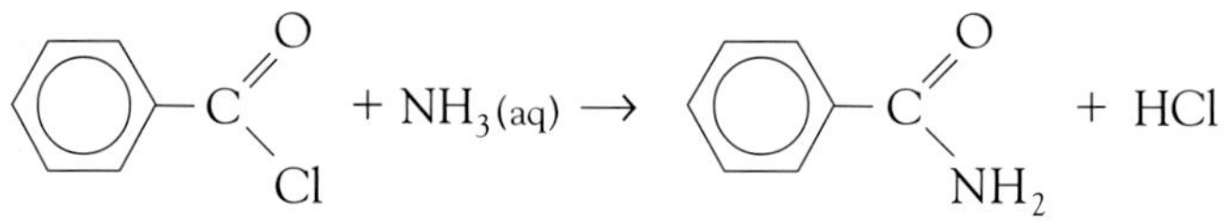

They react with primary amines (see Topic 27), forming substituted amides:

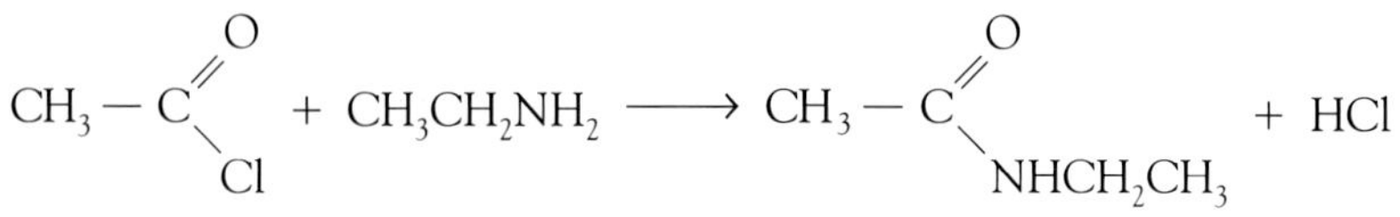

N-ethylethanamide

The hydrogen chloride produced reacts with another molecule of amine or ammonia in an acid–base reaction:

$$R—NH_2 + HCl \rightarrow R—NH_3^+Cl^-$$

Therefore an excess of amine or ammonia is used to ensure complete reaction. The reactions of amides are described in section 27.5.

The mechanism of the reactions of acyl chlorides

The carbonyl group in acyl chlorides can undergo nucleophilic addition in a similar manner to carbonyl compounds:

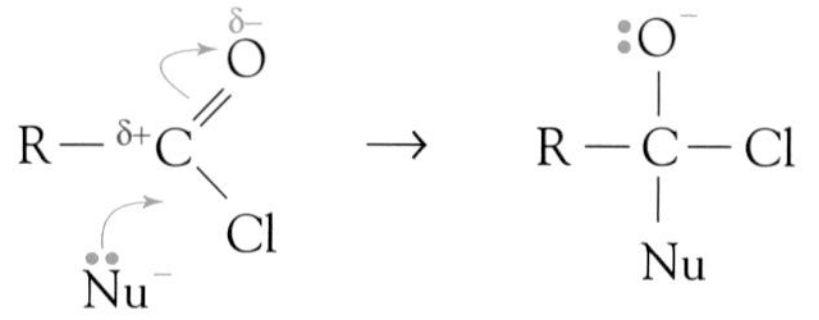

Unlike carbonyl compounds, however, acyl chlorides are provided with an easily removed leaving group, the chloride ion:

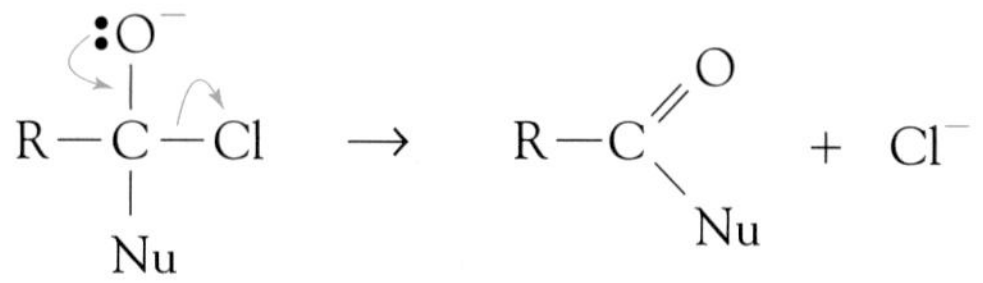

Nucleophilic substitution has taken place, by a mechanism involving addition, followed by elimination.

If the nucleophile is water, the carboxylic acid is formed:

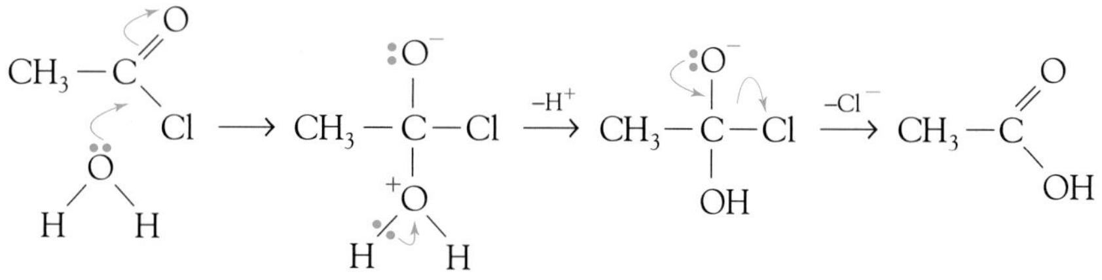

Worked example

Draw the mechanism for the reaction between ammonia and benzoyl chloride:

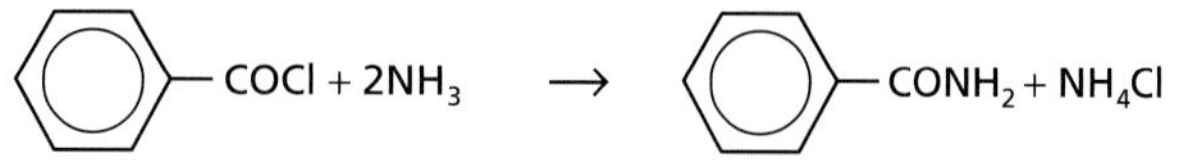

Answer

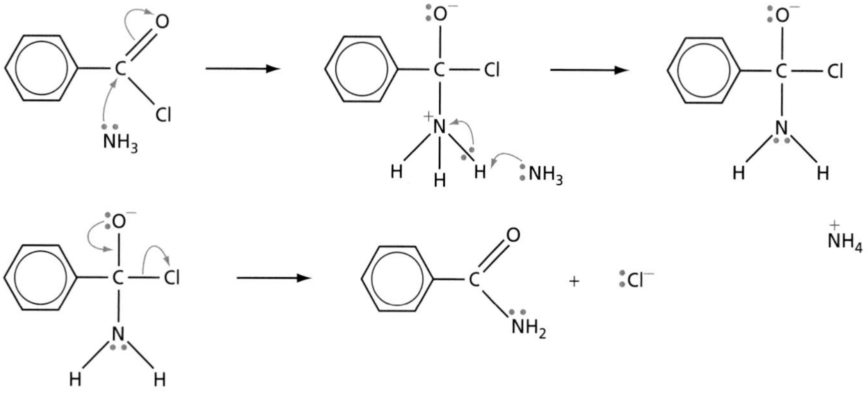

Now try this

Predict the products of, and suggest a mechanism for, the reaction between propan-1-ol and propanoyl chloride.

Explaining the relative reactivity of acyl chlorides, chloroalkanes and chloroarenes

As we saw in section 25.4, chlorobenzene does not undergo hydrolysis with water, or even with concentrated sodium hydroxide solution, due to the overlap of the lone pair on chlorine making the C—Cl bond stronger. Acyl chlorides, on the other hand, react much more readily than chloroalkanes with water or hydroxide. There are two reasons for this.

1 As we saw in the panel on page 445, the unsaturated nature of the C=O bond allows the C—OH bond to be formed before the C—Cl bond has broken. This **addition-elimination** mechanism therefore has a lower activation energy than the straight S_N2 reaction.

2 The carbon atom in acyl chlorides is attached to **two** electronegative atoms (oxygen and chlorine), and so is polarised $\delta+$ to a greater extent than the carbon atom in chloroalkanes. This allows it to attract nucleophiles more strongly.

26.4 Polyesters

Formation of polyesters

When a diol is esterified with a diacid (for example, by heating with hydrogen chloride gas), or reacted with a diacyl chloride, a polyester is produced. Figure 26.4 shows the general reaction for esterification with a diacyl chloride. Here the yellow rectangles represent the rest of the diol molecule and the green circles represent the benzene ring in the diacyl chloride.

Figure 26.4 Formation of a polyester from a diacyl chloride

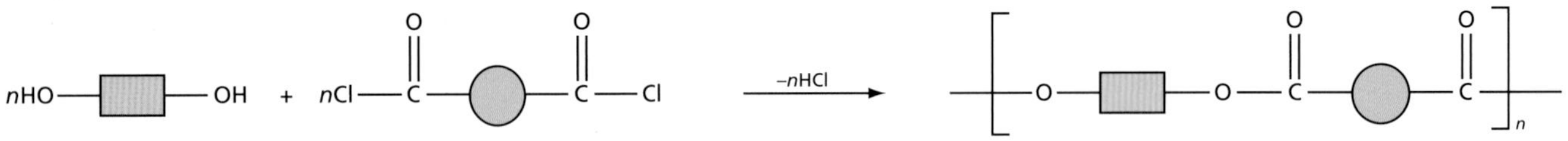

Notice that the direction of the ester bonds

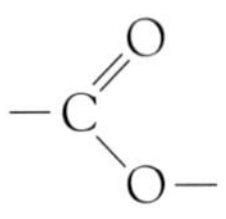

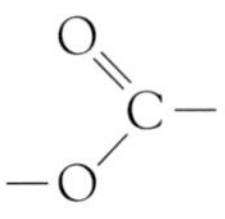

alternates along the chain.

Polyesters are examples of a class of polymers known as **condensation polymers**.

When monomers join together to form a **condensation polymer**, a small molecule such as H_2O, HCl or NH_3 is also produced.

The most commonly used polyester is Terylene, produced by polymerising ethane-1,2-diol and benzene 1,4-dicarboxylic acid, as shown below.

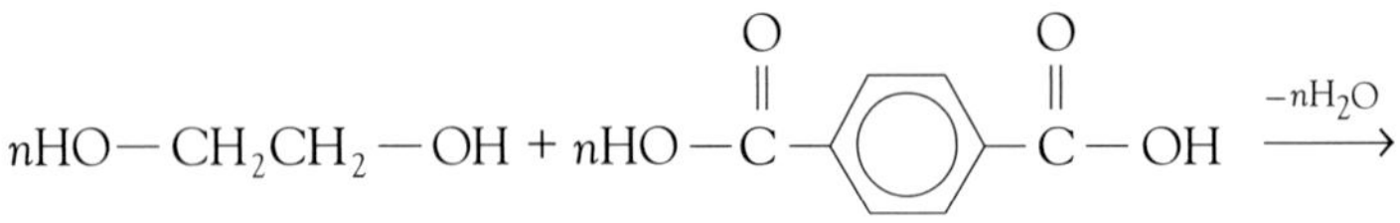

ethane-1,2-diol   benzene-1,4-dicarboxylic acid

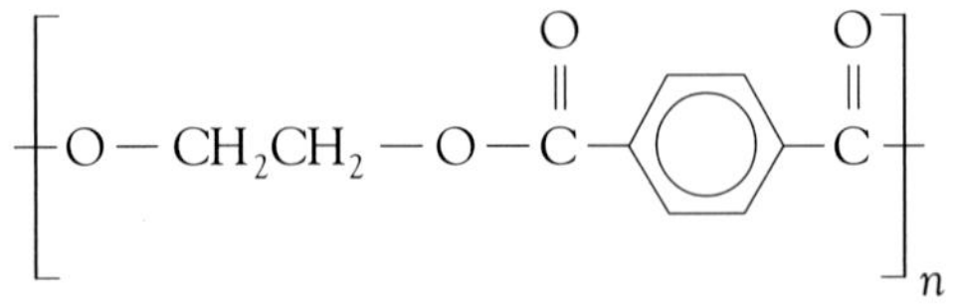

Terylene, a polyester

The polymer can be drawn into fine fibres and spun into yarn. Jointly woven with cotton or wool, it is a component of many everyday textiles. Material made from Terylene is hard-wearing and strong, but under strongly alkaline or acidic conditions the ester bonds can be hydrolysed, causing the fabric to break up.

Figure 26.5 Some products made from polyesters

Further condensation polymers such as nylon and proteins are discussed in Topic 28.

Summary

- The acid strength of **carboxylic acids**, RCO_2H, depends on their structures. In particular, electron-withdrawing chlorine atoms on the alkyl chain near to the CO_2H group increase its acid strength.
- Methanoic and ethanedoic acids can be oxidised further to CO_2.
- **Acyl chlorides** are useful intermediates, forming **esters** with alcohols or phenols, and **amides** with amines.
- Acyl chlorides are much more easily hydrolysed than chloroalkanes.
- **Polyesters** are formed by the condensation of a diol with either a dicarboxylic acid or a diacyl chloride.

Key reactions you should know
- Acyl chlorides:
  $RCOCl + H_2O \rightarrow RCO_2H + HCl$
  $RCOCl + R'OH \rightarrow RCO_2R' + HCl$ (R' = alkyl or aryl)
  $RCOCl + 2R'NH_2 \rightarrow RCONHR' + R'NH_3Cl$
- Condensation polymerisation:
  $$n\text{HO}-\blacksquare-\text{OH} + n\text{HO}_2\text{C}-\bullet-\text{CO}_2\text{H} \rightarrow \text{+}\!\!-\text{O}-\blacksquare-\text{OCO}-\bullet-\text{CO}\!\!-\!\!\text{+}_n$$

Examination practice questions

Please see the data section of the CD for any A_r values you may need.

1 a Describe and explain how the acidities of $CHCl_2CO_2H$ and $CH_2Cl\ CO_2H$ compare to each other, and to the acidity of ethanoic acid. [3]

b For each of the following pairs of compounds, suggest one chemical test (reagents and conditions) that would distinguish between them. State the observations you would make with each compound, writing 'none' if appropriate.

first compound	second compound	test (reagents and conditions)	observation with first compound	observation with second compound
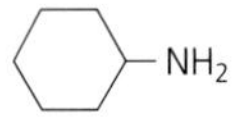 ⬡—NH₂	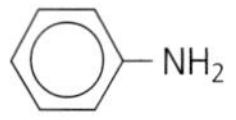 ⌬—NH₂			
CH₃CH₂COCl	CH₃COCH₂Cl			
CH₃CH₂CHO	CH₃COCH₃			

[7]

[Cambridge International AS & A Level Chemistry 9701, Paper 42 Q5 a & b June 2012]

2 a Polyvinyl acetate, PVA, is a useful adhesive for gluing together articles made from wood, paper or cardboard. The monomer of PVA is ethenyl ethanoate, B.

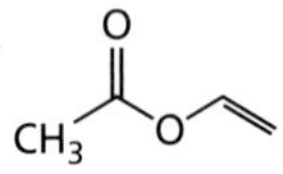

B

PVA is formed from **B** by the process of addition polymerisation.

i Draw a section of the PVA molecule containing at least 2 monomer molecules, and identify clearly the repeat unit.

ii The ester **B** can be hydrolysed in the usual way, according to the following equation.

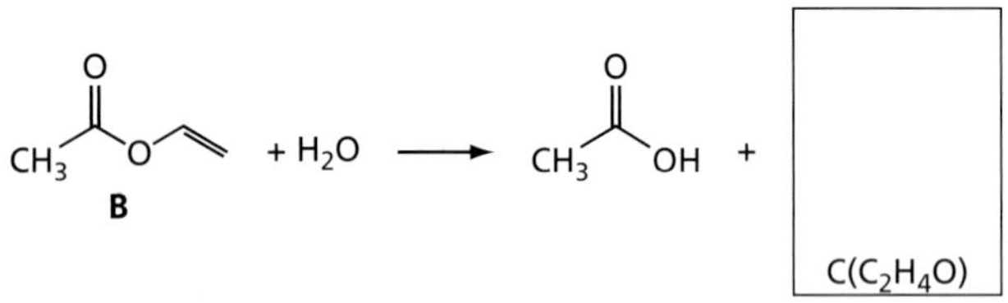

B + H₂O ⟶ CH₃—C(=O)OH + [C(C₂H₄O)]

Use this information to suggest a possible structure for **C** and draw it

iii When substance **C** is extracted from the product mixture, it is found that it **does** not decolourise $Br_2(aq)$, but it does form a pale yellow precipitate with alkaline aqueous iodine.
Suggest a structure for **C** that fits this new information.

iv Suggest a confirmatory test for the functional group in the structure you have drawn in **iii**. Your answer should include the reagent you would use and the observation you would make. [6]

b The following diagram represents a section of another polymer.

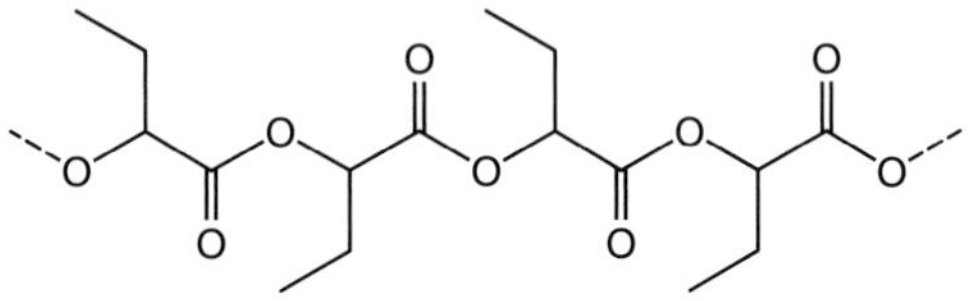

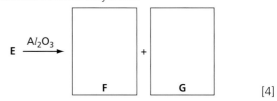

D

i Copy and draw brackets, [ ], around the atoms that make up the repeat unit of this polymer.

ii Name the functional group in polymer **D**.

iii Suggest and draw the structure of the monomer, **E**, that could form this polymer.

iv What *type of polymerisation* is involved in making polymer **D** from its monomer?

v What is the relationship between the repeat unit of polymer **D** and the repeat unit of PVA? [5]

c Monomer **E** exists as two stereoisomers. Heating either isomer with Al_2O_3 gives a mixture of two unsaturated carboxylic acids **F** and **G**, which are stereoisomers of each other.

i Name the *type of stereoisomerism* shown by compound **E**.

ii Suggest structures for **F** and **G**, and name the type of stereoisomerism they show.

E —$\xrightarrow{Al_2O_3}$→ [F] + [G]

[4]

[Cambridge International AS & A Level Chemistry 9701, Paper 42 Q4 June 2011]

27 Amines, amides and amino acids

Functional groups:

amine   amide

amino acid

Amines are organic bases. They react with acids to form salts, and amines that are soluble in water form alkaline solutions. The reaction of aryl amines with nitrous acid is the first step in making dyes.

Amides are neutral compounds formed when ammonia or an amine reacts with an acyl chloride.

Amino acids contain both the basic —NH₂ group and the acidic —CO₂H group; they are therefore amphoteric, reacting with both acids and bases. Amino acids are the building blocks of the important class of biological polymers known as the polypeptides, or proteins.

The topic concludes with a look at synthetic polyamides, including the various types of nylon.

Learning outcomes

By the end of this topic you should be able to:

14.1a) interpret and use the general, structural, displayed and skeletal formulae of the amides and amino acids

20.1a) describe the formation of alkyl amines such as ethylamine (by the reaction of ammonia with halogenoalkanes; the reduction of amides with $LiAlH_4$; the reduction of nitriles with $LiAlH_4$ or H_2/Ni) and of phenylamine (by the reduction of nitrobenzene with tin/concentrated HCl)

20.1b) describe and explain the basicity of amines

20.1c) explain the relative basicities of ammonia, ethylamine and phenylamine in terms of their structures

20.1d) describe the reaction of phenylamine with aqueous bromine, and with nitrous acid to give the diazonium salt and phenol

20.1e) describe the coupling of benzenediazonium chloride and phenol

20.2a) describe the formation of amides from the reaction between NH_3 or RNH_2 and R′COCl

20.2b) recognise that amides are neutral

20.2c) describe amide hydrolysis on treatment with aqueous alkali or acid, and describe the reduction of amides with $LiAlH_4$

20.3a) describe the acid/base properties of amino acids and the formation of zwitterions

20.3b) describe the formation of peptide bonds between amino acids to give di- and tri-peptides

20.3c) describe simply the process of electrophoresis and the effect of pH, using peptides and amino acids as examples.

27.1 Introduction

From prehistoric times, people have been aware that the leaves or seeds of certain plants were pharmacologically active, in alleviating pain, reducing fever or causing euphoria. In the early days of organic chemistry, when chemists were extracting, purifying and attempting to identify the hundreds of compounds that occur in natural organisms, they discovered that many of these compounds were insoluble in water, but dissolved in dilute acids. They called these the **alkaloids** ('alkali-like'). Many of the alkaloids extracted from plants were found to have strong physiological and psychological effects on humans and other animals. These included hypnotics such as morphine, narcotics such as cocaine, and cardiac poisons such as strychnine. All alkaloids contained at least one nitrogen atom, and all belonged to the class of compounds we now call **amines**. Figure 27.1 illustrates some of their structures.

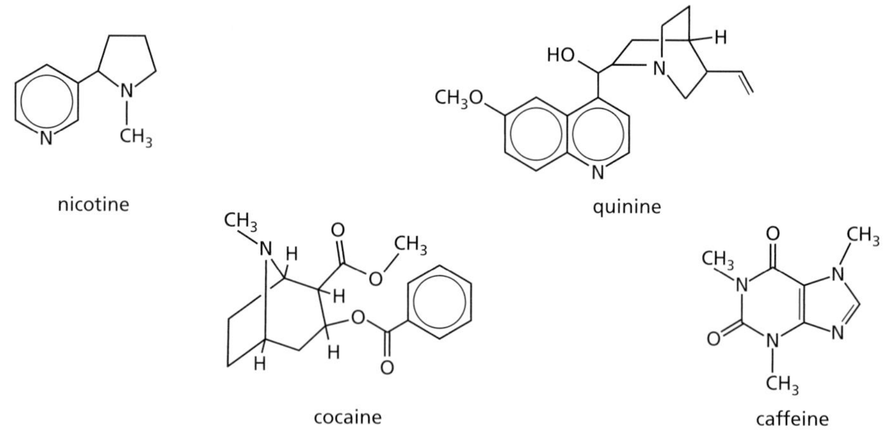

Figure 27.1 The structures of some naturally occurring alkaloids

nicotine

quinine

cocaine

caffeine

Now try this

1 Apart from the nitrogen atoms, what functional groups are contained in
   a the quinine molecule
   b the cocaine molecule?
2 Work out the molecular formula of:
   a quinine
   b caffeine
3 The molecule of nicotine is chiral. Draw out the structure, showing which atom is the chiral one.

Most alkaloids have a bitter taste, but are too involatile (that is, they produce too little vapour) to have an odour. Those amines that are volatile have distinctive smells. The short-chain amines have an astringent ammonia-like smell, but this is replaced by a strong fishy odour in butylamine and longer-chain amines. Aryl amines such as phenylamine have a more 'oily' smell.

The amines can be derived from ammonia by replacing one or more of the hydrogen atoms in NH_3 by organic groups; replacing the amine hydrogen atoms can often be carried out in the laboratory.

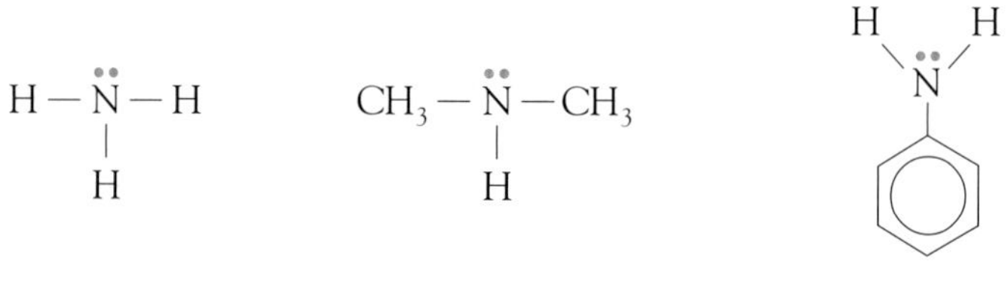

$$H-\overset{..}{N}-H$$
$$|$$
$$H$$

ammonia

$$CH_3-\overset{..}{N}-CH_3$$
$$|$$
$$H$$

dimethylamine

phenylamine

The carbon–nitrogen bond, although quite strongly polarised, is fairly inert and it is not so easily broken as the C—O bond in alcohols (see Topic 16, page 282). There are few reactions in which the $C^{\delta+}$ atom takes part. The key point of reactivity in the amines is the lone pair of electrons on the nitrogen atom, which is more readily donated than a lone pair on oxygen. This lone pair is nucleophilic towards $C^{\delta+}$ and basic towards $H^{\delta+}$. Amines form strong hydrogen bonds to hydrogen-donor molecules such as water. Those amines that contain N—H bonds also form strong intermolecular hydrogen bonds (see Figure 27.2).

The short-chain amines are therefore water soluble and their boiling points are higher than those of the corresponding alkanes (see Table 27.1 and Figure 27.3). The effect of hydrogen bonding on boiling point is not so pronounced as it is in the alcohols, however (compare Table 27.1 with Table 16.1, page 282).

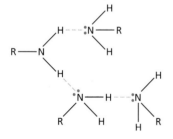

Figure 27.2 Intermolecular hydrogen bonding in amines

Table 27.1 Boiling points of some alkanes and corresponding amines

Number of electrons in the molecule	Alkane		Amine		Increase in boiling point/°C
	Formula	Boiling point/°C	Formula	Boiling point/°C	
18	C_2H_6	−88	CH_3NH_2	−8	80
26	C_3H_8	−42	$C_2H_5NH_2$	17	59
34	C_4H_{10}	0	$C_3H_7NH_2$	49	49
42	C_5H_{12}	36	$C_4H_9NH_2$	78	42
50	C_6H_{14}	69	$C_5H_{11}NH_2$	104	35
58	C_7H_{16}	98	$C_6H_{13}NH_2$	130	32

Figure 27.3 Boiling points of some alkanes and corresponding amines

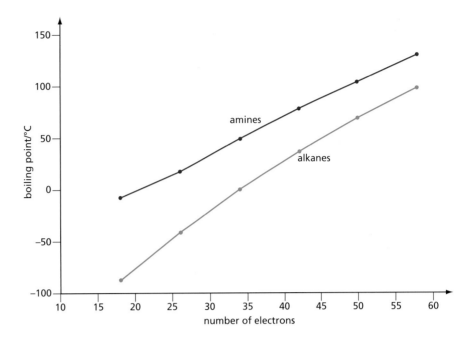

27.2 Isomerism and nomenclature

The most important feature of the amine molecule is the nitrogen atom. This is reflected in the nomenclature of amines. The simpler ones, such as dimethylamine and phenylamine shown opposite, are named as though they are derived from ammonia. Up to three hydrogen atoms in ammonia can be replaced by organic groups. Successive replacement forms **primary**, **secondary** and **tertiary amines** (see Table 27.2). Note that it is the branching that takes place at the nitrogen atom, rather than at the carbon atom attached to it, that determines whether an amine is primary, secondary or tertiary. This is not the same as in the case of the alcohols:

$$
\begin{array}{ccc}
CH_3 & CH_3 & CH_3CH_2CH_2 \\
\diagdown & \diagdown & \diagdown \\
CH-OH & CH-NH_2 & N-H \\
\diagup & \diagup & \diagup \\
CH_3 & CH_3 & CH_3CH_2CH_2
\end{array}
$$

propan-2-ol
a secondary alcohol

2-aminopropane
a **primary** amine

dipropylamine
a **secondary** amine

Table 27.2 Some primary, secondary and tertiary amines

Primary amines	Secondary amines	Tertiary amine
$CH_3CH_2NH_2$ ethylamine	$(CH_3CH_2)_2NH$ diethylamine	$(CH_3CH_2)_3N$ triethylamine
CH₃—⬡—NH₂ 4-methylphenylamine	⬡—N(H)—⬡ diphenylamine	⬡—N(CH₃)₂ N,N-dimethylphenylamine
⬡—NH₂ cyclohexylamine	⬡N—H piperidine	nicotine

An alternative way of naming amines is used when the chain becomes more branched, or when other functional groups are present. This views the —NH_2 group as a substituent, called an **amino group**:

2-aminopentane	4-aminobenzoic acid	1,2-diaminoethane (often abbreviated to 'en')

Apart from the usual structural or positional isomerism that can occur in the carbon chain, amines also demonstrate structural isomerism around the nitrogen atom, through the formation of secondary and tertiary amines as well as primary amines.

Worked example

How many isomers are there with the formula C_3H_9N?

Answer

There are four isomers. Structural isomerism in the propyl group allows two primary amines:

$$CH_3CH_2CH_2—NH_2 \qquad (CH_3)_2CH—NH_2$$

propylamine (1-aminopropane)   2-aminopropane

In addition to these, we can split the three-carbon chain into two or three chains, to form a secondary amine:

$$CH_3—CH_2—NH—CH_3$$

ethylmethylamine

and a tertiary amine:

$$(CH_3)_3N$$

trimethylamine

Now try this

Fourteen of the isomers with the formula $C_8H_{11}N$ contain a nitrogen atom directly bonded to a benzene ring. Draw their structures, and state which are primary, secondary and tertiary amines.

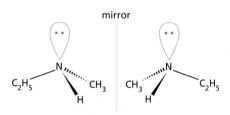

Figure 27.4 Theoretical enantiomers of ethylmethylamine

The nitrogen atom in amines is pyramidal, as it is in ammonia. Although the secondary amine ethylmethylamine could exist as a mirror-image pair of compounds (see Figure 27.4), in practice the nitrogen atom undergoes a rapid inversion, like an umbrella turning inside out, causing a 50:50 mixture of the two forms to exist at room temperature (see Figure 27.5).

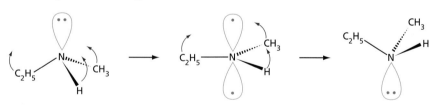

Figure 27.5 Inversion at the nitrogen atom produces a racemic mixture.

27.3 Reactions of amines

Basicity

Amines are basic. They react with acids to form salts:

$$CH_3CH_2\ddot{N}H_2 + HCl \rightarrow CH_3CH_2 - \overset{+}{N}\!\!\diagup^{H}_{\diagdown H}\!-H + :Cl^-$$

The alkylammonium chlorides, sulfates and nitrates are white crystalline solids, soluble in water, but insoluble in organic solvents.

Those amines that are soluble in water form weakly alkaline solutions, just as ammonia does, due to partial reaction with the solvent, producing OH^- ions:

$$CH_3CH_2NH_2 + H_2O \rightleftharpoons CH_3CH_2NH_3^+ + OH^-$$

Table 27.3 lists some amines, with values of their dissociation equilibrium constants, K_b, for the following equilibrium:

$$R_3N + H_2O \rightleftharpoons R_3NH^+ + OH^-$$

$$K_b = \frac{[R_3NH^+][OH^-]}{[R_3N]}$$

The larger K_b is, the stronger is the base. Ammonia is included for comparison.

From Table 27.3 we can see that electron-donating alkyl groups attached to the nitrogen atom increase the basicity of amines. This is expected, since the basicity depends on the availability of the lone pair of electrons on nitrogen to form a dative bond with a proton (see Figure 27.6). Electron donation from an alkyl group will encourage dative bond formation.

Table 27.3 The basicity of some amines

Amine	Formula	K_b /mol dm^{-3}
phenylamine	—NH$_2$	4.2×10^{-10}
ammonia	NH_3	1.8×10^{-5}
ethylamine	$CH_3CH_2NH_2$	5.1×10^{-4}
diethylamine	$(CH_3CH_2)_2NH$	1.0×10^{-3}

Figure 27.6 When an amine acts as a base, the nitrogen lone pair forms a dative bond with a proton.

The most dramatic difference in basicities to be seen in Table 27.3 is between that of phenylamine ($K_b \approx 10^{-10}$) and the alkyl amines ($K_b \approx 10^{-3}$). Taking two compounds of about the same relative molecular mass and shape, we see that phenylamine is about a million times less basic than cyclohexylamine:

phenylamine
$K_b = 4.2 \times 10^{-10}$

cyclohexylamine
$K_b = 3.3 \times 10^{-4}$

This is because in phenylamine, the lone pair of electrons on the nitrogen atom is delocalised over the benzene ring. The bonds around the nitrogen atom can take up a planar arrangement, with the nitrogen's lone pair in a p orbital, so that extra stability can be gained by overlapping this p orbital with the delocalised π bond of the benzene ring (see Figure 27.7).

Figure 27.7 Delocalisation of the nitrogen lone pair in phenylamine

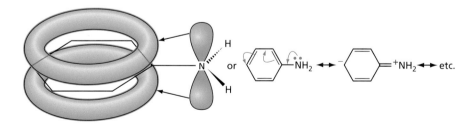

This overlap, causing a drift of electron density from nitrogen to the ring, has two effects on the reactivity of phenylamine.

- It causes the lone pair to be much less basic (see above) and also much less nucleophilic.
- It causes the ring to be more electron rich, and so to undergo electrophilic substitution reactions much more readily than benzene. The enhanced reactivity of phenylamine in this regard is similar to that of phenol (see Topic 25, page 434), an

example being the ease with which phenylamine decolorises bromine water:

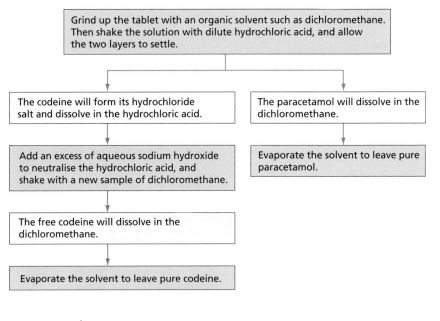

white precipitate

We shall see a similar drift of electron density from nitrogen when we look at amides in section 27.5.

Only the shorter-chained amines (with five or fewer carbon atoms) are soluble in water, but the ionic nature of their salts allows *all* amines to dissolve in dilute aqueous acids. This is a very useful way of purifying amines from a mixture with non-basic compounds. The amines can be regenerated by adding an excess of aqueous sodium hydroxide to the solution, after the non-basic impurity has been extracted with an organic solvent (see the experiment below).

Experiment

The separation of codeine and paracetamol from a medicine tablet

This separation makes use of the fact that codeine is basic, forming water-soluble salts with acids, but paracetamol is slightly acidic, and is insoluble in water (see Figures 27.8 and 27.9).

Figure 27.8 Flow chart for the experimental procedure to be followed

> Grind up the tablet with an organic solvent such as dichloromethane. Then shake the solution with dilute hydrochloric acid, and allow the two layers to settle.

> The codeine will form its hydrochloride salt and dissolve in the hydrochloric acid.

> The paracetamol will dissolve in the dichloromethane.

> Add an excess of aqueous sodium hydroxide to neutralise the hydrochloric acid, and shake with a new sample of dichloromethane.

> Evaporate the solvent to leave pure paracetamol.

> The free codeine will dissolve in the dichloromethane.

> Evaporate the solvent to leave pure codeine.

Figure 27.9 Stages in the separation: **a** the first separation removes the paracetamol in the organic layer, **b** sodium hydroxide neutralises the aqueous layer, and **c** the un-ionised codeine is re-formed and dissolves in the organic layer.

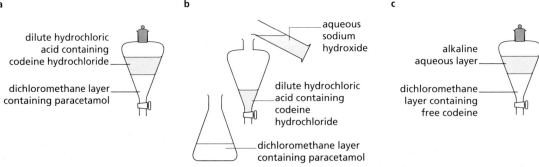

Reactions as nucleophiles

Another similarity between amines and ammonia is their nucleophilicity. The lone pair on the nitrogen atom of amines makes them good nucleophiles. They react with alkyl and acyl halides (see page 274 in Topic 15 and page 444 in Topic 26).

$$CH_3CH_2NH_2 + CH_3COCl \rightarrow CH_3C\overset{O}{\underset{NHCH_2CH_3}{}} + HCl$$

$$\text{C}_6\text{H}_5-NH_2 + CH_3Br \rightarrow \text{C}_6\text{H}_5-NHCH_3 + HBr$$

In the presence of an excess of a bromoalkane, amines can be successively alkylated, first to secondary and then to tertiary amines. These are still strongly nucleophilic, and are readily alkylated to quaternary ammonium salts:

$$CH_3CH_2NH_2 + CH_3CH_2Br \rightarrow (CH_3CH_2)_2NH + HBr$$

$$(CH_3CH_2)_2NH + CH_3CH_2Br \rightarrow (CH_3CH_2)_3N + HBr$$

$$(CH_3CH_2)_3N + CH_3CH_2Br \rightarrow (CH_3CH_2)_4N^+Br^-$$

Quaternary ammonium salts are water-soluble solids, with no basic character at all, because there is no lone pair of electrons on the nitrogen atom. An important naturally occurring ammonium salt is choline. Phosphatidylcholine is a key phospholipid component of cell membranes. Acetylcholine is an important neurotransmitter, allowing a nerve impulse to pass from the end of one nerve to the start of the next one (see Figures 27.10 and 27.11).

Figure 27.10 Some important naturally occurring quaternary ammonium salts

choline chloride

acetylcholine chloride

phosphatidyl choline

Figure 27.11 a Acetylcholine transmits a nerve impulse from one end of a nerve to another at a synapse or to a muscle at a motor end plate. **b** Curare, a deadly poison used by South American Indians on their arrow tips, blocks the action of acetylcholine.

a

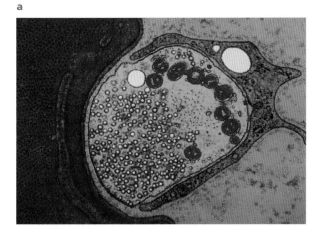

b

Reaction with nitrous acid (nitric(III) acid)

Nitrous acid, HNO_2, is unstable, and has to be made as required by reacting together sodium nitrite and hydrochloric acid:

$$NaNO_2 + HCl \rightarrow NaCl + HNO_2$$

Aromatic and aliphatic amines differ markedly in their reactions with nitrous acid.

Primary aliphatic amines react in warm aqueous acidic solution to give nitrogen gas and alcohols:

$$R-NH_2 + HNO_2 \xrightarrow[\text{30 °C in water}]{\text{NaNO}_2 + \text{HCl}} R-OH + N_2 + H_2O$$

Aryl amines, however, form fairly stable diazonium salts at low temperatures. Under these conditions nitrogen is not evolved, but a solution of the diazonium salt is formed:

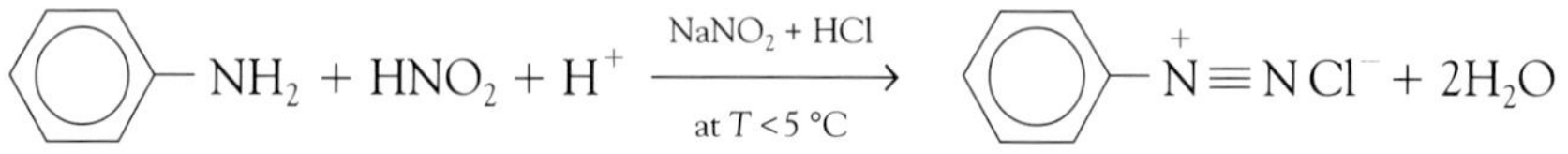

phenyldiazonium chloride

Aryl diazonium salts are unstable and explosive when dry, but can be kept for several days in solution in a refrigerator.

Reactions of diazonium salts

1 Adding a solution of a diazonium salt to hot water causes decomposition, and produces phenol:

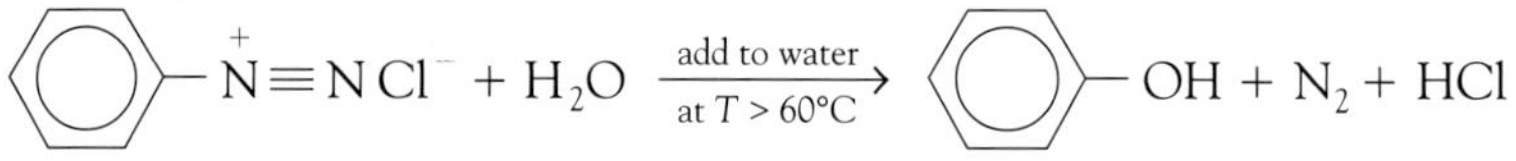

phenol

This is the analogous product to the alcohol produced when an aliphatic primary amine reacts with nitrous acid.

2 The most important reactions of diazonium salts are their use in the formation of **azo dyes** (see panel opposite). When a solution of a diazonium salt is added to an alkaline solution of a phenol, an electrophilic substitution reaction (known as a **coupling reaction**) takes place:

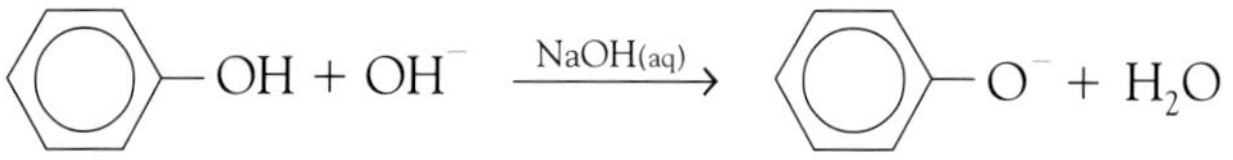

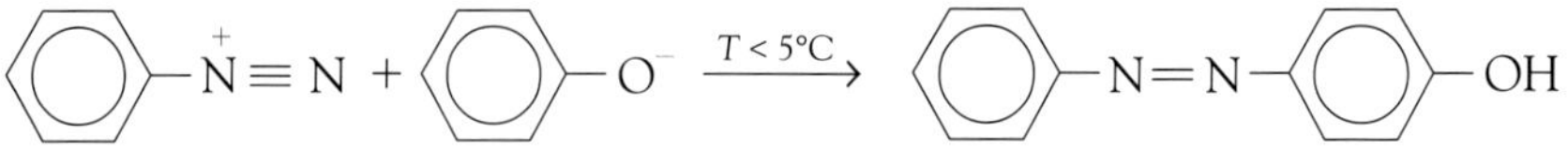

an azo compound

27.4 Preparing amines

The two main methods of preparing alkyl amines both start with halogenoalkanes.

● By nucleophilic substitution with ammonia (see Topic 15, page 273):

$$CH_3CH_2Br + 2NH_3 \xrightarrow[\text{under pressure}]{\text{heat in ethanol}} CH_3CH_2NH_2 + NH_4Br$$

As long as an excess of ammonia is used, further substitutions giving secondary and tertiary amines can be avoided (see page 444).

Azo dyes

A **chromophore** is a group which, especially when joined to other unsaturated groups, causes a compound to absorb visible light, and so become coloured.

The azo group, —N=N—, is called a **chromophore**. Compounds containing this group are highly coloured. Their colours range from yellow and orange to red, blue and green, depending on what other groups are attached to the benzene rings.

The common acid–base indicator methyl orange is an azo compound, made by the coupling reaction shown in Figure 27.12.

Figure 27.12 Methyl orange is an azo compound formed by a coupling reaction.

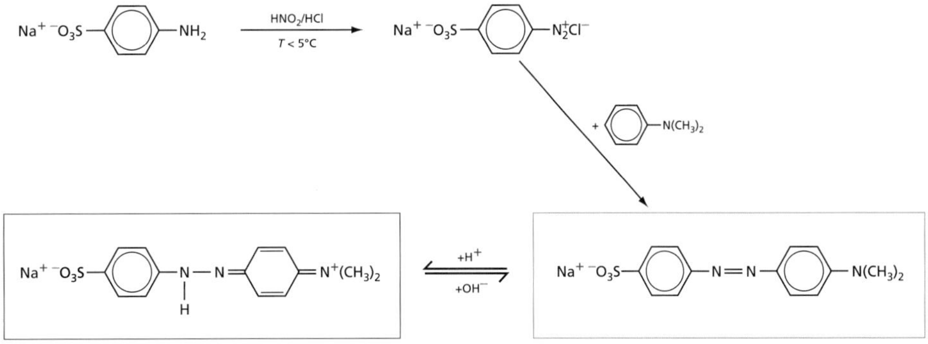

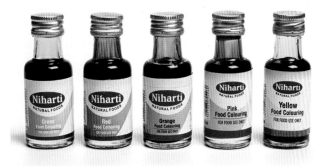

(red, acid form) methyl orange (orange, base form)

Many dye molecules used for dyeing clothes do not easily stick to the fibres of the material by themselves. This is especially the case if their main method of intermolecular bonding (for example, van der Waals', hydrogen bonding, ionic forces) does not match that of the molecules that make up the material. Mordants are often used to help the dye molecules stick. A **mordant** is a polyvalent metal ion, such as Al^{3+} or Fe^{3+}, which can form co-ordination complexes both with the dye molecule and with —OH, —CO or —NH groups on the molecules that make up the fibres of the material. By this means the dye molecule and the fibre molecule are permanently held together by the metal ion. A great number of dyes for clothes, colour printing and food colouring (Figure 27.13) are azo dyes. Figure 27.14 shows two examples.

Figure 27.13 Food colours often contain azo dyes.

Figure 27.14 Some azo dyes used commercially

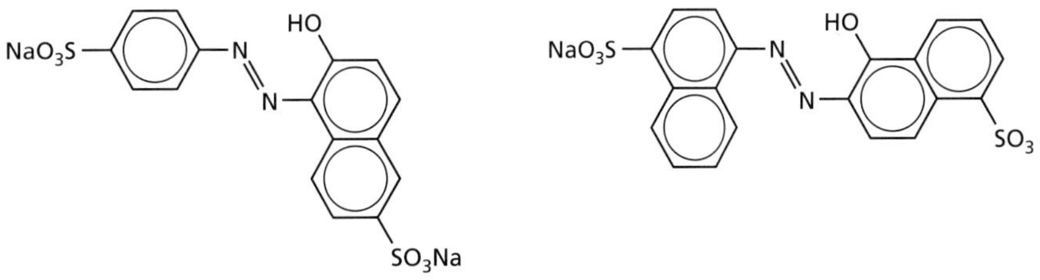

Sunset Yellow Carmoisine (red)

Now try this

1 Draw the structures of the two compounds from which Sunset Yellow can be made by a coupling reaction.
2 What are the structures of the two compounds that couple to form Carmoisine?

- By nucleophilic substitution with sodium cyanide, followed by reduction using either lithium tetrahydridoaluminate(III) (lithium aluminium hydride) or hydrogen over a nickel catalyst (see Topic 15, page 274):

$$CH_3CH_2Br + NaCN \xrightarrow{\text{heat in ethanol}} CH_3CH_2CN \xrightarrow{H_2 + Ni} CH_3CH_2CH_2NH_2$$

Notice that during this reaction the carbon chain length has been extended by one carbon atom.

Amines are also produced when amides are reduced by lithium tetrahydridoaluminate(III) (catalytic hydrogenation only succeeds under conditions of high pressure and temperature). This allows amines to be synthesised from carboxylic acids:

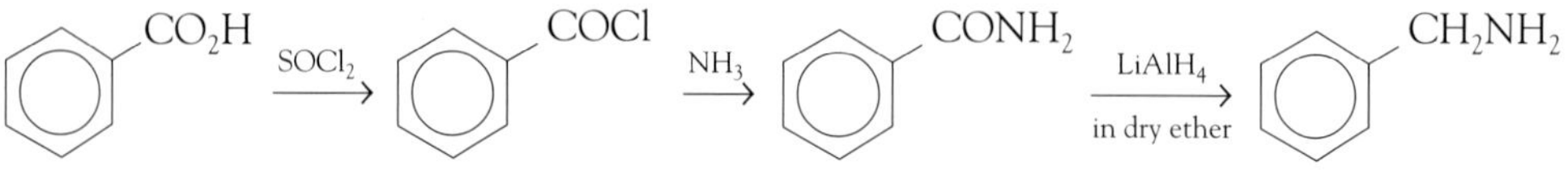

benzoic acid   phenylmethylamine

Aryl amines are most commonly prepared by the reduction of aromatic nitro compounds (see Topic 25, page 424):

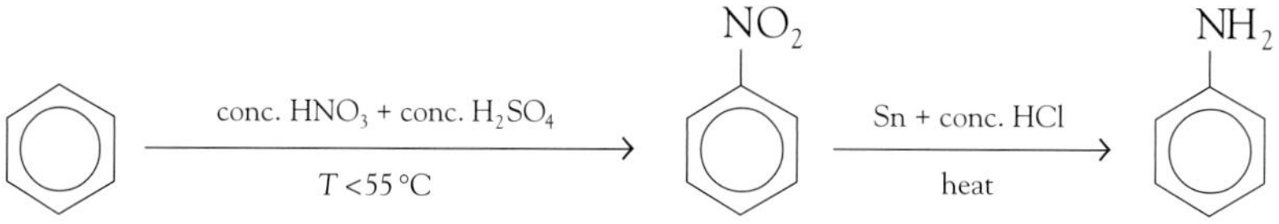

To produce the free amine, excess sodium hydroxide must be added after the reduction is completed.

Worked example

The 'good feeling' factor in chocolate has been identified as 2-phenylethylamine, $C_6H_5CH_2CH_2NH_2$. Suggest a synthesis of this compound from chloromethylbenzene, $C_6H_5CH_2Cl$.

Answer

The target amine has one more carbon atom than the suggested starting material, so the cyanide route is required:

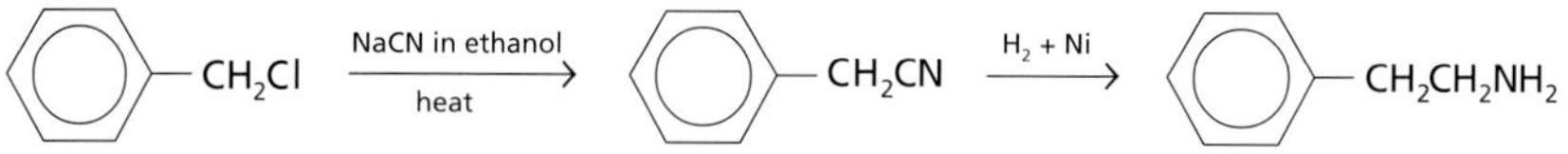

27.5 Amides

Properties of amides

Amides have the functional group:

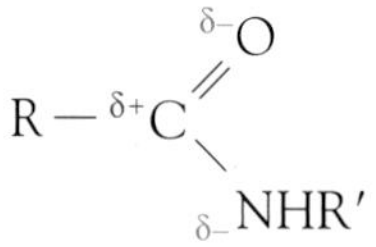

R and R′ can be alkyl, aryl or hydrogen.

Amides are extensively hydrogen bonded, having both $H^{\delta+}$ atoms (on nitrogen) and lone pairs of electrons (on oxygen and nitrogen). Most amides are solids at room temperature, and quite a number are soluble in water.

Unlike amines, amides form neutral solutions in water, and can be protonated only by strong acids. The site of protonation is unusual, and explains why amides are such weak bases.

Now try this

Suggest **two** ways of making butylamine ($CH_3CH_2CH_2CH_2NH_2$), each method starting from a compound containing a different number of carbon atoms.

Now try this

Draw diagrams to show the hydrogen bonding between

a two molecules of ethanamide, CH_3CONH_2
b a molecule of ethanamide and two molecules of water.

Figure 27.15 Delocalisation of the nitrogen lone pair in the amide group

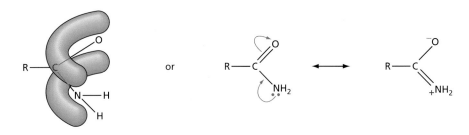

The lone pair on the nitrogen atom in amides is in a p orbital, and can overlap with the π orbital of the adjacent carbonyl group; the electrons are attracted to the electronegative oxygen atom (see Figure 27.15). This confers considerable double-bond characteristics to the C—N bond, including restricted rotation about it (compare with the alkenes, see Topic 14 page 252). This has great significance for the stereochemistry of polypeptide chains in proteins, which we shall meet later in Topic 28. Polypeptide chains contain secondary amide groups:

The properties of synthetic polyamides such as nylon are covered in section 28.2.

When amides react with strong acids, it is the oxygen atom, rather than the nitrogen atom, that is protonated:

Protonation of the nitrogen atom would result in a positively charged nitrogen adjacent to the δ+ carbon of the carbonyl group – an unfavourable situation. On the other hand, protonation of the oxygen atom allows the positive charge on the cation to be delocalised over three atoms – an energetically favourable situation:

So, although amides are more basic (through their oxygen atom) than other carbonyl compounds (because of the electron donation from nitrogen), they are still much less basic than conventional nitrogen bases such as amines.

Preparation and hydrolysis of amides

Amides are most readily prepared by reacting acyl chlorides (see section 26.3) with ammonia or amines.

$$CH_3-C\begin{smallmatrix}O\\\\Cl\end{smallmatrix} \ + \ CH_3NH_2 \longrightarrow CH_3-C\begin{smallmatrix}O\\\\NHCH_3\end{smallmatrix} \ + \ HCl$$

N-methylethanamide

The HCl formed reacts with an excess of amine to produce a salt.

$$CH_3NH_2 + HCl \rightarrow CH_3NH_3{}^+Cl^-$$

Reactions of amides

With water

Because of the high degree of positive charge on the carbon atom in protonated amides, they are susceptible to nucleophilic attack. Hydrolysis is usually carried out in dilute sulfuric acid. It is still quite a slow reaction, however – heating under reflux for several hours is usually required:

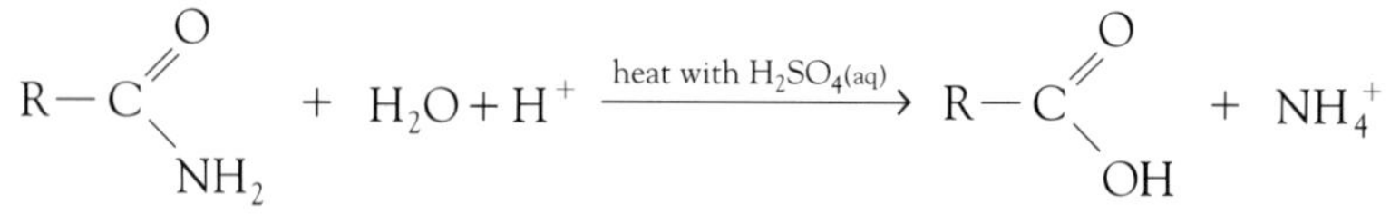

$$R-C\underset{NH_2}{\overset{O}{|}} + H_2O + H^+ \xrightarrow{\text{heat with } H_2SO_{4(aq)}} R-C\underset{OH}{\overset{O}{|}} + NH_4^+$$

Amides can also be hydrolysed under basic conditions, by heating with NaOH(aq).

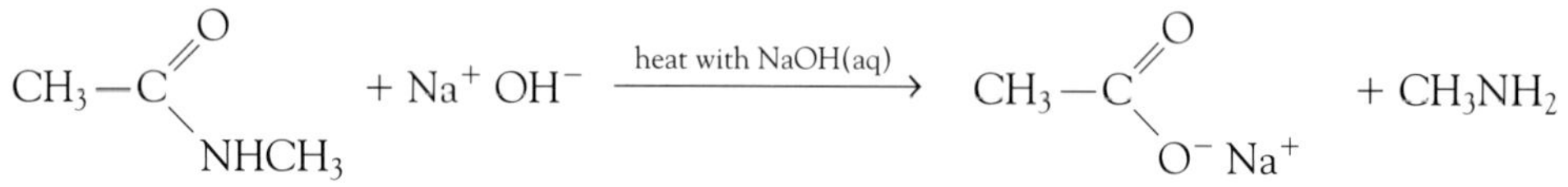

$$CH_3-C\underset{NHCH_3}{\overset{O}{|}} + Na^+ OH^- \xrightarrow{\text{heat with NaOH(aq)}} CH_3-C\underset{O^- Na^+}{\overset{O}{|}} + CH_3NH_2$$

The mechanism of the hydrolysis of amides

The mechanism is very similar to that of the hydrolysis of esters, described in section 18.3. The initial reaction is the protonation of the amide by the mineral acid (this occurs on the oxygen atom, as described above). The protonated amide is then attacked nucleophilically by a water molecule. Protons rearrange themselves amongst the oxygen and nitrogen atoms, and eventually the C—N bond cleaves, and a proton is then transferred to NH_3 (see Figure 27.16).

Figure 27.16 The hydrolysis of an amide

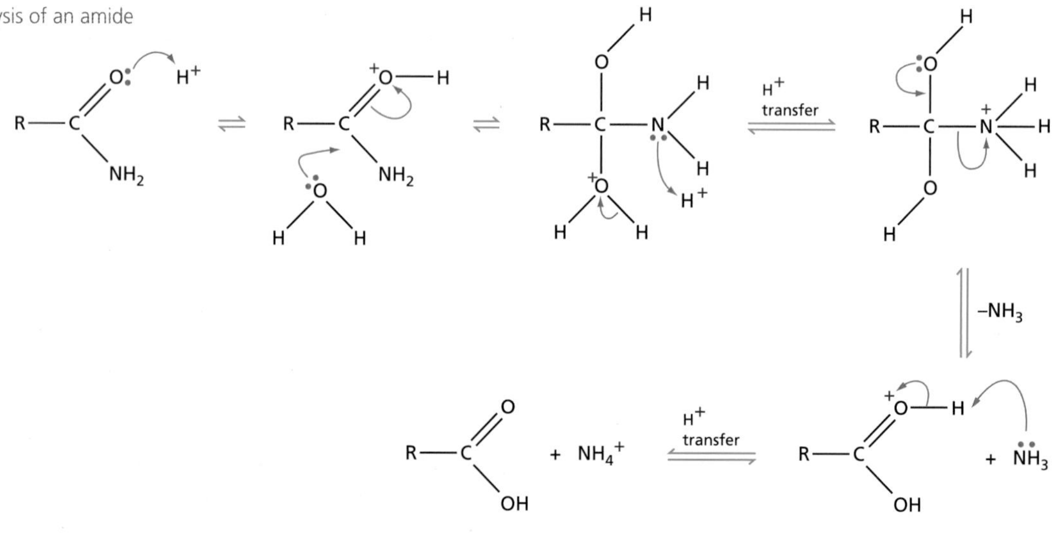

Amides are intermediates in the hydrolysis of nitriles to carboxylic acids:

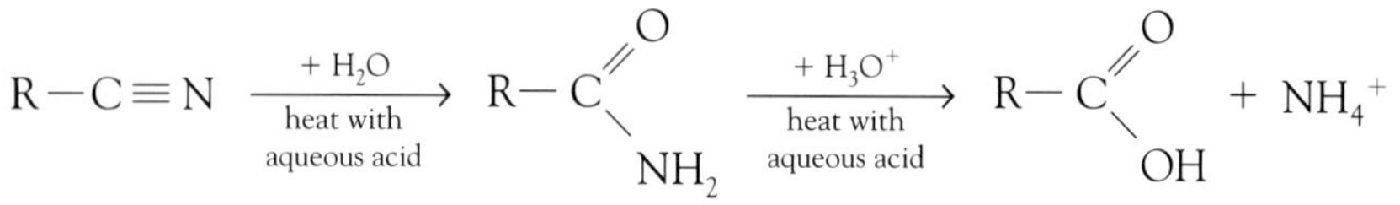

$$R-C\equiv N \xrightarrow[\text{heat with aqueous acid}]{+ H_2O} R-C\underset{NH_2}{\overset{O}{|}} \xrightarrow[\text{heat with aqueous acid}]{+ H_3O^+} R-C\underset{OH}{\overset{O}{|}} + NH_4^+$$

Reduction to amines

Amides can be reduced to amines by lithium tetrahydridoaluminate(III) (lithium aluminium hydride):

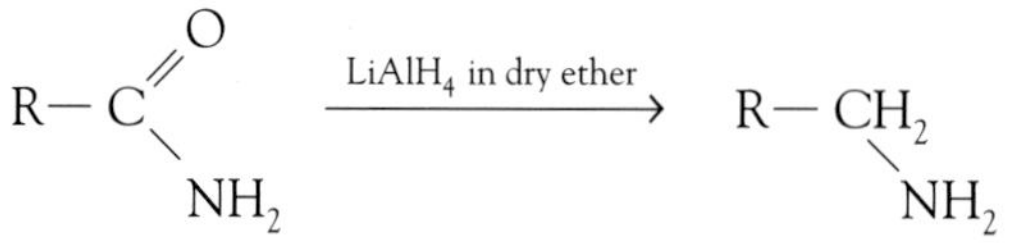

$$R-C\underset{NH_2}{\overset{O}{|}} \xrightarrow{\text{LiAlH}_4 \text{ in dry ether}} R-CH_2\underset{NH_2}{|}$$

Now try this

1 Write a balanced equation for the conversion of ethanenitrile, CH_3CN, into ethanoic acid by heating with hydrochloric acid.

2 (Difficult!) Draw out the mechanism of the hydrolysis of a nitrile to an amide. (Hint: the nitrogen atom has to be protonated twice, in two separate steps of the mechanism.)

Worked example

Suggest products of the following reactions.

a

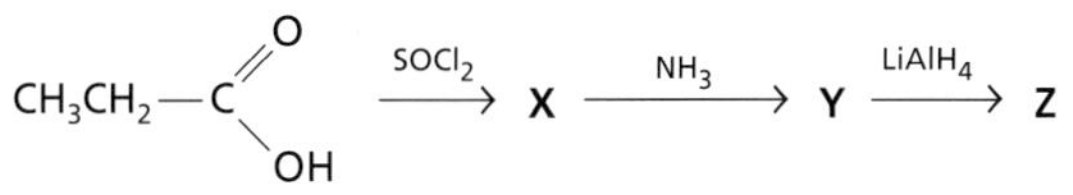

$$CH_3CH_2 - C \overset{O}{\underset{OH}{<}} \quad \xrightarrow{SOCl_2} \quad X \quad \xrightarrow{NH_3} \quad Y \quad \xrightarrow{LiAlH_4} \quad Z$$

b

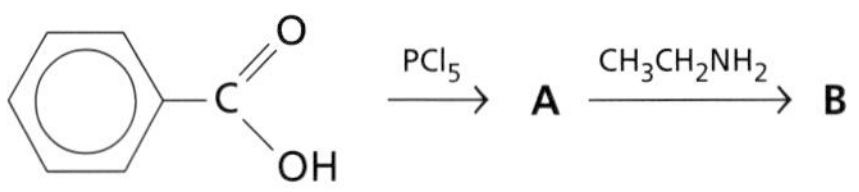

$$\xrightarrow{PCl_5} \quad A \quad \xrightarrow{CH_3CH_2NH_2} \quad B$$

Answer

a **X** is CH_3CH_2COCl **Y** is $CH_3CH_2CONH_2$ **Z** is $CH_3CH_2CH_2NH_2$

b **A** is ⬡—COCl   **B** is ⬡—CONHC$_2$H$_5$

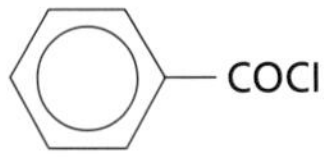

Now try this

Suggest two-stage syntheses of the following compounds, from the stated starting materials.

1   ⬡—CH$_2$NH$_2$ from   ⬡—COCl

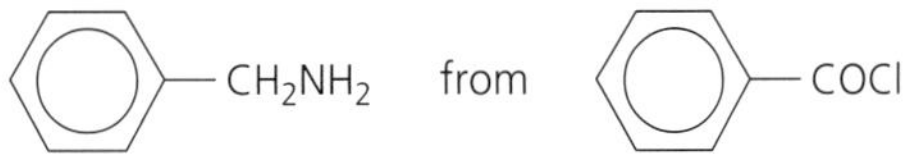

2 $(CH_3)_2CH-CH_2NH_2$ from $(CH_3)CH-Br$

27.6   Amino acids

Amino acids have two functional groups:

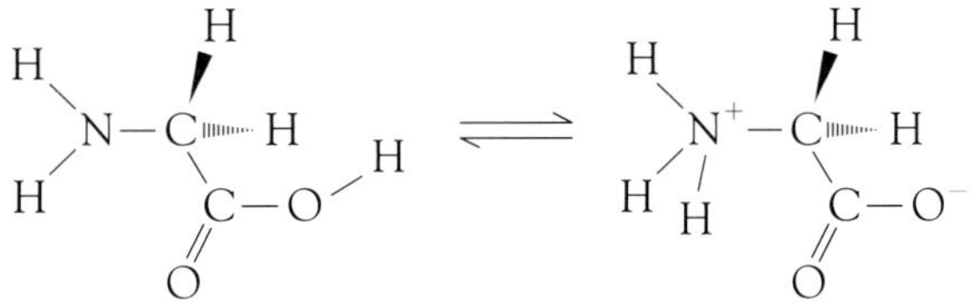

The 2-amino acids (α-amino acids) form an interesting class of compounds, apart from their great importance as the building blocks of proteins. In many compounds containing two or more functional groups, the reactions of one group can be considered independently of those of the other group. They are often widely separated, and dissimilar to each other. In 2-amino acids, however, the two groups are near to each other. What is more, they are of opposite chemical types: the —NH$_2$ group is basic, whilst the —CO$_2$H group is acidic. Interaction between the two is inevitable.

Physical and chemical properties

Whilst alkyl amines and the small-chain aliphatic carboxylic acids are liquids, the amino acids are all solids. They have high melting points (often decomposing before they can be heated to a sufficiently high temperature to melt them), and are soluble in water. In another contrast to amines and carboxylic acids, they are insoluble in organic solvents such as methylbenzene.

Table 27.4 A comparison of some properties of simple amino acids (glycine and alanine) with those of a simple amine and a simple carboxylic acid

Compound	Formula	Melting point/°C	K_a/mol dm^{-3}	K_b/mol dm^{-3}
glycine	$H_2NCH_2CO_2H$	233	1.4×10^{-10}	2.4×10^{-12}
alanine	$H_2NCH(CH_3)CO_2H$	297	9.8×10^{-11}	2.2×10^{-12}
ethylamine	$H_2NCH_2CH_3$	−81	—	5.1×10^{-4}
propanoic acid	$CH_3CH_2CO_2H$	−21	1.3×10^{-5}	—

Table 27.4 compares the properties of the two simplest amino acids with those of a simple amine and a simple carboxylic acid. Apart from their extremely high melting points, another clear difference between the amino acids and their mono-functional counterparts is their much reduced basicity and acidity. Glycine is 200 million times less basic than ethylamine:

$$H_2NCH_2CO_2H + H_2O \rightleftharpoons H_3N^+CH_2CO_2H + OH^- \qquad K_b = 2.4 \times 10^{-12}\,mol\,dm^{-3}$$

$$CH_3CH_2NH_2 + H_2O \rightleftharpoons CH_3CH_2NH_3^+ + OH^- \qquad K_b = 5.1 \times 10^{-4}\,mol\,dm^{-3}$$

Likewise, alanine is over 100 000 times less acidic than propanoic acid:

$$H_2NCH(CH_3)CO_2H + H_2O \rightleftharpoons H_2NCH(CH_3)CO_2^- + H_3O^+$$
$$K_a = 9.8 \times 10^{-11}\,mol\,dm^{-3}$$

$$CH_3CH_2CO_2H + H_2O \rightleftharpoons CH_3CH_2CO_2^- + H_3O^+ \qquad K_a = 1.3 \times 10^{-5}\,mol\,dm^{-3}$$

This large reduction in basicity and acidity is readily explained by the idea that the acidic and basic groups within an amino acid molecule have already reacted with each other:

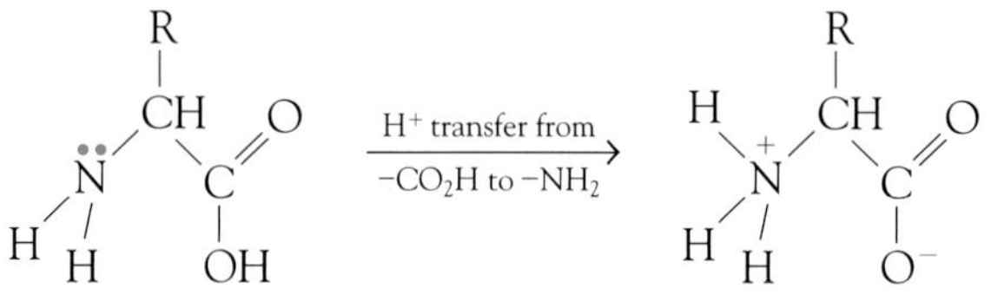

a zwitterion

The product is an 'internal salt', called a **zwitterion** (from the German word *Zwitter*, meaning 'hybrid' – zwitterionic structures were first proposed by the German chemist Küster in 1897).

A **zwitterion** is a molecule that contains both a cationic group and an anionic group.

The salt-like nature of zwitterions also explains the high melting points and the solubility characteristics of amino acids. It is thought that all amino acids exist as zwitterions (or dipolar ions) in solution and in the solid lattice. Their low acidity is therefore due to proton donation from the ammonium group rather than from the carboxylic acid group:

$$H_3N^+CH_2CO_2^- + H_2O \rightleftharpoons NH_2CH_2CO_2^- + H_3O^+ \quad K_a = 1.4 \times 10^{-10}\,mol\,dm^{-3}$$

This value is directly comparable with the ionisation of other ammonium salts:

$$NH_4^+ + H_2O \rightleftharpoons NH_3 + H_3O^+ \qquad\qquad K_a = 5.6 \times 10^{-10}\,mol\,dm^{-3}$$

One further characteristic property of 2-amino acids is their optical activity. All except glycine possess a chiral carbon atom, having four different atoms or groups attached to it. There are two distinct and different ways in which these four groups can arrange themselves around the central carbon atom, as shown in Figure 27.17 (see also page 224). All amino acids derived from the hydrolysis of proteins have the '*l*' configuration.

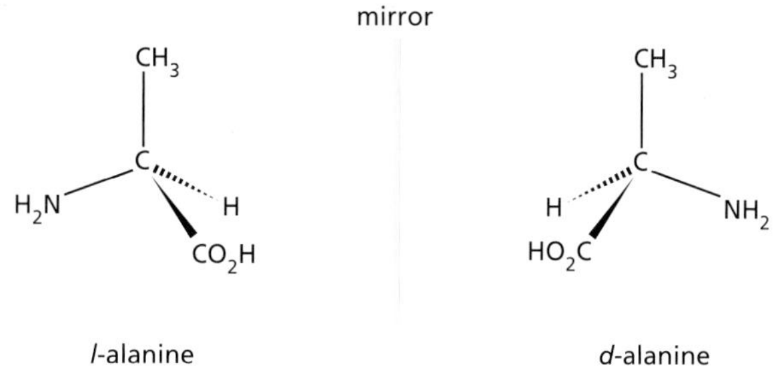

Figure 27.17 *l*- and *d*-alanine

mirror

l-alanine   *d*-alanine

Reactions of amino acids

Because the dipolar form of amino acids is in equilibrium with a small amount of the un-ionised form, amino acids show many of the typical reactions of amines and carboxylic acids.

1 The amino group can be acylated (see page 455) and reacts with nitrous acid (see page 456):

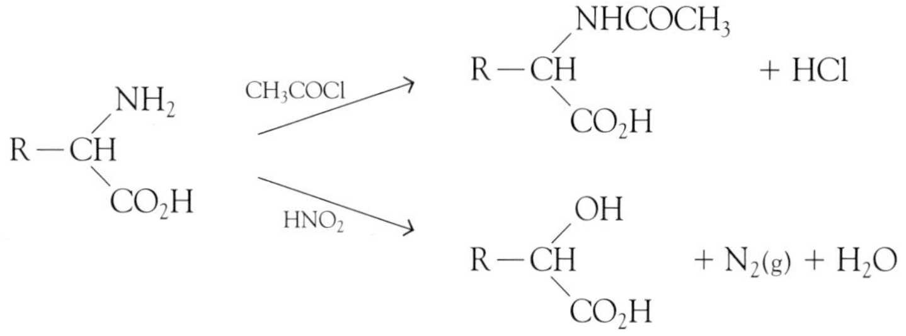

2 The carboxylic acid group can be esterified:

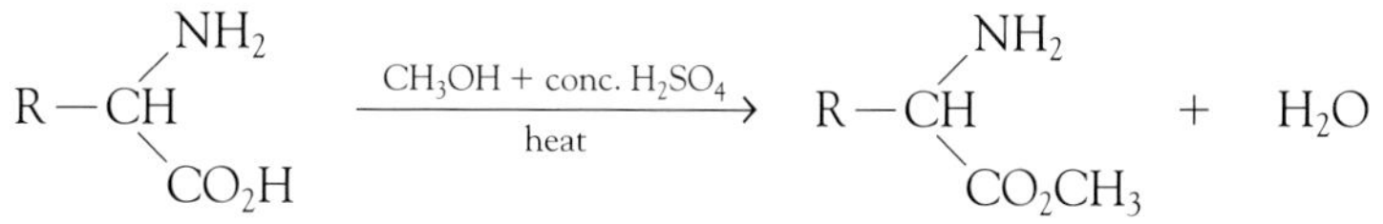

3 Amino acids act as buffers, stabilising the pH of a solution if excess acid or alkali is added (see section 22.3).

On adding acid:

$$NH_2{-}CH_2{-}CO_2H + H^+ \rightarrow {}^+NH_3{-}CH_2{-}CO_2H$$

On adding alkali:

$$NH_2{-}CH_2{-}CO_2H + OH^- \rightarrow NH_2{-}CH_2{-}CO_2^- + H_2O$$

4 In addition to the reactions of the $-NH_2$ and the $-CO_2H$ groups, amino acids with side chains that contain functional groups also show the reactions of that group. Figure 27.18 shows the side chains of some amino acids. For example, serine reacts with phosphorus(V) chloride to form a chloroalkyl side chain, and tyrosine reacts with bromine water just as phenol does (see Topic 25, page 435).

Now try this

Draw the structure of the product of each of the following reactions (see Figure 27.18).

a Tyr with Br_2(aq)
b Lys with an excess of hydrochloric acid
c Glu when heated with an excess of CH_3OH and a trace of concentrated H_2SO_4
d Ser with an excess of CH_3COCl

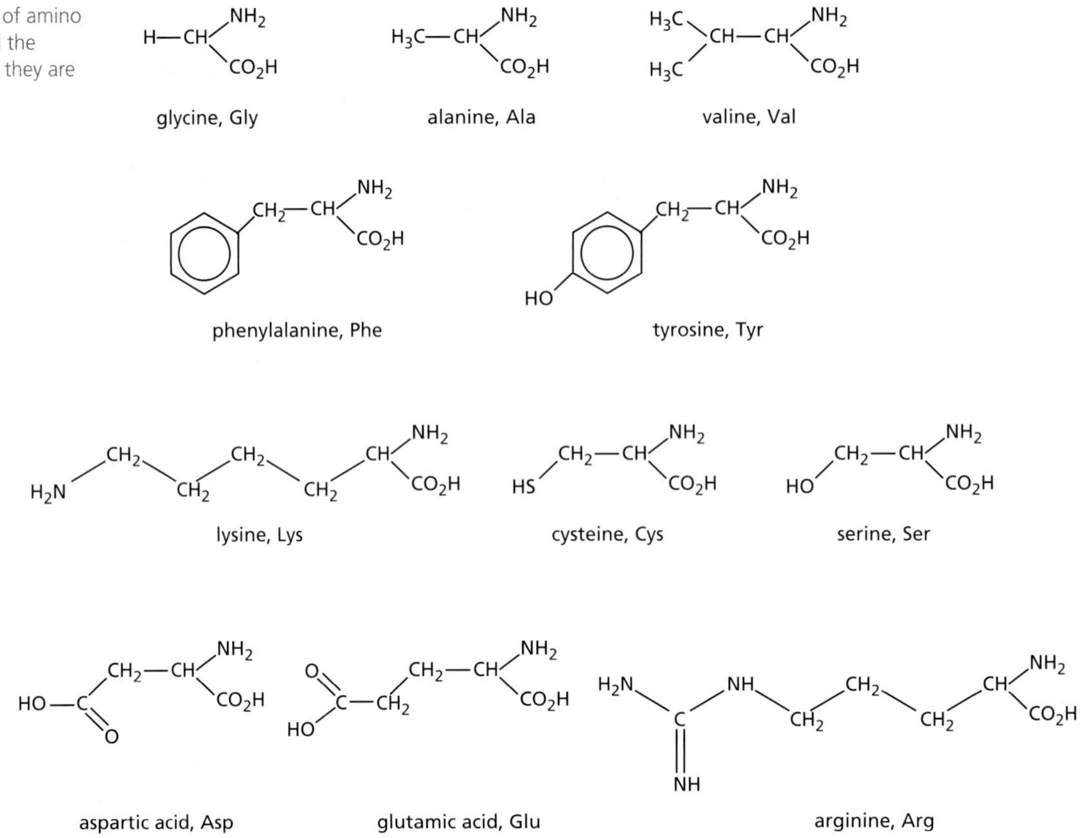

Figure 27.18 A selection of amino acids, with their names and the three-letter codes by which they are known

glycine, Gly

alanine, Ala

valine, Val

phenylalanine, Phe

tyrosine, Tyr

lysine, Lys

cysteine, Cys

serine, Ser

aspartic acid, Asp

glutamic acid, Glu

arginine, Arg

5 Peptide bond formation–two or more amino acids can undergo condensation reactions between themselves to form amides, which, if formed from amino acids, are called **peptides**. If the peptide is composed of two amino acids it is called a dipeptide.

$$H_2NCH_2CO_2H + H_2NCH(CH_3)CO_2H \rightarrow H_2NCH_2CONHCH(CH_3)CO_2H + H_2O$$
$$\text{the dipeptide Gly-Ala}$$

Notice that the dipeptide Ala–Gly is not the same as Gly–Ala: it is a structural isomer of it.

$$H_2NCH(CH_3)CONHCH_2CO_2H$$
$$\text{the dipeptide Ala-Gly}$$

If three amino acids join together, a tripeptide is formed.

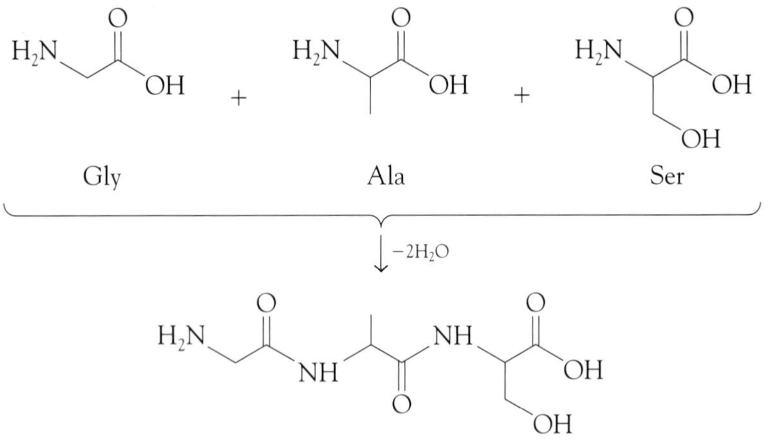

Gly   Ala   Ser

$-2H_2O$

the tripeptide Gly − Ala − Ser

Notice that the prefixes *di-* and *tri-* refer to the number of amino acids in the peptide, and not the number of peptide bonds.

Worked example

How many possible tripeptides can be formed from the three amino acids Gly, Ala and Ser, if each tripeptide contains all three amino acids?

Answer

There are six isomeric tripeptides: Gly–Ala–Ser, Gly–Ser–Ala, Ala–Ser–Gly, Ala–Gly–Ser, Ser–Gly–Ala and Ser–Ala–Gly.

We shall meet peptides again in Topic 28, in the description of proteins.

27.7 Electrophoresis of amino acids and peptides

The principle of electrophoresis is best explained by looking first at how the simple amino acid, glycine, behaves at different pH values. This is because of the amphoteric properties of amino acids. If we have a solution of an amino acid in water, the average charge on the many molecules of amino acid in the solution depends on the pH of the solution. The pH at which the net overall charge is zero is called the **isoelectric point** of that amino acid. For glycine, this is pH 6.07.

$$\underset{\text{at pH} < 6.07}{^+NH_3CH_2CO_2H} \overset{+\,H^+}{\underset{}{\rightleftharpoons}} \underset{\text{at pH} = 6.07}{^+NH_3CH_2\,CO_2^-} \overset{-\,H^+}{\underset{}{\rightleftharpoons}} \underset{\text{at pH} > 6.07}{NH_2CH_2CO_2^-}$$

In solution at pH < 6.07, the average charge on glycine molecules becomes positive. At pH > 6.07, the molecules become negatively charged.

Depending on the side groups, different amino acids have different tendencies to form cations or anions in solution, and so their isoelectric points will occur at different pH values. The more basic ones such as lysine, which have a tendency to form cations if dissolved in water, will have their isoelectric points at a more alkaline pH than most. On the other hand, the acidic amino acids such as aspartic acid will have their isoelectric points at a more acidic pH than most (see Table 27.5).

Table 27.5 The isoelectric points of some amino acids

Amino acid	R group in $R-CH(NH_2)CO_2H$	Isoelectric point (IEP)
glycine	$-H$	6.07
lysine	$-(CH_2)_4NH_2$	9.74
aspartic acid	$-CH_2CO_2H$	2.98

If all three of these amino acids were dissolved together in the same buffer solution kept at pH 6.07, on average, the molecules of glycine would be electrically neutral,

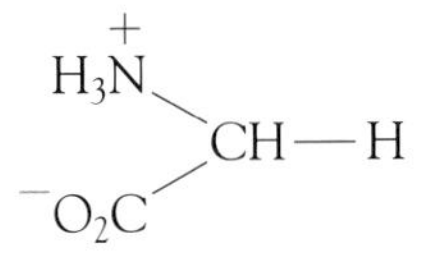

$$\underset{^-O_2C}{\overset{^+H_3N}{\diagdown}} CH-H$$

the molecules of lysine would be positively charged,

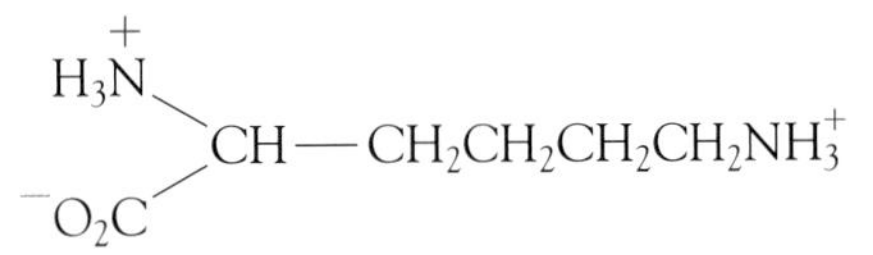

$$\underset{^-O_2C}{\overset{^+H_3N}{\diagdown}} CH-CH_2CH_2CH_2CH_2NH_3^+$$

and the molecules of aspartic acid would be negatively charged.

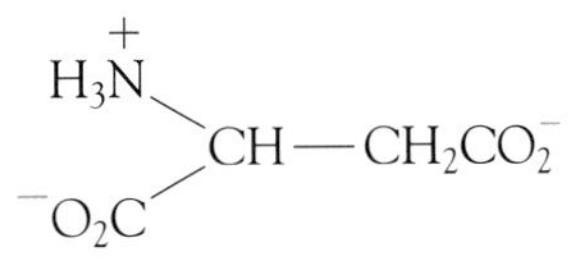

$$\underset{^-O_2C}{\overset{^+H_3N}{\diagdown}} CH-CH_2CO_2^-$$

Figure 27.19 Electrophoresis separates lysine, aspartic acid and glycine.

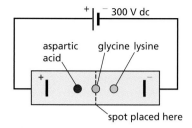

If a drop of the buffer solution at pH 6.07 containing these three amino acids was spotted onto a gel-coated plate, immersed in a conducting liquid, and a potential difference applied, the lysine would move to the cathode, the aspartic acid would move to the anode, and the glycine, having no overall electrical charge, would not move at all (see Figure 27.19).

This separation of a mixture in an electric field is called **electrophoresis**. The electrophoretic mobility depends not only on average charge but also on the size and shape of a molecule:

$$v = \frac{EZ}{F}$$

where v = velocity, E = electric field (voltage applied), Z = average charge and F = frictional resistance.

The frictional resistance is composed of various factors:

- the size of the pores in the gel support (the smaller the pores, the slower the sample moves)
- the shape of the molecule (large, spiky molecules travel slower than small, spherical ones)
- the size/molecular mass of the molecule (larger molecules travel slower than smaller ones).

Other factors that can affect the speed at which the components of a sample move during electrophoresis are:

- the pH of the buffer solution (the average number of charged groups on an amphoteric molecule depends on pH; as the pH increases, the average charge becomes more negative/less positive)
- the temperature of the apparatus (the speed of movement increases with temperature).

Electrophoresis is used regularly to separate and identify not only mixtures of amino acids, but mixtures of peptides obtained from proteins, and even mixtures of proteins themselves (see Figure 27.20). By this means, abnormal proteins can be identified in patients suffering from genetic disorders, and possible treatments monitored.

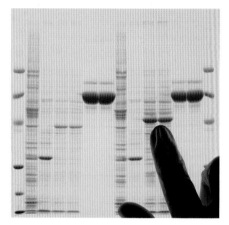

Figure 27.20 Electrophoresis of proteins

27.8 Synthetic polyamides (nylons)

Synthetic polyamides (nylons) are an important class of manufactured polymer. They are condensation polymers (see section 28.2), usually formed by the reaction between a diamine and a dicarboxylic acid:

$$H_2N-(CH_2)_n-NH_2 \quad + \quad HO_2C-(CH_2)_m-CO_2H$$

$$\downarrow -H_2O$$

Values of n and m vary from 4 to 10.

Diacyl chlorides can also be used, in a reaction similar to the preparation of simple amides (see section 26.3).

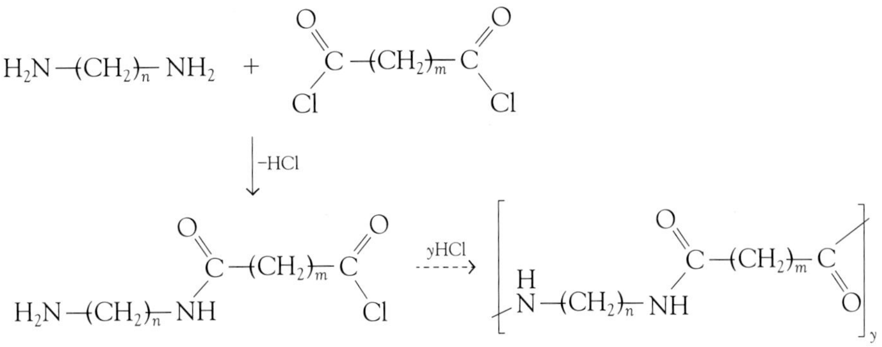

Further information about polyamides can be found in Topic 28.

Summary

- **Amines** are weak organic bases, reacting with acids to form salts, and dissolving in water to give alkaline solutions.
- There are three types of amine – **primary**, RNH_2, **secondary**, R_2NH, and **tertiary**, R_3N.
- Aryl amines react with nitric(III) acid (nitrous acid) to form **diazonium salts**, from which many useful **azo dyes** are manufactured.
- **Amino acids** contain the —NH_2 and —CO_2H functional groups adjacent to each other. In solution and in the solid they exist as **zwitterions**, $^+NH_3CHRCO_2^-$.
- Most amino acids are chiral. Those isolated from natural proteins have the '*l*' configuration.
- Electrophoresis can be used to separate amino acids and peptides.
- **Synthetic polyamides** such as nylon are made by condensing diamines with dicarboxylic acids.

Key reactions you should know
(R = alkyl or aryl unless otherwise stated.)
- Amines:
  - Alkylation:
    $R—NH_2 + CH_3Br \rightarrow R—NHCH_3 + HBr$
  - Acylation:
    $R—NH_2 + CH_3COCl \rightarrow R—NHCOCH_3 + HCl$
  - With nitrous acid (nitric(III) acid):

    alkyl—NH_2 + HNO_2 $\xrightarrow{\text{at room temperature}}$ alkyl—OH + $N_2(g)$ + H_2O

    aryl—NH_2 + HNO_2 + H^+ $\xrightarrow{\text{at } T < 5\,°C}$ aryl—N_2^+ + $2H_2O$
- Amides:
  - Hydrolysis:

    $RCONH_2 + H_2O \xrightarrow{\text{heat with } H^+} RCO_2H + NH_4^+$

    $RCONH_2 + H_2O \xrightarrow{\text{heat with } OH^-} RCO_2^- + NH_3$
  - Reduction:

    $RCONH_2 \xrightarrow{\text{LiAlH}_4 \text{ in ether}} RCH_2NH_2$
- Polymerisation:

  $nH_2N — \bullet — NH_2 + nH_2OC —\blacksquare— CO_2H \longrightarrow {\text{---}[NH—\bullet— NHCO —\blacksquare— CO]}_n$

Examination practice questions

Please see the data section of the CD for any A_r values you may need.

1 Phenol and chlorobenzene are less reactive towards certain reagents than similar non-aromatic compounds. Thus hexan-1-ol can be converted into hexylamine by the following two reactions,

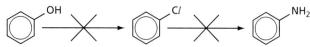

$$CH_3(CH_2)_5OH \xrightarrow{\quad I \quad} CH_3(CH_2)_5Cl \xrightarrow{\quad II \quad} CH_3(CH_2)_5NH_2$$
hexan-1-ol   1-chlorohexane   hexylamine

whereas neither of the following two reactions takes place.

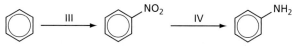

a i Suggest reagents and conditions for
   reaction I
   reaction II
 ii What type of reaction is reaction II?
 iii Suggest a reason why chlorobenzene is much less reactive than 1-chlorohexane. [4]
b Phenylamine can be made from benzene by the following two reactions.

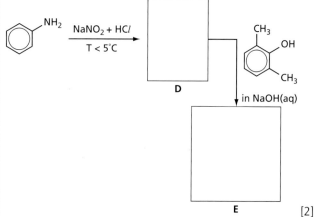

 i Suggest reagents and conditions for
   reaction III
   reaction IV
 ii State the type of reaction for
   reaction III
   reaction IV [5]
c Suggest a reagent that could be used to distinguish phenylamine from hexylamine.
   reagent and conditions
   observation with phenylamine
   observation with hexylamine [2]
d Phenylamine is used to make azo dyes. Draw the structural formula of the intermediate **D** and of the azo dye **E**.

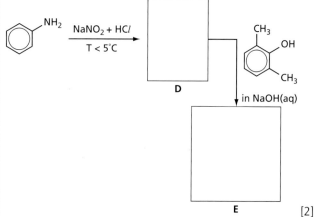

E [2]

[Cambridge International AS & A Level Chemistry 9701, Paper 4 Q6 June 2009]

2 a Describe the reagents and conditions required to form a nitro compound from the following.
 i methylbenzene

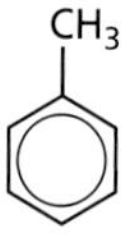

 ii phenol

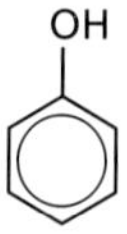

[3]

b Draw the structure of the intermediate organic ion formed during the nitration of benzene. [1]
c Write the reagents needed to convert nitrobenzene into phenylamine. [1]
d Phenylamine can be converted into the organic compounds **A** and **B**.
 i Suggest the structural formulae of **A** and **B**.
 ii Suggest suitable reagents and conditions for step 1.

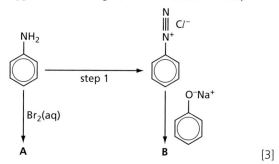

[3]

e When phenylamine is treated with propanoyl chloride a white crystalline compound, **C**, $C_9H_{11}NO$, is formed.
 i Name the functional group formed in this reaction.
 ii Calculate the percentage by mass of nitrogen in C.
 iii Draw the structural formula of **C**. [3]

[Cambridge International AS & A Level Chemistry 9701, Paper 42 Q3 June 2013]

A Level
Organic chemistry

28 Addition and condensation polymers

Functional groups:

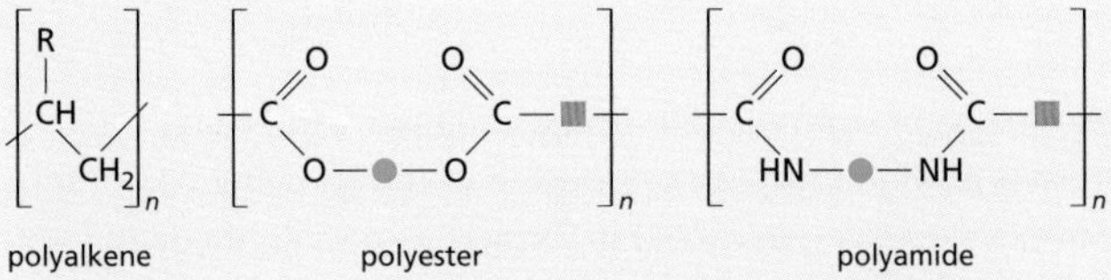

polyalkene   polyester   polyamide

This topic takes a more detailed look at the addition polymerisation of alkenes, first met in Topic 14, and at condensation polymerisation to form polyesters and polyamides, met in Topics 26 and 27 respectively.

Learning outcomes

By the end of this topic you should be able to:

21.1a) describe the formation of polyesters and polyamides

21.1b) describe the characteristics of condensation polymerisation in polyesters as exemplified by Terylene, in polyamides as exemplified by polypeptides, proteins, nylon 6, nylon 6,6 and Kevlar

21.1c) deduce the repeat unit of a condensation polymer obtained from a given monomer or pair of monomers

21.1d) identify the monomer(s) present in a given section of a condensation polymer molecule

21.2a) predict the type of polymerisation reaction for a given monomer or pair of monomers

21.2b) deduce the type of polymerisation reaction which produces a given section of a polymer molecule

21.3a) discuss the properties and structure of polymers based on their methods of formation (addition or condensation, see also Topic 14)

21.3b) discuss how the presence of side-chains and intermolecular forces affect the properties of polymeric materials (e.g. polyalkenes, PTFE (Teflon), Kevlar)

21.3c) explain the significance of hydrogen-bonding in the pairing of bases in DNA in relation to the replication of genetic information

21.3d) distinguish between the primary, secondary (α-helix and β-sheet) and tertiary structures of proteins and explain the stabilisation of secondary structure (through hydrogen bonding between $C=O$ and $N-H$ bonds of peptide groups) and tertiary structure (through interactions between R-groups)

21.3e) describe how polymers have been designed to act as non-solvent based adhesives, e.g. epoxy resins and superglues, and as conducting polymers, e.g. polyacetylene

21.4a) recognise that polyalkenes are chemically inert and can therefore be difficult to biodegrade

21.4b) recognise that a number of polymers can be degraded by the action of light

21.4c) recognise that polyesters and polyamides are biodegradable by hydrolysis

21.4d) describe the hydrolysis of proteins.

28.1 Addition polymers

We saw in Topic 14 how ethene could be polymerised in different ways to produce low-density poly(ethene), LDPE, and high-density poly(ethene), HDPE. Addition polymers with a variety of properties can be made by polymerising other compounds containing double bonds.

Propene (obtained by catalytic cracking) can be polymerised in a similar way to ethene, using the Ziegler–Natta catalyst (see section 14.4):

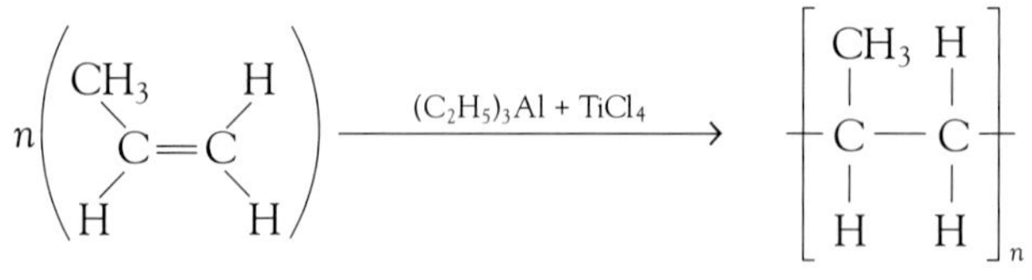

poly(propene)

During polymerisation, the propene subunits line up in a regular head-to-tail arrangement, because as each propene molecule is added to the growing chain, the more stable secondary carbocation (see page 248) is formed as an intermediate (see Figure 28.1). Note that only two of the three carbon atoms in propene become part of the chain: the third one remains as a methyl side group at the 'tail'. The triethyl aluminium is an electrophile, acting as a homogeneous catalyst.

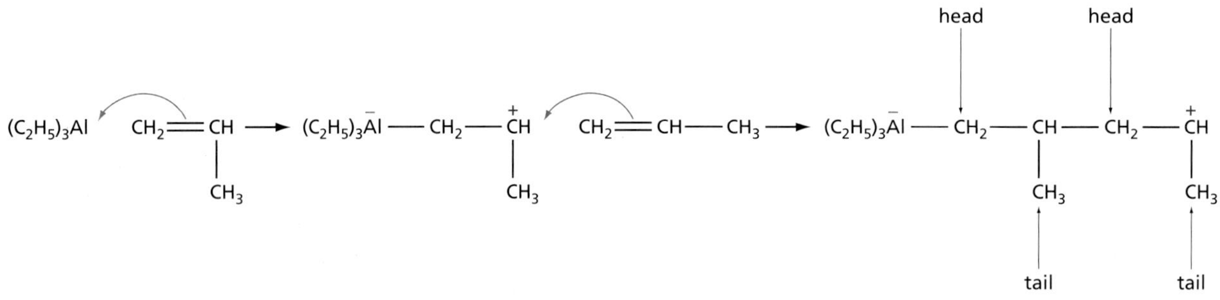

Figure 28.1 The propene units line up head to tail.

Phenylethene can be polymerised to form poly(phenylethene), otherwise known as polystyrene:

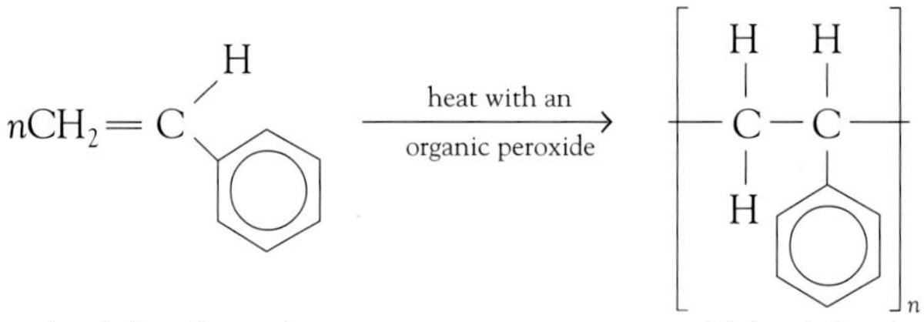

phenylethene (styrene)

poly(phenylethene) (polystyrene)

Phenylethene is obtained commercially by reacting together ethene and benzene followed by dehydrogenation. It is an important component of the co-polymers SBR and ABS (see the panel on pages 471–472).

The mechanism of this polymerisation is similar to that of the polymerisation of propene, but it involves radicals as intermediates, rather than cations. The more stable secondary radical is formed at each stage and the polymer forms in a regular head-to-tail chain (see Figure 28.2). The initial alkoxy radical is formed when the weak O—O bond in the peroxide is broken by heat.

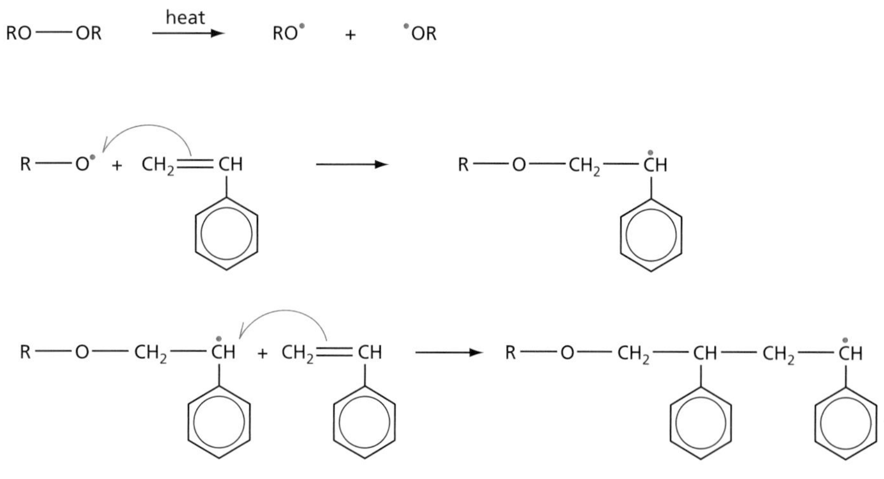

Figure 28.2 In phenylethene, the monomer units line up head to tail.

$$RO — OR \xrightarrow{heat} RO^\bullet \ + \ ^\bullet OR$$

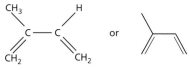

Poly(phenylethene) is used for packaging (for example, egg boxes), and model kits (because when it is moulded it can depict fine details exactly). When an inert gas is bubbled through the phenylethene as it is being polymerised, the familiar white, very light, solid known as **expanded polystyrene** is formed. This is used as a protective packaging, and for sound and heat insulation.

Rubber – natural and synthetic co-polymers

The rubber tree is indigenous to South America, but large plantations have been established in India and Malaysia. When its bark is stripped, it oozes a white sticky liquid called latex, which is an emulsion of rubber in water. Purification of this produces rubber. In its natural state, rubber is not particularly useful. It has a low melting point, is sticky and has a low tensile strength. In 1839, Charles Goodyear discovered the process of **vulcanisation**, which involves heating natural rubber with sulfur. This produces a substance with a higher melting point and of greater strength. Rubber is an addition polymer of 2-methylbutadiene (isoprene) (see Figure 28.3).

Figure 28.3 Isoprene undergoes addition polymerisation to form rubber.

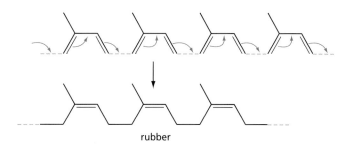

isoprene
(2-methylbutadiene)

rubber

Figure 28.4 A rubber tree being tapped. The white latex is converted to rubber suitable for tyres by the process of vulcanisation.

Note that when a diene undergoes addition polymerisation, a double bond is still present in the product. This double bond may be *cis* or *trans*. In natural rubber, the long chain is *cis* across all the double bonds. From different trees, a naturally occurring substance called gutta percha can be obtained; this is an isomer of rubber in which all double bonds are *trans*. Gutta percha is harder than rubber, and in the 19th century was used as a coating for golf balls.

The presence of the double bond in rubber allows further addition reactions to take place. If liquid rubber is heated with sulfur, sulfur atoms add across some of the double bonds in different chains, cross-linking the rubber molecules (see Figure 28.5). This is what happens during the vulcanisation process. It stops the chains from moving past each other, and gives the material more rigidity.

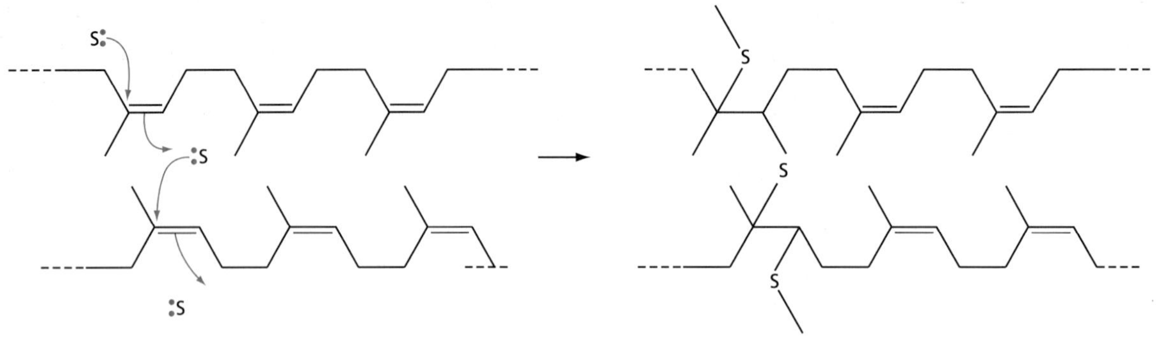

Figure 28.5 Vulcanisation of rubber

Not all double bonds have sulfur added to them. That would make the substance too hard. About 5% sulfur by mass is adequate to give the desired properties. Many millions of tonnes of vulcanised rubber are made each year for the manufacture of car tyres.

Synthetic rubber-like polymers were developed when rubber was in short supply during the Second World War. The one most commonly used today is a co-polymer of phenylethene and butadiene, called **SBR** (styrene–butadiene rubber).

A **co-polymer** is formed when two or more different alkenes are polymerised together. Even if the ratio of monomers is 50:50, there is no guarantee that the monomer fragments will alternate along the chain. The order is fairly random (in contrast to condensation co-polymerisation – see sections 28.2 and 28.3).

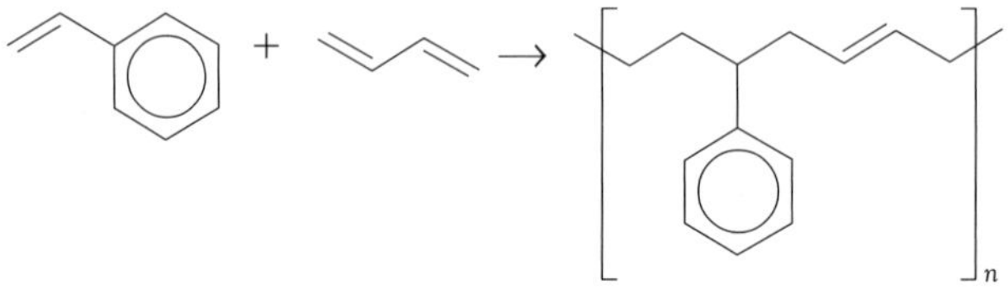

styrene   butadiene   SBR

The product SBR still contains a double bond, and so can be vulcanised just like natural rubber.

Another co-polymer involving phenylethene and butadiene as reagents, along with acrylonitrile, is the tough, rigid plastic **ABS**, or acrylonitrile–butadiene–styrene:

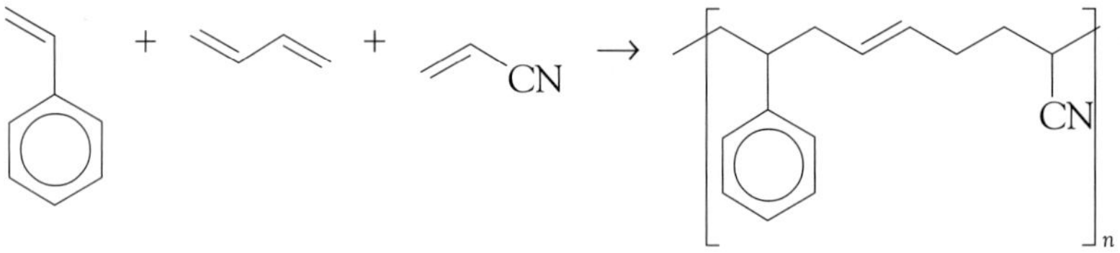

phenylethene   butadiene   acrylonitrile   ABS

Once again, the monomers join together in a fairly random manner, so the drawing of a particular 'repeat unit' does not imply a regular order. ABS is used for suitcases, telephones and other objects that need to be strong and hard, but not too brittle.

Figure 28.6 Products made from rubber, SBR and ABS

Addition polymerisation of other ethene derivatives

Most compounds containing the $>C=C<$ group will undergo polymerisation. Several important polymers are made from monomers in which some or all of the hydrogen atoms in ethene have been replaced by other atoms or groups. Some of them, with their uses, are listed in Table 28.1.

Table 28.1 Structures and uses of some addition polymers

Monomer	Polymer	Uses
chloroethene $CH_2=CHCl$	poly(chloroethene) (polyvinylchloride, PVC)	guttering, water pipes, windows, floor coverings
tetrafluoroethene $CF_2=CF_2$	poly(tetrafluoroethene) (Teflon, PTFE)	non-stick cookware, bridge bearings
methyl 2-methylpropenoate $CH_2=C(CH_3)CO_2CH_3$	poly(methyl 2-methylpropenoate) (Perspex)	protective 'glass', car rear lights, shop signs
methyl 2-cyanopropenoate $CH_2=C(CN)CO_2CH_3$	poly(methyl 2-cyanopropenoate) (cyanoacrylate)	instant 'superglue'
cyanoethene $CH_2=CH-CN$	poly(cyanoethene) (acrylic)	co-polymerised with components of 'acrylic' fibres, ABS

Worked example 1

Draw **two** repeat units of the polymer formed from each of the following monomers or pair of monomers.

a $CH_2=CH-OCOCH_3$
b $CH_2=CCl-CH=CH_2$
c $CH_2=CHCl$ and $CH_2=CH-CH_3$

Answer

a
$$\begin{array}{c} CH_2 \quad\quad CH_2 \\ \quad CH \quad\quad CH \\ OCOCH_3 \quad OCOCH_3 \end{array}$$

b
$$\begin{array}{c} Cl \quad\quad\quad Cl \\ C \quad CH_2 \quad C \quad CH_2 \\ CH_2 \quad CH \quad CH_2 \quad CH \end{array}$$

c

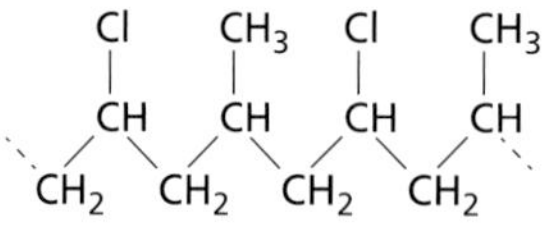

Worked example 2

The following are sections of addition polymers. Draw the monomer(s) from which they are formed.

a $\cdots CH_2-CH(CN)-CH_2-CH(CN)-CH_2\cdots$
b $\cdots CH_2-C(CN)=CH-CH(CN)-CH_2-C(CN)=CH-CH(CN)-CH_2\cdots$

Answer

a $CH_2=CH(CN)$ b $CH_2=C(CN)-CH=CH(CN)$

Now try this

Suggest the monomer(s) from which the following polymers have been made.

1

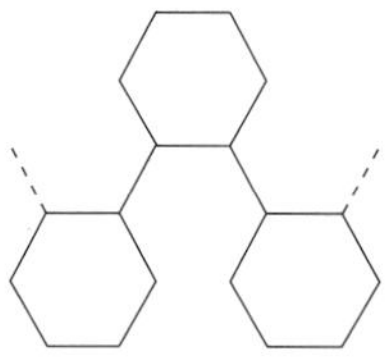

2

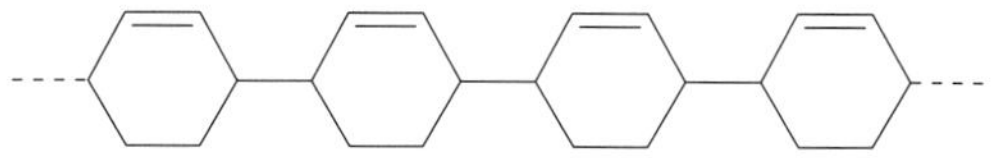

28.2 Condensation polymers

There are two strategies for making condensation polymers. Each strategy results in a different type of polymer.

The first strategy, producing **type I** polymers, starts with a monomer containing two different functional groups that can condense together to form either ester or amide groups (see Figure 28.7).

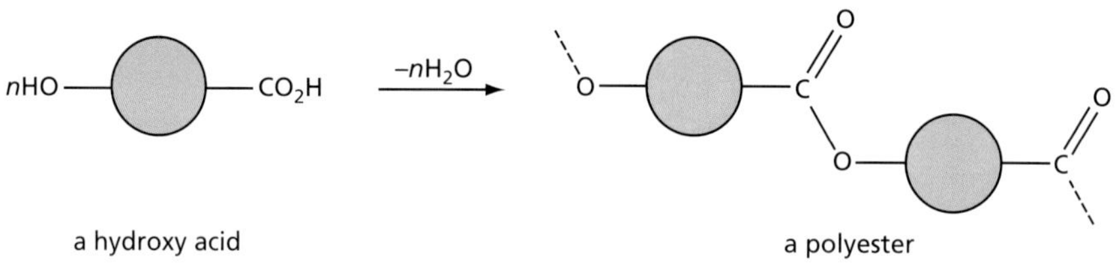

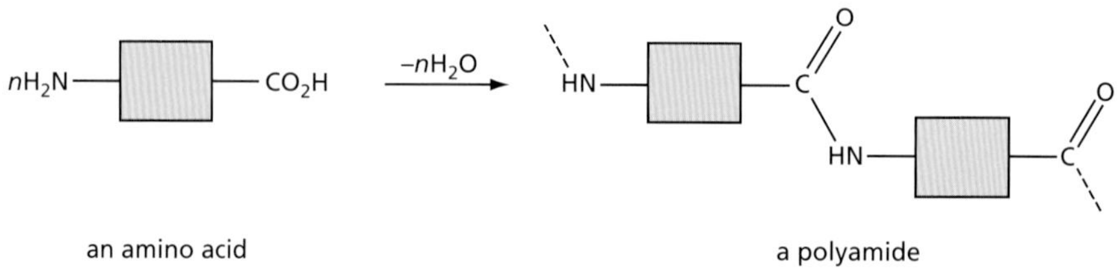

Figure 28.7 Formation of type I condensation polymers

The second strategy, forming **type II** polymers, starts with two different monomers, one containing two carboxylic acid groups or two acyl chloride groups, and the other containing two alcohol groups or two amine groups (see Figure 28.8).

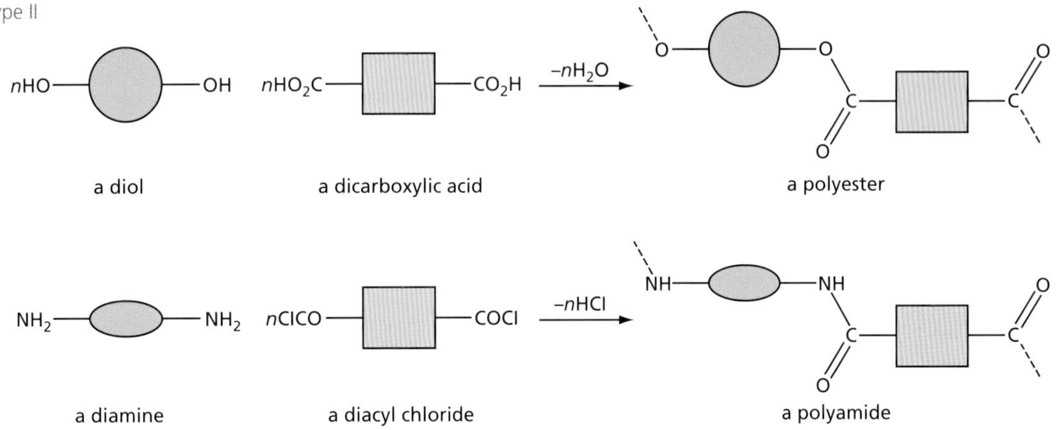

Figure 28.8 Formation of type II condensation polymers

a diol   a dicarboxylic acid   a polyester

a diamine   a diacyl chloride   a polyamide

The two types of polymer differ in the direction of successive functional groups along the chain. In type I polymers, the direction of each ester or amide linkage is the same (see Figure 28.9a), whereas in type II polymers, the direction of the ester or amide linkages alternates (see Figure 28.9b).

Table 28.2 lists some common condensation polymers. The different nylons (polyamides) are named according to the number of carbon atoms contained within each monomer: the first number is the number of carbon atoms in the diamine, and the second is the number of carbon atoms in the diacid.

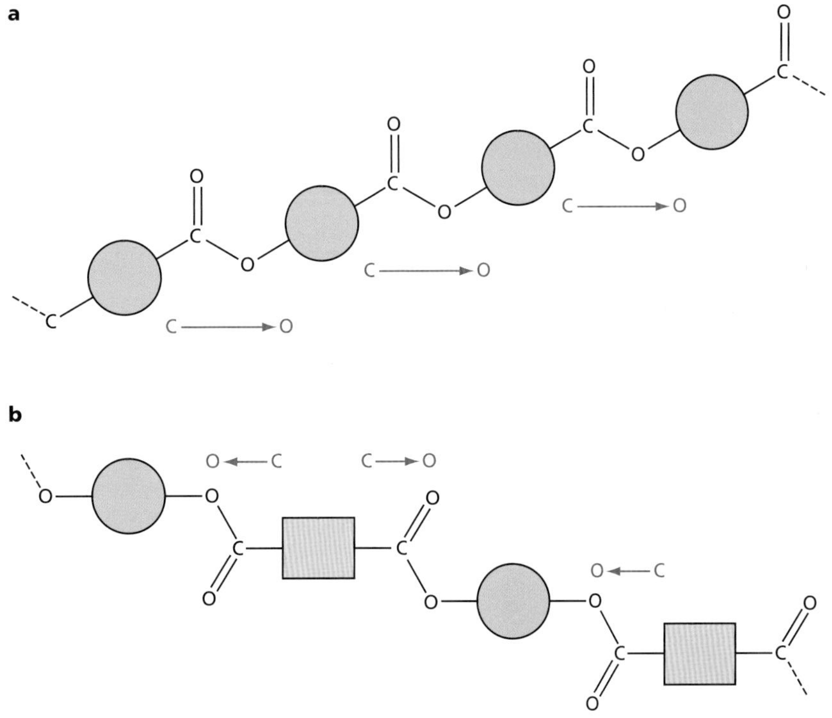

Figure 28.9 The orientation of adjacent ester linkages in **a** a type I polyester and **b** a type II polyester

a

b

Table 28.2 The monomers of some condensation polymers

Name	Formula(e) of monomer(s)	Type of polymer
Polyesters		
Terylene	$HOCH_2CH_2OH$ $HO_2C-C_6H_4-CO_2H$	type II
polylactic acid (PLA)	$HOCH(CH_3)CO_2H$	type I
Polyamides		
nylon 6,6	$NH_2-(CH_2)_6-NH_2$ $HO_2C-(CH_2)_4-CO_2H$	type II
nylon 6,10	$NH_2-(CH_2)_6-NH_2$ $HO_2C-(CH_2)_8-CO_2H$	type II
nylon 6	$NH_2-(CH_2)_5-CO_2H$	type I
polypeptides (proteins)	$NH_2-CHR-CO_2H$ (R = various side chains)	type I

Worked example

Draw the repeat unit of the chain of:

a nylon 6,10
b nylon 6.

Answer

The repeat unit is the smallest unit from which the polymer chain can be built up by repetition. That of nylon 6,10 will include one molecule of each of the two monomers, while that of nylon 6 will contain just the one monomer unit:

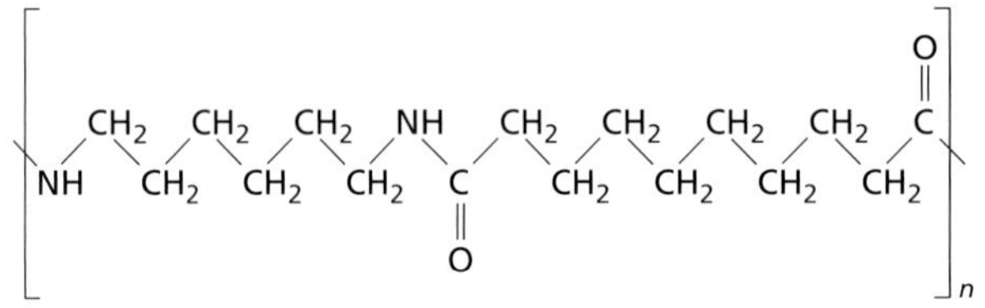

nylon 6,10

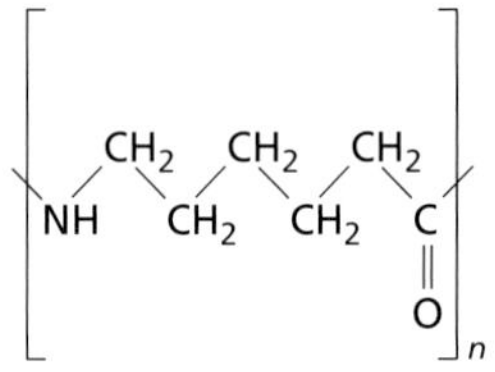

nylon 6

Now try this

1 Kevlar is a polymer which is five times stronger than steel on an equal-weight basis. It is used in bicycle tyres, racing sails and body armour because of its high tensile strength-to-weight ratio.

Kevlar is made from the following two monomers.

a Draw the repeat unit of the polymer Kevlar.
b Decide whether Kevlar or a type I or a type II polymer.

2 A biodegradable co-polymer can be made from lactic acid, $CH_3CH(OH)CO_2H$, and 3-hydroxypropanoic acid, $HOCH_2CH_2CO_2H$. Draw the repeat unit of this co-polymer.

Like simple esters and amides, polyesters and polyamides can be hydrolysed by strong acids or alkalis. Many common fibres used for clothing, such as nylon, polyesters, wool and silk (both of which are proteins), can be dissolved accidentally by careless splashes of laboratory acids or alkalis.

Figure 28.10 Some uses of nylon

28.3 Further polymers and their properties

Kevlar

Kevlar owes its strength to the many regular hydrogen bonds between its chains. Because of the 1,4-disubstituted arrangement of the functional groups in the two aryl rings, the long chains of Kevlar are linear in shape. They can readily align side-by-side and form strong interchain hydrogen bonds (see Figure 28.8).

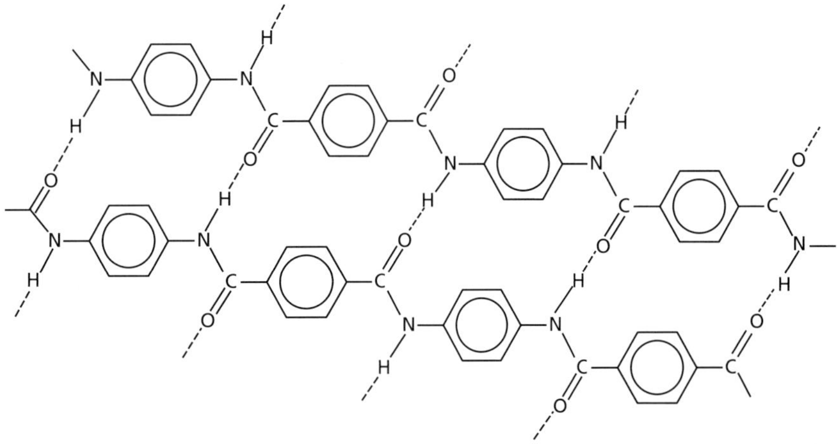

Figure 28.11 Hydrogen bonding between adjacent chains of Kevlar

The tensile strength of Kevlar is five times that of steel, on a weight-for-weight basis (its density is small: about $1.5\,g\,cm^{-3}$), yet it is quite flexible. It finds uses in tyres for cars and bicycles, bullet-proof vests and the 'skins' on modern drumheads.

Now try this

Nomex is a related polymer, made from 1,3-diaminobenzene and benzene-1,3-dicarbonyl chloride, which is used in fire-proof clothing.

a Draw the structure of part of a chain of Nomex, showing **two** repeat units, and suggest how the strength of the interchain hydrogen bonds might differ from that in Kevlar. Explain your reasoning.

b How might the flexibility of Nomex compare to that of Kevlar?

A dramatic use of Kevlar, together with glass-reinforced polyester, is illustrated by the footbridge at Aberfeldy in Scotland (see Figure 28.12). This was the first bridge to use composite materials and was built in 1992. It was much lighter than the possible alternatives made from steel or concrete, and was built for only a third of their cost.

Figure 28.12 The Aberfeldy footbridge is made from glass-reinforced polyester. Its cables are made from Kevlar

Cross-linked polymers

Earlier in this topic we saw how rubber was vulcanised through the cross-linking of the chains of the rubber molecules with sulfur atoms. This made the rubber harder, stronger, less flexible and less 'sticky'. The interchain hydrogen bonding in Kevlar has a similar effect in that Kevlar is more rigid and stronger than polyamides such as nylon 6,6 or nylon 6,10.

The interchain bonding in Kevlar produces two-dimensional sheets, which can bend and slide over each other. More rigidity can be produced if the polymer chains are cross-linked by covalent bonds into a three-dimensional lattice. One of the first polymers to have these properties was **bakelite**, developed in 1907 in New York by the Belgian chemist Leo Baekeland. Bakelite was the world's first synthetic plastic, made from phenol and methanol, and its manufacture marked the start of the plastics age. It is a **thermosetting** condensation polymer: as the monomers are heated together, the mixture becomes more and more viscous and eventually sets solid. Further heating does not melt the polymer, but causes it to char. Bakelite is quite brittle, but strong enough to be used in many applications where a hard, electrically insulating, heat-resistant material is required.

The main molecular requirement for cross-linking is for the polymer chain to contain side groups that can link to other chains (see Figures 28.13 and 28.14).

As can be seen in Figure 28.13, the acid-catalysed reaction between phenol and methanal produces intermediate **A**, which has three —CH$_2$OH groups on the phenolic benzene ring in the 2-, 4- and 6-positions (a similar orientation to that found during the bromination of phenol, see Topic 25). Each of these is able to undergo a condensation reaction with another phenol molecule (for simplicity, only one group is shown in the figure as reacting) to give intermediate **B**. The newly attached phenol ring can then react with more methanal, giving intermediate **C**, which can condense with another phenol, giving intermediate **D**. Reaction of **D** with two more methanal molecules produces intermediate **E**. Compound **E** contains five —CH$_2$OH groups, two of which are needed to continue the linear chain, leaving three available for condensation with three new phenol molecules (see Figure 28.14).

Figure 28.13 The building-up of the linear chains in bakelite

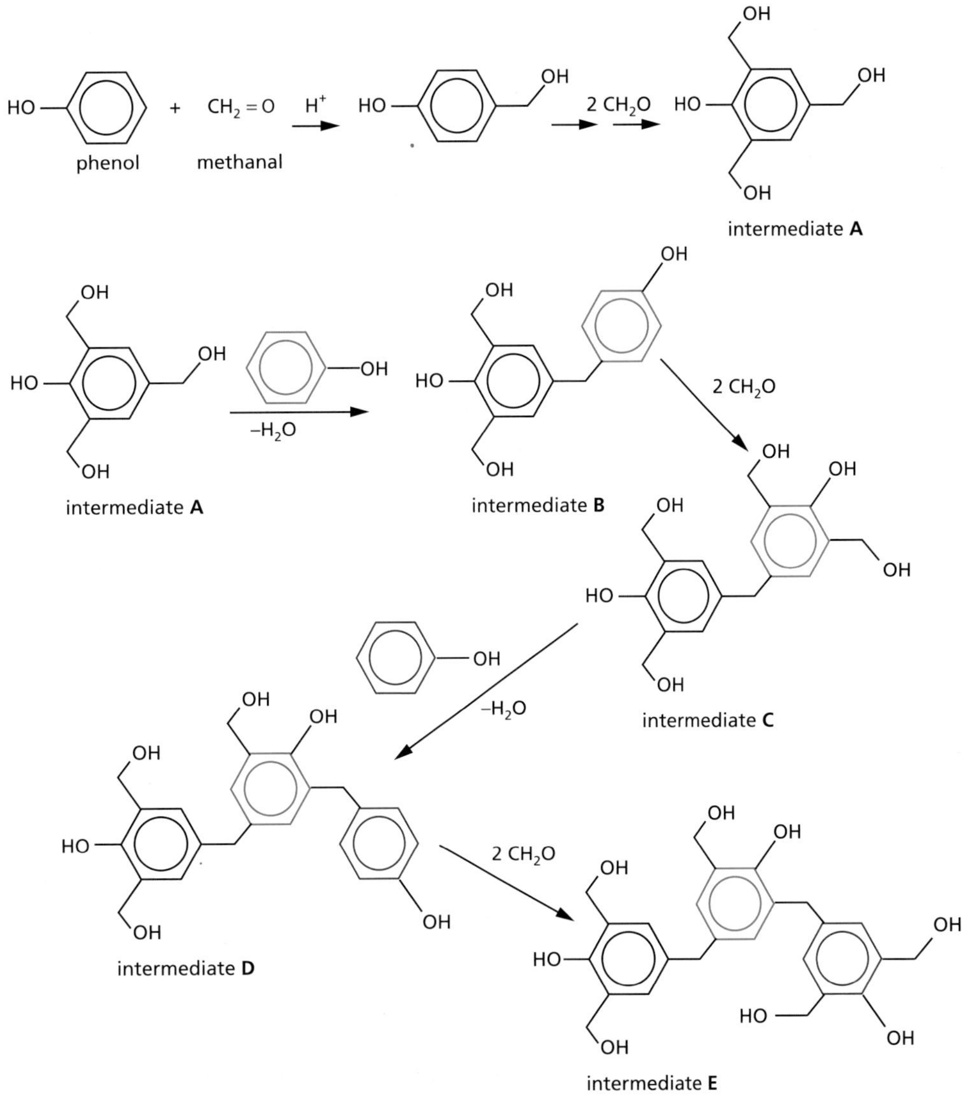

Figure 28.14 The formation of cross-links in bakelite

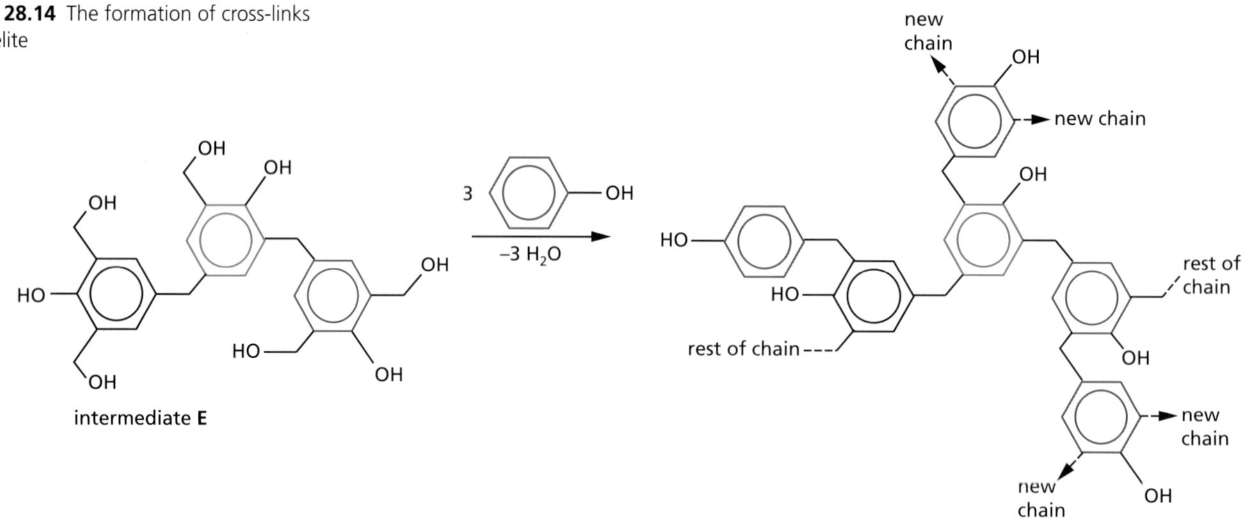

Each of these new phenolic benzene rings can now start two chains, independently of the original one. You can see how complicated matters can become! Although the structure of bakelite can become very involved, it is relatively easy to work out the ratio of phenol to methanal that is needed to make it.

Worked example

a Calculate the phenol:methanal ratio in bakelite.
b State the assumptions you made in your calculation.

Answer

a One way would be to write out the formula of a section of the bakelite molecule, and count the number of phenol rings and methanal —CH_2— groups. An easier way is to recognise that each phenol is joined to three —CH_2— groups derived from methanal, and each —CH_2— group is joined to two phenolic rings. So the phenol:methanal ratio is 2 : 3.

b The main assumption is that each phenol reacts with the maximum number of methanal molecules that it can (three). It is also assumed that the sample of bakelite is so large that the 'boundary effects', of phenolic rings at the edges of the polymer not being joined to other —CH_2— groups, are very small.

Cross-linking of polymers does not have to result in such an inflexible solid as bakelite, but cross-linking is often introduced to provide added strength to a flexible polymer. For example, if a small amount of a tri- or tetra-hydroxy alcohol (see Figure 28.15) is added to the monomer mix for a polyester, a stronger cross-linked polymer can form (see Figure 28.16).

Figure 28.15 Two polyhydroxy alcohols

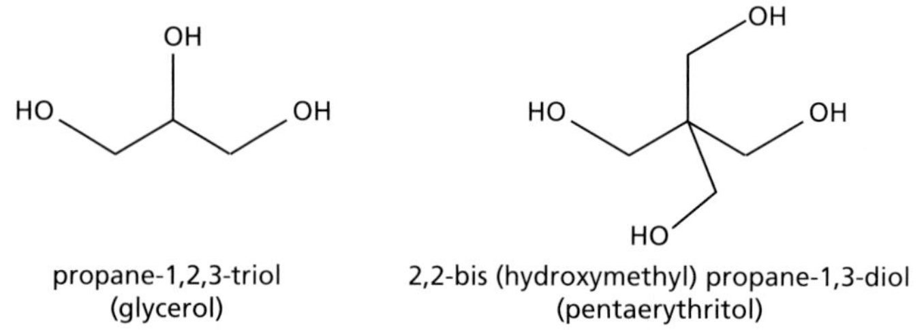

propane-1,2,3-triol
(glycerol)

2,2-bis (hydroxymethyl) propane-1,3-diol
(pentaerythritol)

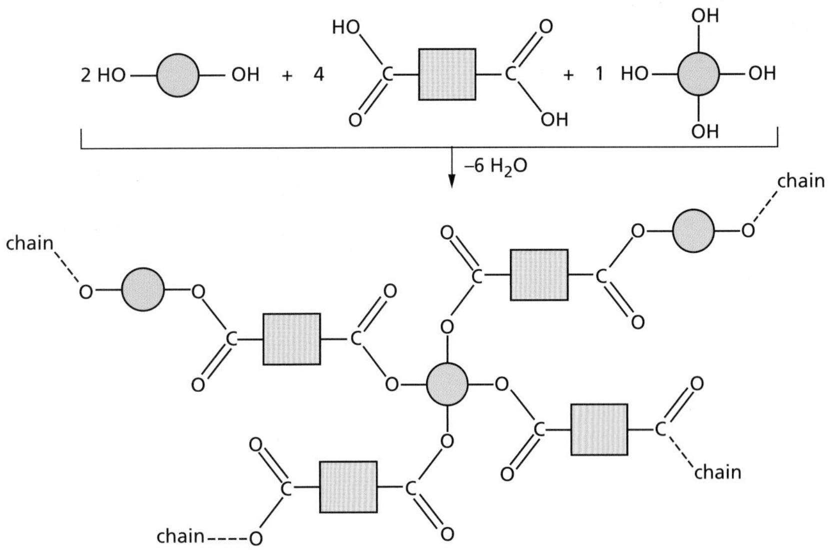

Figure 28.16 Cross-linking in condensation polymers

PTFE

The opposite effect is noticed with polytetrafluoroethene, PTFE (see Topic 14). Here, all of the hydrogen atoms in the polymer have been replaced by fluorine atoms. Fluorine is the most electronegative element in the Periodic Table, and strongly attracts the electrons in covalent bonds attached to it. This means that the electrons are held firmly and are not easily polarised. This results in very weak van der Waals' forces between PTFE molecules and each other and, more importantly, between PTFE molecules and other molecules. It is the intermolecular forces that cause molecules to stick together, and so PTFE is used as a non-stick surface on cookware, and also for the bearings at the ends of bridges that might expand and contract with changes in the environment's temperature.

'Ecopolymers'

Until recently, the raw materials for most polymers came from the products of the refining of crude oil, which is a finite, non-renewable and diminishing resource. Also, the disposal of polymeric materials after their useful life has caused problems of pollution, since many are non-biodegradable. Both of these problems have been partially overcome by an increase in the recycling of polymers: thermal cracking of polyethene can convert it into its monomer, ethene, which can then be reused; the polyethylene-terephthalate (PET) from plastic bottles can be converted into a wool-like material for fleeces and other clothing.

An alternative solution is to use renewable plant material to make polymers that are also biodegradable. The most commonly used is poly(2-hydroxypropanoic acid) (poly(lactic acid), PLA).

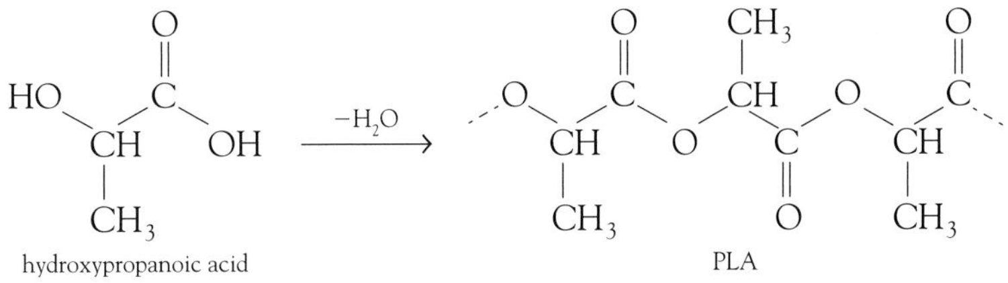

Hydroxypropanoic acid is a naturally occurring compound, which is made industrially by the fermentation of corn starch or sugar. PLA is easily hydrolysed, either chemically or by esterase enzymes, to re-form hydroxypropanoic acid, which is readily metabolised by all living matter into CO_2 and water.

Now try this

The repeat unit of another biodegradable polymer is as follows:

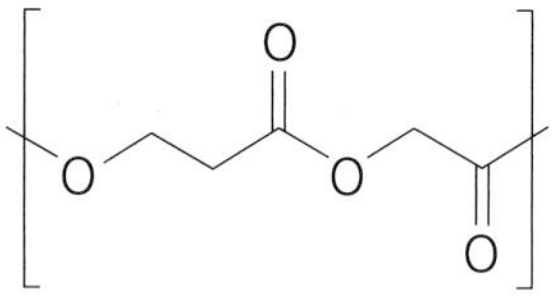

Draw the structures of the products of the hydrolysis of this monomer.

The uses of PLA include replacing conventional plastics in the manufacture of bottles, disposable cups and textiles, but it is also used in applications where its slow biodegradability is an advantage. For example, in medicine it has found uses as a component of self-dissolving stitches and for implants in the form of screws and pins to keep bones in place whilst they heal (see Figure 28.17).

Figure 28.17 A biodegradable screw made of PLA

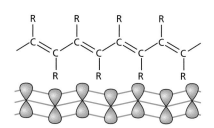

Figure 28.18 The structure of a 'molecular wire'

Conducting polymers

Graphite is a conductor of electricity because within each of its layers of carbon atoms there are delocalised electrons covering every atom within the layer. Thus, when a potential difference is applied at opposite ends of the layer, the delocalised electrons will flow from one end to the other.

Some long-chain polymer molecules act as 'molecular wires'. They contain **conjugated double bonds**, which means that the π bonds on adjacent alkene units overlap, and the π electrons become delocalised throughout the whole length of the polymer chain (see Figure 28.18).

Just as with the three-dimensional delocalised electrons in metals, or the two-dimensional delocalised electrons in the sheets of graphite or graphene, these one-dimensional delocalised electrons can conduct a current when a potential difference is applied to the ends of the polymer. An eventual aim is to construct micro-miniature circuits that can lead electrical signals to 'molecular microchips' in the heart of computing devices that will be thousands of times smaller than today's silicon-based units.

Three main classes of conducting polymers are the polyacetylenes, the polyphenylene vinylenes and the polythiophenes (see Figure 28.19).

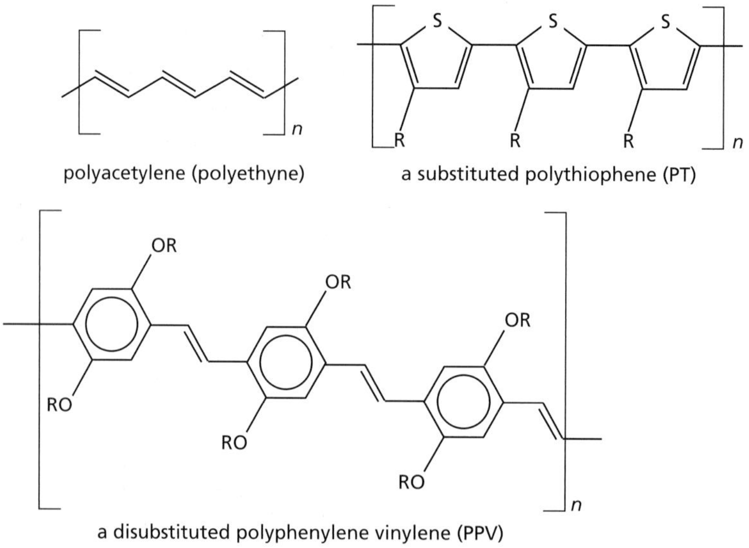

Figure 28.19 Three types of conducting polymer

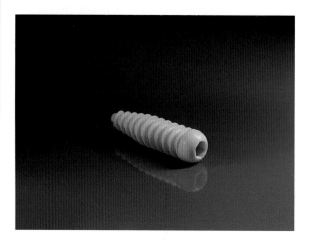

polyacetylene (polyethyne)   a substituted polythiophene (PT)

a disubstituted polyphenylene vinylene (PPV)

These polymers are not used as ordinary conductors – their conductivity is less than that of copper, for example – but as semiconductors, for use in organic polymer light-emitting diodes (OP-LEDs) and photovoltaic (solar) panels, for converting electrical energy into light, or light into electrical energy. For their use in OP-LEDs, by altering the nature of the side groups R and OR, the molecules can be made to electroluminesce (that is, emit light when an electrode potential is applied) in any part of the spectrum, from red, through yellow and green, to blue. An everyday use of OP-LEDs is for traffic lights (see Figure 28.20).

Figure 28.20 Red, yellow and green OP-LED used in traffic lights

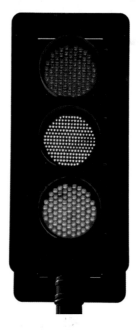

Polymer glues

Most glues achieve their stickiness through the van der Waals' forces that exist between their molecules and the molecules of the two surfaces they stick together. To be an effective and strong glue, the other property they need is for their own molecules to stick to each other strongly. If they are polymeric, consisting of long molecules, this last quality is fairly well assured. There are two important types of polymer glue; one is an addition polymer and the other is a condensation polymer. In each case, the setting of the glue is a result of its continued polymerisation 'in situ', in other words whilst it is in contact with the surfaces being stuck.

Addition polymer adhesives

A common addition polymer adhesive has already been mentioned in Table 28.1, poly(cyanoacrylate), or 'superglue'. Because of its two electron-withdrawing groups at one end of the ethene $C=C$, it has the unusual property of its polymerisation being catalysed by nucleophiles, such as water (this is why it is so dangerous if it is in contact with the skin or the eyes). The glue is applied as the low-M_r low-viscosity monomer, which spreads well into uneven surfaces. Once the two surfaces are brought together, the polymerisation of the monomer is catalysed either by nucleophilic groups on the surfaces (for example, —OH groups) or by moisture in the air (see Figure 28.21).

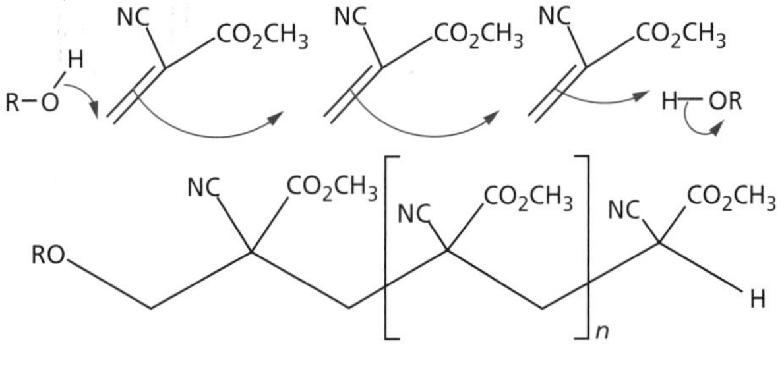

Figure 28.21 The polymerisation of methyl cyanoacrylate to form 'superglue'

Condensation polymer adhesives

A common group of condensation polymer adhesives are the epoxy resins. These are supplied as separate tubes of the two components, which are thoroughly mixed immediately prior to use. One component consists of a mixture of a medium-length polymer containing epoxy groups (compound **P** in Figure 28.22), and the other component (called the 'hardener') contains a diamine which opens up the epoxy rings by nucleophilic attack, and allows cross-linking to take place. The result is an inert thermosetting adhesive of tremendous strength. Figure 28.22 shows a simplified outline of this process.

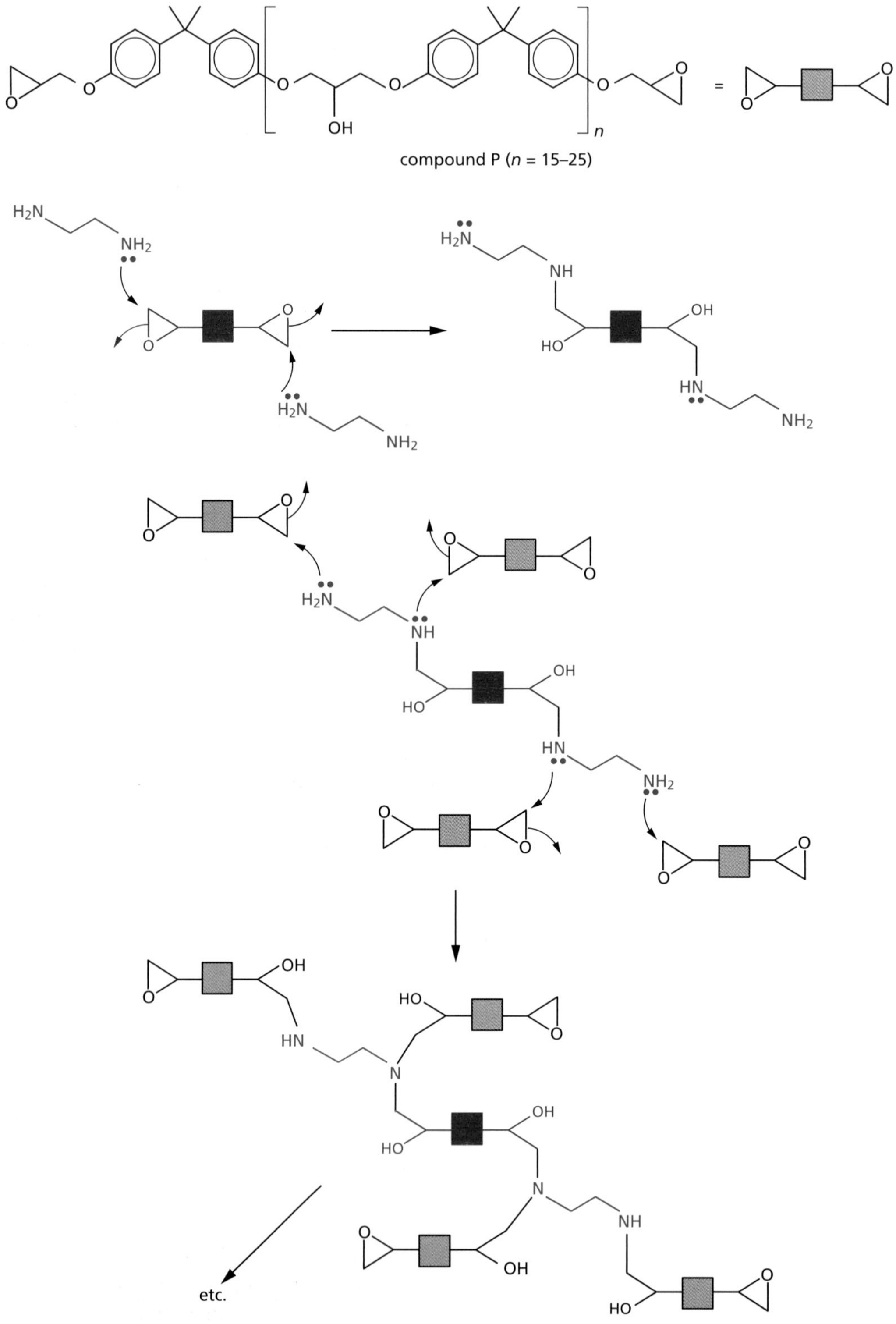

Figure 28.22 An outline of the cross-linking of epoxy compounds by amines to form a hard epoxy-resin adhesive

compound P (n = 15–25)

28.4 Proteins

There are three main classes of polymers that occur in nature: the proteins, the nucleic acids, and the polysaccharides such as starch and cellulose. In this section and the following section we shall look at some of the properties of the first two of these.

Figure 28.23 Some important protein-containing foods

Proteins make up about 16% of the human body. They not only occur as components of blood, skin, bone, nails, hair, nerves, muscles and tendons, but also are responsible for the smooth workings of the body in their roles as enzymes, receptors and hormones. Protein-containing foods, such as meat, beans, cheese, eggs and milk, are therefore important constituents in our diet (see Figure 28.23). Their importance to us is emphasised by their name, which is derived from the Greek *proteios*, meaning foremost or primary.

The structure of proteins

The backbone of a protein is one or more chains made up of polymers of amino acids. Some proteins also have non-amino-acid groups called **prosthetic groups** (for example, the haem group in haemoglobin). The shape and physical and biochemical properties of a protein are the direct result of the number, type and sequence of the amino acids that make it up. This amino acid sequence is determined by the **gene**, the sequence of organic bases in the length of DNA that codes for the protein. Amino acids do not join together to form proteins randomly, but join in a well-ordered and predetermined sequence.

The amino acids are joined by amide bonds between the amino group of one acid and the carboxyl group of another (see Figure 28.24). Proteins are therefore composed of polyamides, and are condensation polymers. As we saw in section 27.6, when an amide bond is formed between two amino acids, it is called a **peptide bond**, and simple proteins (those that contain just one chain, and no prosthetic groups) are also known as **polypeptides**.

Figure 28.24 Structure of a polypeptide

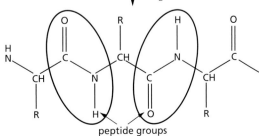

The formulae of peptides are often described by the following shorthand. The customary three-letter abbreviations for all the amino acids in the chain are written down, starting with the **N terminal** (the end of the chain that has a free —NH$_2$ group) on the left, and finishing with the **C terminal** (the end that has a free —CO$_2$H group) on the right.

Worked example

Convert the following structural formula into the correct shorthand:

$$NH_2—CH_2—CO—NH—CH(CH_3)—CO_2H$$

Answer

This structure contains one peptide bond in the middle, so it is a **dipeptide**, containing two amino acid residues. The residue to the left of the peptide bond is glycine (Gly) and that to the right is alanine (Ala) (see Figure 27.18). The shorthand formula is therefore:

$$Gly—Ala$$

Note that this is a different compound from Ala—Gly, because of the left-to-right direction of the peptide bond. The structural formula of Ala—Gly is:

$$NH_2—CH(CH_3)—CO—NH—CH_2—CO_2H$$

Now try this

1 Draw the structural formulae of the following two tripeptides (use Figure 27.18 to help you):

   a Val—Ser—Tyr

   b Tyr—Ala—Gly

2 Use the customary three-letter abbreviations to describe the following tetrapeptide:

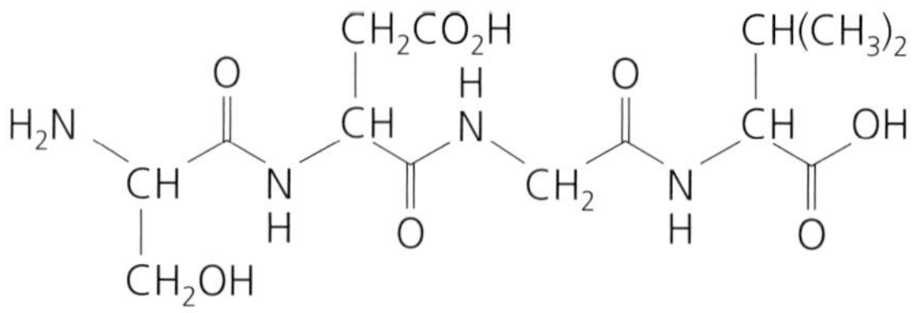

The amide (peptide) bonds in proteins can be hydrolysed by the same reagents as hydrolyse simple amides: hot aqueous acid or hot aqueous alkali. They can also be hydrolysed by a variety of proteolytic enzymes. Many of these (for example, trypsin and pepsin) are found in the digestive system, where they help to break down the proteins in our food into smaller peptides and amino acids, which can then readily be absorbed through the intestinal wall.

   Biochemists recognise various levels of protein structure.

Primary structure

The primary structure of proteins is the sequence of amino acids, covalently joined by peptide bonds, making up the polypeptide chain, as described above. This sequence is determined by the arrangement of organic bases in the section of DNA that codes for the protein.

Secondary structure

The amino acid chain that constitutes a polypeptide is quite flexible, with two-thirds of the bonds within the chain being single bonds, around which free rotation is allowed. The other third of the bonds in the chain are the C—N bonds within peptide groups. Because of the overlap between the lone pair of electrons on the nitrogen atom (in a p orbital) and the π bond of the carbonyl group next to it, the peptide group is rigid and planar. It exists in the *trans* arrangement (see Figure 28.25).

Figure 28.25 Overlap of p orbitals makes
the peptide group planar

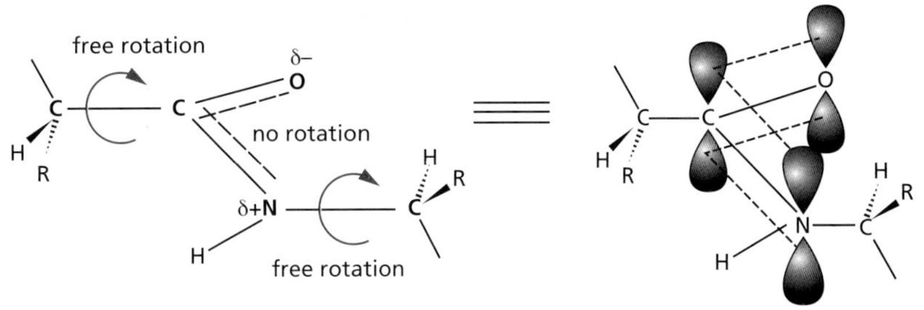

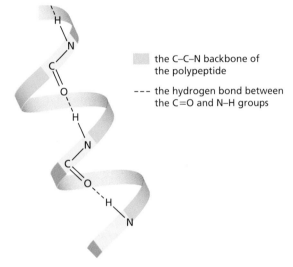

This inflexibility of the peptide group allows parts of the polypeptide chain to take
up an **α-helical arrangement** (see Figure 28.26). This lends strength and rigidity to
some protein structures. Strong hydrogen bonds can form between the N—H of one
peptide group and the C=O of another one three/four amino acids along the chain.

Figure 28.26 The α helix in proteins

the C–C–N backbone of
the polypeptide

--- the hydrogen bond between
the C=O and N–H groups

The α-helix confers **one-dimensional** rigidity to the polypeptide chain. Another type
of secondary structure is the **β-pleated sheet**. This is formed by the C=O group
of one chain hydrogen bonding to the N—H group in the same chain, or another
chain, but separated by many amino acid residues. The resulting sheet of anti-parallel
hydrogen-bonded chains confers a degree of **two-dimensional** rigidity to the
polypeptide chain. The chains are described as *anti*-parallel because adjacent chains
run in opposite directions, as shown in Figure 28.27.

Figure 28.27 The β-pleated sheet in proteins
(The tetrahedral arrangement of bonds around
the sp³ carbon in the chain means that the R
groups stick out above or below the planar
sheet.)

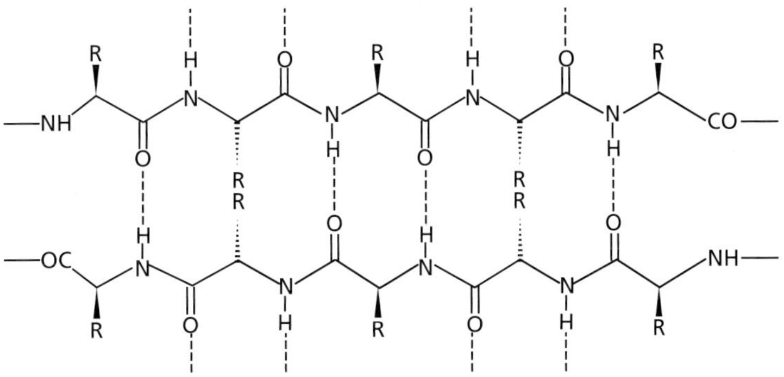

Tertiary structure

Even if a particular polypeptide chain contains regions of α-helices or β-pleated
sheets, there will be many amino acids not involved in these regions. Other
interactions can occur between the side groups of these amino acids, which cause the
chain to twist and fold.

We can recognise four different interactions between side chains, which are
associated with their different structures.

• Van der Waals' attractions occur between nonpolar side chains, for example, glycine, alanine, valine and phenylalanine.

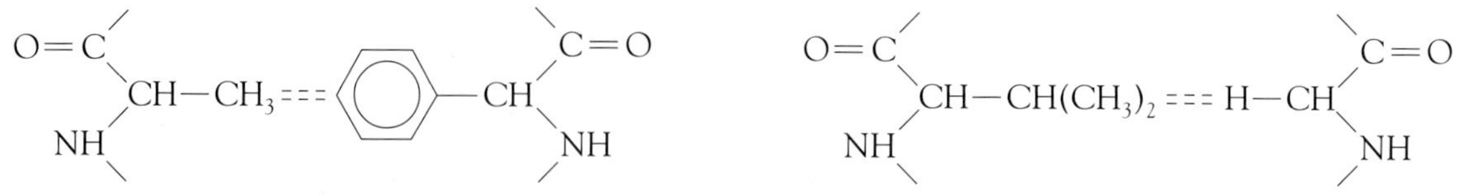

Ala   Phe   Val   Gly

• Hydrogen bonding occurs between side chains that contain —OH (for example, serine and tyrosine), $-NH_2/-NH_3^+$ (for example, lysine) and $-CO_2H/-CO_2^-$ (for example, aspartic acid and glutamic acid).

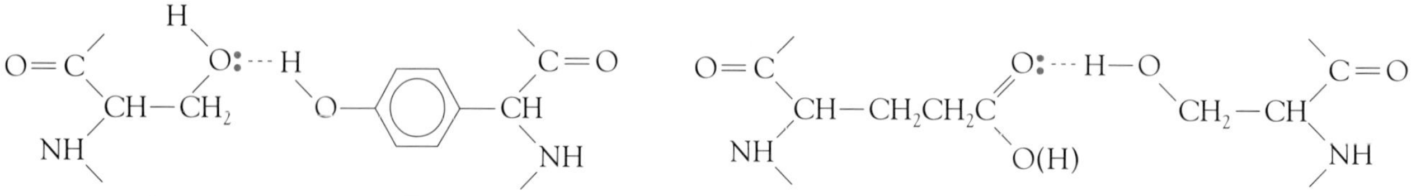

Ser   Tyr   Glu   Ser

• Ionic attractions occur between side chains containing functional groups that exist in their anionic form at pH7 (for example, aspartic acid and glutamic acid) and those that exist in their cationic form (for example, lysine and arginine).

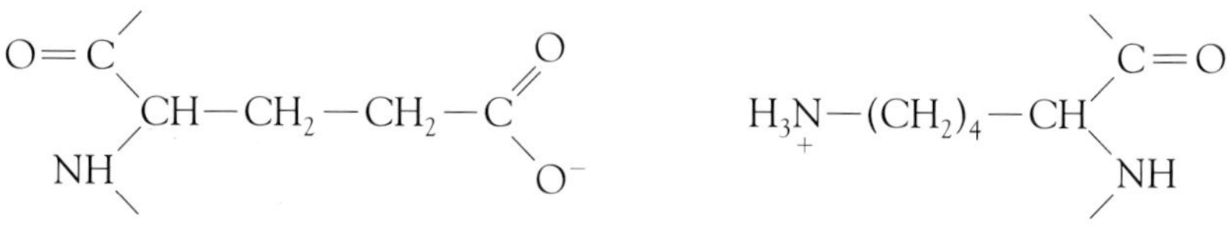

Glu   Lys

• Disulfide bridges can form between the side chains of two cysteine residues.

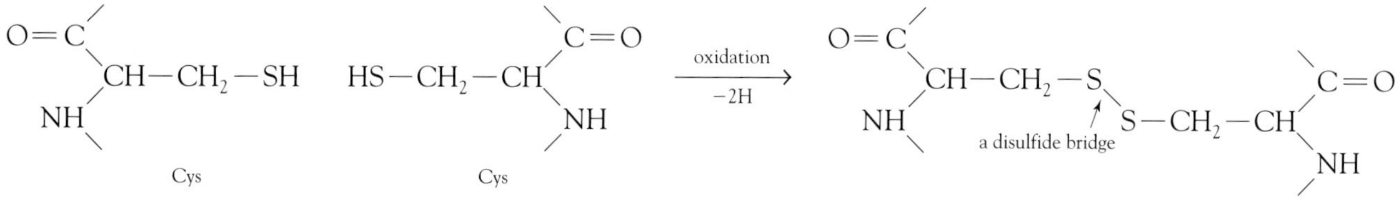

Cys   Cys

The **tertiary structure** confers some **three-dimensional** rigidity to the polypeptide, and is important in determining both the overall shape and the function of the protein. For example, many enzymes are globular (fairly spherical) proteins that are soluble in water. In these, the non-polar side chains are often pointing towards the central hydrophobic part of the molecule, whereas the hydrogen-bonding and ionic groups are on the outside, where they can be solvated by water molecules.

The strongest of the tertiary interactions is the disulfide bridge. Once these are formed, by an oxidation reaction, the various parts of the chain are drawn together, and this allows the other interactions to play their part in creating the overall three-dimensional structure.

28.5 DNA

The genes of all organisms are composed of chains of deoxyribonucleic acid, DNA. It stores the information needed to make specific proteins in the cell.

Like proteins, DNA consists of a primary structure, which comprises the covalent bonds joining the monomers together, and a secondary structure, which is formed by hydrogen bonding between the chains.

The monomer units of DNA are called nucleoside phosphates. These consist of the sugar molecule deoxyribose, which has a cyclic organic base bonded to carbon atom 1, and a phosphate group bonded to carbon atom 5. All DNA molecules contain just four different bases. In a particular DNA molecule, these are arranged in a specific order along the chains.

Figure 28.28 shows the structures of the four bases and Figure 28.29 shows how adenine forms the monomer unit adenosine phosphate.

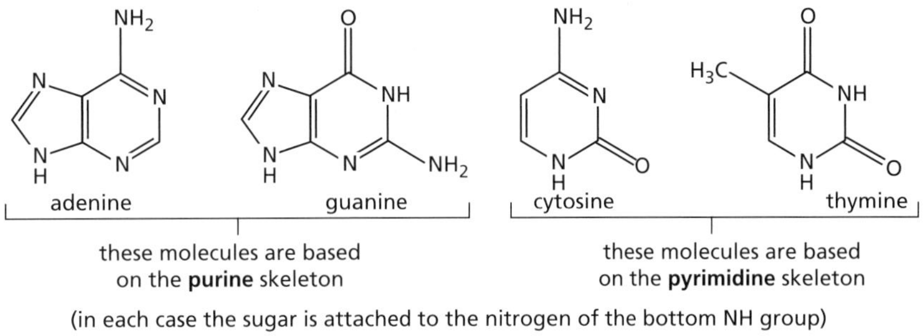
adenine
guanine
cytosine
thymine
these molecules are based on the **purine** skeleton
these molecules are based on the **pyrimidine** skeleton
(in each case the sugar is attached to the nitrogen of the bottom NH group)

Figure 28.28 The four bases of DNA

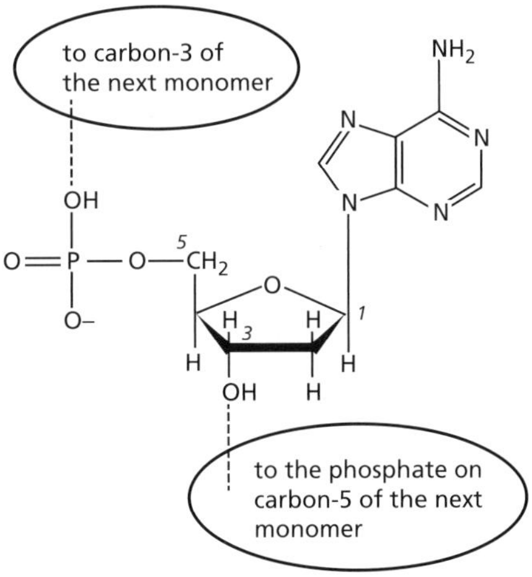
to carbon-3 of the next monomer
to the phosphate on carbon-5 of the next monomer

Figure 28.29 The monomer unit adenosine phosphate

Many thousands of monomer units join up in a specific sequence of bases to form a linear condensation polymer.

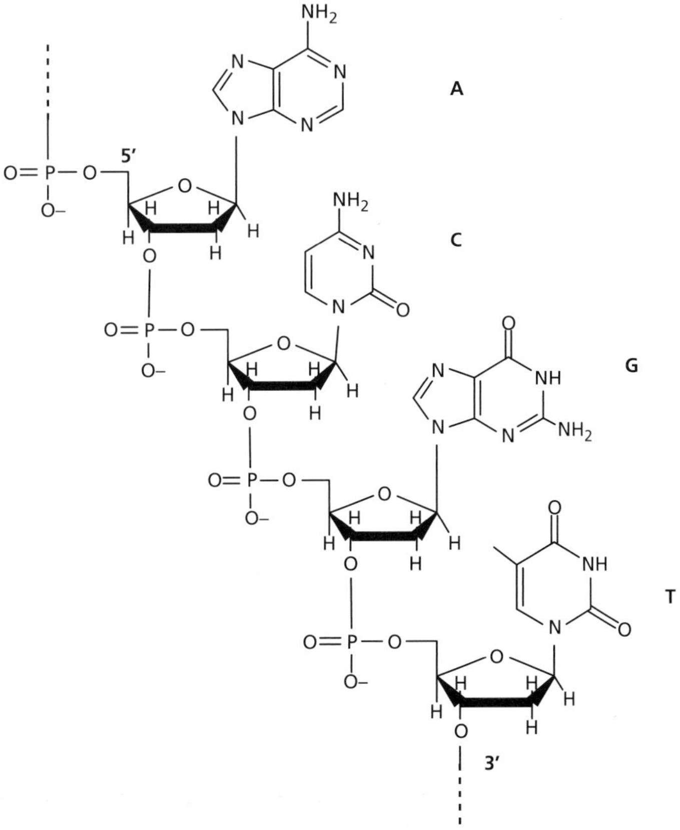
A
C
G
T
5′
3′

Figure 28.30 Part of a single strand of DNA showing the four bases A, C, G and T

Just as with proteins, DNA has a hydrogen-bonded secondary structure. A DNA molecule consists of a double helix of two sugar-phosphate strands coiled around each other, with the purine and pyrimidine bases (both of which are planar molecules) arranged in hydrogen-bonded pairs, one from each chain, stacked above each other on the inside of the double helix.

The pairing of the bases is unique: adenine (A) always pairs with thymine (T), and guanine (G) always pairs with cytosine (C).

As can be seen from their molecular structures, a purine molecule, with its two rings of atoms, is larger than a single-ring pyrimidine molecule. To allow the distance between the sugar-phosphate chains to be constant, the base pairs must consist of one purine and one pyrimidine (see Figure 28.28): there is not enough space between the chains to accommodate two purines, and two pyrimidines would not be close enough to hydrogen-bond with each other.

But why should adenine pair with thymine and not with cytosine, or guanine pair with cytosine and not with thymine? The answer lies in the number of hydrogen bonds that can form between the bases, and the directional sense of those hydrogen bonds. Figure 28.31 shows the hydrogen bonding that occurs.

Figure 28.31 The hydrogen bonded base pairs T—A and C—G (R represents the deoxyribose ring to which the base is joined.)

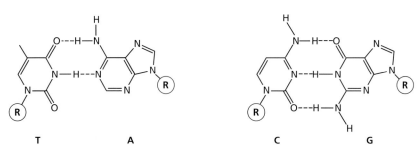

The overall structure is shown in Figure 28.32.

28.6 Distinguishing between addition and condensation polymers

We can recognise which type of polymer will be formed from given monomers by looking at their structures.

- If each of the monomers contains a C=C group, they are likely to produce an addition polymer (even if they also contain acid, ester or amide groups).
- If the monomers contain two different functional groups (carboxylic acid or acyl chloride, amine or alcohol) at the end of a chain of carbon atoms, they are likely to produce a condensation polymer.

We can recognise what type of polymerisation has produced a polymer by looking at a section of the backbone chain.

- If the *chain* contains only carbon atoms, the polymer is an addition polymer (even if any side chains contain ester or amide groups).
- If the chain contains hetero atoms (that is atoms that are not carbon – usually oxygen or nitrogen), the polymer is a condensation polymer.

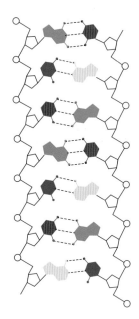

Figure 28.32 A section of the double-stranded DNA molecule

Worked example 1

Predict the type of polymerisation that each of the following pairs of molecules will undergo, and draw the structure of the repeat unit.

a CH_2=CH—$CONH_2$ and CH_3OCH=CH_2
b $NH_2CH_2CH_2NH_2$ and $ClCOCH_2CH(CH_3)COCl$

Answer

a These two monomers will form an addition polymer, because each monomer contains a C=C group. The repeat unit is either

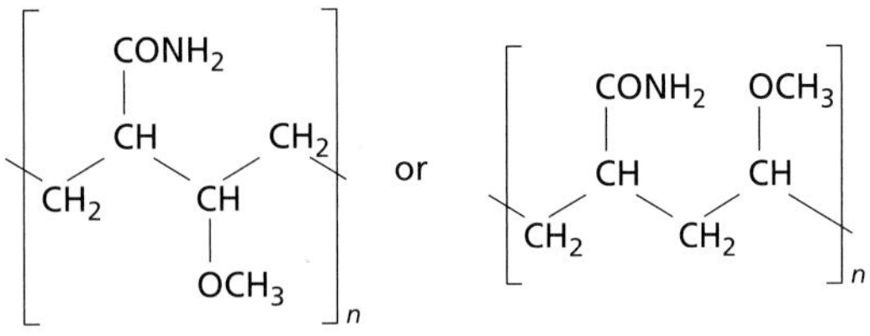

or

Which one is formed depends on the orientation of the monomers.

b These two monomers will form a condensation polymer, because each monomer contains two functional groups. The repeat unit is:

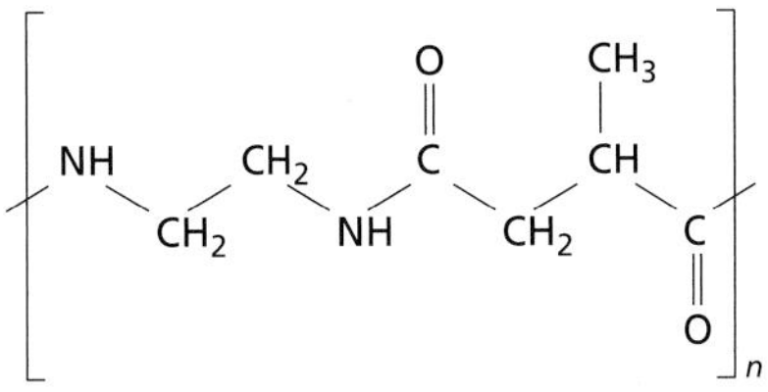

Identifying the monomers that have produced a condensation polymer is relatively straightforward: the chain is broken at the ester or amide bonds, an H atom is added to the —O or —NH groups so formed, and an —OH is added to the —CO groups.

Worked example 2

Draw the structures of the monomers used to make the following section of a co-polymer:

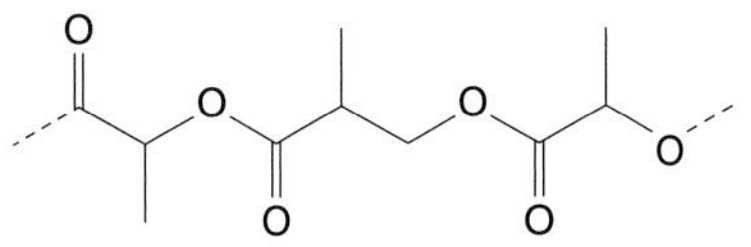

Answer

Splitting the co-polymer section at the ester groups produces three hydroxyacid molecules, of which two are identical, so the monomers are 2-hydroxypropanoic acid and 3-hydroxy-2-methylpropanoic acid:

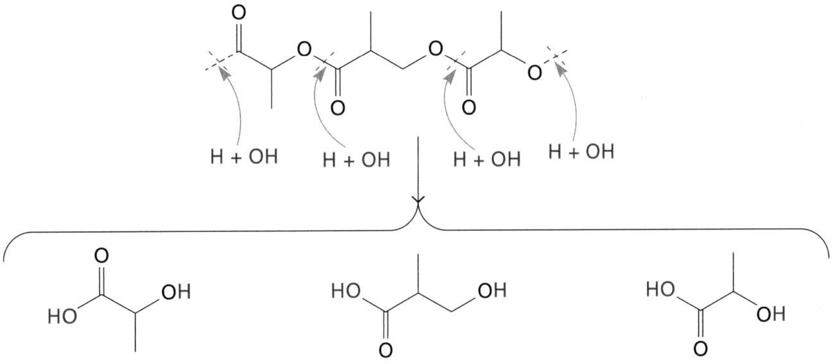

2-hydroxypropanoic acid   3-hydroxy-2-methylpropanoic acid   2-hydroxypropanoic acid

Identifying the monomers that have produced an addition polymer can be a little more difficult. If there are no C=C bonds in the chain, the carbon chain is split into two-carbon sections. Each of these will have been formed from a substituted ethene as monomer. If, however, the carbon chain contains C=C bonds, these are likely to have come from butadiene-like monomers, so the chain is split into a four-carbon unit at this point, with the C=C bond in the middle of the four-carbon unit.

Worked example 3

Draw the structures of the monomers used to make the following section of a co-polymer:

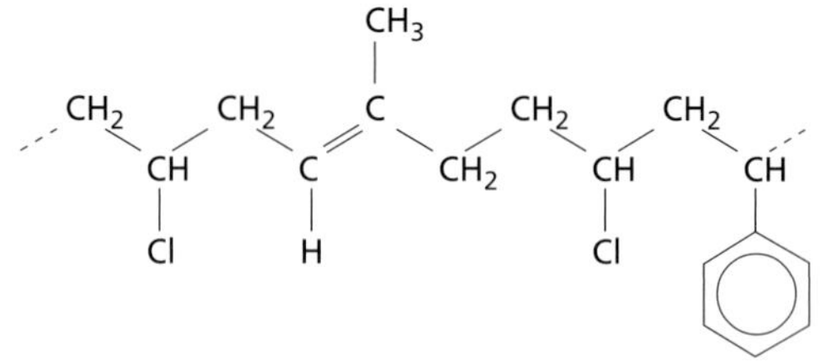

Answer

The polymer contains a double bond, so it is split into a four-carbon unit at this point. All single-bonded lengths are split into two-carbon units:

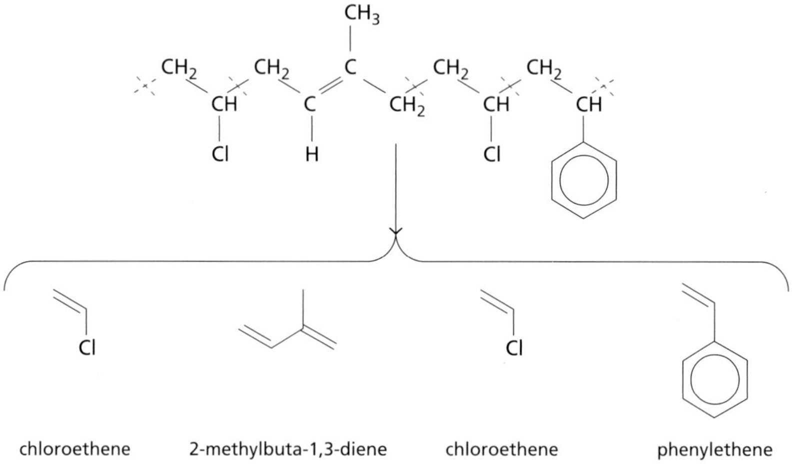

| chloroethene | 2-methylbuta-1,3-diene | chloroethene | phenylethene |

Thus the monomers are chloroethene, phenylethene and 2-methylbuta-1,3-diene.

Now try this

Identify the monomers used to make the following sections of polymer molecules.

1

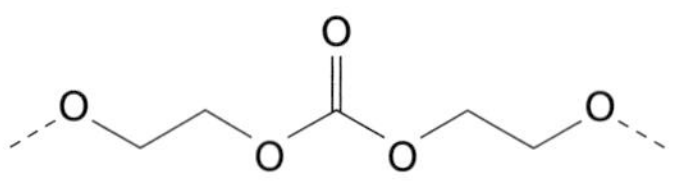

2

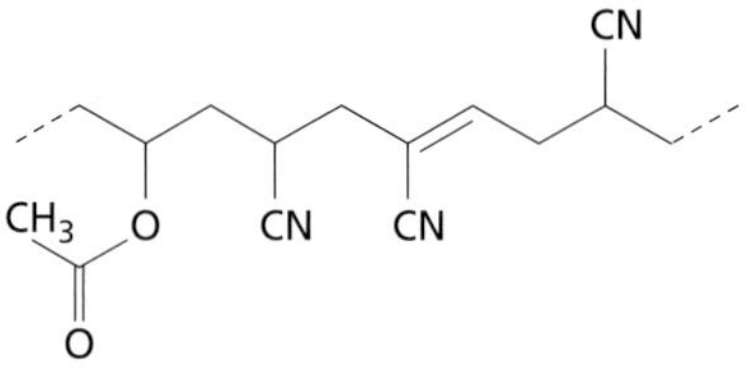

28.7 The disposal of polymers

We saw in section 14.4 how polyalkenes can be disposed of, by incineration, recycling, depolymerisation or bacterial fermentation. Some polymers can also be broken down by ultraviolet rays or strong sunlight. Because of the radical-stabilising effect of alkyl groups, the bond between a hydrogen atom and a tertiary carbon atom is weaker than the usual C—H bond (bond energy $420\,\text{kJ mol}^{-1}$).

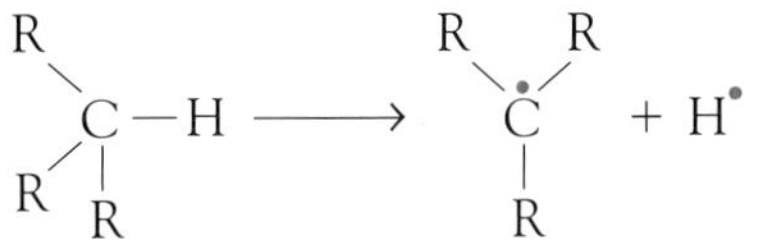

$$\underset{\substack{R \\ R}}{R}\!\!-\!C\!-\!H \longrightarrow \underset{\substack{| \\ R}}{R}\!\!\overset{\bullet}{C} + H^{\bullet} \quad \Delta H = 404\,\text{kJ mol}^{-1}$$

Oxygen molecules (which are di-radicals) can readily react with tertiary C—H groups.

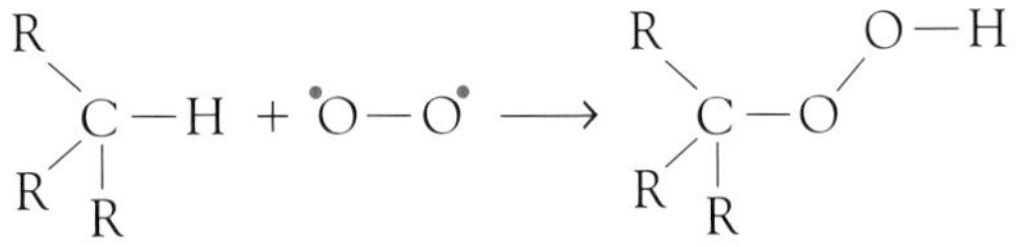

$$\underset{\substack{R \\ R}}{R}\!\!-\!C\!-\!H + {}^{\bullet}O\!-\!O^{\bullet} \longrightarrow \underset{\substack{| \\ R}}{R}\!\!\overset{O-H}{C}\!\!-\!O \qquad \Delta H = -90\,\text{kJ mol}^{-1}$$

The hydroperoxides formed by this reaction can undergo further bond breaking and molecular rearrangements, resulting in smaller molecules that can be more easily metabolised by bacteria in the environment.

The main way that condensation polymers are broken down in the environment is by the hydrolysis of their ester or amide bonds.

$$-R\!-\!CO\!-\!OR'\!- + H_2O \rightarrow -R\!-\!CO_2H + HO\!-\!R'\!-$$

$$-R\!-\!CO\!-\!NHR'\!- + H_2O \rightarrow -R\!-\!CO_2H + H_2N\!-\!R'\!-$$

Esterase and peptidase enzymes in soil bacteria can carry out these transformations.

Summary

- Many different **addition polymers** can be made by polymerising ethene molecules in which one or more hydrogen atoms have been replaced by other atoms or groups.
- **Co-polymers** are polymers made from more than one type of monomer.
- **Condensation polymers** are usually either polyesters or polyamides.
- Condensation polymers can be divided into two categories – **type I** and **type II** – depending on the relative orientation of adjacent functional groups within the chain.
- The properties and uses of various polymers depend on their chemical structures, and the bonding between the chains.
- Proteins and nucleic acids are two important natural examples of condensation polymers.
- Addition and condensation polymers can be distinguished, and the structures of their monomers determined, by studying the make-up of the backbone chain.

Key reactions you should know
- Addition polymerisation:

  $n\text{CH}_2\!=\!\text{CH}\!-\!\text{R} + n\text{CH}_2\!=\!\text{CH}\!-\!\text{R}' \rightarrow [-\text{CH}_2\!-\!\text{CHR}\!-\!\text{CH}_2\!-\!\text{CHR}'\!-]n$

  (The proportions of $\text{CH}_2\!=\!\text{CHR}$ and $\text{CH}_2\!=\!\text{CHR}'$ can be varied.)
- Condensation polymerisation to give polyesters:

  $n\text{HO}-\blacksquare-\text{OH} + n\text{HO}_2\text{C}-\bullet-\text{CO}_2\text{H} \rightarrow \{\!-\text{O}-\blacksquare-\text{OCO}-\bullet-\text{CO}\!-\}_n$
- Condensation polymerisation to give polyamides:

  $n\text{H}_2\text{N}-\bullet-\text{NH}_2 + n\text{H}_2\text{OC}-\blacksquare-\text{CO}_2\text{H} \rightarrow \{\!-\text{NH}-\bullet-\text{NHCO}-\blacksquare-\text{CO}\!-\}_n$

Examination practice questions

Please see the data section of the CD for any A_r values you may need.

1 Proteins exist in an enormous variety of sizes and structures in living organisms. They have a wide range of functions which are dependent upon their structures. The structure and properties of an individual protein are a result of the primary structure – the sequence of amino acids that form the protein.

a Proteins are described as condensation polymers.
  i Write a balanced equation for the condensation reaction between two glycine molecules, $H_2NCH_2CO_2H$.
  ii Draw the skeletal formula for the organic product. [2]

b X-ray analysis has shown that in many proteins there are regions with a regular arrangement within the polypeptide chain. This is called the secondary structure and exists in two main forms.
  i State the two forms of secondary structure found in proteins.
  ii Draw a diagram to illustrate **one** form of secondary structure. [4]

c There are around 20 different common amino acids found in humans most of which have the same general structure.

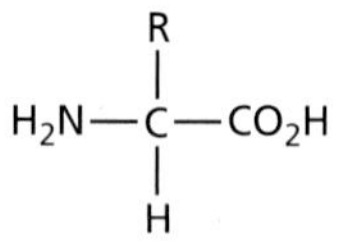

$$H_2N-\underset{\underset{H}{|}}{\overset{\overset{R}{|}}{C}}-CO_2H$$

The nature of the group R affects which bonds are formed as the secondary structure of the protein is further folded to give the tertiary structure.

Copy and complete the table indicating the type of **tertiary** bonding that each pair of the amino acid residues is likely to produce.

residue 1	residue 2	type of tertiary bonding
–HNCH(CH$_2$CH$_2$CH$_2$CH$_2$NH$_2$)CO–	–HNCH(CH$_2$CH$_2$CO$_2$H)CO–	
–HNCH(CH$_3$)CO–	–HNCH(CH$_3$)CO–	
–HNCH(CH$_2$SH)CO–	–HNCH(CH$_2$SH)CO–	
–HNCH(CH$_2$OH)CO–	–HNCH(CH$_2$CO$_2$H)CO–	

[4]

[Cambridge International AS & A Level Chemistry 9701, Paper 41 Q6 November 2011]

2 In recent years there has been considerable interest in a range of polymers known as 'hydrogels'. These polymers are hydrophilic and can absorb large quantities of water.

a The diagram shows part of the structure of a hydrogel.

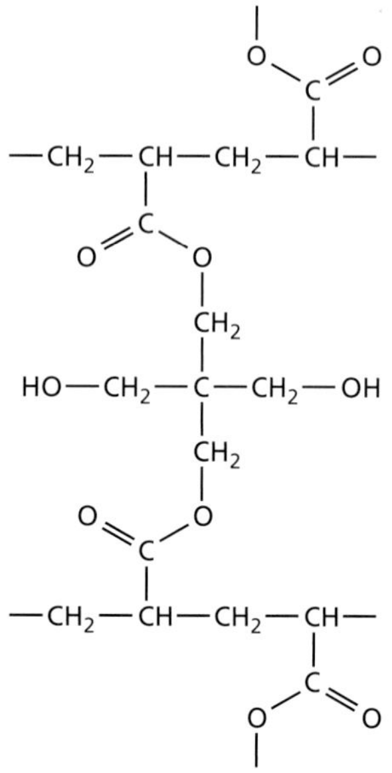

The hydrogel is formed from chains of one polymer which are cross-linked using another molecule.
  i Draw the structure of the monomer used in the polymer chains.
  ii State the type of polymerisation used to form these chains.
  iii Draw the structure of the molecule used to cross-link the polymer chains.
  iv During the cross-linking, a small molecule is formed as a by-product. Identify this molecule. [5]

b Once a hydrogel has absorbed water, it can be dried and re-used many times. Explain why this is possible, referring to the structure above. [2]

c Not every available side chain in the polymer is cross-linked, and the amount of cross-linking affects the properties of the hydrogel.
  i The amount of cross-linking has little effect on the ability of the gel to absorb water. Suggest why this is the case.
  ii Suggest **one** property of the hydrogel that will change if more cross-linking takes place. Explain how the increased cross-linking brings about this change. [3]

[Cambridge International AS & A Level Chemistry 9701, Paper 42 Q8 June 2013]

29 Techniques of analysis

In this topic we look at various methods used to separate mixtures of compounds and identify their components. We also explain the principles and applications of three major analytical techniques used to investigate the structures of molecules: mass spectrometry, infrared spectroscopy and nuclear magnetic resonance spectroscopy.

Learning outcomes

By the end of this topic you should be able to:

22.1a) explain and use the terms R_f value in thin layer chromatography and retention time in gas–liquid chromatography from chromatograms

22.1b) interpret gas–liquid chromatograms in terms of the percentage composition of a mixture

22.2a) analyse an infrared spectrum of a simple molecule to identify functional groups

22.3a) deduce the molecular mass of an organic molecule from the molecular ion peak in a mass spectrum

22.3b) deduce the number of carbon atoms in a compound using the M+1 peak

22.3c) deduce the presence of bromine and chlorine atoms in a compound using the M+2 peak

22.3d) suggest the identity of molecules formed by simple fragmentation in a given mass spectrum

22.4a) analyse a carbon-13 NMR spectrum of a simple molecule to deduce the different environments of the carbon atoms present, and the possible structures for the molecule

22.4b) predict the number of peaks in a carbon-13 NMR spectrum for a given molecule

22.5a) analyse and interpret a proton NMR spectrum of a simple molecule to deduce the different types of proton present using chemical shift values, the relative numbers of each type of proton present from relative peak areas, the number of non-equivalent protons adjacent to a given proton from the splitting pattern, using the $n + 1$ rule, and the possible structures for the molecule

22.5b) predict the chemical shifts and splitting patterns of the protons in a given molecule

22.5c) describe the use of tetramethylsilane, TMS, as the standard for chemical shift measurements

22.5d) state the need for deuterated solvents, e.g. $CDCl_3$, when obtaining an NMR spectrum

22.5e) describe the identification of O—H and N—H protons by proton exchange using D_2O.

29.1 Chromatography

The name 'chromatography', coming from two Greek words meaning 'colour image', was invented by the Russian scientist Mikhail Tsvet. He developed the technique around the start of the twentieth century to separate the carotenes, chlorophyll and xanthophylls that are the coloured pigments in plants. Today, however, most chromatography is carried out to separate colourless compounds, which can then be identified by a variety of methods.

We shall look at three types of chromatography:

- paper chromatography (PC)
- thin layer chromatography (TLC)
- gas chromatography (GC), otherwise known as gas–liquid chromatography (GLC).

Each type of chromatography relies on the same overall principle for separating the components of a mixture: the compounds are distributed between a *mobile* (or *moving*) *phase* (a liquid or a gas) and a *stationary phase*. The stationary phase may be a solid which adsorbs the mixture solutes, or a thin film of liquid on the surface of an inert solid.

Paper chromatography (PC)

Although it may appear dry to the touch, chromatography paper (which is like smooth filter paper but with an accurate and constant thickness) contains water molecules hydrogen-bonded to the OH groups on its cellulose molecules. This layer of water molecules is the stationary phase. The moving phase is chosen to be less polar than water. It is usually an organic solvent, or a mixture of solvents: ethanol or an ethanol–water mixture is often used.

Figure 29.1 Separation by partition

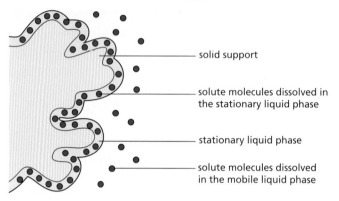

A spot of the solution of the mixture to be analysed is placed about 1 cm from the edge of a rectangular sheet of chromatography paper (often about 20 cm × 20 cm), along with other spots of 'reference' compounds at the same distance from the edge. After the mixture solvent has evaporated, the edge of the sheet is placed in a chromatography 'tank' and immersed in the moving phase solvent, making sure the spots are above the surface of the solvent. The solvent is drawn up the sheet by capillary action. As it passes the point where the spots have been applied, the compounds in the mixture are partitioned between the stationary water layer and the moving solvent, depending on their polarity (Figure 29.1). Although the process is continually changing and never reaches equilibrium because fresh solvent is constantly sweeping up the paper, how quickly the spots move up the paper is determined by the values of their partition coefficients between the solvent and water. The more polar a compound is, the smaller is its partition coefficient. This means it will spend more time dissolved in the water layer on the cellulose than in the moving solvent, and it will progress more slowly up the paper.

When the solvent has almost reached the end of the sheet of paper, the paper is taken out of the tank and allowed to dry. As the solvent evaporates, the solutes will stay at the places they had reached on the paper. If they are not already coloured, they can be made visible by spraying the paper with a specific developing agent, which reacts with the compounds in the spots to form a coloured product. Specific agents have been developed for particular classes of compound: ninhydrin for amino

Figure 29.2 A paper chromatogram of an amino acid mixture with five 'references' spotted alongside, before and after development with ninhydrin

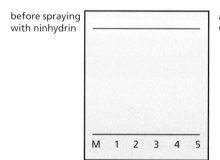

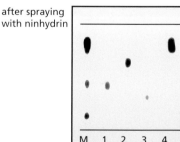

acids (Figure 29.2); Tollens' or Fehling's reagents for reducing sugars. A general visualising agent for most organic compounds is iodine: when the paper is placed in a tank containing iodine vapour, the iodine is absorbed preferentially by the organic compounds in the spots, turning them brown.

The usual way of identifying the compounds that make up the various spots on a chromatography sheet is to measure their **retardation factor** R_f values (Figure 29.3). These are compared with the R_f values of known 'reference' compounds, spots of which were applied to the sheet at the same time as the mixture.

Figure 29.3 The retardation factor, R_f

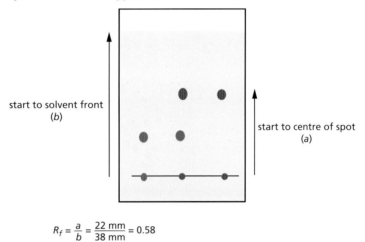

$$R_f = \frac{a}{b} = \frac{22 \text{ mm}}{38 \text{ mm}} = 0.58$$

The retardation factor, R_f, is defined by the equation:

$$R_f = \frac{a}{b} = \frac{\text{distance moved by solute}}{\text{distance moved by solvent}}$$

Each compound has a characteristic R_f value for a given solvent, but occasionally different compounds can have very similar R_f values in a particular solvent, and so it is not easy to separate them. If a new solvent – one with a different polarity – is used, the compounds are likely to have different R_f values to each other, so they can now be separated.

This is applied in **two-dimensional chromatography**. This technique uses two moving phase solvents of different polarities. In this technique a spot of the mixture of compounds is placed at a corner of a square sheet of chromatography paper. The sheet is placed in a tank containing the first solvent, and is left until the solvent front reaches the far edge of the sheet. The sheet is removed from the tank, and allowed to dry thoroughly. It is then turned through 90° and placed in a different tank containing the second solvent, so that the spots that have been partially separated by the first solvent lie along the bottom of the sheet, just above the level of the second solvent in the tank. When the second solvent front has reached the end of the sheet, it is removed, dried and sprayed with the developing agent (Figure 29.4). The R_f values in each solvent can be measured, and compared with those of reference compounds. This type of chromatography has been used to identify the amino acids obtained from the hydrolysis of proteins.

Figure 29.4 Two-dimensional chromatography

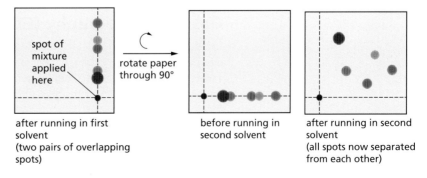

Thin layer chromatography (TLC)

The technique of thin layer chromatography is similar to that of paper chromatography, but the theory behind it is rather different. TLC relies on the fact that the attractive forces that cause different compounds to be adsorbed onto a solid surface differ from one compound to another (Figure 29.5). The solid used is usually powdered silica or alumina, with the small size of the particles providing a large surface area per gram, and it is coated onto a glass, plastic or thin aluminium plate.

Figure 29.5 Separation by adsorption

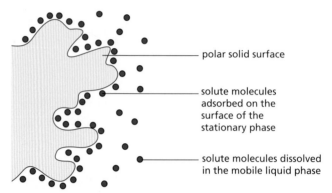

polar solid surface

solute molecules adsorbed on the surface of the stationary phase

solute molecules dissolved in the mobile liquid phase

The method used is almost identical to that of paper chromatography. Spots of the mixture to be analysed, along with any reference compounds, are applied close to the lower edge of the TLC plate and allowed to dry. The plate is then placed in a tank containing the solvent, making sure that the solvent level is below the spot.

As the solvent ascends the plate by capillary action, and passes over the compounds which have been adsorbed onto the solid particles, the compounds will partially dissolve in the solvent. How readily this occurs depends on both how soluble a particular compound is in the solvent, and how strong the attraction is between the compound and the solid support. This is how separation is achieved. Once the solvent front is near the top of the plate, the plate is removed, allowed to dry, and the spots can then be made visible. The ways in which spots of colourless compounds on a TLC plate are made visible can be similar to those used for paper chromatography, but an additional technique is often used for compounds containing aromatic rings, or other systems that absorb ultraviolet (UV) radiation. The solid support is pre-impregnated with an insoluble compound that is fluorescent: it absorbs UV light and re-emits it as visible light. When placed under a lamp which shines only UV light, the plate emits a bright white light unless a particular spot contains a compound that absorbs UV light. These spots will show up as dark areas.

TLC has several advantages over paper chromatography: the plates are quicker to run; better separation of the components of a mixture can be achieved; and its results are usually more reproducible. This is because the plates can be made very precisely, with constant very small particle size and moisture content. It is the technique of choice for synthetic chemists wanting to observe the progress of their reactions in real time.

Gas chromatography (GC)

In gas chromatography (GC) the mobile phase is an inert gas such as helium or nitrogen. This is passed through a column made of glass or metal, which is between 1 m and 3 m in length. The column is packed with fine solid particles (brick dust is often used), coated with oil that has a high boiling point (Figure 29.6).

Figure 29.6 Diagram of gas chromatography apparatus

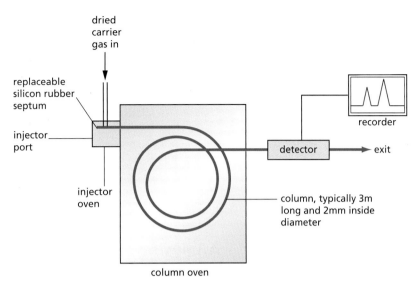

The components of mixtures that are separated by gas chromatography must have a reasonable vapour pressure at the temperature of the oven, which can be up to about 250 °C. The more volatile components spend more time in the vapour phase, and so travel through the column faster: they have shorter **retention times**. Less volatile compounds spend more of their time dissolved in the oil in the column, and so take longer to be carried through the column by the flowing gas: their retention times are longer.

The mixture to be analysed is first injected through a self-sealing disc (a rubber septum) into a small chamber. It is heated to maybe 50 °C above the temperature of the oven, where it is vaporised. The mixture then passes through the column, and separation occurs. The gas emerging from the column passes through a detector, which records the presence and amount of each component on a chart recorder or computer.

Use of gas chromatography in analysis

After a mixture has been separated by gas chromatography its components must be identified. If a sample is being analysed for the presence of known or suspected compounds, such as illegal recreational or performance-enhancing drugs, or the presence of known contaminants in foodstuffs, reference samples of these compounds will be used. If all variables are kept constant throughout the sampling of the mixture and the running of the reference compounds through the same machine, a simple comparison of retention times will allow an identification to be made.

The variables that must be kept the same are:

- the flow rate of the mobile solvent or the carrier gas
- the temperature of the oven
- the length and diameter of the column
- the chemical make-up of the solvent
- the polarity of the stationary phase.

This technique can routinely be coupled to an in-line mass spectrometer. The mass spectrum of each component in a sample can then be found, which enables further verification of the identity of the component. If, on the other hand, the sample is an unknown compound, the mass spectrum can allow its identification. Other physical methods of analysis that can be coupled to gas chromatography machines are infrared spectroscopy and ultraviolet-visible spectroscopy.

Examples of the use of this technique are:

- testing athletes for residues of performance-enhancing drugs in their blood or urine
- detection of explosive residues on skin or clothing
- comparing caffeine contents of various natural and decaffeinated coffees
- detection of pesticide residues in fresh and processed foodstuffs.

29.2 Mass spectrometry

In Topic 2 it was explained how a mass spectrometer can be used to determine the isotopic masses of individual atoms.

Mass spectrometry is used in four main ways to determine the structures of organic compounds:

1 finding the molecular formula of a compound by measuring the mass of its molecular ion to a high degree of accuracy
2 finding the number of carbon atoms in a molecule by measuring the abundance ratio of its molecular ion (M) peak and the M+1 peak
3 finding whether a compound contains chlorine or bromine atoms, and if so, how many of each, by measuring the abundance ratios of the M+2, M+4 and M+6 peaks.
4 working out the structure of a molecule by looking at the fragments produced when an ion decomposes inside a mass spectrometer.

Analysing the molecular ion

If we vaporise an organic molecule and subject it to the ionising conditions inside a mass spectrometer, the mass/charge ratio (m/e) for the molecular ion can be measured, and hence the relative molecular mass can be found.

For example, one of the non-bonding electrons on the oxygen atom of propanone can be removed by electron bombardment, to give an ionised molecule:

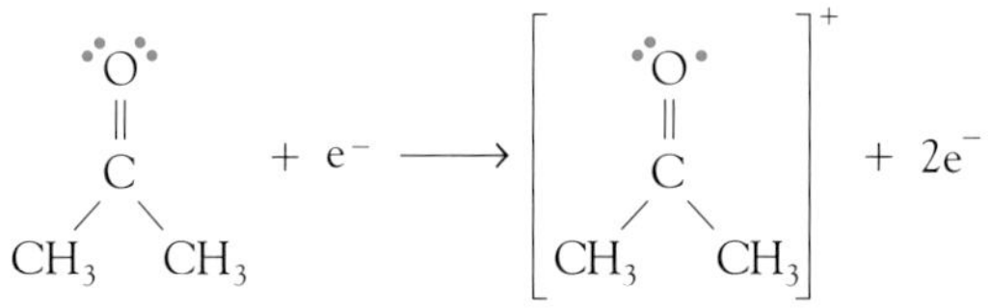

The m/e ratio for the resulting molecular ion is $(3 \times 12 + 6 \times 1 + 16):1$, which is 58.

Using **very high resolution mass spectrometry**, we can measure m/e ratios to an accuracy of five significant figures (1 part in 100 000). By this means, it is not only possible to measure the M_r value of a compound (its relative molecular mass), but also to determine its molecular formula. We can do this because the accurate atomic masses of individual atoms are not exact whole numbers.

Worked example

The three compounds in Table 29.1 all have an approximate M_r of 70.

Table 29.1 Three compounds with M_r of approximately 70

Name	Structure	Molecular formula
pentene	$CH_3CH_2CH_2CH{=}CH_2$	C_5H_{10}
2-aminopropanenitrile	$CH_3CH(NH_2)CN$	$C_3H_6N_2$
but-1-en-3-one	$CH_2{=}CHCOCH_3$	C_4H_6O

Use the following accurate relative atomic masses to calculate their accurate M_r values, and decide how sensitive the mass spectrometer needs to be in order to distinguish between them:
H = 1.0078
C = 12.000
N = 14.003
O = 15.995

Answer

The accurate M_r values are as follows:

C_5H_{10} = 5 × 12.000 + 10 × 1.0078   = 70.078
$C_3H_6N_2$ = 3 × 12.000 + 6 × 1.0078 + 2 × 14.003 = 70.053
C_4H_6O = 4 × 12.000 + 6 × 1.0078 + 15.995   = 70.042

The last two are quite close together. They differ by 11 parts in 70 000, or about 0.16%. However, this is well within the capabilities of a high resolution mass spectrometer.

Now try this

A compound has an accurate M_r of 60.068. Use the accurate relative atomic masses given above to decide whether the compound is 1,2-diaminoethane, $H_2NCH_2CH_2NH_2$, or ethanoic acid, CH_3CO_2H.

The M+1 peak

There are two stable isotopes of carbon, ^{12}C and ^{13}C. Their relative abundances are 98.9% for ^{12}C and 1.1% for ^{13}C. This means that out of every 100 methane (CH_4) molecules, CH_4, about 99 molecules will be $^{12}CH_4$ and just one molecule will be $^{13}CH_4$. For ethane, C_2H_6, the chances of a molecule containing one ^{13}C atom will have increased to about 2 in 100, because each C atom has a chance of 1 in 100 to be ^{13}C, and there are two of them. By measuring the ratio of the M to M+1 peaks, we can thus work out the number of carbon atoms the molecule contains. The formula relating the (M+1)/M ratio to the number of carbon atoms is:

$$n = \frac{100}{1.1} \times \frac{A_{M+1}}{A_M}$$

where n = number of carbon atoms

A_{M+1} = the abundance of the M+1 peak, and

A_M = the abundance of the molecular ion peak.

The abundances of the M and M+1 peaks are sometimes quoted as percentages, and sometimes as the actual heights of the two peaks on a printout of the mass spectrum, in arbitrary units. The units do not matter, however, as it is only the *ratio* that is important.

Worked example

The molecular ion peak of a compound has an *m/e* value of 136, with a relative abundance of 17%, and an M+1 peak at *m/e* 137 where the relative abundance is 1.5%. How many carbon atoms are in the molecule?

Answer

$$n = \frac{100}{1.1} \times \frac{1.5}{17} = 8.02$$

A molecule of the compound therefore contains 8 carbon atoms.

Now try this

A compound contains C, H and O atoms. Its mass spectrum has a peak at *m/e* 132 with a relative abundance of 43.9 and a peak at *m/e* 133 with a relative abundance of 2.9.

Calculate the number of carbon atoms in each molecule, and suggest its molecular formula.

The M+2 and M+4 peaks

Although fluorine and iodine each have only one stable isotope, chlorine and bromine have two. Their natural percentage abundances are shown in Table 29.2.

Table 29.2 The abundances of the isotopes of chlorine and bromine

Element	Isotope	Natural abundance	Approximate ratio
chlorine	^{35}Cl	75.5	3 : 1
	^{37}Cl	24.5	
bromine	^{79}Br	50.5	1 : 1
	^{81}Br	49.5	

Any compound containing one chlorine atom, therefore, will have two 'molecular ion' peaks, one due to molecules containing ^{35}Cl and the other due to molecules containing ^{37}Cl. For example, the mass spectrum of chloromethane, CH_3Cl, will have peaks at *m/e* 50 (12 + 3 + 35 = 50) and at *m/e* 52 (12 + 3 + 37 = 52), corresponding to the species $CH_3{}^{35}Cl^+$ and $CH_3{}^{37}Cl^+$. The relative abundances of the two peaks will be in the ratio 3 : 1 which is the ratio of the two Cl isotopes.

A similar situation occurs with bromine, although in this case the two molecular ion peaks will be of equal heights, since the isotopic abundance ratio is near to 1 : 1.

Mass spectra are slightly more complicated when the molecule contains more than one halogen. The simplest situation is that for two bromine atoms. Take the molecule 1,2-dibromoethane, $C_2H_4Br_2$. Each carbon can be attached to either a ^{79}Br or a ^{81}Br atom, and there is a (roughly) equal chance of either. We therefore arrive at the possibilities in Table 29.3, each of which is equally likely.

Table 29.3

Formula	*m/e* value	Peak
$^{79}BrCH_2CH_2{}^{79}Br$	186	M
$^{79}BrCH_2CH_2{}^{81}Br$	188	M + 2
$^{81}BrCH_2CH_2{}^{79}Br$	188	M + 2
$^{81}BrCH_2CH_2{}^{81}Br$	190	M + 4

There will thus be three molecular ion peaks, with relative abundances of $1:2:1$.

Worked example

Work out the *m/e* values and the relative abundances of the molecular ion peaks for dichloromethane, CH_2Cl_2.

Answer

Just as with dibromoethane above, there will be four possible formulae. Their *m/e* values are given in the following table.

Formula	Ion reference	*m/e* value
$CH_2{}^{35}Cl^{35}Cl$	a	84
$CH_2{}^{35}Cl^{37}Cl$	b	86
$CH_2{}^{37}Cl^{35}Cl$	c	86
$CH_2{}^{37}Cl^{37}Cl$	d	88

The abundance ratio of ion **a**: ion **b** is $3:1$ (because of the $^{35}Cl:{}^{37}Cl$ ratio)
The abundance ratio of ion **c**: ion **d** is also $3:1$
The abundance ratio of ion **a**: ion **d** is $9:1$, because each of the two chlorine atoms has 3 times the probability of being ^{35}Cl rather than ^{37}Cl.
The overall probabilities are therefore:

Mass number	84	86	88
individual probabilities	3	1	
	3	1	
	3	1	
		3	1
total sum of probabilities	9	6	1

Analysing molecular fragments

If the ionising electron beam in a mass spectrometer has enough energy, the molecular ions formed by the loss of an electron can undergo bond fission, and molecular fragments are formed (see Figure 29.7). Some of these will carry the positive charge, and therefore appear as further peaks in the mass spectrum.

Figure 29.7 Ionic fragments formed from propanone

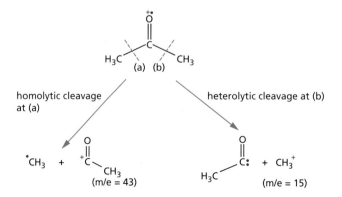

We therefore expect the mass spectrum of propanone to contain peaks at $m/e = 15$ and 43, as well as the molecular ion peak at 58 (see Figure 29.8).

Figure 29.8 Mass spectrum of propanone

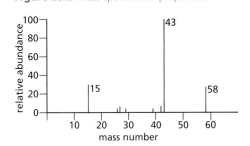

Figure 29.9 Mass spectrum of propanal

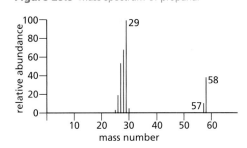

The fragmentation pattern can readily distinguish between isomers. Compare Figure 29.8 with Figure 29.9, which shows the mass spectrum of propanal. Here there is no peak at $m/e = 15$, nor one at $m/e = 43$. Instead, there is a peak at $m/e = 57$ and several from $m/e = 26$ to 29. This is readily explained by the fragmentations shown in Figure 29.10.

Figure 29.10 Ionic fragments formed from propanal

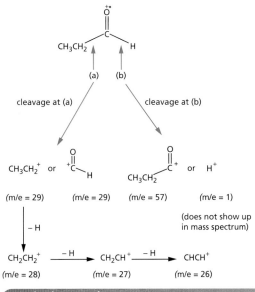

Worked example

Use the following atomic mass data to calculate the accurate M_r values for the two ionic fragments at m/e 29 in the mass spectrum of propanal. Would a mass spectrometer with a sensitivity of 1 part in 10 000 be able to distinguish between them?
H = 1.0078   C = 12.000   O = 15.995

Answer
C_2H_5 is $2 \times 12.000 + 5 \times 1.0078$ = 29.039
CHO is $12.000 + 1.0078 + 15.995$ = 29.003
These masses differ by 36 in 29 003 or 1 part in 8056, so this (fairly inaccurate) spectrometer would just be able to distinguish between them.

Now try this

Suggest the formulae of the ions at m/e values 26, 27 and 28 in the mass spectrum of propanal, and suggest an explanation of how they might arise.

Depending on what type of cleavage occurs at (a) and (b) in Figure 29.10, one or other or both of each pair of ion fragments may appear. The peaks of highest abundance in the mass spectra of organic compounds are associated with particularly stable cations, such as acylium and tertiary carbocations:

$$R - \overset{+}{C} = O \longleftrightarrow R - C \equiv \overset{+}{O}$$

the acylium ion

a tertiary carbocation

The interpretation of the fragmentation pattern in the mass spectra of organic compounds is therefore an important tool in the elucidation of their structures. A further example will show the power of the technique.

Worked example

Figure 29.11 shows the mass spectra of two compounds with the molecular formula $C_2H_4O_2$. One is methyl methanoate, and the other is ethanoic acid. Decide which is which by assigning structures to the major fragments whose m/e values are indicated.

Figure 29.11 Mass spectra of methyl methanoate and ethanoic acid. Which is which?

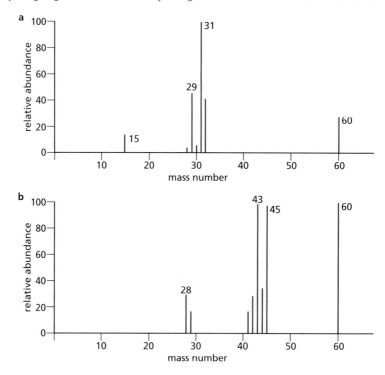

Answer
Apart from the molecular ion at $m/e = 60$, the major peaks in spectrum **a** are at m/e values of 15, 29 and 31. These could be due to:
CH_3^+ ($m/e = 15$)
$C_2H_5^+$ or CHO^+ ($m/e = 29$)
CH_3O^+ ($m/e = 31$)
This fits with the structure of methyl methanoate (see Figure 29.12).

Figure 29.12 Ionic fragments formed from methyl methanoate

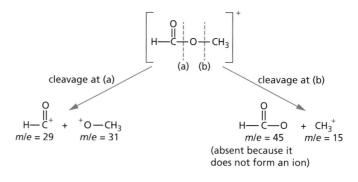

The peak at $m/e = 29$ can come *only* from methyl methanoate, and not from ethanoic acid. The major peaks in Figure 29.11b, apart from the molecular ion at $m/e = 60$, are at m/e values of 28, 43 and 45. These could be due to:

CO^+ ($m/e = 28$)
CH_3CO^+ ($m/e = 43$)
CO_2H^+ ($m/e = 45$)
These arise from the fragmentations shown in Figure 29.13.

Figure 29.13 Ionic fragments formed from ethanoic acid

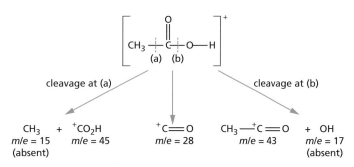

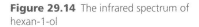
Now try this

A compound of molecular formula $C_3H_6O_2$ has major peaks at $m/e = 27, 28, 29, 45, 57, 73$ and 74. Suggest formulae for these fragments, and a structure for the compound.

The peak at $m/e = 43$ can come only from ethanoic acid, and not from methyl methanoate.

29.3 Infrared spectroscopy

Most organic molecules absorb infrared radiation. The frequencies that are absorbed depend on the stiffness of their bonds and the masses of the atoms at each end, and the intensity of an absorption depends on the change in dipole moment as the bond vibrates. So bonds to electronegative atoms such as oxygen, that are found in alcohols (O—H) and carbonyl compounds (C=O) show very strong absorptions. Although an infrared spectrum shows a series of absorptions (see Figure 29.14), these are always referred to as peaks rather than troughs.

Figure 29.14 The infrared spectrum of hexan-1-ol

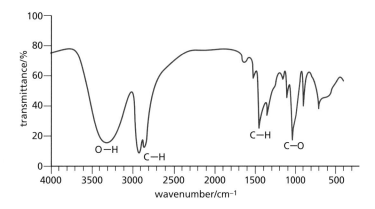

The process is important in that it is small molecules in the atmosphere (especially CO_2, CH_4, H_2O and CFCs) that are responsible for the greenhouse effect: they absorb infrared radiation that is emitted from the surface of the Earth, thus preventing it from being lost to space. In consequence, the amount of heat lost is less than that gained from solar radiation, and the Earth warms up (see the greenhouse effect, page 241).

The infrared spectrum of a compound can allow us to identify the functional groups it contains, as each functional group has a characteristic absorption frequency or range of frequencies. The data are listed in Table 29.4.

Table 29.4 Some infrared (IR) absorption frequencies for organic groups
(The exact frequency of absorption of a group depends on its molecular environment, and can be 50 cm^{-1} or so higher or lower than the frequencies given here.)

Type of bond	Bond	Frequency of absorption (wavenumber)/cm^{-1}
bonds to hydrogen	O—H	3600
	O—H (hydrogen bonded)	3200–3500
	N—H	3400
	O—H in RCO$_2$H (strongly hydrogen bonded)	2500–3300
	C—H	2800–2900
triple bonds	–C≡C– or –C≡N	2200
double bonds to oxygen	C=O in RCOCl	1800
	C=O in RCO$_2$R	1740
	C=O in RCHO	1730
	C=O in RCO$_2$H	1720
	C=O in R$_2$CO	1715
C=C double bonds	C=C in alkenes	1650
	C=C in arenes	1600 and 1500
single bonds	C—O	1100–1250

Worked example

Compounds **T** and **U** are isomers with the molecular formula C$_3$H$_6$O$_2$. Suggest their structures based on the spectra shown in Figures 29.15 and 29.16.

Figure 29.15 Infrared (IR) spectrum of T

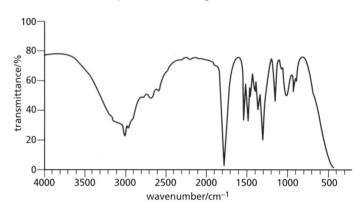

Figure 29.16 IR spectrum of U

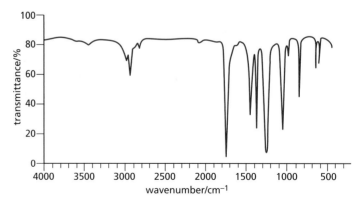

Answer

Both **T** and **U** show a C=O absorption in their spectrum at about 1700–1800 cm^{-1}, and a C—O absorption at about 1250 cm^{-1}. **T** shows a broad hydrogen-bonded O—H band from 3300 to 2500 cm^{-1}, whilst **U** shows no O—H band at all.

So **T** is **CH$_3$CH$_2$CO$_2$H** (propanoic acid) and **U** could be either the ester **CH$_3$CO$_2$CH$_3$** (methyl ethanoate) or the ester **HCO$_2$CH$_2$CH$_3$** (ethyl methanoate).

> **Now try this**
>
> Compound **V** (C_3H_6O) gives a silver mirror when warmed with Tollens' reagent. It can be converted to compound **W** by reagent **X**. Use the spectra in Figures 29.17 and 29.18 to identify the functional groups present in **V** and **W**, and suggest the identity of reagent **X**.
>
>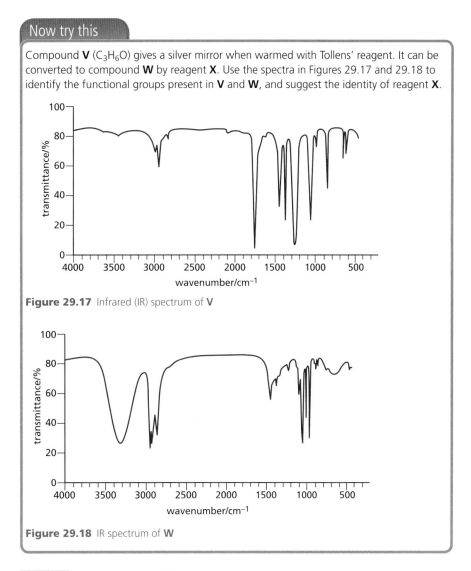
>
> **Figure 29.17** Infrared (IR) spectrum of **V**
>
> **Figure 29.18** IR spectrum of **W**

29.4 Proton (^{1}H) nuclear magnetic resonance (NMR) spectroscopy

The basis of NMR spectroscopy

We cannot see molecules with our naked eyes, but some of the most direct evidence for their structures and shapes comes to us from nuclear magnetic resonance (NMR) spectroscopy. Every aspect of an NMR spectrum demands an exact interpretation, and there is usually a unique molecular structure that gives rise to a particular spectrum. NMR is potentially the most powerful technique at the disposal of the structural organic chemist.

Like electrons, nucleons (protons and neutrons) have spin. If an atom has an even number of nucleons, the spins cancel out and there is no overall magnetic moment. If, however, an atom has an odd number of nucleons, there is an overall magnetic moment. As a result, the nucleus can take up one of two orientations. In the absence of a magnetic field, the energies of the two orientations are the same, but in the presence of an external magnetic field one orientation has a slightly higher energy than the other. This splitting into two energy levels forms the basis of NMR spectroscopy: nucleons can be persuaded to flip from the lower energy spin state to the higher spin state (i.e. to resonate) by irradiating a sample with electromagnetic radiation of the right frequency. The absorption of this frequency is detected by the NMR spectrometer.

The extent of the splitting is proportional to the strength of the external field: to increase the splitting, very large external fields are used. The strength of the applied magnetic field is measured in **tesla**, T; one tesla is about 10000 times as strong as the Earth's magnetic field. Many machines use a field strength of 9.4 T.

For a hydrogen atom, this creates an energy difference of 0.16 J. The Planck equation, $E = hf$, shows that this corresponds to a frequency of 400 MHz, which is in the UHF (ultra high frequency radiowaves) region of the electromagnetic spectrum. An NMR spectrometer therefore detects the absorption of UHF radiation by a sample, in a similar fashion to any other spectrometer. The principal difference is that the sample is also subjected to a strong magnetic field. Compounds containing hydrogen atoms therefore show a nuclear magnetic resonance absorption band at 400 MHz. The frequencies absorbed by other common atoms that have an odd number of nucleons are shown in Table 29.5.

Table 29.5 The absorption frequencies, in an external field of 9.4 T, for some common atoms studied with nuclear magnetic resonance (NMR). In this topic we are looking at ^{1}H and ^{13}C magnetic resonance.

Nucleus	Absorption frequency/MHz
^{1}H	400
^{13}C	101
^{19}F	377
^{31}P	162

Figure 29.19 This NMR spectrometer measures the absorbance of UHF radiofrequency radiation by hydrogen atoms (which have an odd number of nucleons – just one proton) in an external magnetic field. The absorption frequency varies slightly depending on the chemical environment of the hydrogen atom, allowing identification of hydrogen-containing compounds.

The absorption is very weak, because the populations of atoms at each of the two energy levels are almost the same. At room temperature for a hydrogen atom, the population of the upper level differs from the lower end by only 1 part in 30000, so that absorption is nearly always cancelled out by re-emission. The situation is even worse for carbon-13, because carbon contains only 1% of this isotope. In order to make the absorbance as intense as possible:

- the sample is cooled, which increases the population difference
- as large a magnetic field as possible is used, to increase the splitting
- the absorption is measured many times and the results are averaged by computer.

NMR is often used to detect the hydrogen atoms in water, and the analysis of water in the human body forms the basis of magnetic resonance imaging (MRI) (see the panel on page 509).

Magnetic resonance imaging (MRI)

Because the human body is made up mostly of water, it responds to nuclear magnetic resonance. By suitable scanning, an image of the water distribution in the body can be built up, which is invaluable in the diagnosis of various illnesses, in particular brain disorders. The word 'nuclear' has been dropped from the name of the technique magnetic resonance imaging (MRI), to avoid suggesting to patients that nuclear radiation is involved.

The MRI scanner is large, because the magnetic field must pass through the human body. A fine beam of radiation is applied, giving the absorption pattern of a cross-section of the body about 1 cm thick. Within this cross-section, water molecules can be studied in different parts of the body, as follows:

● The magnetic field is not uniform but varies from one side to another. As the frequency of the radiation is changed, water at different depths inside the body responds to the signal. This enables a one-dimensional picture to be built up.
● The radiation beam is rotated through 360°. This enables a computer to produce a two-dimensional image – a 'slice'.

A typical brain scan containing 20–30 slices can be obtained in less than 10 minutes. More subtle analysis of the data makes it possible to distinguish between water held, for example, in grey or white tissue or in cancerous or normal cells. The technique is invaluable in the diagnosis of brain tumours or Alzheimer's disease. It has the great advantage of being non-invasive, and the UHF radiation is much safer than X-rays, which are used in alternative techniques.

A refinement of MRI is to detect phosphorus-31 rather than protons. Areas of the brain that are actively in use require ATP (adenosine triphosphate) for their biochemical reactions. It is therefore possible to locate the regions of the brain that are most actively involved when different mental processes (for example, sight, reasoning or spatial work) are being carried out.

Figure 29.20 A magnetic resonance imaging (MRI) scanner gives a three-dimensional picture of the inside of the body, allowing non-invasive diagnosis of many diseases.

Figure 29.21 Images from an MRI scan of a human head

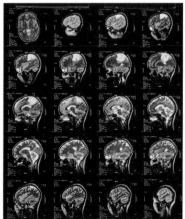

Analysing organic molecules

Both ^{1}H and ^{13}C nuclear magnetic resonance (NMR) spectroscopy are important in the analysis of organic compounds. We shall look first at ^{1}H NMR.

Magnetic resonance spectroscopy would be of little value in chemical analysis if all hydrogen atoms absorbed the same frequency of radiation. The electron cloud around the hydrogen atom partially screens the nucleus from the external magnetic field through the 'diamagnetic effect'.

The electrons within molecules – both bonded and non-bonded electrons – are usually 'paired' (that is, they occur as pairs of electrons spinning in opposite directions). When a molecule is placed in an external field, the electron pairs rotate in their orbits in such a way that they produce a magnetic field which opposes the external field. This phenomenon is called **diamagnetism** (see Figure 29.22). The effect is to shield nearby protons from the external field. This in turn reduces the frequency at which they absorb energy when they flip from their lower to their higher energy state.

Figure 29.22 In an external magnetic field, the electron pairs rotate in such a way that they produce an opposing magnetic field.

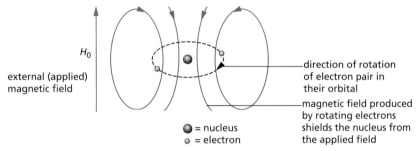

When, however, a proton is near an electronegative atom (or group) within a molecule, the bonding electrons are drawn away from the proton to the electronegative atom. The proton is less shielded from the external magnetic field, and hence it absorbs radiation at a higher frequency. The effect is very pronounced if the proton is attached to a benzene ring. In this situation the mobile delocalised π electrons in the ring can create a strong diamagnetic effect, opposing the external field. This has the effect of strengthening the magnetic field in the vicinity of the protons (see Figure 29.23).

Figure 29.23 In benzene, the field created by the rotation of the π electrons reinforces the applied field.

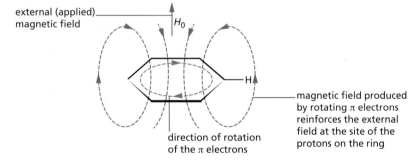

The extent to which a proton is de-shielded from the external magnetic field is measured by its chemical shift.

Chemical shift

For convenience, the hydrogen atoms in the compound tetramethylsilane, $(CH_3)_4Si$ (known as TMS), are used as a reference. The sample being investigated is mixed with a drop of TMS and the frequency of the absorption, f, is measured relative to that of TMS, f_{TMS}. There are a number of reasons why this compound is used.

- All the hydrogen atoms in TMS are equivalent, so it gives a single absorption peak.
- Most other groups absorb at higher frequencies than TMS. This is because the protons in TMS are near the electropositive silicon atom, which does not draw the electrons away from the hydrogen atoms as much as carbon atoms.

The extent of the difference in frequency of absorption, is called the **chemical shift**, symbol δ (delta). It is defined as:

$$\delta = 10^6 \times \frac{(f - f_{TMS})}{f_{TMS}}$$

Values of δ are quoted in parts per million (ppm). Because chemical shifts are very small, the magnetic field must be identical throughout the sample. To achieve this, the superconducting magnets used in modern machines are very carefully constructed.

As mentioned above, the chemical shift increases as the electronegativity of the atom attached to the hydrogen increases. For example, in a halogenoalkane, a carbon atom attached to a fluorine atom is more electron-withdrawing than a carbon atom attached to an iodine atom. This is due to the inductive effect of the halogen atom, which in turn depends on its electronegativity (see section 3.10). The more electron-withdrawing carbon atom has a greater effect in drawing the electron cloud away from an attached hydrogen atom, reducing the screening and increasing the absorption frequency (see Table 29.6).

Table 29.6 Chemical shifts vary with the partial charge on the carbon atom attached to the hydrogen. A carbon atom attached to a highly electronegative atom such as fluorine has a larger partial charge than a carbon atom attached only to hydrogen atoms.

Group	Electronegativity of atom attached to carbon	δ/ppm
F — CH_3	4.0	4.5
Cl — CH_3	3.2	2.9
Br — CH_3	3.0	2.5
I — CH_3	2.7	2.0
H — CH_3	2.2	0.8

The values of some chemical shifts for hydrogen atoms within different organic groups are given in Table 29.7.

Table 29.7 Some chemical shifts. Protons associated with particular groups absorb in a *region* of the spectrum rather than at a definite frequency.

Proton type	Groups	δ/ppm	Range
$Si(CH_3)_4$	tetramethylsilane (TMS)	0.0	0
C—CH_3	end of alkyl chain	0.9	±0.5
C—CH_2—C	middle of alkyl chain	1.4	±0.5
=C—CH—	adjacent to C=C	1.9	±0.5
—C(O) —CH—	adjacent to C=O (ketones, esters, acids)	2.3	±0.5
C_6H_5 — CH —	adjacent to arene ring	2.5	±0.5
O—CH—	adjacent to oxygen (alcohols, ethers, esters)	3.3	±0.5
=C—H	alkenyl	5.5	±1
C_6H_5—H (benzene)	aryl	7.5	±1.5
—C(O) —H	aldehydic	9.0	±1.5
—O—H	alcohols	3	±2*
Ar—O—H	phenols	5.5	±1*
—CO—O—H	carboxylic acids	11.0	±2*

* These δ values are very dependent on the nature and acidity of the solvent and the extent of hydrogen bonding between molecules and the solvent.

Low and high resolution NMR spectroscopy: splitting patterns

When the ^1H NMR spectrum of ethyl ethanoate is scanned at low resolution (Figure 29.24), three peaks are observed, corresponding to the three different chemical environments of the protons in the molecule.

Figure 29.24 Nuclear magnetic resonance (NMR) spectrum of ethyl ethanoate at low resolution. Note that the δ scale, by convention, has its zero on the right of the x-axis.

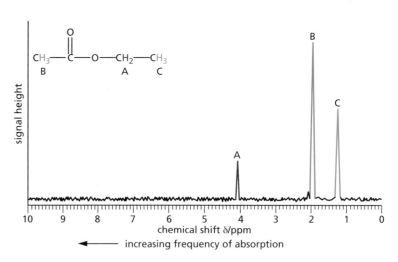

Figure 29.25 NMR spectrum of ethyl ethanoate at high resolution

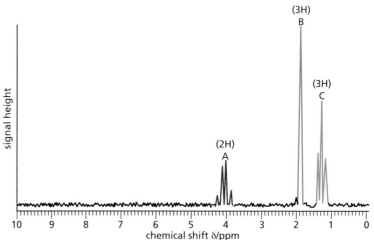

The areas under the three peaks are in the ratio 2:3:3, being proportional to the number of protons at each chemical environment. On published NMR spectra, the relative area under each peak is sometimes shown by a number above the peak as can be seen in Figure 29.25. At higher resolution (see Figure 29.25), peaks A and C are seen to be multiple peaks, although B remains a single peak. This is because the nuclear spins of the protons in the ethyl group, responsible for peaks A and C, interact with each other. This is called **spin–spin coupling**, and it is a general phenomenon observed whenever protons on adjacent carbon atoms are in different chemical environments. The splitting of the peak arises because the magnetic field experienced by a proton is slightly altered due to the orientation of the magnetic moments (the spin states) of the protons on the adjacent carbon atom. Consider the protons in the CH₃ group of the ethyl chain. The field they experience will depend on the orientation of the magnetic moments of the —CH₂— protons, as shown in Figure 29.26

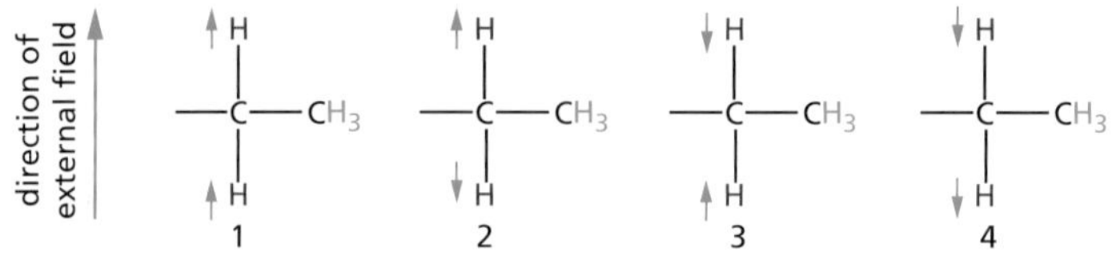

Figure 29.26 The directions of the magnetic moments of the —CH₂— protons have an effect on the —CH₃ protons.

In situations 2 and 3, the magnetic moments of the two —CH₂— protons cancel each other out, so the field experienced by the —CH₃ group protons will be the same as the applied field. In situations 1 and 4, however, the magnetic moments of the —CH₂— protons reinforce each other. Consequently the field experienced by the protons of the —CH₃ group will be, respectively, higher and lower than the applied field. There should therefore be a total of three frequencies at which these —CH₃ protons absorb. What is more, the probabilities of the four states 1 to 4 are equal, so overall there is twice the chance of the —CH$_3$ protons experiencing no change in field (situations 2 and 3) as there is for the —CH$_3$ protons to experience either an enhanced field (situation 1) or a reduced field (situation 4). We therefore expect the intensities of the lines in the triplet of lines to be in the ratio $1:2:1$.

A similar argument can be applied to the modification of the magnetic field experienced by the —CH$_2$— group protons, by the protons in the adjacent —CH$_3$ group. In this case we expect a quartet of lines, in the ratio of $1:3:3:1$.

Worked example

By considering the different combinations of ↑ and ↓ magnetic moments of the —CH$_3$ protons, explain how the ratio $1:3:3:1$ arises.

Answer
The possible combinations of the three —CH$_3$ protons are shown in Figure 29.27.

Figure 29.27

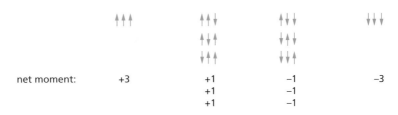

There are three times as many combinations giving a net magnetic moment of +1 or −1, compared with +3 or −3.

The general rules concerning the splitting of the resonance peak of a proton by other protons are as follows.

- Protons in identical chemical environments do not split each other's peaks.
- The peak of a proton adjacent to n protons in a different environment is split into $(n+1)$ lines.
- The relative intensities of the $(n+1)$ lines are in the pattern shown in Table 29.8.
- The interaction between protons separated by more than three bonds is usually too weak to cause any splitting of each other's peaks.

Table 29.8 The splitting of a peak for a proton next to protons in a different chemical environment

Number of protons adjacent to resonating proton	Number of lines in split peaks	Relative intensities of lines
1	2	$1:1$
2	3	$1:2:1$
3	4	$1:3:3:1$
4	5	$1:4:6:4:1$

Peaks that are not split are referred to as *singlets* (*s*). If a peak is split into two lines it is a *doublet* (*d*); three lines is a *triplet* (*t*); four lines a *quartet* (*q*), etc. If there are so many lines that it is difficult to count them, the peak is referred to as a *multiplet* (*m*).

The use of 'heavy water', D$_2$O

Protons directly attached to oxygen or nitrogen atoms can appear almost anywhere in an NMR spectrum. The field strength at which they resonate depends on the acidity and hydrogen-bonding ability of the solution. Because of easy proton exchange with other O—H or N—H protons in the sample, these protons often do not cause the splitting of the peaks of adjacent protons. They can, however, be identified by **deuterium exchange**. If the compound containing them is dissolved in D$_2$O ('heavy water', D$=^2$H), the protons are exchanged with deuterium atoms in the water. The peaks due to the —OH or —NH$_2$ protons disappear (deuterium atoms, having an even number of nucleons, do not resonate):

$$CH_3CH_2-OH + D_2O \rightleftharpoons CH_3CH_2-OD + HDO$$

Now try this

Predict the splitting pattern (the number of lines and the relative intensities of the lines) for a proton adjacent to:
a   one other proton
b   four other protons.

Figure 29.28 NMR spectra of ethanol: **a** showing the —OH peak, and **b** with D_2O added

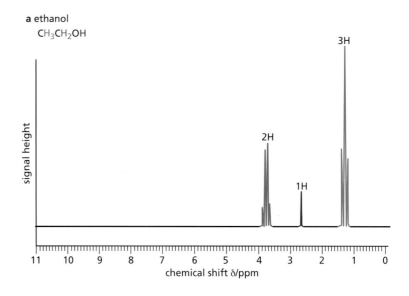

a ethanol
CH_3CH_2OH

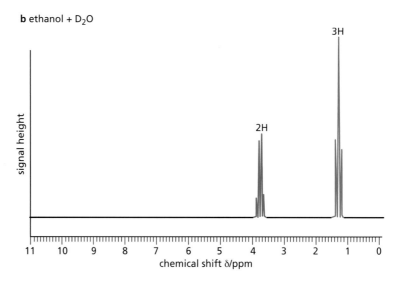

b ethanol + D_2O

Worked example

Figure 29.29 shows the nuclear magnetic resonance (NMR) spectrum of an acid with the molecular formula $C_8H_8O_2$. Work out its structure. (Use the δ values in Table 29.7, page 511 to help you.)

Figure 29.29

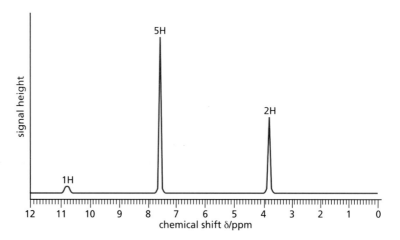

Answer

The high $C:H$ ratio in the molecular formula suggests the presence of a benzene ring, and this is confirmed by the peak at $\delta = 7.6$. The broad peak at $\delta = 10.8$ is typical of the O—H hydrogen of a carboxylic acid. The two-proton single peak at $\delta = 3.7$ is a CH_2 group flanked by both an aryl ring and a CO_2H group, both of which would cause a high-field shift in resonance (by about $1\,\delta$ unit each).

The structure is therefore $C_6H_5—CH_2—CO_2H$.

Now try this

1   Apart from ethyl ethanoate (see Figure 29.25) there are at least 15 other isomers of $C_4H_8O_2$. The ^1H NMR spectra of two of them, **W** and **X**, are shown in Figure 29.30. Explain the splitting patterns seen in these spectra, and use the spectra to suggest the structures of **W** and **X**.

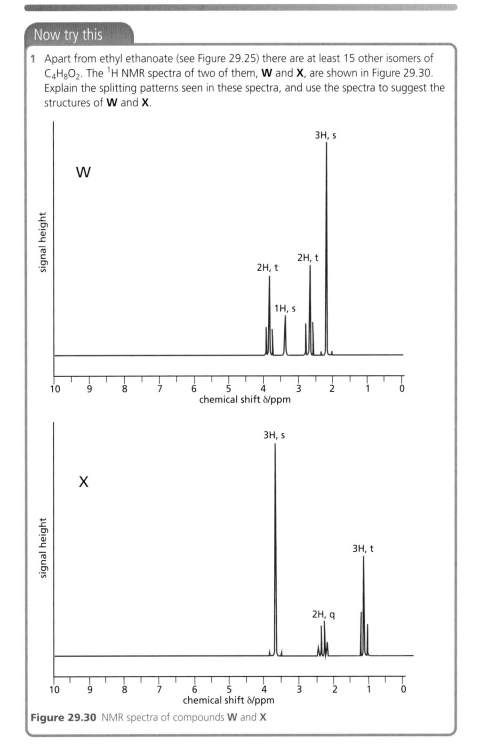

Figure 29.30 NMR spectra of compounds **W** and **X**

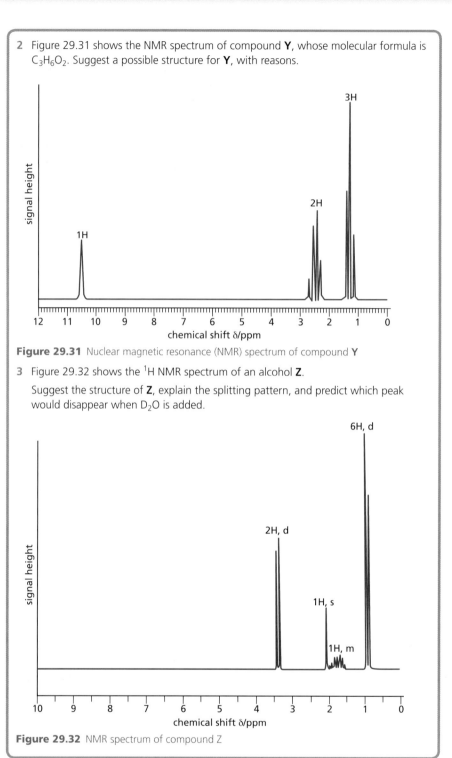

2 Figure 29.31 shows the NMR spectrum of compound **Y**, whose molecular formula is $C_3H_6O_2$. Suggest a possible structure for **Y**, with reasons.

Figure 29.31 Nuclear magnetic resonance (NMR) spectrum of compound **Y**

3 Figure 29.32 shows the ^1H NMR spectrum of an alcohol **Z**.

Suggest the structure of **Z**, explain the splitting pattern, and predict which peak would disappear when D_2O is added.

Figure 29.32 NMR spectrum of compound Z

29.5 ^{13}C Nuclear magnetic resonance (NMR) spectroscopy

A ^{13}C NMR spectrum is simpler than a ^1H spectrum. This is not only because there are usually fewer carbon atoms in a molecule than there are hydrogen atoms, but also because the absorbances in a ^{13}C spectrum usually appear as singlets – they are not split into multiplets by adjacent atoms as are many lines in a ^1H spectrum.

Because of the very small natural abundance of ^{13}C atoms (1.1%), the chances of two adjacent carbon atoms in a molecule both being ^{13}C atoms is only just over 1 in 100, and so the splitting of a peak due to adjacent ^{13}C atoms is very unlikely. Although the spin-interaction between ^{13}C atoms and ^1H atoms is very large, usually when a ^{13}C spectrum is run, the ^1H–^{13}C coupling is removed by irradiating the sample with broad-frequency 'white noise'. Although this greatly simplifies the spectrum, it has the disadvantage that the intensities of the peaks are not dependent on the number of carbon atoms, and so it is not possible to determine the number of carbon atoms associated with a particular absorbance.

Each carbon atom in a different chemical environment produces a single-peak absorbance at a different chemical shift. Table 29.9 shows some chemical shift values for ^{13}C in different chemical environments.

Table 29.9 Some ^{13}C chemical shifts

Environment of ^{13}C	Chemical shift range (ppm from TMS)
C (alkane)	0–30
C (alkene)	110–150
C—N	50–55
C—O	60–65
C (aryl)	110–160
C in —COX (X = O, N)	160–175
C=O	200–220

The ^{13}C NMR spectrum of ethyl ethanoate is illustrated in Figure 29.33.

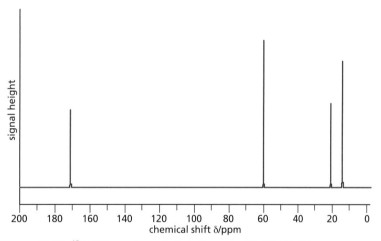

Figure 29.33 ^{13}C NMR spectrum of ethyl ethanoate, $CH_3CO_2CH_2CH_3$

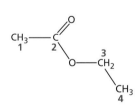

Figure 29.34 Ethyl ethanoate

The spectrum in Figure 29.33 shows four peaks, one for each of the carbon atoms in the molecule. From Table 29.9 we can see that the peak on the left of the spectrum, at 171 ppm, is for carbon **2** in Figure 29.34, and the peak at 60 ppm is for carbon **3** in Figure 29.34.

It is a little more difficult to assign the other two peaks, at 13 and 22 ppm, but because the CH_3 next to the C=O is likely to be more deshielded due to the electron-withdrawing effect of the C=O, we can identify carbon **1** with the peak at 22 ppm, leaving the peak at 13 ppm associated with carbon **4**.

Now try this

Two isomers of ethyl ethanoate are propyl methanoate and prop-2-yl methanoate. Their ^{13}C spectra are shown in Figure 29.35. Decide which compound gives which spectrum.

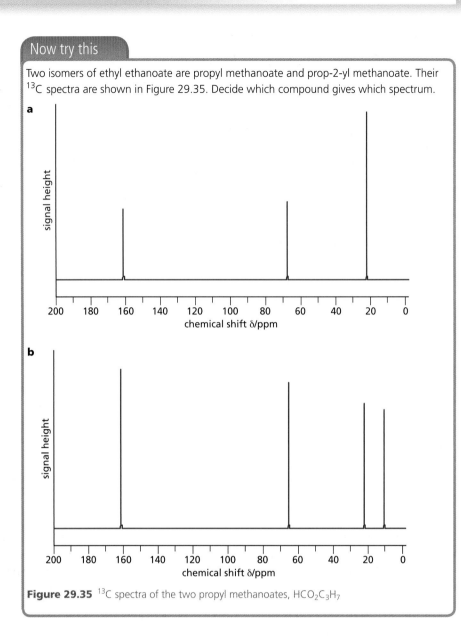

Figure 29.35 ^{13}C spectra of the two propyl methanoates, $HCO_2C_3H_7$

Summary

- **Paper chromatography (PC)** uses the principle of **partition** to separate components of a mixture.
- Components of a mixture are separated in **thin layer chromatography (TLC)** through the principle of **adsorption**.
- Components are separated in **gas chromatography (GC)** by their relative volatilities and attraction to the non-polar coating on the solid support.
- The routine uses of gas chromatography in analysis include the detection of alcohol, drugs, food additives and impurities and explosive residues.
- Measuring the accurate mass of the molecular ion peak in **mass spectrometry** allows us to work out the molecular formula of a compound.
- Use of the M+1 peak in mass spectrometry enables us to determine the number of carbon atoms in a molecule.

- Use of the M+2 peak in mass spectrometry enables us to determine the number of chlorine and/or bromine atoms in a molecule.
- The **fragmentation pattern** in mass spectrometry helps us to determine the structure of molecules.
- **Infrared (IR)** spectroscopy can identify the functional groups within a molecule.
- **Nuclear magnetic resonance (NMR)** spectroscopy can be carried out on compounds that contain atoms such as ^{13}C and ^{1}H, which have magnetic moments that take up orientations with different energies in an external magnetic field.
- Both the **chemical shift** (δ) values and the splitting patterns in a ^{1}H NMR spectrum allow us to determine the structures of molecules.
- The ^{13}C spectrum can allow us to determine the number and environment of carbon atoms within a molecule.

Examination practice questions

Please see the data section of the CD for any A_r values you may need.

1 This question is about the modern techniques of analysis which may be used to determine molecular structures.

 a NMR spectroscopy, in contrast to X-ray crystallography, is frequently used to examine protons in organic molecules.

  i What feature of protons enables their detection by NMR spectroscopy?

  ii The NMR spectrum below was obtained from a compound **X**, $C_xH_yO_z$. In the mass spectrum of the compound, the M : M+1 ratio was found to be 25:2. Determine the values of x, y and z in the formula of X and deduce a possible structure for the compound, explaining how you arrive at your conclusion.

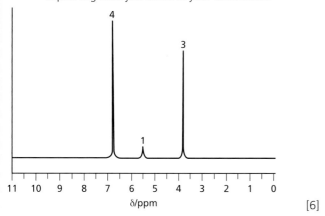

[6]

[Cambridge International AS & A Level Chemistry 9701, Paper 41 Q7 c November 2009]

2 The techniques of mass spectrometry and NMR spectroscopy are useful in determining the structures of organic compounds.

 a The three peaks of highest mass in the mass spectrum of organic compound **L** correspond to masses of 142, 143 and 144. The ratio of the heights of the M : M+1 peaks is 43.3 : 3.35, and the ratio of heights of the M : M+2 peaks is 43.3 : 14.1.

  i Use the data to calculate the number of carbon atoms present in **L**.

  ii Explain what element is indicated by the M+2 peak.

 iii Compound **L** reacts with sodium metal. The NMR spectrum of compound **L** is given below.

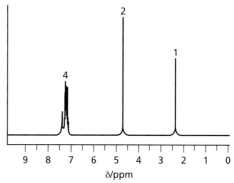

 What does the NMR spectrum tell you about the number of protons in **L** and their chemical environments?

 iv Use the information given and your answers to **i**, **ii** and **iii** to deduce a structure for **L**. **Explain** how you arrive at your answer. [7]

 b The molecular formula C_3H_6 represents the compounds propene and cyclopropane.

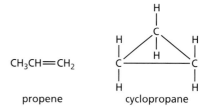

propene   cyclopropane

  i Suggest **one** difference in the fragmentation patterns of the mass spectra of these compounds.

  ii Suggest **two** differences in the NMR spectra of these compounds. [3]

[Cambridge International AS & A Level Chemistry 9701, Paper 42 Q7 June 2013]

30 Organic synthesis and analysis

Many of the commercial applications of organic chemistry involve the synthesis of pharmaceutical compounds. This topic looks at how the reactions that have been introduced in the previous topics can be used to synthesise organic molecules of specific structures; how drug molecules are designed; and how the reactions covered can be used to identify the functional groups in an unknown molecule.

Learning outcomes

By the end of this topic you should be able to:

23.1a) state that most chiral drugs extracted from natural sources often contain only a single optical isomer

23.1b) state reasons why the synthetic preparation of drug molecules often requires the production of a single optical isomer, e.g. better therapeutic activity, fewer side effects

23.2a) for an organic molecule containing several functional groups: identify organic functional groups using the reactions in the syllabus, and predict properties and reactions

23.2b) devise multi-stage synthetic routes for preparing organic molecules using the reactions in the syllabus

23.2c) analyse a given synthetic route in terms of type of reaction and reagents used for each step of it, and possible by-products.

30.1 The synthesis of organic compounds

Simple molecules

Many organic compounds have important uses as pharmaceuticals, pesticides, perfumes and dyes. These compounds often have quite complicated structures, and most of them are manufactured by organic chemists from much simpler starting materials. The science of organic synthesis is rather like building with molecular Lego™: a compound is constructed by combining a sequence of reactions – from two to perhaps 20 or so – which start from known, readily available chemicals and end up with the target molecule.

Figure 30.1 A laboratory where new organic compounds are synthesised

Organic synthesis requires much art and craft as well as science. If you have attempted an organic preparation in the laboratory you will know that the yield and purity of your product will often vary from those given in the 'recipe'. Skill and practice are needed to perfect practical techniques. The reagents and conditions that give an excellent yield with one compound may not be as effective for another compound with an identical functional group.

Nevertheless, it is possible to devise a multi-step method for synthesising a given organic compound by piecing together successive standard organic transformations. We shall illustrate this by making use of the reactions summarised in Charts A–G (see section 30.8), and Table 30.1, pages 535–536).

For example, suppose we needed to devise a synthesis of ethanoic acid, starting from bromoethane:

$$CH_3CH_2Br \longrightarrow CH_3CO_2H$$

The strategy is as follows:

1 Use Chart B (page 536) to work out the structures of all the compounds that can be made from bromoethane in one step.
2 Next we use Chart D (page 537) to work out the structures of all the compounds that could be used to make ethanoic acid.
3 Then we see whether there is a compound that is common between both charts.

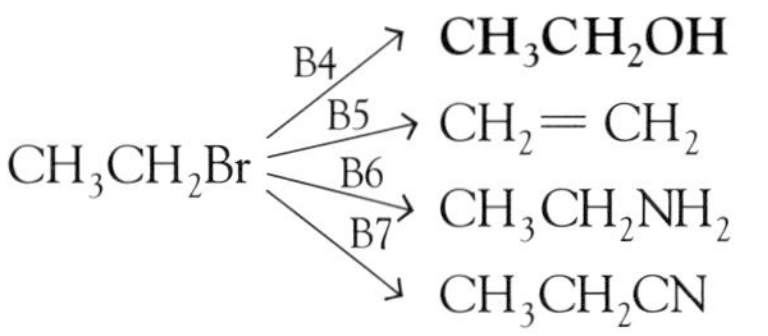

from Chart B

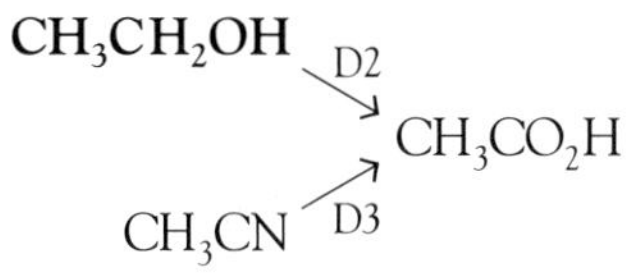

from Chart D

The compound **ethanol** is common to both, so a viable synthetic route would have ethanol as an intermediate:

$$CH_3CH_2Br \longrightarrow CH_3CH_2OH \longrightarrow CH_3CO_2H$$

Finally, from Table 30.1, we find the reagents and conditions required to carry out these two transformations:

$$CH_3CH_2Br \xrightarrow{\text{heat with NaOH(aq)}} CH_3CH_2OH \xrightarrow[\text{Na}_2\text{Cr}_2\text{O}_7+\text{H}_2\text{SO}_4\text{(aq)}]{\text{heat with an excess of}} CH_3CO_2H$$

Steps 1 and 2 above using Charts A–G may need to be repeated if a common compound is not found.

Worked example

Devise a synthesis of propylamine, $CH_3CH_2CH_2NH_2$, from ethene, C_2H_4.

Answer

Chart A (page 536), shows that there are four reactions of ethene that may be useful:

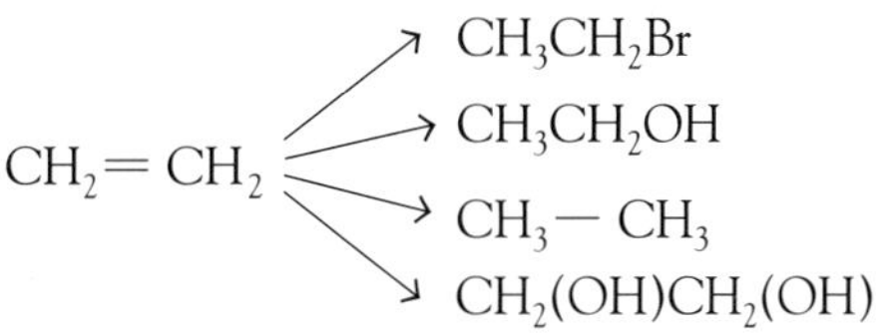

from Chart A

Next, considering the target molecule, we can see from Chart F (page 538) that there are three methods of making amines:

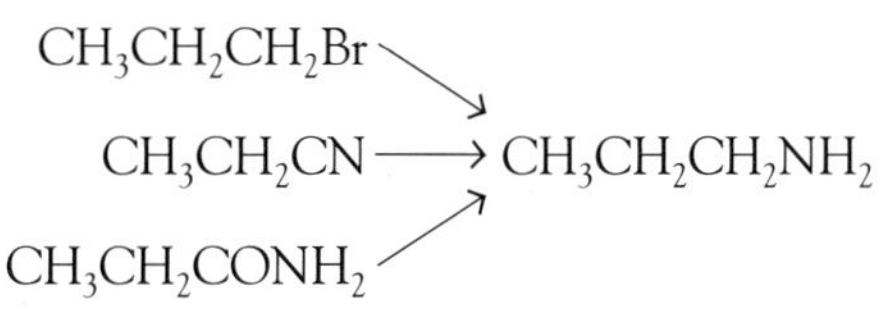

from Chart F

None of the three starting materials for making propylamine is the same as the four products from ethene, so the synthesis will need another step. Therefore we take each of the four products derived from ethene, and use the charts to carry out the same analysis on those products – what compounds can we make from each one in turn?

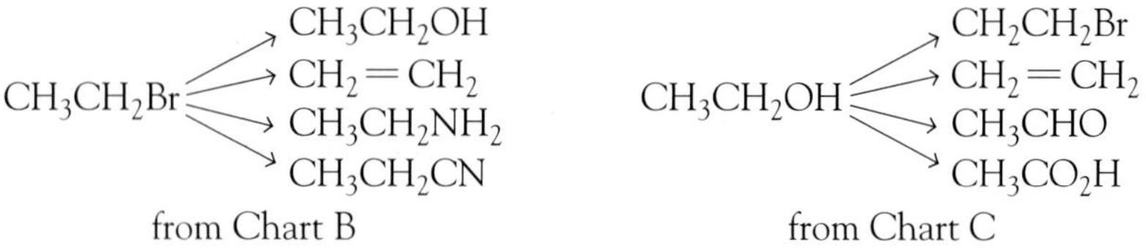

$$CH_3CH_2Br \nearrow \begin{array}{l} CH_3CH_2OH \\ CH_2=CH_2 \\ CH_3CH_2NH_2 \\ CH_3CH_2CN \end{array} \qquad CH_3CH_2OH \nearrow \begin{array}{l} CH_2CH_2Br \\ CH_2=CH_2 \\ CH_3CHO \\ CH_3CO_2H \end{array}$$

from Chart B   from Chart C

We stop immediately at Chart B, as we can see that there is a compound in common between Chart B (page 536) and Chart F (page 538): that is, propanenitrile, CH_3CH_2CN. The synthetic route therefore includes two intermediates:

$$CH_2=CH_2 \longrightarrow CH_3CH_2Br \longrightarrow CH_3CH_2CN \longrightarrow CH_3CH_2CH_2NH_2$$

Adding the reagents and conditions, the whole synthesis can be described as follows:

$$CH_2=CH_2 \xrightarrow{HBr(g)} CH_3CH_2Br \xrightarrow[\text{in ethanol}]{\text{heat with NaCN}} CH_3CH_2CN \xrightarrow{H_2 + Ni} CH_3CH_2CH_2NH_2$$

(Note that a 'shortcut' to the process here is given by the fact that the chain of the product contains one more carbon atom than the starting material: the synthesis must have had a step involving a nitrile.)

Now try this

Devise syntheses for the following compounds, starting with the specified compounds:
1 $CH_3CH(OH)CN$ from CH_3CH_2Br (three steps).
2 $CH_3CO_2CH_3$ from CH_3CN and CH_3OH (two steps).

3 $CH_3CH_2 - C \overset{O-\bigcirc}{\underset{O}{\Big\backslash}}$ from $CH_3CH_2CH_2OH$ and phenol (three steps).

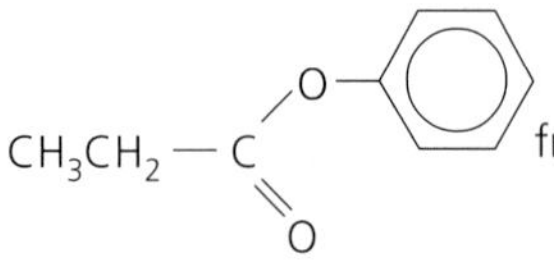

4 $Cl-\bigcirc-CO_2H$ from methylbenzene (two steps).

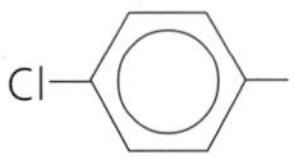

The synthesis of more complicated organic molecules

Many pharmaceutical compounds are made by joining together two or more organic parts, that are first synthesised separately. An example is the compound phenylethanamide that has been used as a pharmaceutical called *antifebrin* (Figure 30.2).

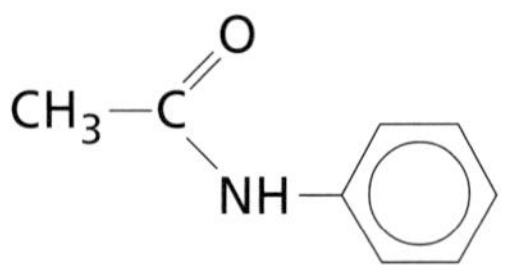

$$CH_3 - C \overset{O}{\underset{NH-\bigcirc}{\Big\backslash}}$$

Figure 30.2 Phenylethanamide

Worked example

Devise a synthesis of phenylethanamide, starting from ethene and benzene. (Both are readily available industrial chemicals that are derived from petroleum.)

Answer
As its name suggests, phenylethanamide is an amide. Amides are formed by reacting together amines and acyl chlorides. Phenylethanamide can be made as follows:

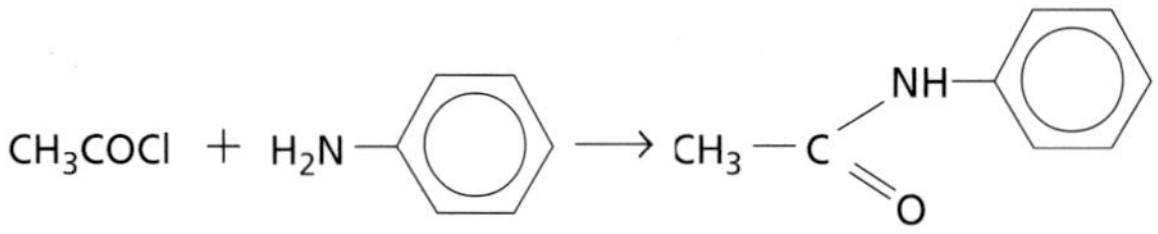

$$CH_3COCl + H_2N-\bigcirc \longrightarrow CH_3 - C \overset{NH-\bigcirc}{\underset{O}{\Big\backslash}}$$

The synthesis is therefore in three parts, as shown in Figure 30.3.

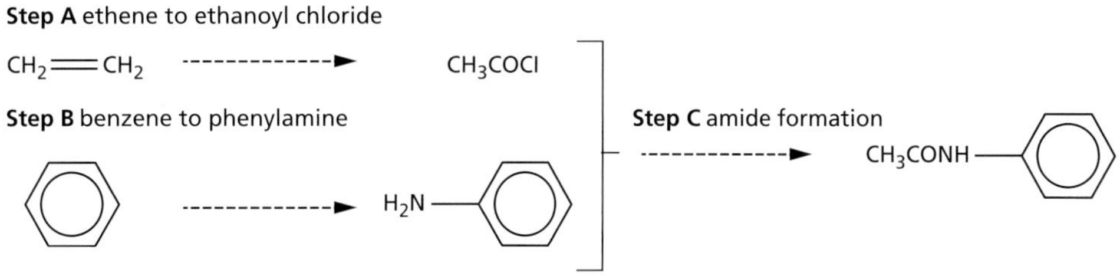

Figure 30.3 Synthesis of phenylethanamide from ethene and benzene

Step A ethene to ethanoyl chloride

$CH_2 = CH_2$ --------→ CH_3COCl

Step B benzene to phenylamine

$H_2N—$⬡

Step C amide formation

--------→ $CH_3CONH—$⬡

Step A Chart D (page 537) shows that acyl chlorides are made from carboxylic acids:

$CH_3CO_2H \longrightarrow CH_3COCl$

Carboxylic acids can be made from alcohols, which in turn can be made from alkenes:

$CH_2 = CH_2 \longrightarrow CH_3CH_2OH \longrightarrow CH_3CO_2H$

So the overall synthesis of ethanoyl chloride from ethene is as follows:

$$CH_2 = CH_2 \xrightarrow{H_2SO_4(aq)\ at\ R.T.} CH_3CH_2OH \xrightarrow[\text{of } Na_2Cr_2O_7 + H^+]{\text{heat with an excess}} CH_3CO_2H \xrightarrow{PCl_5 + heat} CH_3COCl$$

Step B Chart G (page 538) shows a two-step synthesis of phenylamine from benzene:

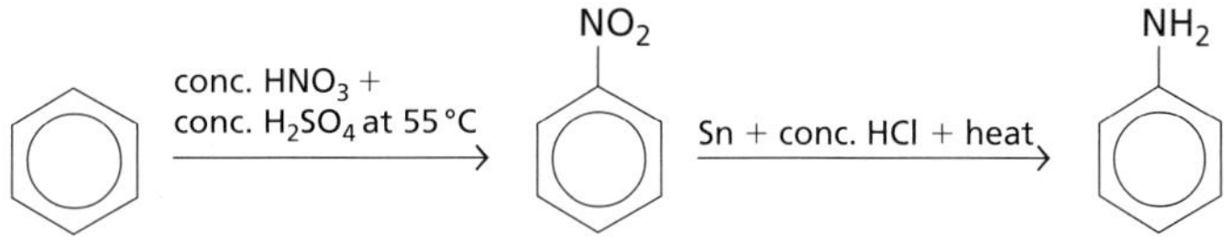

Now try this

1 Suggest a synthesis of the ester prop-2-yl phenylethanoate, $C_6H_5CH_2CO_2CH(CH_3)_2$, from methylbenzene and propene (five steps in all).

2 Suggest a synthesis of ethyl 2-hydroxypropanoate, $CH_3CH(OH)CO_2CH_2CH_3$, from ethanol (four steps).

3 Devise a synthesis of N-phenylmethylbenzamide (Figure 30.4) with all its carbon atoms coming from methylbenzene, $C_6H_5CH_3$ (five steps in all).

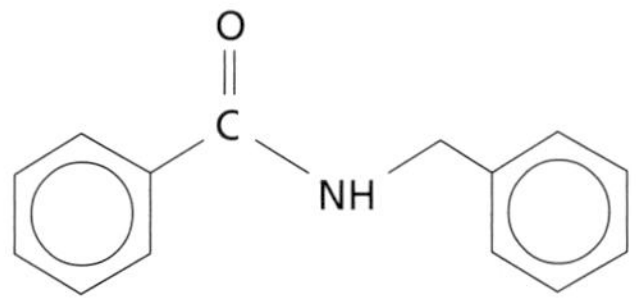

Figure 30.4 N-phenylmethylbenzamide

30.2 Design of drugs

A drug is a chemical substance that interacts with the organs in the body to produce a physiological response. A drug may occur naturally or be synthetic. Naturally occurring drugs may be made within our own bodies by our own metabolism (e.g. hormones, nerve transmitters, endorphins), or may be extracted from some other organism, most often a plant or a fungus.

The effect of a particular drug on the body can be quite varied. In general, we can classify drugs as follows:

1 those aimed (usually lethally) at organisms that are foreign to our bodies, that is, bacteria, viruses and fungi
2 those aimed at cancer cells within our own bodies (again, hopefully, in a lethal manner)
3 those aimed at changing the physiology of our own cells, and hence our physical (and sometimes psychological) well-being.

Most commonly, drugs interact with enzyme and receptor proteins. **Enzymes** are responsible for catalysing most of the chemical reactions that occur within the cell. Little can be done to speed up an enzyme-catalysed reaction (apart from the usual chemical effects of increasing concentration and temperature), but they can be slowed down by using inhibitors. Some drugs have therefore been designed to act as such inhibitors, to treat illnesses that might be caused by an excess of the products of a particular enzyme-catalysed reaction, or to slow down the removal of an important compound from the body. For example, the drug phenelzine (Figure 30.5) inhibits the enzyme monoamine oxidase. This inhibition slows down the metabolism of the nerve transmitter noradrenaline, and thus the concentration of noradrenaline in the nerve synapses can increase. This drug is a successful treatment for depression caused by a deficiency of nerve transmitters.

Figure 30.5 Phenelzine and noradrenaline

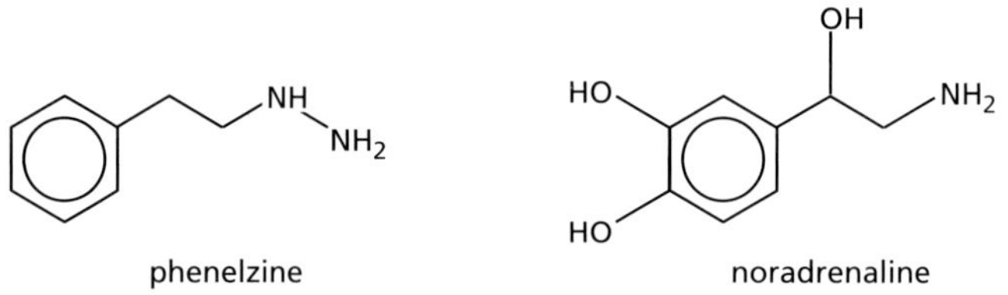

phenelzine

noradrenaline

The other main role of proteins in cells is to act as **receptors**. Receptors are protein molecules that are often found within cell membranes. Much of their influence on the cell's physiology is due to the opposing physicochemical natures of the cell membrane and the cytoplasm. Cell membranes are hydrophobic because of the long alkyl chains in their phospholipid molecules, whereas the cytoplasm is aqueous and hence is hydrophilic (Figure 30.6).

Figure 30.6 Simple diagram of a cell, showing the hydrophobic lipid bilayer membrane, the hydrophilic cytoplasm, and a variety of receptors in the cell wall

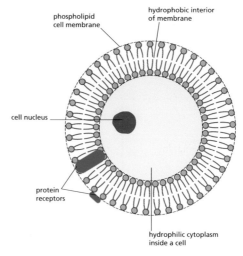

phospholipid cell membrane

hydrophobic interior of membrane

cell nucleus

protein receptors

hydrophilic cytoplasm inside a cell

Unlike enzymes, receptors do not catalyse chemical reactions, but when 'activated' they produce one of the following physiological responses:

- They stimulate a membrane-bound enzyme.
- They cause the release of secondary messengers within the cell's membrane. These migrate to other parts of the membrane, and either activate or inhibit further enzymes.
- They open ion channels through the membrane, allowing hydrophilic ions to pass through the hydrophobic membrane.

Receptors have active sites just like enzymes. When the natural substrate for a receptor binds with its active site, it changes the shape of the whole receptor molecule. It is this change of shape that causes the physiological response. Drugs have been designed to interact with receptors in either of the following two ways:

1 They can mimic the natural substrate – sometimes causing an even stronger physiological response through a larger change in shape. These drugs are called **agonists**.

2 They can bind with the active site – often more strongly than the natural substrate, and hence blocking the binding of the natural substrate - but they do not change the receptor's shape in the right way to cause a physiological response. These drugs are called **antagonists**. They are inhibiting the receptor's normal function.

Various methods are used to decide which molecules may be useful as drugs.

- The stereo-electronic shape of the active site of the enzyme or receptor can be determined by methods such as nuclear magnetic resonance (NMR) and X-ray crystallography. A compound is designed to fit into it.
- The natural substrate can be used to suggest compounds that could mimic its effect.
- If the molecular structure of a traditional remedy, such as the active ingredient of a plant extract, is known, that structure can be used as a basis for further development.

Figure 30.7 shows some examples of drug molecules whose structures have been developed in these ways.

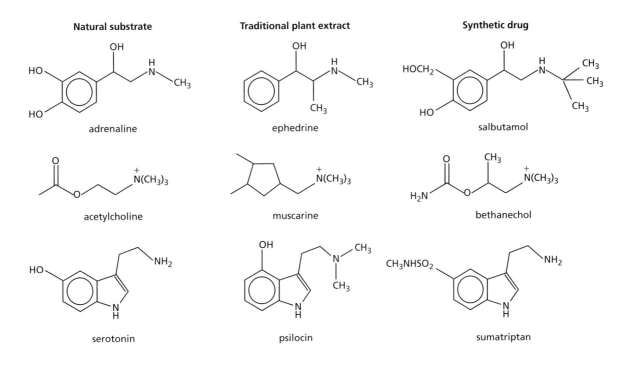

Figure 30.7 The relationship between the chemical structures of the natural substrates, traditional plant extracts and some synthetic drugs.

Now try this

Identify the key structural similarity between the molecules of:
a   adrenaline, ephedrine and salbutamol
b   acetylcholine, muscarine and bethanechol.

When a compound has been identified as having some of the therapeutic properties desired of the drug, pharmaceutical chemists begin to work on modifying its molecular structure to enhance its beneficial properties and minimise undesirable side effects. Many hundreds of compounds may be synthesised before one is found to have the right mix of characteristics to qualify it for further clinical trials.

30.3 Synthesis of chiral molecules

Almost without exception, the active sites of enzymes and receptors are chiral; they are made up from naturally occurring amino acids, all of which (except glycine) are themselves chiral (see Topic 27). Consequently, if a drug molecule which contains a chiral centre interacts with an enzyme or receptor, it is almost certain that only one of its stereoisomers will interact. The other stereoisomer will probably not have the desired physiological effect, or could even have a detrimental effect on another receptor elsewhere in the body.

Apart from any possible harmful side effects the 'other' stereoisomer might have, the cost is another incentive for synthesising just one enantiomer of a chiral compound – if only 50% of the drug is going to be useful pharmacologically, it is wasteful to make twice as much as is needed.

Some examples of pharmaceuticals whose enantiomers have different physiological effects are shown in Figure 30.8 (the chiral centres are circled in red). The modern convention for labelling the two different enantiomers is to use the prefixes **R** and **S**.

Figure 30.8 Different enantiomers can have different physiological effects.

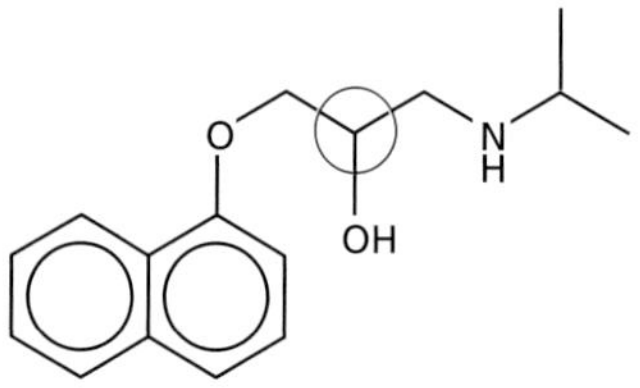

propranolol
R is a beta-blocker for the heart
S is a contraceptive

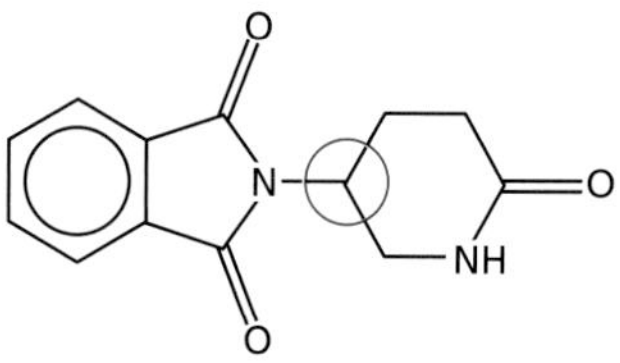

thalidomide
R is mutagenic
S is anti-emetic

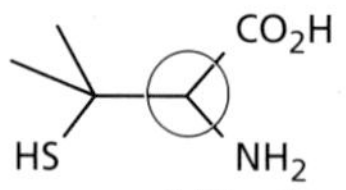

penicillamine
R is anti-arthritic
S is highly toxic

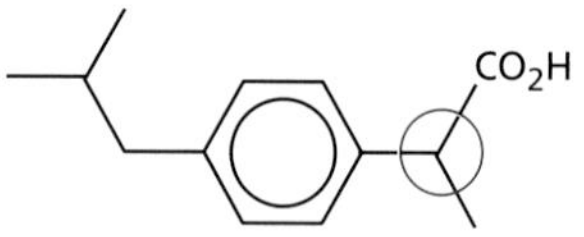

ibuprofen
R is anti-inflammatory
S is inactive

If we synthesise a chiral compound from non-chiral starting materials, we always obtain the **racemic mixture** of enantiomers (see page 224 in section 12.6). It is therefore important to ensure that our target compound, if chiral, is just the one enantiomer that we want. Pharmaceutical chemists have developed ways of preparing drug molecules in a stereoisomerically pure state. These methods can be divided into the following strategies.

1 Start with a chiral compound that is already enantiomerically pure. This will usually be a naturally occurring compound. The synthesis is carefully designed to make sure that the optical activity of the starting material is transferred to each intermediate and to the final target compound, by a process known as **asymmetric induction**.

2 A chiral reagent is used to induce chirality into a molecule at some step during the synthesis. A common method is to use a chiral reducing agent to reduce a symmetric carbonyl compound to just one enantiomer of a chiral alcohol:

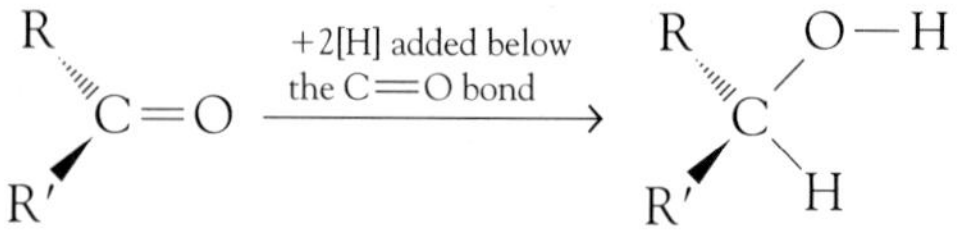

symmetrical   chiral (one isomer)

3 A racemic mixture (either of the final product, or of an intermediate) is **resolved** into its two different enantiomers (see panel on page 527). This often means that the half of the mixture that is the unwanted isomer is wasted, so, economically, the earlier this is done during a long synthesis the better.

4 An enzyme is used to catalyse one of the reactions leading to the target compound. If the product of the enzyme-catalysed reaction is chiral, usually only one isomer will be produced since the enzyme itself is chiral.

5 In a similar (but more wasteful) manner, an enzyme is used to react with (e.g. break down) the unwanted isomer, leaving the desired isomer intact.

Resolving enantiomers

There are two main methods of resolving a mixture of enantiomers.

The first, now used increasingly, is to pass a solution of the racemic mixture through a column of a solid support which is itself chiral, a chiral stationary phase (CSP). Originally cellulose (a polymer of (+)D-glucose) was used, but now many semi-synthetic derivatives of cellulose (e.g. cellulose tribenzoate) or other optically active polymers and resins are used. The column preferentially absorbs one of the isomers, letting the other through at a quicker rate. They are thus separated, and a good yield of each isomer can be recovered. This called **enantioselective chromatography**.

The second method uses an optically active compound to form an adduct with the racemic mixture. The two adducts are no longer enantiomers, but diastereoisomers. They have different physical properties that can be used to separate them. **Diastereoisomers** are stereoisomers that are not mirror images of each other. They usually contain more than one chiral centre.

For example, if the racemic mixture is a carboxylic acid, forming a salt with an optically active natural base, such as strychnine, will produce the following:

$$(+)\text{acid} + (-)\text{acid} + 2\ (+)\text{base} \longrightarrow [(+)\text{acid-}(+)\text{base}] + [(-)\text{acid-}(+)\text{base}]$$

enantiomers   pair of diastereoisomers

The diastereoisomers will often have different solubilities, and so can be separated by fractional crystallisation. The free carboxylic acids can then be liberated from the salts by reaction with dilute HCl(aq).

Other adducts that have been used include carboxylic esters with an optically active alcohol such as menthol, or, if it is the alcohol that required resolution, with an optically active acid such as tartaric acid.

Figure 30.9 (−)-menthol and (+)-tartaric acid

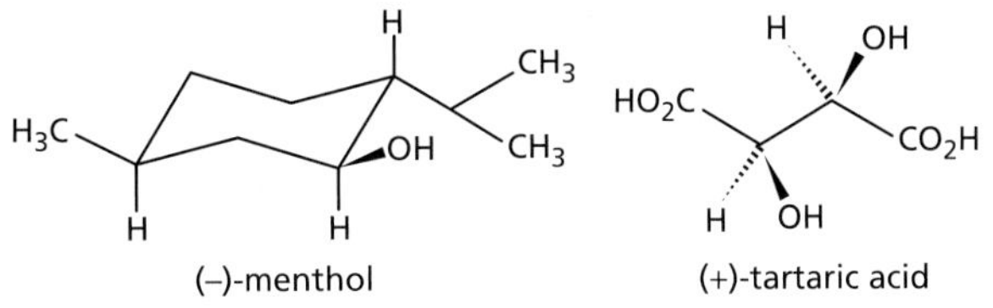

(−)-menthol   (+)-tartaric acid

30.4 The use of prodrugs

Even if a drug molecule has all the right properties to make it effective *in vitro* (in the test tube), it might still not be clinically useful. This is because there are so many obstacles between its point of entry into the body and its intended destination.

For example, if a drug is taken orally, it must be able to survive the strongly acidic conditions in the stomach and the multitude of hydrolytic enzymes in the intestines. To get into the bloodstream, it must be non-polar enough to diffuse through the hydrophobic cell walls of the intestines, but polar enough to dissolve in the blood. If its target is the brain, it must have an extremely non-polar form in order to pass through the blood–brain barrier. On the way it is likely to pass through the liver, where it will encounter many enzymes whose function is to rid the body of foreign molecules as soon as possible.

There are several ways the pharmaceutical chemist can overcome these problems. One is to create a compound, called a **prodrug**, that is inactive by itself but can be broken down into the active drug once inside the body. The prodrug is designed either to be stable to the extreme chemical conditions in the digestive tract, or to be lipophilic enough to pass through the intestinal cell walls, or both.

Worked example

An unwelcome side effect of the non-steroidal anti-inflammatory drug (NSAID) ibuprofen is irritation of the gastro-intestinal tract. This can be reduced by reacting it with phenylmethanol to form a prodrug.

ibuprofen   phenylmethanol

The resulting compound has the added advantage of being more easily transported through the intestinal cell walls into the bloodstream.

a What type of compound will the prodrug be?
b Suggest reagents and conditions for making the prodrug.
c Draw the structure of the resulting compound.
d Suggest why the prodrug is more readily absorbed through the cell walls than is ibuprofen.
e What type of reaction can occur within the target cell to re-form ibuprofen from the prodrug?

Answers

a The prodrug will be an ester.
b Heat ibuprofen and the alcohol in the presence of a small quantity of concentrated H_2SO_4.
c

d The ester is much less likely to hydrogen-bond with water than ibuprofen, which is a carboxylic acid. It is also more lipophilic, because the extra benzene ring allows it to form more van der Waals' attractions to the long alkyl chains within the phospholipid membrane of the cell wall.
e The ester can be **hydrolysed** back to ibuprofen, by an esterase enzyme.

Now try this

Compound **A** (Figure 30.10) has excellent *in vitro* activity as an antibacterial agent, but its *in vivo* (in the body) activity is poor because it is too polar to be absorbed through cell walls.

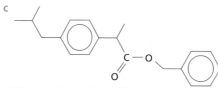

Figure 30.10 Compound **A**

a Name **four** polar functional groups in the molecule of **A**.
b Suggest reagents and conditions for masking each of these polar functional groups to make a prodrug. Draw the resulting structure of the prodrug you suggest.

 ## 30.5 Chemical tests for functional groups

Here we bring together the tests that have been described in the various topics of the organic sections of this book.

Alkanes (Topic 13)

These burn with a non-smoky flame, are immiscible with water, and less dense than water. They do **not** decolorise bromine water or dilute potassium manganate(VII).

Alkenes (Topic 14)

These burn with a slightly smoky flame, are immiscible with water, and less dense than water. They decolorise bromine water and dilute potassium manganate(VII) at room temperature.

Alcohols (Topic 16)

These are neutral to litmus solution and (if insoluble in water) do not dissolve in aqueous sodium hydroxide or sodium carbonate. They effervesce with sodium metal (giving off hydrogen) and with phosphorus(V) chloride (giving off hydrogen chloride). Primary, secondary and tertiary alcohols are most easily distinguished from each other by attempted oxidation with acidified dichromate(VI), followed by distillation and performance of tests on the products (see Figure 30.11).

Figure 30.11 How to work out whether an alcohol is primary, secondary or tertiary. (UI, universal indicator)

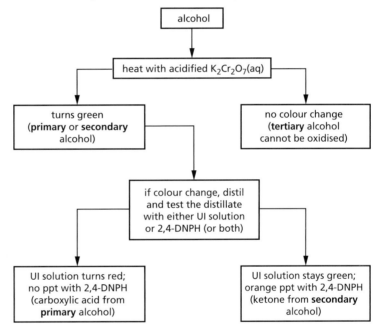

Carbonyl compounds (Topic 17)

Both aldehydes and ketones produce orange precipitates with 2,4-dinitrophenylhydrazine (2,4-DNPH) (Figure 30.12). They may be distinguished from each other by warming with Tollens' reagent (ammoniacal silver nitrate solution) or Fehling's solution (alkaline copper(II) solution). Aldehydes reduce these reagents, to a silver mirror (Figure 30.13), or a red precipitate of copper(I) oxide, respectively, whereas ketones have no effect upon them.

Figure 30.12 Aldehydes and ketones give a precipitate with 2,4-DNPH. **a** Ethanal reacting and, **b** forming an orange precipitate

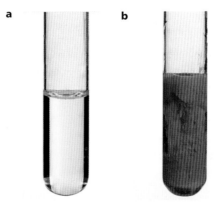

Figure 30.13 Aldehydes give a silver mirror with Tollens' reagent.

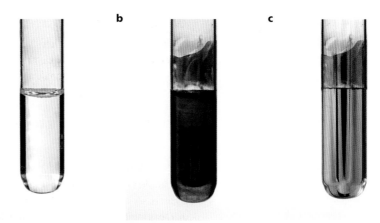

a   b   c

Carboxylic acids (Topic 18)

When a drop of a carboxylic acid is added to a solution of universal indicator, the solution turns red. (Note that some acid derivatives such as acyl chlorides and acid anhydrides will also turn universal indicator red because of their hydrolysis.) Carboxylic acids effervesce with sodium (giving off hydrogen) and with phosphorus(V) chloride (giving off hydrogen chloride). They also dissolve in aqueous sodium hydroxide and sodium carbonate (giving off carbon dioxide in the process).

Phenols (Topic 25)

These are usually insoluble in water, but will dissolve in aqueous sodium hydroxide. They do not dissolve in aqueous sodium carbonate. They effervesce with sodium (giving off hydrogen). When a phenol is added to a dilute neutral solution of iron(III) chloride, a violet coloration is produced. Phenols react with bromine water, decolorising it, and producing a white precipitate of the di- or tri-bromophenol.

Acyl chlorides (Topic 26)

These react with water to produce acidic solutions which turn universal indicator red, often giving off fumes of HCl(g) during the process. They may be distinguished from carboxylic acids by their **not** reacting with sodium metal or phosphorus(V) chloride.

Esters (Topic 18)

These are neutral, sweet-smelling liquids that are immiscible with water. They do not react with sodium metal or phosphorus(V) chloride, or with 2,4-DNPH. On heating with dilute acids or alkalis, however, they are hydrolysed to alcohols and carboxylic acids. These products can be tested for in the usual way.

Amides (Topic 27)

These are neutral, but will evolve the alkaline gas ammonia (test with moist red litmus paper) when boiled with NaOH(aq).

Amines (Topic 27)

These either dissolve in water to give alkaline solutions (if of low M_r), or if insoluble in water, they will dissolve in dilute HCl.

The triiodomethane reaction (Topics 16 and 17)

This is a useful test for the presence of the groups $CH_3CH(OH)-$ or CH_3CO- in a molecule. When alkaline aqueous iodine is added to a compound containing one of these groups (ethanol, ethanal, methyl secondary alcohols or methyl ketones), a pale yellow precipitate of triiodomethane (iodoform) is formed.

30.6 The use of functional-group tests to deduce structures

If the molecular formula of a compound is known, its structure can often be deduced on the basis of the results of various functional-group tests. An example will make this clear.

Worked example 1

Three compounds, **J**, **K** and **L,** are isomers with the molecular formula $C_4H_8O_2$. Use the information below to identify the compounds **J–Q**, and write equations for all reactions that occur.

a Compound **J** is unaffected by hot dilute sulfuric acid, but reacts with sodium metal, Fehling's solution and alkaline aqueous iodine.
b Both **K** and **L** react with hot dilute sulfuric acid. Under these conditions compound **K** gives **M** (CH_2O_2) and **N** (C_3H_8O) and compound **L** gives **P** ($C_2H_4O_2$) and **Q** (C_2H_6O).

- Both **M** and **P** effervesce with sodium carbonate solution.
- **N** and **Q** both react with sodium metal, and are both oxidised by acidified potassium dichromate(VI).
- The oxidation product from **N** gives an orange precipitate with 2,4-dinitrophenylhydrazine, but does not produce a silver mirror when warmed with ammoniacal silver nitrate.
- The oxidation product from **Q** is identical to compound **P**.

Answer

a Compound **J** is not an ester (because it does not react with hot dilute sulfuric acid) but it could be an alcohol or a carboxylic acid (reaction with sodium). Reaction with Fehling's solution suggests an aldehyde, and reaction with alkaline aqueous iodine suggests the group $CH_3CH(OH)-$ or CH_3CO-.

We now consider the second oxygen atom in the formula. Since at least one of the two oxygen atoms is taken up by the aldehyde group, **J** cannot be a carboxylic acid. It must therefore be an alcohol (reaction with sodium). So **J** contains the groups $CH_3CH(OH)-$ and $-CHO$, and must therefore be **$CH_3CH(OH)CH_2CHO$**.
Equations:
$CH_3CH(OH)CH_2CHO + Na \rightarrow CH_3CH(ONa)CH_2CHO + H_2$
$CH_3CH(OH)CH_2CHO + [O] \rightarrow CH_3CH(OH)CH_2CO_2H$
$CH_3CH(OH)CH_2CHO + 4I_2 + 6OH^- \rightarrow CHI_3 + {}^-O_2CCH_2CHO + 5I^- + 5H_2O$

b Both compounds **K** and **L** are esters (reaction with hot dilute sulfuric acid), being hydrolysed to acids (**M** and **P**) and alcohols (**N** and **Q**). **M** and **P** have unique structures for their molecular formulae. **M** is methanoic acid, HCO_2H, and **P** is ethanoic acid, CH_3CO_2H. Both acids will effervesce with aqueous sodium carbonate, giving off carbon dioxide.

The alcohols **N** and **Q** are both oxidised by acidified potassium dichromate(VI) solution, so neither is a tertiary alcohol. **N** produces a ketone on oxidation (orange precipitate with 2,4-DNPH, but no silver mirror with Tollens' reagent). So **N** is the secondary alcohol, propan-2-ol. **Q** must be ethanol, which is oxidised to ethanoic acid (**P**).

Putting all this together shows that **K** must be prop-2-yl methanoate, and **L** must be ethyl ethanoate (Figure 30.14)

Figure 30.14

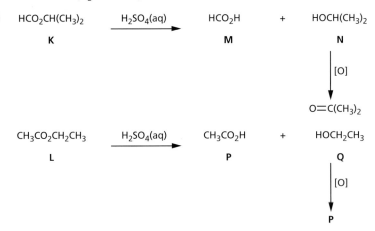

Now try this

Deduce the structures of the following compounds, explaining your reasoning:

1 Compound **R** has the molecular formula C_8H_8O. It effervesces with sodium metal, but not with phosphorus(V) chloride. It decolorises bromine water, giving a white precipitate. It also decolorises dilute aqueous potassium manganate(VII) solution.

2 Compound **S** has the molecular formula $C_6H_{12}O_2$. It is unaffected by hot dilute sulfuric acid and also by hot acidified dichromate, but reacts with both sodium metal and alkaline aqueous iodine. It forms an orange precipitate with 2,4-dinitrophenylhydrazine, but does not react with Fehling's solution.

Functional-group tests are also useful for distinguishing between isomers. The following examples illustrate this.

Worked example 2

For each of the following pairs of isomers, suggest a test that will distinguish between the two compounds.

a $CH_3CH_2CH_2CHO$ and $CH_3CH_2COCH_3$

b $(CH_3)_3COH$ and $(CH_3)_2CHCH_2OH$

c $CH_2{=}CHCH_2OH$ and CH_3CH_2CHO

Answer

a One of the pair is an aldehyde, and the other is a methyl ketone. We could use either Fehling's solution (red precipitate with the aldehyde, no change with the ketone) or alkaline aqueous iodine (yellow precipitate with the methyl ketone, no change with the aldehyde). Note that 2,4-DNPH would not distinguish these: both will give orange precipitates.

b Both compounds are alcohols, so sodium metal or phosphorus(V) chloride would not distinguish between them. One is a tertiary alcohol, so would not be affected by warming with acidified potassium dichromate(VI). The other is a primary alcohol, which would turn acidified dichromate(VI) from orange to green.

c Several tests could be used for this pair. The first compound is an alkene alcohol, so it would react with aqueous bromine, cold potassium manganate(VII) solution, sodium metal or phosphorus(V) chloride. None of these reagents would react with the second compound. Being an aldehyde, however, this second compound would react with 2,4-DNPH, Fehling's solution or Tollens' reagent. Note that both compounds would be oxidised by warm acidified dichromate(VI), so this reagent would not distinguish between them.

Suggest tests that could be carried out on each of the following pairs of isomers that would distinguish between them.

1 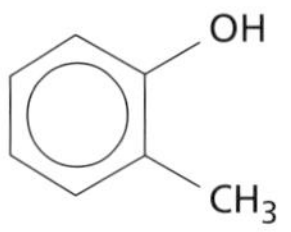 and

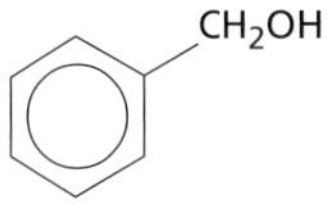

2 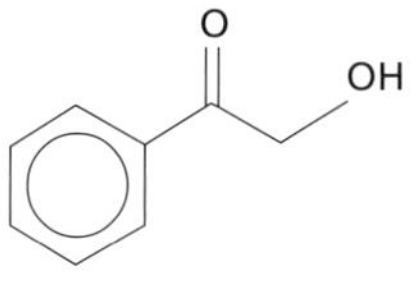 and

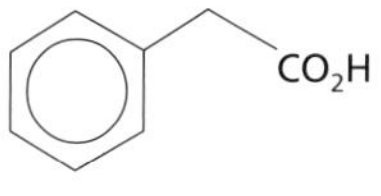

3 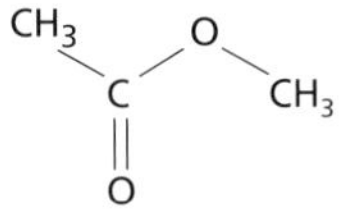 and

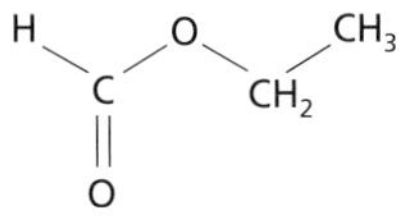

4 CH_3—CH_2—$COCl$ and CH_3—$CHCl$—CHO

30.7 Predicting the reactions of multifunctional compounds

Just as we can deduce the functional groups in a compound from its reactions, so we can work out how a particular multifunctional compound reacts with a particular reagent. The reactions of many multifunctional compounds can be considered to be the sum of the reactions of each of the functional groups they contain, and so a particular compound may react with several different reagents, or a particular reagent may react with several groups in a molecule.

Here are some examples.

Worked example

Predict the product of the reactions between the following compounds and reagents.

a 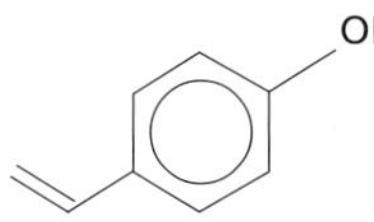 + $Br_2(aq)$

b NH_2 + hot NaOH(aq)

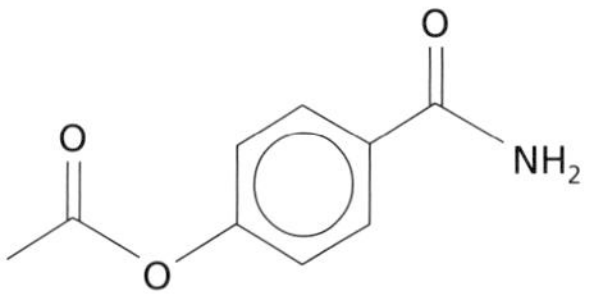

c 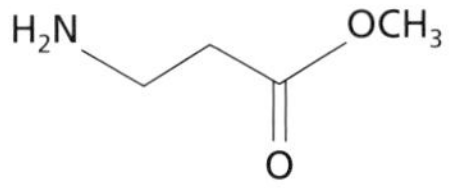 + hot HCl(aq)

d 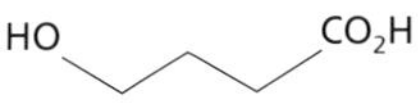 + Na(s)

Answer

a The $Br_2(aq)$ will react with both the alkene and the phenol.

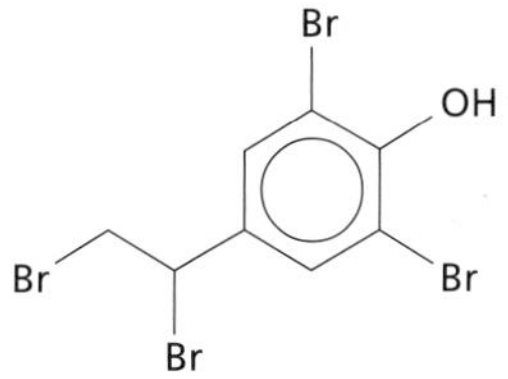

b Hot NaOH will hydrolyse both the ester and the amides, and produce the carboxylate and the phenoxide ions of the product.

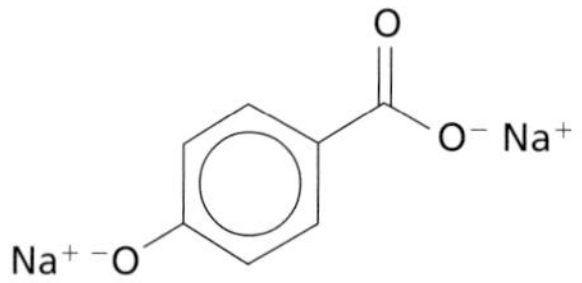

c The HCl(aq) will hydrolyse the ester, but will also form the salt from the amine.

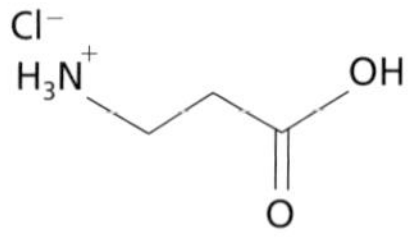

d The sodium metal will react with both the carboxylic acid and the alcohol

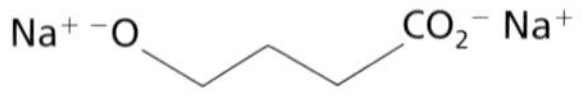

Some functional groups in multifunctional compounds might even react with each other under certain conditions. For example, reacting $HOCH_2CH_2CH_2CO_2H$ with concentrated H_2SO_4 causes the molecule to undergo an internal esterification to produce a cyclic ester (called a lactone).

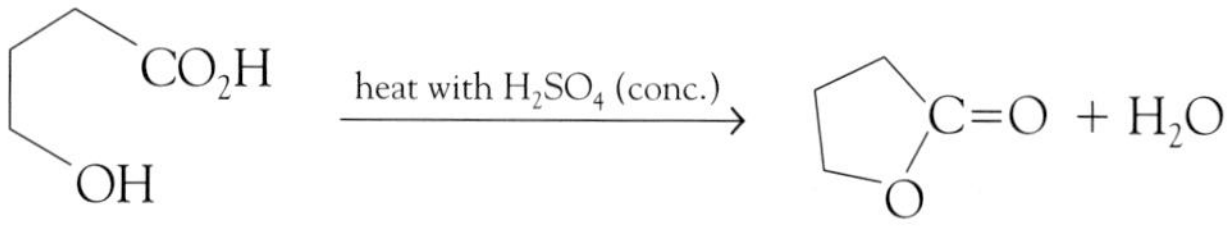

Now try this

Predict the products **A** to **E** of the following reactions.

1

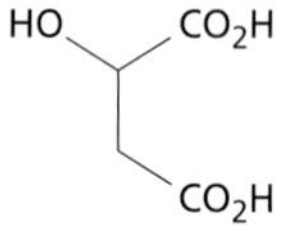

$+ PCl_5 \longrightarrow$ **A**

2

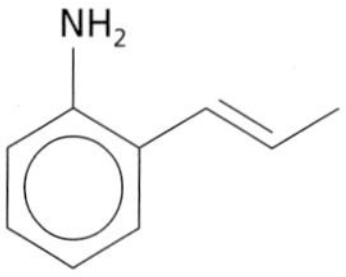

$+ Br_2(aq) \longrightarrow$ **B**

3

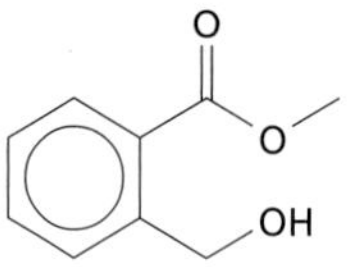

heat with trace of H_2SO_4(conc.) $\longrightarrow$ **C** $(C_8H_6O_2)$

4 $CH_3CH(OH)CO_2H$

heat with trace of H_2SO_4(conc.) $\longrightarrow$ **D** $(C_6H_8O_4)$

5

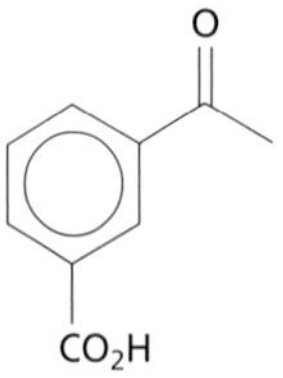

$LiAlH_4 \longrightarrow$ **E**

30.8 Summary of organic transformations

Reagents for organic reactions

Reference (see key below)	Reagents and conditions	Reaction number on chart						
		Chart A Alkenes	Chart B Bromo-alkanes	Chart C Alcohols	Chart D Aldehydes and acids	Chart E Methyl ketones	Chart F Amines	Chart G Benzene
E1	$Cl_2(g) + AlCl_3$							G1
E2	$Br_2(l) + AlBr_3$							G2
E3	HBr(g) at R.T.	A3	B3					
E4	$H_2SO_4(aq)$ at R.T.	A4		C2		E1		
E5	conc. HNO_3 + conc. H_2SO_4 at < 55 °C							G5
E6	R'Cl + $AlCl_3$ + heat							G9
E7	R''COCl + $AlCl_3$ + heat							G10
F1	$X_2(g)$ + light (X═Cl or Br)		B2					G3
N1	NaOH(aq) + heat		B4	C1				
N2	NH_3 in ethanol + heat under pressure		B6			F1		
N3	NaCN in ethanol + heat		B7					
N4	HCN + NaCN in ethanol/water				D5	E5		
N5	HBr(conc.) or NaBr + conc. H_2SO_4 + heat		B1	C4				
N6	$SOCl_2$ or PCl_5 + heat				D8			
N7	R'–OH at R.T.				D9			
N8	R'–OH + conc. H_2SO_4 + heat				D7			
N9	phenol + NaOH (aq) in the cold				D10			G8
N10	$H_2SO_4(aq)$ + heat				D3 D12			
N11	$R'NH_2$ at R.T.				D11			
O1	$Na_2Cr_2O_7(aq)$ + $H_2SO_4(aq)$ + heat			C6 C7	D1 D2 D6	E2		
O2	$KMnO_4(aq)$ at R.T.	A6						
O3	$KMnO_4$(conc.) + $H_2SO_4(aq)$ + heat	A7						
O4	$KMnO_4$(conc.) + $OH^-(aq)$ + heat							G4
R1	$H_2(g)$ + Ni catalyst at R.T.	A5		C3	D4	E3	F2	
R2	$NaBH_4$ in aqueous methanol + heat			C3	D4	E3		
R3	$LiAlH_4$ in dry ether			C3 C9	D4 D13	E3	F2 F3	
R4	Sn + conc. HCl(aq) + heat							G6
X1	NaOH in ethanol + heat	A1	B5					
X2	Al_2O_3 + heat	A2		C5				

Continued

Reference (see key below)	Reagents and conditions	Reaction number on chart						
		Chart A Alkenes	Chart B Bromo-alkanes	Chart C Alcohols	Chart D Aldehydes and acids	Chart E Methyl ketones	Chart F Amines	Chart G Benzene
X3	conc. H_2SO_4 + heat	A2		C5				
M1	I_2(aq) + OH^-(aq) + warm					E4		
M2	HNO_2 (or $NaNO_2$ + HCl(aq)) at < 5°C							G7
M3	R"COCl at R.T.			C8		F5		
M4	$R'CO_2H$ + conc. H_2SO_4 + heat			C8				
M5	R'Br + heat					F4		

Table 30.1 Reagents for organic reactions

E, electrophilic reagents; F, free-radical reagents; N, nucleophilic reagents; O, oxidising agents; R, reducing agents; X, elimination reagents; M, miscellaneous; R.T., room temperature

Charts for organic syntheses

Chart A: Synthetic routes involving alkenes

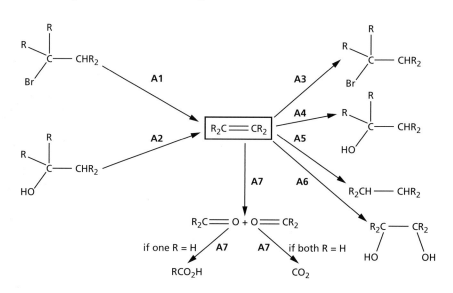

Chart B: Synthetic routes involving bromoalkanes (or chloroalkanes)

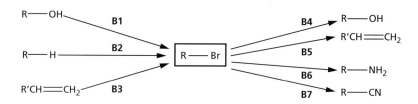

Chart C: Synthetic routes involving alcohols

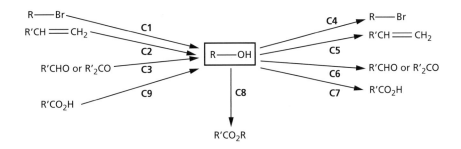

Chart D: Synthetic routes involving aldehydes and carboxylic acids

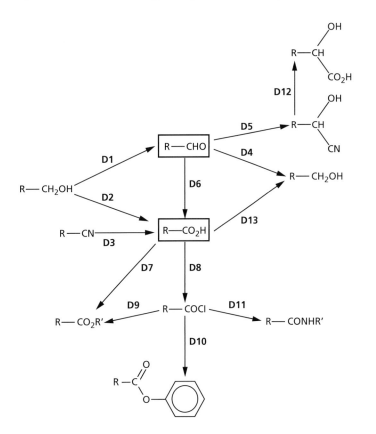

Chart E: Synthetic routes involving methyl ketones

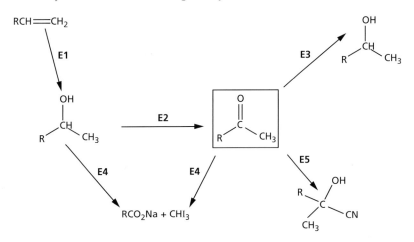

Chart F: Synthetic routes involving amines

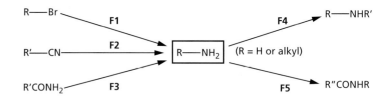

Chart G: Synthetic routes involving arenes

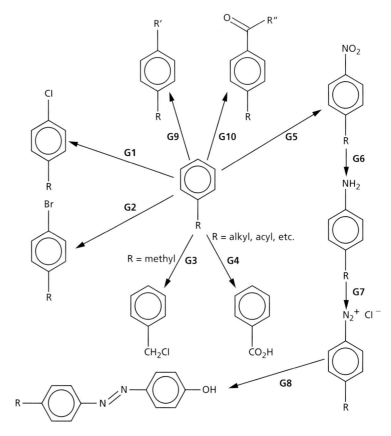

(R = hydrogen or methyl or other alkyl unless stated otherwise)

Summary

- General organic synthetic reactions can be used to make a large variety of organic compounds from simpler, easily available compounds.
- Pharmaceutical molecules can be designed to target specific enzymes and/or receptors, to treat a variety of diseases and illnesses.
- The use of **prodrugs** is a method of protecting drug molecules from degradation before they reach their site of action.

- Various methods are available for the synthesis of **optically pure** compounds for use as drugs.
- Knowledge of the reactions of organic functional groups allows the identities of organic compounds to be determined and isomers to be distinguished.

Examination practice questions

Please see the data section of the CD for any A_r values you may need.

1 a Describe and explain how the acidities of ethanol and phenol compare to that of water. [4]

  b Complete the following equations showing **all** the products of each of these reactions of phenol. Include reaction conditions where appropriate. If no reaction occurs write **no reaction**.

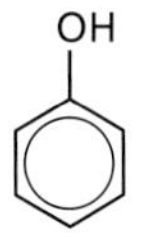

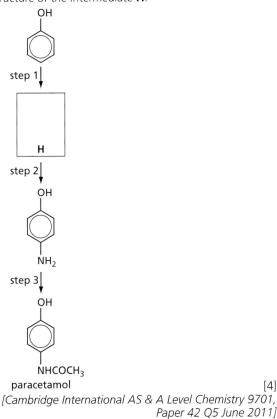

OH + Na ⟶

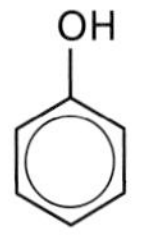

OH + NaOH ⟶

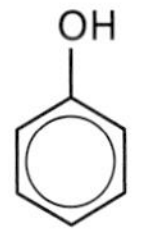

OH + CH$_3$CO$_2$H ⟶

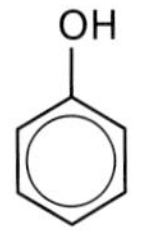

OH + Br$_2$ ⟶ [5]

  c The analgesic drug paracetamol can be synthesised from phenol by the following route. Suggest reagents and conditions for the each of three steps, and suggest the structure of the intermediate **H**.

OH

step 1 ↓

[H]

step 2 ↓

OH

NH$_2$

step 3 ↓

OH

NHCOCH$_3$

paracetamol [4]

[Cambridge International AS & A Level Chemistry 9701, Paper 42 Q5 June 2011]

2 Compound **C** has the molecular formula $C_7H_{14}O$. Treating **C** with hot concentrated acidified $KMnO_4(aq)$ produces two compounds, **D**, C_4H_8O, and **E**, $C_3H_4O_3$. The results of four tests carried out on these three compounds are shown in the following table.

test reagent	result of test with		
	compound C	compound D	compound E
Br$_2$(aq)	decolourises	no reaction	no reaction
Na(s)	fizzes	no reaction	fizzes
I$_2$(aq) + OH$^-$(aq)	no reaction	yellow precipitate	yellow precipitate
2,4-dinitrophenylhydrazine	no reaction	orange precipitate	orange precipitate

  a State the functional groups which the above four reagents test for.
    i Br$_2$(aq)
    ii Na(s)
    iii I$_2$(aq) + OH$^-$(aq)
    iv 2,4-dinitrophenylhydrazine [4]

  b Based upon the results of the above tests, suggest structures for compounds **D** and **E**. [2]

  c Compound **C** exists as two stereoisomers. Draw the structural formula of **each** of the two isomers, and state the type of stereoisomerism involved. [3]

[Cambridge International AS & A Level Chemistry 9701, Paper 41 Q5 November 2011]

Index

Index

Index

Index

Index